Mohd Arif Anuar Mohd Salleh ·
Dewi Suriyani Che Halin · Kamrosni Abdul Razak ·
Mohd Izrul Izwan Ramli
Editors

Proceedings of the Green Materials and Electronic Packaging Interconnect Technology Symposium

EPITS 2022, 14–15 September, Langkawi, Malaysia

Springer

Editors
Mohd Arif Anuar Mohd Salleh
Faculty of Chemical Engineering Technology
Universiti Malaysia Perlis
Arau, Perlis, Malaysia

Kamrosni Abdul Razak
Faculty of Chemical Engineering Technology
Universiti Malaysia Perlis
Arau, Perlis, Malaysia

Dewi Suriyani Che Halin
Faculty of Chemical Engineering Technology
Universiti Malaysia Perlis
Arau, Perlis, Malaysia

Mohd Izrul Izwan Ramli
Faculty of Chemical Engineering Technology
Universiti Malaysia Perlis
Arau, Perlis, Malaysia

ISSN 0930-8989 ISSN 1867-4941 (electronic)
Springer Proceedings in Physics
ISBN 978-981-19-9266-7 ISBN 978-981-19-9267-4 (eBook)
https://doi.org/10.1007/978-981-19-9267-4

This Springer imprint is published by the registered company Springer Nature Singapore Pte Ltd.
The registered company address is: 152 Beach Road, #21-01/04 Gateway East, Singapore 189721, Singapore

Preface

This book highlights an insight into the recent research and cutting-edge technologies, which gains immense interest with the colossal and exuberant presence of adepts, young and brilliant researchers, business delegates and talented student communities. EPITS meeting's goal is to bring together a multi-disciplinary group of scientists and engineers from all over the world to present and exchange break-through ideas relating to the electronic packaging. It promotes top-level research to globalize the quality research in general, thus making discussions, presentations more internationally competitive and focusing attention on the recent outstanding achievements in the field of electronic materials, and future trends and needs.

This book also included papers from various fields in both Geopolymer and Sustainable Material Engineering and Technology. The conference also provides the opportunities for academicians, professionals, practitioners and policymakers in the engineering and computing fields to share their thoughts and empirical works to both those involved in their field or those interested in the subject being researched.

Arau, Malaysia

Mohd Arif Anuar Mohd Salleh
Dewi Suriyani Che Halin
Kamrosni Abdul Razak
Mohd Izrul Izwan Ramli

Contents

Contributors

Hairul Effendy Ab Maulod Fakulti Teknologi Kejuruteraan Mekanikal dan Pembuatan, Universiti Teknikal Malaysia Melaka, Melaka, Malaysia

S. Abbirahmi Environmental Technology Division, School of Industrial Technology, Universiti Sains Malaysia, Penang, Gelugor, Malaysia

Anis Aqilah Abd Ghani Fakulti Kejuruteraan Pembuatan, Universiti Teknikal Malaysia Melaka, Melaka, Malaysia

N. H. Abd Halim EMZI-UiTM Nanoparticles Colloids and Interface Industrial Research Laboratory (NANO-CORE), Centre for Chemical Engineering Studies, Universiti Teknologi MARA, Permatang Pauh, Pulau Pinang, Malaysia

M. I. F. Abd Maruzuki Centre for Electrical Engineering Studies, Universiti Teknologi MARA, Pulau Pinang, Malaysia

Jeefferie Abd Razak Fakulti Kejuruteraan Pembuatan, Universiti Teknikal Malaysia Melaka, Melaka, Malaysia

Nur Umairah Afifah Abd. Wahab Fakulti Kejuruteraan Pembuatan, Universiti Teknikal Malaysia Melaka, Melaka, Malaysia

Fariha Awatif Abdul Aziz Fakulti Teknologi Kejuruteraan Mekanikal and Pembuatan, Universiti Teknikal Malaysia Melaka, Melaka, Malaysia

Mohd Sharizal Abdul Aziz School of Mechanical Engineering, Universiti Sains Malaysia, Nibong Tebal, Seberang Perai Selatan, Penang, Malaysia

B. Abdul Hadi Centre for Civil Engineering Studies, Universiti Teknologi MARA, Pulau Pinang, Malaysia

Sabarina Abdul Hamid Renewable Energy Research Laboratory (RENERAL), Electrical Engineering Section, British Malaysia Institute, Universiti Kuala Lumpur, Gombak, Selangor, Malaysia

N. Abdul Latif School of Civil Engineering, Universiti Teknologi MARA, Shah Alam, Selangor, Malaysia

Q. K. Abdul Rahman EMZI-UiTM Nanoparticles Colloids and Interface Industrial Research Laboratory (NANO-CORE), Centre for Chemical Engineering Studies, Universiti Teknologi MARA, Pulau Pinang, Malaysia

A. K. Abdul Razak EMZI-UiTM Nanoparticles Colloids and Interface Industrial Research Laboratory (NANO-CORE), Centre for Chemical Engineering Studies, Universiti Teknologi MARA, Permatang Pauh, Pulau Pinang, Malaysia

Mohd Mustafa Albakri Abdullah Center of Excellence Geopolymer and Green Technology, Universiti Malaysia Perlis, Perlis, Malaysia

Muhamad Haikal Abdullah Faculty of Chemical Engineering and Technology, Universiti Malaysia Perlis, Arau, Perlis, Malaysia

R. A. Abdullah School of Civil Engineering, Faculty of Engineering, Universiti Teknologi Malaysia, Johor Bahru, Malaysia

S. Z. Abidin College of Engineering, Universiti Tenaga Nasional, Kajang, Selangor, Malaysia

A. Abu Bakar Centre for Electrical Engineering Studies, Cawangan Pulau Pinang, Universiti Teknologi MARA, Permatang Pauh, Pulau Pinang, Malaysia

Maria Abu Bakar Institute of Microengineering and Nanoelectronics (IMEN), Level 4 Research Complex, Universiti Kebangsaan Malaysia, Bangi, Selangor, Malaysia

Nur Maizatul Shima Adzali Industrial Chemical Process Programme, Faculty of Chemical Engineering Technology, Universiti Malaysia Perlis, Perlis, Malaysia

M. I. Ahmad Faculty of Bioengineering and Technology, Universiti Malaysia Kelantan, Jeli, Kelantan, Malaysia

Syazwana Ahmad Faculty of Chemical Engineering Technology, Universiti Malaysia Perlis, Arau, Perlis, Malaysia

Zulkifli Ahmad School of Materials and Mineral Resources Engineering, Universiti Sains Malaysia, Nibong Tebal, Penang, Malaysia

Nur Hidayah Ahmad Zaidi Materials Engineering Programme, Faculty of Chemical Engineering Technology, Universiti Malaysia Perlis, Jejawi, Arau, Perlis, Malaysia

Warid Wazien Ahmad Zailani School of Civil Engineering, College of Engineering, Engineering Complex, Tunku Abdul Halim Muadzam Shah, Universiti Teknologi MARA, Shah Alam, Selangor, Malaysia

Ho Seon Ahn Department of Mechanical Engineering, Incheon National University, Incheon, Republic of Korea

Mohd Suhami A'isyah Faculty of Chemical Engineering Technology (FTKK), Universiti Malaysia Perlis (UniMAP), Jejawi, Arau, Perlis, Malaysia

M. H. Aiman Faculty of Mechanical and Automotive Engineering Technology, Universiti Malaysia Pahang, Pekan, Pahang Darul Makmur, Malaysia

M. S. Akbar Faculty Science, Universiti Teknologi Malaysia, Skudai, Johor, Malaysia

Mohd Mustafa Al Bakri Abdullah Center of Excellence Geopolymer and Green Technology, Universiti Malaysia Perlis, Arau, Perlis, Malaysia

A. J. H. Alansare College of Graduate Studies, Universiti Tenaga Nasional, Kajang, Selangor Darul Ehsan, Malaysia

Azuwa Ali Faculty of Electrical Engineering Technology, Universiti Malaysia Perlis, Kampus Tetap Pauh Putra, Arau, Perlis, Malaysia

Suffi Irni Alias Faculty of Chemical Engineering Technology, Universiti Malaysia Perlis, Kangar, Perlis, Malaysia

Md. Abdul Alim Directorate of Technical Education, Dhaka-1207, Bangladesh

Ayah Almsedeen Faculty of Civil Engineering Technology, Universiti Malaysia Pahang, Kuantan, Malaysia

Rasyidah Alrozi School of Chemical Engineering, College of Engineering, Universiti Teknologi MARA(UiTM), Cawangan Pulau Pinang Kampus, Permatang Pauh, Pulau Pinang, Malaysia

Iqlima Nuril Amini Department of Civil Engineering, Institut Teknologi Sepuluh Nopember, Surabaya, Indonesia

Astuty Amrin Razak Faculty of Technology and Informatics, UTM Kuala Lumpur, Kuala Lumpur, Malaysia

Marvrick Anak Anen Fakulti Kejuruteraan Pembuatan, Universiti Teknikal Malaysia Melaka, Melaka, Malaysia

Le Huynh Tuyet Anh Faculty of Materials Technology, Ho Chi Minh City University of Technology (HCMUT), Ho Chi Minh City, Vietnam;
Vietnam National University Ho Chi Minh City, Linh Trung Ward, Thu Duc District, Ho Chi Minh City, Vietnam

Luong Thi Quynh Anh Department of Metallurgy and Alloys Materials, Faculty of Materials Technology, Ho Chi Minh City University of Technology (HCMUT), Ho Chi Minh City, Vietnam;
Vietnam National University Ho Chi Minh City, Thu Duc District, Ho Chi Minh City, Vietnam

Nguyen Cong Tuan Anh Faculty of Materials Technology, Ho Chi Minh City University of Technology (HCMUT), Ho Chi Minh City, Vietnam;
Vietnam National University Ho Chi Minh City, Linh Trung Ward, Thu Duc District, Ho Chi Minh City, Vietnam

M. R. Aridi College of Engineering, Universiti Tenaga Nasional, Kajang, Selangor, Malaysia

Asliza Aris College of Creative Arts, Universiti Teknologi Mara, Shah Alam, Malaysia

Mohd Fadzil Arshad School of Civil Engineering, College of Engineering, Universiti Teknologi MARA, Shah Alam, Selangor, Malaysia

Noriza Arzain College of Creative Arts, Universiti Teknologi Mara, Shah Alam, Malaysia

Dayang Izzah Nabilah Awang Azman Faculty of Mechanical and Manufacturing Engineering, Universiti Tun Hussein Onn Malaysia, Johor, Malaysia

Azliza Azani Faculty of Chemical Engineering Technology (FTKK), Universiti Malaysia Perlis (UniMAP), Jejawi, Arau, Perlis, Malaysia;
Center of Excellence Geopolymer and Green Technology (CEGeoGTech), Universiti Malaysia Perlis, Jejawi, Arau, Perlis, Malaysia

Muhammad Mokhzaini Azizan Faculty of Engineering and Built Environment, Universiti Sains Islam Malaysia, Nilai, Negeri Sembilan, Malaysia

Siti Aishah Binti Abd Aziz Faculty of Chemical Engineering Technology (FKTK), Universiti Malaysia Perlis (UniMAP), Jejawi, Perlis, Malaysia

M. A. Azmah Hanim Department of Mechanical and Manufacturing Engineering, Faculty of Engineering, Universiti Putra Malaysia, Serdang, Selangor, Malaysia;
Advance Engineering Materials and Composites Research Center, (AEMC), Faculty of Engineering, Universiti Putra Malaysia, Serdang, Selangor, Malaysia

Nur Najwa Umirah Binti Nor Azman College of Engineering, Universiti Tenaga Nasional, Kajang, Selangor, Malaysia

Zuraihana Bachok School of Mechanical Engineering, Universiti Sains Malaysia, Nibong Tebal, Penang, Malaysia

Nur Hanis Badaruddin Faculty of Applied Sciences, Universiti Teknologi Mara, Shah Alam, Malaysia

Muhammad Hazirul Bin Badrul School of Electrical Engineering, College of Engineering, Universiti Teknologi MARA, Shah Alam, Selangor, Malaysia

Sharifah Shahnaz Binti Syed Bakar Faculty of Chemical Engineering Technology (FKTK), Universiti Malaysia Perlis (UniMAP), Jejawi, Perlis, Malaysia

Muhammad Fadlin Hazim Baser Faculty of Chemical Engineering Technology (FTKK), Universiti Malaysia Perlis (UniMAP), Jejawi, Arau, Perlis, Malaysia

N. N. A. Basir Material Science and Technology Program, Faculty of Applied Sciences, Universiti Teknologi MARA, Shah Alam, Selangor, Malaysia

Erwan Basiron Western Digital®, SanDisk Storage Malaysia Sdn.Bhd., Penang, Malaysia

N. A. N. Basri College of Graduate Studies, Universiti Tenaga Nasional, Kajang, Selangor Darul Ehsan, Malaysia;
School College of Graduate Studies, Universiti Tenaga Nasional, Jalan IKRAM-UNITEN, Kajang, Selangor Darul Ehsan, Malaysia

S. Beddu Department of Civil Engineering, Universiti Tenaga Nasional, Jalan Ikram-Uniten, Kajang, Selangor Darul Ehsan, Malaysia

Iszmir Nazmi bin Ismail College of Engineering, Universiti Tenaga Nasional, Kajang, Selangor, Malaysia

Ammar Athallah Budiarto Faculty of Chemical Engineering and Technology, Universiti Malaysia Perlis (UniMAP), Kangar, Perlis, Malaysia

Xuan Viet Cao Department of Polymer Materials, Faculty of Materials Technology, Ho Chi Minh City University of Technology (HCMUT), Ho Chi Minh City, Vietnam;
Vietnam National University Ho Chi Minh City (VNUHCM), Ho Chi Minh City, Vietnam

C. W. Chai Faculty of Mechanical Engineering Technology, Universiti Malaysia Perlis, Arau, Perlis, Malaysia

Umer Masood Chaudry Department of Mechanical Engineering, Incheon National University, Incheon, Republic of Korea

Zuraidawani Che Daud Faculty of Mechanical Engineering Technology, Universiti Malaysia Perlis, Kampus Alam UniMAP, Pauh Putra, Arau, Perlis, Malaysia

Dewi Suriyani Che Halin Faculty of Chemical Engineering Technology, Universiti Malaysia Perlis, Arau, Perlis, Malaysia

Faizul Che Pa Center of Excellence Geopolymer and Green Technology (CEGeoGTech), Universiti Malaysia Perlis, Jejawi, Arau, Perlis, Malaysia;
Faculty of Chemical Engineering and Technology, Universiti Malaysia Perlis, Jejawi, Arau, Perlis, Malaysia

Che Wan Nur Zulaikha Che Wan Adnan School of Civil Engineering, College of Engineering, Universiti Teknologi MARA, Shah Alam, Selangor, Malaysia

T. W. Cheng Institute of Mineral Resources Engineering, National Taipei University of Technology, Taipei, Taiwan, R.O.C.

Norhasliya Mohd Daud Faculty of Engineering, Universiti Pertahanan Nasional Malaysia, Kuala Lumpur, Malaysia

T. T. Dele-Afolabi Department of Mechanical Engineering, Faculty of Engineering, Ajayi Crowther University, Oyo, Oyo State, Nigeria

Mohd Nazree Derman Faculty of Mechanical Engineering Technology, Universiti Malaysia Perlis, Kampus Alam UniMAP, Pauh Putra, Arau, Perlis, Malaysia

Kieu Do Trung Kien Faculty of Materials Technology, Ho Chi Minh City University of Technology (HCMUT), Ho Chi Minh City, Vietnam;
Vietnam National University Ho Chi Minh City, Linh Trung Ward, Ho Chi Minh City, Vietnam

Patrik Dobrovszky Faculty of Materials Science and Technology in Trnava, Slovak University of Technology in Bratislava, Trnava, Slovakia

Abang Annuar Ehsan Institute of Microengineering and Nanoelectronics, Universiti Kebangsaan Malaysia, Bangi, Selangor, Malaysia

Januarti Jaya Ekaputri Department of Civil Engineering, Institut Teknologi Sepuluh Nopember, Surabaya, Indonesia;
Konsorsium Riset Geopolimer Indonesia, NASDEC, Kampus ITS, Sukolilo, Surabaya, Indonesia;
Department of Civil Engineering, Institut Teknologi Sepuluh Nopember, Kota SBY, Indonesia

M. R. Eqwan College of Engineering, Universiti Tenaga Nasional, Kajang, Selangor, Malaysia

F. Ezzah Department of Chemical and Environmental Engineering, Malaysia-Japan International Institute of Technology (MJIIT), Universiti Teknologi Malaysia, Kuala Lumpur, Malaysia

Nur Fatihah Mohd Fadzil Faculty of Chemical Engineering and Technology, Universiti Malaysia Perlis (UniMAP), Arau, Malaysia

Ameeq Farooq Department of Metallurgy and Materials Engineering, Corrosion Control Research Cell, University of the Punjab, Lahore, Pakistan

M. Fathullah School of Manufacturing Engineering, Universiti Malaysia Perlis (UniMAP), Arau, Perlis, Malaysia;
Center of Excellence Geopolymer and Green Technology (CEGeoGTech), Universiti Malaysia Perlis (UniMAP), Kangar, Perlis, Malaysia

Mohd Sharin Ghani Fakulti Kejuruteraan Elektrik, Universiti Teknikal Malaysia Melaka, Melaka, Malaysia

M. J. Ghazali Department of Mechanical and Manufacturing Engineering, Faculty of Engineering and Built Environment, Universiti Kebangsaan Malaysia, Bangi, Selangor, Malaysia

T. L. Goh Department of Earth Sciences and Environment, Faculty of Science and Technology, Universiti Kebangsaan Malaysia, Bangi, Selangor Darul Ehsan, Malaysia

Mahalaksmi Gunasilan Faculty of Mechanical Engineering and Technology, Universiti Malaysia Perlis, Arau, Perlis, Malaysia

Khairul Anwar Abdul Halim Faculty of Chemical Engineering and Technology, Universiti Malaysia Perlis (UniMAP), Arau, Perlis, Malaysia;
Biomedical and Nanotechnology Research Group, Center of Excellence Geopolymer and Green Technology (CEGeoGTech), Universiti Malaysia Perlis (UniMAP), Arau, Perlis, Malaysia

Kotiba Hamad School of Advanced Materials Science and Engineering, Sungkyunkwan University, Suwon, Republic of Korea

Esah Hamzah Faculty of Mechanical Engineering, Universiti Teknologi Malaysia (UTM), Skudai, Johor, Malaysia

Andrie Harmaji Department of Metallurgical Engineering, Institut Teknologi Sains Bandung, Kabupaten Bekasi, Indonesia;
Konsorsium Riset Geopolimer Indonesia, NASDEC, Kampus ITS, Sukolilo, Surabaya, Indonesia

Nur Zalifah Binti Hasmadi Faculty of Chemical Engineering and Technology, Universiti Malaysia Perlis (UniMAP), Arau, Malaysia

Se-yoon Hong Plant Engineering Center, Institute for Advanced Engineering, Cheoin-Gu, Yongin-Si, Gyeonggi-Do, Korea

Ho Cheng How Department of Mechanical, Materials and Manufacturing Engineering, The University of Nottingham Malaysia, Semenyih, Selangor Darul Ehsan, Malaysia

Jakub Hrbal Faculty of Materials Science and Technology in Trnava, Slovak University of Technology in Bratislava, Trnava, Slovakia

C. Y. Huang Institute of Mineral Resources Engineering, National Taipei University of Technology, Taipei, Taiwan, R.O.C.

Ly Cam Hung Hochiminh City University of Natural Resources and Environment, Ho Chi Minh City, Vietnam

M. I. Hussain Faculty of Mechanical Engineering Technology, Universiti Malaysia Perlis, Arau, Perlis, Malaysia

Mohamed Saiful Firdaus Hussin Fakulti Teknologi Kejuruteraan Mekanikal dan Pembuatan, Universiti Teknikal Malaysia Melaka, Melaka, Malaysia

Nur Hazwani Hussin Faculty of Engineering and Built Environment, Universiti Sains Islam Malaysia, Nilai, Negeri Sembilan, Malaysia

Dai Phu Huynh Department of Polymer Materials, Faculty of Materials Technology, Ho Chi Minh City University of Technology (HCMUT), Ho Chi Minh City, Vietnam;
Vietnam National University Ho Chi Minh City (VNUHCM), Ho Chi Minh City, Vietnam;
Polymer Research Center, Ho Chi Minh City University of Technology (HCMUT), Ho Chi Minh City, Vietnam

Nguyen Ngoc Tri Huynh Department of Silicate Materials, Faculty of Materials Technology, Ho Chi Minh City University of Technology (HCMUT), Ho Chi Minh City, Vietnam;
Vietnam National University Ho Chi Minh City (VNU HCM), Linh Trung Ward, Ho Chi Minh City, Vietnam

Thi Ngoc Diem Huynh Department of Polymer Materials, Faculty of Materials Technology, Ho Chi Minh City University of Technology (HCMUT), Ho Chi Minh City, Vietnam;
Vietnam National University Ho Chi Minh City (VNUHCM), Ho Chi Minh City, Vietnam

Azmi Ibrahim School of Civil Engineering, College of Engineering, Universiti Teknologi MARA (UiTM) Shah Alam, Shah Alam, Selangor, Malaysia

I. Ibrahim School of Civil Engineering, Faculty of Engineering, Universiti Teknologi Malaysia, Johor Bahru, Malaysia

Lokman Hakim Ibrahim Faculty of Chemical Engineering and Technology, Universiti Malaysia Perlis (UniMAP), Arau, Malaysia;
Advanced Polymer Research Group, Center of Excellence Geopolymer and Green Technology (CEGeoGTech), Universiti Malaysia Perlis (UniMAP), Arau, Malaysia

W. M. W. Ibrahim Faculty of Mechanical Engineering Technology, Universiti Malaysia Perlis, Arau, Perlis, Malaysia

Azlina Idris School of Electrical Engineering, College of Engineering, Universiti Teknologi MARA, Shah Alam, Selangor, Malaysia

M. R. Isa Department of Mechanical Engineering, College of Engineering, Univerisiti Tenaga Nasional, Jalan IKRAM-UNITEN, Kajang, Selangor, Malaysia

M. Ishak Faculty of Mechanical and Automotive Engineering Technology, Universiti Malaysia Pahang, Pekan, Pahang Darul Makmur, Malaysia

M. H. H. Ishak School of Aerospace Engineering, Universiti Sains Malaysia, Nibong Tebal, Seberang Perai Selatan, Penang, Malaysia

Mohd Fakhrurrazi Ishak Centre for Sustainability of Ecosystem & Earth Resources (Earth Centre), Universiti Malaysia Pahang, Kuantan, Malaysia

Adlil Aizat Ismail Western Digital®, SanDisk Storage Malaysia Sdn.Bhd., Penang, Malaysia;
Institute of Microengineering and Nanoelectronics, Universiti Kebangsaan Malaysia, Bangi, Selangor, Malaysia

M. S. Ismail Faculty of Mechanical Engineering Technology, Universiti Malaysia Perlis, Arau, Perlis, Malaysia

Norshafarina Ismail Radiation Processing Technology Division, Malaysian Nuclear Agency, Dengkil, Selangor, Malaysia

Z. Itam Department of Civil Engineering, Universiti Tenaga Nasional, Jalan Ikram-Uniten, Kajang, Selangor Darul Ehsan, Malaysia

M. F. Jaafar College of Engineering, Universiti Tenaga Nasional, Kajang, Selangor, Malaysia

Azman Jalar Institute of Microengineering and Nanoelectronics (IMEN), Level 4 Research Complex, Universiti Kebangsaan Malaysia, Bangi, Selangor, Malaysia;
Department of Applied Physics, Faculty of Science and Technology, Universiti Kebangsaan Malaysia, Bangi, Selangor, Malaysia

A. A. A. Jaludin School of Aerospace Engineering, Universiti Sains Malaysia, Nibong Tebal, Seberang Perai Selatan, Penang, Malaysia

N. H. Jamadon Department of Mechanical and Manufacturing Engineering, Faculty of Engineering and Built Environment, Universiti Kebangsaan Malaysia, Bangi, Selangor, Malaysia

S. Janasekaran Faculty of Engineering, Built Environment & Information Technology, Center for Advanced Materials and Intelligent Manufacturing, SEGi University, Petaling Jaya, Selangor, Malaysia

N. F. Jasmy EMZI-UiTM Nanoparticles Colloids and Interface Industrial Research Laboratory (NANO- CORE), Centre for Chemical Engineering Studies, Universiti Teknologi MARA, Pulau Pinang, Malaysia

Ramadhansyah Putra Jaya Department of Geotechnics and Infrastructure, Faculty of Civil Engineering and Earth Resources, Universiti Malaysia Pahang, Pahang, Gambang, Malaysia

Banjuraizah Johar Faculty of Chemical Engineering Technology, Universiti Malaysia Perlis, Kangar, Perlis, Malaysia

Iman Nur Sazniza Johari Faculty of Mechanical and Manufacturing Engineering, Universiti Tun Hussein Onn Malaysia, Johor, Malaysia

Ridhwan Jumaidin Fakulti Teknologi Kejuruteraan Mekanikal and Pembuatan, Universiti Teknikal Malaysia Melaka, Melaka, Malaysia

Tea-Sung Jun Department of Mechanical Engineering, Incheon National University, Incheon, Republic of Korea

Hwansoo Jung Industry-Academic Convergence Campus, Hanbat National University, Yuseong-Gu, Daejeon, Korea

M. N. Jusoh School of Civil Engineering, Faculty of Engineering, Universiti Teknologi Malaysia, Johor Bahru, Malaysia

Ros Shilawani S. Abdul Kadir School of Electrical Engineering, College of Engineering, Universiti Teknologi MARA, Shah Alam, Selangor, Malaysia

N. M. Kamal Department of Civil Engineering, Universiti Tenaga Nasional, Kajang, Selangor Darul Ehsan, Malaysia

Mohamad Anuar Kamaruddin Environmental Technology Division, School of Industrial Technology, Universiti Sains Malaysia, Gelugor, Penang, Malaysia;
Center of Excellence Geopolymer and Green Technology, Universiti Malaysia Perlis, Arau, Perlis, Malaysia

Roslan Kamaruddin School of Mechanical Engineering, Universiti Sains Malaysia, Nibong Tebal, Seberang Perai Selatan, Penang, Malaysia

Kamarulzaman Kamarudin School of Mechatronic Engineering, Universiti Malaysia Perlis, Arau, Perlis, Malaysia

K. Khairudin EMZI-UiTM Nanoparticles Colloids and Interface Industrial Research Laboratory (NANO-CORE), Centre for Chemical Engineering Studies, Universiti Teknologi MARA, Permatang Pauh, Pulau Pinang, Malaysia

M. Khalid Graphene and Advanced 2D Materials Research Group (GAMRG), School of Engineering and Technology, Sunway University, Petaling Jaya, Selangor, Malaysia

Shing Fhan Khor Faculty of Electrical Engineering Technology, Universiti Malaysia Perlis, Arau, Perlis, Malaysia

Pham Trung Kien Faculty of Materials Technology, Ho Chi Minh City University of Technology (HCMUT), Ho Chi Minh City, Vietnam;
Vietnam National University Ho Chi Minh City, Linh Trung Ward, Thu Duc District, Ho Chi Minh City, Vietnam;
Polymer Research Center, Ho Chi Minh City University of Technology (HCMUT), Ho Chi Minh City, Vietnam

Marcel Kuruc Faculty of Materials Science and Technology in Trnava, Slovak University of Technology in Bratislava, Trnava, Slovakia

Thi Thai Ha La Department of Polymer Materials, Faculty of Materials Technology, Ho Chi Minh City University of Technology (HCMUT), Ho Chi Minh City, Vietnam;
Vietnam National University Ho Chi Minh City (VNUHCM), Ho Chi Minh City, Vietnam

Elsee Layu Fakulti Teknologi Kejuruteraan Mekanikal and Pembuatan, Universiti Teknikal Malaysia Melaka, Melaka, Malaysia

Jing Rou Lee School of Mechanical Engineering, Universiti Sains Malaysia, Nibong Tebal, Seberang Perai Selatan, Penang, Malaysia

W. H. Lee Institute of Mineral Resources Engineering, National Taipei University of Technology, Taipei, Taiwan, R.O.C.

Wai Keong Leong Center of Excellence Geopolymer and Green Technology (CEGeoGTech), Universiti Malaysia Perlis (UniMAP), Arau, Perlis, Malaysia

Oh Chai Lian School of Civil Engineering, College of Engineering, Universiti Teknologi MARA (UiTM) Shah Alam, Shah Alam, Selangor, Malaysia

C. M. Low Faculty of Chemical Engineering Technology, Universiti Malaysia Perlis (UniMAP), Arau, Perlis, Malaysia

M. M. Mahat Material Science and Technology Program, Faculty of Applied Sciences, Universiti Teknologi MARA, Shah Alam, Selangor, Malaysia

E. M. Mahdi Materials Technology Group, Industrial Technology Division, Malaysia Nuclear Agency, Kajang, Selangor, Malaysia

T. S. A. Manan Institute of Tropical Biodiversity and Sustainable Development, Universiti Malaysia Terengganu (UMT), Terengganu Darul Iman, Malaysia

Sajjad Ali Mangi Department of Civil Engineering, Mehran University of Engineering Technology, SZAB Campus, Sindh, Pakistan

M. H. Mansor College of Engineering, Universiti Tenaga Nasional, Kajang, Selangor, Malaysia

Muhammad Naufal Mansor Faculty of Electrical Engineering Techology, University Malaysia Perlis, Arau, Malaysia

Yuganantheni K. Marippan Environmental Technology Division, School of Industrial Technology, Universiti Sains Malaysia, Gelugor, Penang, Malaysia

M. I. F. Maruzuki Centre for Electrical Engineering Studies, Cawangan Pulau Pinang, Universiti Teknologi MARA, Pulau Pinang, Malaysia

Hazizan Md. Akil School of Materials and Mineral Resources Engineering, Universiti Sains Malaysia, Nibong Tebal, Pulau Pinang, Malaysia

Lim Jia Mei Environmental Technology Division, School of Industrial Technology, Universiti Sains Malaysia, Gelugor, Penang, Malaysia

Jan Milde Faculty of Materials Science and Technology in Trnava, Slovak University of Technology in Bratislava, Trnava, Slovakia

Do Quang Minh Faculty of Materials Technology, Ho Chi Minh City University of Technology (HCMUT), Ho Chi Minh City, Vietnam;

Vietnam National University Ho Chi Minh City, Linh Trung Ward, Ho Chi Minh City, Vietnam

Amiruddin Mishad School of Civil Engineering, College of Engineering, Universiti Teknologi MARA (UiTM) Shah Alam, Shah Alam, Selangor, Malaysia

Noor Aina Misnon Faculty of Engineering, Universiti Pertahanan Nasional Malaysia, Kuala Lumpur, Malaysia

Daud Mohamad Department of Civil Engineering, Universiti Tenaga Nasional, Jalan IKRAM-UNITEN, Kajang, Selangor Darul Ehsan, Malaysia

Noraiham Mohamad Fakulti Kejuruteraan Pembuatan, Universiti Teknikal Malaysia Melaka, Melaka, Malaysia

Hassan Mohamed College of Engineering, Universiti Tenaga Nasional, Kajang, Selangor, Malaysia

Ahmad Azmin Mohammad Advance Soldering Material Group, School of Materials and Mineral Resources Engineering, Universiti Sains Malaysia (USM), Nibong Tebal, Pulau Pinang, Malaysia

Nurul Nazihah Mohamad Nasir Faculty of Chemical Engineering and Technology, Universiti Malaysia Perlis, Jejawi, Arau, Perlis, Malaysia;
Center of Excellence Geopolymer and Green Technology (CEGeoGTech), Universiti Malaysia Perlis, Jejawi, Arau, Perlis, Malaysia

Jariah Mohamad Juoi Fakulti Kejuruteraan Pembuatan, Universiti Teknikal Malaysia Melaka, Melaka, Malaysia

Muhammad Faheem Mohamad Tahir Center of Excellence Geopolymer and Green Technology (CEGeoGTech), Universiti Malaysia Perlis, Jejawi, Arau, Perlis, Malaysia;
Faculty of Chemical Engineering and Technology, Universiti Malaysia Perlis, Jejawi, Arau, Perlis, Malaysia

Nur Syahirah Mohamad Zaimi Center of Excellence Geopolymer and Green Technology (CEGeoGTech), Universiti Malaysia Perlis, Jejawi, Arau, Perlis, Malaysia;
Faculty of Chemical Engineering Technology (FTKK), Universiti Malaysia Perlis (UniMAP), Jejawi, Arau, Perlis, Malaysia

Mohd Amran Mohd Ali Fakulti Kejuruteraan Pembuatan, Universiti Teknikal Malaysia Melaka, Melaka, Malaysia

Nur Majidah Mohd Asri Faculty of Chemical Engineering Technology, Universiti Malaysia Perlis, Jejawi, Arau, Perlis, Malaysia

Mohd Hisbany Mohd Hashim School of Civil Engineering, College of Engineering, Universiti Teknologi MARA (UiTM) Shah Alam, Shah Alam, Selangor, Malaysia

Mohd Hanafiah Mohd Isa Fakulti Kejuruteraan Pembuatan, Universiti Teknikal Malaysia Melaka, Melaka, Malaysia

Nurul Aida Husna Mohd Mahayuddin Faculty of Chemical Engineering and Technology, Universiti Malaysia Perlis (UniMAP), Arau, Perlis, Malaysia

N. B. Mohd Nazeri EMZI-UiTM Nanoparticles Colloids and Interface Industrial Research Laboratory (NANO- CORE), Centre for Chemical Engineering Studies, Universiti Teknologi MARA, Pulau Pinang, Malaysia

Zulkifli Mohd Rosli Fakulti Teknologi Kejuruteraan Mekanikal and Pembuatan, Universiti Teknikal Malaysia Melaka, Melaka, Malaysia

Mohd Arif Anuar Mohd Salleh Center of Excellence Geopolymer & Green Technology (CeGeoGTech), Universiti Malaysia Perlis (UniMAP), Arau, Perlis, Malaysia;
Faculty of Chemical Engineering and Technology, Universiti Malaysia Perlis (UniMAP), Arau, Perlis, Malaysia

Fateh Amera Mohd Yusoff Institute of Microengineering and Nanoelectronics (IMEN), Level 4 Research Complex, Universiti Kebangsaan Malaysia, Bangi, Selangor, Malaysia

D. Mudjono Consultant of Pavement/Material Engineering, PT Aria Jasa Reksatama, Surabaya, Indonesia

Nurmunira Muhammad Faculty of Civil Engineering Technology, Universiti Malaysia Pahang, Kuantan, Malaysia

Nurshamimie Muhammad Fauzi School of Civil Engineering, College of Engineering, Engineering Complex, Tunku Abdul Halim Muadzam Shah, Universiti Teknologi MARA, Shah Alam, Selangor, Malaysia

Miradatul Najwa Muhd Rodhi School of Chemical Engineering, College of Engineering, Universiti Teknologi MARA, Shah Alam, Selangor, Malaysia

Mazidah Mukri School of Civil Engineering, College of Engineering, Engineering Complex, Tunku Abdul Halim Muadzam Shah, Universiti Teknologi MARA, Shah Alam, Selangor, Malaysia

Kaalidass Muniary School of Mechanical Engineering, Universiti Sains Malaysia, Penang, Malaysia

Logendran Murgaya Faculty of Mechanical and Manufacturing Engineering, Universiti Tun Hussein Onn Malaysia, Johor, Malaysia

Nuzaimah Mustafa Fakulti Teknologi Kejuruteraan Mekanikal and Pembuatan, Universiti Teknikal Malaysia Melaka, Melaka, Malaysia

Azrul Affandhi Musthaffa Stare Resources Sdn. Bhd., Kuala Lumpur, Malaysia

Nani Fadzlina Naim School of Electrical Engineering, College of Engineering, Universiti Teknologi MARA, Shah Alam, Selangor, Malaysia

Raihanah Muhammad Navil Department of Mechanical Engineering, College of Engineering, Univerisiti Tenaga Nasional, Jalan IKRAM-UNITEN, Kajang, Selangor, Malaysia

Muhammad Firdaus Mohd Nazeri Center of Excellence Geopolymer and Green Technology (CEGeoGTech), Universiti Malaysia Perlis (UniMAP), Arau, Perlis, Malaysia

N. B. M. Nazeri EMZI-UiTM Nanoparticles Colloids and Interface Industrial Research Laboratory (NANO-CORE), Centre for Chemical Engineering Studies, Universiti Teknologi MARA, Pulau Pinang, Malaysia

Nguyen Vu Uyen Nhi Faculty of Materials Technology, Ho Chi Minh City University of Technology (HCMUT), Ho Chi Minh City, Vietnam;
Vietnam National University Ho Chi Minh City, Linh Trung Ward, Ho Chi Minh City, Vietnam

M. I. Nizam EMZI-UiTM Nanoparticles Colloids and Interface Industrial Research Laboratory (NANO-CORE), Centre for Chemical Engineering Studies, Universiti Teknologi MARA, Pulau Pinang, Malaysia

D. S. Nor Syamsudin EMZI-UiTM Nanoparticles Colloids and Interface Industrial Research Laboratory (NANO- CORE), Centre for Chemical Engineering Studies, Universiti Teknologi MARA, Pulau Pinang, Malaysia

Norazman Mohamad Nor Faculty of Engineering, Universiti Pertahanan Nasional Malaysia, Kuala Lumpur, Malaysia

Razif Muhammed Nordin Department of Chemistry, Faculty of Applied Sciences, Perlis Branch Arau Campus, Universiti Teknologi MARA, Arau, Perlis, Malaysia;
Green and Functional Polymer Research Group, Universiti Teknologi MARA, Shah Alam, Selangor, Malaysia

Noor Izza Farisya Noor Hamdan Faculty of Mechanical and Manufacturing Engineering, Universiti Tun Hussein Onn Malaysia, Johor, Malaysia

Norlin Nosbi Department of Mechanical Engineering, Centre for Corrosion Research (CCR), Institute of Contaminant Management for Oil and Gas (ICM), Universiti Teknologi PETRONAS, Perak, Seri Iskandar, Malaysia

A. Numan Graphene and Advanced 2D Materials Research Group (GAMRG), School of Engineering and Technology, Sunway University, Petaling Jaya, Selangor, Malaysia

Mohd Firdaus Omar Faculty of Chemical Engineering Technology, Universiti Malaysia Perlis, Arau, Perlis, Malaysia;
Geopolymer and Green Technology, Centre of Excellent (CEGeoGTech), Universiti Malaysia Perlis, Arau, Perlis, Malaysia

M. S. Osman EMZI-UiTM Nanoparticles Colloids and Interface Industrial Research Laboratory (NANO-CORE), Centre for Chemical Engineering Studies, Universiti Teknologi MARA, Permatang Pauh, Pulau Pinang, Malaysia;
Center of Excellence Geopolymer and Green Technology (CEGeoGTech), Universiti Malaysia Perlis (UniMAP), Kangar, Perlis, Malaysia

S. Osman Material Science and Technology Program, Faculty of Applied Sciences, Universiti Teknologi MARA, Shah Alam, Selangor, Malaysia

S. H. Osman Fuel Cell Institute, Universiti Kebangsaan Malaysia, Bangi, Malaysia

Saliza Azlina Osman Faculty of Mechanical and Manufacturing Engineering, Universiti Tun Hussein Onn Malaysia, Johor, Malaysia

Fatin Fatini Othman Faculty of Chemical Engineering Technology, Universiti Malaysia Perlis, Kangar, Perlis, Malaysia

Noraziana Parimin Faculty of Chemical Engineering and Technology, Universiti Malaysia Perlis (UniMAP), Arau, Perlis, Malaysia

Jozef Peterka Faculty of Materials Science and Technology in Trnava, Slovak University of Technology in Bratislava, Trnava, Slovakia

Quoc Phu Phan Department of Polymer Materials, Faculty of Materials Technology, Ho Chi Minh City University of Technology (HCMUT), Ho Chi Minh City, Vietnam;
Vietnam National University Ho Chi Minh City (VNUHCM), Ho Chi Minh City, Vietnam

Nguyen Minh Phong Faculty of Materials Technology, Ho Chi Minh City University of Technology (HCMUT), Ho Chi Minh City, Vietnam;
Vietnam National University Ho Chi Minh City, Linh Trung Ward, Thu Duc District, Ho Chi Minh City, Vietnam

Huynh Dai Phu Faculty of Materials Technology, Ho Chi Minh City University of Technology (HCMUT), Ho Chi Minh City, Vietnam;
Vietnam National University Ho Chi Minh City, Linh Trung Ward, Thu Duc District, Ho Chi Minh City, Vietnam;
Polymer Research Center, Ho Chi Minh City University of Technology (HCMUT), Ho Chi Minh City, Vietnam

Nguyen Huu Huy Phuc Faculty of Materials Technology, Ho Chi Minh City University of Technology (HCMUT), Ho Chi Minh City, Vietnam;
Vietnam National University, Ho Chi Minh City, Vietnam

Tran Huu Phuc Faculty of Materials Technology, Ho Chi Minh City University of Technology (HCMUT), Ho Chi Minh City, Vietnam;
Vietnam National University Ho Chi Minh City, Linh Trung Ward, Ho Chi Minh City, Vietnam

Le Thi Kim Phung Vietnam National University Ho Chi Minh City, Ho Chi Minh City, Vietnam;
Faculty of Chemical Engineering, Ho Chi Minh City University of Technology (HCMUT), Ho Chi Minh City, Vietnam

Nguyen Vinh Phuoc SANOMAT Co., Ltd, Dong Nai Province, Vietnam

Tran Ngo Quan Faculty of Materials Technology, Ho Chi Minh City University of Technology (HCMUT), Ho Chi Minh City, Vietnam;
Vietnam National University Ho Chi Minh City, Linh Trung Ward, Thu Duc District, Ho Chi Minh City, Vietnam

A. Rahim School of Civil Engineering, Faculty of Engineering, Universiti Teknologi Malaysia, Johor Bahru, Malaysia

Nor Hidayah Rahim Fakulti Kejuruteraan Elektrik, Universiti Teknikal Malaysia Melaka, Melaka, Malaysia

Nur Liza Rahim Faculty of Civil Engineering and Technology, Universiti Malaysia Perlis, Jejawi, Arau, Perlis, Malaysia

Shayfull Zamree Abd Rahim Geopolymer and Green Technology, Centre of Excellent (CEGeoGTech), Universiti Malaysia Perlis, Arau, Perlis, Malaysia

A. S. A. Rahman School of Civil Engineering, Universiti Teknologi MARA, Shah Alam, Selangor, Malaysia

M. S. Abd Rahman College of Engineering, Universiti Tenaga Nasional, Kajang, Selangor, Malaysia

Shahnurriman Abdul Rahman Faculty of Engineering and Built Environment, Universiti Sains Islam Malaysia, Nilai, Negeri Sembilan, Malaysia

N. A. Rahmat College of Engineering, Universiti Tenaga Nasional, Kajang, Selangor, Malaysia

Norhasyimi Rahmat School of Chemical Engineering, College of Engineering, Universiti Teknologi MARA, Shah Alam, Selangor, Malaysia

Mohd Raizamzamani School of Civil Engineering, College of Engineering, Universiti Teknologi MARA (UiTM) Shah Alam, Shah Alam, Selangor, Malaysia

Raja Izamshah Raja Abdullah Fakulti Kejuruteraan Pembuatan, Universiti Teknikal Malaysia Melaka, Melaka, Malaysia

H. Ramachenran Department of Civil Engineering, Universiti Tenaga Nasional, Jalan IKRAM-UNITEN, Kajang, Selangor Darul Ehsan, Malaysia

K. Ramanathan School of Civil Engineering, Faculty of Engineering, Universiti Teknologi Malaysia, Johor Bahru, Malaysia

Norhidayu Rameli Faculty of Engineering and Built Environment, Universiti Sains Islam Malaysia, Nilai, Negeri Sembilan, Malaysia

M. Z. Ramli College of Engineering, Universiti Tenaga Nasional, Kajang, Selangor, Malaysia

Mohd Izrul Izwan Ramli Faculty of Chemical Engineering Technology (FTKK), Universiti Malaysia Perlis (UniMAP), Jejawi, Arau, Perlis, Malaysia;
Center of Excellence Geopolymer and Green Technology (CEGeoGTech), Universiti Malaysia Perlis, Jejawi, Arau, Perlis, Malaysia

R. Rangappa Department of Mechanical Engineering, BMS Institute of Technology and Management, Bengaluru, Visvesvaraya Technological University, Belagavi, India

A. K. Rasheed Department of New Energy Science and Engineering, School of Energy and Chemical Engineering, Xiamen University Malaysia (XMUM), Sepang, Selangor, Malaysia

A. R. Razali Centre for Electrical Engineering Studies, Cawangan Pulau Pinang, Universiti Teknologi MARA, Permatang Pauh, Pulau Pinang, Malaysia

N. F. M. Razali Material Science and Technology Program, Faculty of Applied Sciences, Universiti Teknologi MARA, Shah Alam, Selangor, Malaysia

Nik AKmar Rejab School of Materials and Mineral Resources Engineering, Universiti Sains Malaysia, Nibong Tebal, Penang, Malaysia

Siti Faqihah Roduan Faculty of Chemical Engineering and Technology, Universiti Malaysia Perlis (UniMAP), Arau, Perlis, Malaysia

Muhammad Faiz Ridhwan Rosli School of Mechanical Engineering, Universiti Sains Malaysia, Nibong Tebal, Seberang Perai Selatan, Penang, Malaysia

Nursyadzatul Tasnim Roslin College of Engineering, Universiti Tenaga Nasional, Kajang, Selangor, Malaysia

R. Rosmamuhamadani Material Science and Technology Program, Faculty of Applied Sciences, Universiti Teknologi MARA, Shah Alam, Selangor, Malaysia

Mohd Syakirin Rusdi School of Mechanical Engineering, Universiti Sains Malaysia, Nibong Tebal, Seberang Perai Selatan, Penang, Malaysia;
School of Aerospace Engineering, Universiti Sains Malaysia, Nibong Tebal, Seberang Perai Selatan, Penang, Malaysia

Mohd Syakirin Bin Rusdi School of Mechanical Engineering, Universiti Sains Malaysia, Nibong Tebal, Seberang Perai Selatan, Penang, Malaysia

Abdullah Aziz Saad School of Mechanical Engineering, Universiti Sains Malaysia, Nibong Tebal, Penang, Malaysia

Shahrizam Saad Faculty of Chemical Engineering and Technology, Universiti Malaysia Perlis (UniMAP), Kangar, Perlis, Malaysia

Ain Najihah Saim Faculty of Chemical Engineering Technology (FTKK), Universiti Malaysia Perlis (UniMAP), Jejawi, Arau, Perlis, Malaysia

Siti Khadijah Salihin Materials Engineering Programme, Faculty of Chemical Engineering Technology, Universiti Malaysia Perlis, Jejawi, Arau, Perlis, Malaysia

S. H. Salleh Faculty of Chemical Engineering and Technology, Universiti Malaysia Perlis, Arau, Perlis, Malaysia;
Centre of Excellence Geopolymer and Green Technology, Universiti Malaysia Perlis, Arau, Perlis, Malaysia

Suzi Seroja Sarnin School of Electrical Engineering, College of Engineering, Universiti Teknologi MARA, Shah Alam, Selangor, Malaysia

N. Saud Faculty of Chemical Engineering Technology, Universiti Malaysia Perlis (UniMAP), Arau, Perlis, Malaysia;
Centre of Excellent On Geopolymer and Green Technology (CEGeoGTech), Universiti Malaysia Perlis, Arau, Perlis, Malaysia

Wan Hafizah Mohd Saufee Center of Excellence Geopolymer and Green Technology (CEGeoGTech), Universiti Malaysia Perlis (UniMAP), Arau, Perlis, Malaysia

A. Z. U. Saufie Faculty of Computer and Mathematical Sciences, Universiti Teknologi MARA, Shah Alam, Selangor, Malaysia

Sian Meng Se San Miguel Yamamura Plastic Films Sdn Bhd, Ayer Keroh, Melaka, Malaysia

S. F. Senin Centre for Civil Engineering Studies, Cawangan Pulau Pinang, Universiti Teknologi MARA, Pulau Pinang, Malaysia

B. H. Setiadji Doctoral Program of Civil Engineering, Faculty Engineering, Diponegoro University, Semarang, Indonesia

S. Setumin Centre for Electrical Engineering Studies, Cawangan Pulau Pinang, Universiti Teknologi MARA, Pulau Pinang, Malaysia

A. I. M. Shaiful Faculty of Mechanical Engineering Technology, Universiti Malaysia Perlis, Arau, Perlis, Malaysia

S. R. Shamsudin Faculty of Mechanical Engineering and Technology, Universiti Malaysia Perlis, Arau, Perlis, Malaysia

Muhamad Syaqir Sha'rani Faculty of Engineering, Universiti Pertahanan Nasional Malaysia, Kuala Lumpur, Malaysia

Goh Wei Shing School of Mechanical Engineering, Universiti Sains Malaysia, Nibong Tebal, Seberang Perai Selatan, Penang, Malaysia

Mohammed Iqbal Shueb Radiation Processing Technology Division, Malaysian Nuclear Agency, Dengkil, Selangor, Malaysia

Norbaya Sidek School of Civil Engineering, College of Engineering, Universiti Teknologi MARA, Shah Alam, Selangor, Malaysia

A. Situmorang Departement of Civil Engineering, Faculty of Engineering, Semarang University, Semarang, Indonesia

F. Somidin Centre of Excellence Geopolymer and Green Technology, Faculty of Chemical Engineering Technology, Universiti Malaysia Perlis (UniMAP), Jejawi, Arau, Perlis, Malaysia

Nguyen Khanh Son Department of Silicate Materials, Faculty of Materials Technology, Ho Chi Minh City University of Technology (HCMUT), Ho Chi Minh City, Vietnam;
Vietnam National University Ho Chi Minh City (VNU HCM), Linh Trung Ward, Ho Chi Minh City, Vietnam

M. S. M. Sufian Department of Civil Engineering, Universiti Tenaga Nasional, Jalan Ikram-Uniten, Kajang, Selangor Darul Ehsan, Malaysia

P. Sukhvinder Department of Civil Engineering, Universiti Tenaga Nasional, Kajang, Selangor Darul Ehsan, Malaysia

Ho Jin Sung Plant Engineering Center, Institute for Advanced Engineering, Cheoin-Gu, Yongin-Si, Gyeonggi-Do, Korea

M. Syafiq Department of Mechanical and Manufacturing Engineering, Faculty of Engineering and Built Environment, Universiti Kebangsaan Malaysia, Bangi, Selangor, Malaysia;
Jabatan Kejuruteraan Mekanikal, Politeknik Banting Selangor, Banting, Selangor, Malaysia

A. Syahmi Department of Mechanical and Manufacturing Engineering, Faculty of Engineering and Built Environment, Universiti Kebangsaan Malaysia, Bangi, Selangor, Malaysia

Fevi Syaifoelida College of Engineering, Universiti Tenaga Nasional, Kajang, Selangor, Malaysia

M. Syarafi Shuhaimi Department of Mechanical Engineering, Universiti Tenaga Nasional, Kajang, Selangor, Malaysia

Syed Nuzul Fadzli Syed Adam Faculty of Mechanicall Engineering Technology, Universiti Malaysia Perlis, Sungai Chuchuh, Padang Besar, Perlis, Malaysia

Sharifah Shahnaz Syed Bakar Faculty of Chemical Engineering and Technology, Universiti Malaysia Perlis (UniMAP), Kangar, Perlis, Malaysia;
Faculty of Chemical Engineering and Technology, Universiti Malaysia Perlis, Arau, Perlis, Malaysia

Mustaffar Ali Azhar Taib Division of Advanced Ceramic Materials Technology (ADTEC), Taiping, Perak, Malaysia

Chi Ying Tan Center of Excellence Geopolymer and Green Technology (CEGeoGTech), Universiti Malaysia Perlis, Jejawi, Arau, Perlis, Malaysia;
Faculty of Chemical Engineering Technology (FTKK), Universiti Malaysia Perlis (UniMAP), Jejawi, Arau, Perlis, Malaysia

Hong Ann Tan Department of Mechanical, Materials and Manufacturing Engineering, The University of Nottingham Malaysia, Semenyih, Selangor Darul Ehsan, Malaysia

Le Nhat Tan Faculty of Materials Technology, Ho Chi Minh City University of Technology, Ho Chi Minh City, Vietnam;
Vietnam National University Ho Chi Minh City, Linh Trung Ward, Ho Chi Minh City, Vietnam

Mohd Nor Mad Tan School of Electrical Engineering, College of Engineering, Universiti Teknologi MARA, Shah Alam, Selangor, Malaysia

Y. P. Tan Centre of Excellence Geopolymer and Green Technology, Faculty of Chemical Engineering Technology, Universiti Malaysia Perlis (UniMAP), Jejawi, Arau, Perlis, Malaysia

T. Tasman Doctoral Program of Civil Engineering, Faculty Engineering, Diponegoro University, Semarang, Indonesia;
Consultant of Pavement/Material Engineering, PT Aria Jasa Reksatama, Surabaya, Indonesia

T. S. A. Tengku Zahidi EMZI-UiTM Nanoparticles Colloids and Interface Industrial Research Laboratory (NANO- CORE), Centre for Chemical Engineering Studies, Universiti Teknologi MARA, Pulau Pinang, Malaysia

Bui Khac Thach Faculty of Materials Technology, Ho Chi Minh City University of Technology, Ho Chi Minh City, Vietnam;
Vietnam National University Ho Chi Minh City, Linh Trung Ward, Ho Chi Minh City, Vietnam

N. Thachnatharen Faculty of Defence Science and Technology, National Defence University of Malaysia, Kuala Lumpur, Malaysia

Sam Sung Ting Geopolymer and Green Technology, Centre of Excellent (CEGeoGTech), Universiti Malaysia Perlis, Arau, Perlis, Malaysia

Phan Dinh Tuan Hochiminh City University of Natural Resources and Environment, Ho Chi Minh City, Vietnam

Tran Anh Tu Vietnam National University, Ho Chi Minh City, Vietnam

Najua Tulos Faculty of Applied Sciences, Universiti Teknologi Mara, Shah Alam, Malaysia

Huynh Khanh Tuong Department of Polymer Materials, Faculty of Materials Technology, Ho Chi Minh City University of Technology (HCMUT), Ho Chi Minh City, Vietnam;
Vietnam National University Ho Chi Minh City, Ho Chi Minh City, Vietnam

Tran Ho Van Anh Department of Silicate Materials, Faculty of Materials Technology, Ho Chi Minh City University of Technology (HCMUT), Ho Chi Minh City, Vietnam;
Vietnam National University Ho Chi Minh City (VNU HCM), Linh Trung Ward, Ho Chi Minh City, Vietnam

Tran Van Khai Faculty of Materials Technology, Ho Chi Minh City University of Technology (HCMUT), Ho Chi Minh City, Vietnam;
Vietnam National University Ho Chi Minh City, Linh Trung Ward, Thu Duc District, Ho Chi Minh City, Vietnam

Bui Van Tien Faculty of Materials Technology, Ho Chi Minh City University of Technology (HCMUT), Ho Chi Minh City, Vietnam;
Vietnam National University Ho Chi Minh City, Linh Trung Ward, Thu Duc District, Ho Chi Minh City, Vietnam;
Polymer Research Center, Ho Chi Minh City University of Technology (HCMUT), Ho Chi Minh City, Vietnam

V. Vicki Wanatasanappan Institute of Power Engineering, Universiti Tenaga Nasional, Kajang, Malaysia

Cao Xuan Viet Department of Polymer Materials, Faculty of Materials Technology, Ho Chi Minh City University of Technology (HCMUT), Ho Chi Minh City, Vietnam;
Vietnam National University Ho Chi Minh City, Ho Chi Minh City, Vietnam

Juyana A. Wahab Faculty of Chemical Engineering and Technology, Universiti Malaysia Perlis (UniMAP), Arau, Perlis, Malaysia

W. M. H. Wan Ahmad Faculty of Chemical Engineering and Technology, Universiti Malaysia Perlis, Arau, Perlis, Malaysia

Wan Amirul Shafiz Wan Zulkifli Fakulti Kejuruteraan Pembuatan, Universiti Teknikal Malaysia Melaka, Melaka, Malaysia

Zahira Yaakob Department of Chemical and Process Engineering, Faculty of Engineering and Built Environment, Universiti Kebangsaan Malaysia, Bangi, Selangor, Malaysia

S. M. Yahaya Applied Chemistry Program, Faculty of Applied Sciences, Universiti Teknologi MARA, Shah Alam, Selangor, Malaysia

Shuhaida Yahud Faculty of Electronic Engineering Technology (FTKEN), Universiti Malaysia Perlis (UniMAP), Pauh Putra Campus, Arau, Perlis, Malaysia

H. M. Yee School of Civil Engineering, University Teknologi MARA (UiTM) Cawangan Pulau Pinang, Pulau Pinang, Malaysia;
School of Engineering, College of Engineering, Universiti Teknologi MARA Cawangan Pulau Pinang, Kampus Permatang Pauh, Permatang Pauh, Pulau Pinang, Malaysia

Chae-Eun Yeo Plant Engineering Center, Institute for Advanced Engineering, Cheoin-Gu, Yongin-Si, Gyeonggi-Do, Korea

Feizal Yusof School of Mechanical Engineering, Universiti Sains Malaysia, Nibong Tebal, Penang, Malaysia

Yusliza Yusuf Fakulti Teknologi Kejuruteraan Mekanikal and Pembuatan, Universiti Teknikal Malaysia Melaka, Melaka, Malaysia

N. M. Zahari College of Engineering, Universiti Tenaga Nasional, Kajang, Selangor, Malaysia;
Department of Civil Engineering, Universiti Tenaga Nasional, Jalan Ikram-Uniten, Kajang, Selangor Darul Ehsan, Malaysia

H. A. Zaharin Department of Mechanical and Manufacturing Engineering, Faculty of Engineering and Built Environment, Universiti Kebangsaan Malaysia, Bangi, Selangor, Malaysia

T. Zaharinie Department of Mechanical Engineering, Faculty of Engineering, University of Malaya, Kuala Lumpur, Malaysia

A. Q. Zaifuddin Faculty of Mechanical and Automotive Engineering Technology, Universiti Malaysia Pahang, Pekan, Pahang Darul Makmur, Malaysia

W. W. A. Zailani School of Civil Engineering, College of Engineering, Universiti Teknologi MARA, Shah Alam, Selangor, Malaysia

Ameer Haqimie Zainal School of Civil Engineering, College of Engineering, Universiti Teknologi MARA (UiTM) Shah Alam, Shah Alam, Selangor, Malaysia

Mohamad Ikhwan Syafiq Zainan Faculty of Engineering, Universiti Pertahanan Nasional Malaysia, Kuala Lumpur, Malaysia

Nur Hazirah Zaini Faculty of Engineering and Built Environment, Universiti Sains Islam Malaysia, Nilai, Negeri Sembilan, Malaysia

M. M. Zainoodin Department of Civil Engineering, Universiti Tenaga Nasional, Jalan Ikram-Uniten, Kajang, Selangor Darul Ehsan, Malaysia

Nasaie Zainuddin Faculty of Applied Sciences, Universiti Teknologi Mara, Shah Alam, Malaysia

A. A. Zakaria College of Engineering, Universiti Tenaga Nasional, Kajang, Selangor, Malaysia

M. N. Zakaria Eco-Technology Program, Faculty of Applied Sciences, Universiti Teknologi MARA, Shah Alam, Malaysia;
Selangor School of Materials and Mineral Resources Engineering, Universiti Sains Malaysia, Nibong Tebal, Penang, Malaysia

Muhammad Hisyam Zakaria College of Creative Arts, Universiti Teknologi Mara, Shah Alam, Malaysia

Muhammad Salihin Zakaria Faculty of Chemical Engineering and Technology, Universiti Malaysia Perlis (UniMAP), Arau, Malaysia;
Biomedical and Nanotechnology Research Group, Center of Excellence Geopolymer and Green Technology (CEGeoGTech), Universiti Malaysia Perlis (UniMAP), Arau, Malaysia

M. I. Zalrusli School of Civil Engineering, Faculty of Engineering, Universiti Teknologi Malaysia, Johor Bahru, Malaysia

M. H. Zawawi College of Engineering, Universiti Tenaga Nasional, Kajang, Selangor, Malaysia

Zol Effendi Zolkefli Western Digital®, SanDisk Storage Malaysia Sdn.Bhd., Penang, Malaysia

Muhammad Syukri Bin Zubir School of Mechanical Engineering, Universiti Sains Malaysia, Nibong Tebal, Seberang Perai Selatan, Penang, Malaysia

Ahmad Wafi Mahmood Zuhdi College of Engineering, Universiti Tenaga Nasional, Kajang, Selangor, Malaysia

Sana Zulfiqar School of Mechanical Engineering, Universiti Sains Malaysia, Nibong Tebal, Penang, Malaysia

Muhammad Nubli Zulkifli Renewable Energy Research Laboratory (RENERAL), Electrical Engineering Section, British Malaysia Institute, Universiti Kuala Lumpur, Gombak, Selangor, Malaysia

Part I
Electronic Materials and Technology

Chapter 1
Effect of Al Addition to the Solidification and Microstructure Formation on Sn–Ag–Cu Solder Alloy

Mohd Izrul Izwan Ramli, Ain Najihah Saim, and Nur Syahirah Mohamad Zaimi

Abstract Due to the use of Pb in electronic packaging which has been prohibited since 2006, the development of Pb-free solder has gained attention due to its comprehensive performance. One of the most prevalent lead-free solder alloys used in industry is Sn–Ag–Cu solder alloy. Aluminum was added to the Sn–3.0Ag–0.5Cu solder alloy in various concentrations ($x = 0.5$ wt%, 1.5 wt%) to observe the influence on solidification and microstructure formation. Because of the formation of large primary intermetallic during the reflow process, the incorporation of aluminum in the lead-free solder alloy has been examined. The formation of new intermetallic, Ag–Al and Cu–Al, has suppressed the formation of primary intermetallic, Cu_6Sn_5 and Ag_3Sn. Since Ag_3Sn intermetallic can reduce the dependability of solder junctions, the new intermetallic formation is preferable. However, when the amount of aluminum added to the Sn–3.0Ag–0.5Cu solder alloy increases, the wettability of the solder alloy on the Cu strip deteriorates.

Keywords Lead-free solder alloy · Solidification phase · Microstructure formation · Aluminum · Wettability

1.1 Introduction

Lead-free tin-silver-copper solder alloys, commonly known as Sn–Ag–Cu solder alloys, are utilized as electronic solder materials. Due to the environmental difficulties associated with the use of Pb solder, the use of Sn–Ag–Cu alloy has been

M. I. I. Ramli (✉) · N. S. Mohamad Zaimi
Center of Excellence Geopolymer and Green Technology (CEGeoGTech), Universiti Malaysia Perlis, Jejawi, Arau, Perlis, Malaysia
e-mail: izrulizwan@unimap.edu.my

M. I. I. Ramli · A. N. Saim · N. S. Mohamad Zaimi
Faculty of Chemical Engineering Technology (FTKK), Universiti Malaysia Perlis (UniMAP), 02600 Jejawi, Arau, Perlis, Malaysia

M. A. A. Mohd Salleh et al. (eds.), *Proceedings of the Green Materials and Electronic Packaging Interconnect Technology Symposium*, Springer Proceedings in Physics 289,
https://doi.org/10.1007/978-981-19-9267-4_1

advocated [1, 2]. The Sn–Ag–Cu solder alloy provides superior properties to Sn–Pb solders [3, 4]. Despite this, the inclusion of the fourth element and nanoparticle appears to improve the solder alloy while also improving the preceding bulk solder alloy [5]. The Sn–Ag–Cu solder alloy may be the next alternative solder in the field of electronic packaging because of these benefits. The addition of aluminum to the Sn–Ag–Cu solder alloy improves the qualities of the lead-free solder alloy by preventing the production of large primary intermetallic in the solder alloy. Due to their low melting point and strong thermal mechanical fatigue resistance, Sn–3.0Ag–0.5Cu and Sn–4.0Ag–0.5Cu solders are the most extensively used [6]. The high Ag concentration in the solder formulations, on the other hand, stimulated the production of massive Ag_3Sn intermetallic inside the bulk solder matrix, which can readily induce fractures. The creation of the intermetallic can be prevented by adding aluminum to the solder alloy [7]. Aluminum is frequently employed in the development of Sn–Ag–Cu solder due to its low density, good electroconductivity, and thermal conductivity [8]. The microstructure of the Sn–Ag–Cu solder alloy may be improved by adding the appropriate quantity of aluminum since the inclusion of aluminum improves the strength of solder joints. The addition of aluminum will enhance the production of Ag–Al and Cu–Al IMC and at the same time suppressed the formation of Ag_3Sn and Cu_6Sn_5. In this project, the amount of aluminum used will be varied from 0 wt%, 0.5 wt% to 1.5 wt% to study the effect on the solidification and microstructure formation of Sn–3.0Ag–0.5Cu.

1.2 Experimental Procedure

1.2.1 Materials

Sn–3.0Ag–0.5Cu, aluminums, and copper strip.

1.2.2 Sample Preparation

Cast a Sn–3.0Ag–0.5Cu solder with aluminum with the percent age of 0.5% and 1.5%. Pour molten metal into a mold and the excess metal on a metal plate. Mount the sample from the mold for the grinding and polishing process. Roll the excess metal into a solder sheet with a thickness of 0.06 mm and punch it into a disk with a diameter of 2.5 mm. Heat the solder sheet on a copper strip with the temperature of 270 and 470 °C.

1.2.3 *Characterization*

The phase transformation and solidification in Sn–3.0Ag–0.5Cu–xAl were determined by ThermoCalc software. The microstructure of the bulk sample was characterized by Scanning Electron Microscope (SEM). The wettability analysis that includes the intermetallic (IMC) thickness, contact angle, and the spreading ratio of the solder sheet on the copper strip was analyzed using an optical microscope.

1.3 Results and Discussion

1.3.1 *Phase Transformation and Solidification*

Figures 1.1 and 1.2 show the phase transformation of Sn–3.0Ag–0.5Cu–0.5Al and Sn–3.0Ag–0.5Cu–1.5Al, respectively. The solidification sequence of Sn–3.0Ag–0.5Cu–1.5Al shows a higher melting temperature than Sn–3.0Ag–0.5Cu–0.5Al which is at 466.78 °C. This is due to the higher concentration of aluminum.

1.3.2 *Microstructure Analysis*

The microstructure of bulk solder samples that contain 0 wt%, 0.5 wt%, and 1.5 wt% of aluminum was analyzed with SEM with a magnification of 1500x. Figure 1.3a shows the microstructure of Sn–3.0Ag–0.5Cu that has β-Sn dendrites that were surrounded by a eutectic area and its primary intermetallic, Cu_6Sn_5. In Fig. 1.3b, the β-Sn dendrites are refined due to the inclusion of aluminum in the composition of Sn–3.0Ag–0.5Cu. New intermetallics are formed after the addition of 0.5 wt% aluminum, which are Cu–Al and Ag–Al intermetallics. The formation of Cu_6Sn_5 is suppressed alongside the addition of aluminum to the composition. Figure 1.3c shows the increasing size of Cu–Al intermetallic and the formation of Cu–Al_2 that occurred due to the high amount of aluminum added to Sn–3.0Ag–0.5Cu composition.

1.3.3 *Wettability Analysis*

For contact angle comparison between the samples that are heated at 270 °C and 470 °C, the sample of Sn–3.0Ag–0.5Cu at 270 °C has the lowest contact angle, 14.642°, which suggests greater wettability. However, the contact angle for Sn–3.0Ag–0.5Cu–1.5Al recorded the greatest value at 32.402° at 270 °C. Figure 1.4 shows the comparison of the contact angles obtained from the samples at both temperatures.

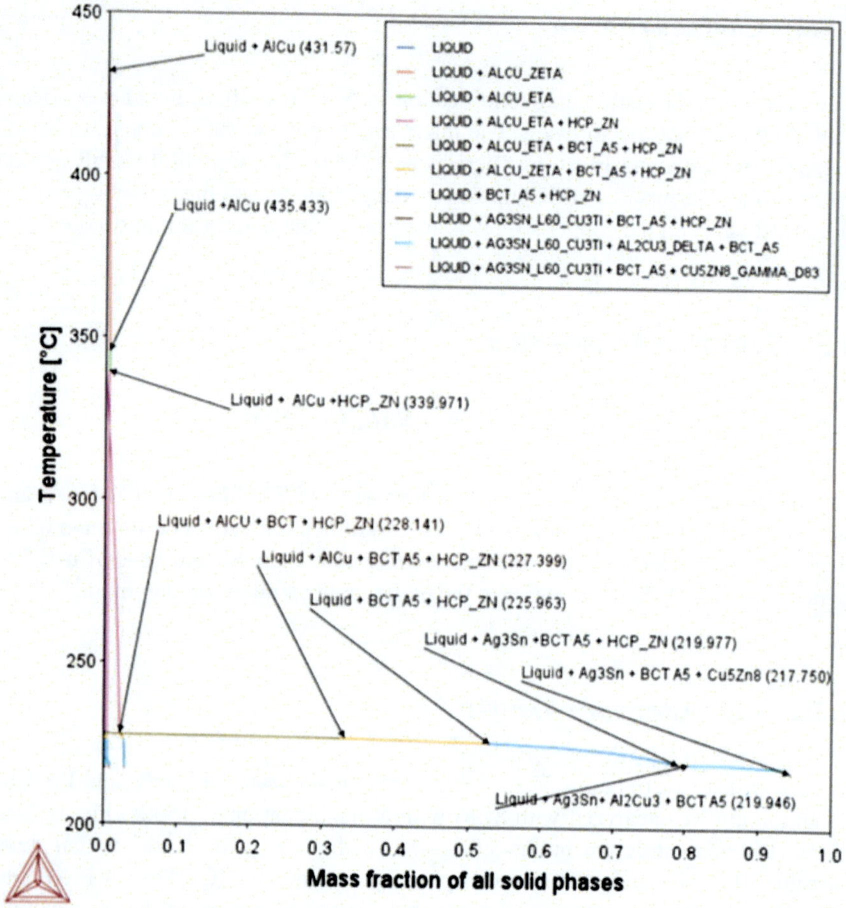

Fig. 1.1 Solidification sequence and mass fraction of Sn–3.0Ag–0.5Cu–0.5Al

For the spreading ratio of the samples at different temperatures, Sn–3.0Ag–0.5Cu has the maximum spreading rate at 270 °C. This outcome is consistent with the contact angle data, which shows that Sn–3.0Ag–0.5Cu has the lowest contact angle at 270 °C, indicating a greater rate of molten solder spreading. As opposed to this, with Sn–3.0Ag–0.5Cu–1.5Al, the spreading rate rises from 51.67% at 270 °C to 77.50% at 470 °C. Figure 1.5 shows the comparison of the spreading rate of the samples at different temperatures. It indicates that the Sn–3.0Ag–0.5Cu has the highest wettability at 270 °C, while the Sn–3.0Ag–0.5Cu–1.5Al has the best wettability at 470 °C.

The intermetallic (IMC) thickness of the samples at different temperatures is also analyzed as a solder junction that contains intermetallic shows that it has effective wetting, but an excessive quantity might weaken the joint owing to its brittleness [9–11]. To find the thickness of interfacial intermetallic, the area of intermetallic

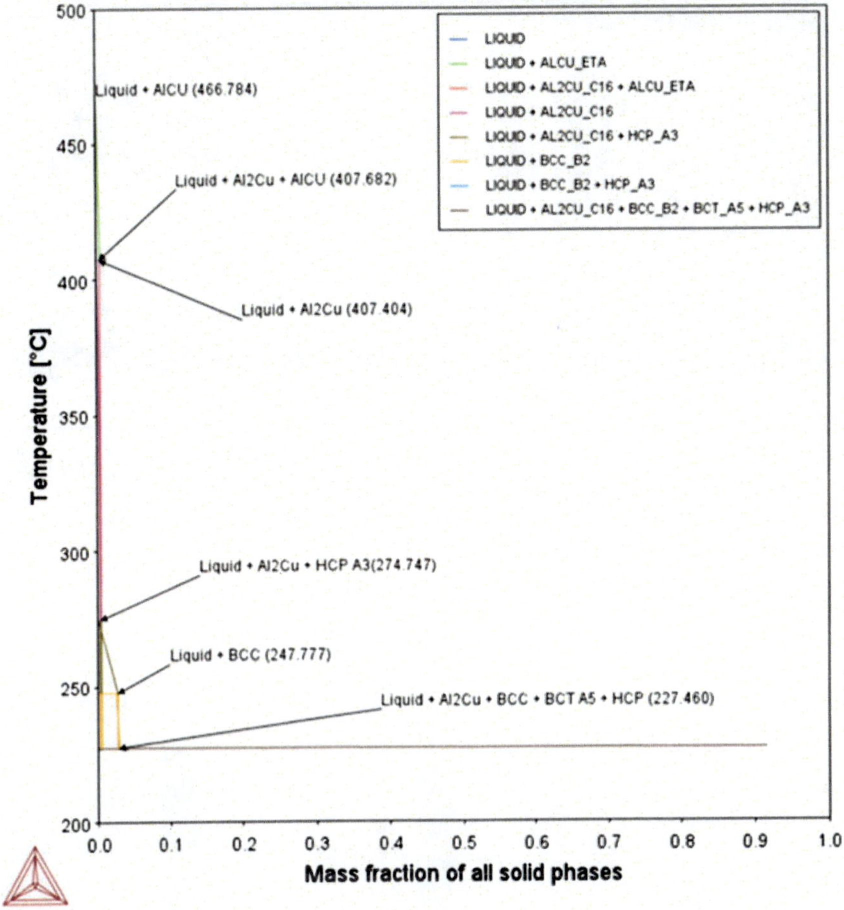

Fig. 1.2 Solidification sequence and mass fraction of Sn–3.0Ag–0.5Cu–1.5Al

will be divided by the length of intermetallic. Based on Fig. 1.6, Sn–3.0Ag–0.5Cu–0.5Al has the thinnest intermetallic layer at both 270 °C and 470 °C, measuring 3.925 μm and 6.743 μm, respectively. However, a large amount of aluminum inclusion causes the interfacial intermetallic thickness to grow for Sn–3.0Ag–0.5Cu–1.5Al as the thickness of the intermetallic layer grows as more aluminum is added to the composition.

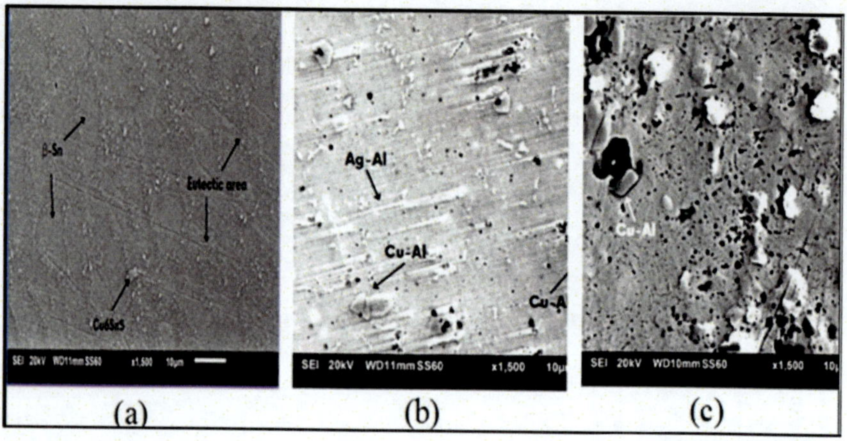

Fig. 1.3 SEM micrographs of **a** Sn–3.0Ag–0.5Cu, **b** Sn–3.0Ag–0.5Cu–0.5Al and **c** Sn–3.0Ag–0.5Cu–1.5Al at magnification of 1500x

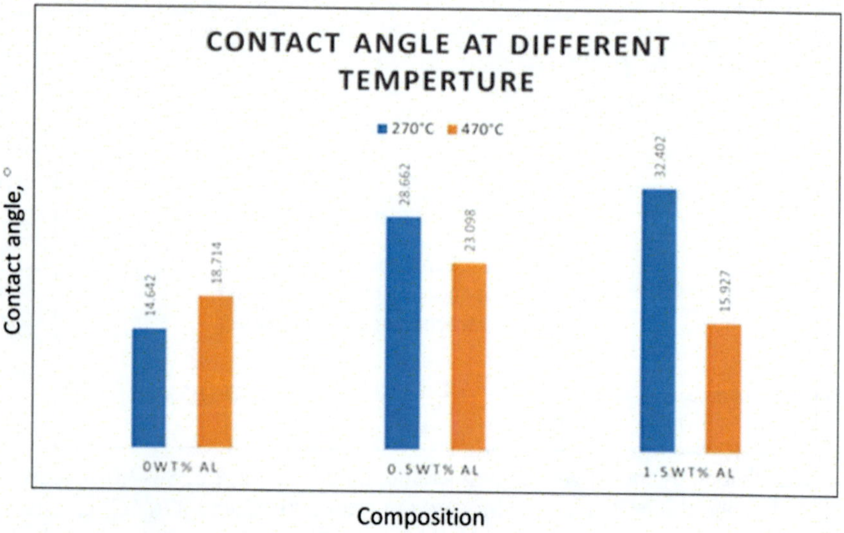

Fig. 1.4 Contact angle of samples at different temperatures

1.4 Conclusion

As a result of the addition of aluminum to the Sn–3.0Ag–0.5Cu so der alloy, the solidification sequence of Sn–3.0Ag–0.5Cu has been altered. Due to the higher melting point of aluminum compared to other solder alloys, the solidification sequence begins at a higher temperature with an increasing amount of aluminum. The formation of Cu–Al and Ag–Al intermetallics in the microstructure of bulk solder has suppressed the

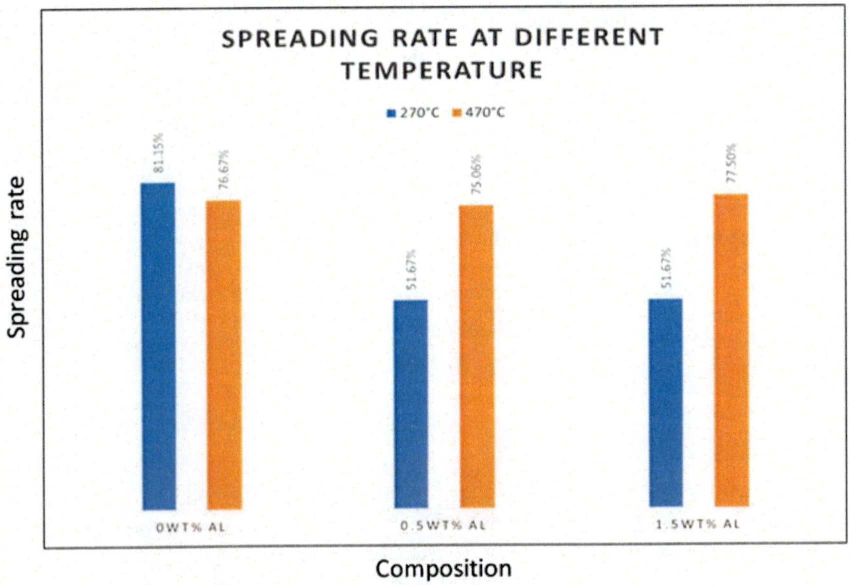

Fig. 1.5 Spreading rate of samples at different temperatures

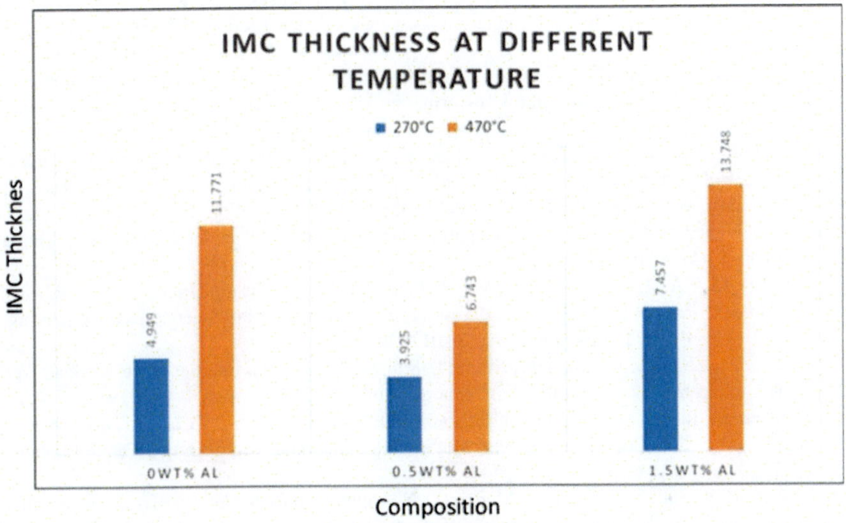

Fig. 1.6 IMC thickness of samples at different temperatures

formation of primary intermetallic, Cu_6Sn_5. With the increasing amount of aluminum added, the Cu_6Sn_5 intermetallic is less likely to be generated. The primary β-Sn dendrites also become more refined with the addition of aluminum to the Sn–3.0Ag–0.5Cu composition. For wettability, Sn–3.0Ag–0.5Cu has the best wettability at the temperature of 270 °C while Sn–3.0Ag–0.5Cu–1.5Al has the best wettability at the temperature of 470°.

Acknowledgements The author is grateful to acknowledge Ministry of Higher Education Malaysia regarding the use of the ISIS Neutron and Muon Source entitle the neutron tomography studies of the geopolymer ceramic used for reinforcement materials in a solder alloy for a robust electric/electronic solder joint under reference no: JPT.S (BPKI)2000/016/018/019(29).

References

1. N. Jiang, L. Zhang, Z.Q. Liu, L. Sun, W.M. Long, P. He, M. Zhao, Reliability issues of lead-free solder joints in electronic devices. Sci. Technol. Adv. Mater. **20**(1), 876–901 (2019)
2. J.W. Xian, M.M. Salleh, S.A. Belyakov, T.C. Su, G. Zeng, K. Nogita, C.M. Gourlay et al., Influence of Ni on the refinement and twinning of primary Cu_6Sn_5 in Sn–0.7 Cu–0.05 Ni. Intermetallics **102**, 34–45 (2018)
3. M. Drienovsky, M. Palcut, P. Priputen, E. Cuninková, O. Bošák, M. Kubliha, L.R. Trnková, Properties of Sn–Ag–Cu solder joints prepared by induction heating. Adv. Mater. Sci. Eng. (2020)
4. F. Somidin, H. Maeno, X.Q. Tran, D. D McDonald, M.A.A. Mohd Salleh, S. Matsumura, K. Nogita, Imaging the polymorphic transformation in a single Cu_6Sn_5 grain in a solder joint. Materials **11**(11), 2229 (2018)
5. A. Kantarcıoğlu, Y. Kalay, Effects of Al and Fe additions on microstructure and mechanical properties of SnAgCu eutectic lead-free solders. Mater. Sci. Eng., A **593**, 79–84 (2014)
6. K. Maslinda, A.S. Anasyida, M.S. Nurulakmal, Effect of Al addition to bulk microstructure, IMC formation, wetting and mechanical properties of low-Ag SAC solder. J. Mater. Sci.: Mater. Electron. **27**(1), 489–502 (2015)
7. C.Y. Tan, M.A.A.M. Salleh, in *The Effect of Aluminium Addition on the Microstructure of Lead-Free Solder Alloys: A Short Review*. IOP Conference Series: Materials Science and Engineering, vol. 701, No. 1 (IOP Publishing, 2019), p. 012026
8. M. Zhao, L. Zhang, Z.Q. Liu, M.Y. Xiong, L. Sun, Structure and properties of Sn–Cu lead-free solders in electronics packaging. Sci. Technol. Adv. Mater. **20**(1), 421–444 (2019)
9. N. Saleh, M. Ramli, M.M. Salleh. in *Effect of Zinc Additions on Sn-0.7 Cu-0.05 Ni Lead-Free Solder Alloy*. IOP Conference Series: Materials Science and Engineering (IOP Publishing, 2017)
10. M.A.A. Mohd Salleh, K. Nogita, S. Mcdonald, Non-metal reinforced lead-free composite solder fabrication methods and its reinforcing effects to the suppression of intermetallic formation: short review. Appl. Mech. Mater. **421**, 260–266 (2013)
11. K. Nogita, M.A.A. Mohd Salleh, E. Tanaka, G. Zeng, S.D. McDonald, S. Matsumura, In situ TEM observations of Cu6Sn5 polymorphic transformations in reaction layers between Sn-0.7 Cu solders and Cu substrates. Jom, 68(11), 2871–2878 (2016)

Chapter 2
Effect of Isothermal Aging on Mechanical Properties of Sn–0.7Cu–xZn Lead-Free Solder

Mohd Izrul Izwan Ramli, Muhammad Fadlin Hazim Baser, Nur Syahirah Mohamad Zaimi, and Chi Ying Tan

Abstract Effect of isothermal aging on mechanical properties of the lead-free solder Sn–0.7Cu with addition of zinc had been studied. All samples were aged at 120, 150 and 180 °C; each temperature would hold for 24, 120 and 240 h in an oven. After aging process, samples for micro-hardness Vickers test undergo a series of process such as grinding and polishing. These samples need to be flat and smooth before indentation. Then, all the samples were done for mechanical testing which is shear strength test and micro-hardness Vickers test. Additional element into lead-free solder (Sn–0.7Cu) has shown improving in mechanical strength, but decreases with increasing aging time. However, addition of zinc was not giving to much impact in mechanical strength before aging, even after addition of high percentage of zinc at 1.5 wt%. Furthermore, after aging process, it improves in shear strength after aging at 24 h.

Keywords Lead-free solder alloy · Isothermal aging · Mechanical

2.1 Introduction

The Developed in recent years of lead-free solder has been rapidly occur to substitute of lead–tin soldering alloys that widely used in electronics parts. The replacement of this lead–tin solder has been set by European Union's (EU) proposal to reduce waste from electrical and electronic equipment (WEEE) and RoHS regulations [1]. In general, several of free-lead solder materials make the solder stronger and have a higher melting point, and they have ability to use it in high-temperature applications

M. I. I. Ramli (✉) · N. S. Mohamad Zaimi · C. Y. Tan
Center of Excellence Geopolymer and Green Technology (CEGeoGTech), Universiti Malaysia Perlis, Jejawi, Arau, Perlis, Malaysia
e-mail: izrulizwan@unimap.edu.my

M. I. I. Ramli · M. F. H. Baser · N. S. Mohamad Zaimi · C. Y. Tan
Faculty of Chemical Engineering Technology (FTKK), Universiti Malaysia Perlis (UniMAP), 02600 Jejawi, Arau, Perlis, Malaysia

M. A. A. Mohd Salleh et al. (eds.), *Proceedings of the Green Materials and Electronic Packaging Interconnect Technology Symposium*, Springer Proceedings in Physics 289,
https://doi.org/10.1007/978-981-19-9267-4_2

such as advanced electronic components. For example, the mixture of copper, tin, and silver has been recognized as the famous lead-free solder due to their eutectic composition of SAC solder alloy has given an optimum strength, low melting temperature, high fatigue resistance, and plasticity. However, there are some problems with this solder composition; high percentage of silver in the SAC357 will increase the cost of these solders. Other than that, SAC has a thick intermetallic and low creep strength compounds which lead to decrease in the reliability of that solders. Thus, to increase the reliability of this solders, one of the methods that can be used to improve the properties of this solders is by adding alloying elements (Ni, Ti, Sb, Bi, Zn, Ni, Ga) and rare earth elements [2].

These elements will react with the solder matrix during preparation or reflow process. These reactions will increase the bonding between the element particles and the solder base and change the physical properties of the phases formed [3]. Adding of these elements also effected on the reaction/growth rate that can be increased or decreased and form another reaction layer at the interface that would normally appear from other reaction products. Another important characteristic is including the structure of intermetallic compounds (IMCs) and their microstructure of the solder [4].

This research focuses on the effect of isothermal aging on mechanical properties of the lead-free solder Sn–0.7Cu with addition of zinc. The addition of bismuth and zinc will play an important role in the enhancement of the structure and mechanical properties of the solder joints. It improves the performance of Sn–0.7Cu solders. From previous studies, with 1 wt% addition of zinc into a Sn–3.5 wt%-Ag eutectic solder matrix, it will eliminate the large β-Sn dendritic grains on the composite solder and improve the solidification microstructure of the Sn–3.5 wt% Ag eutectic solder by a finer and more uniform phase distribution throughout the solder materials [5]. In this study, the materials with different addition of zinc will undergo isothermal aging at a different temperature with different aging time.

2.2 Experimental Procedure

In this study, Sn–0.7Cu ingot is obtained from Nihon Superior, Japan. The material us utilized to supplement the Sn–Cu ed to add into the Sn–Cu were Zinc from Sigma-Aldrich (M) Sdn. Bhd. Firstly, the Sn–0.7Cu ingot was cut into a small size at 150 mm length and put into the crucible. These base solder alloys was then heated at above its melting point and maintained about an hour. The solder alloys were mixed with four different compositions, which were 0, 0.5, 1.0 and 1.5 wt% of Zn. After that, temperature was increased to 550 °C. This heating is hold for one hour, and for every ten minutes, the mixing material was stirred to perform homogeneously mixed and lastly solidified to room temperature. Then, roll mill machine was used to roll the solder material into a sheet form. After weighing process, each material with different composition is formed in a rectangular shape for which material that will conduct shear test while for material that undergoes micro-hardness test, it will be formed in a circle before reflowing it on glass by using reflow oven. In this process, there are three different temperatures of aging which are 120, 150 and 180 °C, and each temperature undergoes three different aging times, 24, 120 and 240 h. After aging process, sample for micro-hardness test was ground on the surface of testing area. During polishing process, the process is carried out using polishing film and it will be divided it into three different stages by using variety of polishing solutions which are 1.0 μm alumina for the first process, then proceed with 0.3 μm alumina and lastly using Oxide Polishing Suspension (OPS). The Vickers hardness investigates the strength the materials can resist the load. Vickers hardness involves indenting the test material in form of a right pyramid with a diamond indenter. The full load is normally applied at 10–15 s for one indention. Indention was done on the flat surface of the specimen. ASTM B933-09 was followed as a standard for the Vickers micro-hardness testing. For this study, six indentions were taken for each solder material and the average for each sample is calculated. The lap shear procedure is normally used to estimate the shear, creep, and thermal fatigue activities of solder joints. This study investigates the effect of solder Sn–0.7Cu with addition of zinc with different composition and thermal aging temperature on their geometrical features. A universal testing machine (UTM) or also known as a universal tester material testing machine was used to test the shear strength of materials. The setup is followed to the ASTM D1002.

2.3 Results and Discussion

2.3.1 Hardness

Figure 2.1 shows the average of micro-hardness results of Sn–0.7Cu with different composition of zinc before aging process; it is observed that pure solder alloy had highest hardness at 9.1HV. However, after adding 0.5 wt% of Zn, the micro-hardness reading is slightly decreased to 8.42HV. Adding 1.0 wt% of Zn shows that the solder alloy becomes more harden at 8.72HV but still lower than the solder without addition of zinc as shown in Fig. 2.2. Besides, after adding 1.5 wt% of Zn, average of micro-hardness reading is dropped to 8.63HV.

The important solder material properties depend on their microstructure, and the grain structure of the solidified alloy depends on the solder composition. However, when 1.0 and 1.5 wt% of Zn was added into the solder alloys, the hardness was slightly decreased; addition of small amount of Zn in the Sn–0.7Cu is shown in Fig. 2.3. Specifically, the motion of dislocation and growth grains effected the strength of the solder alloy, especially in hardness [5, 6]. This lack of performance with addition of zinc has been an issue that by Vaynman et al. [7] where the zinc-containing lead-free solders not suitable to use due to some drawbacks in mechanical properties and reliability database.

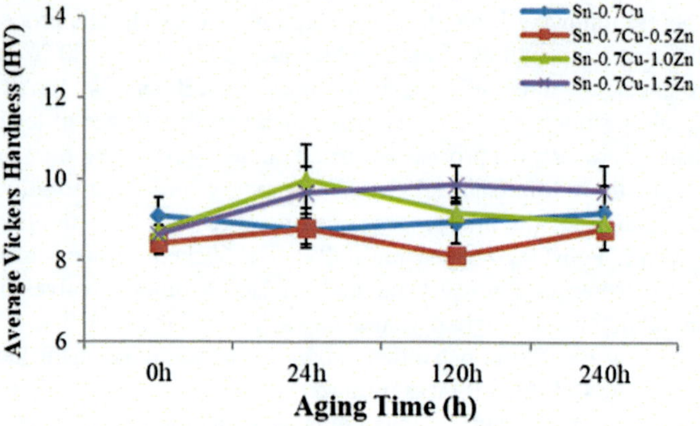

Fig. 2.1 Average Vickers hardness of Sn–0.7Cu with different composition of zinc after aging at 120 °C with different aging time

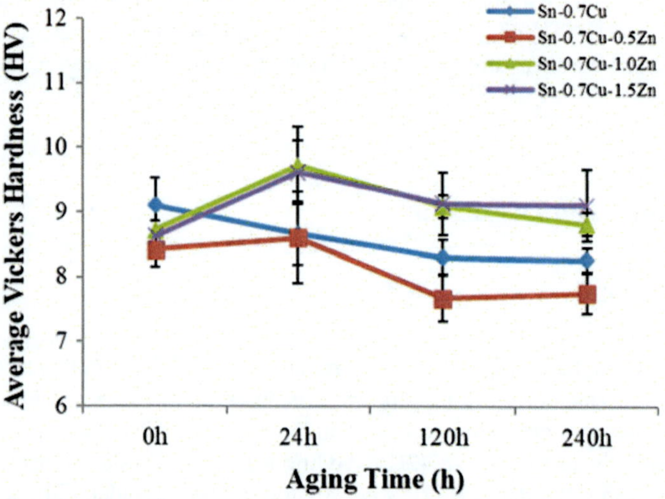

Fig. 2.2 Average Vickers hardness of Sn–0.7Cu with different composition of zinc after aging at 150 °C with different aging time

2.3.2 *Shear Strength*

Figure 2.4 shows decline trend with increasing of aging time at 24, 120 and 240 h. The ultimate tensile strength (UTS) of pure Sn–0.7Cu solder alloy before aging shows at 11.21 MPa, but it slightly decreases with addition of 0.5 wt% of Zn. However, adding high percentage of zinc in lead-free solder of Sn–0.7Cu before aging reduces shear strength of solder due to highly drop in ultimate tensile strength (UTS) which is at 9.66 MPa with addition of 1.5 wt% Zn. However, after aging process at 24 h,

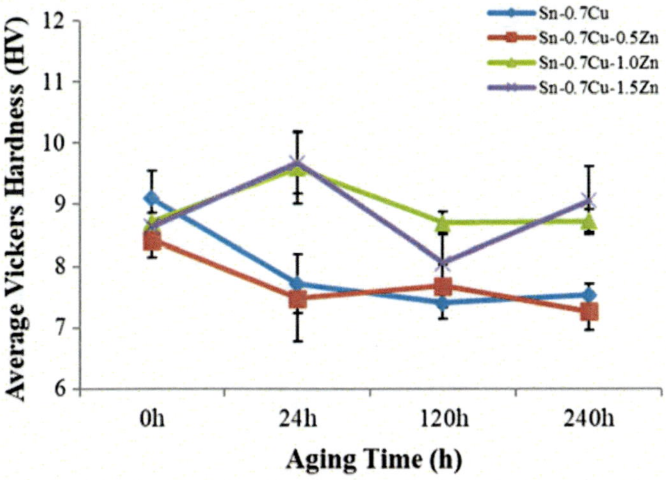

Fig. 2.3 Average Vickers hardness of Sn–0.7Cu with different composition of zinc after aging at 180 °C with different aging time

the solder that contains 0.5 wt% of Zn improves in their shear strength. Then, after increasing the aging time up to 120 h, all the composition of the solder alloy slightly decreases and dramatically drops when aging time increases to 240 h. Therefore, small percentage of zinc at 0.5% added in Sn–0.7Cu will improve shear strength after aging until 120 h at 150 °C as shown in Fig. 2.4. Before aging process, adding of zinc in Sn–0.7Cu solder alloy causes drop in their tensile strength due to formation of Cu_5Zn_8 phase after reflow process. R. Mayappan et al. state that solder material reacts with the Cu substrate and forms interface termed intermetallic compound (IMC). For Sn–Zn solders reacting with Cu, it is expected that the first creating phase is Cu_5Zn_8 primary intermetallic when soldering was done at 250 °C resulting in decrease in the strength [8–10].

Other than that, which contributes in decreasing of ultimate tensile strength (UTS) which are increases is the size of the bulk Cu_6Sn_5 intermetallic compounds as the aging time increase. The structure and the thickness of the IMC layer are more important for its brittleness; this is one of the main factors affecting the occurrence of brittle fractures. C. Yang et al. also reported that when the solder involvements continued thermal aging, the multi-stack grain structure of Cu_3Sn/Cu_6Sn_5 would affect its mechanical property. As a result, brittle fracture would occur inside Cu_3Sn/Cu_6Sn_5 and the apparent toughness becomes lower.

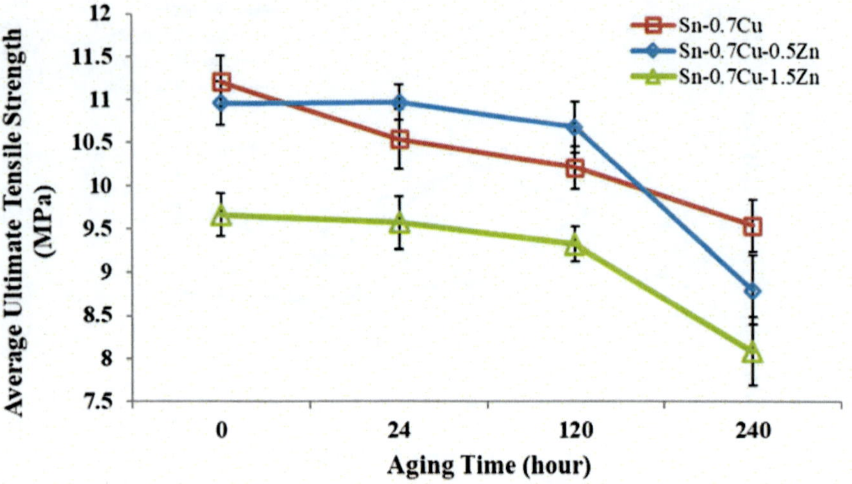

Fig. 2.4 Average ultimate tensile strength of Sn–0.7Cu with different composition of zinc after aging process at 150 °C

2.4 Conclusion

In this study, the isothermal aging that has effect on mechanical properties of the lead-free solder Sn–0.7Cu with zinc addition had been looked into. It was discovered that adding zinc on lead-free solder Sn–0.7Cu gives different performance in mechanical strength. However, after aging at temperature 120 °C and 150 °C with 24 h, all solder alloys with different composition of zinc improve their micro-hardness.

Furthermore, it was found that addition of the zinc on lead-free solder Sn–0.7Cu gives different performance in shear strength, where adding small amount of zinc at 0.5 wt% gives high strength compared to adding 1.5 wt% of zinc in Sn–0.7Cu. However, pure solder alloy shows highest shear strength after reflow process; in addition, after aging at 24 h, the solder which contains zinc maintains their shear strength and just slightly decreases with aging time.

Acknowledgements The authors would like to acknowledge that the study covered in this manuscript was funded by Ministry of Higher Education Malaysia regarding the use of the ISIS Neutron and Muon Source entitle the neutron tomography studies of the geopolymer ceramic used for reinforcement materials in a solder alloy for a robust electric/electronic solder joint under reference no: JPT.S (BPKI)2000/016/018/019(29).

References

1. S.M. Amli, M.M. Salleh, M. Ramli, H. Yasuda, J. Chaiprapa, F. Somidin, Z. Shayfull, K. Nogita, Origin of primary Cu_6Sn_5 in hypoeutectic solder alloys and a method of suppression to improve mechanical properties. J. Electron. Mater. **50**(3), 710–722 (2021)
2. S.A. Musa, M.A.A. Mohd Salleh, N. Saud, Zn–Sn based high temperature solder—a short review. Adv. Mater. Res. **795**, 518–521 (2013)
3. M. Ramli, M.M. Salleh, H. Yasuda, J. Chaiprapa, K. Nogita, The effect of Bi on the microstructure, electrical, wettability and mechanical properties of Sn–0.7 Cu-0.05 Ni alloys for high strength soldering. Mater. Des. **186**, 108281 (2020)
4. R. Sayyadi, H. Naffakh-Moosavy, The role of intermetallic compounds in controlling the microstructural, physical and mechanical properties of Cu–[Sn–Ag–Cu–Bi]–Cu solder joints. Sci. Rep. **9**(1), 1–20 (2019)
5. N. Saleh, M. Ramli, M.M. Salleh, in *Effect of Zinc Additions on Sn-0.7 Cu-0.05 Ni Lead-Free Solder Alloy*, IOP Conference Series: Materials Science and Engineering, (IOP Publishing, 2017)
6. M.A.A. Mohd Salleh, M.H. Hazizi, Z.A. Ahmad, K. Hussin, K.R. Ahmad, Wettability, electrical and mechanical properties of 99.3 Sn–0.7 Cu/Si_3N_4 novel lead-free nanocomposite solder. Adv. Mater. Res. **277**, 106–111 (2011)
7. S. Vaynman, G. Ghosh, Some fundamental issues in the use of Zn-containing lead-free solders for electronic packaging. Mater. Trans. **45**(3), 630–636 (2004)
8. R. Mayappan, Z.A. Ahmad, Effect of Bi addition on the activation energy for the growth of Cu5Zn8 intermetallic in the Sn–Zn lead-free solder. Intermetallics **18**(4), 730–735 (2010)
9. F. Somidin, H. Maeno, M.M. Salleh, X.Q. Tran, S.D. McDonald, S. Matsumura, K. Nogita, Characterising the polymorphic phase transformation at a localised point on a Cu_6Sn_5 grain. Mater. Charact. **138**, 113–119 (2018)
10. J.W. Xian, M.M. Salleh, S.A. Belyakov, T.C. Su, G. Zeng, K. Nogita, C.M. Gourlay et al., Influence of Ni on the refinement and twinning of primary Cu_6Sn_5 in Sn–0.7 Cu–0.05 Ni. Intermetallics **102**, 34–45 (2018)

Chapter 3
Effect of Sb Addition to the Solidification and Microstructure of Sn–Ag–Cu Alloys

Mohd Izrul Izwan Ramli, Mohd Suhami A'isyah, Azliza Azani, and Nur Syahirah Mohamad Zaimi

Abstract Several countries have banned the use of lead (Pb) in electronics due to environmental and public health concerns. Lead-free solder is a key research subject in electronic device packaging. Tin–silver–copper (SAC) has great dependability, remarkable creep resistance, and thermal fatigue qualities. SAC solder has a greater melting point, less wetting ability, and coarser microstructure. Adding a small quantity of antimony (Sb) to SAC alloys will thin the intermetallic layer as well as slowing the reflow process. These intermetallic layers reportedly outperformed SAC alloys. The addition of Sb may improve solder connections in electronics. This study studied the effect of Sb on Sn–3.0Ag–0.5Cu (SAC305) solidification and microstructure. Samples of Sn–3.0Ag–0.5Cu–0.5Sb and Sn–3.0Ag–0.5Cu–1.5Sb solder alloy were prepared and each composition was analyzed for bulk solder, intermetallic compound (IMC), and wettability. Thermo-Calc Software modeled the solidification sequence analysis, while J Image Software analyzed wettability and intermetallic. Scanning electron microscopy (SEM) images of bulk solder and solder junctions were also analyzed. Sb-added SAC305 solder alloys produced $Ag_3(Sn, Sb)$ and $Cu_6(Sn, Sb)_5$ IMCs. Suppressing IMC growth induced SnSb and $Ag_3(Sn, Sb)$ tiny particle pinning along IMC grain boundaries. Original SAC305 IMCs (Ag_3Sn and Cu_6Sn_5) have lack ductility when soldered on a Cu substrate, lowering thermal fatigue lifetime and mechanical characteristics. The new improved solder alloys will increase pinning inhibits IMC development and also element interdiffusion. Therefore, this will improve the solder's mechanical characteristic.

Keywords Solidification · Microstructure · SAC alloys · Wettability · Antimony

M. I. I. Ramli (✉) · M. S. A'isyah · A. Azani · N. S. Mohamad Zaimi
Faculty of Chemical Engineering Technology (FTKK), Universiti Malaysia Perlis (UniMAP), 02600 Jejawi, Arau, Perlis, Malaysia
e-mail: izrulizwan@unimap.edu.my

A. Azani · N. S. Mohamad Zaimi
Center of Excellence Geopolymer and Green Technology (CEGeoGTech), Universiti Malaysia Perlis, Jejawi, Arau, Perlis, Malaysia

M. A. A. Mohd Salleh et al. (eds.), *Proceedings of the Green Materials and Electronic Packaging Interconnect Technology Symposium*, Springer Proceedings in Physics 289,
https://doi.org/10.1007/978-981-19-9267-4_3

3.1 Introduction

Tin–lead (Sn–Pb) solders have been widely used in electronic packaging. However, in response to mounting environmental and public health concerns about lead toxicity, several countries have enacted legislation prohibiting the use of lead in electronic applications [1]. As a result, lead-free solder has established itself as a critical area of research in the field of packaging for electronic devices. To find an alternative to lead-based solder alloys, several lead-free tin-based alloys were developed with elements such as silver (Ag), copper (Cu), zinc (Zn), bismuth (Bi), indium (In), and silver–copper (Ag–Cu) [2–4]. When lead-free solders were introduced initially to the electronics industry in the early 1990s, the tin–silver alloy Sn96.5/3.5Ag was one of the first materials to be studied. The most noticeable difference between this and the industry standard was the significantly higher melting point of 221 °C (Sn63/Pb37) versus 183 °C (Sn63/Pb37). Due to the global legislative requirements noted above, it is necessary to discover viable alternatives to Pb-containing solders for electronic assemblies. The key requirements for an alternative solder alloy are a low melting point. The melting point of the solder joint should be low enough to avoid thermal damage to the assembled assembly being soldered yet high enough to sustain the solder joint's operating temperatures. The solder should keep its mechanical characteristics at these temperatures.

In addition to the aspect of cost, electronic system producers are reluctant to convert to a more expensive solder unless its improved qualities are demonstrated or governmental compulsion to do so. Tin–silver–copper, also known as SAC, has been identified as the most potential lead-free solder currently available for substituting tin–lead solder, given the high reliability, outstanding creep resistance, and thermal fatigue qualities. Numerous issues, however, remain unresolved. For example, SAC solder has a higher melting point, poor wetting properties, and coarser microstructures than conventional solder. Two techniques are used to enhance the qualities of lead-free solder even more. The first method to improve SAC solder wettability is to alloy it with alloying elements such as gallium (Ga). Additionally, rare earth elements can enhance overall performance. Incorporating micro or nanoparticles is another method that could be used to improve performance. Metal particles, organic compounds, ceramic particles, carbon nanotubes, and polymer particles all make up this material. SAC solder's properties vary according to the type and size of particles added. Not only the microstructure of the solder can be altered by the metal particles but the metal particles can also create a new phase within the solder matrix whereby compound or ceramic particles cannot. According to previous findings, adding minor amounts of Sb to SAC alloys led to the formation of intermetallic layers that were thinner and expanded more slowly during consecutive reflows. According to researchers, these intermetallic layers were outstanding to those created by SAC reference alloys. They may improve the performance of solder joints and electronic assemblies in some cases.

3.2 Experimental Procedure

3.2.1 Materials

The material used in this study to prepare solder alloys was developed using the following composition (in weight percent, wt%) which are Sn–3.0Ag–0.5Cu, Sn–3.0Ag–0.5Cu–0.5Sb, and Sn–3.0Ag–0.5Cu–1.5Sb. Three samples were made from each of the compositions for bulk solder, solder ball, and wettability analysis.

3.2.2 Methodology

The first step of this research is the casting process. SAC305 solder and Sb element were deposited into a crucible and melted at 350 °C in a top-loading furnace with stirring in between to ensure homogeneity until the dissolution process is complete. The molten mixture was then poured into a hollow steel mold and the surplus molten mixture onto a plate mold, allowing it to solidify. The molten mixture in the hollow steel mold was used for bulk solder microstructure analysis, while the surplus molten mixture on the plate mold was rolled into sheet metal to form solder balls. This process is repeated for each weight percent.

The surplus molten mixture that has been solidified is rolled with a rolling mill to reduce its thickness and make the thickness uniform. The final product of rolling is in the form of a sheet of material. This process is repeated for each weight percent of the solidified molten mixture. The sheet metal is then subjected to the sheet metal punching process. The sheet metal is punched by hand using a manual punch press to create the desired hole shape to form solder balls. The sheet metal is sandwiched between the punch and the die during the punching process. The punch is pressed onto and through the sheet metal. This process is repeated for each weight percent of sheet metal. The punch-pressed sheet metal is then subjected to a reflow process, resulting in the formation of solder balls. It is placed on a copper substrate, and solder paste flux is applied to the copper substrate. Solder-paste flux act as an adhesive to hold the sheet metal until solder particles undergo reflow.

After mounting was complete, each sample was ground to remove any damaged, deformed surface material, debris and heat from the mounted specimens while minimizing additional surface deformation. Grinding stages include coarse grinding, medium grinding, and fine grinding. The specimen is ground to obtain a flat surface free of previous tool marks and cold working caused by specimen cutting. The debris is washed away with tap water, and the specimen is dried immediately after all of the scratches on the sample are parallel.

After fine grinding, the specimen is carefully rinsed to ensure that debris does not contaminate the next polishing step. Before the polishing process, specimens and hands were thoroughly cleaned. A small amount of Alumina Suspension is applied to the polishing cloth. The specimens are held in place with both hands, and moderate pressure is applied. The sample is then cleaned using soap water and rinsed with distilled water to remove the watermarks. Coating of the sample is done before proceeding with Scanning Electron Microscope testing. Sputter coating of an ultra-thin coating of platinum was applied to the sample.

3.3 Results and Discussion

3.3.1 Solidification

Figure 3.1 illustrates the solidification process. The solidification process for Sn–Ag–Cu with 0.5 wt% Sb addition is as follows: liquid at 200.1 °C, primary β-Sn formed at 220 °C, Ag_3Sn β-Sn binary eutectic structure developed around 218 °C, and finally eutectic Ag_3Sn β-Sn Cu_6Sn_5 nucleated around 215.6 °C.

Figure 3.2 shows the solidification sequence of Sn–Ag–Cu alloys with a 1.5 wt% Sb addition. The solidification process for Sn–Ag–Cu with 1.5 wt% Sb addition is as follows: Primary β-Sn formed at 221.69 °C, Ag_3Sn β-Sn binary eutectic structure developed around 218.456 °C, and finally eutectic Ag_3Sn β-Sn Cu_6Sn_5 nucleated around 216.58 °C.

3.3.2 Microstructure Analysis

Figure 3.3 depicts the microstructures of various Sb concentrations casted with SAC305 solder alloys. Initially, the microstructure of SAC305 solder alloys was constituted of intermetallic compounds (IMCs) such as Ag_3Sn and Cu_6Sn_5. Figure 3.3a illustrates the β-Sn phase and the IMCs phase (Ag_3Sn, Cu_6Sn_5) microstructures in these initial SAC305 solder alloys. Meanwhile, Fig. 3.3b, c, illustrated the formation of new Ag_3(Sn, Sb) and Cu_6(Sn, Sb)$_5$ IMCs due to the addition of Sb elements to SAC305 solder alloys. The addition of Sb also caused the grain size to increase. A larger grain size will reduce the strength and toughness of the material.

3.3.3 Intermetallic Compound (IMC Layer)

Novel Ag_3(Sn, Sb) and Cu_6(Sn, Sb)$_5$ IMCs have been formed after the addition of Sb elements to SAC305 solder alloys as shown in Fig. 3.4. As a result of restraining IMC growth, increasing Sb concentration resulted in Sn–Sb synthesis and an increase in Ag_3(Sn, Sb) small particle pinning along the grain boundary of the IMCs. The morphology of the IMC layer becomes more consistent with smaller scallop-shaped

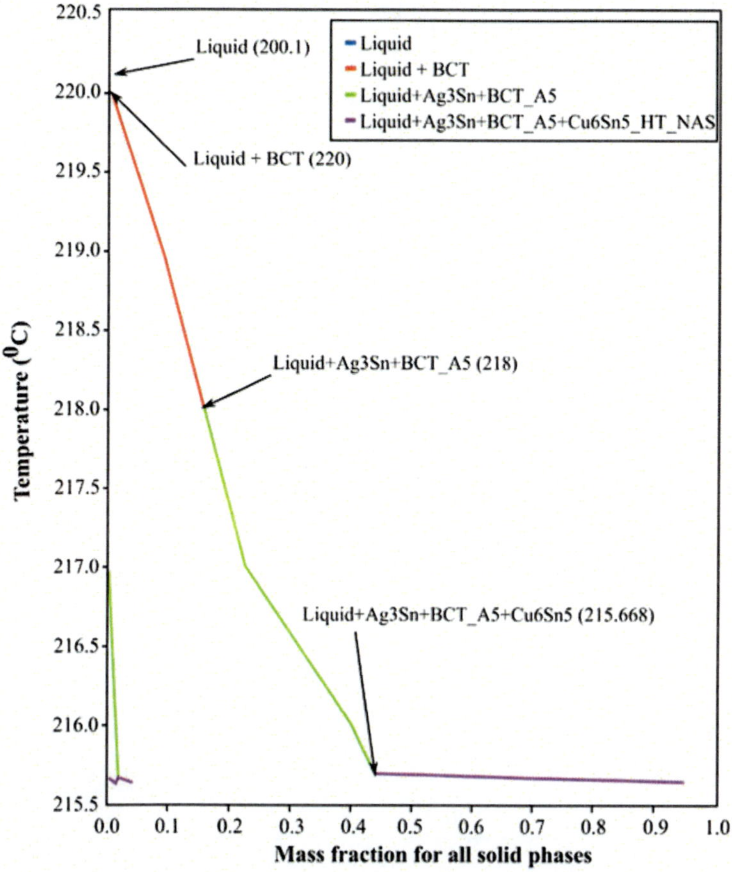

Fig. 3.1 Solidification sequence of Sn–Ag–Cu alloys with 0.5 wt% Sb addition

grain after the Sb element is introduced. The IMC layer also has a better uniform distribution when Sb elements are added to SAC305 alloys. Brittle failure is caused by thicker IMC. However, after incorporating Sb, more $Ag_3(Sn, Sb)$ particles precipitate on the Cu_6Sn_5 layer, causing the hardening effect and improving the strength of the SAC305 solder joint.

The thickness of the intermetallic compound of SAC305 with the addition of the Sb element is given in Table 3.1. The increase in Sb elements has resulted in a reduction in the thickness of the IMC layer. The genuine IMCs of SAC305 solder alloy (Ag_3Sn and Cu_6Sn_5) are brittle and lack ductility when soldered on a Cu substrate, compromising the thermal fatigue life and mechanical qualities of solder connections. By cause of the improved pinning along the grain boundary, the latest produced solder alloy may restrain the growth of IMCs and reduce element interdiffusion [5]. As a consequence, the mechanical properties of solder connections have been improved a lot.

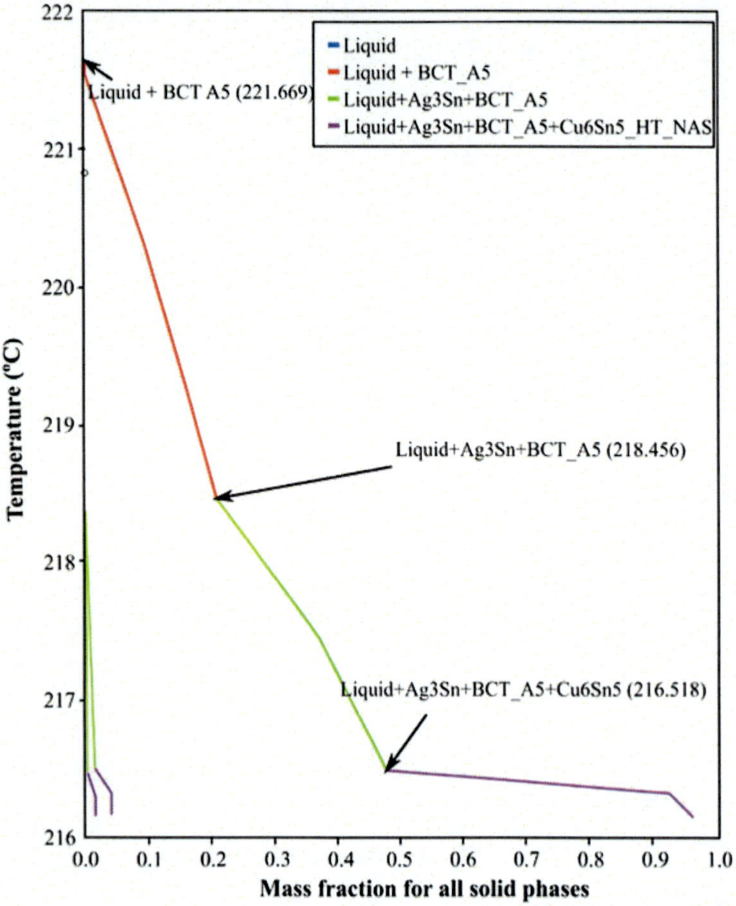

Fig. 3.2 Solidification sequence and mass fraction of Sn–Ag–Cu alloys with 1.5 wt% Sb addition

3.3.4 Wettability

The measurement of the contact angle involves the control of a number of parameters that can affect the wetting behavior of the solder [6]. A perfect metallurgical bond must be established at the interface between the solder (liquid) and the Cu (solid) substrate for good solder wetting.

Table 3.2 shows the contact angle of SAC305 solder alloys with the addition of the Sb element. An increasing trend of the contact angle was observed as the wt% of Sb increased. A minimum contact angle of 30.56° was observed with a 0.5 wt% Sb addition. The results show that a small amount of reinforcement in SAC305 alloys decreases the contact angle compared to pure SAC305. A lower contact angle represents better wettability of the solder joint [7–10]. The increase in

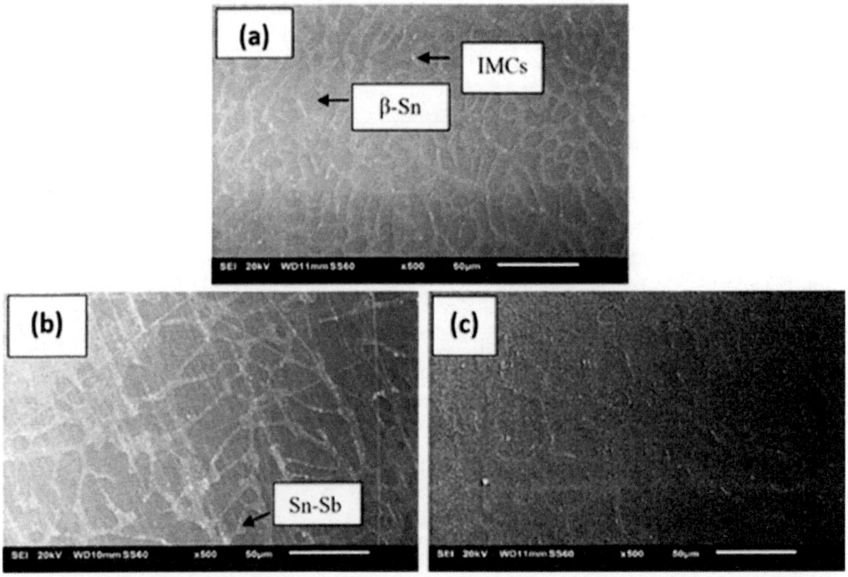

Fig. 3.3 SEM images of SAC305 bulk solder alloys for different wt% of Sb: **a** 0 pct, **b** 0.5 pct, and **c** 1.5 pct at 500 × magnification

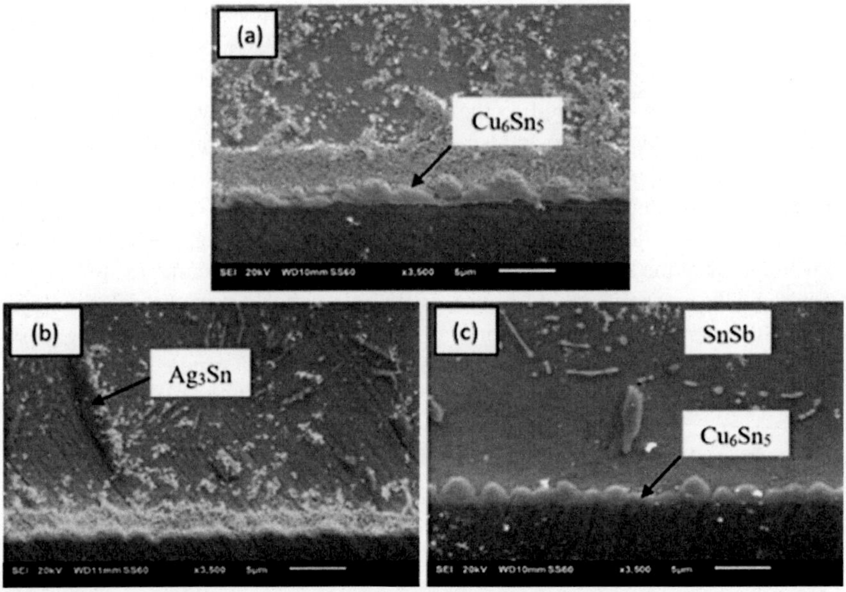

Fig. 3.4 SEM images of intermetallic compounds **a** SAC305, **b** SAC305-0.5Sb, and **c** SAC305-1.5Sb solder joint

Table 3.1 Thickness of intermetallic compounds

Solder alloys	IMC thickness (μm)
SAC305	1.36
SAC305-0.5Sb	1.64
SAC305-1.5Sb	1.33

Table 3.2 Influence of Sb addition on the contact angle in SAC305 alloys

Solder alloys	Average contact angel (θ)
SAC305	31.88
SAC305-0.5Sb	30.56
SAC305-1.5Sb	32.94

contact angle with the increase of Sb shows that a higher concentration of Sb leads to low wettability.

3.4 Conclusion

SAC solder alloys have been casted with different concentrations of antimony (Sb) in an effort to overcome a few of their drawbacks. The effects of Sb addition on the microstructural, IMC, and wettability of SAC305 alloys showed that the addition of Sb have led to improved microstructural of the IMCs whereby a more uniform distribution occurred compared to the conventional SAC alloys. In addition, the formations of new $Ag_3(Sn,Sb)$ and $Cu_6(Sn,Sb)_5$ IMCs on the grain boundary of the β-Sn phase also improve the drawbacks. Besides that, the addition of Sb formed more refine microstructure of the primary β-sn. The highest Sb content of 1.5 wt% resulted in the formation of SnSb and small particle pinning of $Ag_3(Sn,Sb)$ along the grain boundary of the IMCs. The addition of Sb suppresses and refines the formation of intermetallic compounds and improves the microstructure of the solder alloys although the increase in grain size leads to a decrease in the strength and toughness of the material. Aside from that, the increase of Sb elements has resulted in a reduction in IMC layer thickness which improves the wettability of solder alloys.

Acknowledgements The authors acknowledge that the Ministry of Higher Education Malaysia regarding the use of the ISIS Neutron and Muon Source grant funding entitle the neutron tomography studies of the geopolymer ceramic used for reinforcement materials in a solder alloy for a robust electric/electronic solder joint under reference no: JPT.S (BPKI)2000/016/018/019(29).

References

1. W.N. C. *Weng, Evolution of Pb-Free Solders*. IntechOpen (2017)
2. L.M. Lee, A.A. Mohamad, Interfacial reaction of Sn–Ag–Cu lead-free solder alloy on Cu: a review. Adv. Mater. Sci. Eng. **2013**, 1–11 (2013)
3. M. Zhao, L. Zhang, Z.Q. Liu, M.Y. Xiong, L. Sun, Structure and properties of Sn–Cu lead-free solders in electronics packaging. Sci. Technol. Adv. Mater. **20**(1), 421–444 (2019)
4. Q. Guo, Z. Zhao, C. Shen, A comparison study on microstructure and mechanical properties of Sn-10Bi and Sn–Ag–Cu solder alloys and joints. Microelectron. Reliab. **78**, 72–79 (2017)
5. H.T. Lee, H.S. Lin, C.S. Lee, P.W. Chen, Reliability of Sn–Ag–Sb lead-free solder joints. Mater. Sci. Eng., A **407**(1–2), 36–44 (2005)
6. E. Gregerson, Britannica, antimony | Definition, Symbol, Uses, & Facts. Encyclopedia Britannica (2022)
7. P. Sungkhaphaitoon, T. Plookphol, The effects of antimony addition on the microstructural, mechanical, and thermal properties of Sn–3.0Ag–0.5Cu solder alloy. Metal. Mater. Trans. A **49**, 652–660 (2018)
8. M. I. I. Ramli, M.M. Salleh, H. Yasuda, J. Chaiprapa, K. Nogita, The effect of Bi on the microstructure, electrical, wettability and mechanical properties of Sn–0.7 Cu–0.05 Ni alloys for high strength soldering. Mater. Des. **186**, 108281 (2020)
9. G. Zeng, S.D. McDonald, D. Mu, Y. Terada, H. Yasuda, Q. Gu, K. Nogita et al., The influence of ageing on the stabilisation of interfacial (Cu, Ni) 6 (Sn, Zn) 5 and (Cu, Au, Ni) 6Sn5 intermetallics in Pb-free Ball Grid Array (BGA) solder joints. J. Alloy. Compd. **685**, 471–482 (2016)
10. M.I.I. Ramli, M.A.A. Mohd Salleh, F.A. Mohd Sobri, P. Narayanan, K. Sweatman, K. Nogita, Relationship between free solder thickness to the solderability of Sn–0.7 Cu–0.05 Ni solder coating during soldering. J. Mater. Sci. Mater. Electron. **30**(4), 3669–3677 (2019)

Chapter 4
Investigating the Mechanical Performance of New Green Concrete Based on Unglazed Fired Roof Tile Waste

Zulkifli Mohd Rosli, Elsee Layu, Wan Amirul Shafiz Wan Zulkifli, Jariah Mohamad Juoi, Fariha Awatif Abdul Aziz, and Ridhwan Jumaidin

Abstract Concrete is commonly used as building material due to its high compressive strength and durability. Looking forward into the technologies, the use of recycled materials in concrete production has grown in popularity. For example, recycling of ceramic roof tiles waste as an aggregate in concrete will help to alleviate industrial waste disposal issues while also preserving natural aggregate supplies. In this work, unglazed fired roof tile waste (URTW) at 40, 50, and 60 wt% was investigated to potentially replace natural sand fine aggregate in producing new green concrete. The aim of this paper is to report on the mechanical properties of the new green concrete produced. A total of 12 concrete specimens, 100 mm × 100 mm × 100 mm each, were tested. Three concrete specimens were kept as control specimens, 9 specimens with URTW as fine aggregate at 40%, 50%, and 60% weight fraction as fine aggregates were accordingly prepared according to M25 grade of concrete (BS1881-108:1983). Eventually, all specimens were tested for its workability which is slump test (BS 1881:part 102), water absorption test and compressive strength by using compression testing. It was found that the strength gradually increases with the 40, 50, and 60 wt% of URTW as the fine aggregates. The strength achieved by the concrete (41.00 N/mm^2) is above than the target mean strength of M25 (25.00 N/mm^2). However, it is also observed that the use of URTW influence the workability of the concrete mixture, here the workability of concrete mixture with 60 wt% of URTW is reduced compared to 20 and 50 wt%. Thus, it can be concluded that the promising percentage of weight for URTW used as fine aggregate in the concrete mixture is at 50 wt%.

Keywords Unglazed tile waste · Workability · Compressive strength · Concrete · Recycling

Z. Mohd Rosli (✉) · E. Layu · F. A. Abdul Aziz · R. Jumaidin
Fakulti Teknologi Kejuruteraan Mekanikal and Pembuatan, Universiti Teknikal Malaysia Melaka, Hang Tuah Jaya, 76100 Durian Tunggal, Melaka, Malaysia
e-mail: zmr@utem.edu.my

W. A. S. Wan Zulkifli · J. Mohamad Juoi
Fakulti Kejuruteraan Pembuatan, Universiti Teknikal Malaysia Melaka, Hang Tuah Jaya, 76100 Durian Tunggal, Melaka, Malaysia

M. A. A. Mohd Salleh et al. (eds.), *Proceedings of the Green Materials and Electronic Packaging Interconnect Technology Symposium*, Springer Proceedings in Physics 289,
https://doi.org/10.1007/978-981-19-9267-4_4

4.1 Introduction

Presently, the growth of building construction has increased yearly over time. Stated that the estimated concrete manufactured each year is roughly 25 billion tons of concrete worldwide. Concrete is one of the most significant contributors to greenhouse gas emissions. At the same time, concrete has unique properties where the recovery often falls between the standard definition of reuse and recycling [1]. The replacement of aggregate in concrete formulation with recycled material is the step to achieving the concept of green concrete. Green concrete is a type of regular concrete that includes recycled or environmentally friendly materials, lasts longer, or performs better than traditional concrete, decreasing the need for replacement in the future [2]. Green products are typically distinguished by two main goals which is waste reduction and resource efficiency maximization. Commonly, the fabrication route of green concrete is identical to that of standard concrete production [3].

On the other hand, most manufacturing industries face solid waste management, especially in developing countries like Asia. These issues are essential to have extra attention because they could affect the environment badly. At the same time, due to the progress of construction work, these wastes are accumulated from time to time. Somehow, dumping it could risk the environment because the manufactured materials contain other materials, making it an odd substance to the environment [4]. By that, many researchers try to solve these issues by doing a lot of research, studying, and improving hypotheses to develop results that could help overcome environmental threat issues. All these studies show that concrete which is 90% dependent on natural resources had a high impact on the environment. Current progress in work related to green concrete had reported that reduced utilization of natural resources in concrete development is deemed possible by substituting the cementitious material with recycled substances [5, 6]. It is observed that most formulations are mostly able to accommodate at least 20% of recycled materials such as coal bottom ash as fine aggregate in fabricating green concrete [7]. The aim of this article is to report works on the development of green concrete using unglazed roof tile waste for replacing fine aggregates in concrete fabrication.

4.2 Methodology

4.2.1 Portland Composite Cement

Ordinary Portland cement, (CEM II/B-M) grade 32.5, this type of cement is available in 50 kgs bags which complied to MS EN 197-1:2014 CEM II/B-M 32.5R. The initial time setting of this cement is not less than 75 min.

4.2.2 Fine Aggregate

Locally, river sand has been used as fine aggregate in a concrete mix in the experimental work. The standard specification requires compliance with the requirement of BS 882:1992 for aggregate from the natural resource for concrete.

4.2.3 Coarse Aggregate

Uncrushed coarse aggregate available from the local sources was used for the experiment work in the concrete mix design. The maximum size of coarse aggregate is 20 mm. Coarse aggregate used in experimental study was confirming BS 882:1992.

4.2.4 Unglazed Fired Roof Tiles Waste (URTW)

Unglazed roof tile waste has been collected from dump waste at a roof tiles manufacturer. Roof tile was manufactured for the housing development industry. The defective roof tile was taken and crushed to make the required size of fine aggregate.

4.2.5 Water

Tap water has been used for mixing and curing of concrete specimens. The water available from the local source was fit for usage.

4.2.6 Mix Design

The mix design prepared for M25 grade of concrete according to BS1881-108:1983, the proportion is 1:1.50:1.38 (Cement: Coarse aggregate: Fine aggregate) with water to cement ratio (w/c) of 0.5, by weight [8]. The graded aggregate was used in the concrete composition for better slump value results. The fine aggregate (sand) was replaced with unglazed roof tile waste (URTW) at 0, 40, 50, and 60% in the concrete mix (Table 4.1).

Table 4.1 Batching details

Sample	% of replacement with URTW	Sand (kg)	URTW (kg)	Coarse aggregate (kg)	Cement (kg)	Water (kg)	w/c ratio
CM0	0	7.2	0	6.7	4.8	2.4	0.5
URTW40	40	4.3	2.9	6.7	4.8	2.4	0.5
URTW50	50	3.6	3.6	6.7	4.8	2.4	0.5
URTW60	60	2.9	4.3	6.7	4.8	2.4	0.5
Total		18	10.8	26.8	19.2	9.6	0.5

4.2.7 *Workability Test*

Slump value

A slump test was conducted to study the behavior of the concrete containing URTW as fine aggregate in fresh concrete. The slump test was carried out in accordance with BS EN 12,350-2 [9]. The apparatus consists of a slump cone with a base plate. The cone is kept over the base plate and filled with freshly mixed concrete of the desired grade in three layers, and each layer is tamped 25 times with a tamping rod of standard dimensions. After leveling the concrete, the slump cone is lifted upwards carefully and the concrete subsides. The slumped value was calculated as the difference between the initial and end heights of the concrete slumped as shown in Fig. 4.1.

Fig. 4.1 Measurement of slump fall

4.2.8 *Water Absorption Test*

Water absorption test was carried out in this study to calculate the water absorption percentage of hardened concrete in accordance with BS 1881-122. Prior to the testing, the specimen cubes (100 mm × 100 mm × 100 mm) were taken out one hour in advanced from the curing tank at day 28. The specimen was wiped to surface dry condition and weighed to obtain the saturated surface dry weight, W_f, of the specimen. The initial weight W_i is taken before the samples are immersed in the water for curing process. The water absorption percentage of the hardened concrete was calculated by using Eq. (4.1), where W_i is the initial weight of cast concrete samples and W_f is the saturated surface dry weight after curing.

$$\text{Water Absorption (\%)} = \frac{W_f - W_i}{W_i} \times 100\% \tag{4.1}$$

4.2.9 *Compressive Strength*

Compressive strength test is the most common test conducted on concrete, because it is the desirable characteristic properties of concrete are quantitatively related to its compressive strength. Compressive strength was determined by using Compression Testing Machine which is available in CAST Consult Sdn Bhd. The capacity of these machine is 3000 kN, respectively. Constant loading rate between 0.4 and 0.8 N/mm^2 was applied on the hardened cubic concrete specimen [10]. The compressive strength of concrete was tested using 100 mm × 100 mm × 100 mm cube specimens. The test was carried out by placing a specimen between the loading surfaces of a machine, and the load was applied until the specimen fails. Three test specimens were cast for each proportion and used to measure the compressive strength for each test conditions, and average value was considered.

4.3 Results and Discussion

4.3.1 *Slump Value*

The result of workability in terms of slump is given in Table 4.2. It is evident from the table that workability of concrete made using URTW increased with increase in replacement level (40% and 50%) and decreased with 60% replacement level. The decreased of slump value at higher percentage of URTW could be due to the nature of ceramic waste, with pores that contribute to a large surface area, consuming more

Table 4.2 Workability of URTW concrete

Sample	Replacement level of URTW (%)	Slump value, mm	Acceptable slump value, mm (BS 1881:part 102)
CM0	0	80	108–60
URTW40	40	35	
URTW50	50	65	
URTW60	60	35	

Table 4.3 Water absorption of URTW concrete in various replacement level at day 28 water absorption of URTW concrete

Sample	Replacement level of URTW (%)	% absorption water at 28 days	Max % of water absorption of the concrete	Min % of water absorption of the concrete
CM0	0	1.70	3	2
URTW40	40	1.69	3	2
URTW50	50	1.71	3	2
URTW60	60	1.63	3	2

water and hence decreasing the concrete's workability, when present at a higher percentage in the concrete mixing [11].

4.3.2 *Water Absorption*

The average of water absorption for 0, 40, 50, and 60% and of percentage replacement of URTW at day 28 were 1.70, 1.69, 1.71, and 1.63%, respectively, as given in Table 4.3. These results show that the water absorption for all specimens was not exceeding 3% as recommended for most mechanical structures for example guidelines stated by the BS 6349 British standard code for maritime structure.

4.3.3 *Compressive Strength*

Waste aggregate has a significant effect on fresh concrete and hardened concrete behavior. At an early age, the concrete continuously developed compressive strength, even at 28 days. The compressive strength of the URTW concrete has a great bonding with the other cementitious materials. It has been reported that using recycled ceramic aggregate can meliorate distinctly the structure of the concrete, thus improving its strength [12, 13]. According to Table 4.4, the compressive strength of concrete prepared using URTW exhibits comparable performance to that of control concrete.

Table 4.4 Compressive strength of concrete in various replacement level at day 28

Sample No.	Replacement level of URTW (%)	Compressive strength (N/mm^2)
CM0	0	32
URTW40	40	37
URTW50	50	41
URTW60	60	41

4.4 Conclusion

In can be concluded that replacing unglazed roof tile waste with sand as fine aggregate in the concrete mix has improved the compressive strength of the concrete. As stated in BS 812, the unglazed roof tile waste can partially replace the sand as fine aggregate in the concrete mix design. With the increased percentage of the unglazed roof tile waste in the concrete mix, workability decreases. At the replacement of unglazed roof tile for 40%, the compressive strength of the concrete improved 38% from the minimum target strength of 28 days. The strength achieved by the concrete is more than the target mean strength of M25. The compressive strength has improved, but not all the sample workability is within the range. So, the best ratio to use as M25 grade concrete is the proportion of URTW50. So it could be concluded that 50% replacement of unglazed roof tile waste in concrete can safely be used in the concrete composition without considerable loss of compressive strength in construction.

Acknowledgements The author acknowledges the financial support of the UTeM short-term grant (PJP S01782) for this work.

References

1. M.D.H. Doye, Green concrete: efficient & eco-friendly construction materials. IOSR J. Mech. Civ. Eng. **14**(03), 33–35 (2017). https://doi.org/10.9790/1684-1403023335
2. P. Zhang, P. Zhang, J. Wu, Z. Guo, Y. Zhang, Y. Zheng, Mechanical properties and durability of sustainable concrete manufactured using ceramic waste: a review. J. Renew. Mater. **11**(2), 937–974 (2023)
3. K. Nandhini, J. Karthikeyan, Sustainable and greener concrete production by utilizing waste eggshell powder as cementitious material—a review, Constr. Build. Mater. **335**, 127482 (2022)
4. M. Osial, A. Pregowska, S. Wilczewski, W. Urbańska, M. Giersig, Waste Management for green concrete solutions: a concise critical review. Recycling **7**, 37 (2022)
5. K. Kumar, S. Dixit, R. Arora, N.I. Vatin, J. Singh, O.V. Soloveva, S.B. Ilyashenko, V. John, D. Buddhi, Comparative analysis of waste materials for their potential utilization in green concrete applications. Materials **15**, 4180 (2022)
6. C.A. Clear, British Standards and the use of recycled aggregate in concrete Eurocodes European Standards are British Standards, pp. 1–25 (2021)
7. H. Zhou, R. Bhattarai, Y. Li, B. Si, X. Dong, T. Wang, Z. Yao, Towards sustainable coal industry: turning coal bottom ash into wealth. Sci. Total Environ. **804**, 149985 (2022)

8. D.C. Teychenné, R.E. Franklin, H.C. Erntroy, Design of normal concrete mixes. Build. Res. Establ. Ltd, **331**(1), 46, 2010, [Online]. Available: https://epdf.pub/design-of-normal-concrete-mixes-br-331-ci-sfb.html
9. British Standard Institution, Testing fresh concrete—Part 2: slump test. Bs En 12350-2, pp. 5–8 (2009)
10. BS 1881-116, Testing concrete. in *Method for Determination of Compressive Strength of Concrete Cubes* (BSI, London, 1983)
11. A.A. Negm, A. El Nemr, F. Elgabbas, M.A. Khalaf, High and normal strength concrete using grounded vitrified clay pipe (GVCP). Clean. Mater. **5**, 100107 (2022)
12. F. Liu, J. Liu, B. Ma, J. Huang, H. Li, Basic properties of concrete incorporating recycled ceramic aggregate and ultra-fine sand. J. Wuhan Univ. Technol. Mater. Sci. Edn. **30**(2), 352–360 (2015)
13. L.G. Li, Z.Y. Zhuo, J. Zhu, J.J. Chen, A.K.H. Kwan, Reutilizing ceramic polishing waste as powder filler in mortar to reduce cement content by 33% and increase strength by 85%. Powder Technol. **355**, 119–126 (2019)

Chapter 5
One-Pot Fusion-Impregnation Synthesis of Nickel Supported Magnesium Aluminate for Hydrogen Rich Syngas Production

Norhasyimi Rahmat and Zahira Yaakob

Abstract Hydrogen has been identified as one of interesting zero concept emission fuel to power electricity and transportation. The production of hydrogen includes carbon dioxide reforming of methane with steam which is scarcely reported and still has a lot of areas and scopes for investigations. Therefore, the main objective of this study is to investigate the catalytic activity of newly synthesized nickel supported magnesium aluminate using the newly method of one-pot fusion-impregnation and its fundamental behaviour in steam reforming of methane. The calcined catalysts were characterized using X-ray diffraction (XRD), nitrogen adsorption–desorption isotherms, Brunauer–Emmett–Teller (BET) surface area, Barrett, Joyner and Halenda method, high resolution transmission electron microscopy (TEM), temperature programme oxidation (TPO) and field emission electron microscope (SEM). The catalyst synthesized in this study displayed good properties of high surface area and narrow pore size distribution. It was carried out for reforming reaction of carbon dioxide and methane with steam at temperatures ranged between 700 and 900 °C and different Ni loading. The highest value for hydrogen yield and conversion of methane and carbon dioxide was shown at 25 wt% nickel loaded on magnesium aluminate. The higher metal loading did not exhibit an improved pattern which can imply a saturated and agglomerated particle that prohibit oxidation of carbon on catalyst surface for nickel loading of more than 40 wt% in this study.

Keywords One-pot synthesis · Syngas · Hydrogen · Steam reforming · $Ni/MgAl_2O_4$

N. Rahmat (✉)
School of Chemical Engineering, College of Engineering, Universiti Teknologi MARA, 40450 Shah Alam, Selangor, Malaysia
e-mail: norhasyimi@uitm.edu.my

Z. Yaakob
Department of Chemical and Process Engineering, Faculty of Engineering and Built Environment, Universiti Kebangsaan Malaysia, 43600 Bangi, Selangor, Malaysia

M. A. A. Mohd Salleh et al. (eds.), *Proceedings of the Green Materials and Electronic Packaging Interconnect Technology Symposium*, Springer Proceedings in Physics 289,
https://doi.org/10.1007/978-981-19-9267-4_5

5.1 Introduction

The increasing global population growth for the past few decades, which has reached for more than 1.5 billion people, has greatly influenced the ever-changing energy demand in the world. It is of vital importance to ensure energy resources and technologies are available to meet the demand of human development, social well-being, economic growth and environmental sustenance. Currently, the main primary source to empower energy production depends on fossil fuel, nuclear, hydro-energy and other renewables. The fossil fuel has been reported to be remained as the backbone of the energy source production [1], and its reserves are ample for long-term exploitation despite the reports claiming on the depletion of fossil fuel reserves [2]. This implies the longer term of greenhouse gases release, especially CO_2 and CH_4, to the environment from the fossil fuel utilization. Crude oil is contributing the highest percentage of energy source contributor which accounts for 32.8%. This is followed by coal with 27.2, 20.9% natural gas and 20% renewable energy sources, i.e. nuclear power, solar, wind, geothermal, hydroelectric dams, combustible biomass and waste as well as other alternative energy sources [3].

A zero-emission concept fuel should be the substantial replacement for the primary energy fuel and combustible biomass since they cause the emission of anthropogenic CO_2 to the atmosphere. One of the options is hydrogen, which has emerged as a promising and sustainable fuel with zero emission for electricity generation and zero tailpipe emission for fuel vehicles. Hydrogen could be essentially synthesized from any hydrocarbon on the earth. Basically, the renewable hydrocarbon, non-renewable hydrocarbon and renewable energy are the abundant raw source for hydrogen synthesis. The renewable energy source from water [4], wind [5] and geothermal [6] involves high capital cost in order to sustainably and innocuously produce hydrogen. Production of hydrogen from non-renewable hydrocarbon such as coal and natural gas exploits the chemical and thermochemical technologies in which these methods require carbon sequestration to mitigate the carbon release [7]. The technologies involved in the production of hydrogen from non-renewable hydrocarbon are chemical and thermochemical process. Among challenges producing hydrogen from non-renewable hydrocarbon is the uncertain price of fossil fuels which makes the final cost of hydrogen production volatile and unpredictable.

Recently, the production of hydrogen from biogas has been the main interest since it could be an effective effort to mitigate the release of methane and carbon dioxide that pose a potent effect to the greenhouse system. Attainment of established yield and purity of hydrogen is one of the challenges in the process of producing hydrogen from biogas since the composition of CH_4 as the reactant limiter is varying. Besides, the specific reaction mechanisms from different reaction studies have not been thoroughly investigated. Another principal challenge is to minimize the carbon deposits on catalyst surface and other carbonaceous impurities since those compounds could exacerbate the efficiency of chemical reaction and heighten the activation energy.

Expensive noble metals and support material are usually associate with the development of catalyst to improve catalytic activity and stability as well as selectivity in hydrogen production. The use of Rh and Pd on different supports for example La–Al_2O_3, $CeZrO_2$–Al_2O_3 and La_2O_3–CeO_2–ZrO_2 [8] is frequently reported for different synthesis of catalyst developed for steam reforming of biogas for example impregnation [9], co-precipitation [9] and sol–gel [10]. Lanthana, perovskite, silica and spinel are the types of support materials commonly prepared using these methods [11].

Lately, transition non-noble Ni has been comprehensively investigated for steam reforming of biogas. Nevertheless, the fast deactivation of Ni at temperature of more than 700 °C due to sintering and easy formation of coke inhibits the catalytic activity which disrupts the process stability of biogas steam reforming [12]. However, those methods investigated are usually associated with tedious and lengthy steps of catalyst preparation. Moreover, if one method has proven to encounter a problem, i.e. coke deposition but it did not necessarily overcome other shortcomings such as metal sintering and low feed conversion. In order to address all deficiencies at the level of catalyst development and reactions studies, it always has to come with greater cost. Therefore, in this study, it is essential to develop a catalyst system that addresses shortcomings of monotonous catalyst preparation, low cost of metal and support material that promotes minimum carbon deposition at reasonably high feed conversion and high yield of hydrogen. Solid state fusion is another attractive method to prepare support materials; this method has been reported to be the simplest due to its one-pot dry method that skips complex and tedious steps, such as washing, filtration and drying commonly associated with wet chemical syntheses [13]. Therefore in this study, the use of $MgAl_2O_4$ as catalyst support, together with non-noble metal nickel as active catalyst, was synthesized via one-pot fusion-impregnation synthesis to investigate its catalytic activity towards feed conversion, H_2 yield and coke formation with the control of reaction parameters and conditions, instead of adopting and incorporating with expensive promoters and bimetallic catalysts.

5.2 Methodology

5.2.1 Chemicals and Methods

In this study, a new approach was taken to synthesize $MgAl_2O_4$ in a dry medium with only one single step and within short processing time. The approach does not require expensive and hazardous chemicals and reagents. The stoichiometric amounts of magnesium nitrate, aluminium nitrate and citric acid (Mg $(NO_3)_2 \cdot 6H_2O$: Al $(NO_3)_3 \cdot 9H_2O$: $C_6H_8O_7H_2O = 1:2:3$) were rigorously mixed and pulverized using a mortar and a pestle for minimum 1 h until the mixture had the yoghurt-like texture. The mixture was then transferred into an oven at 100 °C for 2 h to ensure the fusion between Al^{3+} and Mg^{2+} precursors. Subsequently, the annealing process was taken place in a furnace under ambient air which was repeated under different annealing temperatures of 600, 700, 800, 900 and 1000 °C. For different annealing durations, the samples were annealed at 700 °C for 4, 5, 6 and 7 h. The metal loaded was varied from 5 to 50 wt%. At a temperature of 40 °C, a 10 g of support powder was cautiously added to a beaker containing 1 L of purified water. Subsequently, at temperature of 60 °C, calculated amount of metal precursors ($(Ni(NO_3)_2 \cdot 6H_2O)$) was accordingly added. At a constant temperature of 90 °C, the mixture was stirred for at least 6 h until it started to be bubble-broth like. At this stage, the stirring rate was slowed down, and the temperature needed to be decreased to below 40 °C. Subsequently, the concentrated mixture was placed in an oven for overnight drying at 100 °C before it was calcined in a muffle furnace at 700 °C for 5 h.

5.2.2 *Characterization*

Brunauer–Emmett–Teller (BET) was used to determine the specific surface areas of freshly synthesized $Ni/MgAl_2O_4$ while Barrett–Joyner–Halenda (BJH) method was applied to compute the pores and volumes by using micropore surface analyser model 3 Flex Micromeritic with sorption gas of N_2. All catalyst samples were degassed prior to measurements at a temperature of 300 °C for 3 h. X-ray diffraction (XRD) of Bruker D8 Focus advanced powder diffractometer was used to characterize the phase evolution of $Ni/MgAl_2O_4$. The setting was set at scanning speed of $4° min^{-1}$ and step of 0.02° with Cu Kα radiation at 2θ degree, from 10° to 80°.

Transmission electron microscope (TEM) was used to characterize the morphology of catalysts using model Philips CM-12 at accelerating voltage of 100 and 120 kV. TEM samples were prepared by dispersing the powders in ethanol. Then, the samples were ultrasonicated for 10 min before it was dropped onto a carbon-coated copper grid, dried and analysed. Thermogravimetric analysis model Mettler Toledo Gas STAR System was used to analyse thermal behaviour of the catalysts at atmospheric air at a heating rate of 10 °C min^{-1} and an air flow rate of 20 ml min^{-1} from room temperature to 900 °C. Micromeritics AutoChem 2920 chemisorption analyser was used to investigate the temperature programmed reduction (TPR) of the fresh catalysts with temperature setting from 30 to 900 °C under a 10% H_2/N_2 gas mixture flow (20 ml min^{-1}) and a heating rate of 10 °C per min. Thermo TPDRO 1100 with He purging assistance was used as CO_2 temperature program desorption from temperature 40 to 1100 °C at a heating rate of 10 °C min^{-1}. In this study, temperature programme oxidation (TPO) from 30 to 1100 °C was used to characterize the oxidation behaviour on catalyst surface.

5.2.3 *Catalytic Activity*

The rig used in this study, namely as a reforming system, consists of four major segments, which are feeding, mixing, reaction and separation segments. The feeding segment mainly consists of piping systems which are connected to gas cylinders and peristaltic pumps, whereas the mixing segment is especially to prepare the feed stream at its respective concentration to be mixed at pre-heated temperature before going into the reaction segment. The reaction system consists of a vertical stainless steel fixed bed reactor which is equipped with three thermocouples for temperature difference recording as well as catalyst holders for catalyst loading. It has an internal diameter of 0.025 m and a length of 0.6 m embedded with an electric muffle furnace for heating. The separator segment consists of a condenser and separation tank particularly to separate steam and product gases. The product gases were collected in gas sampling bags for analysis by gas chromatography. The performance of catalytic activity in this study was evaluated based on CH_4 and CO_2 conversion as well as hydrogen and carbon monoxide yield which can be found from these equations (Fig. 5.1):

$$\text{Methane conversion} = \frac{(CH_4)_{in} - \eta(CH_4)_{out}}{\eta(CH_4)_{in}} \times 100$$

$$\text{Carbon dioxide conversion} = \frac{(CO_2)_{in} - \eta(CO_2)_{out}}{\eta(CO_2)_{in}} \times 100$$

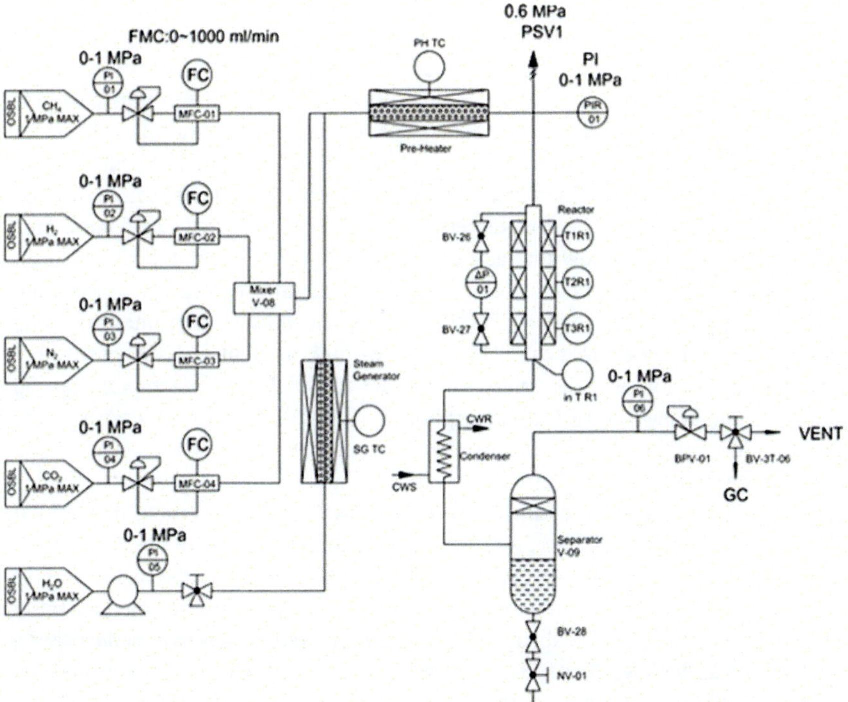

Fig. 5.1 Schematic of piping and instrumentation diagram of reforming system which is segmented into feeding, mixing, reaction and separation

Unreacted and product gases were analysed by Shimadzu Gas Chromatography model GC-2014C with GC Solution software version 2.32. Helium was used as carrier gas which flowed at 40 ml/min to elute gas samples in the capillary column. TDX-01 column was used to separate hydrogen, carbon monoxide, methane, nitrogen and carbon monoxide while thermal conductivity detector (TCD) was used to obtain the chromatogram of those gases. The temperature of the chromatograph injector was set at 150 °C, whereas the initial and final temperatures of the oven were set at 150 and 300 °C. For each sample injection, the total analysis time was 25 min.

5.3 Results and Discussion

The XRD patterns for calcined 25 wt% Ni/$MgAl_2O_4$ and $MgAl_2O_4$ spinel support are illustrated in Fig. 5.2a. $MgAl_2O_4$ spinel has distinction peaks at $2\theta = 19.2, 31.3, 36.9, 44.8, 59.5$ and $65.4°$. $NiAl_2O_4$ is formed in this synthesis method due to high chance reaction between NiO and Al_2O_3 during calcination process at high temperature. The formation of $NiAl_2O_4$ is also shown by strong diffraction peaks at $2\theta = 19.3$, 31.6, 37.3, 44.9, 59.6 and 65.5° at respective diffraction planes of (1 1 1), (2 2 0), (3 1 1), (4 0 0), (5 1 1), (4 4 0) and (5 3 3) with lattice constant of 8.106 Å (COD pattern: 900-53-50). $NiAl_2O_4$ is a spinel type which embraces cubic system with space group of Fd3m. However, the diffraction peaks of 25 wt% Ni/$MgAl_2O_4$ is differed from $MgAl_2O_4$ peaks in terms of peak intensity, crystallite size and the presence of NiO at diffraction peaks of $2\theta = 43.3$ and 62.8°. The reduction behaviour of NiO species on $MgAl_2O_4$ support is depicted in Fig. 5.2b. Two distinct reduction peaks can be observed from the figure, which can be distinguished at temperature range of 400–550 °C and 650–800 °C. The first reduction peak can be attributed to reducible free NiO particles which has low interaction with $MgAl_2O_4$ whereas the second peak can be assigned to NiO species that has strong interaction with spinel support and reduce at temperature above 650 °C. The second peak shows higher intensity value in comparison with the first peak. This indicates high amount of NiO species that has strong interaction with the spinel support are reducible within temperature 650 to 800 °C. The behaviour of CO_2 desorption for 25 wt% Ni/$MgAl_2O_4$ is shown in Fig. 5.2c. A weak basic site was observed for peak at temperature 254 °C while strong basicity was shown at distinguished peak between 400 and 540 °C. In this study, 25 wt% Ni/$MgAl_2O_4$ synthesized from F-IMP method has more than one of basic sites. As displayed in the figure, the weak and basic sites are referring to gas desorption of both phase $MgAl_2O_4$ and Ni in combination due to the metal-support interaction. Profile analysis area indicates that 25%Ni/$MgAl_2O_4$ has the capability to desorb 793 μmol g^{-1}. The distribution of pore size and N_2 adsorption desorption isotherms for 25 wt% Ni/$MgAl_2O_4$ is displayed in Fig. 5.2d. The synthesized catalyst exhibits mesoporous structure with narrow pore size distribution of 10 to 20 nm. This is corroborated with the crystallite size (10–15.4 nm) determined by XRD. The catalyst also shows the H3 hysteresis loop with relative pressure P/P^o in the range of 0.6–0.8. The adsorption and desorption pattern of this sample follows the

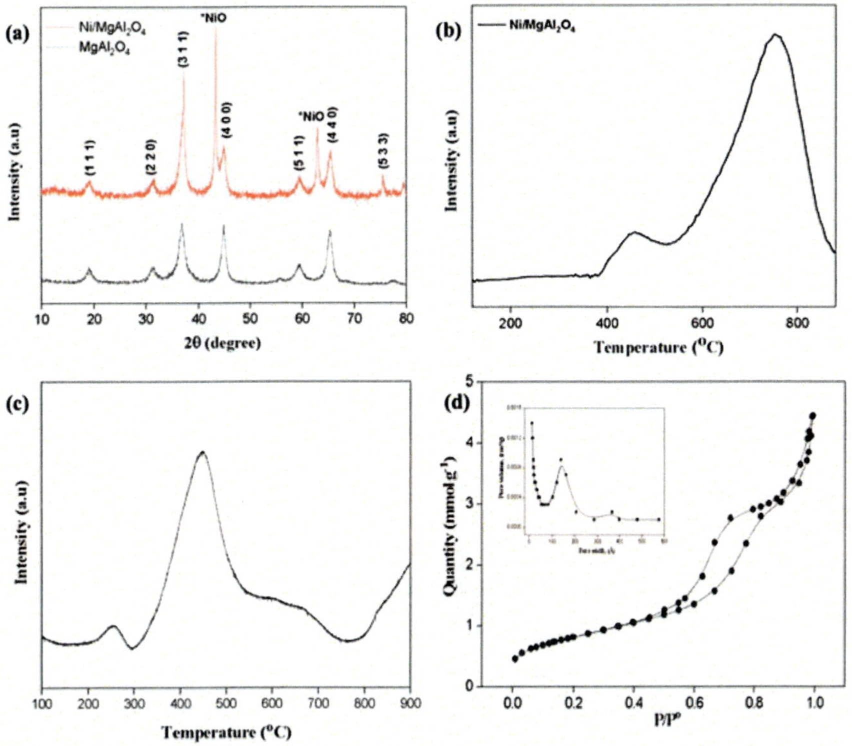

Fig. 5.2 Characterization of freshly synthesized Ni/$MgAl_2O_4$ **a** X-ray diffraction (XRD), **b** temperature programme oxidation (TPO), **c** CO_2-temperature desorption and **d** nitrogen adsorption–desorption isotherm

characteristic of a IV type isotherm which indicates the mesoporous structure has developed during the synthesis of the catalyst [14].

The morphology of prepared 25 wt% Ni/$MgAl_2O_4$ depicted in Fig. 5.3a–c with different magnifications of 200, 100 and 20 nm was to identify the dispersity and segregation of Ni species on catalyst support as well as the lattice structure. A homogeneous distribution of the active phase on $MgAl_2O_4$ is observed in Fig. 5.3a and b. The irregular shape of Ni species substituted in spinel structure is observed to display low particle aggregation which is in good terms with the result of N_2 adsorption–desorption isotherm in Fig. 5.2d. Figure 5.3c exhibits the lattice structure of $MgAl_2O_4$ at higher TEM magnification which indicates good properties for catalytic activity.

Figure 5.4 displayed the effect of wt% loading of Ni towards conversion of CH_4 and CO_2 as well as H_2 yield resulted from CO_2 reforming of methane with steam. From Fig. 5.4a, conversion of CH_4 and CO_2 was observed to be in inclining trend at Ni loading of 5–25 wt%. It achieved the highest conversion of CH_4 and CO_2 at respective 95.8% and 41.1% with 25 wt% Ni loaded $MgAl_2O_4$. A declining trend was exhibited at Ni loading of more than 40 wt%. Increment of Ni active sites at

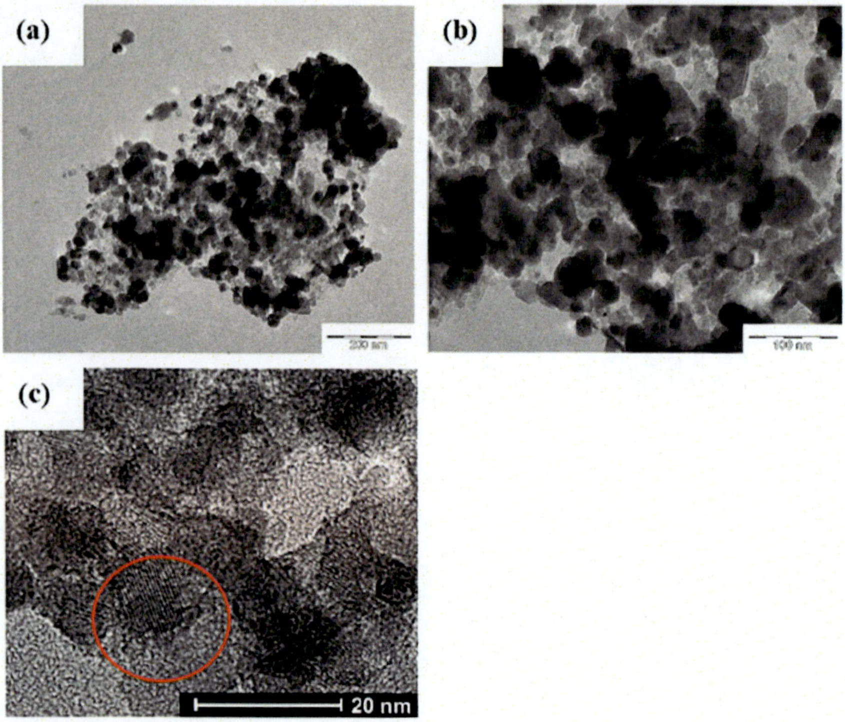

Fig. 5.3 Morphology properties of Ni/$MgAl_2O_4$ at different magnifications of transmission of electron microscopy (TEM). **a** 200 nm, **b** 100 nm and **c** 20 nm

Ni loading between 5 and 25 wt% could attribute to the increment of CH_4 and CO_2 conversion. The dissociation of CH_4 with active metal sites participation in CO_2 and H_2O activation tends to improve oxygen mobility and later will combine with carbon deposited on the catalyst surface. The basicity properties of $MgAl_2O_4$ catalyst support as shown by CO_2-adsorption profile, Fig. 5.2c, provide an enhanced CO_2 adsorption and chemisorption sites. These circumstances of better oxygen mobility and CO_2 adsorption–chemisorption promote better stability of catalytic activity on catalyst surface. The higher loading metal on catalyst surface could also hamper the oxygen mobility within lattice structure of $MgAl_2O_4$. This result corroborates with the findings by An et al. who found that the catalyst with 12 wt% Ni loading has the highest catalytic activity and lowest deposited carbon on catalyst surface [15] on the effect of different wt% of Co loading on MgAl for the ethanol steam reforming and H_2 yield. It was reported that the loading of metal more than 15 wt% could hamper the catalytic activity due to metal agglomeration, increased presence of Co^{3+} species and the decrease in surface area of the support [16]. Other study found that the high loading of metal could promote metal phase segregation and inhibit the interaction between active metal and support [17]. Figure 5.4b exhibits the H_2 yield and carbon

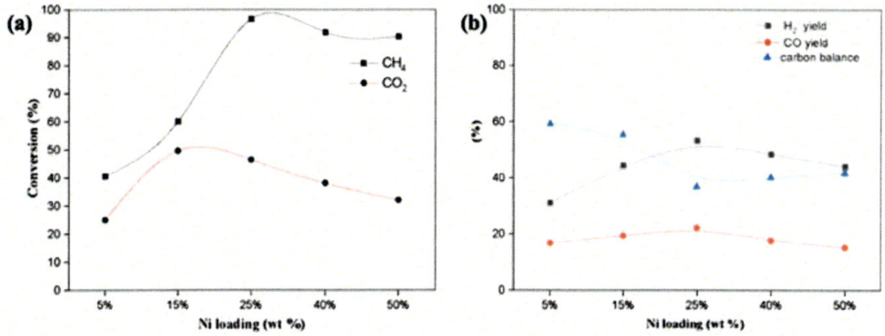

Fig. 5.4 Catalytic activity of $Ni/MgAl_2O_4$ at different wt% Ni content evaluated for **a** CH_4 and CO_2 conversion **b** H_2 and CO yield with carbon balance

formation trend for different Ni wt% loading, which agrees with behaviour pattern shown in Fig. 5.4a.

5.4 Conclusion

CO_2 reforming of CH_4 with steam to produce hydrogen rich syngas was successfully investigated over newly synthesized catalysts namely $Ni/MgAl_2O_4$. The newly developed methods, namely as one-pot fusion-impregnation synthesis follows a safe and harmless steps, circumvent the monotonous, weary and tedious synthesis tread, and more importantly able to convert high CH_4 and CO_2 which is of equivalent quality with other expensive and noble metal catalysts. Various characterization analyses had demonstrated the established properties of $Ni/MgAl_2O_4$. Besides, the synthesis method had successfully incorporated and segregated the active metals on support with minimal agglomeration. The experimental results indicated that 25 wt% $Ni/MgAl_2O_4$ catalyst showed the highest performance in associated with hydrogen yield and feed conversion. It can be postulated from this study that higher metal loading of more than 40 wt% did not show an improved pattern of catalytic activity which could be attributed to an agglomerated and saturated particle condition.

Acknowledgements A research grant from Universiti Teknologi MARA PY/2022/00386 project code 109036220002 and technical support from College of Engineering, Universiti Teknologi MARA is highly acknowledged for supporting this research study.

References

1. L. Cifarelli et al., World energy resources. EPJ Web of Conferences. **98**(2015). https://doi.org/10.1051/epjconf/20159801001
2. M. Höök et al., Depletion of fossil fuels and anthropogenic climate change—a review. Energy Policy **52**, 797–809 (2013). https://doi.org/10.1016/j.enpol.2012.10.046
3. A. Rahman et al., Environmental impact of renewable energy source based electrical power plants: solar, wind, hydroelectric, biomass, geothermal, tidal, ocean, and osmotic. Renewable and Sustainable Energy Reviews **161** (2022). https://doi.org/10.1016/j.rser.2022.112279
4. A. Alkaisi et al., A review of the water desalination systems integrated with renewable energy. Energy Procedia. **110**, 268–274 (2017). https://doi.org/10.1016/j.egypro.2017.03.138
5. U. Singh et al., Wind energy scenario, success and initiatives towards renewable energy in India—a review. Energies. **15**(6) (2022). https://doi.org/10.3390/en15062291
6. E.H.H. Al-Qadami et al., Evaluation of the pavement geothermal energy harvesting technologies towards sustainability and renewable energy. Energies. **15**(3) (2022). https://doi.org/10.3390/en15031201
7. I.B.S. Poblete et al., Sewage-water treatment and sewage-sludge management with power production as bioenergy with carbon capture system: a review. Processes. **10**(4) (2022). https://doi.org/10.3390/pr10040788
8. A. Erdőhelyi, Hydrogenation of carbon dioxide on supported Rh catalysts. Catalysts. **10**(2) (2020). https://doi.org/10.3390/catal10020155
9. T.L. LeValley et al., The progress in water gas shift and steam reforming hydrogen production technologies—a review. Int. J. Hydrogen Energy **39**(30), 16983–17000 (2014). https://doi.org/10.1016/j.ijhydene.2014.08.041
10. H. Arbag et al., Coke minimization during conversion of biogas to syngas by bimetallic tungsten–nickel incorporated mesoporous alumina synthesized by the one-pot route. Ind. Eng. Chem. Res. **54**(8), 2290–2301 (2015). https://doi.org/10.1021/ie504477t
11. K. Wittich et al., Catalytic dry reforming of methane: insights from model systems. ChemCatChem **12**(8), 2130–2147 (2020). https://doi.org/10.1002/cctc.201902142
12. P. Strucks et al., A short review on Ni-catalyzed methanation of CO_2: reaction mechanism, catalyst deactivation, dynamic operation. Chem. Ing. Tec. **93**(10), 1526–1536 (2021). https://doi.org/10.1002/cite.202100049
13. N. Rahmat et al., Renewable hydrogen-rich syngas from CO_2 reforming of CH_4 with steam over Ni/$MgAl_2O_4$ and its process optimization. Int. J. Environ. Sci. Technol. **17**(2), 843–856 (2019). https://doi.org/10.1007/s13762-019-02520-2
14. C. Schlumberger et al., Characterization of hierarchically ordered porous materials by physisorption and mercury porosimetry—a tutorial review. Advanced Materials Interfaces. **8**(4) (2021). https://doi.org/10.1002/admi.202002181
15. L. An et al., The influence of Ni loading on coke formation in steam reforming of acetic acid. Renewable Energy **36**(3), 930–935 (2011). https://doi.org/10.1016/j.renene.2010.08.029
16. M. Espitia-Sibaja et al., Effects of the cobalt content of catalysts prepared from hydrotalcites synthesized by ultrasound-assisted coprecipitation on hydrogen production by oxidative steam reforming of ethanol (OSRE). Fuel **194**, 7–16 (2017). https://doi.org/10.1016/j.fuel.2016.12.086
17. P. Djinović et al., Influence of active metal loading and oxygen mobility on coke-free dry reforming of Ni–Co bimetallic catalysts. Appl. Catal. B **125**, 259–270 (2012). https://doi.org/10.1016/j.apcatb.2012.05.049

Chapter 6
Optimization of Recycled Polypropylene Concrete Aggregates Processing Using Water-Assisted Melt Compounding via Response Surface Methodology

Noraiham Mohamad, Anis Aqilah Abd Ghani, Marvrick Anak Anen, Jeefferie Abd Razak, Raja Izamshah Raja Abdullah, Mohd Amran Mohd Ali, Hairul Effendy Ab Maulod, and Sian Meng Se

Abstract Plastic waste aggregate for concrete is a vast communicating topic nowadays due to overusing and depletion of natural sands and gravels. Despite efforts, plastic waste aggregates (PWA) still weaken concrete due to inadequate particle interaction. Incorporating clay particles into plastic particles is a wise step to mitigate this issue. Water-assisted melt compounding is a promising process for intercalating clay particles into polymer particles. The performance of recycled biaxially-oriented polypropylene composite aggregate (PCA) is highly dependent on the manufacturing processing parameters. In this study, the effect of water-assisted melt compounding process and formulation parameters using a Haake internal mixer on the tensile property of the PCA was investigated. The processing parameters (temperature and duration), clay and water content, called independent variables, were optimized to maximize the response (tensile strength) using response surface methodology (RSM) via a two-level full factorial design. The selected model accurately analysed the interaction between the parameters, with the coefficient of determination approaching a unity of more than 0.9633.

Keywords Concrete · Plastic recycling · Plastic waste aggregates · Optimization · Response surface methodology

---

N. Mohamad (✉) · A. A. Abd Ghani · M. A. Anen · J. Abd Razak · R. I. Raja Abdullah · M. A. Mohd Ali
Fakulti Kejuruteraan Pembuatan, Universiti Teknikal Malaysia Melaka, Hang Tuah Jaya, 76100 Durian Tunggal, Melaka, Malaysia
e-mail: noraiham@utem.edu.my

H. E. Ab Maulod
Fakulti Teknologi Kejuruteraan Mekanikal dan Pembuatan, Universiti Teknikal Malaysia Melaka, Hang Tuah Jaya, 76100 Durian Tunggal, Melaka, Malaysia

S. M. Se
San Miguel Yamamura Plastic Films Sdn Bhd, 75450 Ayer Keroh, Melaka, Malaysia

M. A. Mohd Salleh et al. (eds.), *Proceedings of the Green Materials and Electronic Packaging Interconnect Technology Symposium*, Springer Proceedings in Physics 289,
https://doi.org/10.1007/978-981-19-9267-4_6

6.1 Introduction

Concrete is the world's most commonly used building material and the second-most widely used material after water [1]. Concrete is a composite material consisting of several basic elements. It combines an aggregated matrix with a binder. The Portland cement, or asphalt as a binder, hardens and holds the aggregate together. Global concrete production is expected to exceed 18 billion tonnes annually by 2050 [2], shattering previous records. In most countries, unprocessed gravel, sand pits, river sands or crushed rocks are the primary sources of raw materials and aggregates [3]. Besides, recycled aggregates obtained from various industrial waste products (construction and demolition) are all sources of natural aggregates. The increasing demand for concrete aggregates, mainly natural stone and sand, disrupts the natural equilibrium of the ecosystem. Given the widespread use of aggregates in concrete production, developing a more environmentally friendly aggregate from other highly accumulated waste material is critical.

Plastic products, especially for the packaging industries, are in great demand due to their low cost and ease of production. Hence, plastic waste fast accumulates over time and grows [4]. Numerous studies have investigated the possibility of substituting plastic waste for aggregates in concrete [5] and PWA as an alternative to fine aggregates in concrete production over the last several decades. PWA concrete is generally made by directly substituting the volume of coarse or fine natural aggregates with the plastic waste of the same mass or quantity [6]. Because of its reduced workability and strength, PWA-based concrete still requires improvement to be used in stress-bearing construction structures. As a result, PWA quality and surface properties must be improved. One method is remelting and refabricating the plastic waste into suitable sizes and shapes. When heat-modified PWA was used, the mechanical properties of concrete were improved [7]. Secondly, plastic waste is reformulated into composites by combining natural elements in the aggregates. Yet, different surface properties between polymer and natural elements desire an innovative process method.

Water-assisted compounding is a hybrid process between solution-assisted and traditional melt mixing. This novel approach is widely reported to enhance the mechanical properties of polymer reinforced with clay composite systems [7]. The alleged benefits of water-assisted compounding are in delivering more excellent clay particle dispersion in the polymer matrix. Its capability to modify the composites' properties makes it a promising method for achieving these objectives. This research looked into and highlighted the possibility of producing plastic composite aggregates with improved properties using melt compounding with water-assisted.

Optimization of processing parameters is strongly needed to produce high-quality plastic aggregates. Optimized parameters help to reduce the cost due to rejected products and have proven to be cost-effective [8]. Therefore, this study focuses on the screening methodology using widespread used Response Surface Methodology (RSM) for the optimization of water-assisted melt compounding processing parameters; *A*: clay content (0–10%), *B*: temperature (130–180 °C), *C*: time (5–10 min) and *D*: water content (0–10%).

6.2 Methodology

6.2.1 *Raw Materials*

San Miguel Plastic Films Sdn Bhd supplied recycled biaxially-oriented polypropylene (BOPP) pellets. The pellets have a density of 0.92 g/cm^3. Bentonite clay has a particle size between 0.81 and 174 μm and was supplied by Edutech Sales Sdn Bhd.

6.2.2 *Designing the Processing Parameters Using Response Surface Methodology (RSM)*

Design Expert Software (Statistics Made Easy, version 10) was used to generate the experiment. A 2^4 factorial design with three replications at the centre points was used for four independent variables. As a result, 19 experiments were carried out in this study (Table 6.1). Table 6.2 shows each variable's low, middle and high levels. Based on the results of the experiments, a factorial model is chosen and analysed using analysis of variance (ANOVA) and regression analysis.

6.2.3 *Preparation, Testing and Analysis*

The BOPP plastic waste granules were melt-mixed with clay–water slurry in a Haake internal mixer. First, an aqueous clay slurry was prepared between the clay and water ratio set by the experimental matrix. Then, the slurry was introduced using a separatory funnel into the internal mixer at minute 2 of the compounding. The plastic composite aggregate (PCA) with clay formulation, plastic waste aggregate (PWA) without clay formulation, and process parameters were according to the design matrix as listed in Table 6.1. After compounding, the solidified PCA and PWA compounds were cut into smaller pieces and carefully filled into a mould cavity to undergo the fabrication process using a hot press machine. It was compressed using a GT7014-A hot press machine at the constant temperature and time of 200 °C and 15 minutes. The thickness of the mould's cavity used was 3 mm, suitable for tensile testing.

A tensile test was conducted on samples following the ASTM D638 standard to determine the *TS* of the compounds. The tensile test was performed using UTM (Shimadzu AGS-X Series) at room temperature and 50 mm/min crosshead speed. The physical look of the samples was examined to identify the homogeneity of the produced compounds and used to support the optimization results.

Table 6.1 Design matrix for the experiment and tensile strength experimental results

Standard order	Run order	Factor *A*: %clay	Factor *B*: temperature	Factor *C*: time	Factor *D*: %water	*TS* (Mpa)
1	13	− 1	− 1	− 1	− 1	31.80
2	1	1	− 1	− 1	− 1	22.95
3	8	− 1	1	− 1	− 1	31.63
4	5	1	1	− 1	− 1	25.45
5	12	− 1	− 1	1	− 1	31.41
6	15	1	− 1	1	− 1	26.42
7	3	− 1	1	1	− 1	32.57
8	6	1	1	1	− 1	24.98
9	14	− 1	− 1	− 1	1	33.07
10	10	1	− 1	− 1	1	23.15
11	2	− 1	1	− 1	1	32.92
12	11	1	1	− 1	1	25.45
13	19	− 1	− 1	1	1	30.86
14	18	1	− 1	1	1	26.42
15	16	− 1	1	1	1	32.16
16	4	1	1	1	1	26.13
17	7	0	0	0	0	30.53
18	9	0	0	0	0	30.30
19	17	0	0	0	0	30.05

Table 6.2 Design matrix for the experiment

Factor *A*: %clay	Factor *B*: temperature	Factor *C*: time	Factor *D*: %water
0.0 (−1)	130 (−1)	5.0 (−1)	0.0 (−1)
5.0 (0)	155 (0)	7.5 (0)	5.0 (0)
10.0 (+1)	180 (+1)	10.0 (+1)	10.0 (+1)

6.3 Result and Discussions

6.3.1 Regression Models and R^2 Values of Factor Interaction

Experiments were carried out using data from the two-level factorial design with various clay and water contents, mixing temperature and time combinations. Tensile strength was determined after each sheet was compressed using a hot press set to 200 °C. ANOVA was used to analyse the results. The regression Eq. (6.1) is obtained after the ANOVA is completed. This step generates the tensile strength (*TS*, MPa) of the PCA and PWA by the function of different variables, clay content (*A*, %), mixing

Table 6.3 Regression coefficients and *P*-values of significant terms as calculated from the model

Variables	Coefficient	*P*-value
B_0	+ 28.86	< 0.0001
B_1	− 3.47	< 0.0001
B_2	+ 0.34	0.0029
B_3	+ 0.28	0.0073
B_4	+ 0.18	0.0412
B_{13}	+ 0.59	0.0002
B_{23}	− 0.25	0.0120
B_{123}	− 0.56	0.0002
B_{134}	+ 0.28	0.0078

temperature (*B*, °C), mixing time (*C*, min) and water content (*D*, %). The following Eq. (6.1) includes all terms, regardless of their significance:

$$\begin{aligned} TS = {} & +28.86 - 3.47 * A + 0.34 * B + 0.28 * C + 0.18 * D \\ & + 0.04 * AB + 0.59 * AC - 0.02 * AD - 0.25 * BC \\ & - 0.16 * CD - 0.56 * ABC + 0.28 * ACD \end{aligned} \tag{6.1}$$

The Design Expert® software generates a regression model with one offset, four linear terms and seven interaction terms. Equation (6.1) coefficient values and *P*-value for each term and interaction are listed in Table 6.3. The term *A*, which refers to the clay content and has a sum of square (SS) value of 192.31, has the greatest effect. This is followed by the term B, which stands for mixing temperature. All linear terms, as well as the interaction terms *AC*, *BC*, *ABC* and *ACD*, appear to be significant. When the regression model's significance was tested, the *P*-value obtained was very small, 0.0001 (Table 6.3), compared to the desired significance level of 0.0500 [9]. These figures indicate that the regression model accurately describes or predicts the tensile strength pattern. The R^2 indicates that the independent variables tested to account for 96.38% of the sample variation in tensile strength. It also shows that the model only explains 3.63% of the total variation [10].

6.3.2 Interaction Between Variables for Tensile Strength

The response surface for the variation in tensile as a function of clay content and mixing temperature is shown in Fig. 6.1. The response surface plot makes it simple to locate and perceive the interaction between these variables. The response surface plot shows that increasing the clay content from 0 to 10% decreases the tensile strength of PCAs to 22 MPa. The *TS* value rises when the mixing temperature and water content increase from 130 to 180 °C and 0 to 10%, respectively.

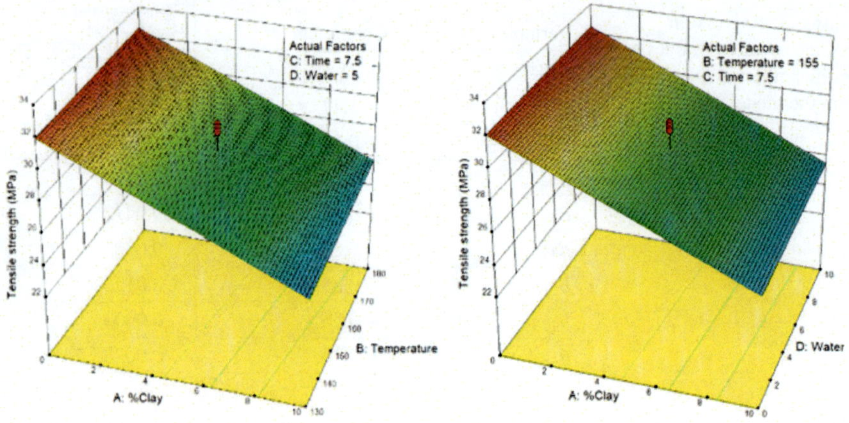

(a) (b)

Fig. 6.1 3D plot of tensile strength for the function of **a** clay content temperature and **b** clay content–water content

The contribution of each factor to the tensile strength is supported by the physical feature of the samples and a Design Expert® perturbation plot, as shown in Fig. 6.2. The physical quality of the pieces comprising the colour evenness and dimensional accuracy depicts the homogeneity of the compounds. The perturbation plot depicts changes in response as the factor moves away from the chosen reference point while all other factors remain constant. This plot demonstrates that as the clay content changes, it has a greater effect than the other variables (mixing temperature, mixing time, and water content) as the reference point changes. However, the increment of clay content harmed the increment of the PCA's tensile strength. Regardless, the interaction between clay content with mixing time (AC) and their interactions with water content (ACD) positively impacted the *TS*. Based on these findings, the maximum *TS* value was obtained for the lowest level of clay content and the highest amount of water, as indicated by Run sample No. 18 (see Table 6.1). The sample showed the highest *TS* of 33.07 MPa. According to the experiment results, too high mixing temperature, too long duration and too much water content caused polymer degradation [11]. It manifested as a yellowish compound from the chain breakage phenomenon where the polymeric chains were cut short by the high temperature and pressure generated during the compounding process.

The optimum values presented by model graphs agree with the numerical optimization generated by the Design Expert® software. For the optimization criteria of clay content targeted to only 1%, the mixing temperature between 130 and 180 °C, mixing time in the range of 5–10 min, and the water content and tensile strength were maximized. The solution given by the software was 32.51 MPa. Therefore, the corresponding optimum A, B, C and D factors were at 1 wt% clay, 180 °C, 5 min and 10 wt%, respectively. The desirability of these criteria was close to unity [12] with a value of 0.981. The findings were in good agreement with the physical features

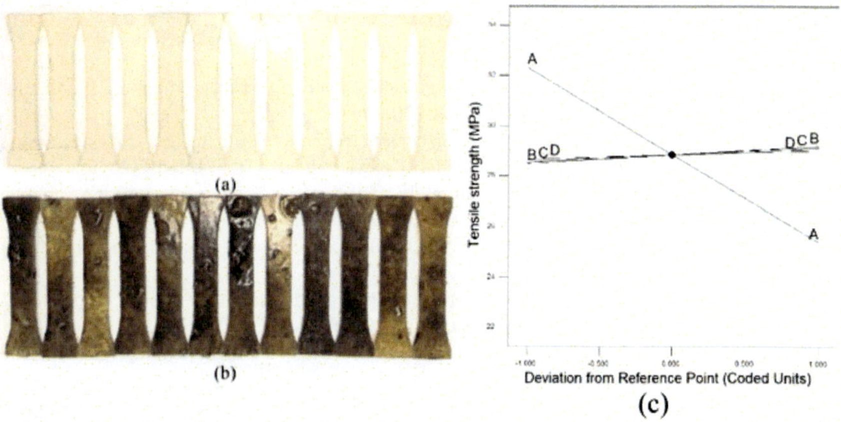

Fig. 6.2 **a** Best sample (Run 14), **b** worst sample (Run 1) and **c** perturbation plot

of the samples (Fig. 6.2a and b). The best sample of BOPP without clay (Fig. 6.2a) was observed to have uniform surface criteria compared to the worst sample with the poorest surface homogeneity (Fig. 6.2b). The worst sample prepared from the combination of BOPP at the highest amount of clay, lowest temperature and lowest time and without water showed inhomogeneous and irregular surface conditions. Therefore, the optimization was in agreement with the samples' physical features. Consequently, the water-assisted process to improve the properties of thermoplastic particles during compounding is validated [7].

6.4 Conclusions

A systematic investigation of the effect of formulation and processing parameters on the tensile strength of BOPP/clay composites was successfully conducted using Design Expert Software via the RSM. All factors investigated using the 2^4 factorial designs significantly altered the PCA's tensile strength. The final combination of factors to achieve the highest tensile strength in the presence of at least 1 wt% clay was 180 °C, 5 min and 10 wt% of water. The screening design model was adequate for presenting and predicting the significant tensile strength, as the coefficient of determination, R^2 0.9638, was close to unity. The optimum parameters could be used to produce the PCA and tested for their potential in reinforcing various concrete mixes.

Acknowledgements The authors are grateful to Universiti Teknikal Malaysia Melaka for the financial support through PJP/2020/FKP/PP/S01780, San Miguel Yamamura Plastic Films Sdn Bhd, Pegasus Polymers, Plastflute Manufacturing Sdn Bhd and Politeknik Melaka for the resources and equipment.

References

1. C.R. Gagg, Cement and concrete as an engineering material: an historic appraisal and case study analysis. Eng. Fail. Anal. **40**, 114–140 (2014)
2. A. Naqi, J.G. Jang, Recent progress in green cement technology utilizing low-carbon emission fuels and raw materials: a review. Sustainability (Switzerland) **11**(537), 1–18 (2019)
3. M. Naderi, A. Kaboudan, Experimental study of the effect of aggregate type on concrete strength and permeability. J. Build. Eng. **37**, 101928 (2021)
4. A.K. Jassim, Recycling of polyethylene waste to produce plastic cement. Procedia Manuf **8**(October 2016), 635–642 (2017)
5. I. Almeshal et al., Use of recycled plastic as fine aggregate in cementitious composites: a review. Constr Build Mater. **253**(119146), 1–27 (2020)
6. L. Gu, T. Ozbakkaloglu, Use of recycled plastics in concrete: a critical review. Waste Manage **51**(July 2019), 19–42 (2019)
7. N. Mohamad et al., Brief review on potential production of plastic waste concrete aggregates using water-assisted melt compounding, in *Intelligent Manufacturing and Mechatronics*, ed. by M.N. Ali Mokhtar, Z. Jamaludin, M.S. Abdul Aziz, M.N. Maslan, J.A. Razak. SympoSIMM 2021. Lecture notes mechanical engineering (Springer, Singapore, 2022). https://doi.org/10.1007/978-981-16-8954-3_50
8. A.J. Kulkarni et al., Introduction to optimization, in *Cohort Intelligence: A Socio-inspired Optimization Method*. Intelligent Systems Reference Library, vol. 114 (Springer, Cham, 2017). https://doi.org/10.1007/978-3-319-44254-9_1
9. S.A. Ghani et al., Application of response surface methodology for optimizing the oxidative stability of natural ester oil using mixed antioxidants. IEEE Trans. Dielectr. Electr. Insul. **24**(2), 974–983 (2017). https://doi.org/10.1109/TDEI.2017.006221
10. I. Elganidi et al., Optimization of reaction parameters for a novel polymeric additives as flow improvers of crude oil using response surface methodology. J. Pet. Explor. Prod. Technol. **12**, 437–449 (2022). https://doi.org/10.1007/s13202-021-01349-1
11. N. Mohamad et al., Epoxidized natural rubber–alumina nanoparticle composites: optimization of mixer parameters via response surface methodology. J. Appl. Polym. Sci. **115**(1), 183–189 (2010). https://doi.org/10.1002/app.31056
12. V.A. Yiga et al., Optimization of tensile strength of PLA/clay/rice husk composites using Box-Behnken design. Biomass Convers. Biorefin. (2021). https://doi.org/10.1007/s13399-021-01971-3

Chapter 7
Recycling of Glazed Roof Tile Waste for Fine Aggregate in Green Concrete

Jariah Mohamad Juoi, Yusliza Yusuf, Zulkifli Mohd Rosli, Nuzaimah Mustafa, Wan Amirul Shafiz Wan Zulkifli, Fariha Awatif Abdul Aziz, and Nur Umairah Afifah Abd. Wahab

Abstract Glazed Roof Tile Waste (GRTW) is identified as having the potential to replace the natural sand aggregate used in cement mixture. This is an innovative effort to produce green concrete. The produced green concrete is intended for building materials application, used for the construction of buildings, drainage, paving, and landscaping. The purpose of this paper is to report on the workability of the concrete mixture, physical characteristics, and compression strength of the green concrete produced. Work involved in this research starts with a preparation of the GRTW thru grinding and milling. The crushed GRTW was sieved and analyzed according to BS 882:1992. Next, the GRTW with different weight fraction (10, 20, and 50%) was mixed with sand, cement, and water to identify its workability thru slump test (BS EN 12350-2). The mixture with accepted workability was casted into concrete specimen (100 mm × 100 mm × 100 mm) and subjected to compression test for 7 and 28 days, and the compressive strength results were compared with Gred M 25 concrete standard. It was found that the workability of concrete mixture gradually increases with the increases of GRTW weight fraction of 10 and 20%, however the workability decrease below the acceptable workability with 50% replacement of GRTW. Hence, the substitution of 20% of GRTW had been proven to have better mechanical performance in term of the slump value achieved, water absorption and compressive strength and can replace sand as fine aggregate in concrete.

Keywords Tile waste · Workability · Compressive strength · Concrete · Recycling · Fine aggregate

---

J. Mohamad Juoi (✉) · W. A. S. Wan Zulkifli · N. U. A. Abd. Wahab
Fakulti Kejuruteraan Pembuatan, Universiti Teknikal Malaysia Melaka, Hang Tuah Jaya, 76100 Durian Tunggal, Melaka, Malaysia
e-mail: jariah@utem.edu.my

Y. Yusuf · Z. Mohd Rosli · N. Mustafa · F. A. Abdul Aziz
Fakulti Teknologi Kejuruteraan Mekanikal and Pembuatan, Universiti Teknikal Malaysia Melaka, Hang Tuah Jaya, Melaka, Malaysia

M. A. A. Mohd Salleh et al. (eds.), *Proceedings of the Green Materials and Electronic Packaging Interconnect Technology Symposium*, Springer Proceedings in Physics 289,
https://doi.org/10.1007/978-981-19-9267-4_7

7.1 Introduction

Ceramic materials contribute to the large amount of wastes within the construction and demolition wastes. This huge amount of production has caused them to be among the most commonly consumed materials in the world [1]. During ceramic production, studies have shown that about 30% of the material goes to wastes [2]. Some of the ceramic waste is also due to a result of destroying constructions. In general, this waste is not beneficially utilized, neglected, and cause environmental and disposal problems [3]. This suggest to the need for exploring innovative ways of re-using ceramic wastes. On the other hand, the high annual aggregate utilization production for concrete production has a significant impact on the environment. Aggregates constitute about 70% of total constituents in concrete production [4]. Utilization of ceramic wastes is possible in a number of ways, either as a full replacement for coarse aggregates, fine aggregates or a partial fraction of both aggregates [3, 5–14]. It was reported that ceramic waste had able to substitute conventional coarse aggregates in the production of nonstructural concrete artefacts, where the primary requirement is tensile strength and resistance to abrasion and not compressive strength. Meanwhile, ceramic waste is also possible to replace crushed stone aggregate, and it was found that the workability of the concrete mixture is good and that the strength characteristics are comparable to that of the conventional concrete [15, 16]. Roof tile waste is a type of ceramic waste produce in construction industry. Roof tile waste aggregates term as porous clay aggregates (PCA) had leads to an increase in compressive strength and the reduction of autogenous shrinkage of concrete [17]. Fundamentally, roof-tile waste aggregate is reported as an effective internal curing material for concrete. It was also found that the roof tile waste can develop the compressive strength and reduce the autogenous shrinkage of high performance concrete with a very low water-to-binder ratio of 0.15 [18]. According Liu et al. [19] it is possible to use recycled ceramic aggregate with a diameter of less than 9.5 mm in concrete as a partial substitute for natural aggregate. Since ordinary concrete has a higher apparent density than recycled ceramic roof tiles concrete, this can help to minimize the self-weight of structure. Since the ultrafine sand has a high mud content, the splitting tensile strength of recycled ceramic roof tiles is weak under similar workability conditions when the replacement rate is less than 20%. Furthermore, when the replacement rate exceeds 40%, the compressive and fracturing tensile strengths of the concrete exceed those of the reference concrete. The use of 100% recycled roof tiles as fine aggregate increases both splitting tensile strength and compressive strength significant. Halicka et al. used ceramic sanitary ware waste as coarse aggregate in concrete mixes [20]. The porosity of the ceramic particles' structure was revealed by scanning electron microscopy. Ceramic aggregates also have a low crushing ratio and a high water absorption, as discovered by studying their properties. It is also reported that ceramic sanitary ware waste aggregate can be used to make high-performance and abrasion-resistant concrete. For example, Medina et al. [14] use ceramic sanitary ware waste as a partial substitute of gravel with replacement levels of 20 and 25%. Here, mixing ceramic aggregate with natural gravel slightly increased the porosity of the fabricated concrete. As the replacement percentage increased, both compressive and tensile splitting strength increased. In comparison to reference concrete physical properties, the recycled ceramic aggregate concrete has a lower slump, lower density, higher water absorption, higher sorptivity, and higher porosity.

7.2 Methodology

7.2.1 Portland Composite Cement

CIMA's Portland Composite Cement is known as NS Composite Cement is used throughout this experiment. It is made by grinding Portland cement clinker and other carefully selected inorganic components, as allowed by MS EN 197-1:2014 and subjected to severe quality control management. It is appropriate for all general-purpose applications. NS Composite Cement can increase workability, minimize bleeding, reduce pollution, improve cohesiveness, improve surface finishing, and provide a richer mix by using the right proportions of water, sand, and aggregates.

7.2.2 Fine Aggregate

River sand has been used as fine aggregate in this experimental work. It is conformed to the overall limit of fine aggregate requirements according to BS 882:1992. It was pass through 4.75 mm size of sieve.

7.2.3 Coarse Aggregate

Coarse aggregate obtained from local supplier. The maximum size of coarse aggregate is 20 mm. Coarse aggregate used in experimental study was confirming BS 882:1992. Specific gravity and water absorption of the coarse aggregate were 2.59 and 1.01%, respectively.

7.2.4 Glazed Roof Tiles Waste (GRTW)

GRTW were obtained from Monier Roofing Tiles Sdn Bhd. GRTW has been used as fine aggregate in this experimental work and crushing of GRTW was carried out by using crusher machines to make sure the size of GRTW as small as sand size. It is conformed to overall limit of fine aggregate according to BS 882:1992.

7.2.5 *Water*

Tap water has been used for mixing and curing of concrete specimens. The water available from the local source was fit for usage.

7.2.6 *Mix Design*

As per the Building Researches Establishment (BRE) Method guideline, M25 grade of mix design concrete was prepared. Water Cement ratio was 0.50 according to the slump test obtain. Thus, the mix proportion obtains for M25 mix design is 1:1.5:1.38 (C: F.A: C.A). Batching details of the concrete mix are presented in Table 7.1, and the results of concrete mix design (kg/m^3) are tabulated in Table 7.2.

7.2.7 *Workability Test*

Workability of concrete made using GRTW was determined at different replacement level. The consistency and degree of workability of fresh concrete are determined using slump test accordance to BS EN 12350-2. The apparatus consists of a slump cone with a base plate. The cone is kept over the base plate and filled with the freshly mixed concrete of desired grade in three layers which each layer is tamped 25 times with a tamping rod of standard dimensions. After leveling the concrete, the slump

Table 7.1 Batching details

Fine	Batch number			
Aggregate	Batch 1	Batch 2	Batch 3	Batch 4
Sand (%)	100	90	80	50
GRTW (%)	0	10	20	50

Table 7.2 Design mix proportion of M25 grade of concrete per m^3

R (%)	C (kg)	F.A (kg)		C.A (kg)	W (l)
		Sand	GRTW		
0	536.2	803.8	0	739.7	268.1
10		723.4	80.4		
20		643.0	160.8		
Ratio		1.5		1.38	0.50

R = Replacement of GRTW, C = Cement, F.A = Fine Aggregate, C.A = Coarse Aggregate, W = Water cone is lifted upwards carefully and the concrete subsides. This depth of subsidence is termed as slump.

7.2.8 Water Absorption Test

Water absorption test was carried out in this study to calculate the water absorption percentage of hardened concrete in accordance with BS 1881-122. Prior to the testing, the specimen cubes (100 mm × 100 mm × 100 mm) were taken out one hour in advanced from the curing tank at day 28. The specimen was wiped to surface-dry condition and weighed to obtain the saturated surface-dry weight, W_f of the specimen. The initial weight W_i is taken before the samples are immersed in the water for curing process. The water absorption percentage of the hardened concrete was calculated by using Eq. (7.1).

$$\text{Water Absorption (\%)} = \frac{W_f - W_i}{W_i} \times 100\% \tag{7.1}$$

7.2.9 Compressive Strength

Compressive strength was determined by using compression testing machine which is available in CASTConsult Sdn Bhd. The capacity of these machine is 3000kN, respectively. Constant loading rate between 0.4 and 0.8 N/mm^2 was applied on the hardened cubic concrete specimen of 100 mm × 100 mm × 100 mm. The test was carried out by placing a specimen between the loading surfaces and the load was applied until the specimen fails. Three test specimens for each level of GRTW replacement level were utilized during the test, and the average value was reported.

7.3 Results and Discussion

7.3.1 Slump Value

The result of workability in terms of slump is given in Table 7.3. It is evident from the table that workability of concrete made using GRTW increased with increase in replacement level (10 and 20%) and decreased with 50% replacement level.

Table 7.3 Workability (slump) of concrete at different replacement level at day 28

Sample No.	Replacement level GRTW (%)	Slup (mm)
1	0	80
2	10	100
3	20	110
4	50	30

Table 7.4 Water absorption of concrete in various replacement level at day 28

Sample No.	Replacement level GRTW (%)	Water absorption (%)
1	0	1.72
2	10	1.71
3	20	1.61
4	50	1.15

7.3.2 *Water Absorption*

The maritime code BS 6349 specifies that water absorption should not exceed 3, or 2% in critical conditions. The average of water absorption for 0, 10, 20, and 50 of percentage replacement of GRTW at day 28 were 1.72, 1.71, 1.61, and 1.15%, respectively, as shown in Table 7.4. These results show that the water absorption for all specimens was not exceeding 3%.

7.3.3 *Compressive Strength*

To obtain optimum compressive strength in concrete, a design mix proportion has been prepared. Compressive strength of different batch of concrete was tested for 28 days according to BS EN 12350-1:2000 through compressive testing machine. Each batch of mix had three samples used to obtain an average. According to Table 7.5, the compressive strength of concrete prepared using GRTW exhibits comparable performance to that of fresh concrete. Here, it is found that concrete with GRTW of 50% had the highest compressive strength of 45.5 N/mm^2; however, it should be noted that the workability (slump, 30 mm) of the concrete mixture is below the acceptable value of slump, 60 mm.

Table 7.5 Compressive strength of concrete in various replacement level at day 28

Sample no.	Replacement level GRTW (%)	Compressive strength (N/mm^2)
1	0	31.2
2	10	27.8
3	20	31.6
4	50	45.5

7.4 Conclusion

The following conclusion may be derived from the aforementioned study:

- The test results demonstrate that GRTW can be utilized as a substitute for sand in concrete.
- While utilizing GRTW as a partial substitute for fine aggregate, the workability improved as the replacement level grew (10 and 20%). At 50% replacement, the workability is 30 mm below the acceptable range of 60 mm workability.
- It has been determined that the compressive strength of concrete prepared using GRTW is equivalent to that of new concrete.
- The optimal replacement level of GRTW in concrete is 20%, given that its slump test, water absorption test, and compressive strength match British Standard requirements.

Acknowledgements PJP/2020/FKP/PP/S01781.

References

1. A. Agrawal, International Journal of Engineering Research **4**(3) (2016) (May–June)
2. A.B. Ceesay, S. Miyazawa, Recycling, **4**, 29 (2019)
3. P.O. Awoyera, J.O. Akinmusuru, A.R. Dawson, J.M. Ndambuki, N.H. Thom, Green concrete production with ceramic waste and laterite. Construction and Bulding Materials **117**, 29–36 (2016)
4. F. Andreola, L. Barbieri, I. Lancellotti, C. Leonelli, T. Manfredini, Recycling of industrial wastes in ceramic manufacturing: state of art and glass case studies. Ceram. Int. **42**, 13333–13338 (2016)
5. C. Medina, M.S. De Rojas, M. Frías, Properties of recycled ceramic aggregate concretes: water resistance. Cement&Concrete Composites **40**, 21–29 (2013)
6. F. Pacheco-Torgal, S. Jalali, Reusing ceramic waste in concrete. Constr. Build. Mater. **24**, 832–838 (2010)
7. R.M. Senthamarai, P.D. Manoharan, Concrete with ceramic waste aggregate. Cem. Concr. Cemp. **27**(9–10), 910–913 (2005)
8. M. Senthamarai, P.D. Manoharan, D. Gobinath, Concrete made from ceramic industry waste: durability properties. Construction Build Material **25**, 2413–2419 (2011)
9. O. Zimbili, W. Salim, M. Ndambuki, A review on the usage of ceramic waste in concrete production. Journal of Civil and Environmental Engineering **8**, 91–95 (2014)

10. A. Juan, C. Medina, M. I Guerra, J.M. Moran, P. J. Aguado, O. Rodrigues, Re-use of ceramic waste in construction. Ceramic Materials, 197–211 (2012)
11. J. Garcia-Gonzalez et al., Ceramic ware waste as coarse aggregate for structural concrete production. Environ. Technol. **36**(23), 1–10 (2015)
12. F.P. Torgal, S. Jalali, Compressive strength and durability properties of roof tiles waste based concrete. Materials and structure **44**, 155–167 (2011)
13. A.M.M. Al Bakri, M.N. Norazian, H. Kamarudin, The potential of recycled ceramic waste as coarse aggregates for concrete, in *Proceedings of MUCET* (2008)
14. C. Medina, M. Frías, M.I. Sánchez De Rojas, Microstructure and properties of recycled concretes using ceramic sanitary ware industry waste as coarse aggregate. Constr. Build. Mater. **31**, 112–118 (2012). https://doi.org/10.1016/j.conbuildmat.2011.12.075
15. M. Daniyal, S. Ahmad, Application of waste ceramic tile aggregates in concrete. Int. J. Innovative Res. Sci. Eng. Technol. **4**(12), 2319–8753 (2015)
16. Y. Ogawa, P.T. Bu, K. Kawai, R. Sato, Constr. Build. Mater. **236**, 117462 (2020)
17. M. Suzuki, M.S. Meddah, R. Sato, Use of porous ceramic waste aggregates for internal curing of high-performance concrete. Cem. Concr. Res. **39**, 373–381 (2009)
18. R. Sato, A. Shigematsu, T. Nukushina, M. Kimura, Improvement of properties of Portland blast furnace cement type B concrete by internal curing using ceramic roof material waste. J. Mater. Civ. Eng. **23**, 777–782 (2010)
19. F. Liu, J. Liu, B. Ma, J. Huang, H. Li, Basic properties of concrete incorporating recycled ceramic aggregate and ultra-fine sand. Journal of Wuhan University of Technology **30**(2) (2015)
20. A. Halicka, P. Ogrodnik, B. Zegardło, Using ceramic sanitary ware waste as concrete aggregate. Construction and Building Materials **48**, 295–305 (2013)

Chapter 8
A Short Review: Reliability Issues of Lead-Free Sn-Based Alloys for Superconducting Applications

Y. P. Tan and F. Somidin

Abstract As the trends of electronic devices are moving to miniaturization which requires high-density electronic packaging and high-speed performance, superconducting solder alloy has attracted considerable interest to fulfill the requirements of advanced electronic packaging as it provides sufficient superconductivity which minimizing the loss of current densities carried by superconductors. In the past few decades, leaded superconducting solder alloy were widely used in the industry due to its satisfied performance. However, development of lead-free solder alloy for superconducting applications is initiated to replace leaded solder alloy due to its toxicity. Sn-based solder alloy is the most popular candidate for replacing leaded solder alloy in the industry, but the reliability of Sn-based solder alloy is concerned. This paper reviewed the reliability issues of lead-free Sn-based superconducting interconnects when subjected to high current stressing. Electromigration, thermo-migration, and the subsequence issues caused by high current stressing are covered. This article also summarized the studies done on minimizing the reliability issues of potential Sn-based superconducting solder alloy caused by high current stressing.

Keywords Superconducting · Electromigration · Lead-free solder

8.1 Introduction

As the electronic devices moving toward miniaturization, the electronic packaging techniques used had drawn a lot of attention to fulfill the requirements of high performance and minimized power consumption with a more compact form compared to traditional electronic packaging techniques. More advanced electronic packaging techniques are required to fulfill the specifications of miniaturized electronic devices

Y. P. Tan (✉) · F. Somidin
Centre of Excellence Geopolymer and Green Technology, Faculty of Chemical Engineering Technology, Universiti Malaysia Perlis (UniMAP), Taman Muhibbah, 02600 Jejawi, Arau, Perlis, Malaysia
e-mail: yipeitan97@gmail.com

M. A. A. Mohd Salleh et al. (eds.), *Proceedings of the Green Materials and Electronic Packaging Interconnect Technology Symposium*, Springer Proceedings in Physics 289,
https://doi.org/10.1007/978-981-19-9267-4_8

to overcome the limitation of Moore's law. 2.5D or 3D electronic packaging techniques such as 2.5D system in package (SiP) and system on package (SoP) involved the stacking of multi-chip modules (MCM) vertically which reduced the critical interconnect path effectively. Although the stacking of MCM improved the performance of the electronic devices, the choice of solder joints is crucial to reduce the loss of current densities during the operation of the electronic devices, which initiated the research on superconducting solder joint. However, the usage of traditional superconducting solders such as Pb–Sn and Pb–Bi are restricted by Restriction of Hazardous Substances (RoHS) and Waste Electrical and Electronic Equipment (WEEE) in 2003 due to its toxicity. This initiated the development of lead-free solder which are mostly Sn-based to replace the usage of leaded solders in the industry.

Generally, the reflow processing temperature of Sn-based solders is higher than that of leaded solders, which not only causes part and board warpage, but also the degradation of superconducting filaments as they are highly thermal-sensitive if apply in the superconducting industry [1]. In addition, the reliability issues of Sn-based solder such as electromigration (EM) and thermo-migration (TM) which caused by high-density current crowding and thermal gradient had drawn concerns as these issues are dominated by Sn grain orientation [2]. The stacking of MCM increased the temperature gradient within the modules which initiated TM process and led to asymmetrical growth of intermetallic compounds (IMCs) at the both hot end and cold end of the interconnects which directly affect the mechanical performance of the solder joints [3]. Therefore, the reliability issues of the Sn-based solder caused by EM and TM requires high attention to prolong the lifespan of the solder joint.

In this paper, the EM-induced and TM-induced reliability issues of the Sn-based superconducting solder such as Sn-58Bi, Sn-52In, and Sn-9Zn were reviewed. In addition, the efforts reported to improve the atomic migration-induced reliability issues of the Sn-based solder such as alloying were covered as well.

8.2 Electromigration

Under current stress, the momentum transfer from the conduction electron produced an electron wind force (F_{wind}), which has an opposite flow direction of electrostatic force (F_{field}). A directed atomic diffusion process occurs from cathode to anode if the resultant force in the direction of the electron wind exceeds the activation energy (E_a) of the atom [4]. The continually diffusion of cathodic atoms promoted the nucleation of voids and cracks at the current crowding cathode, led to the nucleation of structural defects such as voids and cracks at the current crowding cathode, which eventually resulting in failure of short circuit as the defects propagate [5, 6]. Thus, it is important to study the electromigration mechanism of different solder joints.

In Cu/Sn/Cu solder joints, the Cu atomic flux is generally composed of the EM-induced diffusion flux (J_{EM}) and chemical potential gradient-induced flux (J_{Chem}) as shown in Eqs. (8.1) and (8.2) [7]:

$$J_{\text{EM}} = C\frac{D}{kT}Z^{*}e\rho j \tag{8.1}$$

$$J_{\text{Chem}} = -D_{\text{Cu/Cu-Sn}}\left(\frac{\text{d}C}{\text{d}x}\right) \tag{8.2}$$

In Eq. (8.1), C represents Cu concentration, D is the Cu diffusivity in Sn, k represents Boltzmann's constant, T is the absolute temperature, Z^* is the effective charge number, e is the electron charge, ρ is the resistivity of Sn and j is the current density. In Eq. (8.2), the diffusivity of Cu in Sn-Cu IMCs and the concentration gradient of Cu are denoted as $D_{\text{Cu/Cu-Sn}}$ and $\text{d}C/\text{d}x$, respectively. The direction of both the J_{EM} and J_{Chem} at the cathode are the same as the chemical gradient of Cu atoms is always from Cu substrate to the solder joint. However, at the anode, the direction of J_{EM} and J_{Chem} are opposite. Hence, the thickness of IMCs layer formed at the anode is greater than that at the cathode. According to Chen et al. [8], the thickness of IMC layer at the anode was increased proportionally to the $t^{1/2}$ law, while at the cathode, the thickness of the IMC layer was first increased then decreased during the EM process. A growth model of the cathodic IMC layer was proposed and expressed as Eq. (8.3) [8]:

$$J = J_{\text{EM}} + J_{\text{Chem}} - J_{\text{Dis}} \tag{8.3}$$

where J_{Dis} is the Cu flux resulted from the dissolution of Cu–Sn IMC into Sn matrix. The J_{Dis} value determined the thickness of the IMC layer at the cathode. Initially, the thickness of interfacial layer is insufficient, so J_{Dis} value is negligible. However, as the IMC thickness at the cathode increased to a critical thickness, the J_{Dis} value increased and eventually led to decrease in IMC thickness at the cathode.

In Cu/Sn-58Bi/Cu solder joints, Yue et al. [9] and Chen et al. [10] reported that Bi atom is dominant to diffuse from cathode to anode which to Bi-rich phase segregation and also nucleation of microcracks near the cathode in the solder joint. As the EM testing time prolonged, the coarsening of Bi-rich and Sn-rich phase and propagation of microcracks occurred with increasing ambient temperature due to the continually diffusion of Bi atom to the anode, which accelerates the failure of the solder joint due to the thickening of brittle Bi-rich phase. In Cu/In-48Sn/Cu solder joint, Li et al. [11] found that both of the Sn and In atom migrated in opposite directions, which resulted in Sn-rich layer formed at the anode and In-rich layer formed at the cathode. The Sn-rich layer was significantly thicker than that of the In-rich layer, and the consumption of Cu pad occurred, which caused the spalling characteristics of cathodic IMC. However, in Cu/Sn-9Zn/Cu, Zhang et al. [12] reported that the cathodic IMC layer was thicker than anodic IMC layer. Zhang et al. proposed that the reverse polarity effect may be due to the back stress produced by the migration of Sn atoms drove Zn atom migration from anode to cathode continuously, but Huang et al. [13] believed that the uncommon phenomenon was due to the positive value of effective charge number (Z^*) because there was no back stress in the liquid solder. In general, the EM-induced atomic migration required more attention as it promotes vacancies propagation, and the consumption of Cu pad highly reduced the mechanical performance and the lifespan of the solder joints.

8.3 Thermo-migration

Thermal gradient is easily form between the chip, solder, and the substrate of a module as each of them has different heating conditions and rate of heat dissipation. Generally, the TM process of Sn atom in Sn-based solder joints involved the migration in the direction of cold end to hot end, while Cu, Ni, and Ag atoms typically migrate toward the cold end as shown in Fig. 8.1 [14–16]. TM process not only causes asymmetric growth of the IMCs, but also changes the composition and voids formation of the interfacial layers. Other than thermal gradient, the chemical potential gradient and back stress produced during the EM-induced migration process also triggers TM process. As the temperature gradient and atomic migration rate in 3D ICs increased, TM become the most concerned reliability issues of advanced electronic packaging which limited its applications in the industry.

In Cu/Sn/Cu solder joints, Wei et al. [17] performed a TM test under a temperature gradient of 1046 °C/min and reported that the coarsening of Cu_6Sn_5 occurred at the cold end of the solder joint. As the TM testing prolonged, voids formed in the coarsened Cu_6Sn_5 IMC layer. At the hot end, the IMC layer appeared thinner than that at the cold end which indicated the diffusion of IMC at the hot end into the solder joint which provide sufficient Cu atom for IMC growth at the cold end and thus, resulting asymmetric thickness of the IMC. The Cu atomic flux (J_{TM}) from the hot end to cold end which developed from temperature gradient can be expressed as Eq. (8.4) [18]:

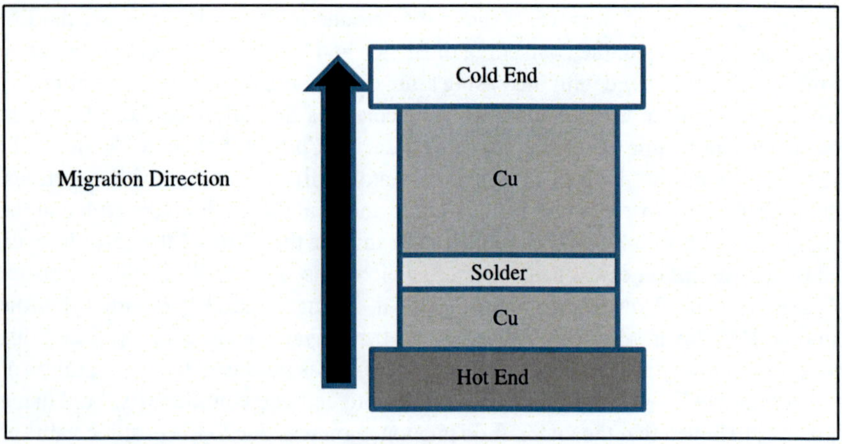

Fig. 8.1 Thermo-migration in solder joint

$$J_{TM} = C\frac{D}{kT}\frac{Q^*}{T} - \frac{\partial T}{\partial x} \quad (8.4)$$

where Q^* is the transport heat of Cu atom in Sn, and $\partial T/\partial x$ is the temperature gradient applied. At the hot end, the direction of J_{TM} and J_{Chem} is the same which accelerates the migration of Cu atom, but opposite at the cold end, thus reduced the risk of Cu substrate consumption at the cold end.

In Cu/Sn-58Bi/Cu solder joints, extremely high activation energy is required for Cu atoms to diffuse from the hot end to the cold end of the solder joint as the amount of grain boundaries in Sn–Bi solder is relatively higher compared to other lead-free Sn-based solder due to its highly brittle property. Therefore, the asymmetric growth of the IMC in Sn-Bi solder is relatively lesser compared to other lead-free solder under TM testing. During TM test, Shen et al. [19] reported that Bi-rich phases were formed at the cold end, grain coarsening and phase aggregation occurred as the temperature gradient increased. Furthermore, Ding et al. [20] reported that Kirkendall voids formed in both end but higher amount of voids at the hot end. This phenomenon was explained by Shen et al. [21] that Bi atoms as the main migration species and diffuse along the direction of heat flux which squeezes the Sn atoms at the hot end during TM testing. In Cu/Sn-9Zn/Cu solder joint, the IMC formation at the cold end was promoted by the diffusion of both Zn atoms and Cu atoms, which resulting Cu pad consumption at the hot end [22]. In summary, TM-induced defects in the solder joints are highly dependent on the temperature gradient exerted to the solder joints and the thermal stability of the dominant migration species in the solder joints.

8.4 Effects of Alloying on Reliability Issues

As the microstructure stability of the IMCs plays an important role in atomic migration-induced reliability issues, it is important to study the Sn grain orientation in the solder joints which is usually the main atomic diffusion species, and the microstructure of the IMCs formed. Several studies were reported on the effect of Sn grain orientation and the IMCs, Cu_6Sn_5 phase transformation by using advanced material characterization technique such as Transmission Electron Microscopy (TEM) [2, 23–25]. For EM performed at 120 °C, the diffusivity of Cu in Sn along the *c*-axis is nearly 61 times higher than that along the *a*-axis. Besides, the diffusivity of Ni atom in Sn along the *c*-axis is around 70,000 times higher than that along the *a*-axis, which facing more severe Ni pad consumption issues compared to Cu substrate. This indicates that the Sn grain orientation has huge effect on the EM and TM in Sn-rich solder joints [2].

The stability of IMCs phase can be modified by alloying, aging process, and altering the fabrication process of the solder joints [26–28]. Researchers had put their efforts in modifying the microstructure and phase of IMCs in lead-free superconducting solder joints by alloying [29–40]. According to the et al. [29], the EM resistance of Sn-58Bi improved when micro-alloyed with only 0.1 wt% of rare earth

(RE) as the RE elements adsorbed at the grains boundaries and reduced the boundary energies which restricted the sliding of the boundaries and thus suppressed the migration of Sn and Bi atoms. In lead-free solder industry, silver (Ag) is one of the popular alloying elements chosen for microstructure refinement. When 0.4 wt% of Ag was added to eutectic Sn-Bi solder, Sun et al. [30] reported that the segregation of Bi-phase was minimized due to the uniform distribution of Ag_3Sn IMCs. The addition of Ag refined the grains of the solder joint which improved the EM resistance. Other than Ag addition, Liu et al. [34] studied the effect of Bi addition on the electromigration of Sn-Zn solder joints with Cu substrate. Based on his results, the IMC thickness formed at the anode is similar to that of the cathode because the main migrating species in Sn-Zn system is Zn, and he claims that Zn has higher diffusivity than Sn in Sn which led to void formation at the anode after further current stressing. In Sn-In-6Zn, Wang et al. [36] stated that refinement of β-phase and enrichment of γ-phase at the cathode and anode, respectively, due to the addition of Zn improved the microhardness of the solder joints by 30%. No significant migration of Zn in Sn–In–Zn/Cu solder joints was observed, and the main migrating species are Sn and In Ref. [36]. However, there are insufficient studies reported on the atomic migration resistance of Sn-In and Sn-Zn superconducting solder.

8.5 Conclusion

In this review, the reliability challenges of Sn-based solder joint under high current stress and high temperature gradient were discussed. The EM-induced and TM-induced issues were covered as well as the studies done on improving the atomic migration resistance of the solder joints. However, there are still insufficient studies reported on the effects of alloying on the reliability of low-temperature superconducting solder joints for improving the atomic diffusion resistance. In addition, the practical operation environment of the solder joints is usually involved loading and stress other than current stress and temperature gradient, such as mechanical stress and magnetic fields. The solder joints may also be applied in extreme operation conditions such as extremely low temperature and rapid changes in temperature. Hence, more efforts can be made on low-temperature superconducting solder joints and also the coupling effect of different stress on the reliability of solder joints.

Acknowledgements The authors would like to acknowledge the financial support from the collaboration between Nihon Superior Co Ltd. and Universiti Malaysia Perlis, UniMAP. The authors are truly appreciated the support from Center of Excellence Geopolymer & Green Technology (CEGeoGTech), Universiti Malaysia Perlis, UniMAP.

References

1. Y. Tsui, E. Surrey, D. Hampshire, Soldered joints—An essential component of demountable high temperature superconducting fusion magnets. Supercond. Sci. Technol. **29**(7) (May 2016). https://doi.org/10.1088/0953-2048/29/7/075005
2. Y.A. Shen, J.A. Wu, *Effect of Sn Grain Orientation on Reliability Issues of Sn-Rich Solder Joints*, pp. 1–20 (2022)
3. T.Y. Lin, F.Y. Ouyang, Effect of temperature on microstructural evolution of solder alloys under thermomigration. J. Appl. Phys. **125**(7), 075103 (2019). https://doi.org/10.1063/1.5081087
4. J. Lienig, M. Thiele, *Fundamentals of Electromigration-Aware Integrated Circuit Design* (2018)
5. Y.W. Chang et al., Study of electromigration-induced formation of discrete voids in flip-chip solder joints by in-situ 3D laminography observation and finite-element modeling. Acta Mater. **117**, 100–110 (2016). https://doi.org/10.1016/J.ACTAMAT.2016.06.059
6. Z. Huang, R.E. Jones, A. Jain, Experimental investigation of electromigration failure in Cu–Sn–Cu micropads in 3D integrated circuits. Microelectron. Eng. **122**, 46–51 (2014). https://doi.org/10.1016/J.MEE.2014.03.003
7. C.E. Ho, P.T. Lee, C.N. Chen, C.H. Yang, Electromigration in 3D-IC scale Cu/Sn/Cu solder joints. J. Alloys Compd. **676**, 361–368 (2016). https://doi.org/10.1016/J.JALLCOM.2016.03.134
8. L.D. Chen, M.L. Huang, S.M. Zhou, Effect of electromigration on intermetallic compound formation in line-type Cu/Sn/Cu interconnect. J. Alloys Compd. **504**(2), 535–541 (2010). https://doi.org/10.1016/J.JALLCOM.2010.05.158
9. W. Yue, H.B. Qin, M.B. Zhou, X. Ma, X.P. Zhang, Electromigration induced microstructure evolution and damage in asymmetric Cu/Sn-58Bi/Cu solder interconnect under current stressing. Trans. Nonferrous Met. Soc. China (English Ed.) **24**(5), 1619–1628 (2014). https://doi.org/10.1016/S1003-6326(14)63233-7
10. C.M. Chen, L.T. Chen, Y.S. Lin, Electromigration-induced Bi segregation in eutectic SnBi solder joint. J. Electron. Mater. **36**(2), 168–172 (2007). https://doi.org/10.1007/S11664-006-0025-0
11. Y. Li, Y.C. Chan, F. Wu, A comparison study of electromigration in In-48Sn solder interconnects with Cu and Au/Ni/Cu pads, in *2016 International Conference on Electronics Packaging. ICEP 2016*, pp. 689–692 (2016). https://doi.org/10.1109/ICEP.2016.7486920
12. X.F. Zhang, J.D. Guo, J.K. Shang, Abnormal polarity effect of electromigration on intermetallic compound formation in Sn–9Zn solder interconnect. Scr. Mater. **57**(6), 513–516 (2007). https://doi.org/10.1016/J.SCRIPTAMAT.2007.05.023
13. M.L. Huang, Q. Zhou, N. Zhao, Z.J. Zhang, Abnormal diffusion behavior of Zn in Cu/Sn-9 wt.%Zn/Cu interconnects during liquid–solid electromigration. J. Electron. Mater. 2013 4210 **42**(10), 2975–2982 (2013). https://doi.org/10.1007/S11664-013-2690-0
14. T.L. Yang et al., Full intermetallic joints for chip stacking by using thermal gradient bonding. Acta Mater. **113**, 90–97 (2016). https://doi.org/10.1016/J.ACTAMAT.2016.04.046
15. Y.S. Yang, C.J. Yang, F.Y. Ouyang, Interfacial reaction of Ni3Sn4 intermetallic compound in Ni/SnAg solder/Ni system under thermomigration. J. Alloys Compd. **674**, 331–340 (2016). https://doi.org/10.1016/J.JALLCOM.2016.03.020
16. F.-Y. Ouyang, Y.-P. Su, Growth kinetic of Ag3Sn intermetallic compound in micro-scale Pb-free solder alloys under a temperature gradient. J. Alloys Compd. **655**, 155–164 (2016). https://doi.org/10.1016/j.jallcom.2015.09.151
17. G.Q. Wei, L.C. Du, Y.P. Jia, L. Qi, Effect of thermomigration on evolution of interfacial intermetallic compounds in Cu/Sn/Cu and Cu/Sn0.7Cu/Cu solder joints. J. Mater. Sci. Mater. Electron. **26**(6), 4313–4317 (2015). https://doi.org/10.1007/S10854-015-2984-5/FIGURES/7
18. Y. Zhong, N. Zhao, H.T. Ma, W. Dong, M.L. Huang, Retardation of thermomigration-induced Cu substrate consumption in Pb-free solder joints by Zn addition. J. Alloys Compd. **695**, 1436–1443 (2017). https://doi.org/10.1016/J.JALLCOM.2016.10.280

19. Y.A. Shen, S. Zhou, J. Li, K.N. Tu, H. Nishikawa, Microstructure and property changes in Cu/Sn-58Bi/Cu solder joints during thermomigration, in *Proceedings—Electronic Components and Technology Conference*, vol. 2019-May, pp. 2003–2008 (2019). https://doi.org/10.1109/ECTC.2019.00307
20. L. Ding, Y. Tao, Y. Wu, Thermomigration in Sn58Bi solder joints at low ambient temperature, in *ICEPT-HDP 2011 Proceedings—2011 International Conference Electronic Packaging Technology and High Density Packaging*, pp. 944–948 (2011). https://doi.org/10.1109/ICEPT.2011.6066985
21. Y.A. Shen, S. Zhou, J. Li, K.N. Tu, H. Nishikawa, Thermomigration induced microstructure and property changes in Sn-58Bi solders. Mater. Des. **166**, 107619 (2019). https://doi.org/10.1016/J.MATDES.2019.107619
22. N. Zhao, Y. Zhong, M.L. Huang, W. Dong, H.T. Ma, Y.P. Wang, In situ study on interfacial reactions of Cu/Sn–9Zn/Cu solder joints under temperature gradient. J. Alloys Compd. **682**, 1–6 (2016). https://doi.org/10.1016/J.JALLCOM.2016.04.282
23. F. Somidin et al., Characterising the polymorphic phase transformation at a localised point on a Cu6Sn5 grain. Mater. Charact. **138**(February), 113–119 (2018). https://doi.org/10.1016/j.matchar.2018.02.006
24. F. Somidin et al., Imaging the polymorphic transformation in a single Cu6Sn5 grain in a solder joint. Materials (Basel) **11**(11), 9–13 (2018). https://doi.org/10.3390/ma11112229
25. K. Nogita, M.A.A. Mohd Salleh, E. Tanaka, G. Zeng, S.D. McDonald, S. Matsumura, In Situ TEM Observations of Cu6Sn5 Polymorphic Transformations in Reaction Layers Between Sn-0.7Cu Solders and Cu Substrates. Jom **68**(11), 2871–2878 (2016). https://doi.org/10.1007/s11837-016-2020-0
26. M.I. Ahmad, et al., Investigations of infrared desktop reflow oven with fpcb substrate during reflow soldering process. Metals (Basel) **11**(8) (2021). https://doi.org/10.3390/met11081155
27. G. Zeng et al., The influence of ageing on the stabilisation of interfacial (Cu, Ni)6(Sn, Zn)5 and (Cu, Au, Ni)6Sn5 intermetallics in Pb-free Ball Grid Array (BGA) solder joints. J. Alloys Compd. **685**, 471–482 (2016). https://doi.org/10.1016/j.jallcom.2016.05.263
28. M.I.I. Ramli, M.A.A. Mohd Salleh, F.A. Mohd Sobri, P. Narayanan, K. Sweatman, K. Nogita, Relationship between free solder thickness to the solderability of Sn–0.7Cu–0.05Ni solder coating during soldering. J. Mater. Sci. Mater. Electron. **30**(4), 3669–3677 (2019). https://doi.org/10.1007/s10854-018-00647-5
29. H. Sun, Y.C. Chan, F. Wu, Influence of the aggregated Ag3Sn on the improvement of electromigration phenomenon in the doped Sn58Bi solder joints. J. Mater. Sci. Mater. Electron. **26**(7), 5129–5134 (2015). https://doi.org/10.1007/S10854-015-3040-1/FIGURES/5
30. H. He, G. Xu, F. Guo, Effect of small amount of rare earth addition on electromigration in eutectic SnBi solder reaction couple. J. Mater. Sci. **44**(8), 2089–2096 (2009). https://doi.org/10.1007/S10853-009-3276-3/FIGURES/5
31. M.A.A. Mohd Salleh, S.D. McDonald, Y. Terada, H. Yasuda, K. Nogita, Development of a microwave sintered TiO2 reinforced Sn-0.7wt%Cu-0.05wt%Ni alloy. Mater. Des. **82**, 136–147 (2015). https://doi.org/10.1016/j.matdes.2015.05.077
32. M.I.I. Ramli, M.A.A. Mohd Salleh, H. Yasuda, J. Chaiprapa, K. Nogita, The effect of Bi on the microstructure, electrical, wettability and mechanical properties of Sn-0.7Cu-0.05Ni alloys for high strength soldering. Mater. Des. **186**, 108281 (2020). https://doi.org/10.1016/j.matdes.2019.108281
33. W.N. Hsu, F.Y. Ouyang, Effect of Ag3Sn: Effective suppression of thermomigration-induced Cu dissolution in micro-scale Pb-free interconnects. Mater. Chem. Phys. **165**, 66–71 (2015). https://doi.org/10.1016/J.MATCHEMPHYS.2015.08.052
34. L. Liu, W. Zhou, H. Zhang, B. Li, P. Wu, Electromigration behavior in Cu/Sn-8Zn-3Bi/Cu solder joint. Microelectron. Reliab. **50**(2), 251–257 (2010). https://doi.org/10.1016/j.microrel.2009.09.018
35. Z. Gao, C. Dong, S. Shang, M. Huang, H. Ma, Y. Wang, Pronounced electromigration of GaInSn/Cu interconnects under super low critical current density. Mater. Lett. **300**, 130137 (2021). https://doi.org/10.1016/j.matlet.2021.130137

36. J. Wang, et al., Electromigration mechanism of Indium-44Tin-6Zinc alloy. J. Electron. Mater. 2019 4810 **48**(10), 6849–6856 (2019). https://doi.org/10.1007/S11664-019-07471-5
37. M.M.A. Mohd Salleh, S. Mcdonald, K. Nogita, Non-metal reinforced lead-free composite solder fabrication methods and its reinforcing effects to the suppression of intermetallic formation: Short review. Appl. Mech. Mater. **421**, 260–266 (2013). https://doi.org/10.4028/www.scientific.net/AMM.421.260
38. M.A.A. Mohd Salleh, M.H.Z. Hazizi, Z.A. Ahmad, K. Hussin, K.R. Ahmad, Wettability, electrical and mechanical properties of 99.3Sn-0.7Cu/Si 3N4 novel lead-free nanocomposite solder. Adv. Mater. Res. **277**, 106–111 (2011). https://doi.org/10.4028/www.scientific.net/AMR.277.106
39. J.W. Xian et al., Influence of Ni on the refinement and twinning of primary Cu6Sn5 in Sn-0.7Cu-0.05Ni. Intermetallics **102**(June), 34–45 (2018). https://doi.org/10.1016/j.intermet.2018.08.002
40. K. Nogita, M.A.A. Mohd Salleh, E. Tanaka, G. Zeng, S.D. McDonald, S. Matsumura, In situ TEM observations of Cu6Sn5 polymorphic transformations in reaction layers between Sn-0.7 Cu solders and Cu substrates. Jom 68(11), 2871–2878 (2016)

Chapter 9
Influence of Epoxy Viscosity on Led Encapsulation Process

A. A. A. Jaludin, Mohd Sharizal Abdul Aziz, Goh Wei Shing, M. H. H. Ishak, Mohd Syakirin Rusdi, and M. I. Ahmad

Abstract There are a few issues experienced in high-power light emitting diode (LED) applications that severely influence the reliability of the LEDs. One of the most problems of the LED failure is wire deformation or wire sweep during the encapsulation process that can indirectly affect the LED life expectancy. The goal of this study was to examine the encapsulation process of the LED and to correlate the impact of viscosity of the epoxy molding compound (EMC) during the LED encapsulation process. The advanced simulation tool, ANSYS Fluent, is utilized to complete an investigation on the fluid–structure interaction (FSI) phenomena between the gold wire bonding and the EMC. The FSI modeling was used to measure the stresses indirectly applied to the gold wire bonding during the encapsulation process. Besides, the volume of the EMC being dispense onto the LED substrate will be simulated by utilizing the volume of fluid (VOF) method. The simulation can be conducted repeatedly by changing the viscosity of the EMC (0.248 kg/m.s, 0.448 kg/m.s, and 0.648 kg/m.s). Parameters such as injection speed, injection time, and inlet contact angle are fixed in this study. The results demonstrated that the wire deformation, stress and strain distributed on the wire bonding increased with increasing viscosity of the EMC. The high viscosity of the EMC would yield higher deformation, stress, and strain distributions of the gold wire bonding.

Keywords LED encapsulation · EMC viscosity · FSI

A. A. A. Jaludin · M. H. H. Ishak (✉) · M. S. Rusdi
School of Aerospace Engineering, Universiti Sains Malaysia, Engineering Campus, 14300 Nibong Tebal, Seberang Perai Selatan, Penang, Malaysia
e-mail: mhafifihafiz@usm.my

M. S. Abdul Aziz · G. W. Shing
School of Mechanical Engineering, Universiti Sains Malaysia, Engineering Campus, 14300 Nibong Tebal, Seberang Perai Selatan, Penang, Malaysia

M. I. Ahmad
Faculty of Bioengineering and Technology, Universiti Malaysia Kelantan, Jeli Campus, Locked Bag No. 100, 17600 Jeli, Kelantan, Malaysia

M. A. A. Mohd Salleh et al. (eds.), *Proceedings of the Green Materials and Electronic Packaging Interconnect Technology Symposium*, Springer Proceedings in Physics 289,
https://doi.org/10.1007/978-981-19-9267-4_9

9.1 Introduction

Light-emitting diode (LED) is the most development innovation in lighting industry. Recently, light-emitting diodes (LEDs) were rapidly developed in various applications such as lighting hardware, traffic lights, camera glimmers, and large area displays because of its high efficiency, low energy or power consumption, and high strength [1]. Eventually, the theoretical viewpoint of LEDs is that they can work as thousands of hours as possible. However, with the rapid growth of the LEDs, the reliability of these high-power LEDs has become major challenge in this current scenario. For instance, material degradation, wire sweep, and structural destruction due to electrical, thermal and mechanical stress will prompt lumen debasement, shading variety just as early glitch of the LEDs gadget [2].

Encapsulation process is a process that adopted in the packaging of the LEDs. Epoxy resins, as elite thermosetting polymer materials have been broadly utilized for the current LEDs encapsulation process due to the benefits like high adhesion, extraordinary machinability, and excellent resistance to chemicals [3]. In fact, improper selection of the rheological properties of the epoxy resins such as viscosity can straightforwardly impact the wire deformation and the stress that acts on the wire structure during encapsulation process. Hence, the wire sweep issue can be occurred during the encapsulation process, and this may lead to the failure of the LEDs. Besides, it is discovered that when the viscosity flow of the encapsulant and the pressure distribution are high, the gold wires are expected to have large deformation when contrasted with those at a low viscosity or pressure [4]. Consequently, full understanding of the effect of viscosity change of the encapsulation material is critical to be concerned in the present paper.

The most popular numerical techniques for simulating the LED encapsulation process are the finite element method (FEM), finite volume method (FVM) [5–7], and fluid structure interaction (FSI) [8–12]. Most studies conducted structural analyses using the FEM method, whereas fluid analyses were conducted using the FVM method [13]. Although silicone polymers have significantly improved the performance of modified epoxy resins, the modified epoxy systems still have flaws. They are not very vulnerable to damage, and they have weak mechanical and adhesive qualities [14]. The use of silicone resins reduces the internal stress of the epoxy, which reduces interface delamination [15]. In this paper, the fluid–structure interaction (FSI) phenomena between the gold wire bonding and the EMC are compared in terms of wire deformation, stress, and strain distributed on the wire bonding.

9.2 Methodology

An experiment setup is prepared to validate the results obtained from the simulation under the actual conditions. The apparatus and material needed to conduct the experiment are a syringe, a needle tip with an inner diameter of 1.194 mm, a high-power light-emitting diode (LED), epoxy molding compound (EMC), and a curing oven. Figure 9.1 shows that the apparatus and experiment setup. The distance between dispensing needle and top surface of the LED substrate is set to 3.24 mm. The syringe must be pressed consistently to maintain the same pressure exerted on it. A digital camera is set in front of the LED package in order to the observation and data collection can be done. Table 9.1 shows the material properties of EMC used in the experiment.

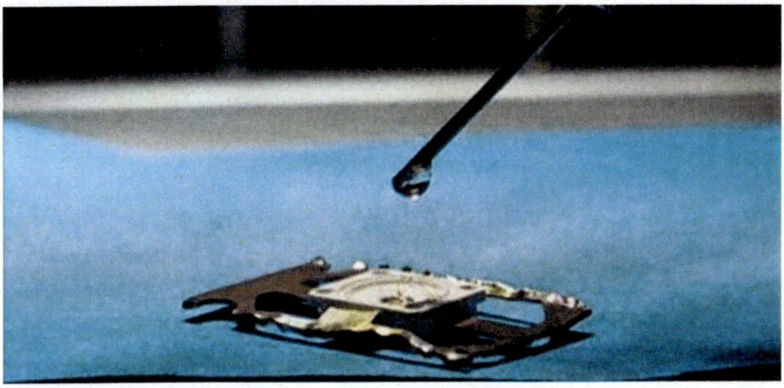

Fig. 9.1 Apparatus and experiment setup

Table 9.1 Material properties of EMC used in experiment

Test	Method	Specification		Result
		Min	Max	
Colour & clarity, visual				
Resin		Blue	Transparent	Pass
Hardener	PEN 10	Blue	Transparent	Pass
Viscosity, Cp, S-01 (K)				
Resin		1000	2000	1180
Hardener	PEN 44	100	250	185
Mix		300	800	448
Refractive Index, 25 °C (K)				
Resin		1.500	1.600	1.534
Hardener	PEN 44	1.400	1.550	1.481
Degassing time, Min, 10 g -30 in Hg vacuum, 30 °C	PEN 13	15	.	5
Gel time, 87 °C, 5 mg, min	PEN 45	5.0	9.0	5.2
Hardness, Shore D	PEN 29	80	90	84

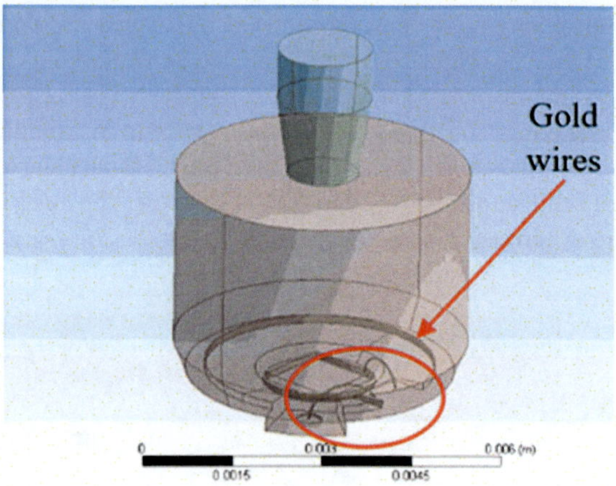

Fig. 9.2 Reflow oven geometry

9.2.1 Numerical Methods

Figure 9.2 depicts the 3D fluid assembly that was modeled using ANSYS 2021 R2. A LED, wire bonds, a substrate, and an LED enclosure make up the 3D model. The main concept is to put up a numerical approach utilizing the FVM-based software ANSYS Fluent to model fluid structure interaction (FSI) and forecast the flow of EMC during the dispensing process onto the LED substrate.

For the statistic of the meshing, the total elements and nodes for the fluid domain are 255,629 and 51,129, respectively, as shown in Fig. 9.3. For the solid body (gold wire diameter = 0.03 mm), the total elements and nodes are 37,408 and 46,230, respectively. For the quality of the mesh, all mesh is set to high smoothing.

In the fluent setup, the simulation was run in pressure-based, absolute velocity formulation, transient time and gravity set to $y = 9.81m/s^2$. In this simulation, the effect of the turbulence flow on the encapsulation dispensing process is neglected. This is because the process is very stable without any obstruction or sharp corner that creates turbulence by imparting velocities perpendicular to the flow, as well as the flow speed of epoxy molding compound (EMC) is low. Therefore, the flow is assumed to be laminar. An implicit volume of fluid (VOF) scheme and transient formulation were applied in each time step of the volume fraction.

9.3 Results and Discussion

9.3.1 Grid Independence Test

Four distinct models with the diameter of the gold wire, $D = 0.030$ mm were examined: Mesh-1 (0.15 mm), Mesh-2 (0.20 mm), Mesh-3 (0.25 mm), Mesh-4

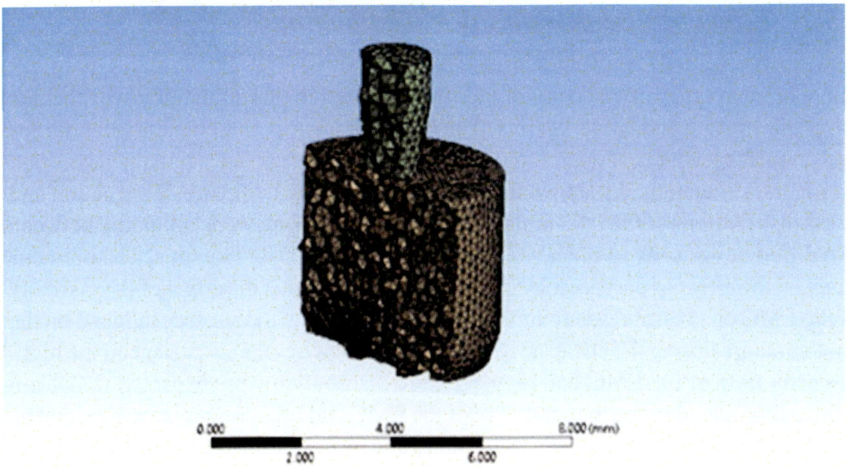

Fig. 9.3 Meshing of the LED encapsulation domain

(0.30 mm), Mesh-5 (0.40 mm), and Mesh-6 (0.45 mm) are established to choose the right mesh element size for the simulation analysis. Figure 9.4 summarizes the number of elements and nodes for each model with the diameter of the gold wire, $D = 0.030$ mm is used. Based on Fig. 9.4, it is observed that the mesh with 255,629 elements gives the least deviation when compared to others. Hence, mesh with 51,129 nodes and 255,629 elements is chosen as the optimal case in terms of accuracy and computational costs.

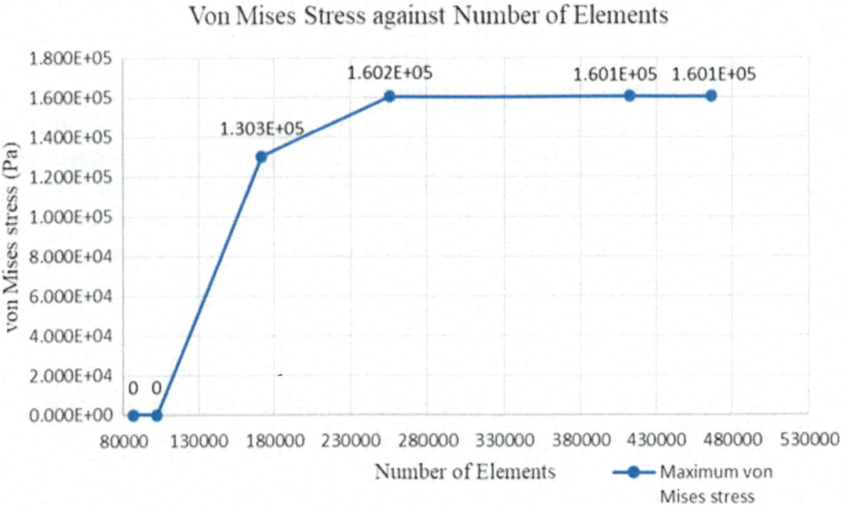

Fig. 9.4 Grid independence test

9.3.2 Effect of Viscosity on the Wire Bonding

There are two types of viscosity of the EMC utilized in this simulation work which are 0.248 kg/m.s and 0.448 kg/m.s. The other parameters and boundary conditions such as inlet speed, inlet contact angle, and injection time will have remained constant during the simulation. As shown in Figs. 9.5 and 9.6, the total wire deformation and stress distributions of the 0.03 mm diameter of the gold wire were compared between two different viscosity of the EMC which are 0.248 and 0.448 kg/m.s. The previous study on the effect of the rheology on wire sweep shows that the effect of the viscosity of the EMC on pressure was found to indirectly influence the stresses induced on the wire structure during the encapsulation process. The wires that experienced the high-viscosity flow of the EMC and high-pressure distribution were expected to deform more as compared to the low viscosity of the EMC and low-pressure distribution [4]. The form of the relation between shear stress and rate of strain depends on a fluid, and most common fluids obey Newton's law of viscosity, which states that the shear stress is proportional to the strain rate. In this present work, the maximum wire deformation, Von Mises stress, and elastic equivalent strain increasing with increasing viscosity same go to the minimum wire deformation and Von Mises stress.

9.4 Conclusion

In this present work, fluid–structure interaction (FSI) technique is utilized to examine the interaction between gold wire bonding and encapsulant in a high-power LED package. ANSYS 2021 R2 is used for the simulations of various viscosity of the epoxy molding compound (EMC) during the LED encapsulation process. The results shown that the maximum wire deformation and von Mises stress distributed on the wire bonding increased with increasing viscosity of the EMC. Since the stress exerted on the gold wire bond is relatively small, hence it does not make enough contribution to the wire deformation or does not show any significant changes on the wire structure. The analysis of the LED encapsulation process in the present work is considered only some of the variables are investigated and discussed. The other factors that will affect the structure of the EMC during the dispensing process and the deformation of the gold wire such as injection speed, injection pressure, and thermal stress as well as the distance between the needle tip and target area are recommended to be examined in the future.

Viscosity (kg/m. s)	Total Deformation (m)
0.248	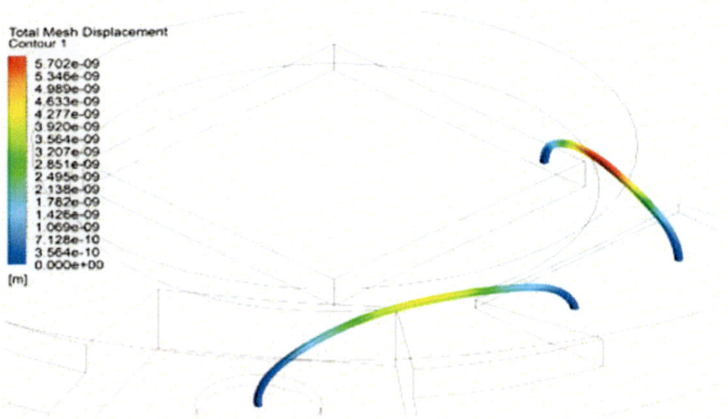
0.448	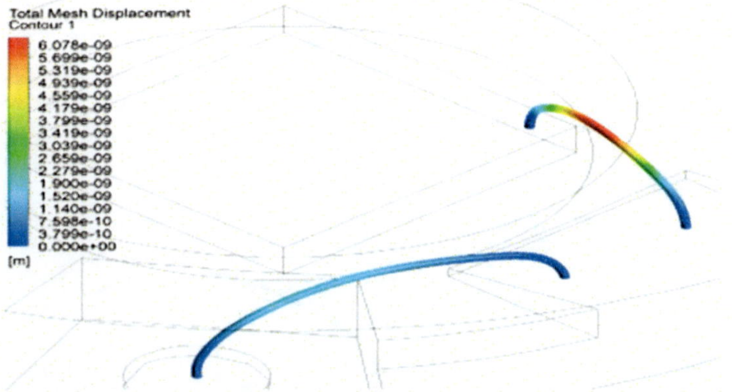

Fig. 9.5 Stress distributions of the gold wires with 0.03 mm in diameter

Viscosity (kg/m. s) | **Von Mises Stress (Pa)**

0.248

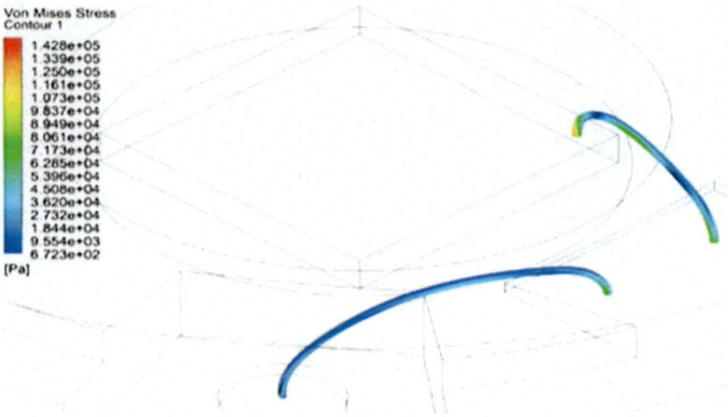

0.448

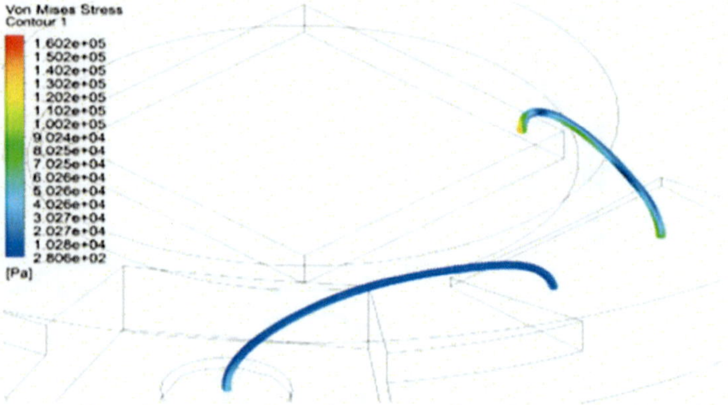

Fig. 9.6 Stress distributions of the gold wires with 0.03 mm in diameter

Acknowledgements The work is financially supported by Ministry of Higher Education under Fundamental Research Grant Scheme, (Grant number FRGS/1/2022/TK10/USM/03/11) and University Science Malaysia Short Term Grant, 304/PAERO/6315555. The authors would also like to thank University Science Malaysia for providing technical support.

References

1. R. Wen, J. Huo, J. Lv, Z. Liu, Y. Yingfeng, Effect of silicone resin modification on the performance of epoxy materials for LED encapsulation. J. Mater. Sci. Mater. Electron. **28**(19), 14522–14535 (2017)
2. M. Buffolo, A. Caria, F. Piva, N. Roccato, C. Casu, C. De Santi, N. Trivellin, G. Meneghesso, E. Zanoni, M. Meneghini, Defects and reliability of GaN-based LEDs: review and perspectives. Physica Status Solidi (a) **219**(8) (2022)
3. X. Shan, Y. Chen, Experimental and modeling study on viscosity of encapsulant for electronic packaging. Microelectron. Reliab. **80**, 42–46 (2018)
4. D. Ramdan, Z.M. Abdullah, M.A. Mujeebu, W.K. Loh, C.K. Ooi, R.C. Ooi, FSI simulation of wire sweep PBGA encapsulation process considering rheology effect. IEEE Trans. Compon. Packag. Manuf. Technol. **2**(4), 593–603 (2011)
5. D.C. Whalley, A simplified reflow soldering process model. J. Mater. Process. Technol. **150**(1–2), 134–144 (2004). https://doi.org/10.1016/j.jmatprotec.2004.01.029
6. Y.S. Son, J.Y. Shin, Thermal response of electronic assemblies during forced convection-infrared reflow soldering in an oven with air injection. JSME Int. J. Ser. B Fluids Therm. Eng. **48**(4), 865–873 (2005). https://doi.org/10.1299/jsmeb.48.865
7. A.M. Najib, M.Z. Abdullah, C.Y. Khor, A.A. Saad, Experimental and numerical investigation of 3D gas flow temperature field in infrared heating reflow oven with circulating fan. Int. J. Heat Mass Transf. **87**, 49–58 (2015). https://doi.org/10.1016/j.ijheatmasstransfer.2015.03.075
8. C. Srivalli, M.Z. Abdullah, C.Y. Khor, Numerical investigations on the effects of different cooling periods in reflow-soldering process. Heat Mass Transf. **51**(10), 1413–1423 (2015). https://doi.org/10.1007/s00231-015-1506-6
9. M.S. Abdul Aziz, M.Z. Abdullah, C.Y. Khor, Thermal fluid-structure interaction of PCB configurations during the wave soldering process. Solder. Surf. Mt. Technol. **27**(1), 31–44 (2015). https://doi.org/10.1108/SSMT-07-2014-0013
10. M.S. Abdul Aziz, M.Z. Abdullah, C.Y. Khor, Influence of PTH offset angle in wave soldering with thermal-coupling method. Solder. Surf. Mt. Technol. **26**(3), 97–109 (2014). https://doi.org/10.1108/SSMT-08-2013-0021
11. M.S. Abdul Aziz, et al., Finite volume-based simulation of the wave soldering process: influence of the conveyor angle on pin-through-hole capillary flow. Numer. Heat Transf. Part A Appl. **69**(3), 295–310 (2016). https://doi.org/10.1080/10407782.2015.1069675
12. M.S.A. Aziz, M.Z. Abdullah, C.Y. Khor, I.A. Azid, A. Jalar, F.C. Che Ani, Influence of printed circuit board thickness in wave soldering. Sci. Iran. **24**(6), 2963–2976 (2017). https://doi.org/10.24200/sci.2017.4311
13. J.R. Lee, M.S. Abdul Aziz, M.H.H. Ishak, C.Y. Khor, A review on numerical approach of reflow soldering process for copper pillar technology, Int. J. Adv. Manuf. Technol. 7–8 (2022)
14. J.-W. Bae, W. Kim, S.-H. Cho, S.-H. Lee, The properties of AlN-filled epoxy molding compounds by the effects of filler size distribution. J. Mater. Sci. **35**(23), 5907–5913 (2000)
15. I. Khalilullah, T. Reza, L. Chen, A.K.M. Monayem, H. Mazumder, J. Fan, C. Qian, G. Zhang, X. Fan, In-situ characterization of moisture absorption and hygroscopic swelling of silicone/phosphor composite film and epoxy mold compound in LED packaging, in *2017 18th International Conference on Thermal, Mechanical and Multi-Physics Simulation and Experiments in Microelectronics and Microsystems (EuroSimE)* (2017), pp. 1–9

Chapter 10
Investigation of Thermal Reflow Profile for Copper Pillar Technology

Jing Rou Lee, Mohd Sharizal Abdul Aziz, Muhammad Faiz Ridhwan Rosli, Mohd Syakirin Rusdi, Roslan Kamaruddin, M. H. H. Ishak, and Mohd Arif Anuar Mohd Salleh

Abstract The reflow soldering process is crucial in flip chip applications for forming a high-quality interconnection joint. This paper aims to investigate the effect of bump diameters of solder bump and copper pillar bump on the reflow temperature distribution during the reflow soldering process. A simplified reflow oven is developed and the virtual reflow process is established using computational fluid dynamics (CFD) software. The simulation study is validated with the experiment result. The numerical findings show that the temperature distribution is uniform in the copper pillar bumps with different diameters but uneven for the solder bumps. This study provides a foundation and insights into the effects of copper pillar bump structure on the reflow temperature profile during the reflow soldering process.

Keywords Copper pillar technology · Infrared-convection reflow oven · Computational fluid dynamics · Surface mount technology

10.1 Introduction

As the demand for high-performance mobile devices rises, so does the demand for semiconductor packaging solutions that can achieve increasingly complex performance, cost-cutting requirements and miniaturization. For older technologies, the traditional solder bump with range size of 75–150 microns was used in most flip

J. R. Lee · M. S. Abdul Aziz (✉) · M. F. R. Rosli · M. S. Rusdi · R. Kamaruddin
School of Mechanical Engineering, Universiti Sains Malaysia, Engineering Campus, 14300 Nibong Tebal, Seberang Perai Selatan, Penang, Malaysia
e-mail: msharizal@usm.my

M. H. H. Ishak
School of Aerospace Engineering, Universiti Sains Malaysia, Engineering Campus, 14300 Nibong Tebal, Seberang Perai Selatan, Penang, Malaysia

M. A. A. Mohd Salleh
Center of Excellence Geopolymer & Green Technology (CeGeoGTech), Universiti Malaysia Perlis (UniMAP), 01000 Arau, Perlis, Malaysia

M. A. A. Mohd Salleh et al. (eds.), *Proceedings of the Green Materials and Electronic Packaging Interconnect Technology Symposium*, Springer Proceedings in Physics 289,
https://doi.org/10.1007/978-981-19-9267-4_10

chips. However, the solder bump has reached its limitation due to the shrink in size of the gadgets. Hence, the copper (Cu) pillar technology with range size 50–100 microns [1] is introduced due to its possibility to solve the new features and sizes of flip chips. The thermal and electrical performances of Cu pillar bump are better than the solder bumps and able to provide a greater standoff height during the reflow soldering process [2, 3].

In the reflow oven, the electronic assembly will be undergone different stages, which are preheating, soaking, reflow and cooling stages [4]. Besides, the restriction of hazardous substances (RoHs) [5] and the environmental and health concerns have accelerated the use of the lead-free solder paste [6]. Thus, the monitoring of the temperature in the reflow oven is important as a proper control of temperature can prevent the solder defects like solder bridge and tombstone from happening.

There are several numerical approaches that can be used to simulate the reflow soldering process; the most common approaches are Finite Element Method (FEM), Finite Volume Method (FVM) [7–9] and Fluid Structure Interaction (FSI) [6, 10–13]. Most researchers used FEM for the structural analysis whereas the FVM for the fluid analysis [14]. By using FVM, Son and Shin [8] studied the effect of conveyor speed on the oven heat transfer rate, and Najib et al. [9] found that fan speed affects the temperature distribution in the solder bumps. Ahmad et al. [15] concluded that the temperature distribution of solder bumps is influenced by fan speed, board position and board thickness. With the aid of FVM, the molten solder flow during the reflow soldering process can be simulated [16, 17]. For Cu pillar bump, most analysis has been done using FEM and stress distribution was predicted for the crack formation. Through FEM, Lau et al. [18] studied the effect of solder bump structures on the stress distribution, while Che et al. [19], Shih and Hong [20], Long et al. [21] and Sun et al. [22] investigated the influence of Cu pillar bump structures on the stress distribution. However, the study of thermal reflow performance of Cu pillar bumps is rarely reported. Therefore, this paper aims to study the effects of bump diameters on the temperature distribution of the Cu pillar bumps within the assembly in the desktop reflow oven. Also, this paper attempts to provide the industry an insight into the application of Cu pillar technology. In the industrial application, smaller bump size is preferable to assemble the microchip. Thus, the thermal reflow performances of both solder bumps and Cu pillar bumps with different bump diameters are compared in terms of reflow peak temperatures and times.

10.2 Methodology

10.2.1 Numerical Methods

Figure 10.1 shows that the 3D assembly of fluid domain is modeled using the ANSYS 2021 R2. It consists of a reflow oven, a ball grid array (BGA) chip, a substrate and the interconnection joints. The basic idea is to simulate a virtual reflow oven and

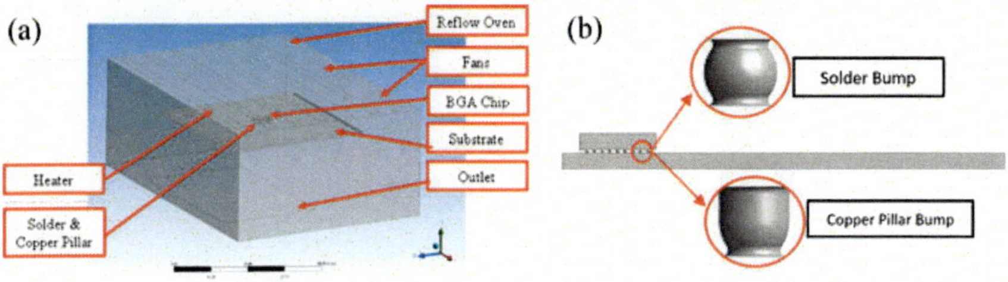

Fig. 10.1 **a** Reflow oven setup, **b** solder bump and Cu pillar bump

analyze the temperature profiles of the solder bumps and Cu pillar bumps. Meshing was performed in the ANSYS fluent (with fluent meshing) using poly-hexcore.

In the fluent setup, the simulation was run in pressure-based as the fluid flow in the reflow oven is assumed incompressible [15]. Absolute velocity formulation is applied as the fan in the fluid domain is fixed. Transient time is set due to the time-dependent characteristic of the fluid and the gravity is set to $y = 9.81$ m/s^2 according to the coordinate of the oven. Then, the energy equation was on, viscous was set to RNG *k*-epsilon, standard wall function and swirl [9]; and radiation set to discrete ordinates (DO) [6]. Besides, the boundary conditions such as the pressure outlet and other wall conditions are set based on their materials and functions.

10.3 Result and Discussions

10.3.1 Grid Independence Test

There are five mesh cases, with the mesh numbers of 377,707, 416,251, 458,961, 500,365 and 539,649, respectively. Time step size of 1 was used because it can reduce the time taken to run and obtain the results for the test. The model with finer mesh size yields higher accuracy, but it spends more computational time. It will simulate a thermal reflow profile which is similar to the actual thermal reflow profile. Hence, case 5 which has the finest size acts as the benchmark, and the percentage differences for other cases are defined. Figure 10.2 shows that the case 4 (500,365) gives the least deviation. Hence, it is chosen as the optimal case in terms of accuracy and computational costs.

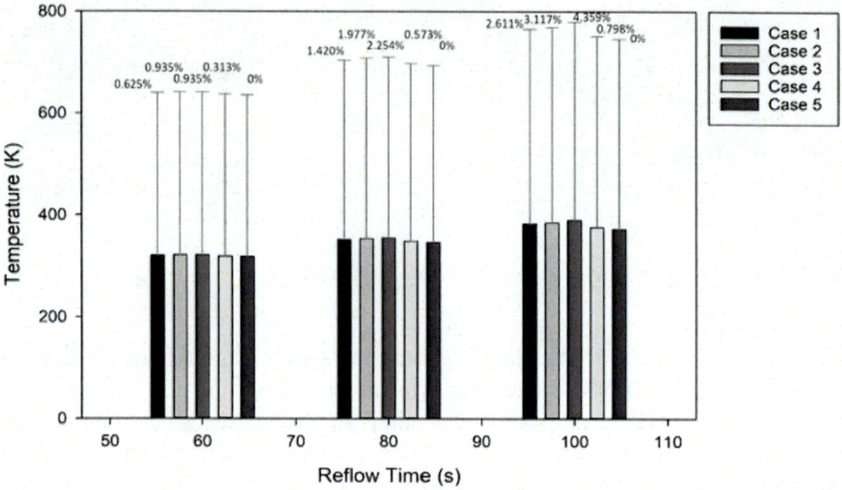

Fig. 10.2 Grid independence test

10.3.2 Experimental Validation

Figure 10.3 shows the validation study for the fluid domain in the preheating and soaking stages (0 ~ 160 s). The discrepancy between experimental and simulation values was below 10%, and this shows that the simulation results are in a good agreement with the experimental data.

10.3.3 Effect of Diameter on the Temperature Profile

Figure 10.4 shows the point 1 (2.0x, 5.5y, −2.0z) which is selected as the critical point in this paper. The critical point 1 has been chosen for studying the peak temperature and the corresponding reflow time during the reflow soldering process. The point 1 is located at the BGA chip's edge, which is also its outermost area and near to the fan. The rapid circulation of the fan might increases the assembly surface temperature at that position [15]. Hence, the monitoring of reflow temperature profile is important to prevent the temperature-sensitive electronic components from overheating and burning.

In the reflow soldering process, the heat generated will pass through the substrate with BGA chip, and the Cu pillar bumps slowly absorb the heat and begin to melt [6]. Table 10.1 shows the comparison of peak temperatures and times at point 1; the Cu pillar bumps give a more consistent peak temperature and reflow time when compared to solder bumps. The peak temperatures of Cu pillar bumps with various

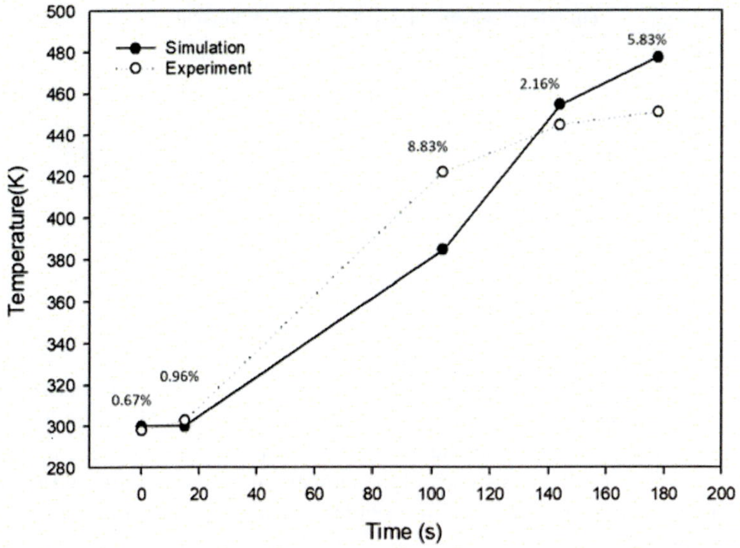

Fig. 10.3 Experimental and simulation results

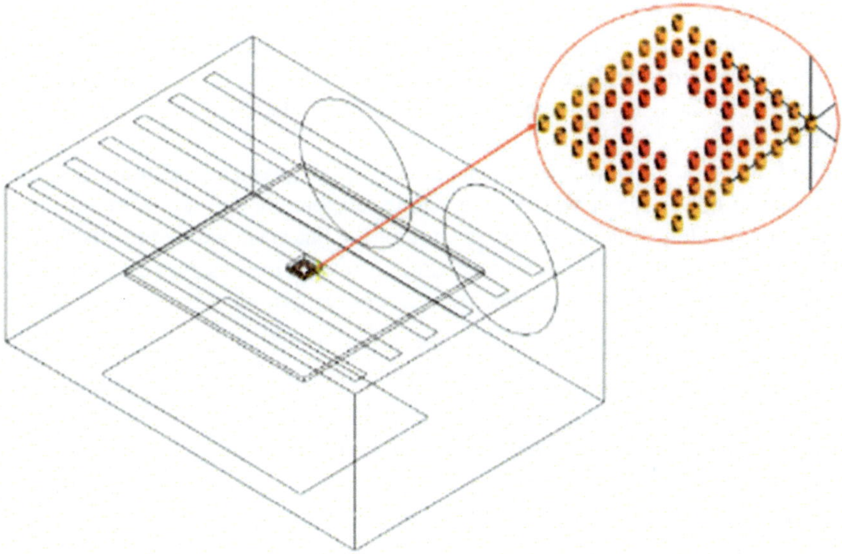

Fig. 10.4 Point 1

Table 10.1 Comparison of reflow temperatures and times for different diameters at Point 1

Diameter (mm)	Solder bump		Cu pillar bump	
	Peak temperature (K)	At peak time (s)	Peak temperature (K)	At peak time (s)
0.15	645.573	257	636.322	**270**
0.20	642.648	258	635.973	**270**
0.25	637.642	256	635.914	**270**
0.30	641.216	259	639.686	**270**
0.35	628.489	272	639.262	**270**

diameters are between 635 and 640 K whereas for solder bumps are between 630 and 645 K. The reflow time required for Cu pillar bumps with different diameters to ramp up to peak is 270 s, whereas for solder bumps, the time needed is varying from 256 to 272 s. The inconsistency in the temperature profile of the solder bumps presents that the temperature distribution in the assembly is not equally distributed [23, 24] when different diameters are considered.

When the diameter of solder bumps and Cu pillar bumps increases from 0.15 to 0.35 mm, the peak temperatures at point 1 presents a fluctuating trend. As the diameter increases from 0.15 to 0.25 mm, the peak temperature presents an upward trend; but, when the diameter increases from 0.25 to 0.35 mm, the peak temperature shows a downward trend. The changes in the peak temperature of the Cu pillar bumps with different diameters are small; this may be due to the thermal characteristic of the Cu.

From the simulation, the peak temperatures of the interconnection joints are in the range of 635 ~ 645 K which is very high temperature and not recommended. According to Ahmad, the peak temperature of the board should be targeted in the range of 523 ~ 533 K. This occurrence might be due to the inaccurate descriptions of boundary conditions of the reflow oven and/or inappropriate time step size used [14].

10.3.4 Temperature Contour for Cu Pillar Bump

The temperature contours of Cu pillar bumps with different diameters at 50, 160 and 270 s are shown in Table 10.2. It can be seen that all the Cu pillar bumps present a similar pattern. At 50 s, the Cu pillar bumps are undergone preheating stage, the outer layers are orange and greenish whereas when reflow time rises to 160 s and then 270 s, the Cu pillar bumps turn red as they are fully soaked and begin to reflow. The similar pattern shows that all the Cu pillar bumps absorb the heat energy at the same rate during the reflow soldering process. The consistency of temperature distribution of Cu pillar bumps shows that it is heated up evenly and the possibility of joint failure is reduced [24].

Table 10.2 Temperature contours of **Cu** pillar bumps with different diameters

<table>
<tr><th rowspan="2">Diameter (mm)</th><th colspan="3">Reflow time (s)</th></tr>
<tr><th>50</th><th>160</th><th>270</th></tr>
<tr><td>0.15</td>
<td>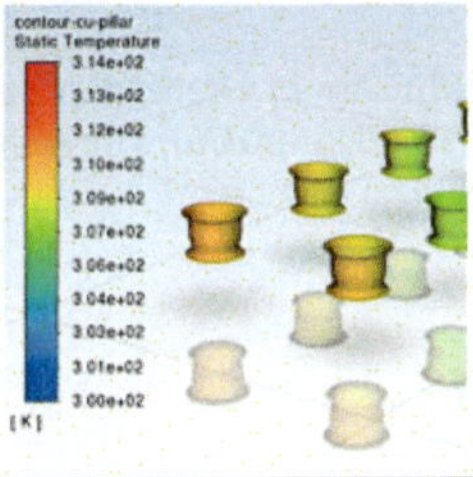
</td>
<td>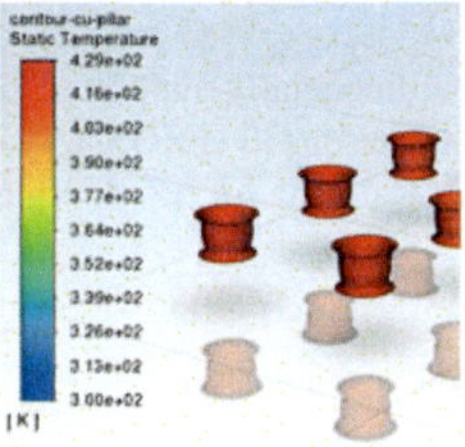
</td>
<td>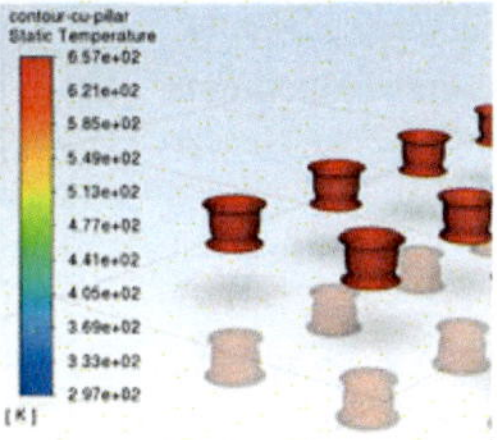
</td></tr>
<tr><td>0.25</td>
<td>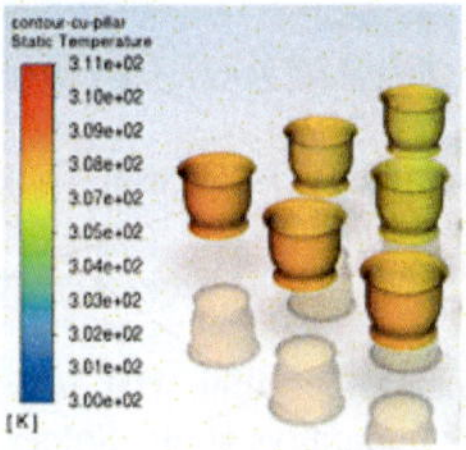
</td>
<td>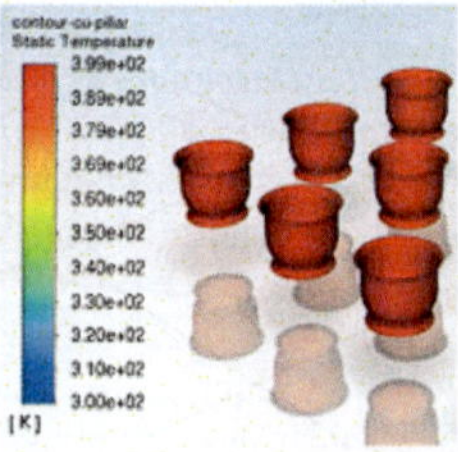
</td>
<td>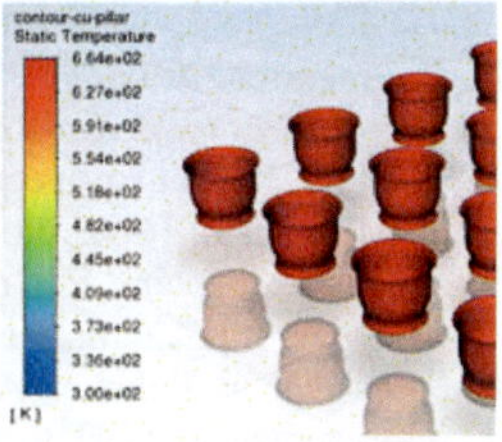
</td></tr>
<tr><td>0.35</td>
<td>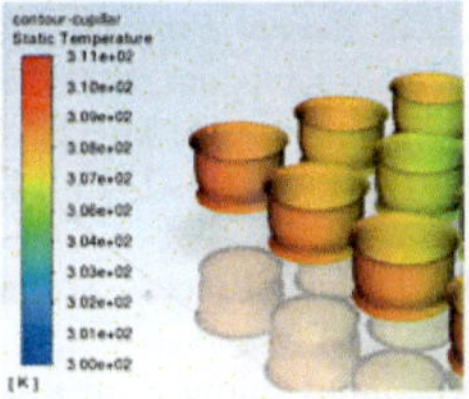
</td>
<td>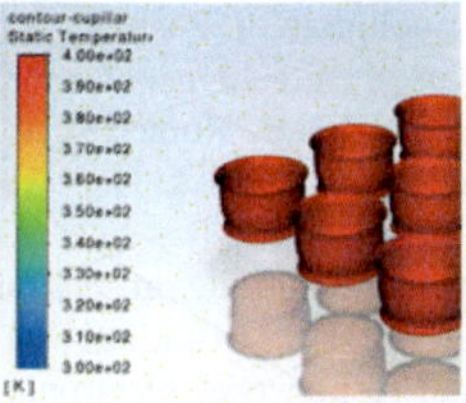
</td>
<td>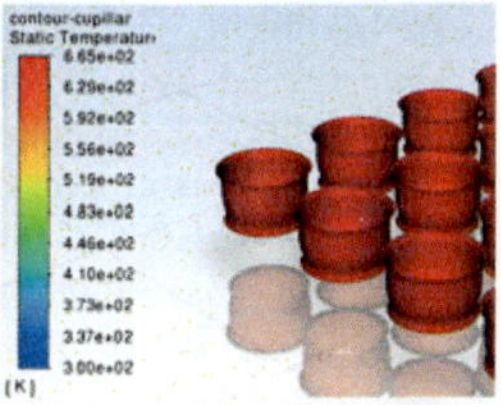
</td></tr>
</table>

10.4 Conclusion

In this paper, the numerical approach has been applied as a prediction tool to study the effect of bump diameters on the interconnection joint performance during the reflow soldering process. ANSYS 2021 R2 is used for the simulations of various diameters of the solder bumps and Cu pillar bumps. The experiment is conducted for validating the simulation results. The simulation results are well validated with the experimental data with the percentage error less than 10% at the preheating and soaking stages. From the simulations, the peak reflow temperature and time for the solder and Cu pillar bumps are compared; the temperature profiles of the Cu pillar bumps with various diameters present a consistent trend. At point 1, the peak temperatures of Cu pillar bumps are in the range of 635 ~ 640 K at the peak time of 270 s. For the solder bumps, the peak temperatures are in the range of 630 ~ 645 K with the peak time varied from 256 to 272 s. In present work, the change in Cu pillar bump diameters has minimal effect on the peak temperature and time. Also, the peak temperatures in the study are higher than 533 K and this may cause overheating and failure in the electronic components. For future work, the simulation setting will be improved to generate more accurate results. More bump diameters or larger diameter differences will be included to study the effect of diameters on the thermal reflow profile.

Acknowledgements The work is financially supported by Ministry of Higher Education under Fundamental Research Grant Scheme, FRGS (FRGS/1/2020/TK0/USM/03/6). The authors would also like to thank Universiti Sains Malaysia for providing technical support.

References

1. T. Gregorich, A. Gu, *Accelerate the Development of Advanced IC Packages Using 3D X-ray Microscopes to Measure and Characterize Buried Features* (2019)
2. J.H. Lau, Recent advances and new trends in flip chip technology. J. Electron. Packag. Trans. ASME **138**(3), 16–22 (2016). https://doi.org/10.1115/1.4034037
3. J.H. Lau, Chapter 2 Flip chip technology versus FOWLP, in *Fan-Out Wafer-Level Packaging* (Springer, Singapore, 2018), pp. 21–68
4. R. Asghar, F. Rehman, A. Aman, K. Iqbal, A.A. Nawaz, Defect minimization and process improvement in SMT lead-free solder paste printing: a comparative study. Solder. Surf. Mt. Technol. **32**(1), 1–9 (2020). https://doi.org/10.1108/SSMT-05-2019-0019
5. W.N.C. Weng, Chapter 5 Evolution of Pb-free solders, in *Recent Progress in Soldering Materials*, ed. by A.A. Mohamad (IntechOpen, London, United Kingdom, 2017), pp. 91–108
6. C. Srivalli, M.Z. Abdullah, C.Y. Khor, Numerical investigations on the effects of different cooling periods in reflow-soldering process. Heat Mass Transf. **51**(10), 1413–1423 (2015). https://doi.org/10.1007/s00231-015-1506-6
7. D.C. Whalley, A simplified reflow soldering process model. J. Mater. Process. Technol. **150**(1–2), 134–144 (2004). https://doi.org/10.1016/j.jmatprotec.2004.01.029
8. Y.S. Son, J.Y. Shin, Thermal response of electronic assemblies during forced convection-infrared reflow soldering in an oven with air injection. JSME Int. Journal, Ser. B Fluids Therm. Eng. **48**(4), 865–873 (2005). https://doi.org/10.1299/jsmeb.48.865
9. A.M. Najib, M.Z. Abdullah, C.Y. Khor, A.A. Saad, Experimental and numerical investigation of 3D gas flow temperature field in infrared heating reflow oven with circulating fan. Int. J. Heat Mass Transf. **87**, 49–58 (2015). https://doi.org/10.1016/j.ijheatmasstransfer.2015.03.075
10. M.S. Abdul Aziz, M.Z. Abdullah, C.Y. Khor, Thermal fluid-structure interaction of PCB configurations during the wave soldering process. Solder. Surf. Mt. Technol. **27**(1), 31–44 (2015). https://doi.org/10.1108/SSMT-07-2014-0013
11. M.S. Abdul Aziz, M.Z. Abdullah, C.Y. Khor, Influence of PTH offset angle in wave soldering with thermal-coupling method. Solder. Surf. Mt. Technol. **26**(3), 97–109 (2014). https://doi.org/10.1108/SSMT-08-2013-0021
12. M.S. Abdul Aziz, et al., Finite volume-based simulation of the wave soldering process: influence of the conveyor angle on pin-through-hole capillary flow. Numer. Heat Transf. Part A Appl. **69**(3), 295–310 (2016). https://doi.org/10.1080/10407782.2015.1069675
13. M.S.A. Aziz, M.Z. Abdullah, C.Y. Khorc, I.A. Azid, A. Jalar, F.C. Che Ani, Influence of printed circuit board thickness in wave soldering. Sci. Iran. **24**(6), 2963–2976 (2017). https://doi.org/10.24200/sci.2017.4311
14. J.R. Lee, M.S. Abdul Aziz, M.H.H. Ishak, C.Y. Khor, A review on numerical approach of reflow soldering process for copper pillar technology. Int. J. Adv. Manuf. Technol. (7–8) (2022)
15. M.I. Ahmad et al., Investigations of infrared desktop reflow oven with FPCB substrate during reflow soldering process. Metals (Basel) **11**(8), 1155 (2021). https://doi.org/10.3390/met11081155

16. M.S. Abdul Aziz, M.Z. Abdullah, C.Y. Khor, A. Jalar, F. Che Ani, CFD modeling of pin shape effects on capillary flow during wave soldering. Int. J. Heat Mass Transf. 72, 400–410 (2014). https://doi.org/10.1016/j.ijheatmasstransfer.2014.01.037
17. M.S. Abdul Aziz, M.Z. Abdullah, C.Y. Khor, F. Che Ani, N.H. Adam, Effects of temperature on the wave soldering of printed circuit boards: CFD modeling approach. J. Appl. Fluid Mech. **9**(4), 2053–2062 (2016). https://doi.org/10.18869/acadpub.jafm.68.235.23709
18. C.S. Lau, M.Z. Abdullah, M. Abdul Mujeebu, N. Md. Yusop, Finite element analysis on the effect of solder joint geometry or the reliability of ball grid array assembly with flexible and rigid PCBS. J. Eng. Sci. Technol. **9**(1), 47–63 (2014)
19. F.X. Che, L.C. Wai, X. Zhang, T.C. Chai, Characterization and modeling of fine-pitch copper ball bonding on a Cu/Low-k chip. J. Electron. Mater. **44**(2), 688–698 (2015). https://doi.org/10.1007/s11664-014-3532-4
20. M.K. Shih, P.C. Hong, Structural design guideline for Cu pillar bump reliability in system in packages module, in *2015 IEEE 17th Electronics Packaging and Technology Conference (EPTC)* (2016), pp. 1–4. https://doi.org/10.1109/EPTC.2015.7412345
21. X.J. Long, J.T. Shang, L. Zhang, Design optimization of pillar bump structure for minimizing the stress in brittle low K dielectric material layer. Acta Metall. Sin. **33**(4), 583–594 (2020). https://doi.org/10.1007/s40195-019-00948-6
22. H. Sun, B. Gao, J. Zhao, Thermal-mechanical reliability analysis of WLP with fine-pitch copper post bumps. Solder. Surf. Mt. Technol. **33**(3), 178–186 (2020). https://doi.org/10.1108/SSMT-06-2020-0027
23. X.Q. Tang, S.J. Zhao, C.Y. Huang, L.K. Lu, Thermal stress-strain simulation analysis of BGA solder joint reflow soldering process, in *2018 19th International Conference on Electronic Packaging Technology (ICEPT)* (2018), pp. 981–986. https://doi.org/10.1109/ICEPT.2018.8480615
24. J.R. Lee, M.S. Abdul Aziz, M.H.H. Ishak, Study on copper pillar bump in flip chip technology using computational fluid dynamics, in *International Invention & Innovative Competition (InIIC)*, (2020), pp. 1–7

Chapter 11
The Effect of Solder Geometry and Intermetallic Compound on Lead-Free Solder Joint Reliability

Hong Ann Tan and Ho Cheng How

Abstract Finite element simulation was performed on 132 pin fleXBGATM package assembly using ANSYS Workbench 21.1 to investigate the integrated effect of solder geometry and intermetallic compound on the thermal fatigue reliability of lead-free solder joint. Various solder geometries represented by shape factor are simulated with intermetallic compound layer thickness of 0, 3 and 6 μm. Global model simulation was firstly performed to identify the most critically affected joint in the package assembly subjected to thermal cycling test based on highest accumulated inelastic strain. Subsequently, sub-modelling analysis technique was applied on the critical joint to enhance the computational efficiency. The results showed that the location of the critical joint is consistent irrespective to solder geometry and intermetallic compound. Hourglass shape generally enhances the reliability of the solder joint, and the optimum performance is found on shape factor of 0.604. Besides, intermetallic compound layer thickness degrades the reliability of solder joint with shape factor above 1 but improves the reliability of solder joint with shape factor below 1.

11.1 Introduction

Driven primarily by the mismatch of coefficient of thermal expansion (CTE) between constituents of package assembly, thermally induced shape deformation or dynamic warpage often results when subjected to temperature profile [1]. The peak temperature experienced by lead-free Ball Grid Array (BGA) package during reflow soldering process associated with board assembly could reach as high as 260 °C which results in the assembly warping dynamically with the reflow temperature profile [1]. Past studies have reported various types of BGA solder joint failure resulting from warpage such as bridging, Head-on-Pillow (HoP) and stretched joint [1]. In this

H. A. Tan · H. C. How (✉)
Department of Mechanical, Materials and Manufacturing Engineering, The University of Nottingham Malaysia, Jalan Broga, 43500 Semenyih, Selangor Darul Ehsan, Malaysia
e-mail: HoCheng.How@nottingham.edu.my

M. A. A. Mohd Salleh et al. (eds.), *Proceedings of the Green Materials and Electronic Packaging Interconnect Technology Symposium*, Springer Proceedings in Physics 289,
https://doi.org/10.1007/978-981-19-9267-4_11

context, less attention had been given to stretch joint resulting from warped package assembly.

With the warpage measurements, Rayasam et al. [2] had developed a computational model to predict the solder joint shape in the presence of PCB and package warpage. However, the implication behind each of the solder shape predicted in terms of the reliability of the package assembly is not discussed. Lau et al. [3] had adopted finite element (FE) sub-modelling technique to investigate the effect of solder joint geometry on the reliability of BGA assembly subjected to accelerated thermal cycling (ATC), but the study was performed using lead solder material and neither the bonding pad nor the intermetallic compound (IMC) layer were considered which could significantly affect the results. Besides, due to the low melting temperature of lead-free solder material, solder joint often suffers from high homologous temperature ($T/T_m > 0.4$) when subjected to cyclic thermal profile. Thermal-induced creep fatigue mechanism is therefore expected to dominate the deformation kinetics [4].

Previous studies had shown that reflow conditions such as temperature, humidity and ageing time have strong correlation with IMC growth [5]. According to Qin et al. [6], the strength of solder joint drops with greater IMC thickness. Lee et al. [7] had experimentally observed the failure location of barrel and hourglass solder geometry in the presence of IMC across various lead-free solder material. It was determined that for solder with Ag content, crack mostly initiates and propagates along IMC/solder interface for barrel shape joint while failure usually occurs at bulk solder near midpoint for hourglass shape joint. However, the effect of IMC thickness is not discussed in the paper. Che and Pang [5] had further verified the effect of IMC thickness on thermal fatigue reliability of solder joint using numerical FE approach. IMC layer was modelled with thickness ranging from 1 to 6 μm to mimic IMC growth. The findings had indicated that thicker IMC results in lower fatigue life specifically for barrel shape solder joint. Findings by Che et al. [5] had also shown that the obtained life prediction of the solder would be significantly overestimated if IMC layer is ignored in numerical simulation in which addition of IMC layer into FE simulation can reduce solder joint fatigue life by more than 8% compared to cases without IMC.

While there are many previous research studies on the effect of solder geometry [3, 7] and IMC layer [5, 6], respectively, none has been done on investigating the integrated effect of solder geometry and IMC thickness on solder joint reliability. Furthermore, most of the research papers support the general statement that 'the thicker the IMC layer, the lower the strength of the solder joint', but this is only a phenomenological statement without considering the solder geometry at the same time. As it is commonly reported that the growth of IMC layer plays a degrading role in the mechanical strength of solder joints specifically for barrel shape solder joint [5, 6], the role of IMC layer in other joint geometries such as column and hourglass shape is still not clear. Hence, the objective of this study is to investigate the integrated effect of IMC layer thickness across various solder geometries on thermal fatigue reliability under ATC.

11.2 Methodology

Figure 11.1 shows the CAD model of fleXBGATM assembly [3, 8]. 3D quarter model of the assembly was created based on its geometrical symmetry for higher computational efficiency. The whole package consists of 132 lead-free solder joints and 264 copper pads on PCB and substrate without underfill. The whole assembly is comprised of moulding compound, substrate, silicon die, solder joints, copper pads and PCB.

11.2.1 Material Properties and Constitutive Model

All materials except solder joints were modelled as linear isotropic elastic materials. Solder joints were simulated using Anand's viscoplastic model [5]. All materials including solder joint were assumed homogenous. Only the material behaviour of solder joint is nonlinear and temperature dependent [4]. The reference temperature of stress-free state of the assembly was specified to 298 K [3]. At this temperature, primary stresses from previous fabricating processes such as reflow soldering were neglected [4]. The material properties including Young's Modulus, Poisson's Ratio and CTE for each component in the assembly are summarized in Table 11.1. Anand's viscoplastic model is used to accurately capture the constitutive relations of solders at high homologous temperature ($T/Tm > 0.4$) as it unifies both plasticity and creep effect simultaneously in the material. Anand's constants which were obtained experimentally [4] are shown in Table 11.2 for SAC 387 solder material.

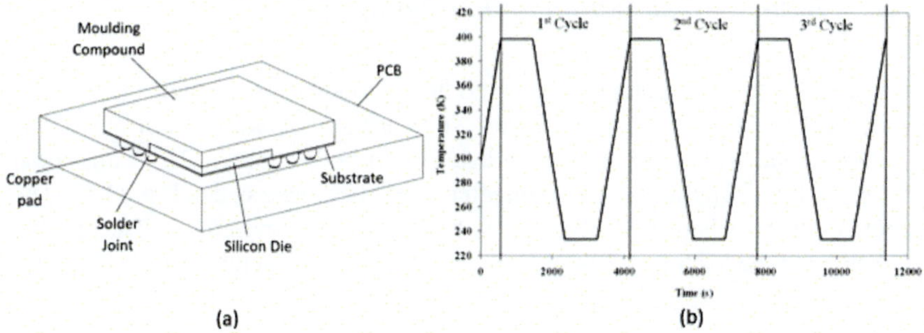

Fig. 11.1 **a** Quarter model of fleXBGATM assembly that includes package substrate (6 mm × 6 mm × 0.05 mm), PCB (10 mm × 10 mm × 1.6 mm) and solder joint pitch of 0.8 mm [3, 8]; **b** The temperature cycle profile [3]

Table 11.1 Material properties of the components in the assembly [3, 4]

Material	Young's modulus (MPa)	Poisson's ratio	CTE (ppm/K)
PCB	18,200	0.25	15
Substrate	26,000	0.39	15
Silicon die	130,360	0.28	2.5
Mould	16,520	0.25	14.8
Copper pad	129,000	0.34	17
Solder (SAC387)	45,000	0.36	17.6

Table 11.2 Anand's parameters for SAC 387 solder material [4]

Anand parameters	SAC 387
Initial deformation resistance, s_0 (MPa)	39.5
Activation energy Q/Universal gas constant R	8710
Pre-exponential factor, A (s^{-1})	24,300
Multiplier of stress, ξ	5.8
Strain rate sensitivity of stress, m	0.183
Hardening/softening constant, h_0 (MPa)	35,412
Coefficient for deformation resistance saturation, $\hat{s}$ (MPa)	65.3
Strain rate sensitivity of saturation, n	0.019
Strain rate sensitivity of hardening, a	1.9

11.2.2 Boundary Condition and Loading

Symmetrical boundary conditions (frictionless support) are applied on the two symmetrical faces of the quarter model. Node at the diagonal bottom of PCB is also constrained with fixed support to avoid rigid body motion [3]. The thermal cycling conditions applied are based on reference [3] as illustrated in Fig. 11.1b. The temperature ranges between 233 and 398 K at a rate of 1 cycle/hour. The starting temperature of the loading is 298 K. Each dwell time in the thermal profile is 15 min and the ramp rate is 11 °C/min.

11.2.2.1 FE Modelling

The main purpose of global model is to identify the critical solder joint in which failure is mostly likely to occur first in the assembly. The critical solder joint is determined based on highest accumulated inelastic strain (NLEPEQ) which represents the solder fatigue damage. Furthermore, the global model is also important to provide

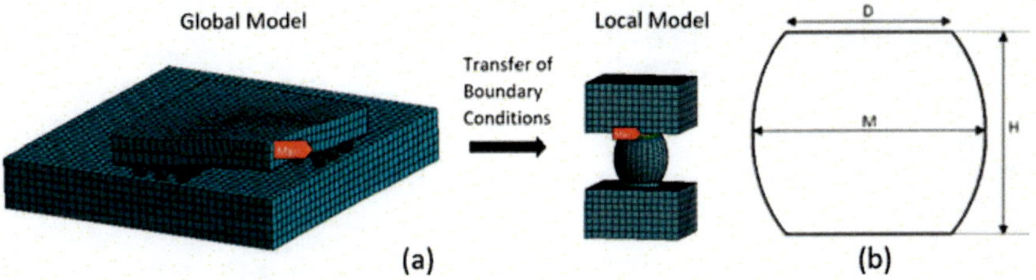

Fig. 11.2 **a** Transfer of the global model of the BGA package assembly to the local model of the critical joint; **b** solder joint shape specifications

driving displacement boundary condition to the sub-model of the critical joint at subsequent stage [3]. An illustration of global model is shown in Fig. 11.2a.

The local model contains the most critical joint located in the package where maximum NLEPEQ occurs as shown in Fig. 11.2a [3]. To create the local model, the critical joint was sliced off from the rest of the assembly. The cut boundary is ensured to be far away from the region of interest. Transfer of displacement boundary condition from global to local model was performed by sub-modelling technique. Local model allows higher mesh density to be used compared to global model while being more computationally efficient and accurate [3]. Accumulated inelastic strain (NLEPEQ) from the local model of each joint geometry and IMC were obtained as this quantity is the fatigue damage parameters described in Darveaux Fatigue law [3, 8]. Reliability of solder joints across various solder geometries and IMC thickness are compared using the fatigue damage parameters.

11.2.2.2 Modelling of Solder Joint Geometries

Figure 11.2b shows the solder joint shape specification where H refers to the standoff height of the solder joint, D is the pad diameter (0.33 mm), M is the midpoint diameter of solder joint. The shape factor (SF) is a non-dimensional parameter and is defined in Eq. 11.1 [9]:

$$\text{Shape Factor, SF} = \frac{M}{D} \tag{11.1}$$

SF can be tailored to form different geometries. According to Eq. 11.1, solder joint is barrel shape when SF > 1, column shape when SF = 1 and hourglass shape when SF < 1.

11.2.2.3 Modelling of Intermetallic Compound (IMC)

Various IMC thicknesses (0, 3 μm, 6 μm) are modelled to mimic the IMC growth. Assumption made is that the addition of IMC layer does not consume any solder or

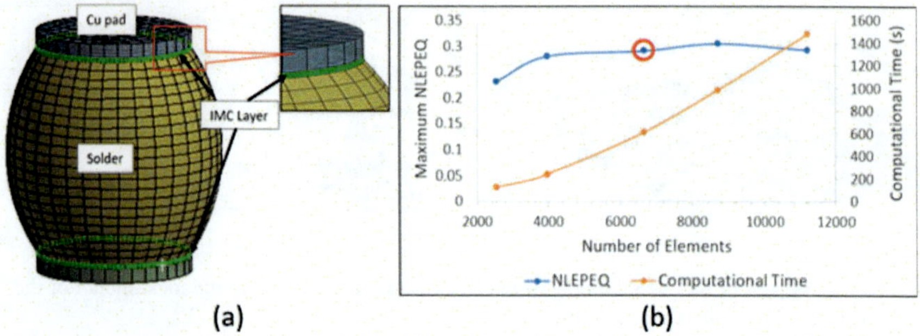

Fig. 11.3 **a** IMC layer (Young's modulus = 86 GPa, CTE = 16.3 ppm/°C, Poisson's ratio = 0.25) between bulk solder and Cu pad; **b** plot of maximum NLEPEQ and computational time against number of mesh elements for barrel shape local model

pad material as shown in Fig. 11.3a. Furthermore, IMC layer is modelled to be flat without roughness and only one IMC species exists in the layer for each case [5]. The material properties of Cu_6Sn_5 IMC are used in the simulation [5].

11.2.2.4 Grid Independence Test

The focus of grid independence test is on the local model where the simulation results are obtained. Quadratic elements were applied for both global and local model. Since only displacement is transferred from the global model to local model [3], it is more independent of the grid size, and hence, a percentage difference with successive mesh of below 10% is acceptable. Furthermore, mesh of local model is declared independent once the percentage difference with successive mesh achieves below 5%. Figure 11.3b shows the plot of maximum NLEPEQ and computational time against number of elements with selected mesh (6660 elements, 28,453 nodes) highlighted for barrel shape local model. Same procedure applies for other geometries [3].

11.3 Results and Discussion

11.3.1 Location of Critical Solder Joint

Failure is most likely to occur first at the solder joint having the highest NLEPEQ at the end of the thermal cycle. Therefore, critical solder joint is defined as the joint that has the highest NLEPEQ. The location of the critical solder joint in the package assembly is consistent across all geometries and IMC thickness simulated in the current study. The critical joint is found to be located at inner row closest to the silicon die as shown in Fig. 11.4. This is due to largest CTE mismatch between silicon die and PCB. As quarter model assumption is adopted in the global simulation, there

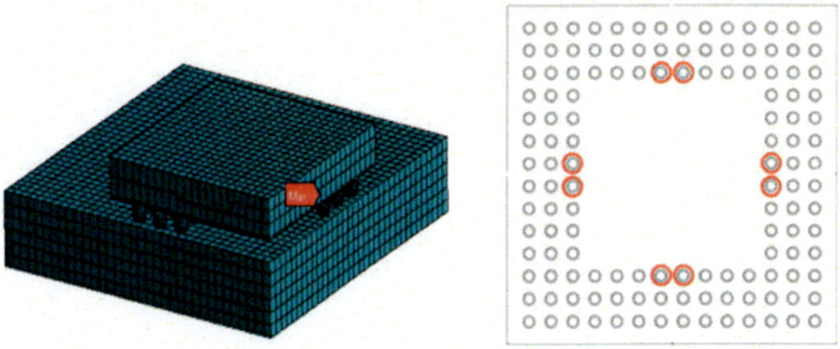

(a) (b)

Fig. 11.4 **a** Location of the critical joint in the quarter model; **b** the location of the critical joints in the full package assembly

are a total of 8 critical joints in the full model of the package assembly. This explains previous findings in which the row of joints nearest to the die edge firstly failed in most of the thermal cycle tests [5].

11.3.2 Effect of Solder Joint Geometries

Various solder joint geometries are simulated without IMC. The distribution of NLEPEQ of each solder geometries in their local models was observed. The failure location of the solder joint of each geometry matches well with the experimental findings [9] as shown in Fig. 11.5a. As copper pads were included in the simulation, maximum NLEPEQ tends to concentrate at the pad/solder interface due to bi-material bonding for barrel shape solder [3]. However, maximum NLEPEQ shifts from the pad/solder interface to the midpoint region as the SF of the solder geometry decreases due to necking which can be clearly seen in hourglass shape solder.

Comparisons are made on the maximum NLEPEQ on different geometries as shown in Fig. 11.5b. Barrel shape has the highest fatigue damage compared to the rest of the geometries. Crack initiates at the edges of interface between the solder and pad due to high stress concentration at that region. Hourglass 2 shape is the most optimal geometry as it has the least fatigue damage among all. This implies that hourglass shape is capable of relieving the stress singularity at the pad/solder interface by shifting the stress distribution towards the midpoint region. However, the reliability performance worsens as the neck of the solder becomes too narrow in hourglass 3. This shows that while decreasing SF can improve the reliability performance, it is countered by the risk of midpoint failure once a critical point is exceeded. Similar trend is observed in numerical study by Lau et al. [3]. However, hourglass 1 shape was found to be the most optimal geometry. The discrepancy might be due to exclusion of copper pad, leaded solder material and different constitutive model applied in their FE simulation.

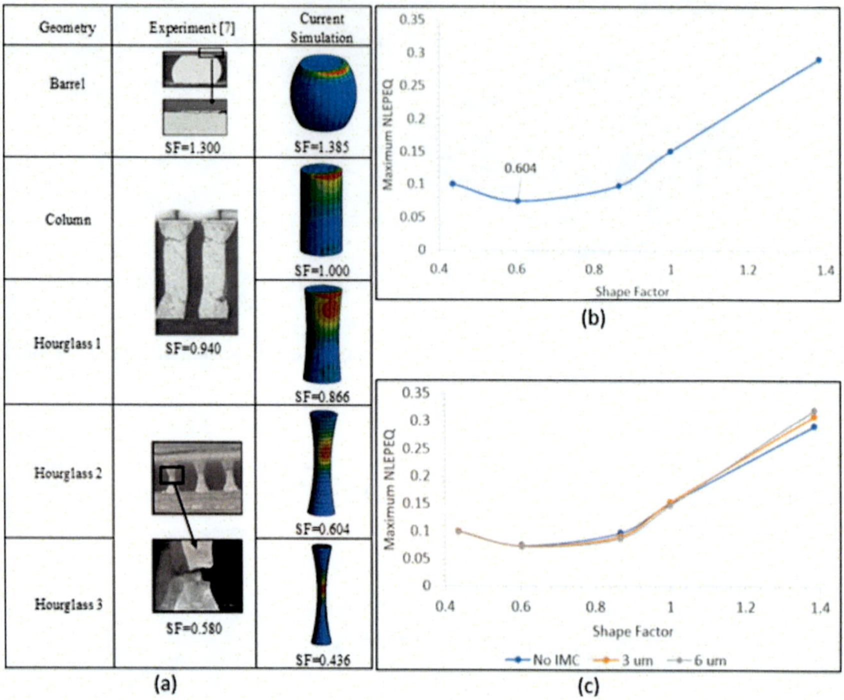

Fig. 11.5 **a** Comparisons between failure location observed in experimental findings and current simulation of various geometries; **b** maximum NLEPEQ against various solder geometries; **c** maximum NLEPEQ against various geometries with different IMC thicknesses

Figure 11.5b shows the plot of maximum NLEPEQ against SF. It can be seen that the maximum NLEPEQ declines as the SF decreases up to the critical point with lowest NLEPEQ. The critical point is the most optimal geometry having SF equal to 0.604. When SF is greater than 0.604, maximum NLEPEQ decreases as the midpoint diameter of the solder joint reduces which is beneficial for thermal fatigue performance of solder joint. Upon the critical point, when SF is lower than 0.604, further decrease in midpoint diameter will lead to higher NLEPEQ due to the risk of midpoint failure as a result of necking which will deteriorate the fatigue performance of the solder joint. Therefore, it is important to ensure that the final solder geometry of the package assembly is having SF greater than and close to 0.604 for better reliability.

11.3.3 Integrated Effect of IMC and Solder Geometries

Addition of Cu_6Sn_5 IMC layers with increasing thickness of 0, 3 and 6 μm results in different effect depending on the solder geometry as shown in Fig. 11.5c. The

NLEPEQ distribution in solder joints with IMC is similar to those shown previously in Fig. 11.5a. Based on Fig. 11.5c, higher IMC thickness would lead to higher maximum NLEPEQ when SF is above 1. Effect of IMC thickness is significant in this region as increase in IMC thickness would lead to degradation of the thermal fatigue performance of the solder joint. Such trend can be clearly observed in barrel shape solder (SF = 1.385) in which stress concentration tends to occur at the IMC/solder interface as a result of bi-material bonding. Brittle nature of IMC also imposes more constraint on the solder joint deformation which eventually leads to higher fatigue damage [6]. Similar failure mode and degrading effect are reported by the literatures that studied effect of IMC thickness specifically on barrel shape joint [5, 6]. A transition point of the trend is observed when SF is equal to 1. At the transition point, IMC thickness on column shape joint does not give rise to significant influence on the reliability performance of the solder joint. Beyond the transition point when SF is below 1, further decrease in SF results in an opposite trend such that the higher IMC thickness, the lower the maximum NLEPEQ. In this region, the stress concentration effect at the IMC/solder interface had been alleviated and maximum NLEPEQ had shifted towards the midpoint region of the solder.

11.4 Conclusions

3D, FE analysis of a quarter-symmetry BGA package assembly model was performed to investigate the effect of solder geometry and IMC on the thermal fatigue reliability of solder joints. It was found that the location of the critically affected joint is consistently at the inner row closest to the silicon die irrespective to solder geometry and IMC. Furthermore, it was determined that the reliability of the solder joint improves with lower SF up to a critical point. Hourglass 2 shape with SF = 0.604 is the most optimal geometry. Further decrease in SF beyond the critical point results in poorer reliability. Besides, IMC thickness plays a degradation role in solder geometries when SF > 1 but it improves the reliability when SF < 1.

References

1. W.K. Loh et al., in *Impact of Low Temperature Solder on Electronic Package Dynamic Warpage Behavior and Requirement*, 2019 IEEE 69th Electronic Components and Technology Conference (ECTC), pp. 318–324 (2019)
2. M. Rayasam et al., A model for assessing the shape of solder joints in the presence of PCB and package warpage. J. Electron. Packag. **128**(3), 184–191 (2006)
3. C.-S. Lau et al., Finite element analysis on the effect of solder joint geometry for the reliability of ball grid array assembly with flexible and rigid PCBS. J. Eng. Sci. Technol. **9**, 47–63 (2014)
4. J.A. Depiver, S. Mallik, E.H. Amalu, Thermal fatigue life of ball grid array (BGA) solder joints made from different alloy compositions. Eng. Fail. Analy. 125 (2021)
5. F.X. Che, J.H.L. Pang, Characterization of IMC layer and its effect on thermomechanical fatigue life of Sn–3.8Ag–0.7Cu solder joints. J. Alloys Compounds **541**, 6–13 (2012)

6. T. An, F. Qin, Effects of the intermetallic compound microstructure on the tensile behavior of Sn3.0Ag0.5Cu/Cu solder joint under various strain rates. Microelectron. Reliability **54**(5), 932–938 (2014)
7. H.-T. Lee, K.-C. Huang, Effect of solder-joint geometry on the low-cycle fatigue behavior of Sn-xAg-0.7Cu. J. Electron. Mater. **45**(12), 6102–6112 (2016)
8. G. Chen, C. Bailey, X. Chen, Finite element analysis of fleXBGA reliability. Soldering Surf. Mount Technol. **18**(2), 46–53 (2006)
9. R. Rajoo et al., Development of stretch solder interconnections for wafer level packaging. IEEE Trans. Adv. Packag. **31**(2), 377–385 (2008)

Chapter 12
Thermal Management on the Solder Joints of Adjacent Ball Grid Array (BGA) Rework Components Using Laser Soldering

Adlil Aizat Ismail, Maria Abu Bakar, Abang Annuar Ehsan, Azman Jalar, Zol Effendi Zolkefli, and Erwan Basiron

Abstract Temperatures for hot air and laser reflow soldering were evaluated for a ball grid array component (BGA) throughout the rework procedure to mitigate the influence of mechanical and thermal-induced damage on the solder joints quality and reliability of the adjacent BGA components on both the printed circuit board assembly (PCBA) sides. Sample H received convection heat from hot air, while sample L received radiation heat from a laser beam. Temperature measurements were obtained with thermocouple wires and analysed to better understand the link between reflow soldering heat source types and heat dissipation on adjacent rework components. The findings were displayed using individual value plots and standard deviation charts. With a radiation heat source, laser reflow soldering is capable to decrease the highest temperatures on the adjacent components during the rework procedure up till 10%. It may also maintain the peak temperatures of the adjacent BGA components below 207 and 204 °C, respectively, which is below the lead-free solder melting point, improving PCBA reliability by reducing solder joint damage.

Keywords Rework · Laser soldering · Adjacent components · Ball grid array (BGA) · Solder joint

---

A. A. Ismail · Z. E. Zolkefli · E. Basiron
Western Digital®, SanDisk Storage Malaysia Sdn.Bhd., Plot 301A, Persiaran Cassia Selatan 1, Taman Perindustrian Batu Kawan, MK13, Batu Kawan, Seberang Perai Selatan, 14100 Penang, Malaysia

A. A. Ismail · M. Abu Bakar (✉) · A. A. Ehsan · A. Jalar
Institute of Microengineering and Nanoelectronics, Universiti Kebangsaan Malaysia, 43600 Bangi, Selangor, Malaysia
e-mail: maria@ukm.edu.my

A. Jalar
Department of Applied Physics, Faculty of Science and Technology, Universiti Kebangsaan Malaysia, 43600 Bangi, Selangor, Malaysia

M. A. A. Mohd Salleh et al. (eds.), *Proceedings of the Green Materials and Electronic Packaging Interconnect Technology Symposium*, Springer Proceedings in Physics 289,
https://doi.org/10.1007/978-981-19-9267-4_12

12.1 Introduction

In the manufacturing industry, rework on printed circuit board assembly (PCBA) is commonly used as a cost-effective way to decrease waste and increase overall income. Rework on PCBA is critical in times when getting components is difficult and product development cycles are short [1, 2]. The major benefit of reworking a PCBA is that it may be done faster than replacing it, depending on the amount of damage [3]. Convection heat via hot air is the common method for reflow soldering a BGA component throughout the rework procedure [4]. During rework, the adjacent components surrounding the rework area are exposed to thermal reflows for the high-density PCBA designs with BGA components population [5]. Reflowing adjacent component solder joints unintentionally can result in component damage and solder joint cracks [6]. The cracks would influence the reliability and quality of the solder joint which affects the PCBA performance [7].

Radiation heat source via laser soldering has been used for the reflow soldering method for surface mount devices on the printed circuit board (PCB) since it was introduced in the 1980s [8]. The key benefits of laser soldering include precise local heat input, non-contact and regulated processing, and excellent solder joint quality with less thermal and mechanical stress [9]. However, the application of laser soldering in the ball grid array (BGA) component rework procedure is still uncommonly used compared to hot air or infrared (IR) reflow soldering [10]. Not many studies have examined the effect of rework heat source type on the adjacent BGA components of high-density PCBA designs with component population rework location solder joints on both sides of the PCBA. Muonio and Stadem [11] stated that laser reballing for BGA due to rework can improve the component reliability. Albert et al. [12] analysed the quality of laser reworked solder joints on the targeted component. Burke and Dai [13] discussed the overall benefits of laser soldering for rework.

The primary goal of this research is to study the effectiveness of heat source types for the rework procedure in avoiding damage from mechanical forces and thermal damage to the solder joint and surrounding components at the rework area on both PCBA sides. For this paper, (a) temperature measurements through thermocouple (TC) wires were used to confirm the solder joint array's thermal distribution; (b) the dye and pull (DP) tests were performed to identify cracks in solder joints after rework; and (c) the DP test findings, as well as temperature data, were examined using both qualitative and quantitative analysis to learn more about the connection between the types of heat source used in the rework procedure and solder joint degradation on adjacent components.

12.2 Methodology

12.2.1 Materials and Sample Description

Two variables served as test subjects, including sample H: rework using convection heat (hot air reflow soldering) and sample L: rework using radiation heat (laser reflow soldering). For each sample, a total of twelve BGA components were monitored set on both sides of the PCBA. On each side of the PCBA, six BGA components mirrored each other, and the gap between the BGA components can be found in Fig. 12.1. U3 BGA component located on the top side of the PCBA was selected for the rework experiment for each sample. According to ASME Y14.44-2008, the letter "U" was chosen as the common reference designation for an integrated circuit component [14]. The BGA component solder joints architecture consisted of 132 ball-type solder joints with an unpopulated column in the middle. Before the rework procedure, the solder paste used for the BGA component assembly was a lead-free type with an alloy composition of Sn96.5, Ag3.0, and Cu0.5 wt% (SAC305). The solder ball diameter of the BGA component was 0.49 ± 0.5 mm. The 14 layers printed circuit board (PCB) with solder mask-defined lands was finished with organic solderability preservative (OSP). The reflow profile was created using a temperature profile for a lead-free rework procedure, which requires the PCBA to be preheated between 100 °C and 190 °C. The temperature of the soak or preheat activation was 220 °C for 90 s. The maximum ramp rate of the component was 4 °C per second. For 80 s, the temperature for reflow dwell time at liquidus must be greater than 220 °C. The peak temperature was 245 °C for 15 s [15].

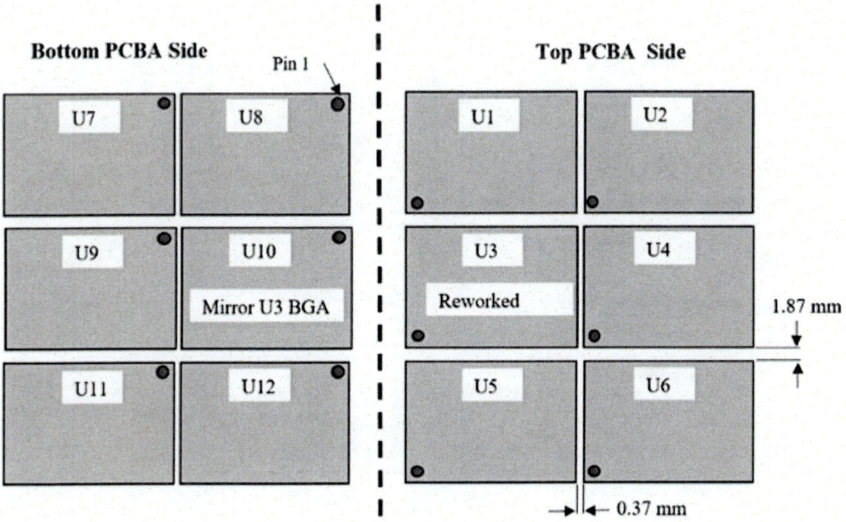

Fig. 12.1 Schematic of the reworked component location (U3) and BGA components' location

12.2.2 *Profiling the Rework Temperature*

TC wires were used to profile the temperature during the rework procedure. Drilling was done on the PCBA sample to ensure that the TC bead and wires were properly placed on the BGA component's solder joint that needed to be monitored, as shown in Fig. 12.2. Epoxy resin was used to fill the hole left by the drilling process. Figure 12.3 shows the placement of the TC wires placement on the rework, mirror rework, and five adjacent components on both the PCBA sides. The TC wires were placed according to IPC-7095D-WAM1 requirement, representing areas with the smallest to biggest thermal mass [16]. Only the temperatures of surrounding BGA components on both sides of the PCBA were documented and examined for this experiment.

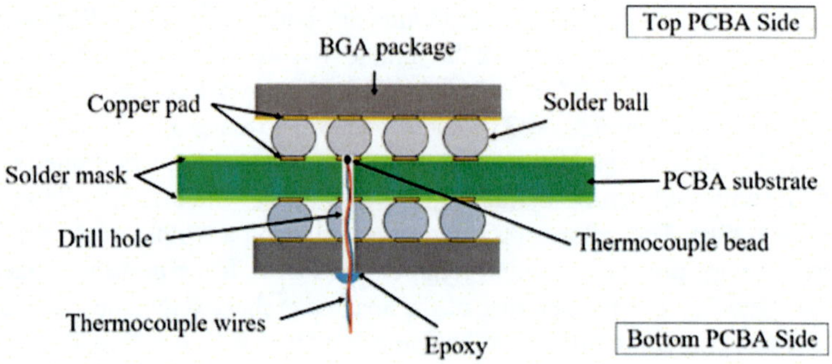

Fig. 12.2 Schematic of TC wire placement on the BGA solder ball for profiling the rework temperature

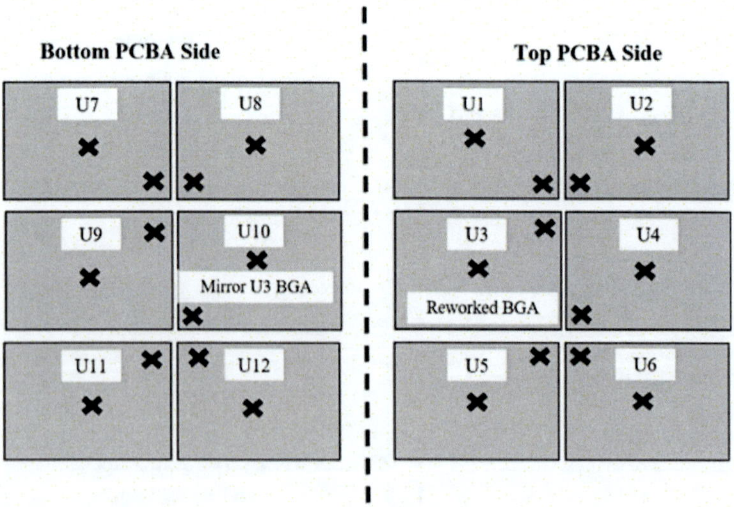

Fig. 12.3 TC wires placement locations on BGA components sample schematic

12.2.3 Rework Method

Both PCBA samples H and L were baked at 125 °C for 9 h in an oven to eliminate moisture from the PCBA and avoid thermal shock on the PCB due to the excess moisture [17]. For monitoring the temperature, a TC wire was attached to the PCBA surface. To achieve a good metallurgical detachment and attachment between base metals of PCB pads and solder alloys during component detachment and attachment, all solder joints must meet the melting point of lead-free alloys in the range of 217–220 °C [18]. The reworked BGA components were attached by applying paste flux to the solder joints of the BGA components and then using heat to reflow via the rework machine [19]. Using the hot air vent and vacuum suction, the reworked component U3 was detached from the sample H PCBA. Solder residues on the PCBA were detached using a soldering iron tip and pre-fluxed copper braid after applying paste flux to the area. A cleaning solution was then used to clean the flux residue. The new component was attached by using hot air reflow via the hot air vent. The hot air rework machine's top vent and the bottom hot air heater provide hot air for detachment and attachment [20]. The reworked component U3 on sample L was detached by using a laser beam controlled by galvanometric mirrors coming from the laser head. The rework component was then lifted by the pickup module. Solder residues on the PCBA were removed using the same method as in hot air rework. The new component was attached to the PCBA using the laser soldering procedure. The laser rework machine uses radiation heat, a laser beam on the top and an IR heater on the bottom for detachment and attachment.

12.2.4 Dye and Pull Test and Dye Penetration Coverage Inspection

IPC-TM-650 Test Methods Manual, 2.4.53, Dye and Pull Test Method was used to look for cracks in solder joints indicative of rework and adjacent BGA components [21]. PCBA samples were subjected to optical and X-ray inspections for the physical defect on reworked and surrounding components [22]. An epoxy compound was used to adhere the tee nut to the package surface of the reworked and adjacent BGA components. The BGA components were separated from the PCBA sample using a pull tester, and the PCB pads and BGA components were inspected for dye penetration via an optical microscope [23]. The dye penetration coverage was analysed using the IPC-TM-650 Test Methods Manual, 2.4.53, Dye and Pull Test Method [24].

12.2.5 Quantitative Analysis

Minitab and JMP software were used in tandem for quantitative analysis and computations. The peak temperature variance, mean standard deviation, and the number of the affected solder joints results were derived using the rework-induced damage data [25].

12.3 Results and Discussion

Figures 12.4 and 12.5 show DP test results for post-hot air rework and post-laser rework, respectively. The area shown was taken from the top corner of the U1 BGA package which is located on the adjacent U3 rework location. This area focuses on nine out of the 132 solder joints overall on the BGA component. According to the observations, the dye penetration coverage represented the affected solder joints due to the crack occurrence. Eight out of nine BGA solder joints were affected due to the hot air rework, whereas only four out of nine solder joints were affected on the laser rework sample.

The plot presented in Fig. 12.6a shows the peak temperature results on samples H and L for the adjacent BGA components on the top PCBA side. Figure 12.6b illustrates the peak temperature results for mirror BGA component positions on the PCBA's bottom side. Furthermore, the peak temperature data for both the PCBA side components for sample L show a decreasing trend against sample H. The use of a radiation heat source via laser soldering during the rework procedure reduced the peak temperature results for the PCBA side components. The plot in Fig. 12.6a, b shows that throughout the rework procedure, there was a correlation between the

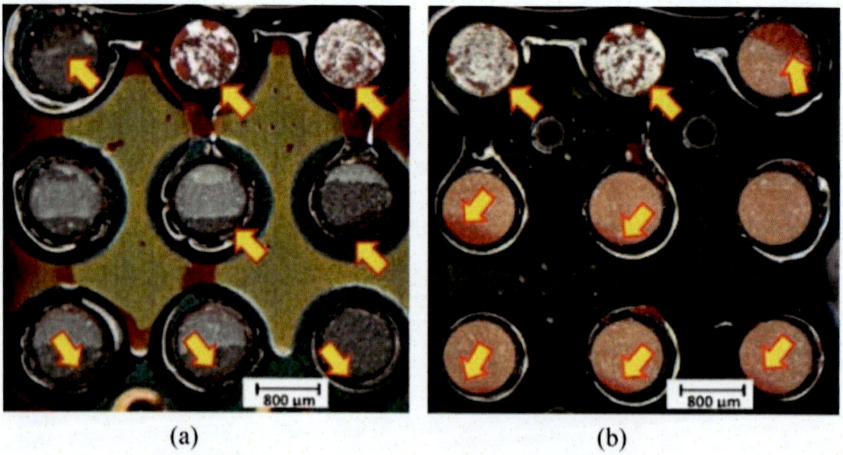

Fig. 12.4 DP test results post-hot air rework: **a** PCB side, **b** component side

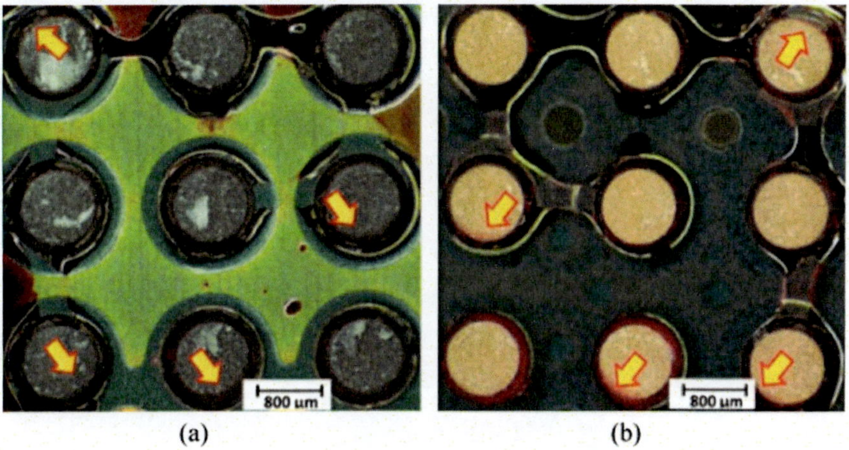

Fig. 12.5 DP test results post-laser rework: **a** PCB side, **b** component side

type of heat source and the peak temperature for both the centre and side of the BGA components on both the PCBA sides. The highest peak temperature was spotted on sample H which was reworked using a convection heat source.

In comparison with sample H, reworking using a radiation heat source resulted in much lower temperatures in the centre and sides of the surrounding components for sample L. In all samples, the range of side peak temperature was smaller than the range of centre peak temperature on both the PCB sides. This result matched with Sommerer et al. [26]. experiment, in which the distance between the TC wires and the heat source location was proportional to the degree of heat retention For both the PCBA sides, sample L had fewer variances in the centre and side peak temperature. The mean for the peak temperatures presented using the square shapes showed a descending trend, indicating that the temperature was reduced throughout the rework procedure by using the radiation heat source. This is in line with Wang et al. [27] and works in laser soldering; the laser heating location is precisely controlled, avoiding the heat-sensitive adjacent components and keeping them at a safe low temperature.

The standard deviation in Fig. 12.7 is the most important number for comparison in this study because it represents the quantitative analytical assessment of the variance from the mean in the peak temperature range between the samples in Fig. 12.6a, b. A substantial standard deviation indicates that there is a significant difference between the data and the mean and that the heat source employed for the rework was ineffective. A low standard deviation suggests that the data is relatively near to the mean, indicating that the heat source used for the rework was effective. Zeng et al. [28] have stated that laser soldering has the advantage of accurately and precisely directing the energy beam onto a target location without heating the adjacent components.

In this paper, the experimental results have been obtained, in which the adjacent BGA components' temperature throughout the rework procedure can be lowered by using a radiation heat source from laser soldering. Laser soldering has the most

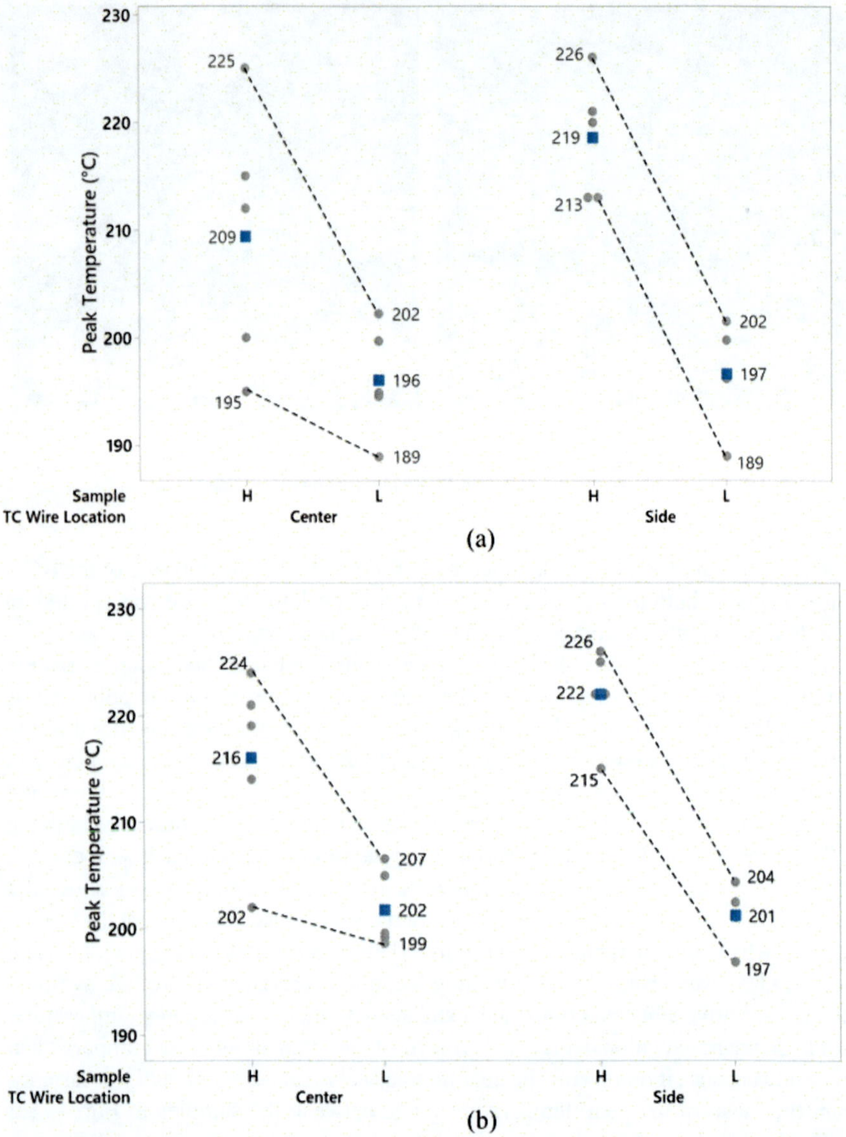

Fig. 12.6 Individual value plot for centre and side of BGAs' peak temperature for samples H and L **a** top PCB side **b** bottom PCB side

efficient heat depletion on the adjacent components where the mean peak temperature on the top PCBA side for the BGA component centre and side was reduced by 6.22% and 10.05% correspondingly. As for the bottom PCBA side, the heat depletion for the BGA component means the peak temperature was 6.48% in the centre and 8.37% on the side. The number of solder joints affected caused by the solder joint crack

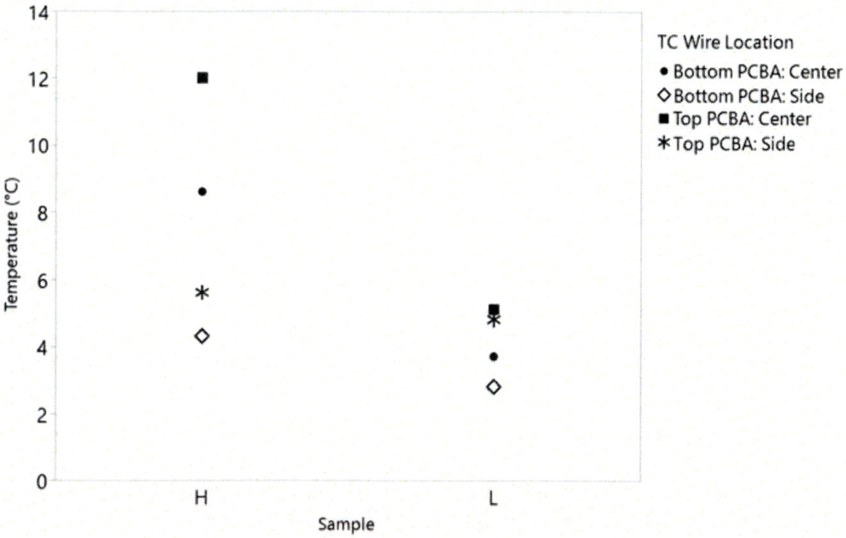

Fig. 12.7 Standard deviation for the peak temperature is based on the TC wire locations on all samples

can be reduced if the adjacent component temperature can be kept below 207 and 204 °C for the centre and side of the BGA component correspondingly throughout the rework procedure. This observation is matched with Chen et al. [29] because reducing the solder joint's temperature exposure will lessen the impact of reliability concerns like the increase of the intermetallic compound (IMC) layer, which impacts the solder joint's shear strength.

12.4 Conclusion

This research delivers a technique to report the issue of the heat source types during reworking high-density PCBA designs with BGA components populated on both sides of the PCBA. Thermal management throughout the rework procedure has been handled effectively by using a radiation heat source via laser soldering. The adjacent components of the rework location were less impacted throughout the rework procedure using laser soldering. This technique can be effectively applied for reworking high-density PCBA designs with BGA components populated on both sides of the PCBA. Laser soldering can reduce up to 10.05% of peak temperatures on the adjacent components. It also can maintain the peak temperature below 207 and 204 °C for the centre and side of the adjacent BGA components correspondingly to enhance PCBA quality and reliability by minimizing solder joint damage as observed in the DP test results.

Acknowledgements The authors would like to thank Western Digital® through SanDisk Storage Malaysia Sdn. Bhd for financial support (RR-2020-004) and collaboration with Universiti Kebangsaan Malaysia.

References

1. X. Wu, C. Zhang, W. Du, An analysis on the crisis of "chips shortage" in automobile industry-based on the double influence of COVID-19 and trade friction. J. Phys. Conf. Ser. **1971**(1), 012100 (2021). https://doi.org/10.1088/1742-6596/1971/1/012100
2. G. Marinova, A. Bitri, Challenges and opportunities for semiconductor and electronic design automation industry in post-Covid-19 years. IOP Conf. Ser.: Mater. Sci. Eng. **1208**(1), 012036 (2021). https://doi.org/10.1088/1757-899x/1208/1/012036
3. G. Fontana et al., Precision handling of electronic components for PCB rework. IFIP Adv. Inf. Commun. Technol. **435**, 52–60 (2014). https://doi.org/10.1007/978-3-662-45586-9_8
4. J. Du, Z. Raz, A reliability qualification process for BGA rework limit. Int. Conf. Control Autom. Robot. 776–781 (2019). https://doi.org/10.1109/iccar.2019.8813504
5. S. Patel et al., Solder immersion process of ceramic column grid array package assembly for space applications. IEEE Trans. Compon. Packag. Manuf. Technol. **10**(4), 717–722 (2020). https://doi.org/10.1109/tcpmt.2019.2961424
6. S. Zain et al., Effect of moisture content on crack formation during reflow soldering of ball grid array (BGA) component. Adv. Robot. Auto. Data. Anal. 309–314 (2021). https://doi.org/10.1007/978-3-030-70917-4_29
7. M. Suhaimi et al., Thermal cycling effect on the crack formation of solder joint in ball grid array package. J. Phys. Conf. Ser. **2169**(1), 012006 (2022). https://doi.org/10.1088/1742-6596/2169/1/012006
8. C. Lea, Laser soldering—production and microstructural benefits for SMT. Solder. Surf. Mt. Technol. **1**(2), 13–21 (1989). https://doi.org/10.1108/eb037672
9. G. Jeong et al., Interfacial reactions and mechanical properties of Sn–58Bi solder joints with Ag nanoparticles prepared using ultra-fast laser bonding. Materials **14**(2), 335 (2021). https://doi.org/10.3390/ma14020335
10. Y. Yost, F. Hosking, D. Frear, *The Mechanics of Solder Alloy Wetting and Spreading*. (Springer, New York, 2012), p. 246
11. J. Muonio, R. Stadem, Solder ball attachment using laser soldering. SMT Circuits Assembly **19**(10), 28–31 (2008)
12. F. Albert, I. Mys, M. Schmidt, in *Laser-Based Rework in Electronics Production*, Proc. SPIE, Laser-based Micro- and Nanopackaging and Assembly, vol. 6459, no. 645905 (2007). https://doi.org/10.1117/12.703125
13. J. Burke, A. Dai, Laser rework process for BGA: a new method for PCBA rework instead of hot air/infrared heating, in *Rosemont*. ed. by S.M.T.A. International (USA, Illinois, 2018), pp.763–768
14. R. Hanifan, Electrical reference designations. SpringerBriefs Appl. Sci. Technol. 59–67 (2014). https://doi.org/10.1007/978-3-319-06983-8_6
15. IPC-7095D-WAM1, *Design and Assembly Process Implementation for Ball Grid Arrays (BGAs)* (2019)
16. P. Ciszewski et al., A comparative analysis of printed circuit drying methods for the reliability of assembly process. Microelectron. Reliab. **129**, 114478 (2022). https://doi.org/10.1016/j.microrel.2022.114478
17. O. Thomas, C. Hunt, M. Wickham, Finite difference modelling of moisture diffusion in printed circuit boards with ground planes. Microelectron. Reliab. **52**(1), 253–261 (2012). https://doi.org/10.1016/j.microrel.2011.08.014

18. M. Abu Bakar et al., Directional growth behaviour of intermetallic compound of Sn3.0Ag0.5Cu/ImSn subjected to thermal cycling. Mater. Sci. Forum. **857**, 36–39 (2016). https://doi.org/10.4028/www.scientific.net/msf.857.36
19. S. Chou et al., Test method to evaluate a robust ball grid array (BGA) ball mount flux. IEEE Electr. Packag. Technol. Conf. 623–628 (2016) https://doi.org/10.1109/eptc.2016.7861555
20. A.A. Ismail et al., Effect of heat shield locations on rework-induced thermal management in ball grid array solder joint. Sci. Rep. **12**, 15118 (2022). https://doi.org/10.1038/s41598-022-19436-6
21. V. Reddy et al., Evaluation of the quality of BGA solder balls in FCBGA packages subjected to thermal cycling reliability test using laser Ultrasonic inspection technique. IEEE Trans. Compon. Packag. Manuf. Technol. **11**(4), 589–597 (2021). https://doi.org/10.1109/tcpmt.2021.3065958
22. A.A. Ismail et al., Cardinal and ordinal directions approach in investigating arrayed solder joints crack propagation of ball grid array semiconductor packages. IEEE Trans. Compon. Packag. Manuf. Technol. (2022). https://doi.org/10.1109/TCPMT.2022.3198406
23. A. Jalar, M. Bakar, R. Ismail, Temperature dependence of elastic-plastic properties of fine-pitch SAC 0307 solder joint using nanoindentation approach. Metall. Mater. Trans. A. **51**(3), 1221–1228 (2020). https://doi.org/10.1007/s11661-019-05614-1
24. C. Chen et al., The failure mode study of the polymer ball interconnected IC package under board level thermal mechanical stress. Int. Microsyst. Packag. Assy. Circuits Technol. Confer. 424–427 (2016). https://doi.org/10.1109/impact.2016.7800064
25. A. El-Sharkawy, A. Uddin, Development of a transient thermal analysis model for engine mounts. SAE Int. J. Mater. Manuf. **9**(2), 268–275 (2016). https://doi.org/10.4271/2016-01-0192
26. Y. Sommerer et al., Uncertainty quantification of thermocouple air temperature measurement in highly radiative environment: application to turbofan engine compartment. Proc. ASME: Turbomac. Tech. Confer. Expo. **5** (2016). https://doi.org/10.1115/gt2016-57938
27. L. Wang et al., 59-2: laser reflow soldering technique for mini/microled displays. SID Symp. Dig. Tech. Pap. **52**(1), 830–832 (2021). https://doi.org/10.1002/sdtp.14811
28. Z. Zeng et al., Numerical modeling and optimization of laser soldering for Micro-USB electric connector. Int. J. Numer. Model.: Electron. Netw. Devices Fields **28**(2), 175–188 (2014). https://doi.org/10.1002/jnm.1995
29. C. Chen et al., Microstructure evolution and shear strength of the Cu/Au80Sn20/Cu solder joints with multiple reflow temperatures. Materials **15**(3), 780 (2022). https://doi.org/10.3390/ma15030780

Chapter 13
Electrical Characterization of Ultrasonic Aluminum Bond on Molybdenum Back Contact of the Thin-Film Solar Module Using Micro-Ohmmeter

Muhammad Nubli Zulkifli and Sabarina Abdul Hamid

Abstract In this paper, the accuracy and suitability of applying the micro-ohmmeter for the electrical characterization of aluminum (Al) bonds on molybdenum back contact of the thin-film solar cell were examined. The standard error of volume resistance for each bond on the solar cell's module was calculated to identify the accuracy of this technique. The results were also compared with contact resistance by the transmission line method (TLM) measurements to identify the suitability of this method. The range of volume resistance was 94.5 mΩ to 154.5 mΩ with minimum and maximum accuracies of 4.4 mΩ and 8.3 mΩ, respectively. The result of contact resistance of TLM also shows the same trend as compared to micro-ohmmeter. The advantages of using micro-ohmmeter are the lower cost and mobile experimental setup as compared to the TLM measurement.

Keywords Ultrasonic bonds · Four-wire probe · Volume resistance · Micro-ohmmeter · Electrical characterization

13.1 Introduction

Ultrasonic bonding is one of the interconnection methods that have been applied as the connection means in the production of thin-film solar cells. The electrical characterization is conducted to evaluate the joint quality and module reliability. The most often techniques applied for resistance measurements are contact resistance and volume resistivity. The transmission line method (TLM) is the common technique used by researchers to evaluate the contact resistance value of the bonds, which is inversely related to the efficiency of the interconnection [1–4], while volume resistivity is typically applied to assess the joint or material properties [5–7].

M. N. Zulkifli (✉) · S. Abdul Hamid
Renewable Energy Research Laboratory (RENERAL), Electrical Engineering Section, British Malaysia Institute, Universiti Kuala Lumpur, Jalan Sungai Pusu, 53100 Gombak, Selangor, Malaysia
e-mail: mnubliz@unikl.edu.my

M. A. A. Mohd Salleh et al. (eds.), *Proceedings of the Green Materials and Electronic Packaging Interconnect Technology Symposium*, Springer Proceedings in Physics 289,
https://doi.org/10.1007/978-981-19-9267-4_13

The contact resistance and volume resistivity were typically measured by a four-point probe with a precise current meter [1, 5–7]. Four-wire or Kelvin method is frequently used for low resistance measurements that can reduce the effect of test lead resistance. The measurement can be made by applying digital multimeter, micro-ohmmeter, or separate current source and voltmeter. Among the low resistance measurement applications are contact resistance, superconductor resistance, and resistivity measurement of conductor [8].

This study is a continuity of a preliminary study on contact resistance evaluation by TLM method using a four-point probe conducted by Basher et al. [1]. The equipment used for the lab-scale test of TLM method is expensive and non-portable besides consuming time to set up. Thus, the measurement of volume resistance using micro-ohmmeter was conducted to identify the simpler and affordable method yet give less error for electrical resistance measurement of Al bond.

13.2 Methodology

The resistance measurement of ultrasonic aluminum (Al) bond on molybdenum (Mo) back contact layer of thin-film solar module was conducted to identify the volume resistance of the bonds. The dimension of the module of 5 cm and 10 cm is shown in Fig. 13.1. The electrical characterization of this study applies the micro-ohmmeter with a kelvin probe (UT620C) to measure the volume resistance between the bonds. The resolution and precision of the micro-ohmmeter are 1 μΩ and + 0.1%, respectively. The two alligator clips from the micro-ohmmeter were connected at the center of the first bond and second bond as shown in Fig. 13.2. The distance between the bond is 1 cm while the width is 0.2 cm. The measurements were conducted on seven modules with five aluminum ribbons on each module. Each aluminum ribbon line is connected by five to ten ultrasonic bonds on the molybdenum layer.

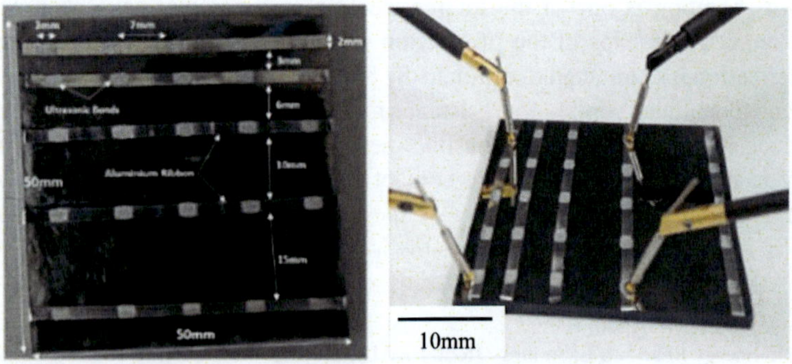

(a) (b)

Fig. 13.1 Dimension of the solar module for **a** 5 cm and **b** 10 cm module

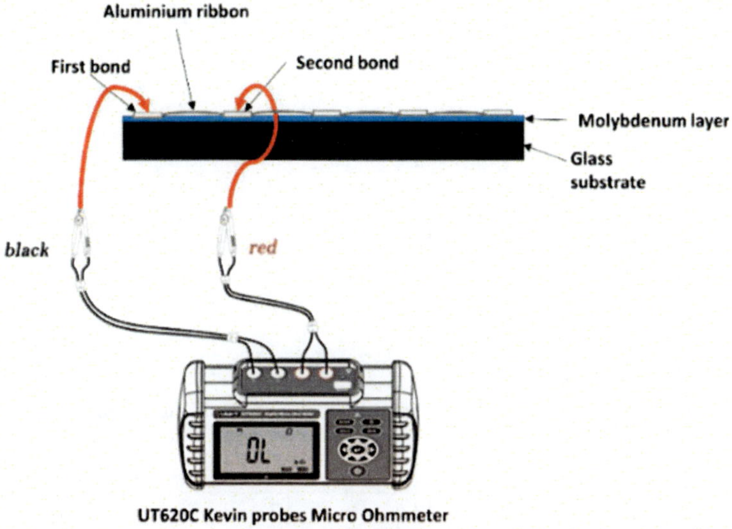

Fig. 13.2 Volume resistance measurement using micro-ohmmeter between two bonds

After the measurement of volume resistance for each module was conducted, the accuracy of the measurements was calculated using the standard error equation. The standard error was calculated by the following Eq. (13.1):

$$SE_x = \frac{s}{\sqrt{n}}, \tag{13.1}$$

where s is the standard deviation and n is the sample size. The average value of volume resistance for each module with the standard error value was tabulated to identify the accuracy of each module.

The electrical characterization conducted for this study is compared with the preliminary study conducted by Basher et al. [1] which evaluated the contact resistance by the TLM method using a four-point probe. The TLM method determines the resistance between bonds at different ribbon distances.

As shown in Fig. 13.1, the distances between ribbons for 5 cm modules are 3, 6, 9, and 15 mm while 10 cm modules are spaced by 5, 10, 20, and 40 mm ribbons. The contact resistance was calculated based on the intercept line from plotting a straight-line graph of resistance versus distance. The suitability of the micro-ohmmeter with the kelvin probe was compared with the TLM measurement.

Twelve modules were evaluated with different ultrasonic parameter settings, namely pressure in bar, amplitude in µm, and maximum energy in Ws as shown in Tables 13.1 and 13.2. The results of resistance measurement were compared in the tabulated graph of variation.

Table 13.1 Bonding parameter for 5 cm module

Module	Pressure (bar)	Amplitude (μm)	Maximum energy (Ws)
I	2	8.4	20
II	1.5	7.7	20
III	3.5	8.4	20
IV	1	9.1	20
V	2.5	9.8	20
VI	1	9.8	20
VII	3	7.7	20

Table 13.2 Bonding parameter for 10 cm module

Module	Pressure (bar)	Amplitude (μm)	Maximum energy (Ws)
1	1.9	9.1	15
2	1.6	7	20
3	1.6	7	17
4	2.5	8.4	20
5	1.6	7	15

13.3 Results and Discussions

The variation of volume resistance for seven modules of Al bond with different parameter windows is shown in Fig. 13.3. It is essential to have a low volume resistance in between bonds to achieve high conductivity or efficiency of the solar module [9]. Based on the measurement, the range of volume resistance is 94.5 to 154.5 mΩ with accuracy or standard error of 4.4 to 8.3 mΩ. The accuracy of each module indicates the consistency of the volume resistance measurement [10]. Although Module IV shows a low volume resistance value, the standard error indicates the inconsistency of the measurement.

Figure 13.4 shows the comparison of standard error between the ribbon for Module IV and Module VI. Module VI shows more consistent with accuracy for all ribbons compared to Module IV. The high standard error of Module IV comes from the high variation of volume resistance in ribbon (R1). The physical condition of the bonds appears to be in a uniform condition without any wavy ribbon. Further analysis of the microstructure will be conducted in the near future to evaluate any void or growth of the crack at the bonding [11].

Figure 13.5 shows the comparison of contact resistance values for TLM and the micro-ohmmeter technique for the 10 cm module. The variation ranges of both techniques were 80.49–120.37 mΩ where the least variation was at module A while the highest variation was at module E. From the physical inspection, it is noted that the module D has wavy ribbon and non-uniform bonds while module E has the non-uniform bond with a small hole at ribbon 3 that increases the contact resistance

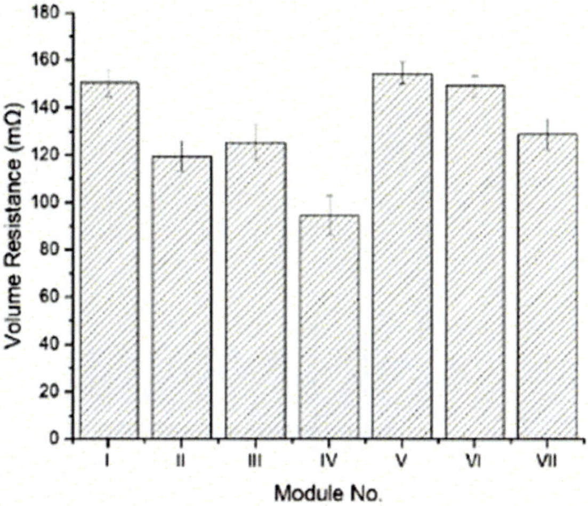

Fig. 13.3 Average volume resistance with a standard error of seven modules

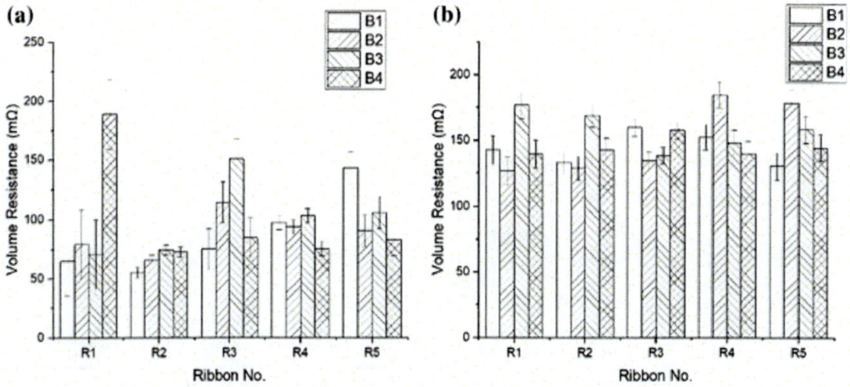

Fig. 13.4 Comparison of standard error within the ribbon for **a** module IV and **b** module VI

variation. This condition may be due to the handling process during transportation or the experimental process [12].

Moreover, Fig. 13.6 depicts the comparison of contact resistance values for TLM and micro-ohmmeter techniques for 5 cm module. The variation ranges of both techniques are 45.63.35–91.93 mΩ. The lowest variation was at Module IV while extreme variation was at Module I. However, based on Figs. 13.5 and 13.6, the variation was decreased for the smaller size of the module.

The TLM technique that used the lab-scale equipment applied a constant current that produces a more precise measurement instead of a train of pulses or direct current (DC) by micro-ohmmeter [13, 14]. The accuracy and resolution of the current meter

Fig. 13.5 Variation of contact resistance value for 10 cm module

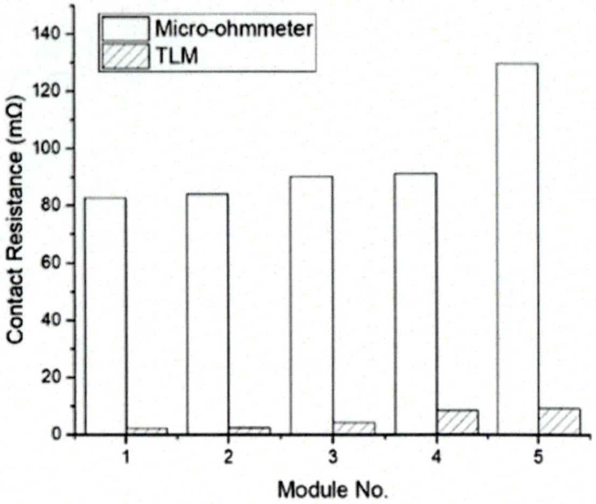

Fig. 13.6 Variation of contact resistance value for 5 cm module

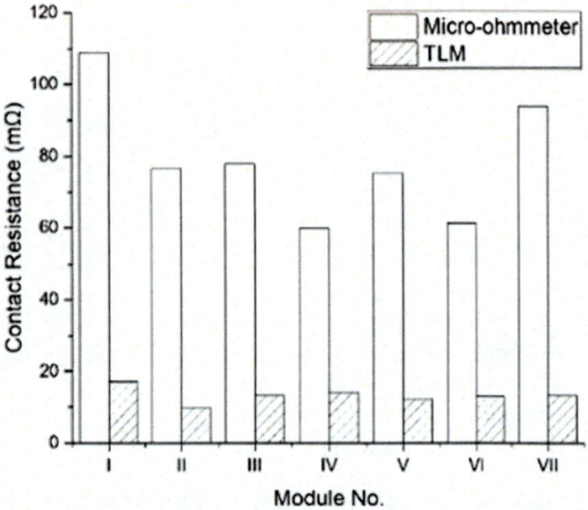

(SourceMeter 2400) used for TLM measurement were 0.012% and 6½-digit resolution when compared to the micro-ohmmeter of $+0.1\%$ accuracy and 1 $\mu\Omega$ resolution. Besides, different experiment setup of TLM gives a large variation in contact resistance value. The probe location and type of probe used will affect the sensitivity of the measurement [15]. Common lab-scale measurement will use probe stations and nanopositioners for resistance measurement. This equipment will consume time for setup and need extra caution in handling [16]. Moreover, the controlled environment, such as room temperature and under dark conditions during measurement for TLM, can minimize the effect of the environment on resistance measurement. The controlled environment can minimize the variation in resistance [15].

The advantages of this technique are the low cost and mobile experimental setup as compared to the TLM measurement. This technique is also simpler in terms of the experimental setup and consumes less time.

13.4 Conclusion

Although the micro-ohmmeter technique is not precise and sensitive as compared to a lab-scale source meter unit used for TLM, based on this study, this technique is a plausible method for low resistance measurement with low cost and is easier to set up. This technique can be applied to identify the category or trend of the low resistance measurement, especially for the Al bonds on the thin-film solar cells.

Acknowledgements This research is sponsored by the Ministry of Higher Education of Malaysia under the Fundamental Research Grant Scheme (FRGS/1/ 2020/TKO/UNIKL/02/10).

References

1. H. Basher, M.N. Zulkifli, A. Jalar, M. Daenen, Effect of ultrasonic bonding parameters on the contact resistance and bondability performances of CIGS thin film photovoltaic solar panel. IEEE J. Photovoltaics 1–9 (2021). https://doi.org/10.1109/JPHOTOV.2020.3047295
2. T. Xu, O. Valentin, C. Luechinger, Reliable metallic tape connection on CIGS solar cells by ultrasonic bonding.in *Thin Film Solar Technology II*, vol. 7771 (2010), p. 77710R. https://doi.org/10.1117/12.860962
3. M. Heimann et al., Ultrasonic bonding of aluminum ribbons to interconnect high-efficiency crystalline-silicon solar cells. Energy Procedia **27**, 670–675 (2012). https://doi.org/10.1016/j.egypro.2012.07.127
4. T. Geipel et al., *Industrialization of Ribbon Interconnection for Silicon Heterojunction Solar Cells with Electrically Conductive Adhesives* (2019)
5. T. Geipel, M. Meinert, A. Kraft, U. Eitner, Optimization of electrically conductive adhesive bonds in photovoltaic modules. IEEE J. Photovoltaics **8**(4), 1074–1081 (2018). https://doi.org/10.1109/JPHOTOV.2018.2828829
6. H.A. Jaffery, M.F. Mohd Sabri, S. Rozali, M.H. Mahdavifard, D.A. Shnawah, Effect of temperature and alloying elements (Fe and Bi) on the electrical resistivity of Sn-0.7Cu solder alloy. RSC Adv. **6**(63), 58010–58019 (2016). https://doi.org/10.1039/c6ra08706j
7. N.A.A.M. Amin, D.A. Shnawah, S.M. Said, M.F.M. Sabri, H. Arof, Effect of Ag content and the minor alloying element Fe on the electrical resistivity of Sn-Ag-Cu solder alloy. J. Alloy. Compd. **599**, 114–120 (2014). https://doi.org/10.1016/j.jallcom.2014.02.100
8. Keithley, *Low Level Measurements Handbook-Precision DC Current, Voltage and Resistance Measurements*, p. 239 (2004)
9. N. Ismail, A. Jalar, A. Afdzaluddin, M.A. Bakar, Electrical resistivity of Sn–3.0Ag–0.5Cu solder joint with the incorporation of carbon nanotubes. Nanomater. Nanotechnol. **11**, 1–9 (2021). https://doi.org/10.1177/1847980421996539
10. S. Guo, G. Gregory, A.M. Gabor, W.V. Schoenfeld, K.O. Davis, Detailed investigation of TLM contact resistance measurements on crystalline silicon solar cells. *Solar Energy* **151**, 163–172 (2017). https://doi.org/10.1016/j.solener.2017.05.015

11. H. Basher, M.N. Zulkifli, M.K. Rahmat, M.G.A. Rahman, A. Jalar, M. Daenen, Bondability of ultrasonic Aluminum bonds on the molybdenum (de)selenide and molybdenum of back contact layer of copper indium gallium (de)selenide CIGS thin film photovoltaic solar panel. Sol. Energy **228**(September), 516–522 (2021). https://doi.org/10.1016/j.solener.2021.09.082
12. Institute of Electrical and Electronics Engineers, Electron Devices Society, Colo. IEEE Photovoltaic Specialists Conference 40 2014.06.08-13 Denver, and Colo. PVSC 40 2014.06.08-13 Denver, "Design and Characterization of an Adhesion Strength Tester for Evaluating Metal Contacts on Silicon Solar Cells," in *IEEE 40th Photovoltaic Specialist[s] Conference (PVSC), 2014 8–13 June 2014, Denver, Colorado*, (2014), pp. 2550–2553
13. J. Deng, W. Yan, Q. Yang, A micro-resistance measurement based design approach of digital micro-ohmmeter. Adv. Mater. Res. **339**(1), 36–42 (2011). https://doi.org/10.4028/www.scientific.net/AMR.339.36
14. T.W. Giants, Aging effects on the electrical properties of silver-filled epoxy adhesives. J. Adhes. Sci. Technol. **12**(6), 593–613 (1998). https://doi.org/10.1163/156856198X00812
15. M.M. Ghorbani, R. Taherian, Methods of measuring electrical properties of material, in *Electrical Conductivity in Polymer-Based Composites: Experiments, Modelling, and Applications* (Elsevier Inc., 2018), pp. 365–394. https://doi.org/10.1016/B978-0-12-812541-0.00012-4
16. T. Lecklider, Nano-measurements need mega care. EE-Eval. Eng. **47**(12), 26–31 (2008)

Chapter 14
Effect of Different Epoxy Materials During LED Wire Bonding Encapsulation Process Using CFD Approach

Muhammad Syukri Bin Zubir, Mohd Syakirin Bin Rusdi, Mohd Sharizal Abdul Aziz, Roslan Kamaruddin, M. H. H. Ishak, and Mohd Arif Anuar Mohd Salleh

Abstract There are several issues concerning high-power LED applications that affect the functionality and reliability of the LED. The most common problem that occurred in LEDs during the encapsulation process is wire deformation which can affect the life expectancy of the LED. The aim of this present study is to analyze the influence of different epoxy materials during the LED encapsulation process. In this study, ANSYS Fluent is utilized to simulate the volume of the epoxy materials being dispensed onto the LED by utilizing the volume of fluid (VOF) strategy. The simulation is conducted by varying the types of epoxy materials (ERL-4221, EMC, and D.E.R.-331). An experimental test was conducted to validate the final structure of the epoxy materials obtained from the simulation setup. The final form of the EMC and ERL-4221 resembled a hemisphere, consistent with the experimental finding. However, D.E.R.-331 is an exception since its final construction was not perfectly hemispherical.

Keywords Epoxy · Encapsulation process · Computational fluid dynamics · Wire bonding

M. S. B. Zubir · M. S. B. Rusdi (✉) · M. S. Abdul Aziz · R. Kamaruddin
School of Mechanical Engineering, Universiti Sains Malaysia, Engineering Campus, 14300 Nibong Tebal, Seberang Perai Selatan, Penang, Malaysia
e-mail: syakirin@usm.my

M. H. H. Ishak
School of Aerospace Engineering, Universiti Sains Malaysia, Engineering Campus, 14300 Nibong Tebal, Seberang Perai Selatan, Penang, Malaysia

M. A. A. Mohd Salleh
Center of Excellence Geopolymer & Green Technology (CeGeoGTech), University Malaysia Perlis (UniMAP), 02600 Taman Muhibbah, Perlis, Malaysia

M. A. A. Mohd Salleh et al. (eds.), *Proceedings of the Green Materials and Electronic Packaging Interconnect Technology Symposium*, Springer Proceedings in Physics 289,
https://doi.org/10.1007/978-981-19-9267-4_14

14.1 Introduction

Light-emitting diode (LED) devices with good quality must have desirable properties such as low power consumption and high efficiency, reliability, and life expectancy. However, it is quite difficult to obtain the LED with the desirable properties as there were several possible factors that could affect the functionality of the device. Alim et al. [1] conducted a study on the encapsulation process in LED packaging. They listed several challenges in LED packaging such as the shape of encapsulation, thermal conductivity, packaging material degradation, and strength of bonding. Therefore, the failure rates of the LED device can often be attributed to the following factors.

Packaging materials is one of the aspects where a variety of failures could occur. For example, heat from the PN junction of the chip combined with the external surrounding can cause the packaging material of the LED to change in color and crack [2]. Epoxy resins, as top thermosetting polymer materials, have been widely used in the present LED encapsulation process due to advantages such as strong adhesion, exceptional machinability, and great chemical resistance [3]. Hence, three different types of epoxy resins used for the LED encapsulation process will be studied in this paper.

Wire deformation or wire sweep is an example where failures can occur at the wire bonding. When the amount of stress applied to the wire bonding is too large, it will cause the wire to break and cause the LED device to fail [4]. There are several factors that can be the cause of wire sweep formation during the encapsulation process. Therefore, the effect of the gold wire number on LED will be a concern study in this paper. Currently, there are different types of LED encapsulation structures available on the market. The type of encapsulation structure used during the encapsulation process is depending on the type of LED itself [5]. For this study, a typical LED encapsulation structure will be used which is a semisphere.

EMC is widely used in the electronics industry as packaging for encapsulating semiconductor materials. During the curing process of this material, residual stress will be induced due to both cure and thermal shrinkage which can lead to product failure [6]. Sadeghinia et al. [6] conducted research to investigate the changes in both elastic modulus and viscoelastic behavior during the curing process. The experimental shear setup was done by using Dynamic Mechanical Analyzer DMAQ800 instrument to measure the changes in the mechanical properties of the EMC.

For the viscoelastic behavior of the EMC, it was determined during an intermittent cure test with the same DMA apparatus. During the test, the shear modulus and viscoelastic shear master curve are extracted to be analyzed. Based on the analysis, it is found that by curing the EMC partially, the rubbery shear modulus will increase and the viscoelastic master curve shifts to a higher time domain. The limitation of this paper is that it only focuses on the factor that will affect the properties of the EMC. However, this paper still contributes by emphasizing the properties of the EMC which is useful for the research of the present paper as it uses EMC as one of the materials for the encapsulant. In this paper, the influence of three types of epoxy resins on the LED encapsulation process will be studied both computationally and experimentally to validate the results acquired.

14.2 Experiment Procedures

The aim of the LED encapsulation experiment is to validate the results obtained from the simulation with the actual results. For the experiment, the material and apparatus needed are a syringe, a needle with a tip of 1 mm in diameter, a high-power LED, a fluid, and a vacuum chamber. Figure 14.1 shows the experimental setup for the experiment. Figure 14.2 depicts the schematic diagram. A digital camera is set in front of the needle and an LED for the data recording. First, the LED package is placed on a flat surface inside the vacuum chamber to prevent the results from being affected by any external factors. Then, the syringe containing a certain volume of EMC A is injected onto the LED surface. The mold cavity of the LED substrate allows the fluid to form a hemisphere shape. The fluid dispensing process is observed, recorded, and then compared with the results obtained from the simulation.

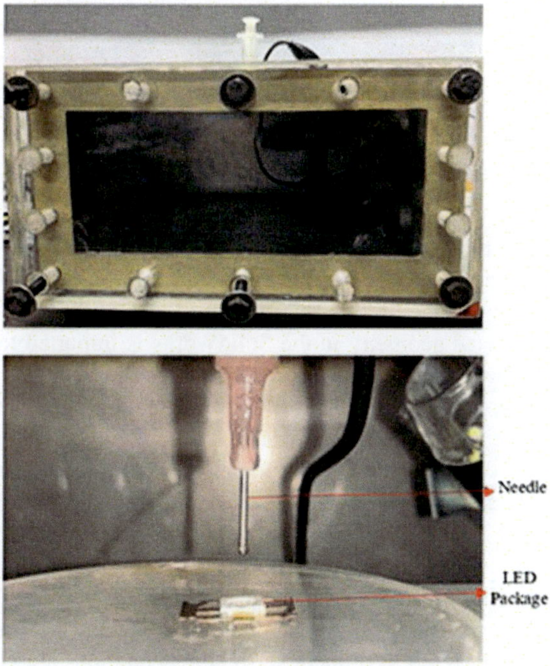

Fig. 14.1 Experimental setup for LED encapsulation process

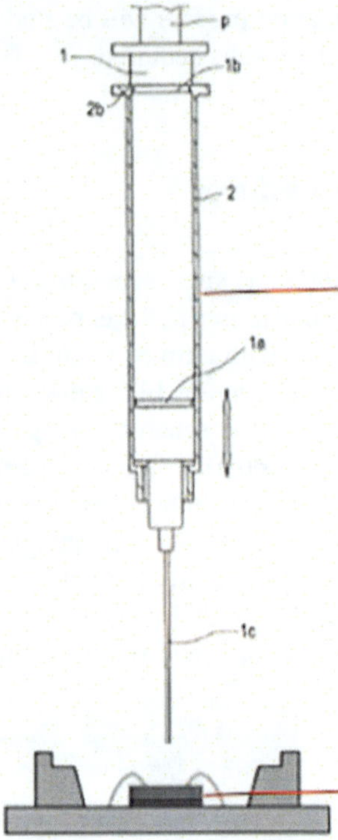

Fig. 14.2 Schematic diagram of experimental setup

14.3 Numerical Methods

In the modeling of the high-power LED package, the complete geometry is drawn by using SolidWorks software as shown in Fig. 14.2. The dimension for the geometry is obtained by measuring the LED prepared for the experimental setup. The completed geometry in the SolidWorks is imported to the ANSYS Workbench so that the simulation can be run. Before the setup for the simulation is run, part like the chip is subtracted by using "Boolean" function in the Design Modeler. The dispensing needle and the domain are defined as the fluid body.

In the simulation setup, the effect of the turbulence flow on the encapsulation dispensing process is neglected [5]. The Reynolds' number for the ERL-4221, EMC, and D.E.R.-331 is 1.57, 1.21, and 0.39, respectively. Therefore, the flow is assumed to be laminar. This is due to the absence of obstruction or sharp corner that can create turbulences which makes the process very stable. For each time step of the volume fraction, volume of fluid (VOF) scheme and transient-based formulation are applied. In the VOF method, the flow equations are going to be summed up in the volume of a single equation set directly, and the interface will be tracked with a phase indicator function. The indicator is used to track the interface between two phases which in this case are air and epoxy resin that has a value of 1 or 0 when a control volume is fully filled. In the setup, air is defined as phase 1 with value

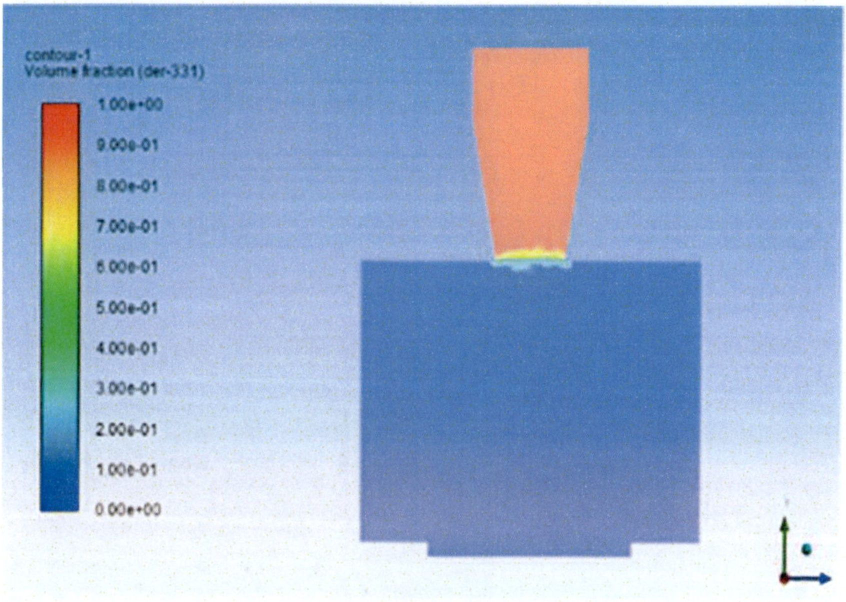

Fig. 14.3 Reflow oven geometry

0, while epoxy resin is defined as phase 2 with value 1. Figure 14.3 shows that the red part is the patched region of phase 2, while the blue part is the region of phase 1.

Meshing was performed in the ANSYS Fluent (with Fluent meshing) using poly-hexcore as shown in Fig. 14.4. The mesh has total nodes of 364,120 nodes, meshed nodes of 243,190 nodes, and 458,961 cells. For the simulation part, the selection of the mesh type and size is based on the complexity, convergence, accuracy requirements, and computational time required to solve the calculation for 1e-5 times step size. A grid-independent test was conducted for mesh sizes between 0.1 and 0.3 mm. From the results of the grid-independent test, mesh with a 0.15 mm element size is selected as the standard for the simulation in this study. This is because it can generate a similar result with minimal percentage error which is less than 1% compared to 0.2 mm mesh with 14.44% which is a big percentage error. Furthermore, the higher the number of cells, the higher the computation power of the computer is required. Table 14.1 shows the material properties for ERL-4221, EMC, and D.E.R.-331 that were used in the simulation.

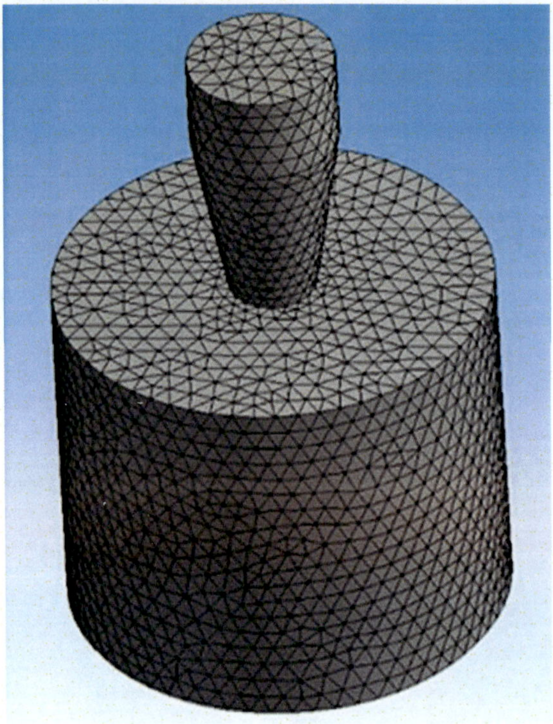

Fig. 14.4 Meshing of the encapsulation process

Table 14.1 Material properties of EMC

Properties	Epoxy molding compound	D.E.R.-331	ERL-4221
Density (kg/m^3)	1800	1159	1173
Viscosity (kg/m.s)	0.448	0.896	0.224
Surface tension (N/m)	0.005	0.005	0.005
Reference	Roslan et al. [5]	Huang et al. (2004)	Huang et al. (2004)

14.4 Result and Discussion

14.4.1 Experimental Validation

In Fig. 14.5, EMC A was used in the experiment as it has the same viscosity with EMC. From the results, the encapsulant structure for the EMC is the same as the experimental structure and the percentage difference is only 5%. This shows that

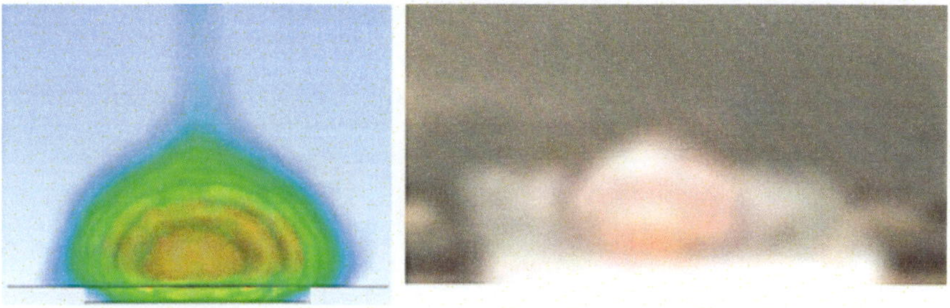

Fig. 14.5 Comparison of epoxy shape between simulation and experiment

the simulation setup for the EMC is almost accurate and can be used for the other epoxy materials to compare the final structure of the encapsulant. The experiment was conducted at 27 °C and in a closed chamber.

14.4.2 Effect of Epoxy on Filling

Based on the results obtained from the simulation vs. experimental section, the setup used for the EMC simulation is used to compare the structure of the other epoxy resin as the EMC structure obtained is the same with actual results. Table 14.2 shows the comparison between the structures of different epoxy materials.

In this simulation, three types of epoxy resins were used which were ERL-4221, EMC, and D.E.R.-331. The parameters and boundary condition in the entire simulation such as inlet speed, contact angle, and injection time were set as constant for the three epoxy resins. Both results from the simulation and experiment are examined, and the comparison between the structures of the epoxy resins during the encapsulation process is illustrated in Table 14.2. The objective of this simulation was to analyze the effect of three different epoxy resins on the final structure of the encapsulant that covers the base of the LED which is approximately 3 mm in diameter. The final structure of the simulation result for the three epoxy resins was expected to be the same as the actual result.

Based on the results shown in Table 14.2, the structure for all three epoxy resins was similar from a time duration of 0–25%. However, starting from time duration of 60% until 100%, the structures for the three epoxy resins started to differ. At the 60% time duration, ERL-4221 structure has already finished dropping off the epoxy resin, while the structure for EMC just started to drop off the epoxy resin. D.E.R.-331 was the slowest compared to the other two as at the 60% time duration, the structure still did not drop off the epoxy resin completely. At 100% time duration which was the final structure, both ERL-4221 and EMC structures appeared like the hemisphere shape as the actual result, while for D.E.R.-331, the structure was not fully hemisphere in shape.

Table 14.2 Comparison between three epoxy materials

Time (%)	Simulation (Fluent)		
	ERL-4221	EMC	D.E.R.-331
0	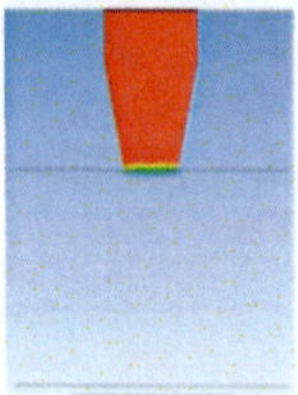	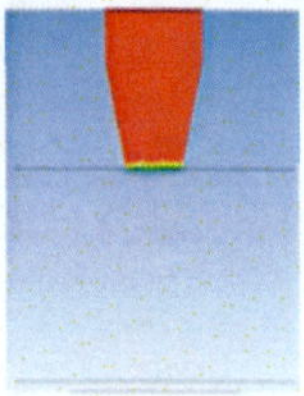	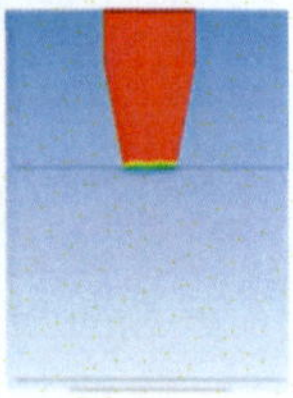
10	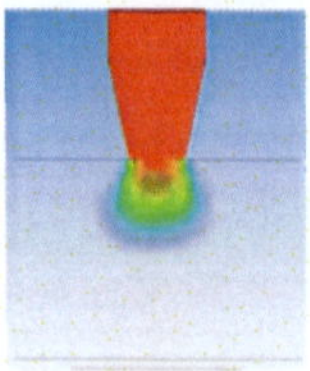	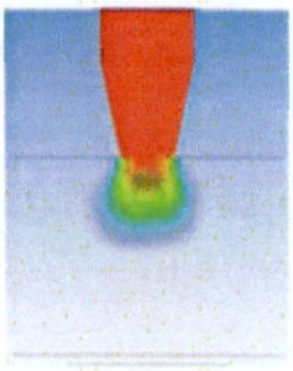	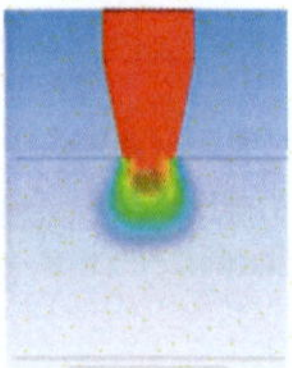
25	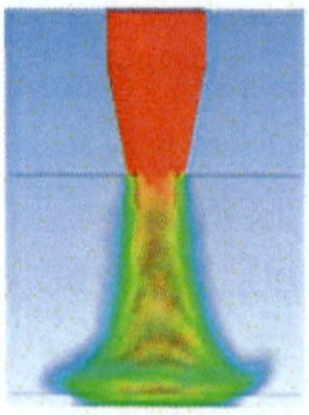	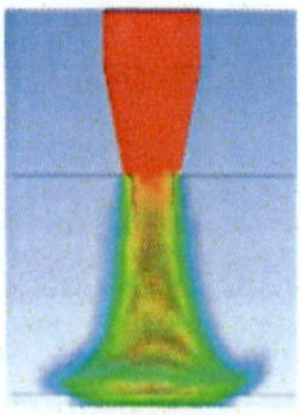	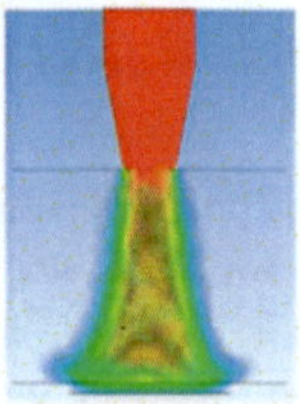
60	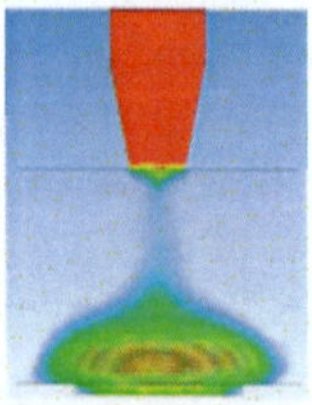	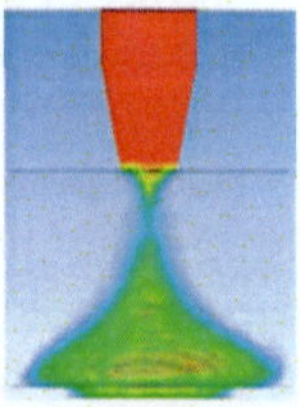	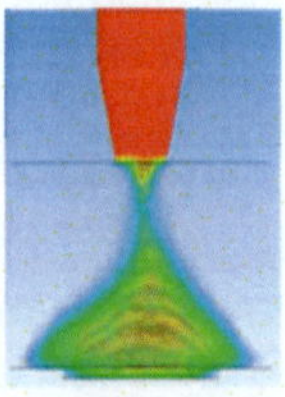
100	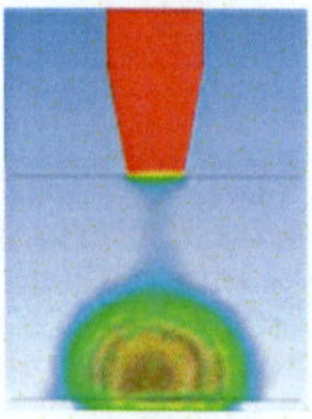	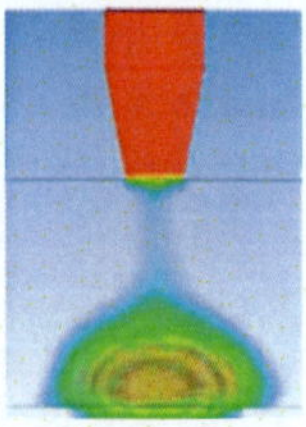	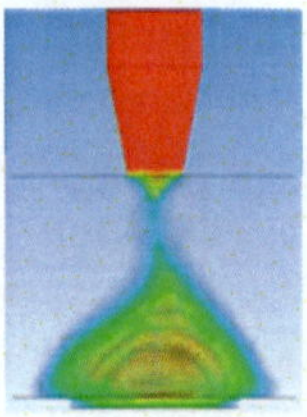
Volume of Epoxy Resin (m^{-3})	8.51143×10^{-8}	8.51143×10^{-8}	8.51143×10^{-8}

From the results obtained, ERL-4221 tends to flow quickly compared to EMC and D.E.R.-331 as it has the lowest viscosity. As the material with a lower viscosity has a shorter gel time, it can flow further while in liquid form compared to the epoxy with higher viscosity [7]. Meanwhile, D.E.R.-331 which has the highest viscosity tends to move sluggishly, thus taking a longer time than epoxy with lower viscosity. So, some modification to the simulation setup was required such as increasing the length of time to complete the dispensing process for a high viscosity epoxy. The simulation result shows that the approximate volume for the epoxy resins to complete the encapsulation process is 8.51143×10^{-8} m^3.

14.5 Conclusion

Based on the simulation results, it can be said that the final structure for EMC tends to be more accurate and precise compared to the other two materials. This conclusion was made after the validation of the epoxy final structure between the simulation and experimental results. The final structure for the EMC and ERL-4221 appeared like a hemisphere shape as in the experimental result. However, it is different for D.E.R.-331 as the final structure was not fully hemisphere in shape. Thus, further modifications need to be done on the time taken to complete the encapsulation process. The volume of the EMC injected was approximately 8.51143×10^{-8} m^3, the same as ERL-4221 and D.E.R.-331, since the speed and injection time for the simulation were constant. Besides, ERL-4221 which is low in viscosity tends to flow quickly since it has a shorter gel time and low resistance which causes the fluid to flow further while it is still liquid and will stop flowing sooner.

Acknowledgements Acknowledgment to the Ministry of Higher Education Malaysia for Fundamental Research Grant Scheme with Project Code: FRGS/1/2021/TKO/USM/03/9. The authors would also like to thank Universiti Sains Malaysia for providing technical support.

References

1. M.A. Alim, M.Z. Abdullah, M.S.A. Aziz, R. Kamarudin, Die attachment, wire bonding, and encapsulation process in LED packaging: A review, in *Sensors and Actuators, A: Physical*, vol. 329 (Elsevier B.V, 2021). https://doi.org/10.1016/j.sna.2021.112817
2. H. Xu, Y. Tang, J. Wu, B. Peng, Z. Chen, Z. Liu, The study on cracking reasons of LED encapsulation silicone, in *2019 20th International Conference on Electronic Packaging Technology, ICEPT* (2019, August 1). https://doi.org/10.1109/ICEPT47577.2019.245288
3. X. Shan, Y. Chen, Experimental and modeling study on viscosity of encapsulant for electronic packaging. Microelectron. Reliab. **80**, 42–46 (2018). https://doi.org/10.1016/j.microrel.2017.11.011
4. D. Ramdan, Z.M. Abdullah, M.A. Mujeebu, W.K. Loh, C.K. Ooi, R.C. Ooi, FSI simulation of wire sweep PBGA encapsulation process considering rheology effect. IEEE Trans. Compon. Packag. Manuf. Technol. **2**(4), 593–603 (2012). https://doi.org/10.1109/TCPMT.2011.2171513

5. R. Asghar, F. Rehman, A. Aman, K. Iqbal, A.A. Nawaz, Defect minimization and process improvement in SMT lead-free solder paste printing: a comparative study. Solder. Surf. Mt. Technol. **32**(1), 1–9 (2020)
6. M. Sadeghinia, K.M.B. Jansen, L.J. Ernst, Characterization of the viscoelastic properties of an epoxy molding compound during cure. Microelectron. Reliab. **52**(8), 1711–1718 (2012). https://doi.org/10.1016/j.microrel.2012.03.025
7. U. Peanpunga, K. Ugsornrat, P. Thorlor, C. Sumithpibul, The effect of epoxy molding compound floor life to reliability performance and mold ability for QFN package. J. Phys: Conf. Ser. **901**(1) (2017). https://doi.org/10.1088/1742-6596/901/1/012088

Chapter 15
Effect of Thermomechanical Treatment on Microstructural and Localized Micromechanical Properties of Sn–0.7Cu Solder Alloy

Fateh Amera Mohd Yusoff, Maria Abu Bakar, and Azman Jalar

Abstract Thermomechanical treatment is employed with the aim that it can modify the microstructure to improve the mechanical properties. This work utilizes the thermomechanical treatment on the Sn–0.7Cu solder alloy for interconnection materials in electronic packaging applications. The structural integrity of the interconnection is important to serve the miniaturization and reliability of electronic packaging. Therefore, this paper is aimed to investigate the effect of temperature and compression during thermomechanical treatment on microstructural and localized micromechanical properties of the Sn–0.7Cu solder alloy. The Sn–0.7Cu solder alloy with a cube-shaped is subjected to heat treatment at 30 °C and 90 °C for 20 min, followed by a compression process until 40% and 80% thickness reduction. Samples without a compression process will be used as a control sample. An infinite focus microscope (IFM) was used to capture and observed the microstructural changes. Localized micromechanical properties such as hardness and reduced modulus were investigated using the nanoindentation approach. The findings show that the subgrain was formed due to the thermomechanical treatment at 90 °C with thickness reduction of 80%. The localized hardness has decreased from 164 to 156 MPa while the reduced modulus has decreased from 57 to 53 GPa after being subjected to heat treatment at 90 °C. Significant micromechanical properties have been observed due to the thermomechanical treatment as the localized hardness and reduced modulus have increased from 156 to 261 MPa, 53 GPa to 78 GPa at 90 °C with a thickness reduction of 80%, respectively. The findings of this work have shown that the thermomechanical treatment has successfully modified the microstructural and micromechanical properties of the Sn–0.7Cu solder alloy.

F. A. Mohd Yusoff · M. Abu Bakar (✉) · A. Jalar
Institute of Microengineering and Nanoelectronics (IMEN), Level 4 Research Complex, Universiti Kebangsaan Malaysia, 43600 Bangi, Selangor, Malaysia
e-mail: maria@ukm.edu.my

A. Jalar
Department of Applied Physics, Faculty of Science and Technology, Universiti Kebangsaan Malaysia, 43600 Bangi, Selangor, Malaysia

M. A. A. Mohd Salleh et al. (eds.), *Proceedings of the Green Materials and Electronic Packaging Interconnect Technology Symposium*, Springer Proceedings in Physics 289,
https://doi.org/10.1007/978-981-19-9267-4_15

Keywords Hardness · Reduced modulus · Sn–0.7Cu solder alloy · Thermomechanical treatment

15.1 Introduction

Thermomechanical treatment is a metallurgy process that combines heat treatment and mechanical loading, resulting in plastic deformation that modifies the microstructure and improves the characteristics of a material through grain refinement [1–3]. This treatment is frequently applied for the structural materials used in the automotive and construction sectors, which are required to have high levels of mechanical characteristics and reliability over the long term [4]. Past research on mechanical properties of alloys is often using heat treatment. Heat treatment is the process of heating metal without allowing it to reach the molten (or melting) stage, then cooling the metal in a controlled manner to get the appropriate mechanical properties. All heat treatments comprise heating and cooling metals, but there are three major modifications in the process, which are the heating temperature, cooling rates, and quenching types employed to get the desired properties. Two most common heat treatments used by many researchers are annealing and aging [5, 6]. Annealing is frequently performed at a substantially greater temperature than aging [7]. High-temperature heat treatment in the prior study had lowered the material's mechanical properties like strength and hardness. This indicates that heat treatment alone cannot improve the properties of a material, leading to the incorporation of mechanical processes during heat treatment. Altering the temperature of the material while subjecting it to a mechanical load at the same time allows one to control the microstructure of the material. This can result in microstructural changes such as the reduction of grain size and the increase of dislocation density. It has been demonstrated that refining a material's grains can improve its mechanical properties. These improvements can include an increase in the material's hardness and tensile strength. According to the findings of Li et al. (2020), the process of applying hot rolling to Nb-Ti steel resulted in a greater tensile strength than that of the material without the application of hot rolling [8]. Therefore, a related understanding of the relationship between effect of thermomechanical treatment on microstructure and mechanical properties is important in a study because these three components are interrelated with each other.

Solder alloy refers to metal alloys composed of two metallic or metallic and nonmetallic components with a melting temperature (T_m) of less than 475 °C [9, 10]. In the electronic industry, this soft solder alloy is frequently used to join or connect numerous electrical components together in order to assure mechanical integration and electrical connectivity of electronic devices [11, 12]. Tin–lead (Sn–Pb) solder systems have been the foundation of the conventional soldering process for decades due to their unique mix of desirable mechanical properties, including low melting temperature (T_m), decent corrosion resistance, good wettability, and most importantly low cost [13, 14]. Despite these benefits, the toxicity of plumbum (Pb) has been widely discussed as a threat to the environment and human health [15–18]. This issue has led to adoption of Restriction of Hazardous Substances (RoHS) directives in July 2006 and new developments in lead-free solders as substitute of lead-based solder alloys [19, 20]. Lead-free solders have been extensively used in electronic industries, with Sn-Cu being one of the most common due to better mechanical properties, good wettability, low T_m and good solderability compared to the other lead-free solder alloys [21]. The reliability of the Sn–Cu solder alloys for the application as solder joint for electronic device is highly dependent on the mechanical behavior and microstructural evolution that occur interactively during processing [22]. The properties of solder materials continue to improve over time to meet the challenges posed by the continuous advances in electronic packaging technology, which aims at miniaturization and multifunctionality to ensure solder joint reliability [23]. These developments mean that solder joints are not only responsible for ensuring effective electrical current and conductivity connections but also need to have the high mechanical strength to ensure good performance in the long term. Despite the mentioned studies on the mechanical performances of Sn–Cu alloys, limited work has been done for the exploration of detailed deformation mechanisms during processing, which is particularly important in electronic packaging industries.

Small-scale mechanical properties characterization of solder materials is necessary due to the solder joint's miniaturization and understanding of the reliability of electronic devices. Previous studies have used conventional methods to determine the mechanical properties of solder alloys, namely microhardness test, shear test, impact test, Vickers test, bending test, and tensile test [24]. However, this is a conventional method that can only determine the mechanical properties in bulk and alloy which usually have anisotropic behavior after treatment. Therefore, the nanoindentation method is a technique that can determine the mechanical properties locally [25]. Nanoindentation is a widely used method to characterize mechanical properties on small structures without damaging the samples [26, 27]. This method also allows control over the load, depth, and exact test position. Mechanical properties and deformations are provided from the load versus depth curve. For example, the mechanical properties of intermetallic compounds (IMC) at the interface of Sn–3.0Ag–0.5Cu/Cu solder joints were investigated using the nanoindentation method [28]. The results show that the hardness, elastic modulus, and creep properties of IMC Cu_3Sn and Cu_6Sn_5 can be determined. Therefore, the present work aims to explore the microstructural features of Sn-Cu solder alloy after thermomechanical treatment at different temperature and thickness reduction. The microstructure and grain boundaries have been characterized through infinite focus microscope (IFM). The emphasis has been focused on microstructure evolution.

15.2 Methodology

Solder alloy in the form of bar was obtained commercially with composition of 99.3 wt% tin (Sn) and 0.7 wt% copper (Cu). The bar solder was cut into six samples with size of 6 mm (length) × 6 mm (width) × 10 mm (height). The first two samples were denoted as control sample as it only subjected to heat treatment in an oven with one sample at 30 °C and another sample at 90 °C. The heat treatment of samples was conducted for 20 min to ensure thermal stability without microstructure alterations. Thermomechanical treatments were conducted by firstly heat-treated at the same temperature as control sample, but quickly compression using push–pull gauge until the thickness of samples reduced to 40% and 80% from its original height as shown in Fig. 15.1. Each treated sample was quickly quenched in a water medium. After that, the samples underwent cross-section for microstructure and mechanical characterization. Firstly, the samples were clamped with sample holder and placed in the center of the mold container. The cold mounting liquid was prepared with ratio of 2:1, which are 20 g of hardener resin powder and 10 g of liquid epoxy resin. The powder and liquid were mixed up in a polystyrene cup and stirred for 30 s to obtain homogenous solution. Next, the solution was poured gently into the mounting container and let it cure for three to four hours. After that, mounting that contained samples was removed from its container and underwent grinding to expose the samples cross-section. Grinding of samples was conducted by *Buehler* grinding machine starting from coarser silicon carbide (SiC) paper (800 grit) to the finest SiC paper (1000 grit, 1200 grit, 2000 grit, and 4000 grit). Consequently, the samples were then polished through *Struers* polish machine using two DP-Nap polishing clothes with one cloth sprayed with 1 μm diamond and another one sprayed with 0.25 μm. The polishing process was started from the bigger size of diamond to the smallest to obtain mirror-like cross-sectional samples without any scratches when viewed under optical microscope (OM). For microstructure analysis, the samples were characterized through infinite focus microscope (IFM) with magnification of 50 ×. Mechanical properties of samples were characterized by nanoindentation method with maximum load of 10 mN, load–unloading rate was 0.5 mN/s and dwell time was 360 s. From the load versus indentation depth graph, hardness properties were obtained using Oliver–Pharr method as per equation below:

$$H = \frac{P_{\max}}{A_c}. \tag{15.1}$$

From the above equation, H is defined as the hardness value of a material in MPa, $P_{\max}$ is the maximum load applied on the material, and A_c is the contact area or area of indentation on the material's structure. Apart from hardness, reduced modulus also can be obtained as per equation below:

$$\frac{1}{E_r} = \frac{(1 - v_s^2)}{E_s} - \frac{(1 - v_i^2)}{E_i}. \tag{15.2}$$

Based on the reduced modulus equation, E_r is referred as reduced modulus, whereas E_s and v_s are Young's modulus and Poisson's ratio of a material. Meanwhile, E_i and v_i are Young's modulus and Poisson's ratio for the indentation.

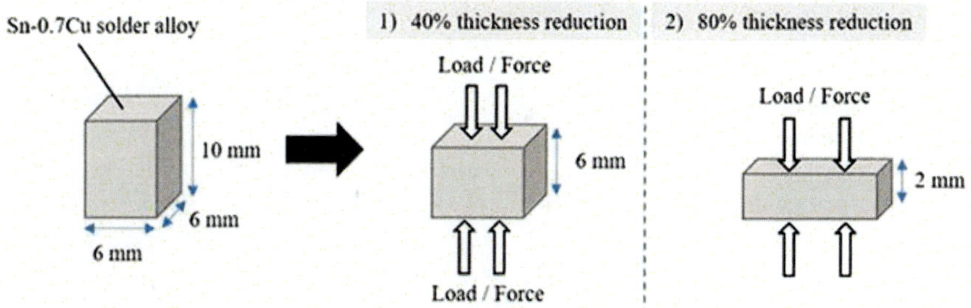

Fig. 15.1 Schematic of sample's height after thermomechanical treatment at various temperatures with 40% and 80% thickness reduction

15.3 Result and Discussions

15.3.1 Microstructure

Figures 15.2 and 15.3 show the morphology, size, and distribution of β-Sn and intermetallic phase in the microstructure of the Sn–0.7Cu solder alloy. Intermetallic phase in Sn–0.7Cu alloy is characterized as Cu_6Sn_5 showing a dark-field image, whereas the bright-filed image is tin-rich (β-Sn) phase or grain [22]. From the figure, it can be seen that increase of temperature up to 90 °C has resulted an increase in size of β-Sn grains or known as grain coarsening. Before heat treatment, the solder alloy at 30 °C shows distinct β-Sn grain and Cu_6Sn_5 that are sparsely distributed between them as shown in Fig. 15.2a. When the sample was heat-treated up to 90 °C, rapid recovery occurred which causes a decrease in Cu_6Sn_5 density due to extinction and rearrangement, some of which merged. This microstructure changes have led to a lower-energy state which resulted in a change in hardness value.

However, introduction of mechanical process such as compression at lower temperature (30 °C) has resulted an increase in dislocation density which accumulated at grain boundaries. After thermomechanical treatment at 30 °C with 40% thickness reduction, the β-Sn grains elongated perpendicular to compression direction. It can also be seen that there are slight formations of subgrains inside β-Sn grains. Increasing the thickness reduction up to 80% has led to increase in subgrains formation. This subgrain acts as an additional barrier for Cu_6Sn_5 movement to across from one grain to another neighbour grain. The difficulty of Cu_6Sn_5 movement and rearrangement has resulted an increase in hardness value. Sample that thermomechanical treated at 90 °C also possessed the same microstructure changes, but the size of β-Sn grains is more finer when compared to thermomechanical treated at 30 °C.

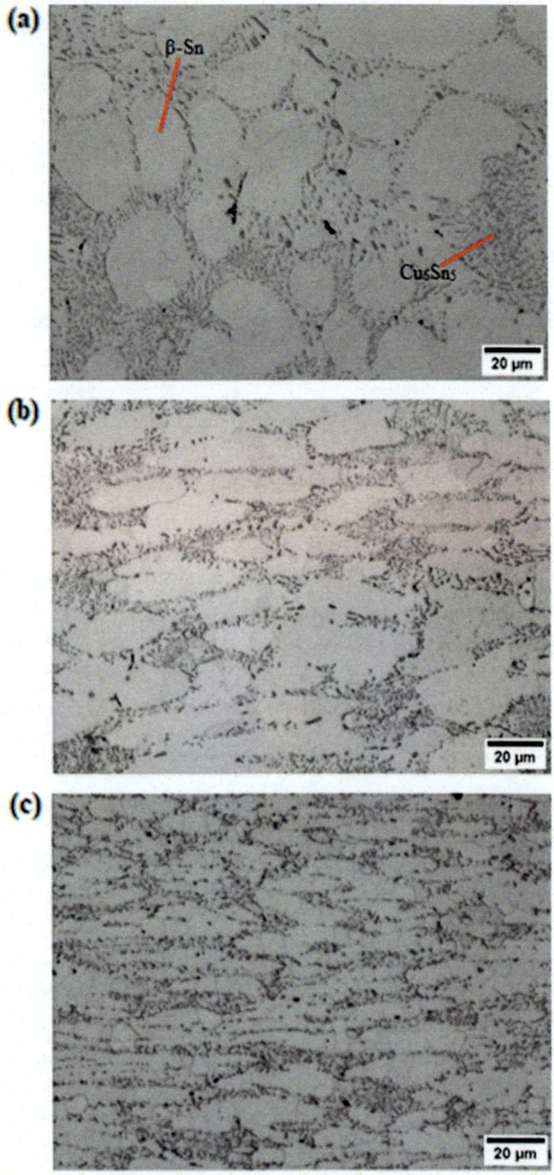

Fig. 15.2 IFM image of Sn–0.7Cu solder alloy subjected to heat treatment at 30 °C with percentage of thickness reduction of **a** 0% (controlled sample), **b** 40%, and **c** 80%

15.3.2 Nanoindentation Test

In microelectronic packaging, most of the reliability researches are focused on failure analysis, especially mechanical properties. A reliability study is required for quality inspection of the solder joint, meaning that if the solder joint fails the inspection, it

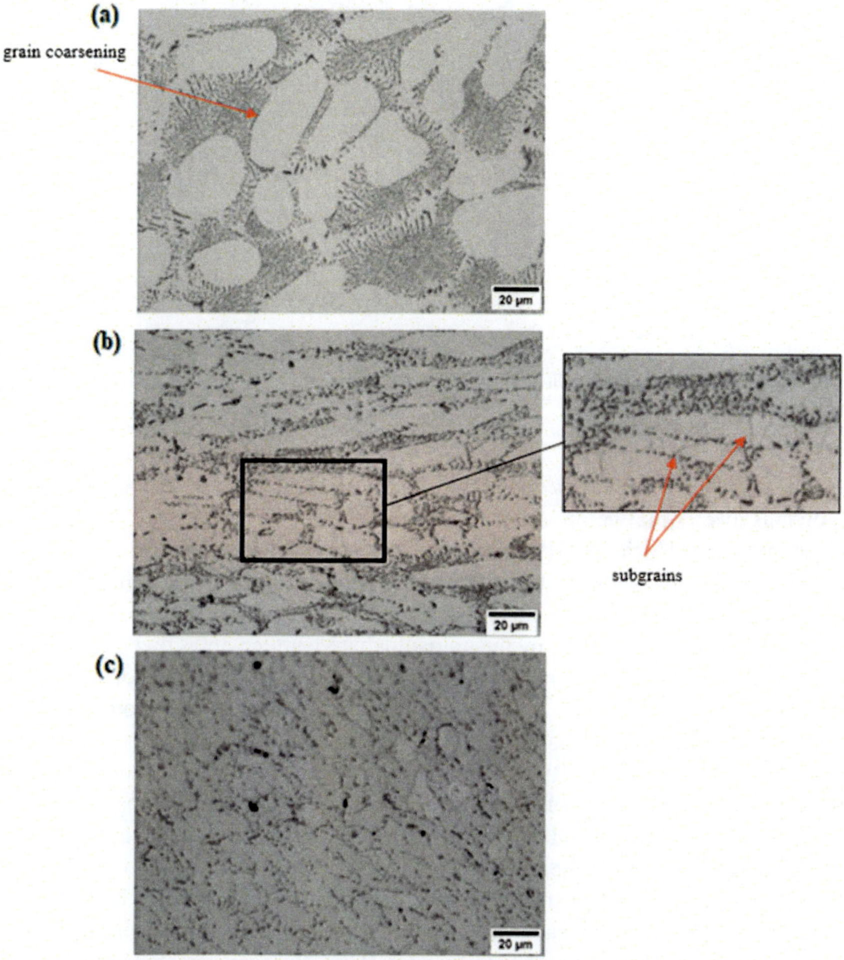

Fig. 15.3 IFM image of Sn–0.7Cu solder alloy subjected to thermomechanical treatment at 30 °C with percentage of thickness reduction of **a** 0% (controlled sample), **b** 40%, and **c** 80%

does not satisfy the specifications and the material becomes unreliable. A study has stated that microstructure evolution will affect the properties of the studied material [29]. In this work, the micromechanical properties of Sn–0.7Cu alloy were obtained through nanoindentation to verify the relationship between microstructure changes and properties. Figure 15.4 depicts load versus depth graph of Sn–0.7Cu solder alloy after treatment at 30 and 90 °C. Based on the graph, it is clearly shown that increase in load has resulted an increase in depth. When a load is applied to the indenter, the indenter's tip started to penetrate from the sample's surface into its structure and therefore increase the depth of penetration. This condition is known as loading, and it continuously penetrates the sample's structure until maximum

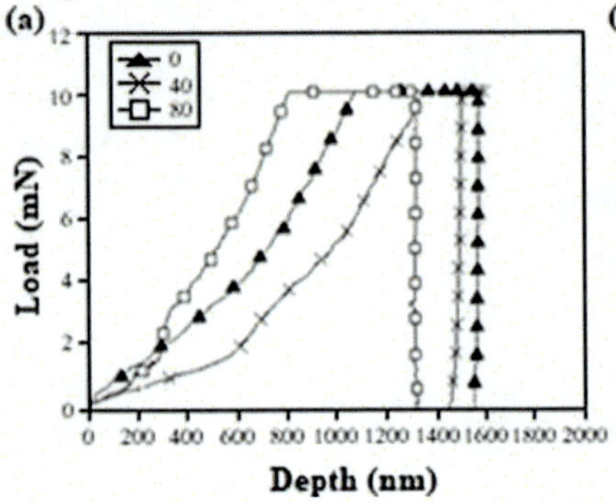

Fig. 15.4 Load versus depth of Sn–0.7Cu solder alloy subjected to treatment at **a** 30 °C and **b** 90 °C

load of 10 mN is reached. At maximum load, the indenter is left static for 360 s (known as dwell time) before the load is released from the indenter. Consequently, the indenter started to move out from sample's structure (known as unloading) making the indentation depth value continuously decrease but at a very low rate. In this work, the nanoindentation test reveals a typical load–depth relationship from the elastic-to-plastic region, as evidenced by the unloading curve after a 360 s dwell time. This indicates that experimental conditions and parameters used during nanoindentation test were optimum for achieving a substantial elastic–plastic deformation [27].

Figure 15.5 shows the relationship between maximum depth and plastic depth of Sn–Cu solder alloy after treatment at different temperatures. Maximum depth refers to the highest ability of indenter to penetrate the structure of sample. From the figures, the different of values between maximum and plastic depth is small. For example, after sample was heat-treated at 90 °C, the maximum depth is 1638 nm, whereas plastic depth is 1606 nm (~ 2% changes). The lower difference between the maximum and plastic depths suggests that the elastic-to-plastic area is expansive [27]. A study on the properties of high entropy alloy (HEA) was investigated by using the nanoindentation method, and it stated that indentation depth has a significant impact on the nanoindentation hardness, with the nanoindentation hardness increasing as the indentation depth decreases [30]. For example, sample that underwent thermomechanical treatment at 90 °C produced a decreasing trend of indentation depth value along with thickness reduction increment (Fig. 15.5b). However, the hardness value for the same sample possessed an increasing value from 216 to 261 MPa as shown in Fig. 15.6. The indentation depth observed in this investigation is consistent with previous research, indicating the data is valid.

Figure 15.6 shows the hardness of solder alloys as measured by nanoindentation testing, showing that the overall hardness properties of solder alloy are directly proportional to temperature and reduction in thickness. During heat treatment from 30 to 90 °C, the solder alloy's hardness decreased from 164 to 156 MPa (~8 MPa reduction). This decrease in solder alloy properties is similar to the findings from studies on alloy heat treatments, such as annealing and aging, which produce a decrease in

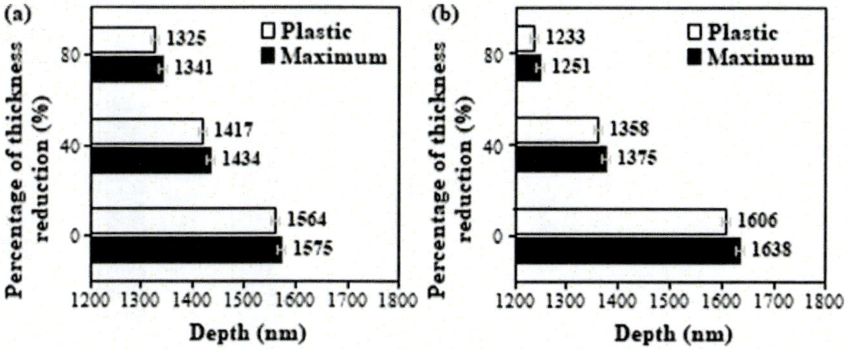

Fig. 15.5 Maximum and plastic depth of Sn–0.7Cu solder alloy subjected to different percentages of thickness reduction at **a** 30 °C and **b** 90 °C

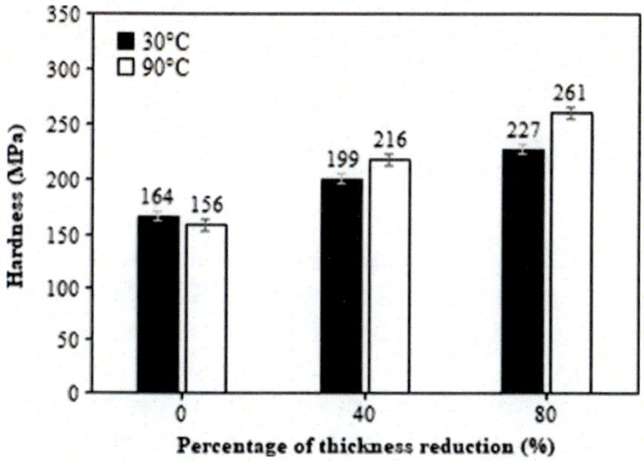

Fig. 15.6 Hardness of Sn–0.7Cu solder alloy subjected to different percentages of thickness reduction at **a** 30 °C and **b** 90 °C

properties as the treatment temperature increases [31–33]. However, when a mechanical process, such as compression, is introduced to the heat treatment, the hardness properties begin to exhibit opposite properties, such as an increase in hardness values compared to heat treatment alone. Thermomechanical processing refers to the introduction of a mechanical process after heat treatment, or the combination of both processes during treatment. At a lower temperature of 30 °C, reducing the solder alloy's thickness to 40% through compression led to an increase in hardness of 199 MPa, which increased to 227 MPa after an additional 80% reduction in thickness. When the temperature of thermomechanical treatment was raised to 90 °C, the hardness value increased slightly to 216 MPa (at 40% thickness reduction) and 261 MPa when the thickness reduction was increased to 80% at the same temperature.

Apart from hardness properties, reduced modulus properties of Sn–0.7Cu solder alloy were obtained as shown in Fig. 15.7. Based on the reduced modulus's figure, the samples show the same trend as hardness properties. Heat-treated samples show decreasing value of reduced modulus starting from 57 GPa at 30 °C to 53 GPa at

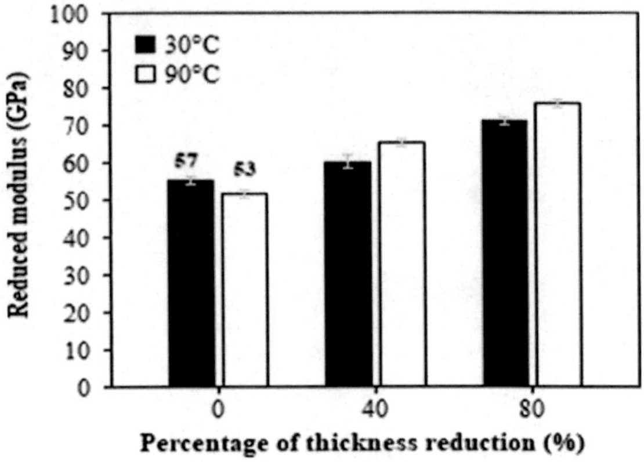

Fig. 15.7 Reduced modulus of Sn–0.7Cu solder alloy subjected to different percentages of thickness reduction at **a** 30 °C and **b** 90 °C

90 °C. However, thermomechanical-treated samples possessed the opposite trend of reduced modulus value as it slightly increased along with temperature increase.

15.4 Conclusion

The effect of thermomechanical treatment on the microstructural and localized micromechanical properties of Sn–0.7Cu solder alloy was successfully investigated. Nanoindentation testing was used to verify the relationship between microstructure changes and micromechanical properties of solder alloy after heat and thermomechanical treatments. From the results, it can be concluded that thermomechanical-treated samples possessed better micromechanical properties compared to heat-treated samples. Due to grain coarsening, the hardness of Sn–0.7Cu solder alloy decreases from 164 to 156 MPa after heat treatment from 30 °C to 90 °C. In contrast, after thermomechanical treatment with a 40% thickness reduction, the hardness of the Sn–0.7Cu solder alloy increased from 199 to 216 MPa. The hardness of the solder alloy increased from 227 to 261 MPa after thermomechanical treatment with a 80% thickness reduction. This occurs as a result of grain refining. In addition to hardness properties, the reduced modulus of Sn–0.7Cu solder alloy follows the same pattern as hardness value. These studies showed that a thermomechanical treatment can improve the mechanical properties of solder alloys for use as interconnection materials in electronic devices.

Acknowledgements The authors would like to acknowledge the financial support of the Ministry of Higher Education, Malaysia (grant number FRGS/1/2019/STG07/UKM/03/1) and The National University of Malaysia (UKM) for the research facilities support.

References

1. I. Abdullah, M.N. Zulkifli, A. Jalar, R. Ismail, Deformation behavior relationship between tensile and nanoindentation tests of SAC305 lead-free solder wire. Solder. Surf. Mt. Technol. **30**, 194–202 (2018). https://doi.org/10.1108/SSMT-07-2017-0020
2. R. Al Adawiyah Ab Rahim, M.N. Zulkifli, A. Jalar, A.M. Afdzaluddin, K.S. Shyong, Effect of isothermal aging and copper substrate roughness on the SAC305 solder joint intermetallic layer growth of high temperature storage (HTS). Sains Malaysiana **49**, 3045–3054 (2020). https://doi.org/10.17576/jsm-2020-4912-16
3. A.M. Afdzaluddin, M.A. Bakar, Effect of coating element on joining stability of Sn–0.3Ag–0.7Cu solder joint due to aging test. Sains Malaysiana **49**, 3045–3054 (2020). https://doi.org/10.17576/jsm-2020-4912-16
4. B. Ali, M.F.M. Sabri, I. Jauhari, N.L. Sukiman, Impact toughness, hardness and shear strength of Fe and Bi added Sn–1Ag–0.5Cu lead-free solders. Microelectron. Reliab. **63**, 224–230 (2016). https://doi.org/10.1016/j.microrel.2016.05.004
5. M. Dada, P. Popoola, N. Mathe, S. Adeosun, S. Pityana, Investigating the elastic modulus and hardness properties of a high entropy alloy coating using nanoindentation. Int. J. Light. Mater. Manuf. **4**, 339–345 (2021). https://doi.org/10.1016/j.ijlmm.2021.04.002
6. J.A. Depiver, S. Mallik, D. Harmanto, Solder joint failures under thermo-mechanical loading conditions—a review. Adv. Mater. Process. Technol. **4**, 1–26 (2021). https://doi.org/10.1080/2374068X.2020.1751514
7. M.A. Fazal, N.K. Liyana, S. Rubaiee, A. Anas, A critical review on performance, microstructure and corrosion resistance of Pb-free solders. Meas. J. Int. Meas. Confed. **134**, 897–907 (2019). https://doi.org/10.1016/j.measurement.2018.12.051
8. M. Hasnine, N. Vahora, Microstructural and mechanical behavior of SnCu–Ge solder alloy subjected to high temperature storage. J. Mater. Sci. Mater. Electron. **29**, 8904–8913 (2018). https://doi.org/10.1007/s10854-018-8908-4
9. J. Hui, Z. Feng, W. Fan, X. Yuan, The influence of power spinning and annealing temperature on microstructures and properties of Cu-Sn alloy. Mater. Charact. **144**, 611–620 (2018). https://doi.org/10.1016/j.matchar.2018.08.015
10. N. Ismail, A. Jalar, M. Abu Bakar, R. Ismail, N.S. Safee, A.G. Ismail, N.S. Ibrahim, Effect of isothermal aging on microhardness properties of Sn–Ag–Cu/CNT/Cu using Nanoindentation. Sains Malaysiana 48, 1267–1272 (2019). https://doi.org/10.17576/jsm-2019-4806-14
11. A. Jalar, M.A. Bakar, R. Ismail, Temperature dependence of elastic-plastic properties of fine-pitch SAC 0307 solder joint using nanoindentation approach. Metall. Mater. Trans. A. **51**, 1221–1228 (2020). https://doi.org/10.1007/s11661-019-05614-1
12. H.S. Joo, S.K. Hwang, Y.-T. Im, Effect of thermomechanical treatment on mechanical and electrical properties of Cu–Cr–Zr alloy in continuous hybrid process. Procedia Manuf. **15**, 1525–1532 (2018). https://doi.org/10.1016/j.promfg.2018.07.325
13. A. Kudryashova, V. Sheremetyev, K. Lukashevich, V. Cheverikin, K. Inaekyan, S. Galkin, S. Prokoshkin, V. Brailovski, Effect of a combined thermomechanical treatment on the microstructure, texture and superelastic properties of Ti-18Zr-14Nb alloy for orthopedic implants. J. Alloys Compd. **843**, 156066 (2020). https://doi.org/10.1016/j.jallcom.2020.156066
14. H. Li, M. Gong, T. Li, Z. Wang, G. Wang, Effects of hot-core heavy reduction rolling during continuous casting on microstructures and mechanical properties of hot-rolled plates. J. Mater. Process. Technol. **283**, 116708 (2020). https://doi.org/10.1016/j.jmatprotec.2020.116708
15. G. Liu, S. Ji, Microstructure, dynamic restoration and recrystallization texture of Sn-Cu after rolling at room temperature. Mater. Charact. **150**, 174–183 (2019). https://doi.org/10.1016/j.matchar.2019.02.032
16. K. Lodo, C. Dalgleish, M. Patel, M. Veitch, A novel public health threat—high lead solder in stainless steel rainwater tanks in Tasmania. Aust. N. Z. J. Public Health. **42**, 77–82 (2018). https://doi.org/10.1111/1753-6405.12723

17. D. Marinho Filizzola, T. da Silva Santos, A. Gomes de Miranda, J.C. Martins da Costa, N. Reis do Nascimento, M. Dantas dos Santos, R. Hoel Bello, G. Garcia del Pino, J. Costa de Macêdo Neto, Annealing effect on the microstructure and mechanical properties of AA 5182 aluminum alloy. Mater. Res. **24**(4) (2021). https://doi.org/10.1590/1980-5373-mr-2020-0545
18. D.M. Mattox, *The "Good" Vacuum (Low Pressure) Processing Environment* (2010). https://doi.org/10.1016/b978-0-8155-2037-5.00003-4
19. C. Morando, O. Fornaro, Influence of aging on microstructure and hardness of lead-free solder alloys. Solder. Surf. Mt. Technol. **33**, 57–64 (2021). https://doi.org/10.1108/SSMT-03-2020-0013
20. Z. Nasiri, S. Ghaemifar, M. Naghizadeh, H. Mirzadeh, Thermal mechanisms of grain refinement in steels: a review. Met. Mater. Int. **27**(7), 2078–2094 (2021). https://doi.org/10.1007/s12540-020-00700-1
21. D. Qu, C. Li, L. Bao, Z. Kong, Y. Duan, Structural, electronic, and elastic properties of orthorhombic, hexagonal, and cubic Cu3Sn intermetallic compounds in Sn–Cu lead-free solder. J. Phys. Chem. Solids. **138**, 109253 (2020). https://doi.org/10.1016/j.jpcs.2019.109253
22. Y. Román-Ochoa, G.T. Choque Delgado, T.R. Tejada, H.R. Yucra, A.E. Durand, B.R. Hamaker, Heavy metal contamination and health risk assessment in grains and grain-based processed food in Arequipa region of Peru. Chemosphere **274** (2021). https://doi.org/10.1016/j.chemosphere.2021.129792
23. B. Santosa, Evaluation of anemia in the residents of tambaklorok exposed to plumbum. Maced. J. Med. Sci. **9**, 831–835 (2021). https://doi.org/10.3889/oamjms.2021.6430
24. P.D. Sonawane, V.K. Bupesh Raja, M. Gupta, Mechanical properties and corrosion analysis of lead-free Sn-0.7Cu solder CSI joints on Cu substrate. Mater. Today Proc. **46**, 1101–1105 (2021). https://doi.org/10.1016/j.matpr.2021.01.521
25. P.D. Sonawwanay, V.K.B. Raja, Advances in lead-free solders. Int. J. Mech. Eng. Technol. **10**, 520–526 (2019)
26. Z. Tang, F. Jiang, M. Long, J. Jiang, H. Liu, M. Tong, Effect of annealing temperature on microstructure, mechanical properties and corrosion behavior of Al-Mg-Mn-Sc-Zr alloy. Appl. Surf. Sci. **514** (2020). https://doi.org/10.1016/j.apsusc.2020.146081
27. M.C. Tanzi, S. Farè, G. Candiani, Organization, structure, and properties of materials. Found. Biomater. Eng. 3–103 (2019). https://doi.org/10.1016/B978-0-08-101034-1.00001-3
28. Z. Wang, A.M. Korsunsky, Effect of temperature on shape memory materials. Encycl. Smart Mater. 239–253 (2022). https://doi.org/10.1016/B978-0-12-803581-8.11793-X
29. K.K.T.A.S. Wardoyo, Plumbum (Pb) in rainwater in West Kalimantan: Impact of plumbum (Pb) in community blood. Nat. Environ. Pollut. Technol. **18**, 1423–1427 (2019)
30. G. Xiao, X. Yang, G. Yuan, Z. Li, X. Shu, Mechanical properties of intermetallic compounds at the Sn–3.0Ag–0.5Cu/Cu joint interface using nanoindentation. Mater. Des. **88**, 520–527 (2015). https://doi.org/10.1016/j.matdes.2015.09.059
31. M.Z. Yahaya, N.A. Salleh, S. Kheawhom, B. Illes, M.F. Mohd Nazeri, A.A. Mohamad, Selective etching and hardness properties of quenched SAC305 solder joints. Solder. Surf. Mt. Technol. **32**, 225–233 (2020). https://doi.org/10.1108/SSMT-01-2020-0001
32. H. Zeng, H. Sui, S. Wu, J. Liu, H. Wang, J. Zhang, B. Yang, Evolution of the microstructure and properties of a Cu–Cr-(Mg) Alloy upon thermomechanical treatment. J. Alloys Compd. **857**, 157582 (2021). https://doi.org/10.1016/j.jallcom.2020.157582
33. P. Zhang, S. Xue, J. Wang, New challenges of miniaturization of electronic devices: electromigration and thermomigration in lead-free solder joints. Mater. Des. **192**, 108726 (2020). https://doi.org/10.1016/j.matdes.2020.108726

Chapter 16
Effects of Heat Treatment on the Properties of SS440C for Blades Applications

Nur Maizatul Shima Adzali, Siti Khadijah Salihin, and Nur Hidayah Ahmad Zaidi

Abstract SS440C steel is commonly used for knife blades, bearings, valve parts, and medical equipment. The composition of SS440C steels is designed to increase hardness especially in blade applications. The effect of quenching and tempering heat treatment on the properties of SS440C was investigated in this study. Quenching heat treatment is done at 1000 °C, followed by tempering at 150 and 500 °C in a muffle furnace. Microstructure of SS440C samples were studied using an optical microscope (OM) and scanning electron microscope (SEM). Properties of SS440C after heat treatment have been investigated using the Rockwell hardness test and tensile test. It was found that the sample quenched at 1000 °C (without temper) had the highest hardness with 58.4HRC, while the as-received annealed sample had 11.4HRC followed by sample tempered at 150 °C with 57.5HRC and 500 °C with 54.1HRC. Tensile testing reveals that quenching and tempering at 500 °C result in the highest maximum stress compared to other samples. Through optical microscopy observation, a sample tempered at 500 °C has larger size of carbide precipitate than sample that quenched and tempered at a 150 °C. Insufficient carbide dissolution or a more abrasive reaction is revealed by larger carbide sizes. In conclusion, SS440C temper at 500 °C reflects that it has better properties than the other.

Keywords SS440C · Quenching · Tempering

N. M. S. Adzali (✉)
Industrial Chemical Process Programme, Faculty of Chemical Engineering Technology, Universiti Malaysia Perlis, Kampus UniCITI Sg. Chuchuh, Padang Besar, 02100 Perlis, Malaysia
e-mail: shima@unimap.edu.my

S. K. Salihin · N. H. Ahmad Zaidi
Materials Engineering Programme, Faculty of Chemical Engineering Technology, Universiti Malaysia Perlis, 02600 Jejawi, Arau, Perlis, Malaysia

M. A. A. Mohd Salleh et al. (eds.), *Proceedings of the Green Materials and Electronic Packaging Interconnect Technology Symposium*, Springer Proceedings in Physics 289,
https://doi.org/10.1007/978-981-19-9267-4_16

16.1 Introduction

Nowadays, steel extends every aspect of our life's applications. Over time, steel forms expanded and supported other building materials like wood, stone, and concrete. Among all steels, stainless steel is one that is used the most. AISI 440C high-carbon martensitic stainless steel (MSS) is a form of nickel-free chrome steel. Approximately 1.1 wt% of it is carbon, which gives it very good mechanical strength and hardness followed by strong wear resistance and mild corrosion resistance. The "C" in 440C stainless steel stands for a higher carbon content, and the increased hardness of this steel makes it perfect for tool and die applications in the food processing industry [1]. It is more cost effective and significantly stronger than carbon steels, with superior corrosion resistance. It has superior wear resistance, toughness, and an extremely high finish value, which is crucial but rarely discussed.

A crucial step in modifying the microstructure and enhancing the properties of martensitic stainless steels, including SS440C, is heat treatment. The heat treatment conditions can be carefully managed to give the steels the ideal balance of high strength and tolerable toughness [2]. The martensitic grades can be quenched and tempered to produce high hardness, unlike other stainless steels. The mechanical properties of quenched and tempered microstructures are significantly influenced by the quantity and type of carbides [3].

Numerous studies have revealed that thermal treatments like normalizing and tempering significantly affect the microstructure and mechanical properties of reversed austenite produced during the quenching process [4, 5]. Due to limited research, it is difficult to prove that heat treatment can improve the properties of SS440C. Moreover, martensitic stainless steel has a higher strength than other types of stainless steel [1]. However, martensitic stainless steel is brittle and has low toughness when compared to other types of stainless steel. In its as-quenched martensitic state, the steel is hard and brittle, with pockets of retained austenite.

Thus, further stainless steel research should be advocated. Quenching and tempering heat treatments to SS440C have been proposed to change the characteristics for better properties and generate qualities required for its use. In order to solve the aforementioned issues, the effects of quenching and tempering temperatures of 1000, 150, and 500 °C on property changes were examined.

16.2 Methodology

16.2.1 Materials

A readily annealed SS440C sample plate (38 cm × 7.6 cm) was cut into different sizes (2 cm × 2 cm and 1 cm × 1 cm) prior to the different heat treatment process. The readily annealed sample was used as the control sample and was not subjected

Table 16.1 Compositional analysis of SS440C by optical emission spectroscopy (OES)

	C (%)	Si (%)	Mn (%)	Cr (%)	Mo (%)	Fe (%)
OES	0.799	0.454	0.250	16.42	0.451	80.97

to any heat treatment, while other samples were quenched at 1000 °C, followed by tempering at two different temperatures (150 and 500 °C).

The as-received material was a SS440C plate. The chemical composition for the material acquired using arc–spark spectrometers is given in Table 16.1.

16.2.2 Experimental Procedures

The as-received (readily annealed) sample served as the control sample and was not subjected to any heat treatment, while other samples were quenched at 1000 °C, followed by tempering at two different temperatures (150 and 500 °C).

Figure 16.1a shows the heating profile of the quenching process, while Fig. 16.1b shows the heating profile of the tempering process. After heat treatment, all samples were grinded and polished before microscopic analysis.

16.3 Characterization Method

After heat treatment, the microhardness of the SS440C was measured using a Rockwell test, followed by tensile tests which used to investigate the mechanical properties of samples after heat treatment. The microstructure changes of every sample were observed by using an optical microscope (OM) and scanning electron microscope (SEM) after heat treatment.

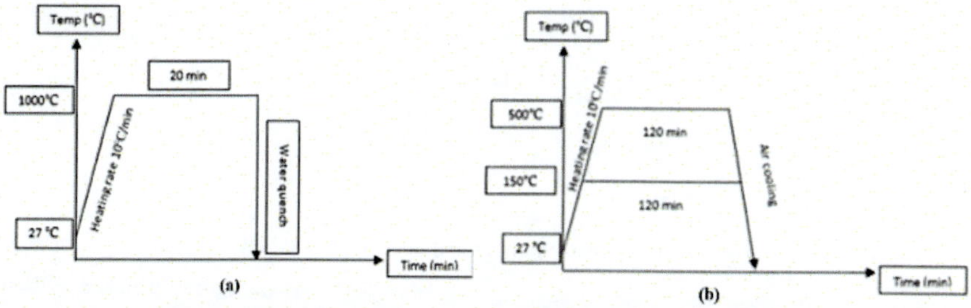

Fig. 16.1 Heating profiles for **a** quenching and **b** tempering process at 150 and 500 °C for SS440C

16.4 Results and Discussion

16.4.1 Hardness

Figure 16.2 shows the hardness of SS440C samples before (readily annealed sample) and after heat treatment. Annealed samples show the lowest hardness value which is 11.4 HRC. This confirms the sample's softness and the presence of austenite. Meanwhile, for the sample quenched at 1000 °C contribute the highest hardness value (58.4 HRC). When all precipitates have been dissolved and grain growth has been minimized, austenitization at 1000 °C for 20 min results in the highest hardness of the quenched sample.

Ebrahimi et al. [6] revealed that at a given solution time, hardness increases to a maximum before suddenly decreasing. The gradual dissolution of $M_{23}C_6$ carbides in austenite is responsible for the primary increase in hardness. These $M_{23}C_6$ carbides are Cr-rich fine secondary carbides fully distributed inside the matrix which appear during tempering. Beyond the peak, however, because there is no pinning force from carbide particles on grain boundaries, grain growth, whether normal or abnormal, leads to larger martensitic packets with lower quenched hardness. It is important to note that the highest hardness of a quenched martensite is obtained by austenitizing at 1000 °C for 60 min when all precipitates have been dissolved and grain growth has been lowered.

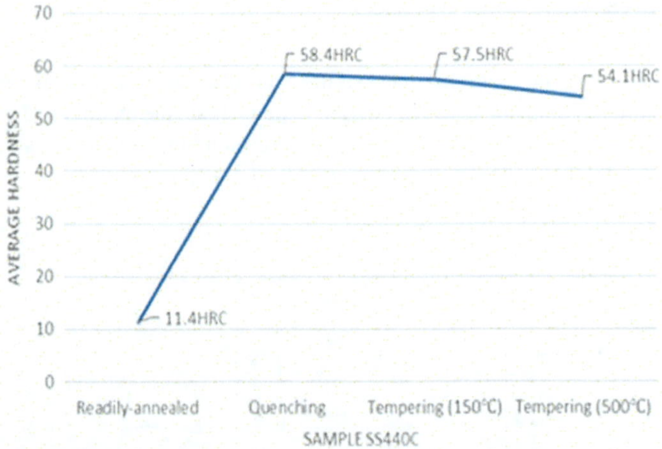

Fig. 16.2 Hardness (HRC) values of ready sample and heat-treated sample of SS440C

16.4.2 Tensile

Tensile testing was performed on the as-received annealed, quenching, and tempering (150 and 500 °C) samples. Figure 16.3 shows a tensile test result for a ready sample and heat-treated sample of SS440C. The total elongation of the as-received annealed SS440C sample was higher than the other three samples. It is because the sample is ductile. The greater the area under the curve, the greater the toughness, as it indicates that the tough material can withstand a greater level of load before fracture or failure. Meanwhile, quenching sample and tempering 150 °C fractures before yield point. That is because the two samples are highly brittle. These two samples have no elongation plastic deformation. Furthermore, sample tempered at 500 °C does not show high elongation but has high stress value, which is 987.6 N/mm^2 (Fig. 16.3). This indicates that the tempering process of SS440C samples at 500 °C has high ductility. Hence, this trend matched the phenomenon obtained in the hardness test (which has been discussed in Fig. 16.2). Due to the specimen's highly brittle martensite structure (quenched sample), the maximum stress value of the quenched sample is the lowest (275.3 N/mm^2). Due to the fact that the elongation sample decreases as hardness increases. Brittle fractures absorb only a limited amount of energy that is absorbed through small plastic deformation regions.

Naveed et al. [7] reported that as temperature of heat treatment increased, the yield strength, ultimate tensile strength, and failure strength all decreased. Because of the increased plastic flow at higher temperatures, the ductility parameters improved. However, the tempering time had no effect on these properties. SEM analysis of the fracture surfaces revealed reduced cracking and dimpled microstructures, indicating improved ductility at higher testing temperatures. Subbiah et al. [4] discussed that the tensile strength and hardness dropped fast when tempering temperature increased from 550 to 600 °C.

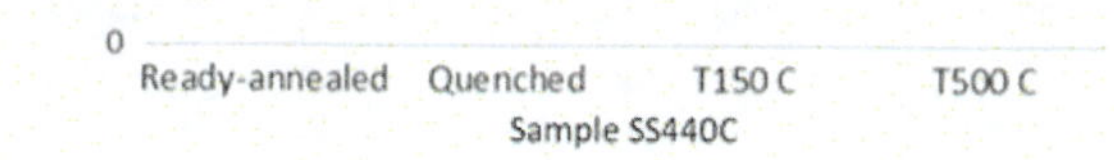

Fig. 16.3 Tensile test graph for heat-treated sample of SS440C

16.4.3 Microstructural Observation

The microstructure of SS440C is revealed using an OM and SEM. Figure 16.4b–d shows OM images of quenching at 1000 °C (b), tempering at 150 °C (c), and 500 °C (d), while Fig. 16.4a shows SEM image for readily sample. Those figures illustrate precipitation of varying sizes. These precipitates are to be iron-chromium carbides, which are present in martensitic stainless steels. The size of the particles after tempering at 500 °C (Fig. 16.4d) has a combination of smaller and bigger particles but more in bigger carbide particles. Furthermore, the size particle at tempering 150 °C (Fig. 16.4c) shows a mix of smaller particle carbide size. This is caused by the precipitation and growth of carbides. A larger carbide size indicates insufficient carbide dissolution or a coarser reaction. A quenching temperature of 1000 °C (Fig. 16.4b) is sufficient for carbides to dissolve optimally. The larger particle will be classified as primary carbide, while the smaller particle will be classified as secondary carbide.

As reported by Li et al. [8], carbides are classified into two types based on their size: primary carbides (> 5 μm) and secondary carbides (< 5 μm). Small-sized primary carbides are generally classified as secondary carbides, but the number of such carbides is insignificant. Furthermore, when compared all sample, which was tempered at a lower temperature, contains a greater amount of secondary carbides.

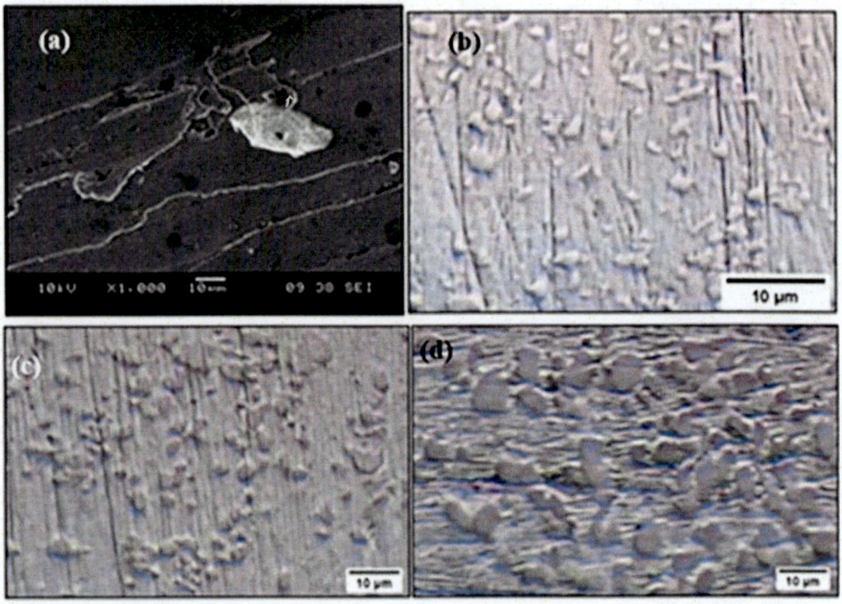

Fig. 16.4 **a** SEM images of annealed (readily sample), **b** OM images of quenching sample, **c** OM image of tempering at 150 °C, **d** OM image for tempering at 500 °C of SS440C

The results of the tensile testing are supported by the SEM images of the fracture surfaces which are shown in Fig. 16.5a–d. The average size of dimples, which roughly indicates the amount of absorbed energy during fracture, clearly decreases from image in Fig. 16.5a–c. The fracture surface from Fig. 16.5a–c has very small dimples that indicate a very low absorbed energy or nearly brittle fracture. Figure 16.5 (tempering at 150 °C) and Fig. 16.5 (quenching) show an intergranular crack. The oxidation process darkens the crack surface. Intergranular fracture, as seen in images in Fig. 16.5b,c, results in low impact toughness values when quenched and tempered at the same time. Tempering at 500 °C for another 2 h (Fig. 16.5d) resulted in a reduction in intergranular fracture and its complete elimination. At high temperatures, the failure morphology was characterized by increased plastic deformation. On the fracture surfaces, reduced cracking and dimpled microstructures were observed.

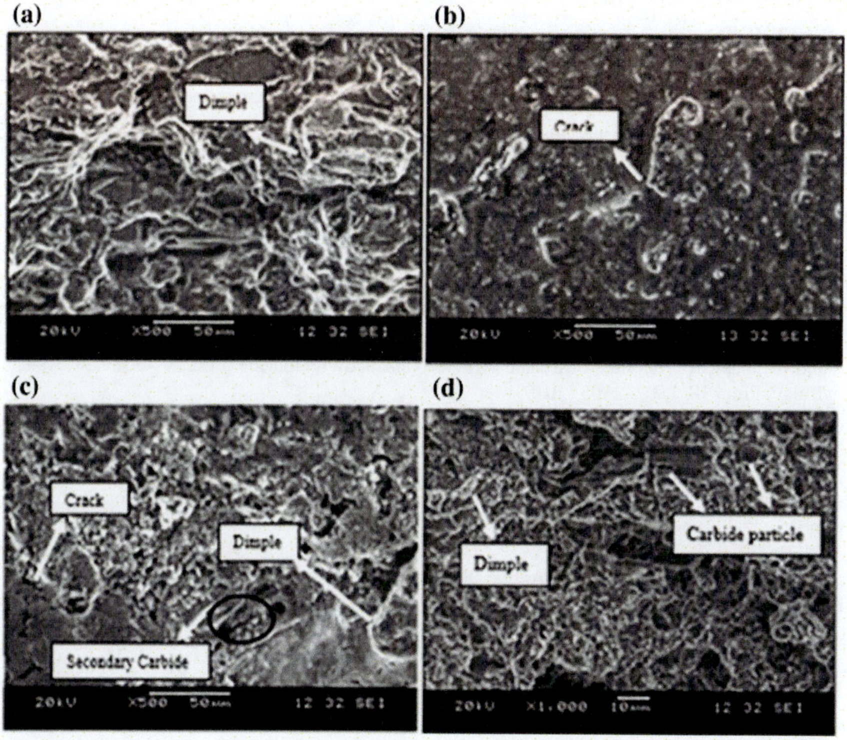

Fig. 16.5 **a** SEM images of fracture surfaces of annealed (readily sample), **b** sample tempered at 150 °C, **c** sample quenched at 1000 °C, **d** sample tempered at 500 °C

16.5 Conclusions

The investigation on the microstructure and mechanical properties of SS440C as-received annealed, quenching, and tempering sample has led to some conclusions. Firstly, the hardness test showed the highest result for quenching sample but the lowest for annealed sample. The hardness of tempering 500 °C is slightly lower than tempering at 150 °C because of the carbide precipitation growth which lowers the hardness of the sample, but its maximum stress is the highest among others. Sample tempering at 500 °C had the best combination of both microstructural and mechanical properties for blade application, having high hardness, 54.1HRC, and maximum stress.

References

1. L. Pan, C.T. Kwok, K.H. Lo, Friction-stir processing of AISI 440C high-carbon martensitic stainless steel for improving hardness and corrosion resistance. J. Mater. Process. Technol. **277** (2020)
2. Y. Liu, T. Wang, Z. Yu Li, J. Xun Zhang, Heat treatment for microstructure and mechanical properties improvement of powder plasma arc melted 17Cr-2Ni steel containing boron. Surf. Coatings Technol. **427** (2021)
3. K.T. Huang, S.H. Chang, P.C. Hsieh, Microstructure, mechanical properties and corrosion behavior of NbC modified AISI 440C stainless steel by vacuum sintering and heat treatments. J. Alloys Compd. **712**, 760–767 (2017)
4. R. Subbiah, K. Lokesh, S.K. Singh, S. Chatterjee, D. Eswaraiah, Investigation on microstructure and mechanical properties of treated AISI 440 steels by tempering process—a review. Mater. Today: Proc. **18**, 2802–2805 (2019)
5. A.A. Salih, M.Z. Omar, J. Syarif, Z. Sajuri, An investigation on the microstructure and mechanical properties of quenched and tempered SS440C martensitic stainless steel. Int. J. Mech. Mater. Eng. **7**(2), 119–123 (2012)
6. G.R. Ebrahimi, H. Keshmiri, A. Momeni, Effect of heat treatment variables on microstructure and mechanical properties of 15Cr-4Ni-0.08C martensitic stainless steel. Ironmaking Steelmaking **38**(2), 123–128 (2011)
7. A. Naveed, R. Ahmad, Tanveer Akhtar, R. Ayub, I.M. Ghauri, Annealing and test temperature dependence of tensile properties of UNS N04400 alloy. J. Mater. Eng. Performance (2013)
8. S. Li, X. Xi, Y. Luo, M. Mao, X. Shi, J Guo., H. Guo, Carbide precipitation during tempering and its effect on the wear loss of a high-carbon 8 Mass% Cr tool steel. Materials **11**(12) (2018)

Chapter 17
The Effect of the Epoxy Curing Method on the Encapsulation of Led

Kaalidass Muniary, Mohd Syakirin Rusdi, Mohd Sharizal Abdul Aziz, Roslan Kamaruddin, M. H. H. Ishak, Md. Abdul Alim, and Mohd Arif Anuar Mohd Salleh

Abstract The electronic sector is developing novel products that are smaller and faster, with better performance, lighter containers, and are more cost effective. An integrated circuit can be used to link several LEDs with different functions to electronic devices packaged. The space between the cover lens and the LED chip is filled with a non-conductive polymeric substance, like epoxy or transparent silicone encapsulant. The goal of encapsulation is to protect the device from physical forces that could weaken the connection between the chip and the substrate. The research is focusing on the experimental investigation of the effect of epoxy on the encapsulation process in LED. The different parameters that affect the encapsulation process, such as needle size, type of epoxy resin, and method of curing, have been studied. The contact angle and contact surface area of each type of epoxy were measured after the curing process. Using a syringe and five different needle sizes, the three types of epoxy resin and hardener are manually injected onto the wafer substrate. ImageJ was used to measure the contact angle and contact area. The results of the experiment demonstrated that the 1:1 self-curing epoxy resin is the most effective type of epoxy resin, as its contact angle and surface area of contact have optimal values. The 21G needle size produces acceptable results in the encapsulation procedure. The contact angle and surface area of contact measurements for this project will determine the appropriate epoxy to use.

K. Muniary · M. S. Rusdi (✉) · M. S. Abdul Aziz · R. Kamaruddin
School of Mechanical Engineering, Universiti Sains Malaysia, Engineering Campus, 14300 Nibong Tebal, Seberang Perai Selatan, Penang, Malaysia
e-mail: syakirin@usm.my

M. H. H. Ishak
School of Aerospace Engineering, Universiti Sains Malaysia, Engineering Campus, 14300 Nibong Tebal, Seberang Perai Selatan, Penang, Malaysia

Md. A. Alim
Directorate of Technical Education, F-4/B, Agargaon, Dhaka-1207, Bangladesh

M. A. A. Mohd Salleh
Center of Excellence Geopolymer & Green Technology (CeGeoGTech), University Malaysia Perlis (UniMAP), Taman Muhibbah, 02600 Perlis, Malaysia

M. A. A. Mohd Salleh et al. (eds.), *Proceedings of the Green Materials and Electronic Packaging Interconnect Technology Symposium*, Springer Proceedings in Physics 289,
https://doi.org/10.1007/978-981-19-9267-4_17

Keywords Epoxy · Encapsulation process · Contact angle · Self-curing resin

17.1 Introduction

The light-emitting diode (LED) is the most advanced technology in the lighting industry. High-power LED rapidly expands in terms of usage since it can last long up to 11 years. In recent years, LED brightness has improved tremendously due to advancements in LED die in terms of luminous efficiency and its ability to be driven under higher-powered conditions [1]. Aside from that, as compared to other light sources, the LED has far higher efficiency. However, obtaining an LED with the desirable qualities is rather challenging because there are several potential conditions that could impact the device's operation. An investigation into the encapsulating procedure in LED packaging was done by Alim et al. [2]. They highlighted a variety of issues with LED packaging, including encapsulation form, heat conductivity, material deterioration, and bonding strength. As a result, the following reasons frequently contribute to LED device failure rates. In the packing of lamp-LEDs, encapsulation is commonly used. The encapsulation technique involves pouring liquid epoxy into the LED wafer first, then placing the substrate in the oven, where the epoxy hardens, and the LED is formed. Because photons emitted by the LED p–n junction are non-directional, meaning they have the same chance of launching in all directions, not all of the light produced by the chip can be emitted. The quality of the semiconductor material, chip structure, geometric shape, encapsulation internal material, and packaging materials influence the amount of light emitted. Therefore, for the LED encapsulation, the right encapsulation method should be selected according to the size of the chip and power. Regardless of the processing method employed, factors such as injection pressure, encapsulant ingredients, resin velocity, and needle nozzle position will all affect the finished part's output. It is possible to influence one or more of the other parameters by changing one of the processing parameters. It is only by understanding how these components interact for a certain process that high-quality parts may be successfully produced.

Some potential points of failure involve the packaging materials themselves. The LED's packaging material, for instance, may undergo a colour change and break due to heat generated by the p–n junction of the chip in conjunction with its external environment [3]. Strong adhesion, remarkable machinability, and great chemical resistance [4] are just a few of the reasons why epoxy resins, as top thermosetting polymer materials, have been widely employed in the current LED encapsulation process. This research will thus investigate three distinct epoxy resins often employed in the LED encapsulation technique.

In order to protect semiconductor components during shipping, the electronics industry frequently use epoxy moulding compound (EMC) as a kind of packaging. Residual stress will be created during curing as a result of cure and thermal shrinkage, which might cause failure of the product. The elastic modulus and viscoelastic behaviour of a material were studied by Sadeghinia et al. [5] throughout the curing process. An equipment called a Dynamic Mechanical Analyzer DMAQ800 was used to monitor the mechanical characteristics of the EMC as shear loads were applied and removed.

From Yang et al. [6], as a hardener and catalyst, the epoxy methyl-hexahydro phthalic anhydride (MHHPA) and tetrabutylphosphium methanesulfonate (TBPM) were added to the oligosiloxane resins, respectively. The molar ratio of MHHPA and cycloaliphatic epoxy oligosiloxane resins was 1.0:0.9. The amount of TBPM in MHHPA was set at 0.5 mol per cent. The weight ratio of the synthesized resins and p-xylene is 1.0–0.1. The resins were kept at 50 °C under a vacuum for 24 h to remove bubbles and solvent. The samples were thermally cured for 2 h at 120 °C in an oven. After the mould was removed, they were thermally cured in a vacuum for 12 h at 175 °C. The curing agent Epikure W and EPON-862 (Diglycidyle Ether of Bisphenol-F) were mixed at a weight ratio of 100:26.4, and the curing cycle was 4 h at 120 °C followed by another 4 h at room temperature [7]. Based on their optical, thermal, and mechanical qualities, cycloaliphatic epoxy hybrids are excellent for LED encapsulation.

The epoxy EO1080 resin is strongly packed with silica-based filler and carbon additives for greater microwave absorption. The exact composition of the fillers is confidential, according to Johnston et al. [8], but measurements of dielectric characteristics at various temperatures can be used to compare with other systems. EO1080 was treated for 2 h at 110 °C or 20 min at 150 °C. The average net input power is 1.833W, and the microwave net energy input is 1200 J. The samples are heated at a constant pace to the chosen curing temperature (120, 150, or 180 °C) for varying periods of time. Thermal curing was accomplished using a thermal cycle that was quite similar to the microwave curing cycle. In the apparatus, the samples were heated to 30 °C and calibrated. The samples were then heated at a pace of 100 °C/min to the optimum curing temperatures, in order to reduce the amount of unrecorded heat that would accumulate before reaching the curing temperature. Results reveal that up to 70% conversion, microwave curing is faster than thermal curing. The findings indicate that the impact of microwave radiation on the curing process is more nuanced than just a rise in temperature.

Zhang et al. [9] conducted a study on a resin that was created by combining DGEBA-based epoxy (E-51 and E-20 from Wuxi resin plant, respectively, the epoxy equivalents of 196 and 500) with DAMI as a curing agent. The mass ratio of E-51/E-20/DAMI is 1:1:0.922. The liquid was heated to 130 °C for 2 min while constantly stirring, then poured into a 12 cm diameter cylindrical glass mould. The resin sample was placed in a cylindrical glass mould and placed in an oven with a pre-set temperature profile that was heated by air. A thermocouple was used to measure the temperature profiles within the epoxy casting part, which were then recorded on a computer. Shore D-type hardness treatment was used to determine the hardness value. Hardness is shown to be dependent on both the degree of cure and the thermal history throughout the curing process by comparing the estimated temperature profile and degree of cure with the Shore hardness of a number of interior sites.

Microwave equipment and Gurit's unidirectional (UD) out-of-autoclave (OoA) carbon fibre reinforced epoxy with PAN (polyacrylonitrile)-based carbon fibre were used. Four piles were laminated together to form 2.4-mm-thick laminates. In a vacuum table, two piles were first debulked for 30 min, and then the two halves were debulked for an additional hour. Although epoxy tape was used to cover the edges of the laminates during microwave curing to eliminate exposed carbon fibres, this was not essential during conventional curing. The utilization of a microwave frequency of 2.45 GHz has resulted in a highly homogenous microwave energy dispersion. In the available microwave frequencies, it provides an excellent balance of microwave heating and dielectric material penetration depth for industrial, scientific, and medical applications [10].

Chaowasakoo and Sombatsompop [11] evaluated a fly ash/epoxy composite containing 3-aminomethyl-3,5,5-trimethyl cyclohexyl-amine. Using a weight-to-weight ratio of 100:60, the resin and hardener were combined. The mixture was thoroughly agitated to prevent air bubbles from forming in the polymer. The liquid was then gradually poured into the mould and let to cure in a preheated oven. Using a differential scanning calorimeter, the optimal curing temperature and time were established (DSC). All of the composite samples were carefully put at the edge of the microwave oven's constantly moving glass plate to get a good pattern of how the microwave field spread out. The curing times were set to 18 min, and the DSC was used to determine the composites' degree of curing. Different microwave power levels were discovered to affect the curing properties of composite materials.

From Tanrattanakul and Jaroendee [12], a general-purpose grade of diglycidylether of bisphenol A (DGEBA) with an epoxy equivalent weight of 0.15 was employed as the epoxy resin. Hardeners included MTHPA (methyl tetrahydrophthalic anhydride) and MTHPA (methyl hexahydrophthalic anhydride) (MHHPA). As an accelerator, tris-2,4,6-dimethyl aminomethyl phenol was utilized (DMP-30). An 80:100 anhydride/epoxy ratio was utilized since the epoxy/anhydride stoichiometric ratio is around 80–90 wt per cent. After thorough mixing, air bubbles were forced out of the resin before it was poured into a mould. A Memmert U500 oven was used for thermal curing. In a Sanyo EM-X412 commercial microwave oven, microwave curing was done at a frequency of 2.45 Hz. The microwave oven has a rotator to prevent hot spots from forming due to uneven heating. The curing samples were quite large to avoid the non-uniformity of the microwave field in the microwave oven. These results demonstrate that multistep heating in a microwave oven is essential for curing glass-reinforced epoxy composites, and that microwave oven-curing results are comparable to, if not superior to, those obtained using a thermal oven.

As both the healing agent and the matrix component, an epoxy resin (diglycidyl ether of bisphenol A, 0.41–0.47 eq/100 g) was utilized Liu et al. [13]. Ethylenediamine was utilized as a curing agent for epoxy monomer to create shell material. Polyamide resin (diethylenetriamine condensate, 180–220 mg KOH/g) is the curing ingredient in the coating matrix. In the coating solution, emulsifiers Tween-80 and Arabic gum were used, with anhydrous alcohol as the solvent. The generated microcapsules were disseminated in a conventional epoxy matrix at room temperature to develop self-healing coatings. A stoichiometric quantity of polyamide hardener was used for curing. The homogenized slurry was then uniformly brushed onto the pre-treated carbon steel plates. After three days of air drying, solid coatings with an average thickness of roughly 150 m were produced. The effect of microcapsule concentration on coating self-healing behaviour was discovered for the optimal self-healing system. For comparison, a covering without microcapsules was made at 10%, 15%, 20%, 25%, 30%, and 35%.

Although many articles and journals have been published about the encapsulation process in the past few years, there is no specific article showing the contact angle and surface area of the contact test studied for this process. Furthermore, the selection between the self-cured epoxy resin and oven-cured epoxy resin is still in progress as the suitable epoxy resin that must be used is still unknown. The encapsulation process study is crucial as it helps the light to be dispersed evenly from the surface of the wafer substrate.

17.2 Experiment Procedures

Encapsulation generally involves injecting epoxy resin with hardener onto a wafer substrate. Oven-curing and self-curing epoxy resins exist. The oven-curing epoxy resin is baked at 150 °C for one hour. Before filling a syringe, epoxy resin and hardener are combined for one minute. The experiment uses needles 16G, 18G, 21G, 22G, and 23G. Single and double drops of solution are injected. This is to examine if the drop technique affects the contact angle and contact area when we require more volume. The contact angle and contact surface area test is performed by putting the substrate on the imaging platform. The platform has top, side, and led-lit cameras. Each sample's photo is kept. ImageJ is used to determine the samples' contact angles and surface areas. Figure 17.1 shows the camera-equipped platform.

The epoxy resin dispensing operation depends on silicone processing conditions to functionalize the resin in the LED assembly with minimal voids. Most shapes are hemispheres. It is preferable to preserve hemispheric shape and manufacturing pace. Methods establish hemispheric processing characteristics. This project uses three epoxy resins and hardeners. Before filling the syringe, the epoxy and hardener are combined for two minutes. Table 17.1 classifies epoxies by characteristics. This project uses different gauge needles for various outcomes. The encapsulation procedure uses five-gauge needles for comparison. Table 17.2 lists gauge numbers, outer/inner diameter, and wall thickness.

According to Yang et al. [14], a perfect injection droplet can be formed only when the needle impact velocity is between 0.15 m/s and 0.4 m/s. The syringe is placed about 2–2.5 cm above the wafer substrate during the droplet injection process. The diameter of one gauge needle will vary from that of another, the process of injecting epoxy resin will, of course, be carried out differently for each type of gauge needle. During each injection of epoxy into the substrate, the syringe is subjected to the same amount of force and pressure as was delivered initially.

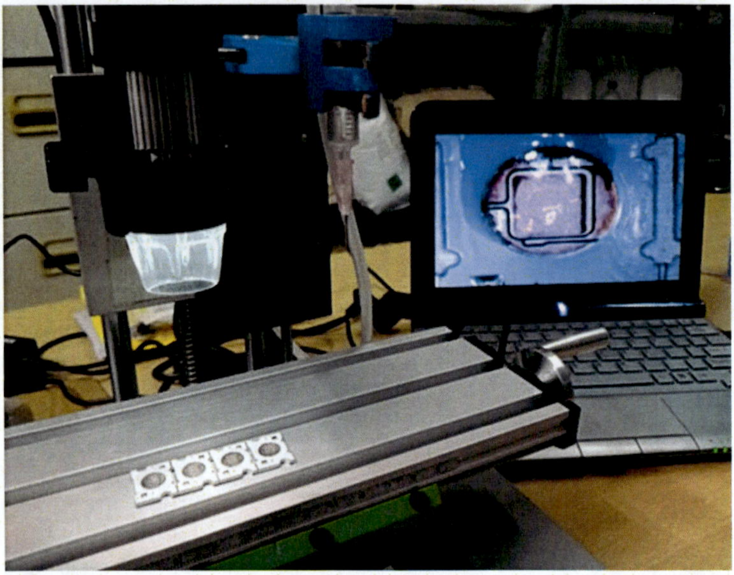

Fig. 17.1 Platform with top camera and image captured

Table 17.1 Material properties of EMC

Epoxy	Type	Ratio epoxy with hardener	Temperature of curing (°C)	Time to cure (h)
A	Oven curing	1:1	150	1
B	Self-curing	2:1	30	24
C	Self-curing	1:1	30	24

Table 17.2 Type of needle gauge and their diameters

Gauge number	Needle outer diameter (mm)	Needle internal diameter (mm)	Needle wall thickness (mm)	Needle dead volume (μL/25.4)
16 gauge needle	1.651	1.194	0.229	28.444
18 gauge needle	1.270	0.838	0.216	14.011
21 gauge needle	0.819	0.514	0.152	5.270
22 gauge needle	0.718	0.413	0.152	3.403
23 gauge needle	0.642	0.337	0.152	2.266

17.3 Result and Discussion

17.3.1 Contact Angle

Each needle is used to inject a single drop and double drop on the wafer substrate. Figure 17.2 shows the graph of the contact angle of 1:1 self-cured epoxy versus needle size. Figure 17.3 shows the graph of the contact angle of 2:1 self-cured epoxy versus needle size. Figure 17.4 shows the graph of the contact angle of 1:1 oven-cured epoxy versus needle size.

From the results in Figs. 17.2, 17.3, and 17.4, the contact angle of the epoxy increases with the needle size overall but not all of them. The size of the needle increasing indicates the diameter of the needle's nozzle decreases. When the diameter of the needle's needle is reduced, less epoxy will be dispensed from the needle onto the wafer substrate. This will result from the epoxy forming a hemisphere shape more likely on the substrate. On the contrary, the large diameter of the nozzle will dispense more epoxy onto the substrate and cause the epoxy to spread evenly on the substrate, reducing the contact angle. The contact angle provides an immediate indicator of the solid's wettability. The solid is said to have poor wetting and is called hydrophobic if the measured contact angle is greater than 90 degrees. The term hydrophilic is used when the contact angle is less than 90 degrees. So, from our results, all the contact angles measured fall under the hydrophilic, which is under 90 degrees.

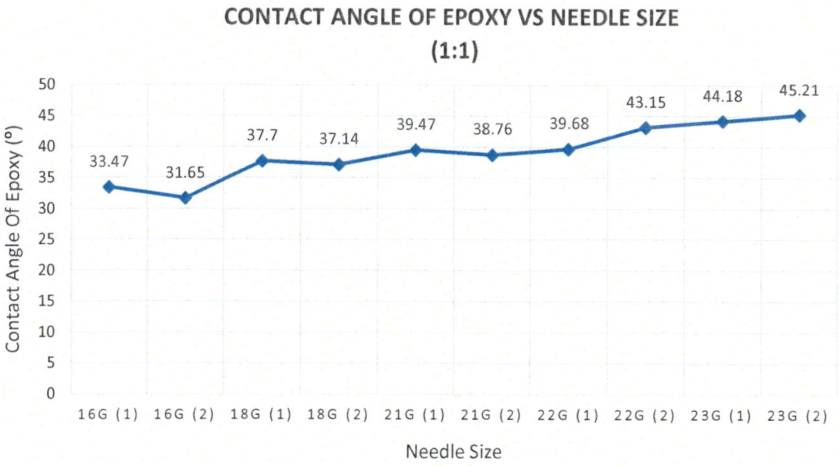

Fig. 17.2 Graph of contact angle of 1:1 self-cured epoxy versus needle size

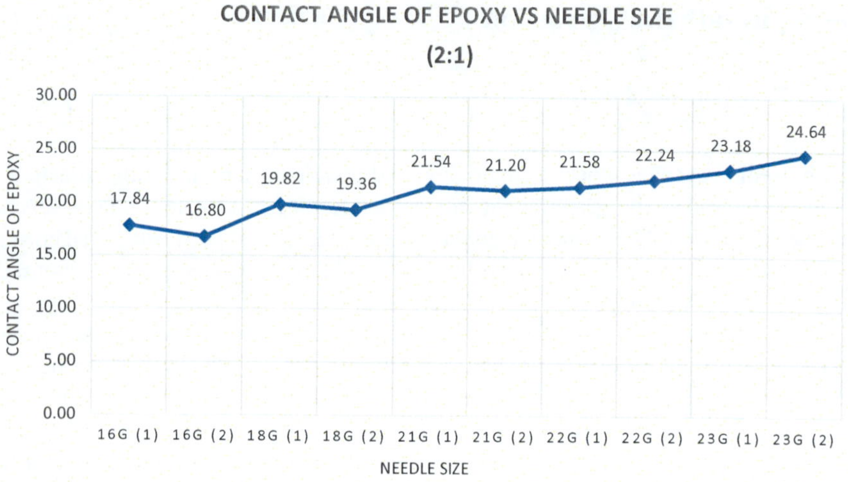

Fig. 17.3 Graph of contact angle of 2:1 self-cured epoxy versus needle size

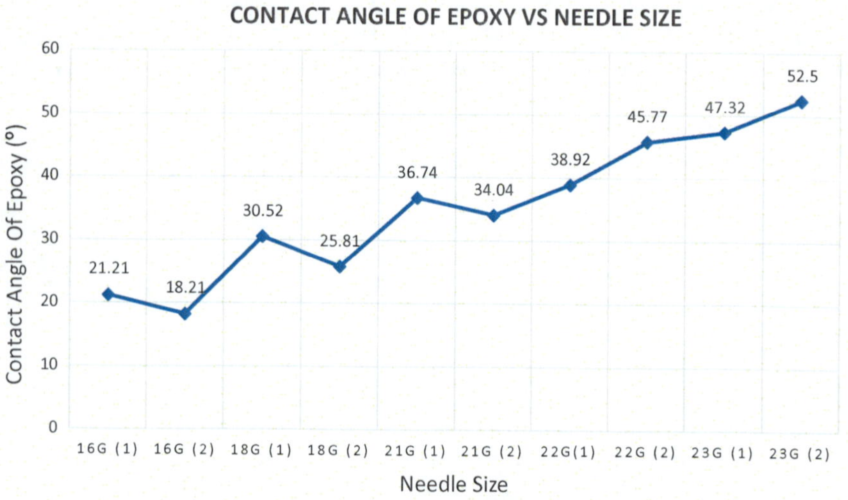

Fig. 17.4 Graph of the contact angle of 1:1 oven-cured epoxy versus needle size

17.3.2 The Surface Area of Contact

When calculating the surface area of contact, the needle gauge size is taken into consideration. On the wafer substrate, a single drop and a double drop are injected into it using each specific needle. Figure 17.5 shows the graph of the surface area of contact of 1:1 self-cured epoxy versus needle size. Figure 17.6 shows the graph of the surface area of contact of 2:1 self-cured epoxy versus needle size. Figure 17.7

shows the graph of the surface area of contact of 1:1 oven-cured epoxy versus needle size.

From the results in Figs. 17.5, 17.6, and 17.7, the surface area of contact of the epoxy is decreasing linearly as the size of the needle size increases as well. The size of the needle increasing indicates the diameter of the needle's nozzle decreases. When the diameter of the needle's nozzle is reduced, less epoxy will be dispensed onto the wafer substrate. This will result in the epoxy distributing lesser on the substrate.

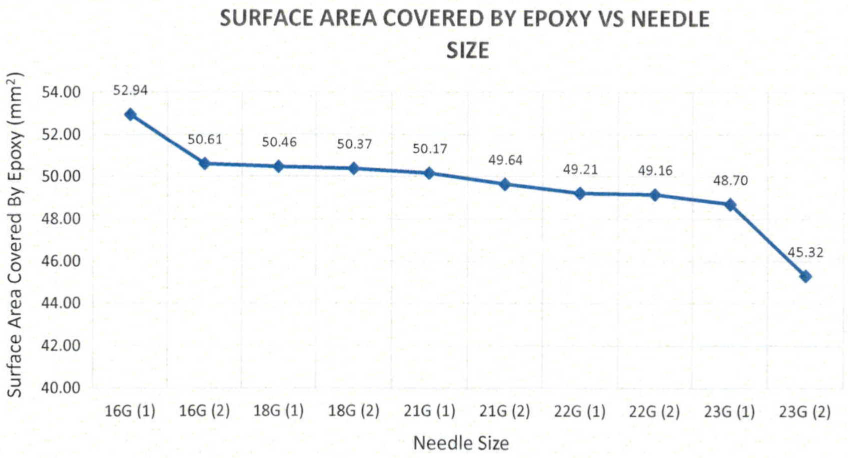

Fig. 17.5 Graph of the surface area of contact of 1:1 self-cured epoxy versus needle size

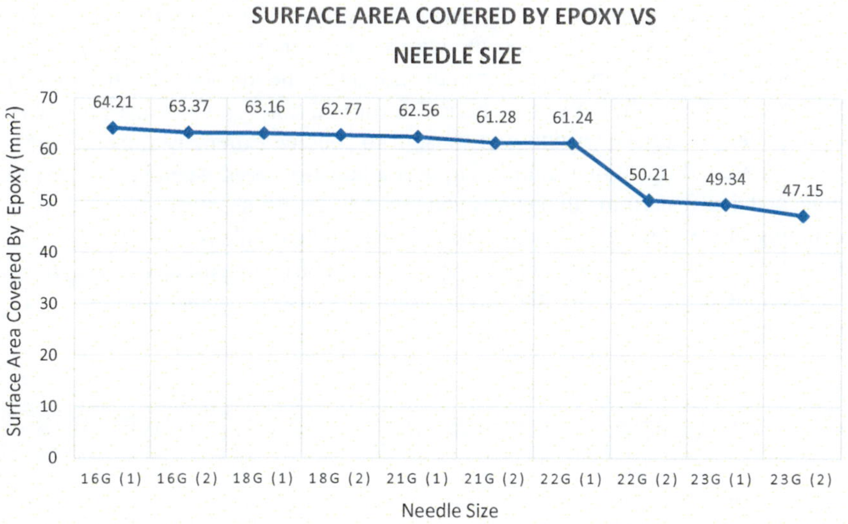

Fig. 17.6 Graph of the surface area of contact of 2:1 self-cured epoxy versus needle size

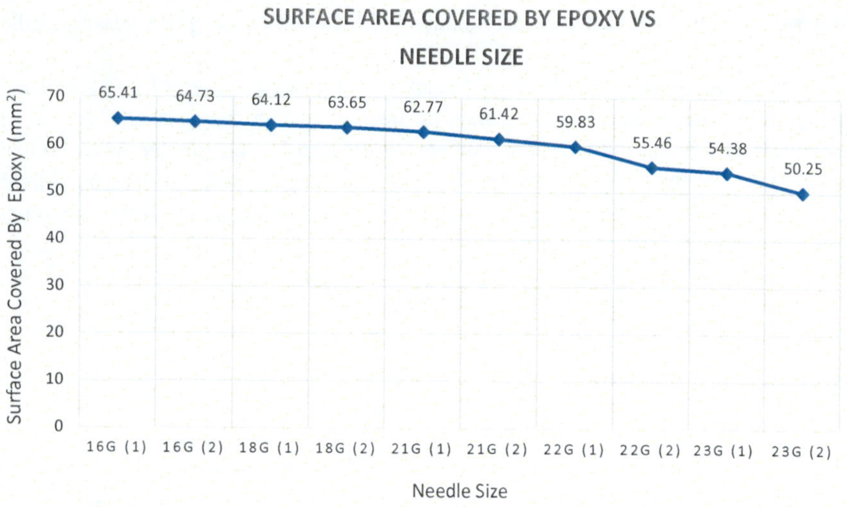

Fig. 17.7 Graph of the surface area of contact of 1:1 oven-cured epoxy versus needle size

On the contrary, the large diameter of the nozzle will dispense more epoxy onto the substrate and cause the epoxy to spread evenly on the substrate, covering a large surface area on it.

The dispensing process makes use of both the needle and the speed at which it moves. When the syringe is pressed, the motion may be separated into three stages: moving, oscillating, and the final or stop stage. Epoxy is dispensed by the needle's motion, reaches the substrate, and bridges the gap between the two. There will be thinning after that related to surface tension and gravity. When surface tension and gravity exceed the viscosity force, the epoxy thread splits down the middle. The top half of the thread retracts back into the nozzle, while the bottom half falls into the substrate. Soon after, a semi-spherical form will emerge. Once the procedure is complete, the surface tension will cause it to become extremely smooth, and it will stick around for further doling. There is a boost in speed when you are really going, and then it starts to fall down as you settle. When the needle is squeezed at the beginning of the moving stage, the driving force on the needle is significantly more than the resistance force formed by the epoxy and the friction pushed by the syringe and speeding down. However, the driving force diminishes together with the epoxy dispensing force.

17.4 Conclusion

The testing results show that, of the three epoxies tested, the 1:1 self-curing epoxy resin provides the best contact angles. The contact angles for this 1:1 self-curing epoxy resin range from 31° to 45°, which is the highest value among the three different types of epoxies. This 1:1 epoxy resin provides superior coverage on the wafer substrate as compared to other epoxies, which range in contact area from 45 to 53 mm^2. The contact angle range is 16° to 24° for the 2:1 self-curing epoxy resin. The amount of material that makes contact with the needle is directly related to its size. More epoxy will be sprayed onto the wafer substrate when the needle's nozzle is enlarged. The results of the trial indicate that a needle size of 21G is ideal. This needle size is ideal for producing reliable values for contact angle and contact area. Choosing the right needle size is critical because it determines the final shape of the wafer after it has been deposited on the substrate. Self-curing epoxy produces satisfactory results, although it takes nearly a day to cure entirely. It just requires an hour in the oven to cure, which is far faster than other methods. As a result of our work, the contact angle test and the contact surface area test are all that are required to choose the proper epoxy. Since the curing process for self-cured epoxy takes time, the technique of curing will be taken into account in the production line based on its requirement. Oven-curing is an efficient process since it can cure the substrates of multiple wafers in a relatively short amount of time. However, there are a wide variety of epoxy resins available, each with its own unique resin-to-hardener ratio. The epoxys intended function and application might also play a role.

Acknowledgements An acknowledgement to the Ministry of Higher Education Malaysia for the Fundamental Research Grant Scheme with the Project Code of FRGS/1/2021/TK0/USM/03/9. The authors would also like to express their gratitude to Universiti Sains Malaysia for extending their assistance in a technical capacity.

References

1. H. Roslan, M.S. Abdul Aziz, M.Z. Abdullah et al., Analysis of LED wire bonding during encapsulation process. IOP Conf. Series: Mater. Sci. Eng. **1007** (2020). https://doi.org/10.1088/1757-899X/1007/1/012173
2. M.A. Alim, M.Z. Abdullah, M.S.A. Aziz, R. Kamarudin, Die attachment, wire bonding, and encapsulation process in LED packaging: a review. Sens. Actuators A Phys. 329 (2021)
3. H. Xu, Y. Tang, J Wu et al., in *The Study on Cracking Reasons of LED Encapsulation Silicone*. 2019 20th International Conference on Electronic Packaging Technology, ICEPT 2019. Institute of Electrical and Electronics Engineers Inc (2019)
4. X. Shan, Y. Chen, Experimental and modeling study on viscosity of encapsulant for electronic packaging. Microelectron. Reliab. **80**, 42–46 (2018). https://doi.org/10.1016/j.microrel.2017.11.011
5. M. Sadeghinia, K.M.B. Jansen, L.J. Ernst, Characterization of the viscoelastic properties of an epoxy molding compound during cure. Microelectron. Reliab. **52**, 1711–1718 (2012). https://doi.org/10.1016/j.microrel.2012.03.025

6. S. Yang, J.S. Kim, J. Jin et al., Cycloaliphatic epoxy oligosiloxane-derived hybrid materials for a high-refractive index LED encapsulant. J. Appl. Polym. Sci. **122**, 2478–2485 (2011)
7. V.K. Rangari, M.S. Bhuyan, S. Jeelani, Microwave curing of CNFs/EPON-862 nanocomposites and their thermal and mechanical properties. Compos. A Appl. Sci. Manuf. **42**, 849–858 (2011). https://doi.org/10.1016/j.compositesa.2011.03.014
8. K. Johnston, S.K. Pavuluri, M.T. Leonard et al., Microwave and thermal curing of an epoxy resin for microelectronic applications. Thermochim. Acta **616**, 100–109 (2015). https://doi.org/10.1016/j.tca.2015.08.010
9. J. Zhang, Y.C. Xu, P. Huang, Effect of cure cycle on curing process and hardness for epoxy resin. Express Polym Lett **3**, 534–541 (2009). https://doi.org/10.3144/expresspolymlett.2009.67
10. M. Kwak, P. Robinson, A. Bismarck, R. Wise, Microwave curing of carbon-epoxy composites: penetration depth and material characterisation. Compos. A Appl. Sci. Manuf. **75**, 18–27 (2015). https://doi.org/10.1016/j.compositesa.2015.04.007
11. T. Chaowasakoo, N. Sombatsompop, Mechanical and morphological properties of fly ash/epoxy composites using conventional thermal and microwave curing methods. Compos. Sci. Technol. **67**, 2282–2291 (2007). https://doi.org/10.1016/j.compscitech.2007.01.016
12. V. Tanrattanakul, D. Jaroendee, Comparison between microwave and thermal curing of glass fiber-epoxy composites: effect of microwave-heating cycle on mechanical properties. J. Appl. Polym. Sci. **102**, 1059–1070 (2006)
13. X. Liu, H. Zhang, J. Wang et al., Preparation of epoxy microcapsule based self-healing coatings and their behavior. Surf. Coat. Technol. **206**, 4976–4980 (2012). https://doi.org/10.1016/j.surfcoat.2012.05.133
14. Y. Yang, S. Gu, Q. Lv et al., Influence of needle impact velocity on the jetting effect of a piezoelectric needle-collision jetting dispenser. AIP Adv. **9**. (2019) https://doi.org/10.1063/1.5086258 dispenser. Micromachines **9**. https://doi.org/10.3390/mi9070330

Chapter 19
Materials Modification of Lead-Free Solder Alloys with Different Reinforcing Components

M. A. Azmah Hanim and T. T. Dele-Afolabi

Abstract As a result of growing technological advancement, the increase in input and output terminals in electronic packaging has increased significantly. This signifies the commensurate rise in the number of solder joint interconnections. Hence, for the purpose of complying with the mission and vision of the electronic industry including the manufacturing of robust, efficient, and miniaturized devices, it is essential to incorporate reinforcing components into the existing traditional lead-free solders for the requisite materials modification. This short review article documents the progress made so far in the improvement of lead-free solders with selected reinforcement. The article covers the evaluation of the reliability level of solders reinforced with three targeted categories of materials such as carbon-based nanomaterials, interlayer components, and agricultural waste.

Keywords Lead-free solder · Reinforcement · Carbon-based nanomaterials · Interlayer components · Agricultural waste · Reliability

19.1 Introduction

The present trend towards miniaturization of electronic gadgets is motivated by the demand for lighter gadgets with advanced functions [1]. Miniaturization entails reducing solder joint size in electronic packages without necessarily affecting its reliability. While electronic devices are evolving towards miniaturized size and enhanced

M. A. Azmah Hanim (✉)
Department of Mechanical and Manufacturing Engineering, Faculty of Engineering, Universiti Putra Malaysia, 43400 Serdang, Selangor, Malaysia
e-mail: azmah@upm.edu.my

Advance Engineering Materials and Composites Research Center, (AEMC), Faculty of Engineering, Universiti Putra Malaysia, 43400 Serdang, Selangor, Malaysia

T. T. Dele-Afolabi
Department of Mechanical Engineering, Faculty of Engineering, Ajayi Crowther University, P.M.B. 1066, Oyo, Oyo State, Nigeria

M. A. A. Mohd Salleh et al. (eds.), *Proceedings of the Green Materials and Electronic Packaging Interconnect Technology Symposium*, Springer Proceedings in Physics 289,
https://doi.org/10.1007/978-981-19-9267-4_19

power, dense packaging technology also faces interconnection challenges at low and moderate temperatures [2]. Environmental considerations and international legislation against the use of lead-based solders with high toxicity levels have made lead-free solders as substitutes for Sn lead solder. Nevertheless, the immoderate evolution of intermetallic (IMC) at the interfacial bonding layer and the transition of temperate Cu_6Sn_5 into deleterious Cu_3Sn at elevated operating temperature pose serious concerns on the bonding reliability.

Recently, the introduction of reinforcing materials is becoming a suitable alternative to revamp commercially available lead-free solders. Xu et al. [3] noticed the ability of graphene nanosheets (GNSs) in suppressing interfacial IMC growth of Sn–Ag–Cu (SAC) solders by decreasing the diffusion coefficients to 0.6×10^{-12} m^2/s from 1.8×10^{-12} m^2/s for the composite solder, as compared to the monolithic SAC solder. Mayappan et al. [4] showcased the significant influence of carbon nanotubes (CNTs) in retarding the concurrent growth of Cu_6Sn_5 and Cu_3Sn IMCs and, as a result, enhancing the shear strength property of CNTs-reinforced Sn–3.5Ag solder. Liu et al. [5] noticed that the introduction of porous Cu into Sn–Bi–Ag solder refined the solder's microstructure and enhanced the solder joint strength integrity.

With a view to presenting recent developments in the production of lead-free solders, this article discusses three targeted categories of reinforcing materials as highly viable revamping materials to enhance the reliability of existing lead-free solders. The scope of discussion covers IMC layer evolution, mechanical strength, and artificial intelligence modelling to provide some suggestions for future reliability research on the design and selection of reinforcing materials for lead-free solder modification.

19.2 Reinforcing Components

This section presents the latest findings on the effect of three targeted reinforcing materials on the reliability of lead-free solders which includes carbon-based nanomaterials, interlayer components, and agricultural wastes.

19.2.1 Carbon-Based Nanomaterials

A study was conducted to evaluate reinforcement of multi-walled carbon nanotubes (MWCNTs) in Sn–5Sb solder from the perspective of IMC layer growth kinetics [6]. It was noticed that the addition of MWCNTs was able to suppress the diffusion paths pivotal to IMC layer growth due to the density disparity between MWCNTs and the solder. The reaction includes (i) immediate reaction between Cu and Sn atoms from the substrate and solder and (ii) reaction between Sn and Cu atoms from Cu_6Sn_5 floating IMC and solder matrix/IMC layer. The reduction in melting temperature of the Sn–5Sb solder with addition of MWCNTs reinforcement since the superior surface free energy capacity of the MWCNTs excited the surface imbalance in the Sn matrix [7].

The addition of MWCNTs also influences the Sn–5Sb solder shear strength by remarkably increasing in value for both the as-reflowed and isothermally aged solder joints [8]. This finding was substantiated by the increase in mechanical properties resulting from the IMC layer growth retardation actuated by MWCNTs' presence in the bulk solder. Nevertheless, the agglomeration effect associated with higher weight fractions of MWCNTs promoted the dysfunctionality and poor distribution of MWCNTs during solder preparation. Hence, the shear strength deterioration was observed at higher addition of the carbon nanotubes.

Despite the noticeable IMC layer thickness increase with increasing isothermal ageing temperature between 120 °C and 170 °C, the MWCNTs were highly effective in suppressing IMC growth in the composite solder (Fig. 19.1) [9, 10]. As expected, MWCNTs addition enhanced the shear strength and hardness properties in aged composite solders relative to its plain counterpart.

Artificial intelligence was deployed using artificial neural network (ANN) to develop a model for numerical investigation of MWCNTs-doped Sn–5Sb solder interconnects in terms of IMC formation and shear strength [11]. Based on the R^2 and the root mean square error (RMSE), the established ANN model accurately predicted the IMC thickness ($R^2 = 0.9913$; RMSE = 0.0234) and the shear strength ($R^2 = 0.9798$; RMSE = 0.0314) of the MWCNTs-reinforced Sn–5Sb interconnects.

Performing the soldering operation using laser as the heat source showed that the dominant IMCs including the Cu_6Sn_5 and Cu_3Sn appeared at the Sn–0.7Cu–xMWCNT/Cu substrate interface with the MWCNTs dispersing homogeneously in the bulk solder [12]. Similar to the remarkable reliability exhibited by the

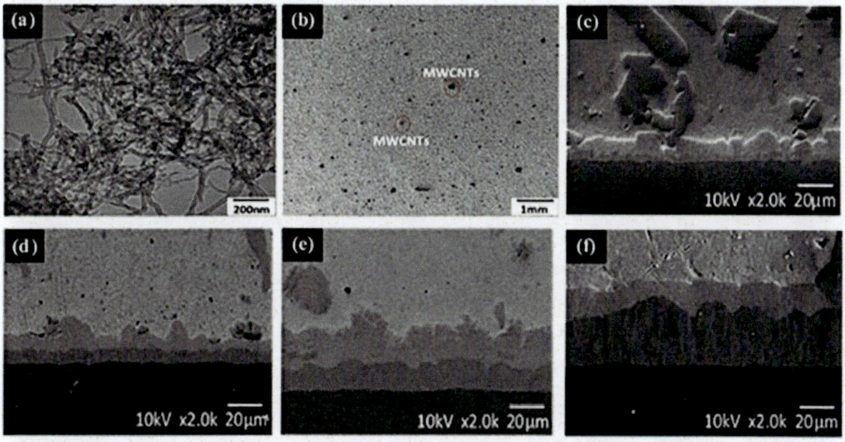

Fig. 19.1 **a** TEM microstructure of MWCNTs, **b** FESEM microstructure of the composite solder, **c** FESEM image of as-soldered Sn–5Sb–0.05CNT and FESEM images of Sn–5Sb–0.05CNT aged for 1500 h at **d** 120 °C, **e** 150 °C, and **f** 170 °C [10]

MWCNTs, the incorporation of the graphene nanosheets (GNS) into the Sn–1.0Ag–0.5Cu (SAC305) solder matrix promoted suppression of the IMC layer growth at the interfaces between SAC305–xGNS/Cu substrate (5.23–3.35 μm) and SAC305–xGNS/ENIAg surface finish (2.96–2.53 μm) [13]. Furthermore, the superior strength of the GNS was showcased in the strength enhancement demonstrated by both solder joint grades.

19.2.2 Interlayer Components

The IMC growth and the solder shear strength property in two types of MWCNTs-doped Sn–3Ag–0.5Cu composite solder loaded with porous Cu interlayers (PCI) (15, 25, and 50 pores per inch (PPI); see Fig. 19.2) were investigated [14]. Before performing the soldering process, the porous Cu components were rolled using a pasta machine to obtain a uniform thin layer in order to decrease the gap created in the resultant joint. The porous Cu components were more efficient in retarding the IMC growth at both solder/Cu and solder/porous Cu interfaces for SAC-0.01MWCNT solders compared to the SAC-0.04MWCNT counterpart.

Improved MWCNTs distribution was noticed in the SAC-0.01CNT-xPCI solder type compared to SAC-0.04CNT-xPCI type which resulted in retardation enhancement of the Sn atoms and subsequent dissolution of Cu atoms at the solder/Cu and solder/porous Cu interfaces to form the IMC layer. Furthermore, the PCI materials were highly effective in improving the shear strength of MWCNTs-reinforced SAC 305 solder joints, especially the SAC-0.01MWCNT.

Comparable investigation on the tensile strength of Sn-based solders reinforced with Ni foam, Ni foam coated with Cu, and Cu–Ni alloy foam was conducted [15]. Compared to the tensile strength of Sn solder, it shows that the tensile strength of the Ni foam/Sn, Cu coated Ni foam/Sn, and Cu–Ni alloy foam/Sn solder was enhanced by 161.1%, 200.2%, and 234.4%, respectively. The improved interfacial reaction in this experiment was highly critical in increasing the strength of Sn-based solder.

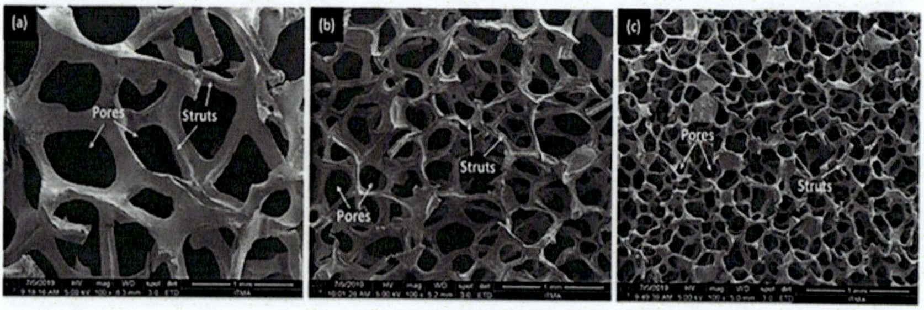

Fig. 19.2 FESEM microstructures of **a** 15 PPI, **b** 25 PPI, and **c** 50 PPI porous Cu interlayers [14]

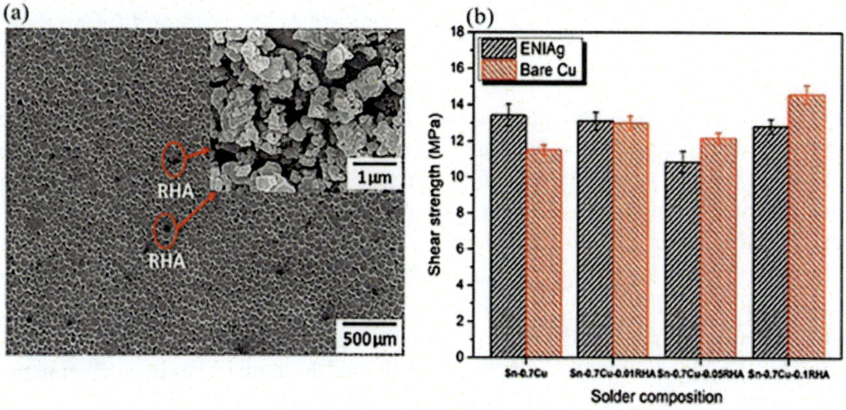

Fig. 19.3 a Microstructure of the compacted composite solder with RHA (inset) and **b** graph of the shear strength versus composite solder composition [16]

19.2.3 Agricultural Wastes

In order to reduce the cost of lead-free solder, low-cost reinforcing material was explored in enhancing the Sn–0.7Cu solder [16]. Rice husk ash (RHA) was chosen (Fig. 19.3a) as a reinforcing material emanating from the high SiO_2 content and matching low density as compared with other available ceramic materials. The microstructural analysis revealed the dominant IMCs observed at the Sn–0.7Cu–xRHA/Cu interface to be Cu_6Sn_5 and Cu_3Sn phases, whereas the $(Cu, Ni)_6Sn_5$ and Ni_3Sn_4 IMCs were formed at the Sn–0.7Cu–xRHA/Electroless Ni Immersion Ag (ENIAg) interface. The strengthening capacity of the RHA was evident, particularly in the Sn–0.7Cu–xRHA/Cu samples where the solder joint having 0.1 wt% RHA reinforcement exhibited the highest shear strength of 14.6 MPa across the board (Fig. 19.3b).

Even though the Sn–0.7Cu solder demonstrated a slightly superior melting temperature (233.62 °C) relative to the Sn–0.7Cu–0.1RHA counterpart (232.82 °C), the latter exhibited a superior liquidus temperature of 248.99 °C as compared with the 247.85 °C exhibited by the former [17]. This finding was attributed to the superior melting temperature and local dissolution of SiO_2 present in the rice husk ash in the liquid solder.

19.3 Conclusion

In the present review, three targeted reinforcing materials which are carbon-based nanomaterials, interlayer components, and agricultural wastes have been successfully utilized as viable materials in revamping existing lead-free solder alloys. Findings from the thermal analysis showed that the existing reflow soldering profiles are applicable in the preparation of composite lead-free solders since the melting temperature difference between them and their plain counterparts was insignificant. More so, the doped solders demonstrated superior mechanical properties when compared with the corresponding plain solders. Significant suppression of IMC layer growth was also noticed in the composite solders which confirms the effectiveness of the reinforcing materials in disrupting the diffusion paths of Cu and Sn atoms responsible for IMC layer growth. Even though the general findings in this review are admissible, to affirm the desired level of performance efficiency and to ensure compliance of the processing method with present industrial regulations and standards, future studies must focus on reducing to the barest minimum, agglomeration of nanoparticle reinforcements which promotes the dysfunctionality of these materials in enhancing the reliability of lead-free solder systems.

Acknowledgements The authors would like to acknowledge that the study covered in this short review was funded by Universiti Putra Malaysia (UPM—Grant Putra; UPM/700-2/1/GPBI/2017/9553600) and (UPM/GP-IPB/2020/9688700).

References

1. X.F. Zhang, H.Y. Liu, J.D. Guo, J.K. Shang, Inhibition of electromigration in eutectic SnBi solder interconnect by plastic prestraining. J. Mater. Sci. Technol. **27**(11), 1072–1076 (2011). https://doi.org/10.1016/S1005-0302(11)60188-6
2. X. Yin, C. Wu, Z. Zhang, W. Yang, C. Xie, X. Yang, Z. Huang, Highly reliable Cusingle bond Cu low temperature bonding using SAC305 solder with rGO interlayer. Microelectron. Reliab. **129**, 114483 (2022). https://doi.org/10.1016/j.microrel.2022.114483
3. L. Xu, L. Wang, H. Jing, X. Liu, J. Wei, Y. Han, Effects of graphene nanosheets on interfacial reaction of Sn–Ag–Cu solder joints. J. Alloys Comp. **650**, 475–481 (2015). https://doi.org/10.1016/j.jallcom.2015.08.018
4. R. Mayappan, A.A. Hassan, N.A. Ab Ghani, I. Yahya, and J. Andas, Improvement in intermetallic thickness and joint strength in carbon nanotube composite Sn–3.5Ag lead-free solder. Mater Today: Proc. **3**(6), 1338–1344 (2016). https://doi.org/10.1016/j.matpr.2016.04.012
5. Y. Liu, B. Ren, Y. Xue, M. Zhou, R. Cao, P. Chen, X. Zeng, Microstructure and mechanical behavior of SnBi-xAg and SnBi-xAg@P-Cu solder joints during isothermal aging. Microelectron. Reliab. **127**, 114388 (2021). https://doi.org/10.1016/j.microrel.2021.114388
6. T.T. Dele-Afolabi, M.A. Hanim, M. Norkhairunnisa, H.M. Yusoff, M.T. Suraya, Growth kinetics of intermetallic layer in lead-free Sn–5Sb solder reinforced with multi-walled carbon nanotubes. J. Mater Sci: Mater Electron. **26**(10), 8249–8259 (2015). https://doi.org/10.1007/s10854-015-3488-z
7. T.T. Dele-Afolabi, M.A. Hanim, M. Norkhairunnisa, M.T. Suraya, H.M. Yusoff, Influence of multi-walled carbon nanotubes on melting temperature and microstructural evolution of Pb-free Sn–5Sb/Cu solder joint. IOP Conference Series: Mater. Sci. Eng. **238**, 012010 (2017)
8. T.T. Dele-Afolabi, M.A. Hanim, M. Norkhairunnisa, H.M. Yusoff, M.T. Suraya, Investigating the effect of isothermal aging on the morphology and shear strength of Sn-5Sb solder reinforced with carbon nanotubes. Alloys Comp. **649**, 368–374 (2015). https://doi.org/10.1016/j.jallcom.2015.07.036
9. T.T. Dele-Afolabi, M.A. Hanim, O.J. Ojo-Kupoluyi, R. Calin, Impact of different isothermal aging conditions on the IMC layer growth and shear strength of MWCNT-reinforced Sn–5Sb

solder composites on Cu substrate. Alloys Comp. **808**, 151714 (2019). https://doi.org/10.1016/j.jallcom.2019.151714

10. T.T. Dele-Afolabi, M.A. Hanim, R. Calin, R.A. Ilyas, Microstructure evolution and hardness of MWCNT-reinforced Sn-5Sb/Cu composite solder joints under different thermal aging conditions. Microelectron. Reliab. **110**, 113681 (2020). https://doi.org/10.1016/j.microrel.2020.113681
11. T.T. Dele-Afolabi, M.A. Hanim, O.J. Ojo-Kupoluyi, D.W. Jung, A.A. Nuraini, A.A. Erameh, Interfacial IMC evolution and shear strength of MWCNTs-reinforced Sn–5Sb composite solder joints: experimental characterization and artificial neural network modelling. J. Mater. Res. Technol. **13**, 1020–1031 (2021). https://doi.org/10.1016/j.jmrt.2021.05.042
12. M.A. Hanim, M.Y.A. Syafiq, T.T. Dele-Afolabi, M.I.S. Ismail, Formation of intermetallic layer with multiwall carbon nanotubes reinforcement in Sn-0.7Cu solders on bare copper surface finish with laser soldering method. AIP Conf. Proc. **2506**, 060001 (2022). https://doi.org/10.1063/5.0083715
13. K. Vidyatharran, M.A. Hanim, T.T. Dele-Afolabi, K.A. Matori, O.S. Azlina, Microstructural and shear strength properties of GNSs-reinforced Sn-1.0Ag-0.5Cu (SAC105) composite solder interconnects on plain Cu and ENIAg surface finish. J. Mater. Res. Technol. **15**, 2497–2506 (2021). https://doi.org/10.1016/j.jmrt.2021.09.067
14. M.A. Hanim, A.B. Dasan, T.T. Dele-Afolabi, T. Ariga, K. Vidyatharran, Influence of porous Cu interlayer on the intermetallic compound layer and shear strength of MWCNT-reinforced SAC 305 composite solder joints. J. Mater Sci: Mater Electron. **32**, 4515–4528 (2021). https://doi.org/10.1007/s10854-020-05194-6
15. H. He, S. Huang, Y. Ye, Y. Xiao, Z. Zhang, M. Li, R. Goodall, Microstructure and mechanical properties of Cu joints soldered with a Sn-based composite solder, reinforced by metal foam. J. Alloys Comp. **845**, 156240 (2020). https://doi.org/10.1016/j.jallcom.2020.156240
16. M.A. Hanim, N.M. Kamil, C.K. Wei, T.T. Dele-Afolabi, O.S. Azlina, Microstructural and shear strength properties of RHA-reinforced Sn–0.7 Cu composite solder joints on bare Cu and ENIAg surface finish. J. Mater Sci: Mater Electron. **31**(11), 8316–8328 (2020). https://doi.org/10.1007/s10854-020-03367-x
17. M.A. Azmah Hanim, C.K. Wei, T.T. Dele-Afolabi, O.S. Azlina, Shear analysis of rice husk ash (RHA) reinforced tin-0.7-copper composite solders on electroless nickel/immersion silver (ENIAg) surfaces. Materialwiss. Werkstofftech. **52**(9), 943–951 (2021). https://doi.org/10.1002/mawe.202000247

Chapter 20
The Effect of Nickel Addition on Lead-Free Solder for High Power Module Devices—Short Review

C. M. Low and N. Saud

Abstract The issue of substituting high lead (Pb) solders in elevated temperature applications such as in high power module devices has been a major concern due to the potential toxicity of lead (Pb) to the environment and human health. A significant increase in health and environmental awareness has paved the way for the development of lead-free solder in the elevated temperature applications. The alloying element plays an important role in Pb-free solder as it relates to the reliability of bonding, as well as the mechanical, thermal and electrical properties of solder joint. The addition of nickel (Ni) tends to enhance the microstructure of the solder and the formation of intermetallic compounds, as well as the properties of the solder joints, yet the experimental data on the effect of the addition of nickel on the high temperature applications is not established and still open for discussion. In this paper, a review on recent research on the effect of nickel addition on the high temperature applications such as in high power module devices is studied.

Keywords Nickel addition · High power module devices · High temperature lead-free solder

20.1 Introduction

In the past few decades, tin–lead solder has been widely used in the elevated temperature applications, for example, high temperature electronic packaging applications, due to its excellent solderability, wettability, and electrical and mechanical properties. However, the use of lead in the electronic packaging industry is restricted

C. M. Low · N. Saud (✉)
Faculty of Chemical Engineering Technology, Universiti Malaysia Perlis (UniMAP), Kompleks Pusat Pengajian Jejawi 2, Kawasan Perindustrian Jejawi, 02600 Arau, Perlis, Malaysia
e-mail: norainiza@unimap.edu.my

N. Saud
Centre of Excellent On Geopolymer and Green Technology (CEGeoGTech), Universiti Malaysia Perlis, Kompleks Pusat Pengajian Jejawi 2, Taman Muhibbah, 02600 Arau, Perlis, Malaysia

M. A. A. Mohd Salleh et al. (eds.), *Proceedings of the Green Materials and Electronic Packaging Interconnect Technology Symposium*, Springer Proceedings in Physics 289,
https://doi.org/10.1007/978-981-19-9267-4_20

by the Waste Electrical and Electronic Equipment (WEEE) and the Restriction of Hazardous Substance (RoHS) directive due to health and environmental concerns [1]. Since the majority of industries are driven toward less usage of lead, lead-free solders have been introduced in order to eliminate the use of lead in electronics once and for all.

The introduction of lead-free solders contributes to the Waste Electrical and Electronic Equipment (WEEE) by reducing the health and environmental risks [2]. High temperature lead-free solder is a solder with a melting point in the range of 250–350 °C, which is used in high power module devices [3]. However, the high soldering temperature of these solders is such a limitation of the high temperature lead-free solder alternatives. This is because high soldering temperature tends to damage the other electrical components [4]. Transient liquid phase (TLP) bonding is a promising bonding approach for elevated temperature electronic packaging applications due to the fact that the transient liquid phase solders are capable of withstanding high temperatures, at the same time only lower bonding temperature is required during the soldering process. According to the principle of TLP bonding, the high melting point element and low melting point element engage in a bonding reaction at a lower bonding temperature, forming a solder joint with a high re-melting temperature [5, 6].

According to studies, the addition of alloying elements, for instance, bismuth (Bi), nickel (Ni), zinc (Zn), aluminum (Al), and indium (In), greatly improves the microstructure and mechanical properties due to their excellent properties, for instance, good thermal and mechanical behavior, as well as corrosion resistance [7–9]. The alloying element plays an important role in lead-free solder due to its effect to the microstructural evolutions and subsequently to the bonding reliability and properties in elevated temperature electronic packaging applications. Recent research has focused substantially on the feasibility of adding alloying elements to lead-free solder to develop more reliable lead-free solder joints, since alloying elements can significantly affect the microstructural stability, phase formation, and phase transformation of solder joints [9–11]. Nickel is selected as a promising alloying element for lead-free solder for high temperature applications because it has been shown to optimize the properties of the typical lead-free solders such as flow characteristic, and improve the mechanical properties such as hardness, strength, and elastic modulus [12, 13]. There are extensive research on the development of high temperature lead-free solder to replace lead solder in high temperature electronic packaging applications. This paper reviews recent research on the effect of nickel addition on typical lead-free solders as well as on high temperature lead-free solders, including transient liquid phase (TLP) solders in high temperature applications.

20.2 The Effect of Nickel Addition on Microstructure

Based on the previous literature, the addition of Ni to typical lead-free solders, which has been applied in high power module applications, tends to refine the microstructure and enhance the properties. The Ni element acts as a nucleation site, causing an acceleration of grain solidification and a reduction in the time required for the growth of intermetallic compounds. This refines the grains and inhibits the growth of intermetallic compounds [8]. In power module applications, the addition of Ni in the Sn–Ag–Cu lead-free solder tends to refine the grain size of the microstructure. With the addition of Ni in the Sn–Ag–Cu lead-free solder alloy, the crack propagation can be impeded by fine grain size due to the intragranular strength of fine grain size is relatively stronger than the coarse grain size, resulting in better solder crack resistance properties [14].

The addition of Ni to the eutectic Sn–3.5Ag solder has a great impact on the solidified microstructure of the solder matrix. When solidification occurs, the dendritic Sn phase is reduced, while the continuous Sn phase is formed with the addition of Ni. Moreover, with Ni addition, a uniform distribution of intermetallic compound occurs in the Sn-rich matrix. In other words, the addition of Ni to the eutectic Sn–3.5Ag refines the particle size of Sn, and the uniform distribution of Ag3Sn intermetallic compound is formed. This is because the high melting temperature of Ni element is considered as a hard inclusion, which implies that the Ni element acts as a refiner and facilitator of the intermetallic compound and a barrier to phase segregation. This provides better mechanical properties for the Sn–Ag solder joints due to the absence of dislocations and microvoids in the formation of intermetallic compound. The addition of 0.1 wt% Ni in the eutectic Sn–Ag lead-free solder promotes the formation of Ag3Sn because the atomic radius of Ni is smaller than Ag, which makes the Ni atoms diffuse faster. With the addition of 3 wt% Ni, the formation of Ag_3Sn is uniformly distributed, but small porosities occur, which may lead to a decrease in the mechanical properties of the solder joints [15].

According to Mu et al., in Sn–Cu lead-free solder, the number of eutectic microstructures increase with the addition of Ni. The Cu_6Sn_5 phases become more rounded and the volume fraction for Sn–dendrites decreases as the Ni content in Sn–Cu–Ni lead-free solder increases as shown in Fig. 20.1. The addition of Ni to Sn-Cu lead-free solder enables the refinement of Cu_6Sn_5 intermetallic compound, which improves the solder reliability. Furthermore, the Ni addition improves the microstructure by converting the hypoeutectic binary composition to a ternary eutectic microstructure. This enhances the fluidity of a Sn–Cu solder alloy [12]. In addition, the Ni content makes the microstructure of the lead-free solder more uniform and stable. In terms of the formation of intermetallic compound in the Sn–Cu lead-free solder, the addition of Ni stabilizes the elevated temperature hexagonal phase. The presence of cracks in the Cu_6Sn_5 layer between the solder alloy and the Cu substrate decreases with the addition of Ni to the Sn–Cu lead-free solder alloy [16].

Adding Ni to high temperature, lead-free solder gives the same trend to the results as typical lead-free solder, which is grain refinement and properties improvement. In high temperature lead-free solder of Sn–5Sb with 240 °C melting temperature, the number of NiSb phases increases and the phases become coarser as the Ni content in the solder increases from 0.05 wt% to 0.5 wt%. This is because the large amount of Ni content reduces the dislocation pinning effect, which develops an extensive grain growth [17]. Besides, the coarsening effect of Sn–Sb compounds formation in

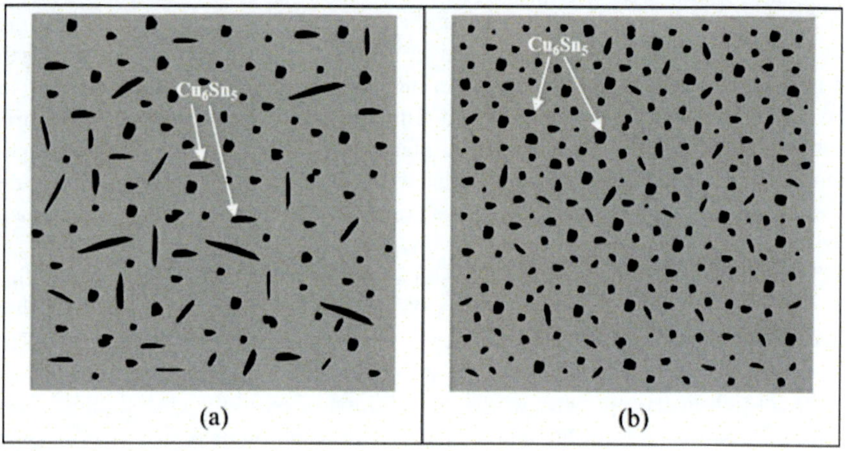

Fig. 20.1 Schematic diagram of microstructure **a** without Ni addition and **b** with Ni addition

high temperature Sn–10Sb lead-free solder is induced primarily by the increase of Ni content from 0.05 wt% to 0.5 wt%, leading to grain growth due to the reduced pinning effect and grain growth obstruction [18].

In Zn–Ni high temperature lead-free solder, which has a melting temperature of 420 °C, with increasing the addition of Ni, the grain size of Zn decreases due to the increased area fraction of intermetallic phase. The intermetallic phase contributes the pinning effect to the Zn grains; therefore, the grain size of Zn is reduced. According to the research conducted by Mallick and the researchers, the intermetallic compound is formed in Zn–Ni high temperature lead-free solder with the presence of Ni_2Zn_7 phase. The formation of Zn–Ni intermetallic compounds increases as the addition of Ni increases. Moreover, the size of the intermetallic phases increases steadily with the volume fraction of intermetallic particles with increasing Ni addition to the solder, resulting in an enhancement in properties of the solder joints [19].

In Sn–Ni high temperature lead-free solder with a melting temperature of 250 °C, the addition of Ni effectively promotes the Sn–Ni reaction and the formation of intermetallic compounds. The number of Ni_3Sn_4 intermetallic compound increases with the addition of Ni, indicating that the intermetallic particles are interconnected. The high re-melting temperature solder joint is formed due to the formation of intermetallic compound composed of Ni_3Sn_4 and residual Ni when the Ni addition is increased to 30 wt%. In the solder joint interlayer, the interval gradually becomes smaller and finally disappears with the increase of Ni addition. Furthermore, with the increase of Ni addition, the Sn–Ni high temperature lead-free solder possesses a good phase transformation resistance, which can prolong the lifetime of the high power module devices [20]. According to the Yoon et al., the Ni_3Sn_4 intermetallic compound and remaining fine Ni particles made up the majority of the bonded layer through TLP sintering at lower temperature and shorter bonding time, showing a very stable solder joint with no new phase formation in the TLP sintered joint during the additional reaction process [21].

The microstructures Zn–4Al high temperature lead-free solder with a melting temperature of 380 °C is composed of intermetallic phase, eutectoid lamellar phase, eutectic lamellar phase and proeutectic β-phase. The presence of Ni initiates the formation of intermetallic compounds. The formation of intermetallic compound in the microstructure increases as the Ni addition increases. In the other words, the addition of Ni is directly proportional to the formation of intermetallic compound. Furthermore, with the addition of Ni, the microstructure shows that the intermetallic compound phase coexists closely with the proeutectic β-phase, and the intermetallic compound in not uniformly distributed, which means that the distribution of intermetallic compounds is not homogeneous throughout the structure. However, the number of proeutectic β-phase reduces with further increasing Ni addition. Besides, as the Ni addition increases, the intermetallic phase changed from irregularly shaped particles to rod-like structures and distributed throughout the matrix [22].

20.3 The Effect of Nickel Addition on Mechanical and Physical Properties

The Sn–5Sb lead-free solder exhibits superior fatigue properties at the Ni content of 0.05 wt% and 0.1 wt%, which is due to the dislocation pinning effect suppressing the dynamic recrystallization, resulting in better fatigue resistance [17]. In Sn–10Sb lead-free solder, the tensile strength increases with the Ni addition increase. However, the tensile strength of Sn–10Sb lead-free solder decreases when the Ni addition exceeds 0.25 wt% due to the coarsening effect of the formation of Sn–Sb compounds. Furthermore, the fine Sn–Sb compounds with 0.05 wt% Ni addition in Sn–10Sb lead-free solder shows stable fatigue properties because the local recrystallization process is suppressed. According to the experimental results of Kobayashi and Shohji, Sn–10Sb high temperature lead-free solder with the addition of 0.10 wt% to 0.25 wt% Ni exhibited excellent mechanical properties, for instance, tensile strength, due to the role of Ni–Sb compound in the strengthening of the dispersion [18].

In Zn–Ni high temperature lead-free solder, with the addition of Ni, the formation of intermetallic compound plays a pinning role in the grain formation process. The increase in intermetallic particles leads to a decrease in the grain size of the Zn phase, which leads to an increase in mechanical properties, for example, hardness and tensile strength. In terms of thermal properties, Mallick et al. reported that the addition of Ni decreased the coefficient of thermal expansion, which may prevent the thermal stress damage to electronic components [19].

Regarding the mechanical properties of the Sn–Ni high temperature lead-free solder, the shear strength increases then decreases with the increase of Ni content. The Sn–Ni high temperature lead-free solder with 20 wt% Ni addition has a combination of ductile and transcrystalline fracture. The Sn–Ni high temperature lead-free solder with 10 wt% Ni exhibits transcrystalline fracture with a dense microstructure, showing improved mechanical properties such as shear strength. The failure mode of high temperature Sn–Ni lead-free solder transforms from ductile fracture to brittle fracture with the addition of 10–24 wt%, demonstrating a decline in shear strength. In addition, the increase of Ni addition in Sn–Ni high temperature lead-free solder results in poor wettability [20]. Furthermore, the residual Ni particles may hinder crack propagation inside the TLP sintering bonded joint, resulting in an improvement in shear strength based on the perspective of joint mechanical reliability [21].

According to Mallick et al., the alteration in microstructure and intermetallic compound significantly affects the mechanical properties of solder alloys. With the increase of Ni addition in Zn–4Al high temperature lead-free solder, the area fraction of intermetallic particles increases, resulting in an increase in hardness. However, the Ni addition exceeding 0.4 mass% tends to reduce the mechanical properties such as hardness and tensile strength of the solder alloy due to the occurrence of pores in the microstructure. Moreover, the coefficient of thermal expansion of the Zn–4Al high temperature lead-free solder decreases with the increase of Ni addition. In terms of melting characteristics, increasing the Ni addition leads to a minor increase in the melting temperature and melting range. Furthermore, as the Ni addition increases, the electrical resistivity also increases. This is probably due to the existence of more intermetallic compounds, a rise in lamellar phases and a drop in the β-phase in the ternary systems [22].

20.4 Conclusion

This study focuses on the effect of nickel (Ni) addition on lead-free solder for high power module devices. Previous studies have shown that the Ni addition affects the microstructure and the formation of intermetallic compound in lead-free solder, thereby improving the properties, stability and reliability of high temperature lead-free solder. Transient liquid phase (TLP) bonding is a good alternative bonding technique for high temperature electronic packaging application because lower bonding temperature is utilized during the soldering process. Therefore, further studies on the effect of nickel addition on high temperature lead-free solder such as transient liquid phase (TLP) solder are required to obtain established experimental data.

Acknowledgements This work was supported by Ministry of Higher Education Malaysia, under Fundamental Research Grant Scheme (FRGS/1/2020/TK0/UNIMAP/02/51) and Faculty of Engineering and Chemical Technology, Universiti Malaysia Perlis.

References

1. D.H. Jung, A. Sharma, M. Mayer, J.P. Jung, A review on recent advances in transient liquid phase (TLP) bonding for thermoelectric power module. Rev. Adv. Mater. Sci. **53**(2), 147–160 (2018). https://doi.org/10.1515/rams-2018-0011
2. C.-M. Huang, A. Raj, M. Osterman, M. Pecht, Assembly options and challenges for electronic products with lead-free exemption. IEEE Access **8**, 134194–134208 (2020). https://doi.org/10.1109/access.2020.3010771
3. M.M. Hasan, A. Sharif, M.A. Gafur, Characteristics of eutectic and near-eutectic Zn-Al Alloys as high-temperature lead-free solders. J. Mater. Sci.: Mater. Electron. **31**(2), 1691–1702 (2019). https://doi.org/10.1007/s10854-019-02687-x
4. S. Menon, E. George, M. Osterman, M. Pecht, High lead solder (over 85 %) solder in the electronics industry: RoHS exemptions and alternatives. J. Mater. Sci.: Mater. Electron. **26**(6), 4021–4030 (2015). https://doi.org/10.1007/s10854-015-2940-4
5. N. Saud, R.M. Said, Transient liquid phase bonding for solder—a short review. IOP Conf. Ser. Mater. Sci. Eng. **701**(1), 012050 (2019). https://doi.org/10.1088/1757-899x/701/1/012050
6. K.E. Aasmundtveit, T.-T. Luu, H.-V. Nguyen, A. Larsson, T.A. Tollefsen, Intermetallic bonding for high temperature microelectronics and microsystems: solid-liquid inter-diffusion bonding. Intermetallic Compd Formation Appl. (2018). https://doi.org/10.5772/intechopen.75139
7. J.H. Lau, N.-C. Lee, Assembly and reliability of lead-free solder joints (2020). https://doi.org/10.1007/978-981-15-3920-6
8. A.E. Hammad, Enhancing the ductility and mechanical behavior of Sn-1.0Ag-0.5cu lead-free solder by adding trace amount of elements Ni and SB. Microelectron. Reliab. **87**, 133–141 (2018). https://doi.org/10.1016/j.microrel.2018.06.015
9. C. Li, Y. Yan, T. Gao, G. Xu, The microstructure, thermal, and mechanical properties of Sn–3.0Ag–0.5Cu–xSb high-temperature lead-free solder. Materials **13**(19), 4443 (2020). https://doi.org/10.3390/ma13194443
10. S.-K. Seo, M. G. Cho, S. K. Kang, J. Chang, H. M. Lee, in *Minor Alloying Effects of Ni or Zn on Microstructure and Microhardness of Pb-Free Solders*. 2011 IEEE 61st Electronic Components and Technology Conference (ECTC) (2011). https://doi.org/10.1109/ectc.2011.5898496
11. G. Zeng, S.D. McDonald, Q. Gu, Y. Terada, K. Uesugi, H. Yasuda, K. Nogita, The influence of Ni and Zn additions on microstructure and phase transformations in Sn–0.7Cu/Cu solder joints. Acta Mater. **83**, 357–371 (2015). https://doi.org/10.1016/j.actamat.2014.10.003
12. D.K. Mu, S.D. McDonald, J. Read, H. Huang, K. Nogita, Critical properties of Cu_6Sn_5 in electronic devices: recent progress and a review. Curr. Opin. Solid State Mater. Sci. **20**(2), 55–76 (2016). https://doi.org/10.1016/j.cossms.2015.08.001
13. T.-T. Chou, R.-W. Song, W.-Y. Chen, J.-G. Duh, Enhancement of the mechanical strength of Sn-3.0Ag-0.5Cu/Ni joints via doping minor ni into solder alloy. Mater. Lett. **235**, 180–183 (2019). https://doi.org/10.1016/j.matlet.2018.10.050
14. S.-W. Guo, Y.-W. Huang, H.-K. Liao, in *High Reliability Lead Free Solder Evaluations in Power Module Application*. 2017 12th International Microsystems, Packaging, Assembly and Circuits Technology Conference (IMPACT) (2017). https://doi.org/10.1109/impact.2017.8255940
15. M.S. Gumaan, R.M. Shalaby, M.K. Mohammed Yousef, E.A.M. Ali, E.E. Abdel-Hady, Nickel effects on the structural and some physical properties of the eutectic Sn-Ag lead-free solder alloy. Soldering Surf. Mount Technol. **31**(1), 40–51 (2019). https://doi.org/10.1108/ssmt-03-2018-0009
16. M.A. Mohd Salleh, I.G. Sandu, M.M. Abdullah, I. Sandu, N.A. Saleh, Nickel (Ni) microalloying additions in Sn-Cu lead-free solder. Short review. IOP Conf. Ser. Mater. Sci. Eng. **209**, 012084 (2017). https://doi.org/10.1088/1757-899x/209/1/012084
17. T. Kobayashi, K. Mitsui, I. Shohji, Effects of Ni addition to Sn–5Sb high-temperature lead-free solder on its microstructure and mechanical properties. Mater. Trans. **60**(6), 888–894 (2019). https://doi.org/10.2320/matertrans.mh201809

18. T. Kobayashi, I. Shohji, Evaluation of microstructures and mechanical properties of Sn-10Sb-Ni lead-free solder alloys with small amount of Ni using miniature size specimens. Metals **9**(12), 1348 (2019). https://doi.org/10.3390/met9121348
19. S. Mallick, M.S. Kabir, A. Sharif, Study on the properties of Zn–xNi high temperature solder alloys. J. Mater. Sci.: Mater. Electron. **27**(4), 3608–3618 (2015). https://doi.org/10.1007/s10854-015-4198-2
20. H. Ji, M. Li, S. Ma, M. Li, Ni_3Sn_4-composed die bonded interface rapidly formed by ultrasonic-assisted soldering of Sn/Ni solder paste for high-temperature power device packaging. Mater. Des. **108**, 590–596 (2016). https://doi.org/10.1016/j.matdes.2016.07.027
21. J.-W. Yoon, Y.-S. Kim, S.-E. Jeong, Nickel–tin transient liquid phase sintering with high bonding strength for high-temperature power applications. J. Mater. Sci.: Mater. Electron. **30**(22), 20205–20212 (2019). https://doi.org/10.1007/s10854-019-02404-8
22. S. Mallick, M.S. Kabir, A. Sharif, Development of Zn-Al-xNi lead-free solders for high-temperature applications. Harsh Environ. Electron. 115–133 (2019). https://doi.org/10.1002/9783527813964.ch5

Chapter 21
Thermal Analysis Simulation Between Hand Soldering and Laser Soldering Process

Logendran Murgaya, Noor Izza Farisya Noor Hamdan, Iman Nur Sazniza Johari, Dayang Izzah Nabilah Awang Azman, and Saliza Azlina Osman

Abstract This study investigates the thermal analysis simulation between hand and laser soldering processes using SolidWorks simulation software. The three-dimensional model design of PCB with components was developed through SolidWorks modelling software, and thermal distribution analysis was conducted by convection. The thermal analysis result shows that the minimum and maximum temperatures recorded for hand soldering were 2.729e+02 °C at node 387 and 2.731e+02 °C at node 25,911, respectively. Meanwhile, for laser soldering are 2.686e+02 °C at node 485 and 2.731e+02 °C at node 22,751. The temperature recorded for hand soldering is higher than laser soldering which is between 269 °C to 272 °C and 230 °C to 263 °C, respectively. In addition, laser soldering reveals 30 times faster than the hand soldering process and shows a minimal thermal effect on PCB component.

Keywords Soldering · Printed circuit board · Thermal analysis · Simulation

21.1 Introduction

With the trends of miniaturization, the lightweight and multi-functions of electronic devices are growing substantially and the size of solder joints is decreasing significantly [1]. Solder joints formed during soldering process are utilized in microelectronics in order to provide electrical, thermal and mechanical continuities in electronic devices and assemblies. The devices and circuit performance are dependent on the strength and reliability of solder joints [2]. Recently, laser soldering has been introduced into the electronic packaging industry and received more attention. Compared with the conventional reflow soldering and hand soldering processes, laser

L. Murgaya · N. I. F. Noor Hamdan · I. N. S. Johari · D. I. N. Awang Azman · S. A. Osman (✉)
Faculty of Mechanical and Manufacturing Engineering, Universiti Tun Hussein Onn Malaysia, Parit Raja, 86400 Johor, Malaysia
e-mail: saliza@uthm.edu.my

M. A. A. Mohd Salleh et al. (eds.), *Proceedings of the Green Materials and Electronic Packaging Interconnect Technology Symposium*, Springer Proceedings in Physics 289,
https://doi.org/10.1007/978-981-19-9267-4_21

soldering reveals obvious advantages for solder bumping in a printed circuit board (PCB), due to its unique properties such as localized and non-contact heating, short processing time including soldering time, fast heating and cooling, good controllability, and high adaptability for various material systems, ease of automation, highly energy efficient and versatile for heat-sensitive materials [3–7]. Because laser soldering is relatively short, the laser process parameters will have a big influence on the soldering process, thus maybe can affecting the soldering quality [8]. Besides that, laser conditions may affect by the circuit board's thermal conductivity. However, by selecting the appropriate laser condition, this problem maybe can be reduced [9]. In contrast, Nishikawa and Iwata [5] stated that intermetallic compound formed by laser soldering process has superior impact reliability than other soldering techniques. However, hand soldering is still considered a reliable soldering process when components need to rework, touch up or replace but some precaution is required. The challenges in reworking lead-free soldering assemblies include spacing between BGA/CSP and lead-frame components. This spacing is necessary to avoid localized secondary reflow of adjacent components during rework operations. Besides that, excessive temperatures and contact times with the soldering tip need to avoid because components cannot tolerate fast ramp rates and are prone to crack or delamination issues [10].

Numerical simulation technology and its analysis results are necessary for a better understanding of relevant researchers and engineers in the electronic industry especially in electronic components for laser soldering [1, 4, 11]. By using simulation, the failure of solder joint can be predicted [12]. In thermal distribution analysis, the heat loss is applied using the node-value information of the model to determine the temperature distribution [13]. A node is a division of one block shaped into several small blocks depending on the shape of the node, where the mathematical modelling will be completed on each small block when performing computational analysis. For example, a solid element is made up of a four-corner rectangular shape. The node is located at the corner of each shape. Although monitoring the localized temperature distribution during a short processing time for laser soldering is generally difficult, many simulation-based methods for predicting the localized temperature distribution have been proposed [1, 4, 12, 13]. Therefore, this study aims to investigate the thermal distribution analysis between hand soldering and laser soldering processes by using SolidWorks simulation software.

21.2 Methodology

The printed circuit board (PCB) was designed with 1.5 mm in thickness. The length and width of the board are 50 mm and 26 mm, respectively. Therefore, the maximum component height on PCB is a battery capacitor with a 7 mm and 8 mm radius. After that, the resistor had a maximum length of 14 mm. At the same time, the resistor chip had 10 mm in length. In addition, the solder leg joint of the components was lying through the PCB with a height of 5.5 mm except for the resistor which had only 5 mm in height. Then, the PCB and its parts must be meshed using solid mesh. The shell

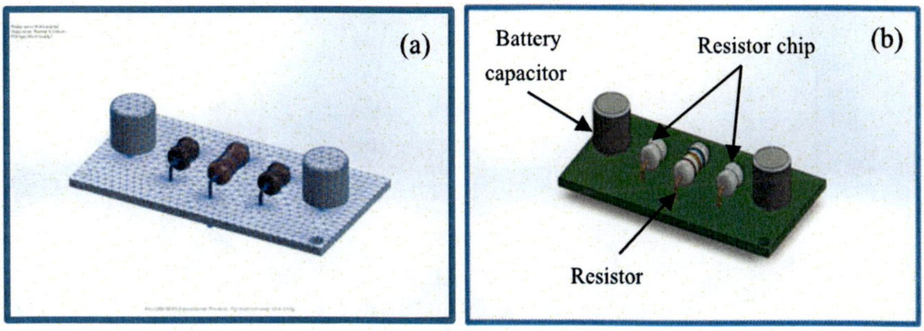

Fig. 21.1 **a** Mesh quality plot of PCB design. **b** 3D model of PCB

Table 21.1 Materials properties setting of PCB

Component	Material	Thermal conductivity W/(m K)	Specific heat J/(kg K)	Mass density kg/m^3
Printed circuit board	FR4	0.29	396	1900
Resistor chip	Copper	390	390	8900
Resistor	Nickel chromium	13	461	8400
Battery capacitor	Aluminium	30	850	3960

mesh is the most useful for thin-walled parts because it provides the best balance between analysis accuracy and computation time as shown in Fig. 21.1 while the materials' properties setting of PCB has been set as represented in Table 21.1. After meshing was successful, the solder node's location (Fig. 21.2) was discovered to start heating the components layer on PCB with two different types of temperatures and soldering times which is 350 °C (30 s) and 360 °C (1 s) for hand soldering and laser soldering during thermal steady-state analysis, respectively. Finally, the thermal simulation was run and recorded.

21.3 Result and Discussions

Hand soldering has almost become an obsolete technique since more precise and robust processes have been created. For hand soldering thermal distribution, the temperature setting was 350 °C to start heating each component on PCB using a soldering tip, as represented in Fig. 21.3. The hand soldering simulation result shows that the minimum and maximum temperatures recorded were 2.729e+02 °C at node 387 and 2.731e+02 °C at node 25,911, respectively. For laser soldering, the temperature setting is 360 °C and the thermal distribution was illustrated in Fig. 21.4. Hence, the minimum and maximum temperatures observed are 2.686e+02 °C at node 485 and

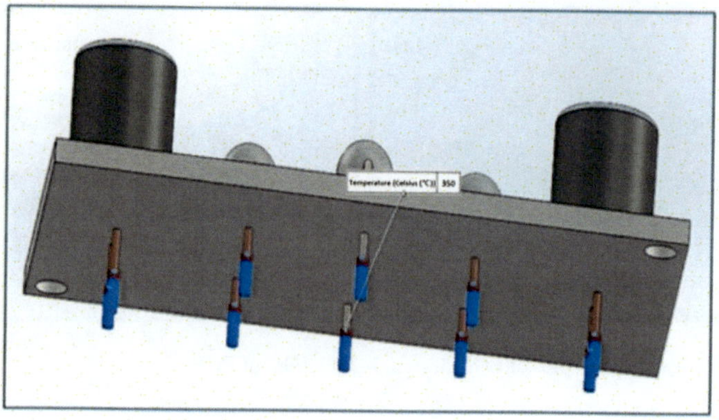

Fig. 21.2 Solder nodes location for hand soldering

2.731e+02 °C at node 22,751, respectively. The result also reveals that the component resistor chips and resistor were exposed to thermal changes during hand soldering process compared to laser soldering. Generally, the melting temperature of lead-free solder (about 220 °C) will melt at a higher temperature than lead-based solder (about 180 °C). The PCB and components should be able to withstand the soldering temperature to avoid damaging components. Thus, the soldering iron needs to remove quickly after the soldering process is finished because the lead-free soldering process requires higher wattage usage to ensure the near components are not affected or less heat exposure.

Name	Type	Min (°C)	Max (°C)
Thermal 1	Temperature at 30 Seconds	2.729e+02 Node: 387	2.731e+02 Node: 25911

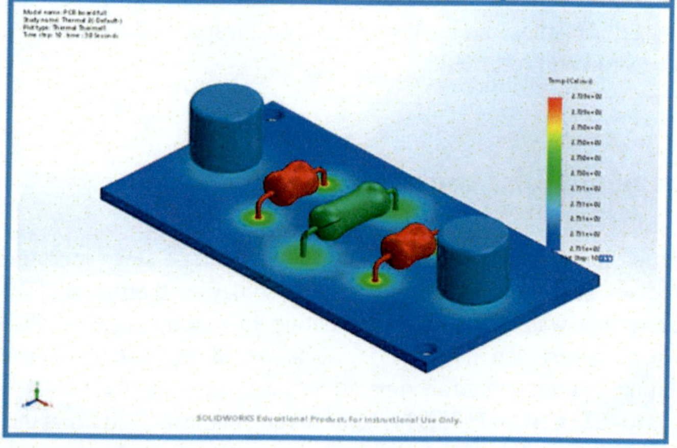

Fig. 21.3 Hand soldering thermal distribution

Name	Type	Min (°C)	Max (°C)
Thermal 2	Temperature at 1 Second	2.686e+02 Node: 485	2.731e+02 Node: 22751

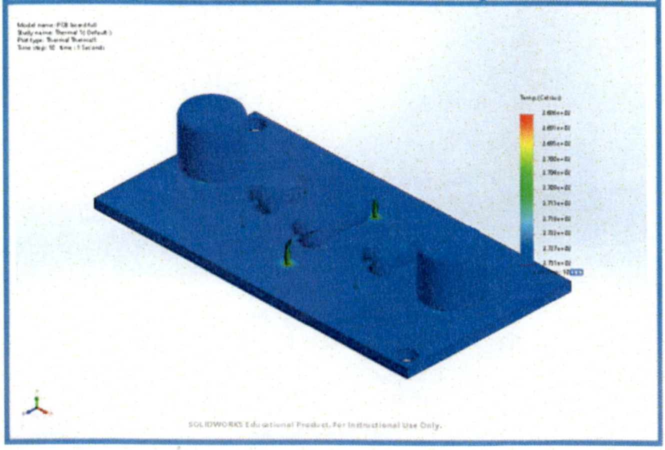

Fig. 21.4 Laser soldering thermal distribution

The thermal distribution of PCB in both hand and laser soldering as shown in Fig. 21.5 is different due to its various temperature range and time. Furthermore, time played a significant role in both soldering methods. Hand soldering process took 3 s to solder the component on PCB while laser soldering takes less than 0.1 s, which is 30 times faster than hand soldering. Besides that, both graphs show the highest temperature values between node #1964 and node #1804 directly at the middle point of the PCB. The components on the middle part of PCB are assigned as copper materials. In terms of nodes temperature, hand soldering was travelled between 269 and 272 °C which is a higher temperature than laser soldering where the temperature of nodes between 230 and 263 °C. However, it was validated that the longest time is taken for soldering, the temperature would be increased on the selected materials gradually.

21.4 Conclusion

In this study, the thermal analysis simulation between hand and laser soldering processes using SolidWorks was investigated. The thermal analysis result shows that the minimum and maximum temperatures recorded for hand soldering were 2.729e+02 °C at node 387 and 2.731e+02 °C at node 25,911, respectively. Meanwhile, for laser soldering are 2.686e+02 °C at node 485 and 2.731e+02 °C at node 22,751. The temperature recorded for hand soldering is higher than laser soldering which is between 269 °C to 272 °C and 230 °C to 263 °C, respectively. In addition,

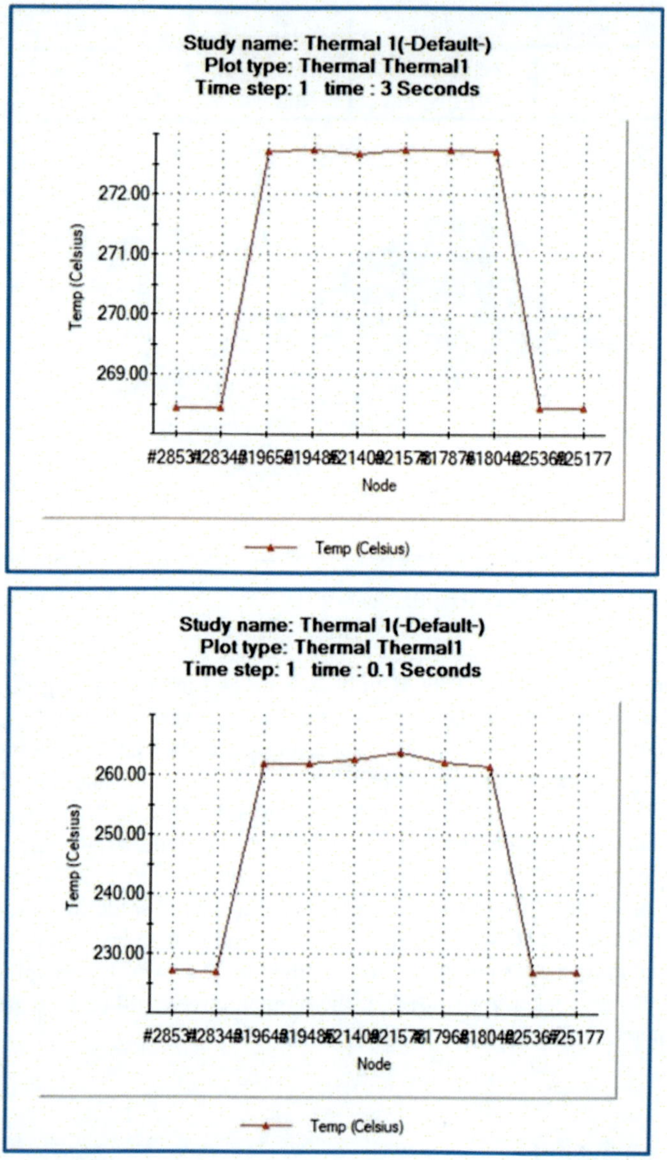

Fig. 21.5 Laser soldering thermal distribution

laser soldering reveals 30 times faster than the hand soldering process and show a minimal thermal effect on PCB component.

Acknowledgements This project was funded by the Ministry of Higher Education (MOHE) through Fundamental Research Grant Scheme (FRGS/1/2019/TK03/UTHM/02/6) and facilities provided by Faculty of Mechanical and Manufacturing Engineering, Universiti Tun Hussein Onn Malaysia.

References

1. Z. Yang, L. Li, W. Chen, X. Jiang, Y. Liu, Numerical and experimental study on laser soldering process of SnAgCu lead-free solder. Mater. Chem. Phys. **273**, 125046 (2021)
2. A. Kunwar, L. An, J. Liu, S. Shang, P. Råback, H. Ma, X. Song, A data-driven framework to predict the morphology of interfacial Cu6Sn5 IMC in SAC/Cu system during laser soldering. J. Mater. Sci. Technol. **50**, 115–127 (2020)
3. B. Liu, Y. Tian, W. Liu, W. Wu, C. Wang, TEM observation of interfacial compounds of SnAgCu/ENIG solder bump after laser soldering and subsequent hot air reflows. Mater. Lett. **163**, 254–257 (2016)
4. H. Tatsumi, S. Kaneshita, Y. Kida, Y. Sato, M. Tsukamoto, H. Nishikawa, Highly efficient soldering of Sn–Ag–Cu solder joints using blue laser. Journal of Manufacturing Processes **82**, 700–707 (2022)
5. H. Nishikawaa, N. Iwata, Formation and growth of intermetallic compound layers at the interface during laser soldering using Sn–Ag Cu solder on a Cu Pad. J. Mater. Process. Technol. **215**, 6–11 (2015)
6. S. Zhao, Z. Tan, H. Wang, M. Gao, Effects of spreading behaviors on dynamic reflectivity in laser soldering. Optics & Laser Technology **155**, 108404 (2022)
7. L. An, H. Ma, L. Qu, J. Wang, J. Liu, M. Huang. The effect of laser-soldering parameters on the Sn-Ag-Cu/Cu interfacial reaction, in The Institute of Electrical and Electronics Engineers (IEEE): 2013 14th International Conference on Electronic Packaging Technology (2013)
8. T. Li, K. Pan, Z. Tan, Y. Cai, T. Teng, Y. Li, Simulation and experimental study of the temperature field of solder ball in the nozzle during laser jet solder ball bonding process, in The Institute of Electrical and Electronics Engineers (IEEE): 2020 21st International Conference on Electronic Packaging Technology (ICEPT) (2020)
9. D. Imai, R. Kibushi, T. Hatakeyama, S. Nakagawa, M. Ishizuka, The effect of thermal conductivity of board to laser condition in laser soldering, in The Institute of Electrical and Electronics Engineers (IEEE): 2014 International Conference on Electronics Packaging (ICEP) (2014)
10. V. Eveloy, S. Ganesan, Y. Fukud, J. Wu, M.G. Pecht, WEEE, RoHS, and what you must do to get ready for lead-free electronics. IEEE Transactions on Components And Packaging Technologies **28**(4), 884–894 (2005); X. Zhang, R. Ma, J. Liu, W. Wu, Morphology and electrical properties of polypropylene/polyamide 6/glass fiber composites with low carbon black loading. J. Polym. Eng. **39**(9), 813–821 (2019). https://doi.org/10.1515/polyeng-2019-0024
11. F.C. Ng, M.A. Abas, Underfill flow in flip-chip encapsulation process: a review. Journal of Electronic Packaging **144**(1), 010803 (1–19) (2022)
12. S.A. Zahiri, M.A. Abas, F.C. Ng, M.F.M. Sharif, F.C. Ani, Numerical simulation of laser soldering process on pin through hole (PTH). Journal of Advanced Research in Fluid Mechanics and Thermal Sciences **98**(1), 137–145 (2022)
13. K. Hollstein, X. Yang, K. Weide-Zaage, Thermal analysis of the design parameters of a QFN package soldered on a PCB using a simulation approach. Microelectron. Reliab. **120**, 114118 (2021)

Chapter 22
Characterization of Al–Mg Alloy by Powder Metallurgy Technique

Zuraidawani Che Daud, Nur Majidah Mohd Asri, and Mohd Nazree Derman

Abstract The powder metallurgy Al has been widely used in the heavy industry, especially in precision technology. Unfortunately, these new materials are problematic in powder metallurgy production and corrosion problems. This research paper aims to study the influence of Mg contents (10, 25, 50, 75, and 90) wt% on microstructure and corrosion behavior on Al–Mg alloy by using powder metallurgy techniques. Al–Mg powder was mixed using a rotation mill with a rotation speed of 120 rpm for 30 min. Then, the mixed powders were compacted at a pressure of 150 MPa. Sintering was done in an argon-controlled atmosphere at a temperature of 500 °C. An optical microscope was used to observe the microstructure of sintered sample; meanwhile, X-ray diffraction (XRD) was used to analyze phase identification. A potentiostat was used to study the corrosion behavior of sintered Al–Mg alloy. The results revealed that Al–90 wt% Mg gives a high corrosion rate.

Keywords Powder metallurgy · Alloy

22.1 Introduction

Precision technology structures such as engine component, screw sealing, and oil seal are made from powder metallurgy aluminum–magnesium because of their lightweight and higher mechanical properties. Powder metallurgy Aluminum–Magnesium alloy (PM Al–Mg), which is composed of aluminum and magnesium, is a kind of lightweight alloy material. It has the advantage of both magnesium and aluminum and has a good application prospect in the special field such as aerospace.

Z. Che Daud (✉) · M. N. Derman
Faculty of Mechanical Engineering Technology, Universiti Malaysia Perlis, Kampus Alam UniMAP, Pauh Putra, 02600 Arau, Perlis, Malaysia
e-mail: zuraidawani@unimap.edu.my

N. M. Mohd Asri
Faculty of Chemical Engineering Technology, Universiti Malaysia Perlis, 02600 Jejawi, Arau, Perlis, Malaysia

M. A. A. Mohd Salleh et al. (eds.), *Proceedings of the Green Materials and Electronic Packaging Interconnect Technology Symposium*, Springer Proceedings in Physics 289,
https://doi.org/10.1007/978-981-19-9267-4_22

Currently, the most used method for preparing aluminum and magnesium were diffusion bonding, roll-bonding, and powder metallurgy. In contrast, the powder metallurgy can easily realize various types of alloys and give full play to the respective characteristics of each component material, which is a low-cost technology for producing high-performance metal alloy. The powder metallurgy process is simple and flexible, and the prepared material has fine microstructure, uniform composition, and easy design of composition [1, 2].

Aluminum (Al) alloy is one of the most common lightweight alloys used in a broad array of applications. The range of its application has been rapidly expanded because of its exhibit of high specific strength and excellent corrosion resistance. Magnesium (Mg) is a material that has the lowest density, 1.74 g/cm^3. Mg is 35% lighter than aluminum (Al) and has a density four times lower than steel (7.86 g/cm^3). The need for lighter materials in the automotive and aviation industries has increased with the rise in oil prices. Mg is a common element added to a wide range of alloys because it can add strength without inhibiting the positive characteristics of the base metal. Using magnesium in the automotive industry particularly attracted great attention, and the use of alloyed magnesium automotive components has a beneficial effect on improving fuel efficiency and reducing greenhouse gas emissions [3–5].

However, magnesium in aluminum will be affected by the strength and corrosion resistant of PM Al–Mg. The strength of this specimen will be increased with the increasing magnesium content but its brittle and interface easily formed between aluminum and magnesium particles. This phenomenon affects to the interfacial bonding quality. Many researchers focus on solving this problem with optimize interface bonding process to reduce the formation of intermetallic Al–Mg.

22.2 Experimental Procedure

Al–Mg alloy was fabricated by powder metallurgical methods. Raw materials of Al powder from Bendosen were mixed with different content of (10, 25, 50, 75, and 90) wt%. Mg powders (from Merck). Then, the mixed powder was cold compacted using a hand press machine with a pressure of 150 MPa. The green samples were sintered at a sintering temperature of 500 °C in a tube furnace for 3 h of sintering time under an argon atmosphere.

The corrosion behavior was tested using a potentiostat Autolab PGSTAT 204. The platinum rode out as the counter electrode and saturated calomel electrode as a reference electrode. These three electrodes were placed in a beaker containing of 3.5% stadium chloride. The test was set up in 100 to with start − 1 V to 1 V with step potential 0.001 V scan rate 0.001 V/s. Corrosion analysis was performed using the Tafel extrapolation method.

The sintered samples were polished according to the standard preparation of the metallographic specimen. For microstructural analysis, all samples were observed under an optical microscope (Optika B383 MET) and X-ray diffraction (XRD) the Bruker D2 Phaser for phase analysis.

Fig. 22.1 Optical micrograph of the samples sintered at 500 °C; **a** Al–90wt% Mg, **b** Al–75wt% Mg, **c** Al–50wt% Mg, **d** Al–25wt% Mg, and **e** Al–10wt% Mg

22.3 Results and Discussions

22.3.1 Microstructural Observation

Metallography examination is performed on all PM Al–Mg. Figure 22.1a–e shows the Al with different Mg contents after sintering at 500 °C. PM Al–90wt% Mg shows more pore compare to other samples. Sample PM Al–10wt% Mg shows lower pore content because of the presence of lower Mg content. It was seen in optical microscopy examinations that there were pores observed on the surface of the sample with irregular shapes and darker colors. The Al samples show the bonding Mg particle have good bonding with Al particles. The pore formation depends on the Al particle shape. Irregular shape of Al particles makes the pore after sintering. The Mg particles fill these pores and made the sample more condense. The increasing of the Mg contents leads to create good particle bonding with the Al and thus decrease pores.

22.3.2 Corrosion Analysis

Tafel extrapolation is used to determine the corrosion rate of the Al samples. E_{corr} in this case is the rest potential, i.e., when there is no external potential applies to the system. At E_{corr}, the oxidation and reduction of species in the solution are taking place at the same rate when the net measurable current are zero. While i_{corr} is the current density at E_{corr}, representing the total of the oxidation current and reduction current that is equilibrium. These two parameters contribute to the Tafel extrapolation. The calculation of corrosion rate has been made from Tafel extrapolation graphs where

Table 22.1 Value of corrosion rate

Samples	Corrosion rate (mpy)
Al–90 wt% Mg	7.934
Al–75 wt% Mg	1.797
Al–50 wt% Mg	0.675
Al–25 wt% Mg	0.537
Al–10 wt% Mg	0.168

corrosion current density, i_{corr} and corrosion potential, E_{corr} are plotted as x-axis and y-axis, respectively. Table 22.1 represents the value of corrosion rate in mpy. The corrosion rate increases as the magnesium content increases. The results in Table 22.1 showed that the sample Al–90wt% Mg has the highest value (7.934 mpy) of corrosion rate and sample Al–10wt% has the lowest value (0.168 mpy) of corrosion rate. Thus, sample Al–10wt% Mg has the highest corrosion resistance.

Figure 22.2 shows the surface morphology after corrosion testing in 3.5% NaCl solution. Figure 22.2a shows pitting on the surface. Figure 22.2b shows less corroded sample compare to sample Fig. 22.2a. Meanwhile, Figs. 22.2c, d, and e show small pitting created on the surface. All samples show corrosion attack. The worst condition is happened on the Al–90wt% Mg. The corrosion attack Mg particle rather than Al particle. Therefore, pitting is generated on the surface due to dissolution of Mg particle. Based on this research, no passivation film is formed on the samples surface. The film formation is retarded by Mg particle. The presence of chlorine ions made the sample surface more active and will destroy passive film on the PM Al surface [6].

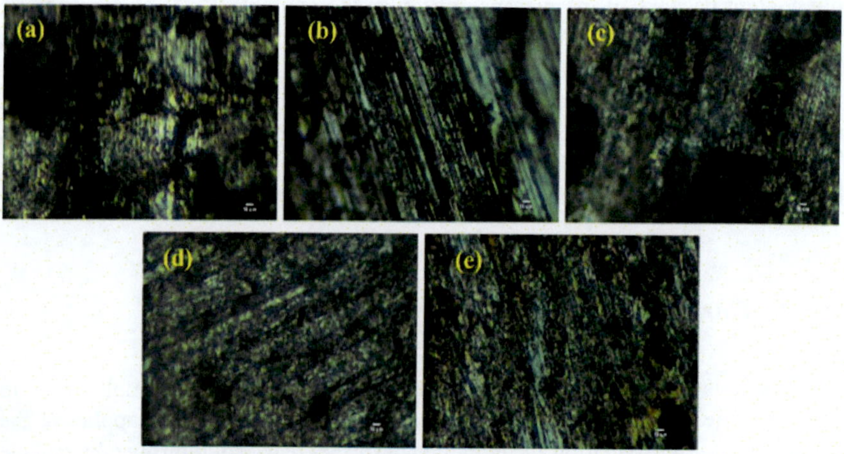

Fig. 22.2 Optical micrograph of the samples after corrosion test; **a** Al–90wt% Mg, **b** Al–75wt% Mg, **c** Al–50wt% Mg, **d** Al–25wt% Mg, and **e** Al–10wt% Mg

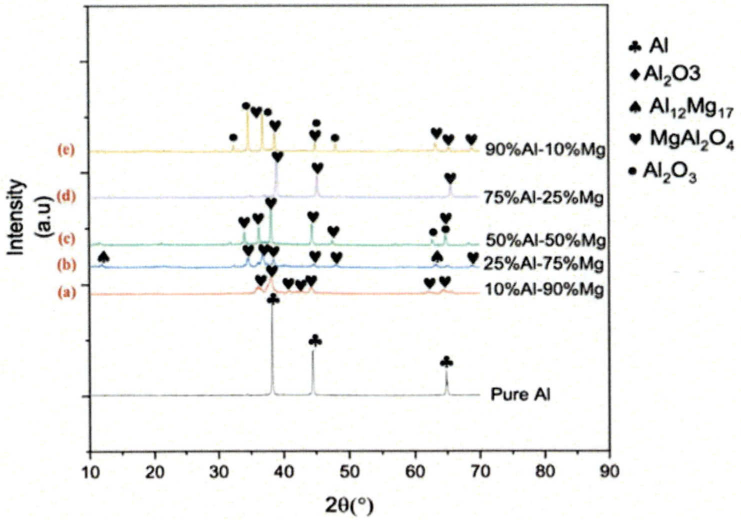

Fig. 22.3 XRD pattern after corrosion test; **a** Al–90wt% Mg, **b** Al–75wt% Mg, **c** Al–50wt% Mg, **d** Al–25wt% Mg, and **e** Al–10wt% Mg

Figure 22.3 shows the XRD pattern for PM Al samples, which is different composition after the corrosion test. The analysis is done when all the samples undergo corrosion testing at 3.5% NaCl. The pure Al as the control sample indicates Al peaks in the XRD pattern. The pure Al shows the aluminum phase and aluminum oxide compounds in the XRD pattern. While, sample Mg shows the magnesium, sodium chloride and sodium oxide found on the Mg surface. For the Al–90wt% Mg sample found Al phase, aluminum magnesium oxide ($MgAl_2O_4$) and Al_2Mg_3 phase exist on the sample surface. Only magnesium aluminum oxide ($MgAl_2O_4$) found on the Al–75wt% Mg. The Al–50wt% Mg found intermetallic compound Al_2Mg_3 and $Al_{12}Mg_{17}$. Al with 75wt% and 90wt% Mg, respectively, shows $MgAl_2O_4$ and Al_2O_3 phases. Intermetallic phase Al_2Mg_3 and $Al_{12}Mg_{17}$ detected due to PM Al–Mg sintering process. All samples detect aluminum oxide phase. The alumina oxide is a product from corrosion PM Al–Mg process [7].

22.4 Conclusion

The percentage of the magnesium content influences the microstructure properties and corrosion behavior of PM Al–Mg alloy. The increasing magnesium content will decrease corrosion resistance of PM Al–Mg alloy. The PM Al–90wt% Mg alloys show the lower corrosion rate.

References

1. R.K. Behera, B.P. Samal, S.C. Panigrahi, K.K. Muduli, Advances in Materials Science and Engineering (2020)
2. C. Randy, H. Rich, D. Ian, D. P. Bishop, Powder Metallurgy **55** (2012)
3. T. Wang, Z. Tang, L. Yang, L. Wu, H. Yan, C. Liu, Y. Ma, W. Liu, Materials Letters (2020)
4. J. Alias, International Journal of Automotive and Mechanical Engineering **17** (2020)
5. P. Jonnard, K.L. Guen, Characterization of Al and Mg Alloys from their X-ray emission bed (2013)
6. H. Parangusan, J. Bhadra, N. Al-Thani, Emergent Mater. **4** (2021)
7. A. Bakkar, V. Neubert, Corrosion Science **49** (2007)

Chapter 23
Investigation on Polyvinyl Chloride (PVC) and Polycaprolactone (PCL) Blend Ratio: Effect on Their Mechanical and Physical Properties

Siti Aishah Binti Abd Aziz, Sharifah Shahnaz Binti Syed Bakar, and Shuhaida Yahud

Abstract The effect of blend ratio for polyvinyl chloride (PVC) and polycaprolactone (PCL) blends on their mechanical and physical properties has been studied in this report. Polymer blends which consists of polyvinyl chloride (PVC) and polycaprolactone (PCL) were prepared by solution casting method with different ratio of PVC/PCL (100/0, 70/30, 50/50, 30/70, 0/100). The tensile and flexural properties of PVC/PCL blends were observed in this work by using universal machine testing. The sample of PVC/PCL blends showed that the tensile and flexural strength of the blend systems increased as the PCL content increased. Hardness test showed that as PCL is absorbed into PVC, the hardness decreased, and this tendency decreased as the amount of PCL is increased. The surface morphology of the PVC/PCL blends was been examined using scanning electron microscopy. The surface morphology resulted that smooth surface can be seen with large pores in the pure PVC while rough surface appeared in the blends due to high content of PCL. The water absorption and density of the PVC/PCL blend is obtained by immersing the sample into the distilled water. The density result showed that the density of blends increased when the content of PCL increased. The water absorption result showed that by increasing the amount of PCL in the blends has decreased the percentage of water absorption.

Keywords Polyvinyl chloride (PVC) · Polycaprolactone (PCL) · Blends · Mechanical properties · Physical properties

S. A. B. A. Aziz (✉) · S. S. B. S. Bakar
Faculty of Chemical Engineering Technology (FKTK), Universiti Malaysia Perlis (UniMAP), 02600 Jejawi, Perlis, Malaysia

S. Yahud
Faculty of Electronic Engineering Technology (FTKEN), Universiti Malaysia Perlis (UniMAP), Pauh Putra Campus, 02600 Arau, Perlis, Malaysia

M. A. A. Mohd Salleh et al. (eds.), *Proceedings of the Green Materials and Electronic Packaging Interconnect Technology Symposium*, Springer Proceedings in Physics 289,
https://doi.org/10.1007/978-981-19-9267-4_23

23.1 Introduction

The use of polymers in the manufacture of medical products has increased in recent years. The benefits of these materials include low-cost manufacturing procedures that can result in a wide range of polymer materials for medical applications such as for cosmetic implants, dental composites, and contact lenses. PVC is the most often used material for storing IV fluids, dialysis solutions, and blood products. Over the last few decades, biodegradable polymers have seen an increase in demand in medical applications. One of these polymers is polycaprolactone (PCL), a polyester that has been widely employed in the tissue engineering area due to its availability, low cost, and ability to be modified [1]. Polymers also have been used in medical devices likes pessary. Pessaries are now commonly composed of inert plastic or silicone in order to minimize smells and absorption of vaginal fluids [2]. A polymer blend is a compound made up of two or more polymers that have been combined to form a new material. Polymer blending has received a lot of interest as a simple and cost-effective way to generate polymeric materials with a wide range of commercial uses. In other words, by correctly selecting the component polymers, the properties of the blends can be changed according to their end application [3]. Polyvinylchloride (PVC) is a high-strength thermoplastic material used in a variety of applications including pipelines and medical equipment. It is the third-most-produced synthetic plastic polymer in the market. Polyvinyl chloride (PVC) contributes for one-third of the volume of polymers used in medical device manufacturing. Intravenous (IV) tubing and blood bags are two common PVC device uses. Polycaprolactone (PCL) is a biodegradable polymer that is widely used in biomedical and ecologically friendly applications. PCL sustained significance in the growing field of biodegradable polymers may be related to a variety of characteristics that make it ideal for medical applications. In PVC, polycaprolactone (PCL) is added as a polymeric plasticizer in order to increase the usability of PVC, apart from that, improve the properties for the PVC/PCL blends [4].

23.2 Experimental Procedure

In this research study, three raw materials were used. First, polyvinyl chloride (PVC) with average weight (M_n) of 47,000 g/mol which was kindly supplied by Faculty of Chemical Engineering Technology, UniMAP. Second, polycaprolactone (PCL) with average weight (M_n) of 80,000 g/mol which was purchased from Sigma-Aldrich. And third, tetrahydrofuran (THF) solvent was purchased from Sigma-Aldrich. Five different blends ratio were prepared in this research study. The polyvinyl chloride/polycaprolactone (PVC/PCL) blends were prepared by using solution blending technique. In tetrahydrofuran (THF), appropriate volumes of PVC and PCL were being dissolved (10% mass fraction of the total polymer). The resultant mixture was stirred at room temperature until it was completely dissolves by using magnetic stirrer for 4 h until it was fully homogenized. The solution was then been transferred to a mold size of 155 mm × 55 mm × 80 mm for casting. Finally, the films were dried at room temperature for 5 days to eliminate the residual solvent.

23.2.1 Tensile and Flexural Test

The tensile and flexural test was performed by using a universal testing machine (Instron, 5567) according to the ASTM D638 and ASTM D790 also the stretching rate was 50 mm/min and 10 mm/min, respectively, with temperature of 25 °C. Rectangular sample bars with dimensions of 100 mm × 10 mm × 4 mm were prepared. Three samples of each blend were tested, and average values of tensile strength, flexural strength, and modulus of elasticity were reported.

23.2.2 Hardness Test

Hardness testing is being performed by using the type of shore A scale durometer. The hardness value was determined by the penetration of the durometer indenter foot into the sample for 10 s.

23.2.3 Water Absorption Test

The water absorption test for PVC/PCL blends, sample with dimensions of 20 mm × 20 mm, was dried at 100 °C temperature in a laboratory oven for 2 h before the film sample been cooled and weighted right away. At room temperature, the dried sample was immersed in 100 mL of distilled water. The sample was submerged in distilled water for 24 h. After a brief immersion, the sample was removed from the water, dried superficially, and weighted. To determine the water absorption (WA), the mass difference between the dry and saturated states in each sample was measured. The water absorption was determined by calculating using Eq. 23.1

$$\mathrm{WA}\% = [\,(m_f - m_i)/m_i] \times 100, \tag{23.1}$$

where m_f = weight of sample after immersed in distilled water for 24 h and m_i = weight of sample before immersed in distilled water.

23.2.4 Density Test

The density of the PVC/PCL blends were determined by calculating the formula of Eq. 23.2

$$\text{Density, } \rho = m/V \tag{23.2}$$

where m = mass of sample and V = volume of sample.

23.2.5 Surface Morphology

The PVC/PCL blends sample were examined by scanning electron microscopy (SEM) to study the surface morphology of the sample. SEM was performed using SEM (JSM-64OLA, JEOL) under magnification of 2000× . Samples were coated beforehand with platinum in this SEM study.

23.3 Results and Discussion

23.3.1 Mechanical Properties

Table 23.1 indicates that pure PVC has the lowest tensile strength, which is 6.79 MPa and pure PCL has the highest tensile strength, which is 14.76 MPa.

The difference between the highest and the lowest tensile strength is 7.97 MPa. The blend ratio of PVC/PCL (100/0) with the lowest tensile strength is then followed by the different blend ratio of PVC/PCL (70/30), (50/50), (30/70), and (0/100) in ascending order with tensile strength 7.94 MPa, 8.23 MPa, 12.02 MPa, and 14.76 MPa, respectively. Thus, the data showed that the PVC/PCL blend film has a better strength than pure PVC up to a PCL content of 70%.

Table 23.1 Results of tensile strength (MPa), flexural strength (MPa), modulus of elasticity (MPa), and hardness for different ratio of PVC/PCL blends

PVC/PCL blend ratio	Tensile strength (Mpa)	Flexural strength (Mpa)	Modulus of elasticity (Mpa)	Hardness shore A
100/0	6.79	2.03	20.42	85.30
70/30	7.94	3.53	49.98	80.00
50/50	8.23	13.33	264.60	91.30
30/70	12.02	15.31	271.41	86.00
0/100	14.76	51.86	723.30	88.00

Table 23.1 also indicates that pure PCL has the highest value of flexural strength while pure PVC has the lowest value of flexural strength compared to another PVC/PCL blend ratio with 2.03 MPa and 51.86 MPa, respectively. The increment material of PCL in the PVC/PCL blend sample will increase the flexure strength value. This shows that the addition of PCL in PVC may contribute to better dispersion and adhesion [5]. Table 23.1 shows that pure PCL has the highest modulus of elasticity compared to pure PVC and other PVC/PCL blends with 723.3 MPa and 20.42 MPa. The decreasing of PVC material in the PVC/PCL blend will increase their modulus of elasticity. The results tell that when the weight percent of PCL increase in the PVC/PCL blends, it will increase the flexure strength and modulus of elasticity of the polymer blends. The presence of PCL in PCV/PCL blends increased the flexural strength demonstrating a reaction between the carbonyl group of the PCL and the HC–Cl group of the PVC [6].

Table 23.1 shows shore A hardness of PVC/PCL blend ratio where the maximum hardness is showed by mixture of 50% PVC and 50% PCL with 91.3 and declined to 86 with increasing the proportion of PCL to 70%. Table 23.1 indicates that pure PCL has the higher value of hardness than pure PVC with 88 and 85.3, respectively. This is corresponding to their crystallinity which PCL is a semi-crystalline material with 37.3% degree of crystallinity while PVC is an amorphous material which the degree of crystallinity (10%) is low [7].

23.3.2 *Physical Properties*

Water absorption increases molecular mobility in the event of swelling and softening. However, the water absorption test shows that there were not much differences in the dimension of the sample. The percent of water absorption also has a small value of increment as given in Table 23.2.

Table 23.2 indicated that pure PVC has highest percentage of water absorption compared to pure PCL has the lowest percentage of water absorption with 0.23% and 0.03%, respectively. Although the highest percentage of water absorption among the blends was 0.23%, it is still considered as small. Therefore, it can be concluded that the polar bonding of the water molecules is weaker than the interfacial adhesion between the polyvinyl chloride (PVC) and polycaprolactone (PCL). As the

Table 23.2 Results of water absorption (%) and density (kg/m^3) for different ratio of PVC/PCL blends

PVC/PCL blend ratio	Water absorption (%)	Density (kg/m^3)
100/0	0.23	157.10
70/30	0.10	116.20
50/50	0.09	181.63
30/70	0.05	158.65
0/100	0.03	160.82

PCL content increases in the blends, percentage of water absorption of blends also decreases. This is due to PCL is made up of nonpolar methylene groups. Methyl group is a carbon atom that joined to three hydrogen atoms. These C–H bonds will be treated as nonpolar covalent bonds in this class. As a result, methyl groups cannot interact with polar substances like water or create hydrogen bonds with them [8].

Table 23.2 indicates that pure PCL has the higher density compared to pure PVC with 160.82 and 157.10 kg/m^3. Table 23.2 showed that the by mixture of 70% PVC and 30% PCL, the density is 116.20 kg/m^3 and escalates to 181.63 kg/m^3 with increasing the proportion of PCL to 50%. Density of a material is directly proportional to a mass of material if the volume remains constant [9]. Thus, from the result tells that pure PVC had the least of mass while PCL had the highest of mass as the volume of material used in this study were constant.

23.3.3 Surface Morphology

Figure 23.1 indicates the surface morphology for fractured surface of PVC/PCL blends at a different ratio with the 2000 × magnification. Figure 23.1a is a pure PVC, which shows that it consists of smooth surface with large pores indicating its brittle fracture. Due to the large of porosity in the blends, caused a low value in tensile strength and flexural strength. Based on Table 23.1, large porosity in the blends caused a low value in tensile strength (6.79 MPa) and flexural strength (2.03 MPa).

Figure 23.1b and c consist the blend ratio of PVC/PCL (70/30) and (50/50) indicate that the circle like structure is the PVC, and the matrix is the PCL. Therefore, Fig. 23.1 describes that the addition amount of PCL will decrease the appearance of pores. The reduction of pores would increase their tensile and flexural strength but lowering the percentage of water absorption.

Figure 23.1d and e consist the blend ratio of PVC/PCL (30/70) and (0/100) respectively, which indicates the most fibrous appearance structure due to high PCL content. As can be seen, pure PCL has rough surface, which indicated that its ductile fracture. Thus, when the pores are stretched, the fibrous look will be clearer.

23.4 Conclusion

A polymer blend is a compound made up of two or more polymers that have been combined to form a new material. Polymer blending has received a lot of interest as a simple and cost-effective way to generate polymeric materials with a wide range of commercial uses. The addition of PCL in the PVC helps in improving the mechanical and physical properties of the polymer blends. In this study, flexural and tensile strength of PVC/PCL blends increased when the weight percentage of PCL increased in the blends. Besides, the addition of PCL content in the blends

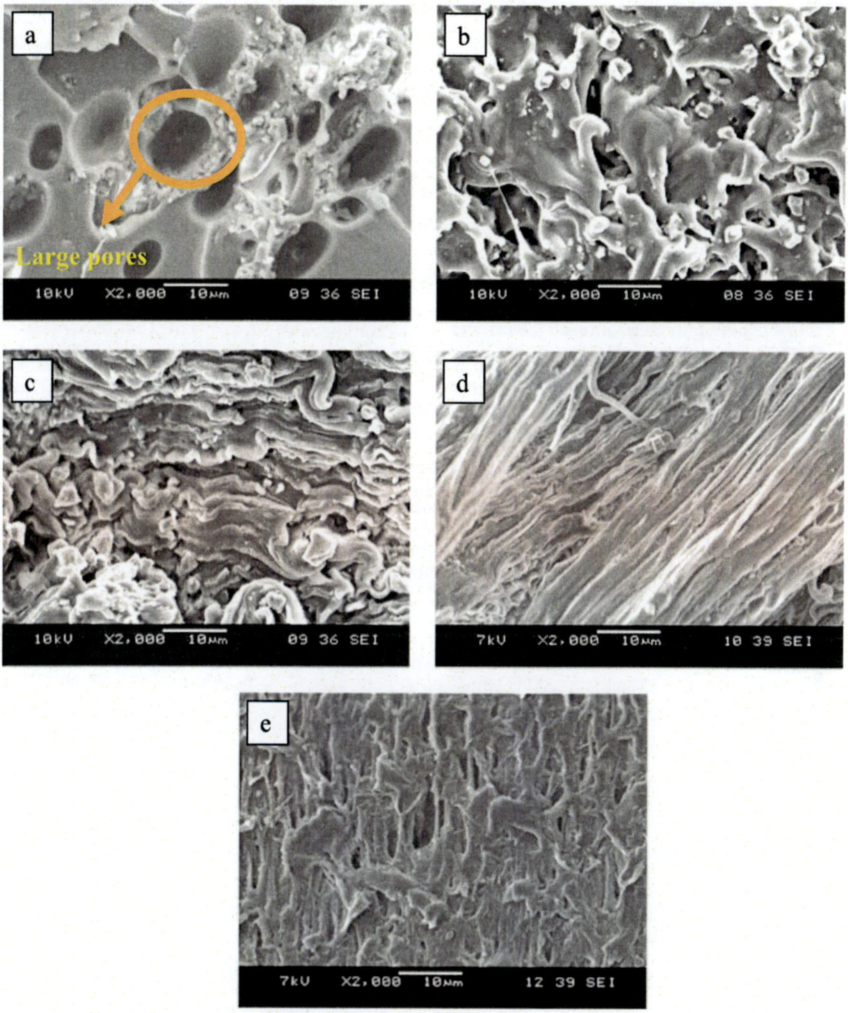

Fig. 23.1 SEM micrograph for different of PVC/PCL blend ratio. **a** 100/0, **b** 70/30, **c** 50/50, **d** 30/70, and **e** 0/100 at magnification of 2000×

also increased their modulus of elasticity. Thus, the lowest modulus elasticity has a flexible property of material which is pure PVC in this study while pure PCL is the stiffer material due to the highest of modulus elasticity. The addition of PCL in the blend has decreasing the percentage water absorption of the blends. In addition, polymer blend decreased when PCL content was at 30% and it increased when PCL content was at 70%. The surface morphology shows that pure PVC consists of a large pore that may cause a lower in tensile and flexural strength. As the pores decrease due to the addition of PCL, a fibrous structure will be seen, thus increasing their mechanical properties.

References

1. E. Malikmammadov, T.E. Tanir, A. Kiziltay, V. Hasirci, N. Hasirci, PCL and PCL-based materials in biomedical applications. J. Biomater. Sci. Polym. Ed. **29**(7–9), 863–893 (2018)
2. G. Al-Shaikh, S. Syed, S. Osman, A. Bogis, A. Al-Badr, Pessary use in stress urinary incontinence: a review of advantages, complications, patient satisfaction, and quality of life. Int. J. Women's Health **10**, 195–201 (2018)
3. J. Parameswaranpillai, S. Thomas, Y. Grohens (2014). Polymer blends: state of the art, new challenges, and opportunities
4. Z. Sun, B. Choi, A. Feng, G. Moad, S.H. Thang, Nonmigratory poly (vinyl chloride)-block-polycaprolactone plasticizers and compatibilizers prepared by sequential RAFT and ring-opening polymerization (RAFT-T-ROP). Macromolecules **52**(4), 1746–1756 (2019)
5. D. Esperanza, S. Iban, B.V. María, In vitro degradation of poly(caprolactone)/nHA composites. Journal of Nanomaterials (2014)
6. A. Campos, J.C. Marconato, S.M. Martins-Franchetti, The influence of soil and landfill leachate microorganisms in the degradation of PVC/PCL films cast from DMF. Polímeros **22**(3), 220–227 (2017)
7. V.B. Carmona, A.C. Corrêa, J.M. Marconcini, L.H.C. Mattoso, Properties of a biodegradable ternary blend of thermoplastic starch (TPS), poly (ε-caprolactone)(PCL) and poly (lactic acid)(PLA). J. Polym. Environ. **23**(1), 83–89 (2015)
8. S. Rahmani, M. Maroufkhani, S. Mohammadzadeh-Komuleh, Z. Khoubi-Arani, Polymer nanocomposites for biomedical applications, in Fundamentals of Bionanomaterials. (Elsevier, 2022) pp. 175–215
9. N. Singh, Relation between mass and density (2017)

Chapter 24
Room Temperature Synthesis and Characterization of HKUST-1, Metal–Organic Frameworks (MOFs)

Syazwana Ahmad, Mohd Firdaus Omar, E. M. Mahdi, Khairul Anwar Abdul Halim, Shayfull Zamree Abd Rahim, Sam Sung Ting, Hazizan Md. Akil, and Norlin Nosbi

Abstract In the present work, HKUST-1 has been synthesized at room temperature with 1:0, 1:1, and 0:1 ratio of ethanol and water. A wide range of reaction conditions were explored in order to understand the effects of solvent and temperature. It was discovered that various precursors yielded products with various BET specific surface areas. The effect of water may therefore be explained by the decrease in reaction rate with an increasing concentration of reactants. The XRD data and SEM analysis showed that both MOFs were very crystalline in the product.

Keywords HKUST-1 · SEM · XRD · Physicochemical characteristics

S. Ahmad · M. F. Omar (✉) · K. A. A. Halim
Faculty of Chemical Engineering Technology, Universiti Malaysia Perlis, Arau, Perlis, Malaysia
e-mail: firdausomar@unimap.edu.my

M. F. Omar · K. A. A. Halim · S. Z. A. Rahim · S. S. Ting
Geopolymer and Green Technology, Centre of Excellent (CEGeoGTech), Universiti Malaysia Perlis, Arau, Perlis, Malaysia

E. M. Mahdi
Materials Technology Group, Industrial Technology Division, Malaysia Nuclear Agency, 43000 Kajang, Selangor, Malaysia

H. Md. Akil
School of Materials and Mineral Resources Engineering, Universiti Sains Malaysia, Engineering Campus, 14300 Nibong Tebal, Pulau Pinang, Malaysia

N. Nosbi
Department of Mechanical Engineering, Centre for Corrosion Research (CCR), Institute of Contaminant Management for Oil and Gas (ICM), Universiti Teknologi PETRONAS, 32610 Perak, Seri Iskandar, Malaysia

M. A. A. Mohd Salleh et al. (eds.), *Proceedings of the Green Materials and Electronic Packaging Interconnect Technology Symposium*, Springer Proceedings in Physics 289,
https://doi.org/10.1007/978-981-19-9267-4_24

24.1 Introduction

Metal–Organic Frameworks (MOFs) are sponge-like materials that are crystalline and very porous and trap, store, and distribute gases. Their structures comprise metal ions, and organic molecules referred to as "linkers," as shown in Fig. 24.1. Beneficial properties, particularly selectivity and capacity, can be altered by selecting the appropriate metal and linker for the MOF [1, 2].

MOFs are coordinated compounds made up of metal ions correlated to organic molecules that are often rigid to produce one-, two-, or three-dimensional structures that can be porous [3, 4]. In the observation, these metals provide metallic coordination environments with various geometries. The development of coordination bonds between the metal ions and the organic linkers can also be reversible, which is possible because of the standard liability of metal complexes.

Chui and a colleague made the initial discovery of HKUST-1 (Hong Kong University of Science and Technology) which is one remarkable type of the MOFs in 1999 [5]. The reported findings concern the synthesis, structure, and preliminary physical and chemical properties of HKUST-1, a highly porous open-framework metal coordination polymer $[Cu_3(TMA)_2(H_2O)_3]$ n. This polymer crystallizes into face-centered cubic crystals with a massive square-shaped network of intersecting three-dimensional (3D) pores (9 Å by 9 Å). The HKUST-1 framework comprises dimeric metal units linked together using benzene-1,3,5-tricarboxylate linker molecules. The paddle wheel unit, also known as the secondary building unit (SBU) of the HKUST-1 structure, is a structural motif often used to define the coordination environment of metal centers [6]. Benzene-1,3,5-tricarboxylate linkers molecules connect two metal centers in the paddle wheel, four molecules. Two metal centers on the paddle wheel unit are linked by one water molecule when hydrated, which is usually found if the material is handled in the air.

It is HKUST-1, which has the chemical formula $Cu_3(BTC)_2(H_2O)$, considered one of the most promising MOFs [7, 8]. HKUST-1 can remove harmful gases such

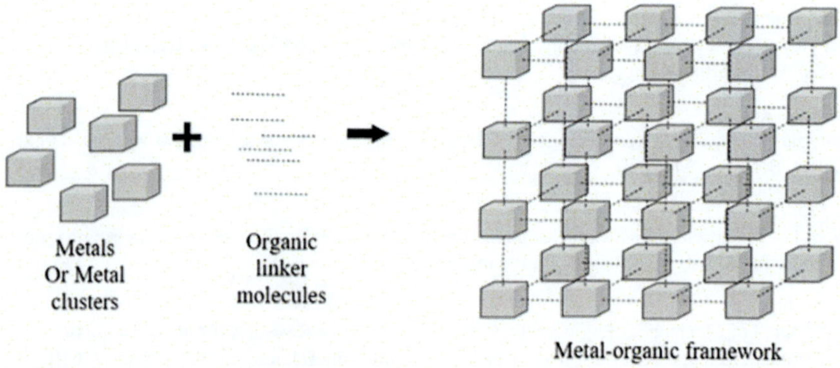

Fig. 24.1 Schematic structure of MOFs

as ammonia and carbon dioxide from contaminated air and flue gases because of the high affinity of its metallic group for NH_3 and CO_2 molecules. In the realms of energy and the environment, HKUST-1 can separate carbon dioxide from other gases such as hydrogen, methane, nitrogen, and oxygen and purify these gases and reduce greenhouse gas emissions. Due to this separation process, the HKUST-1 features promising adsorption characteristics.

The common method that has been used to synthesize MOFs is solvothermal methods, where heat is applied during synthesis. Even though the method used in this study is slower and does not involve heat, it can still be used in materials that are sensitive to heat. David and his colleagues proved that the synthesis reaction without heat is also possible in producing MOFs with very clear crystal. Therefore, by simply mixing organic linker and metal salt at room temperature, we will obtain a good quality of MOF [9].

24.2 Experimental Section

24.2.1 Chemicals

For this research, all materials were supplied from Sigma-Aldrich and Merck. Copper(ii) nitrate trihydrate, 1,3,5-benzenetricarboxylic acid and ethanol were purchased from Sigma-Aldrich. Tetrahydrofuran and *n,n*-dimethylformamide were purchased from Merck, and sodium bicarbonate is from HmBG.

24.2.2 Synthesis of HKSUT-1 with Difference Solvent Ratio

1g of Cu $(NO_3)_2.3H_2O$ 1.5g of sodium bicarbonate and 1g of 1,3,5-benzenetricarboxylic acid were added in a solution of distilled water and stirred for 24 h at room temperature (~25 °C). The solution will turn murky, which confirms the formation of HKUST-1 particles, and the solution will be washed with ethanol, centrifuged, and redispersed in tetrahydrofuran to be stored. All steps above are repeated with a 50:50 ratio of water and ethanol, and HKUST-1 synthesis with water only [10].

24.2.3 Material Characterization

The following characterization methods were chosen for their respective capabilities to determine the morphological properties of HKUST-1. Morphology of the HKUST-1 was examined using the Scanning Electron Microscope (SEM). The powder samples were dispersed and mounted on a stub using adhesives. Each sample was coated with platinum using a sputter coater for a period of 30 s, prior to imaging at 20 kV under high vacuum. The structure of crystallinities was carried out by X-ray diffraction (XRD) operated at 45 kV, and the scanning rate was set to 2° min^{-1} with a continuous scanning range of $2\theta = 5°–40°$. The Brunauer–Emmett–Teller (BET) method was used to estimate a specific surface area since the BET theory is the foundation for this critical analytical methodology. The BET theory aims to explain the physical adsorption of gas molecules on a solid surface. Before each test, the sample was degassed at 150 °C for 12 h for HKUST-1 [11].

24.3 Result and Discussion

24.3.1 Morphologies of HKUST-1

The structures of the synthesized HKUST-1 are illustrated in Fig. 24.2, where a (synthesized with 1:0 ethanol: water ratio), b (synthesized with 1:1 ethanol: water ratio), and c (synthesized with 0:1 ethanol: water ratio). It was observed that in Fig. 2a, HKUST-1 appear in octahedral and cubic. Figure 2b and c show the octahedral, cubic and monoclinic structure, respectively [8, 12, 13]. The grain of the freshly prepared sample emerges from the crystalline crystal with sharp edges and both have an uneven size of particles which are in the range of an average size of 0.5–1.2 μm for HKUST-1. Kareem et al. say that the process of crystallinity slows down when water is present during synthesis. Hence, the structures of b and c have the monoclinic crystal structure, which indicates the crystallization HKUST-1 is decreasing [8, 14]. A study by Kuen et al. shows a low-temperature synthesis yields cubic crystals with sharp edges, while a high-temperature synthesis yields spherical particles [10]. The unreacted chemicals must be thoroughly washed out of the as-synthesized MOF in order for them to be removed; otherwise, they will remain in the crystals.

All the samples were characterized using X-ray diffraction (XRD), and Fig. 24.3 shows the XRD pattern of HKUST-1 (a, b, and c). The diffraction peaks present in between two theta range 5° and 20°; it was found that the highest peak formed at 11. 74°, which is the element of copper (JCPDS card No. 96-230-0382). The intensity of a is higher than b and c because of the increment in crystallinity. This is due to the presence of water during synthesis of b and c, which implies the good crystallinity of these materials.

The maximum Langmuir specific surface area (259 $m^2\ g^{-1}$) was achieved for the HKUST-1. As given in Table 24.1, as the volume of ethanol increased, the Langmuir and BET specific surface areas increased for HKUST-1. This figure is still much below the measurement made by Kuen et al. (1558 $m^2\ g^{-1}$) [10]. Sample activation, or the elimination of amorphous or nonporous impurity phases, must be supervised for this problem. According to Kuen and his colleague, it is possible to post treat a

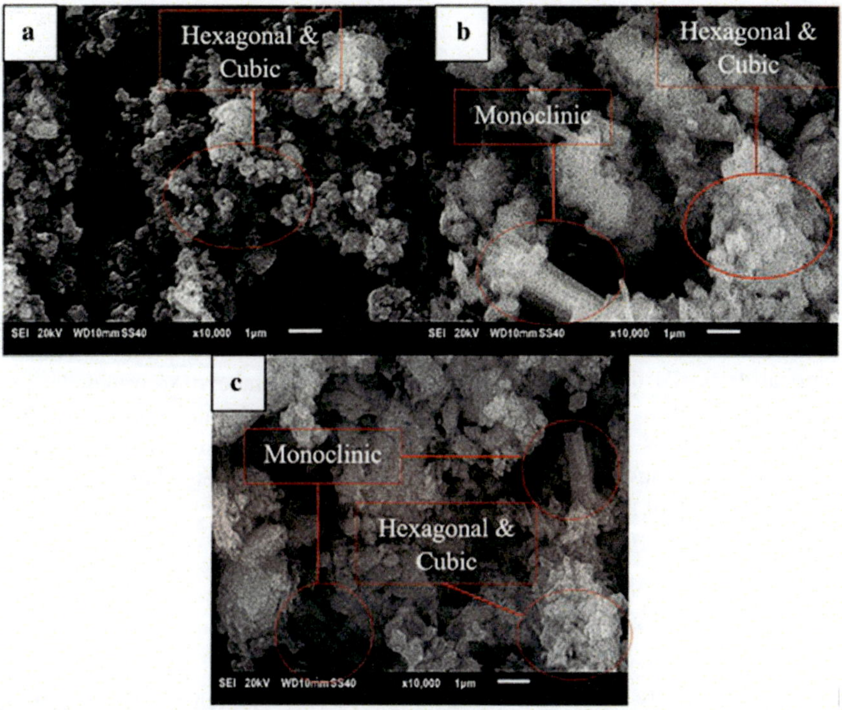

Fig. 24.2 SEM images of HKUST-1 **a–c**

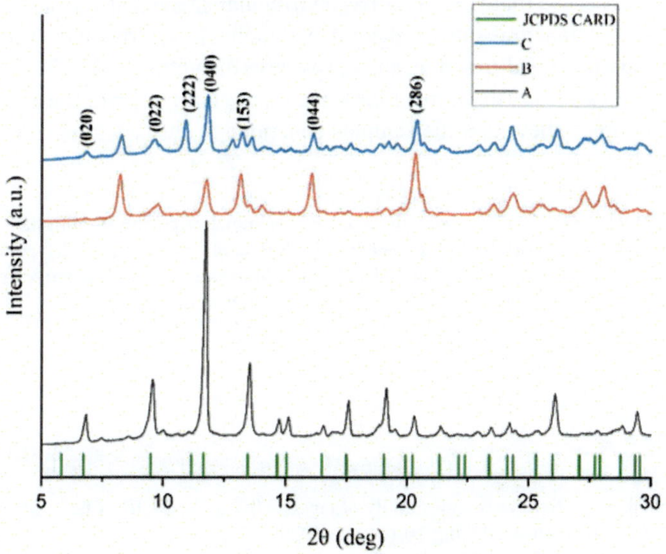

Fig. 24.3 Simulated and observed XRD pattern of HKUST-1 **a–c**

Table 24.1 Summarizes the pore properties of HKUST-1

Sample	S_{BET} (m^2/g)	$S_{Langmuir}$ (m^2/g)	V_p (cm^3/g)	V_m (cm^3/g)
A	259.606	342.918	0.202	0.101
B	5.457	7.360	0.014	0.001
C	25.163	33.670	0.016	0.014

material to get rid of guest molecules that are trapped inside its pores after manufacturing. The following average pore size values for the three samples indicate that the samples created in this study are microporous materials. With increasing ethanol volume, the three HKUST-1 samples' pore textural characteristics became essentially identical. Typically, the produced MOFs include a large number of solvent molecules such as H_2O, ethanol, DMF, and the like. They are employed to manufacture MOF or reactant compounds or by-products within their pores. The pore channels may also get blocked as a result of these impurities [1]. Therefore, a variety of procedures can be used to get rid of those solvent molecules or impurities.

24.4 Conclusion

In the current study, HKUST-1 was synthesized. To comprehend, the effects of solvent, SEM, XRD, and BET analyses were performed to explore the characteristics of it. It was observed that several precursors created a product with diverse surface areas. It was observed that HKUST-1 has an irregular shape and various nanostructures, including octahedral, cub octahedral, and monoclinic structures. The samples were analyzed by X-ray diffraction, and the patterns of synthesized HKUST-1 exhibited and implied good crystallinity of these materials. The results of the average pore size of all samples show that the samples generated in this work are microporous materials.

Acknowledgement The author would like to acknowledge the Malaysian Ministry of Higher Education (MOHE), Fundamental Research Grant (FRGS) (Grant no.: FRGS/1/2020/TK0/UNIMAP/02/58) and University Malaysia Perlis (incentive grant) for sponsoring and providing financial assistance for this research work.

References

1. M. Kandiah et al., Synthesis and stability of tagged UiO-66 Zr-MOFs. Chem. Mater. **22**(24), 6632–6640 (2010). https://doi.org/10.1021/cm102601v
2. P.S. Sharanyakanth, M. Radhakrishnan, Synthesis of metal-organic frameworks (MOFs) and its application in food packaging: a critical review. Trends Food Sci. Technol. **104**(June), 102–116 (2020). https://doi.org/10.1016/j.tifs.2020.08.004

3. L. Asgharnejad, A. Abbasi, M. Najafi, J. Janczak, One-, two- and three-dimensional coordination polymers based on copper paddle-wheel SBUs as selective catalysts for benzyl alcohol oxidation. J. Solid State Chem. **277**(April), 187–194 (2019). https://doi.org/10.1016/j.jssc.2019.06.011
4. H. Furukawa et al., Ultrahigh porosity in metal-organic frameworks. Science **329**(5990), 424–428 (2010). https://doi.org/10.1126/science.1192160
5. Z. Hu et al., Mechanical properties of electrochemically synthesised metal-organic framework thin films. Science **283**(1), 102–116 (2021). https://doi.org/10.1039/x0xx00000x
6. J. Cortés-Súarez et al., Synthesis and characterization of an SWCNT@HKUST-1 composite: enhancing the CO_2 adsorption properties of HKUST-1. ACS Omega **4**(3), 5275–5282 (2019). https://doi.org/10.1021/acsomega.9b00330
7. W.W. Lestari, M. Adreane, C. Purnawan, H. Fansuri, N. Widiastuti, S.B. Rahardjo, Solvothermal and electrochemical synthetic method of HKUST-1 and its methane storage capacity. IOP Conf. Ser. Mater. Sci. Eng. **107**(1) (2016). https://doi.org/10.1088/1757-899X/107/1/012030
8. H.M. Kareem, R.T. Abd Alrubaye, Synthesis and characterization of metal organic frameworks for gas storage. IOP Conf. Ser. Mater. Sci. Eng. **518**(6), 0–7 (2019). https://doi.org/10.1088/1757-899X/518/6/062013
9. D.J. Tranchemontagne, J.R. Hunt, O.M. Yaghi, Room temperature synthesis of metal-organic frameworks: MOF-5, MOF-74, MOF-177, MOF-199, and IRMOF-0. Tetrahedron **64**(36), 8553–8557 (2008). https://doi.org/10.1016/j.tet.2008.06.036
10. K.S. Lin, A.K. Adhikari, C.N. Ku, C.L. Chiang, H. Kuo, Synthesis and characterization of porous HKUST-1 metal organic frameworks for hydrogen storage. Int. J. Hydrogen Energy **37**(18), 13865–13871 (2012). https://doi.org/10.1016/j.ijhydene.2012.04.105
11. Y. Chen, X. Mu, E. Lester, T. Wu, High efficiency synthesis of HKUST-1 under mild conditions with high BET surface area and CO_2 uptake capacity. Prog. Nat. Sci. Mater. Int. **28**(5), 584–589 (2018). https://doi.org/10.1016/j.pnsc.2018.08.002
12. M. Kanno, T. Kitao, T. Ito, K. Terashima, Synthesis of a metal-organic framework by plasma in liquid to increase reduced metal ions and enhance water stability. RSC Adv. **11**(37), 22756–22760 (2021). https://doi.org/10.1039/d1ra00942g
13. X. Chen et al., An efficient modulated synthesis of zirconium metal–organic framework UiO-66. RSC Adv. **12**(10), 6083–6092 (2022). https://doi.org/10.1039/d1ra07848h
14. D. Saha, S. Deng, Hydrogen Adsorption on Metal-Organic Framework MOF-177. Tinshhua Sci. Technol. **15**(4), 363–376 (2010)

Chapter 25
Structural Analysis of Silver-Based Conductive Ink Under Cyclic Loading

Sana Zulfiqar, Abdullah Aziz Saad, Zulkifli Ahmad, Feizal Yusof, and Zuraihana Bachok

Abstract The stretchable conductive ink has become the primary component in the development of stretchable electronic devices. The ink is mainly comprised of a binder or epoxy resin and a conductive filler. Due to the addition of epoxy, the composite, known as conductive polymer, possesses the elastomeric properties. In this study, the silver-based conductive ink was fabricated using PDMS-OH as a binder and silver flakes as filler. To check the dynamic properties of the ink, the cyclic tensile loading was carried out at different strain rates between 0 and 100%. The electrical conductivity of the ink was also evaluated after loading and unloading at each strain rate. The effect of cyclic loading on the ink was checked by using strain and stress controlled mechanisms at 50% and 0.11 MPa, respectively. The behavior of the ink was examined for ten cycles using both the stress–strain sensitivities. Finally, the morphology and elemental analysis of the conductive ink after ten cycles of loading and unloading were studied by scanning electron microscopy (SEM).

Keywords Stretchable electronics · Conductive ink · Cyclic loading · Morphology

25.1 Introduction

Stretchable electronics is considered as an emerging technology that has replaced rigid printed circuit boards (PCBs) by the stretchable and flexible circuits. These circuits mainly comprised of stretchable or flexible polymeric substrate and stretchable polymer–matrix conductive ink. The stretchable circuits have a wide range of applications in the various fields of engineering such as healthcare monitoring system

S. Zulfiqar · A. A. Saad (✉) · F. Yusof · Z. Bachok
School of Mechanical Engineering, Universiti Sains Malaysia, 14300 Nibong Tebal, Penang, Malaysia
e-mail: azizsaad@usm.my

Z. Ahmad
School of Materials and Mineral Resources Engineering, Universiti Sains Malaysia, 14300 Nibong Tebal, Penang, Malaysia

M. A. A. Mohd Salleh et al. (eds.), *Proceedings of the Green Materials and Electronic Packaging Interconnect Technology Symposium*, Springer Proceedings in Physics 289,
https://doi.org/10.1007/978-981-19-9267-4_25

[1], telecommunications [2], strain sensors, wearable devices, e-skin for robots [3], MP3 or DVD players [4], 3D interconnect devices [5–7], and home appliances.

The polymers and their composites exhibit good physical and mechanical properties. Several methods have been used to examine the different mechanical properties of polymers, such as tensile and compression tests, creep, fatigue, hardness test, nano-indentation, and so on [8, 9]. The performance of these materials with respect to cyclic loading is under debate nowadays. The rate sensitivity of any polymer or polymeric composite depends on its viscoelastic nature. Mostly polymers are considered as rate-dependent materials that have three important properties: stress relaxation, hysteresis, and creep [10]. The cyclic loading depends on the stress amplitudes and the number of cycles. The material strength and elasticity can be increased or decreased by changing the values of stress amplitude and maximum applied load [11]. The behavior of polymers in the elastic strain range can be determined by the stress–strain relations, below the yield point, obtained via tension–compression tests and hysteresis loops [12]. The commencement of multiple micro-cracks produced by stress concentration in few microscopic locations causes a broad hysteresis loop in the initial cycle of loading. As a result, the first cycle is characterized by a nonlinear stress–strain relationship. In the next loading cycles, a decrease in quick defect initiation expansion as well as the rapid disappearance of micro-crack nuclei can be seen. However, the continuous increase in the cyclic loading may cause defects at the micro-level, creep effect (time), and cracks propagation at a slower rate [13].

Furthermore, the effect of cyclic loading on different composites of polymers has also been studied [14]. Bociaga [15] investigated the structural behavior and change in mechanical properties of poly-propylene injection molded parts during cyclic loading. Yu [16] examined the viscoelastic properties of an epoxy-based amorphous thermosetting shape memory polymers using tensile testing machine for cyclic tension and DMA for shape memory behavior. Moreover, Rozo Lopez [17] generated a micromechanical model for fiber-reinforced polymers behavior under cyclic loading. Ly [18] studied the cyclic response in fiber-reinforced polymers by considering the effect of energy dissipation on the micro and macroscopic performances of these composites. Also, Zheng [19] used a double cantilever beam (DCB) fracture mechanics testing method to measure the adhesion energy between the silver and PDMS substrate and the mechanical properties under cyclic loading.

Based on the literature review, the performance of stretchable conductive ink is very important in stretchable electronic applications. Many works have been done on the mechanical examination of different conductive inks to study their behavior under tensile or compression loadings. However, very limited research has been conducted on the effect of cyclic loading on conductive polymers. The main objective of this research was to study the behavior of proposed silver-based conductive ink under cyclic tensile loading. The ink was tested using two modes; stress controlled and strain controlled mechanisms. Lastly, the surface morphology and element composition of the ink after cyclic loading were analyzed using finite element scanning electron microscopy (FESEM) and energy disruptive X-ray (EDX) spectroscopy, respectively.

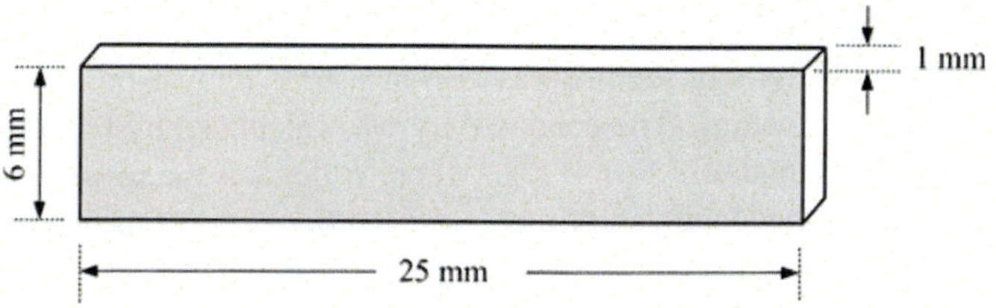

Fig. 25.1 Geometry of sample for cyclic tensile loading

25.2 Methodology

25.2.1 Materials and Fabrication of Ink

Poly(dimethylsiloxane) hydroxy terminated (PDMS-OH) of 110×10^3 g/mol molecular weight and viscosity of 50×10^3 cSt used as a binder, (3-glycidyloxypropyl) trimethoxysilane (ETMS) (236.34 g/mol molecular weight, $\geq$ 98% purity, 1.07 g/ml at 25 °C specific gravity) worked as cross-linking agent, octamethylcyclotetrasiloxane (D4) (0.956 g/ml density) acted as an organic solvent, toluene (92.14 g/mol molecular weight, 99% purity, 0.867 g/ml density) to control the viscosity, silver flakes (2–3.5 μm particle size) performed as a conductive filler and dibutyltin dilaurate (DBDTL) (631.56 g/mol molecular weight, 1.066 g/ml density, 95% purity) and acetic acid (99% purity) operated as catalysts.

A conductive ink was formed by dissolving silver flakes into the base polymer PDMS-OH. Toluene was added into the mixture and allowed it to mix for 24 h at room temperature on a magnetic or mechanical stirrer at 220–280 rpm so that all silver particles dissolved in the binder. After that, organic solvent (D4) and cross-linking agent (ETMS) were added into the mixture and again stirred for 5–10 min. Finally, few drops of acetic acid and DBDTL were poured into the mixture to cure. The conductive ink mixture was then dispensed on a rectangular-shaped mold and allowed it to cure at room temperature for 24 h. Three samples of rectangular shape geometry (Fig. 25.1) were cut from the sheet and proceeded to the experimentation.

25.2.2 Cyclic Tensile Loading

The cyclic tensile loading is also known as a step cycle test which can be conducted on different models of tensile testing machines like universal tensile machine (UTM). Instron 3366 UTM machine was utilized to observe the behavior of silver-based conductive ink under cyclic loading. Three samples of the ink were fabricated and gripped at the gauge length of 25 mm. The test was carried out at room temperature with a minimum load cell of 500 N and loading rate 10 mm/min. Three cyclic tensile tests for ten cycles were conducted on the samples (i) at variable strains (0 to 100%),

(ii) constant strain (50%) level, and (iii) constant stress (0.11 MPa). In addition, the electrical conductivity of the ink was also measured by two-point probe multimeter during tensile loading. The conductivity was calculated by taking reciprocal of resistivity that is equal to $\rho = RWt/L$, where R denotes the resistance in ohms, W is the width, L is the length, and t represents the thickness of the conductive ink.

25.2.3 Scanning Electron Microscopy (SEM)

The morphology of the samples with their surface modifications, crystal alignment, and failure behavior of silver-based conductive ink were investigated by SEM images. The microstructural analysis of the ink was carried out before and after application of the load. The morphology of the samples was observed in a HITACHI S-3400N-II SEM machine under 15 kV voltage depending on the properties of the conductive polymer ink. The magnification level of ×5000 was opted to study the morphology of the silver ink. In addition, the elemental composition of both the samples at different locations were computed and observed the detailed size of silver particles in respective samples.

25.3 Result and Discussions

25.3.1 Cyclic Loading Test Results

The engineering stress–strain curve of the silver-based conductive ink at different strain rates was obtained using simple uniaxial tensile testing machine. The sample was fixed at 25 mm gauge length and moved upward with the loading rate of 10 mm/min and 500 N load cell. The graph between tensile strain and tensile stress of 10 cycles is depicted in Fig. 25.2. It is worth noticing that by increasing tensile strain, mostly the stress of the ink also increased. The ink sample was first stretched to 10% with an initial length of 25 mm and then allowed to come back to its initial length at a loading rate of 10 mm/min. The same procedure was carried out for rest of the cycles with their respective strain rates. The maximum plastic deformation occurred in the first cycle. Also, the mechanical hysteresis and modulus of elasticity decreased as the number of cycles increased. Moreover, by increasing the number of cycles, the elastic recovery increased and the residual strain decreased which demonstrates that the current formulation of silver-based conductive ink is more elastic in nature. The elastic recovery (ER) [20] of each cycle was calculated by taking the ratio of the difference between stretched and recovered lengths to the change in length, as given in Table 25.1.

It can be seen from Table 25.1 that the elastic recovery increases in the first two cycles, while in third cycle, the ink showed less recovery than the previous cycles. The

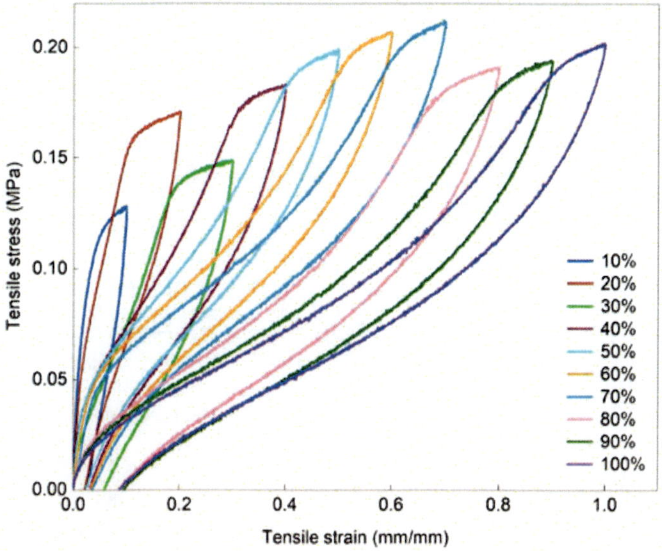

Fig. 25.2 Engineering stress–strain curve of silver ink at different tensile strains

Table 25.1 Elastic recovery of silver-based conductive ink at different strains with the initial length of 25 mm

Strain (%)	10	20	30	40	50	60	70	80	90	100
$L_{stretched}$	27.5	30.0	32.5	35.0	37.5	40.0	42.5	45.0	47.5	50.0
$L_{recovered}$	25.7	25.6	26.5	25.7	25.6	25.8	25.9	27.1	27.3	27.2
ER (%)	72.8	89.0	80.0	92.7	95.2	94.5	94.7	89.9	89.8	91.2

maximum elastic recovery and less residual strain were obtained at 50% strain and after that start decreasing slightly. Also, the conductivity of the ink before stretching was computed as 4.167×10^4 S/m, while after stretching the conductivity values were obtained as 2.778×10^3, 9.69×10^2, 5.411×10^2, 3.255×10^2, 1.743×10^2, and 1.68×10^2 S/m at 10–60% strains, respectively.

Furthermore, two cyclic tensile tests were also performed on two samples of the ink one by one, at a constant strain of 50% and constant stress of 0.11 MPa. The corresponding stress–strain curves of strain and stress controlled mechanisms are illustrated in Fig. 25.3. It can be seen from both the figures that the huge hysteresis curves were formed between the first and second cycles under constant strain and stress conditions, whereas the difference from second cycle onward the tenth cycle was almost negligible. This is due to the fact that the maximum plastic deformation occurred after the first cycle and ink behaved slightly visco-elastically from second cycle to tenth cycle with a small increment in the plastic deformation. The modulus of elasticity of each loading curve of every cycle also increased as the number of cycles increased. In stress controlled cyclic test, no asymmetric stress cycling occurred

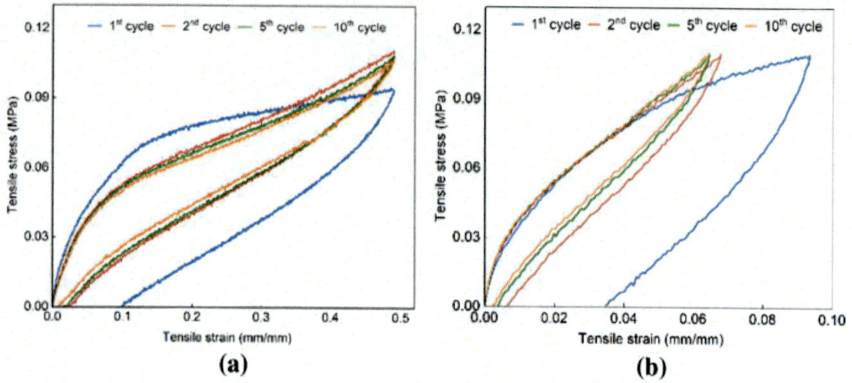

Fig. 25.3 Cyclic loading behavior of silver ink at **a** constant strain and **b** constant stress

which means that the silver conductive ink did not obey ratcheting behavior [21]. Hence, the mechanical properties of the ink changed between the first two cycles.

25.3.2 Morphological Analysis

The surface morphology of silver-based conductive ink was carried out using SEM machine. Two samples of the ink were analyzed, i.e., before and after ten cycles. Different locations of the surfaces of both the samples were examined under ×5000 magnification scale. Figure 25.4 represents the SEM images of silver conductive ink and its composition before loading.

It is found that the silver particles were fully spread homogeneously throughout the sample. Due to this, the conductive ink became denser and gave good hardness value. Moreover, greater than 90% by weight of the silver particles were present in the formulation. Likewise, the SEM images of silver conductive after undergoing ten times cyclic tensile loading were studied at two locations using ×5000 magnification scale. Figure 25.5 shows the microscopic view of conductive ink in terms of shape, size, and bonding.

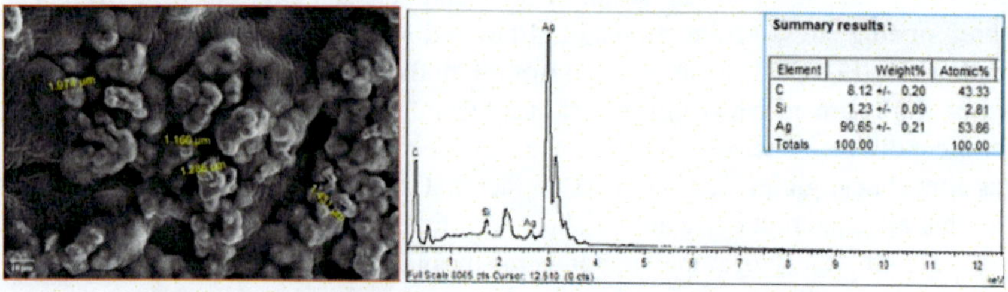

Fig. 25.4 Distribution of silver particles and composition of the ink before loading

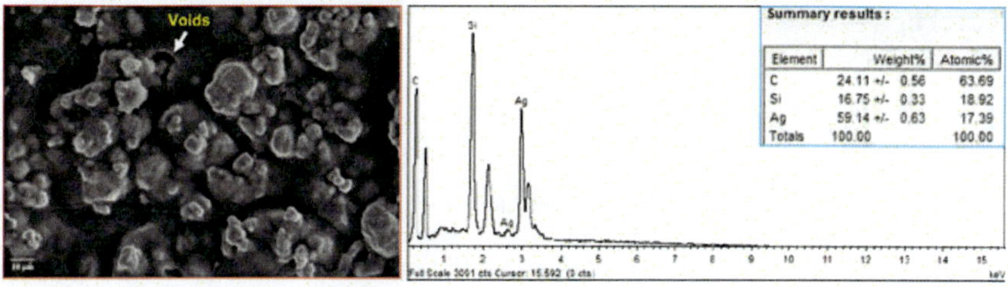

Fig. 25.5 Distribution of silver particles and composition of the ink after cyclic loading

It is also noted that few small circular-shaped voids were formed on the surface of the ink sample after loading and unloading ten times. This is due to the oxidation of carbon atoms which developed new functional groups. These functional groups then create the small pits on the sample and hence the surface roughness of the sample increased. Though, the formulated conductive ink possessed a good conductive path even after ten cycles. Furthermore, the composition of ink after cyclic test was also checked at different spectrums and found that about 60% of the silver particles were still present on the surface to provide conductivity.

25.4 Conclusion

The conductive dispersion was fabricated using PDMS-OH as a base polymer and silver particles as a conductive filler. The mechanical properties of the ink were evaluated under cyclic tensile loading test. Three tests were performed on a simple UTM machine under tensile loading by considering constant strain (mean stress relaxation cycling), constant stress (symmetric stress cycling), and variable deformations, i.e., from 0 to 100%. The behavior of ink under variable deformations was elastic because it exhibited larger elastic recovery than the plastic deformation in each cycle. In constant strain case, the elastic modulus increased as the number of cycles increased. In addition, the silver ink showed symmetric stress cyclic behavior at constant stress and did not undergo ratchetting. Finally, the surface morphology of the formulated ink before and after loading was also examined using SEM. The SEM images shows that the silver particles were homogenously spread throughout the surface and in a good condition even after ten cycles. Approximately more than 90% and 55% by weight of the silver particles were present in the samples before and after loading, respectively.

Acknowledgements The authors would like to acknowledge Universiti Sains Malaysia (USM) for giving the Short Term Grant 304.PMEKANIK.6315494 and the Research University Grant 1001.PMEKANIK.8014067 to complete this project. The authors would also like to thanks Polymer and Rubber labs of School of Materials and Mineral Resources Engineering, USM, for their support in the experimentation.

References

1. W. Dang, L. Manjakkal, W.T. Navaraj, L. Lorenzelli, V. Vinciguerra, R. Dahiya, Stretchable wireless system for sweat pH monitoring. Biosens. Bioelectron. **107**, 192–202 (2018). https://doi.org/10.1016/j.bios.2018.02.025
2. T. Chang et al., A general strategy for stretchable microwave antenna systems using serpentine mesh layouts. Adv. Funct. Mater. **27**(46) (2017). https://doi.org/10.1002/adfm.201703059
3. C.G. Núñez, W.T. Navaraj, E.O. Polat, R. Dahiya, Energy-autonomous, flexible, and transparent tactile skin. Adv. Funct. Mater. **27**(18) (2017). https://doi.org/10.1002/adfm.201606287
4. B. You, Y. Kim, B.K. Ju, J.W. Kim, Highly stretchable and waterproof electroluminescence device based on superstable stretchable transparent electrode. ACS Appl. Mater. Interfaces **9**(6), 5486–5494 (2017). https://doi.org/10.1021/acsami.6b14535
5. S. Zulfiqar, A.A. Saad, M.W. Chek, M.F.M. Sharif, Z. Samsudin, M.Y.T. Ali, Structural and random vibration analysis of LEDs conductive polymer interconnections. IOP Conf. Ser. Mater. Sci. Eng. **815**(1) (2020). https://doi.org/10.1088/1757-899X/815/1/012003
6. S. Zulfiqar et al., Alternative manufacturing process of 3-dimensional interconnect device using thermoforming process. Microelectron. Reliab. **127**, 114373 (2021). https://doi.org/10.1016/j.microrel.2021.114373
7. S. Zulfiqar, A.A. Saad, Z. Ahmad, F. Yusof, Z. Bachok, Structural analysis and material characterization of silver conductive ink for stretchable electronics. Int. J. Integr. Eng. **13**(7), 128–135 (2021). Available: https://publisher.uthm.edu.my/ojs/index.php/ijie/article/view/9536
8. N.A. Aziz, A.A. Saad, Z. Ahmad, S. Zulfiqar, F.C. Ani, Z. Samsudin, Stress analysis of stretchable conductive polymer for electronics circuit application (Chap. 8), in *Handbook of Materials Failure Analysis*, ed. by A.S.H. Makhlouf, M. Aliofkhazraei (Butterworth-Heinemann, 2020), pp. 205–224
9. S. Zulfiqar, A.A. Saad, Z. Ahmad, F. Yusof, K. Fakpan, Analysis and characterization of polydimethylsiloxane (PDMS) substrate by using uniaxial tensile test and Mooney-Rivlin hyperelastic model. J. Adv. Manuf. Technol. **16**(1) (2022). Available: https://jamt.utem.edu.my/jamt/article/view/6280
10. K.K. Chawla, *Composite Materials: Science and Engineering* (Springer International Publishing, 2019)
11. Y. Liu, F. Dai, A review of experimental and theoretical research on the deformation and failure behavior of rocks subjected to cyclic loading. J. Rock Mech. Geotech. Eng. **13**(5), 1203–1230 (2021). https://doi.org/10.1016/j.jrmge.2021.03.012
12. W. Grellmann, S. Seidler, *Polymer Testing* (Hanser Munich, 2013)
13. A. Głuchowski, W. Sas, Long-term cyclic loading impact on the creep deformation mechanism in cohesive materials. Materials (Basel) **13**(17) (2020). https://doi.org/10.3390/ma13173907
14. J. Beter et al., Viscoelastic behavior of glass-fiber-reinforced silicone composites exposed to cyclic loading. Polymers (Basel) **12**(9), 1862 (2020). https://doi.org/10.3390/POLYM12091862
15. E. Bociaga, M. Kula, K. Kwiatkowski, Analysis of structural changes in injection-molded parts due to cyclic loading. Adv. Polym. Technol. **37**(6), 2134–2141 (2018). https://doi.org/10.1002/adv.21872
16. K. Yu, H. Li, A.J.W. McClung, G.P. Tandon, J.W. Baur, H.J. Qi, Cyclic behaviors of amorphous shape memory polymers. Soft Matter **12**(13), 3234–3245 (2016). https://doi.org/10.1039/c5sm02781k
17. N. Rozo Lopez, J. Chen, C. Hopmann, A micromechanical model for loading and unloading behavior of fiber reinforced plastics under cyclic loading. Polym. Compos. **41**(9), 3892–3902 (2020). https://doi.org/10.1002/pc.25684
18. M. Ly, K.A. Khan, A. Muliana, Modeling self-heating under cyclic loading in fiber-reinforced polymer composites. J. Mater. Eng. Perform. **29**(2), 1321–1335 (2020). https://doi.org/10.1007/s11665-020-04663-7

19. X. Zheng, Q. Wang, J. Luan, Y. Li, N. Wang, Patterned metal/polymer composite film with good mechanical stability and repeatability for flexible electronic devices using nanoimprint technology. Micromachines **10**(10) (2019). https://doi.org/10.3390/mi10100651
20. M. Miraftab, *Fatigue Failure of Textile Fibres* (Elsevier Science, 2009)
21. S.K. Paul, A critical review of experimental aspects in ratcheting fatigue: microstructure to specimen to component. J. Mater. Res. Technol. **8**(5), 4894–4914 (2019). https://doi.org/10.1016/j.jmrt.2019.06.014

Chapter 26
Tensile and Dielectric Properties of Tin Dioxide Reinforced Deproteinized Natural Rubber Nanocomposites for Electrical Insulator

Noraiham Mohamad, Hairul Effendy Ab Maulod, Jeefferie Abd Razak, Mohd Sharin Ghani, Nor Hidayah Rahim, Mohd Hanafiah Mohd Isa, Dewi Suriyani Che Halin, Mohammed Iqbal Shueb, and Norshafarina Ismail

Abstract Tin dioxide, SnO_2 nanoparticles combined with deproteinized natural rubber could be an effective electrical insulator that inherits its parents' insulation and conductivity. The deproteinized natural rubber (DPNR) nanocomposites containing uncalcined SnO_2 nanofillers at 0.5, 1.0, 3.0, and 7.0 phr were investigated for tensile and dielectric properties. The nanocomposites were prepared using a Haake internal mixer through a melt compounding method and vulcanized by a semi-EV system. Their properties were explored and compared with the DPNR vulcanizate. The highest tensile strength of about 28 MPa was obtained at 3.0 phr SnO_2 loading. The elongation at break increases with the increase of nanofiller loading up to 3 phr and then decreases. With a rise in SnO_2, the dielectric constant decreased but increased once 7 phr SnO_2 was added. The nanocomposites exhibited the minimum dielectric constant at the optimum SnO_2 loading of 3.0 phr. Therefore, the SnO_2-reinforced DPNR nanocomposites are promising to be further explored for a stretchable electric insulator material.

N. Mohamad (✉) · J. Abd Razak · M. H. Mohd Isa
Fakulti Kejuruteraan Pembuatan, Universiti Teknikal Malaysia Melaka, Hang Tuah Jaya, 76100 Durian Tunggal, Melaka, Malaysia
e-mail: noraiham@utem.edu.my

H. E. Ab Maulod
Fakulti Teknologi Kejuruteraan Mekanikal & Pembuatan, Universiti Teknikal Malaysia Melaka, Hang Tuah Jaya, 76100 Durian Tunggal, Melaka, Malaysia

M. S. Ghani · N. H. Rahim
Fakulti Kejuruteraan Elektrik, Universiti Teknikal Malaysia Melaka, Hang Tuah Jaya, 76100 Durian Tunggal, Melaka, Malaysia

D. S. Che Halin
Faculty of Chemical Engineering Technology, Universiti Malaysia Perlis, Arau, Perlis, Malaysia

M. I. Shueb · N. Ismail
Radiation Processing Technology Division, Malaysian Nuclear Agency, Dengkil, Selangor, Malaysia

M. A. A. Mohd Salleh et al. (eds.), *Proceedings of the Green Materials and Electronic Packaging Interconnect Technology Symposium*, Springer Proceedings in Physics 289,
https://doi.org/10.1007/978-981-19-9267-4_26

Keywords Tensile properties · Dielectric properties · Rubber nanocomposites · Tin dioxide · Electrical insulator

26.1 Introduction

Insulation or packaging technology is crucial to the electrical cable industry. It entails creating effective heat dissipation systems, developing new insulation materials with high dielectric strength, and optimizing stress control techniques in cable joints and terminations [1]. Natural rubber (NR)-based electrical wire insulation dates back to the early days of the electrical industry. Historically, manufacturers discovered that adding various compounding components, such as mineral fillers, improves the performance of rubber. Until the 1930s, the only polymeric material used as a wire and cable dielectric was natural rubber-based electrical insulation [2]. However, because of their unsaturated backbone, NR-based insulation materials are inferior in terms of thermal degradation and ozone resistance. It has low oil and chemical resistance due to its non-polarity [3]. Therefore, many researchers investigate various NR-based blends and composites with other elastomers such as NR-CR [4], NR-EDPM, thermoplastics example, NR-HDPE, biodegradable polymers such as NR-Chitosan [5] and NR-PVA [6], and different fillers for this purposes. The electrical insulator highly desires materials having a low dielectric permittivity (ε') and a low dissipation factor (tan δ). Polymer materials with high tensile strength, low dielectric constant, and low tan δ are appealing for electrical insulation. They are relatively strong, flexible, lightweight, and simple to produce. Sadly, polymer materials often have a low resistance to thermal degradation. Therefore, adding suitable filler material is required to develop a dielectric polymer matrix composite [7] and prolong the service life of polymer-based insulators.

Filler materials, such as micro or nanoscale fibres and particles, have significantly improved the dielectric properties of NR-based materials. The surface and volume resistivity of various organic fibres and compounds (oil palm empty fruit bunch, potato starch nanocrystal) and inorganic fillers [alumina trihydrate and nano-montmorillonite (MMT)] were found to be increased. Furthermore, they can reduce leakage current and increase resistance to tracking and erosion [3]. These nanofillers acted as a barrier to the movement of electrical charges. Meanwhile, nanofillers such as silicon dioxide and titanium dioxide were reported to decrease the partial discharge (PD) pulse number compared to polymers without fillers. On the other hand, the iron oxide–carbon nanotubes and Nickel–Cobalt-Zinc ferrite nanoparticles have been shown to increase PD activities while decreasing electrical resistivity [3]. As a result, highly conductive filler materials are likely to harm NR composites' dielectric properties. Therefore, adding semiconductor fillers to the NR matrix is the better solution for improving dielectric strength and mechanical properties.

The rutile structure SnO_2 is an *n*-type environmentally friendly semiconductor with 3.6 eV bandgap energy [8]. However, nanocrystalline SnO_2 has exceptional optical and electrical properties due to its high surface-to-volume ratio and quantum confinement effects. It offers tremendous advantages for various applications, such as photocatalysis, optoelectronics, and spintronics devices [9]. The SnO_2 nanoparticles dispersed polyethylene oxide (PEO) matrix have also been investigated to prepare high-performance nanocomposite solid polymer electrolytes (NSPEs). It's utilized to make transparent organic resistive memory devices, gas sensors, solar cell electrodes, electrochromic windows, and energy storage/converter electrolytes [10]. Incorporating polymers and other fillers into NR improves its mechanical properties and electrical insulation. Most studies focus on synthetic rubbers such as silicone rubber and styrene butadiene rubber, and only limited research focuses on NR's electrical properties. Therefore, more research is needed to create electrical insulating materials containing NR as blends or composites. This paper is a preliminary work toward the effort. It focuses on the feasibility of producing DPNR reinforced with uncalcined SnO_2 nanocomposites via a melt compounding method. The uncalcined SnO_2 nanofillers varied from 0.5, 1, 3, and 7 in a DPNR matrix. The tensile properties and dielectric characteristics of the DPNR vulcanizate and nanocomposites were investigated. These properties were supported by the morphological analyses conducted on the fractured surfaces by field emission scanning electron microscopy (FESEM).

26.2 Methodology

26.2.1 Raw Materials

The formulation recipe is summarized in Table 26.1. The main materials used in this study were deproteinized natural rubber (DPNR) and tin dioxide (SnO_2). DPNR with Mooney viscosity UML (1 + 8) at 150 °C of 60 ± 6 MU was supplied by Edutech Supply & Services. The commercial ultrafine nano SnO_2 of > 99.9%C was purchased from SAT NANO, China, with a specific surface area of 70 m^2/g and a bulk density of 6.48 g/cm^3. The SnO_2 was used without calcination. Sulfur, zinc oxide, stearic acid, N-cyclohexyl-2-benzothiazole sulfonamide (CBS) and tetramethyl thiuram disulfide (TMTD), N-(1,3-dimethylbutyl)-N′-phenyl-p-phenylenediamine (6PPD) were used as vulcanization agent, activator, primary and secondary accelerators, and antidegradant, respectively. They were supplied by Lembaga Getah Malaysia.

The compounding process was performed per ASTM D-3192 in a Haake internal mixer. During the mixing procedure, DPNR was firstly masticated, and after 0.5 min, compounding chemicals (zinc oxide, stearic acid, and 6PPD) were added before adding SnO_2 at 1 min. After sweeping at 2 min, the compound was dropped onto a two-roll mill machine. Finally, sulfur and accelerators (CBS and TMTD) were added and mixed for 3 min before the compounds were dumped and conditioned at room temperature for 24 h. The compounds were compressions moulded at 160 °C and 110 kg/force pressure at the respective cure times, t_{90}. The step was to produce sheets of nanocomposites with suitable thicknesses for further testing and analyses.

Table 26.1 The formulation for preparation of DPNR nanocomposites

Materials/Chemicals	Loading [a](phr)
DPNR	100
Sulfur	1.5
Zinc oxide	5.0
Stearic acid	2.0
CBS	1.0
TMTD	0.3
6PPD	2.0
SnO_2	0.0, 0.5, 1.0, 3.0, 7.0

[a]Part per hundred rubber

26.2.2 Testing and Analyses

The tensile tests were conducted following ASTM D412 using a Shimadzu Universal Testing Machine (UTM) at a crosshead speed rate of 500 mm/min. The test to measure the dielectric properties was conducted using the transmission-line permittivity measuring method. It was performed using the microwave vector network analyzer (MVNA model Anritsu 32747D Agilent) with a GPC7 coaxial cable. The testing procedure used the S-parameter transmission/reflection method. The measurement frequency range was 500 MHz–18 GHz. Meanwhile, the tensile fractured surfaces of the DPNR nanocomposites were examined under the FESEM Model Gemini SEM 500 from Carl Zeiss at 10,000 × magnifications.

26.3 Result and Discussions

26.3.1 Tensile Properties

Figure 26.1 depicts the tensile characteristics of DPNR vulcanizate (0 phr SnO_2) and DPNR nanocomposites at different nanofiller loadings. The tensile strength was observed to reduce at low SnO_2 nanofiller loadings. But, it dramatically increased as the SnO_2 nanofillers loading increased by more than 1 phr. This finding demonstrated a similar trend to a prior study by Mankar et al. [11] when 2–12 phr nano tin oxide was added to the SBR matrix. Adding 3 phr SnO_2 nanofillers to the DPNR matrix enhanced the average tensile strength by about 10%. However, the tensile strength decreased slightly at too-low or too-high nanofiller loading of 0.5, 1.0, and 7 phr. It could be due to the homogeneity effect and compromised dispersion at these loadings. The limited dispersion particles formed aggregates and agglomerates. They have acted as stress concentrators, causing the nanocomposites to experience premature failure. From Fig. 26.1a, the modulus of the DPNR nanocomposites was

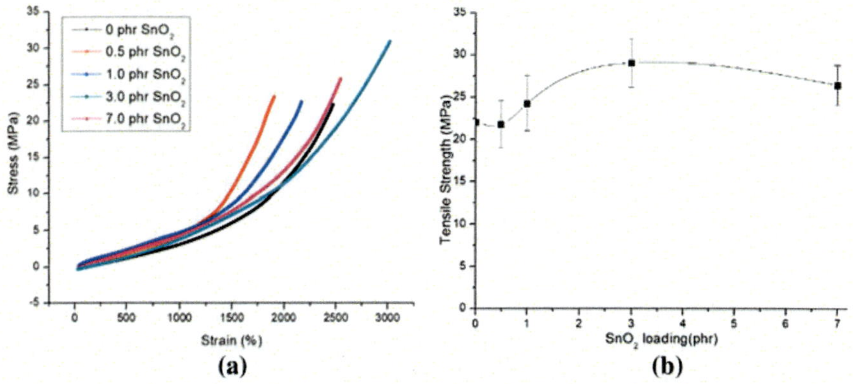

Fig. 26.1 **a** Tensile stress versus strain curve and **b** tensile strength of the SnO_2 reinforced DPNR nanocomposites

higher than the DPNR vulcanizate. The modulus varies with the percentage of strain experienced by the nanocomposites under the tension deformation—the nanocomposites filled with higher SnO_2 nanofillers than 1.0 phr experienced reductions in their modulus. The pattern was observed clearly at strain percentages of more than 1000%. It shows that the presence of SnO_2 nanofillers softens the DPNR chains. It is in good agreement with the highest elongation at break (%EB) demonstrated by the nanocomposite loaded by 3 phr SnO_2. The nanocomposites exhibited a slight reduction of %EB at 7 phr SnO_2 loadings. In contrast, the low nanofiller loadings at 0.5 and 1.0 phr showed a higher modulus at this strain level. It manifested as the stiffness of the materials, where the polymeric chains were harder to deform under stress. This condition corresponds to SnO_2 distribution and dispersion characteristics. The agglomerates impeded the chain mobility and caused premature failure, reducing the composites' elongation. This observation is in good agreement with their morphological characteristics.

26.3.2 Dielectric Properties

Relative permittivity or dielectric constant (ε') describes polymers' ability to store charges. Meanwhile, dielectric loss or loss tangent ($\tan\delta$) quantifies a dielectric substance's inherent electromagnetic energy dissipation. It is also the measure of electromagnetic wave energy absorbed when it passes through a medium. Significant influences on the dielectric characteristics of elastomers are the polymeric macromolecules and their chain characteristics, type of filler, and loading level [12]. Meanwhile, the types of polarization possible in polymeric systems are electronic, atomic, dipole, and interfacial [6]. Figure 26.2 depicts the frequency (f)-dependent real permittivity, which reflects the dielectric constants (ε') of the SnO_2-reinforced

DPNR nanocomposites from 500 to 18,000 MHz (log f of 2.7–4.3). In SnO_2-reinforced DPNR nanocomposites, the frequency-dependent dielectric behaviour exhibited three different regions: (1) at low f of log 2.7–2.9, (2) transition between low to high f at log 3.0–3.9, and (3) at high f of 4.0–4.3. From Fig. 26.2, the dielectric constant is unsteady in the low-frequency zone (up to log f 2.9 Hz). In the transition region, it decreases gradually and reaches its minimum values at log f of 3.7–3.9 MHz, which varies according to different nanofiller loadings. At the high-frequency region (log f 4.0–4.3 MHz), the line is dramatically increased and levelled to its initial value at the low-frequency zone. The greater dielectric constant in the lower frequency range could be interfacial polarization (IP) or Maxwell–Wagner (M-W) polarization. The polarizations happen due to electrically heterogeneous materials, specifically polymer composites and blends [12]. In our study, the SnO_2 nanofillers have higher electrical conductivity than the DPNR particles. Therefore, it created persistent dipoles at the polymer-filler interfaces, improving the polarization capability and enhancing the dielectric constant. As the fraction of SnO_2 nanofillers rises, more active sites are produced, causing the dielectric constant to rise. These dipoles are substantially larger than orientation, ionic, and electronic dipoles.

However, differences in SnO_2 level of distribution and dispersion played an important role in affecting the dielectric constant. It is noted that the low-frequency effect on nanocomposites with a lower amount of SnO_2 loadings has relatively fluctuated. It could result from low polarization due to anisometry in the electrical conductivity of SnO_2 particles, their cluster size, and orientation in the DPNR matrix [12]. The morphological characteristics of the nanocomposites support this finding. At low

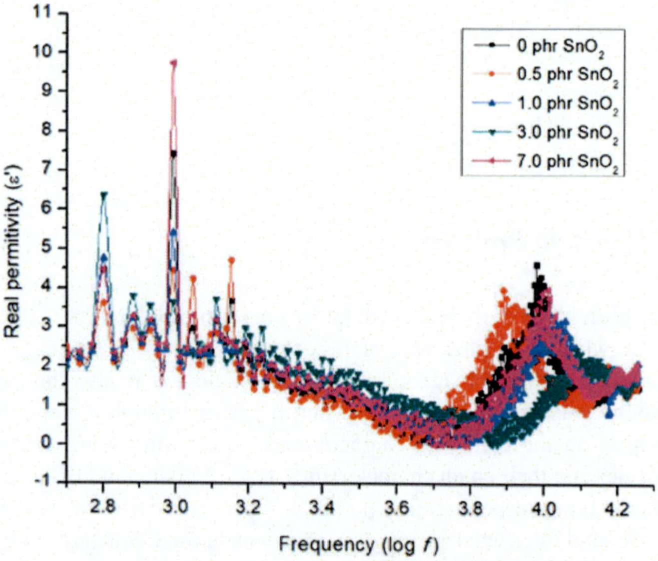

Fig. 26.2 Dielectric constant of SnO_2-reinforced DPNR nanocomposites at log f of 2.7–4.3

Table 26.2 The dielectric constant of the DPNR nanocomposites at the frequency of ~ 8000 MHz

SnO_2 loading (phr)	Dielectric constant, ε'	tan δ
0	2.2248	3.7796
0.5	3.3502	1.0408
1	1.2232	3.2914
3	0.4449	1.8848
7	1.6341	13.7834

SnO_2 loadings, lesser homogeneous distributions of the dispersed aggregates and agglomerates were observed. Meanwhile, in the second region (transition frequency range), the dielectric constant decreases as frequency increases due to the inability of induced dipoles to align in the direction of the applied electric field. The reduction was prominent in the nanocomposites with SnO_2 loadings. The lowest value was obtained by the DPNR nanocomposites reinforced by 3.0 phr SnO_2 at $\log f = 3.90$. The rise in the dielectric constant in the third region (high-frequency zone) might be caused by electronic polarization. The effect of filler loading on the dielectric properties was further characterized at a frequency of 8000 MHz ($\log f = 3.90$) (Table 26.2). At this frequency, the average dielectric constant was the lowest with the variation of nanofiller loading in the DPNR matrix except at 0.5 phr. The energy loss factor was the highest for the nanocomposites loaded with 7 phr SnO_2. Too high a dielectric constant and tan δ are not desirable for the electrical insulative properties. The dielectric behaviour of the composites corresponds to the nanocomposites' morphological characteristics.

Meanwhile, SnO_2 nanofillers at a high loading create an inorganic–organic interface that increases polarization, increasing the dielectric constant. The nano-filler situated closer together in SnO_2 aggregates or agglomerates increases the time needed for space charge polarization. Hence, the DPNR nanocomposite has a more extended relaxation period [13]. Relaxation peaks are the fluctuations observed in the samples. Numerous relaxation peaks caused by the main chain or flexible side group indicate the importance of dielectric property studies of filled polymers [13]. Thus, the dielectric constant of less homogenous SnO_2-reinforced nanocomposites was greater than the rest of the samples. The high dielectric constant of the nanocomposites is also manifested at the frequency of ~ 10,000 MHz ($\log f = 4.0$). Therefore, the composites' dielectric constant depends on their polarization and relaxation mechanisms corresponding to the level of filler loadings [14] and wave frequency.

26.3.3 Morphological Characteristics

Morphological characteristics of DPNR nanocomposites at low and high SnO_2 loadings are depicted in Fig. 26.3. The images show the difference between agglomerates formed at 0.5 (low loading) and 7.0 phr (high loading) SnO_2. The tensile and dielectric properties were highly affected by the filler loadings. The agglomerates may

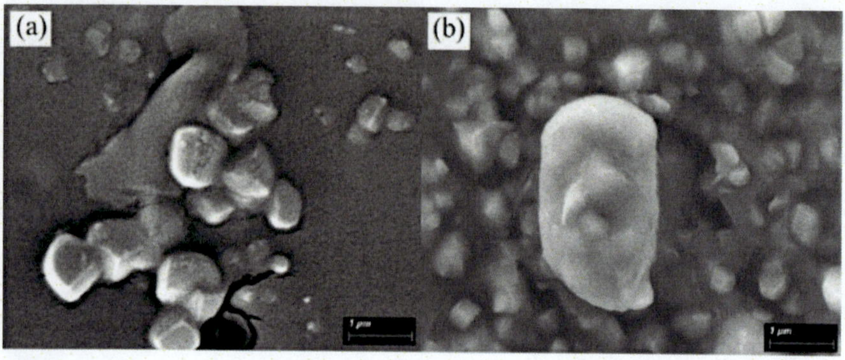

Fig. 26.3 Micrographs showing aggregates and agglomerates of **a** 0.5 phr and **b** 7.0 phr nanofiller loadings at 10,000 × magnifications

be played a role as stress concentrators and correspond to the tensile properties exhibited by the nanocomposites. In a report by Jobish et al. [13], the dielectric constant decreases significantly and becomes practically frequency independent by forming interpenetrating polymer networks (IPN) structures. In Fig. 26.3b, the SnO_2 agglomerates at high loading were dispersed closer together, forming dense filler network structures, which might accelerate the polarization under the electromagnetic wave. Hence, the dielectric constant was increased. The agglomerates were larger, with sizes of up to 5 μm in diameter. In contrast, the agglomerates at low filler loading are smaller, but the distribution was least homogenous. In this study, more uniform nanofiller agglomerates with even sizes situated slightly farther away at optimum loading (between 1.0 and 3.0 phr) resulted in dielectric stability (the image is not included). At this level, the homogenously distributed agglomerate resulted in close interaction between its components [15]. Under the influence of an electric field, the efficient network established by the scattered SnO_2 agglomerates may efficiently disturb the interfacial polarization throughout the nanocomposites. The obtained dielectric properties might be enhanced by the calcination of SnO_2 prior compounding process [16].

26.4 Conclusions

In this investigation, SnO_2-reinforced DPNR nanocomposites were fabricated at nanofiller loadings of 0.5, 1.0, 3.0, and 7.0 phr. The SnO_2 nanofillers efficiently strengthen the DPNR chains at a suitable filler loading of 3.0 phr. At this level, the nanocomposites had a tensile strength of 30 MPa and the greatest elongation at break. The frequency-dependent dielectric constant value of the SnO_2-reinforced DPNR nanocomposites varies with nanofiller loading. The SnO_2-reinforced DPNR nanocomposites exhibited lower dielectric permittivity and energy dissipation than

DPNR vulcanizates. The condition was highly prevailed by composites at optimal nanofiller loadings of 3.0 phr. Therefore, the SnO_2-reinforced DPNR nanocomposites have promising potential to be adopted in an electrical insulation application. Yet, further studies on structure-properties-performance for various factors and service life assessment for a specific type of insulator are required.

Acknowledgements The authors would like to thank Tin Industry (Research and Development) Board, Kementerian Air, Tanah Dan Sumber Asli, Malaysia, for KHAS-TIN/2021/FKP/C00007 grant, Universiti Teknikal Malaysia Melaka (UTeM), and Malaysian Nuclear Agency for facilities and supports.

References

1. J.-U. Lee et al., The status of electrical insulation technology in Korea. IEEE Electr. Insul. Mag. **14**(2), 18–25 (1998)
2. C. Zuidema et al., A short history of rubber cables. IEEE Electr. Insul. Mag. **4**(27), 45–50 (2011)
3. S. Junian et al., Natural rubber as electrical insulator: a review. J. Adv. Res. Mater. Sci. **61**(1), 1–12 (2019)
4. A. Das et al., Evaluation of physical and electrical properties of chloroprene rubber and natural rubber blends. KGK-Kautschuk Gummi Kunststoffe **58**(5), 230–238 (2005)
5. J. Jobish, C. Nakason, Dielectric properties of natural rubber/chitosan blends: effects of blend ratio and compatibilization. J. Non. Cryst. Solids **357**(7), 1816–1821 (2011)
6. J. Jobish et al., Dielectric properties and AC conductivity studies of novel NR/PVA full-interpenetrating polymer networks. J. Non. Cryst. Solids **358**(8), 1113–1119 (2012)
7. K. Silakaew et al., Highly enhanced frequency- and temperature-stability permittivity of three-phase poly(vinylidene-fluoride) nanocomposites with retaining low loss tangent and high permittivity. Results Phys. **26**, 104410 (2021)
8. S. Lei, H. Yi, Research on magnetic property of environmental friendly material SnO_2: Mn, S. E3S Web of Conferences **237**, 01023 (2021)
9. Zulfiqar et al., Zn-doped SnO_2 nanoparticles: Structural, optical, dielectric and magnetic properties. Int. J. Mod. Phys. B **31**, 1750234 (1–11) (2017)
10. P. Dhatarwal et al., Significantly enhanced dielectric properties and chain segmental dynamics of PEO/SnO_2 nanocomposites. Polym. Bull. (2020). https://doi.org/10.1007/s00289-020-03215-2
11. R.V. Mankar et al., Evaluation of thermal and mechanical properties of Styrene-Butadiene Rubber-nanocomposite by using tin oxide as filler. Mater. Today: Proc. **15**(3), 371–379 (2019)
12. K. Ravikumar et al., Dielectric properties of natural rubber composites filled with graphite. Mater. Today: Proc. **16**, 1338–1343 (part 2) (2019)
13. F. Wang et al., Research on the dielectric properties of nano-ZnO/silicone rubber composites. IOP Conf. Ser.: Mater. Sci. Eng. **231**, 012060 (2017)
14. F. Danafar, M. Kalantari, A review of natural rubber nanocomposites based on carbon nanotubes. J. Rubber Res. **21**(4), 293–310 (2018). https://doi.org/10.1007/BF03449176
15. F.K. Saidu, G.V. Thomas, Synthesis, characterization and dielectric studies of poly(1-naphthylamine)–tungsten disulphide nanocomposites. SN Appl. Sci. **2**(1158) (2020). https://doi.org/10.1007/s42452-020-2889-7
16. N.H. Rahim et al., Effects of filler calcination on structure and dielectric properties of polyethylene/silica nanocomposites. IEEE Trans. Dielectr. Electr. Insul. **26**(1), 284–291 (2019). https://doi.org/10.1109/TDEI.2018.007796

Part II
Green Materials and Technology

Chapter 27
FEM on Short-Term Chloride Penetration on Carbon Fibre Reinforced Polymer (CFRP) Strengthened to RC Beam

Amiruddin Mishad, Mohd Hisbany Mohd Hashim, Azmi Ibrahim, Oh Chai Lian, Ameer Haqimie Zainal, and Mohd Raizamzamani

Abstract The strengthening or repairing of reinforced concrete beams using carbon fibre-reinforced polymer (CFRP) has gained extensive attention in the field of structural engineering, realising that the CFRP has several unique mechanical properties such as high corrosion resistance, faster installation, low maintenance cost, lightweight, and high strength compared to the conventional materials. Nowadays, civil engineers are facing problems due to the corrosion of reinforced steel which leads to the deterioration of concrete. The strengthening technique was used in this research is near-surface mounted (NSM) which the CFRP reinforcement was embedded in grooves to protect them against corrosion and high temperatures. The finite element model for RC beams strengthened with NSM-CFRP was developed using LUSAS software version 19.0, a commercially available code, and specific materials behaviour models. The results obtained from FE analysis including the load–deflection relationship, maximum stress, failure mode, and crack pattern behaviour in the concrete at the mid-span section of the beam for all specimen beams have been discussed and analysed. The objectives of this study were to develop FE models of all the specimen beams, to determine the flexure performance of NSM-CFRP under chloride penetration at 28 days, and to compare the results from analytical analysis and experimental tests from previous research with the same dimensions of beams. The beams were tested and loaded in four points bending until the cracks were developed. The specimen beams were installed with CFRP and then compared with the control beam. The behaviour of these beams was simulated in a finite element model and developed in the LUSAS program. Based on the analytical analysis, found that the flexural strength of the beam strengthened with the NSM-CFRP rod has significantly increased the flexural performance by 10% compared to other strengthening materials. It has good flexural performance despite having been exposed to chloride penetration for a short-term duration. The failure modes obtained from the analyses for all the specimen beams are flexural and shear failures, and the

A. Mishad (✉) · M. H. Mohd Hashim · A. Ibrahim · O. C. Lian · A. H. Zainal · M. Raizamzamani
School of Civil Engineering, College of Engineering, Universiti Teknologi MARA (UiTM) Shah Alam, Shah Alam, Selangor, Malaysia
e-mail: amiruddinmishad@uitm.edu.my

M. A. A. Mohd Salleh et al. (eds.), *Proceedings of the Green Materials and Electronic Packaging Interconnect Technology Symposium*, Springer Proceedings in Physics 289,
https://doi.org/10.1007/978-981-19-9267-4_27

crack pattern and maximum stress mostly occurred at the lowest bottom mid-span of the beam.

Keywords CFRP rod · CFRP plate · NSM · CFRP · FEM · Flexural performance of beam

27.1 Introduction

Carbon fibre-reinforced polymer (CFRP) is a composite material made up of carbon fibre and a polymer matrix. The carbon fibre provides strength and stiffness, while the polymer serves as a cohesive matrix to bind the fibres together [1–5]. CFRP has widely been used for repairing and strengthening reinforced concrete structures. CFRP is one of the most reasonable options to replace the use of steel reinforcement bars in the strengthening and repair of new and existing structures. CFRP can not only improve the structure's performance, but it can also extend the structure's life, giving it a longer lifespan. Previous research shows that CFRP has been utilised to strengthen and improve the flexural performance of a reinforced concrete beam [6]. The excessive load applied on the beam and the condition of the steel-reinforced bar within the beam could cause flexural failure. The beam will be unable to support load if the steel reinforcement bar corrodes, causing the strength to decrease.

RC beams can be strengthened in a variety of methods to avoid early failure, extend service life, and improve the strength and load-carrying capacity of the concrete beam. One common method of strengthening RC beams is to use externally bonded sheets or plates with high-strength materials like fibre-reinforced polymer, polyester, steel plates, wire mesh, and textile fabrics. In comparison to other materials, CFRP is commonly used to strengthen RC beams. CFRP has several advantages, including non-corrosiveness, high longitudinal tensile strength, stiffness, strength-to-weight ratio, insect, fungi, and chemical resistance, low heat transmissibility, and easy installation [6–9]. Besides that, CFRP is more durable than steel reinforcement bars because it can withstand the effects of the environment and it has a great saltwater corrosion resistance.

Laboratory testing is required to determine the flexural strength of reinforced concrete beam but requires many samples and hence involves high cost and time consumption. Therefore, another way of determining flexural strength is by undergoing finite element modelling and simulation, which can assist to speed up the process to obtain the data. The main concept behind FEM is to break down the modelling into simpler shapes and geometries known as finite elements [10, 11]. The focus of this research is to develop a finite element model (FEM) for strengthening reinforced beams using CFRP and to analyse the model to see which methods produce the best results in terms of flexural strength [12–18].

This research is to identify which type of CFRP material is suitable was used as strengthening material and to investigate the flexural performance of the reinforced concrete beam under chloride penetration. For this research, finite element analysis is adopted to investigate the behaviour of different types of CFRP materials strengthened to RC beams by using LUSAS software. The results from finite element models from the LUSAS were compared and evaluated with the previous experimental results from previous studies. The data collected consists of the load–deflection relationship, crack pattern, strain in the concrete at the mid-span section of the beam, and failure modes of all types of beams strengthened with CFRP.

27.2 Methodology

Generally, experimental work and analytical research was used to solve the civil engineering problems which can cut cost of material, reduce time production, and provide accurate result. Meanwhile, certain problems can be solved successfully by experimental work only or just by utilizing theory. However, a combination of both techniques which are theoretical modelling and experimental works was required for most problems.

The main topic of this section on the analytical program was used to evaluate the experimental findings and observe the failure modes analytically. All tested beams including control beam specimens were developed the model and analysed by using LUSAS [19] software version 19.0. The analytical work, which involved modelling, meshing, selecting geometry and material types, and assigning the supports and loadings, is shown in Fig. 27.1.

27.3 Result and Discussions

The findings and results on the flexural performance of reinforced concrete specimen beams consisting of a control beam strengthened with a CFRP rod and plate using the near-surface mounted (NSM) strengthening technique. The analytical data for all specimen beams were analysed and evaluated in this chapter. The results comprised the flexural parameters of load, deflection, maximum stress, deformation, failure mode, and cracking. The data was presented and illustrated in graphs, figures, and tables to explain the results effectively and clearly. The findings from all specimen beams were reported in terms of load–deflection curves, stress distribution, failure modes, and cracking behaviour.

27.3.1 Load Deflection Behaviour

Figure 27.2 shows the load versus deflection graph of all specimen beams obtained from finite element analysis, consisting of the control beam, the beam strengthened with the CFRP rod, the CFRP vertical plate, and the CFRP horizontal plate.

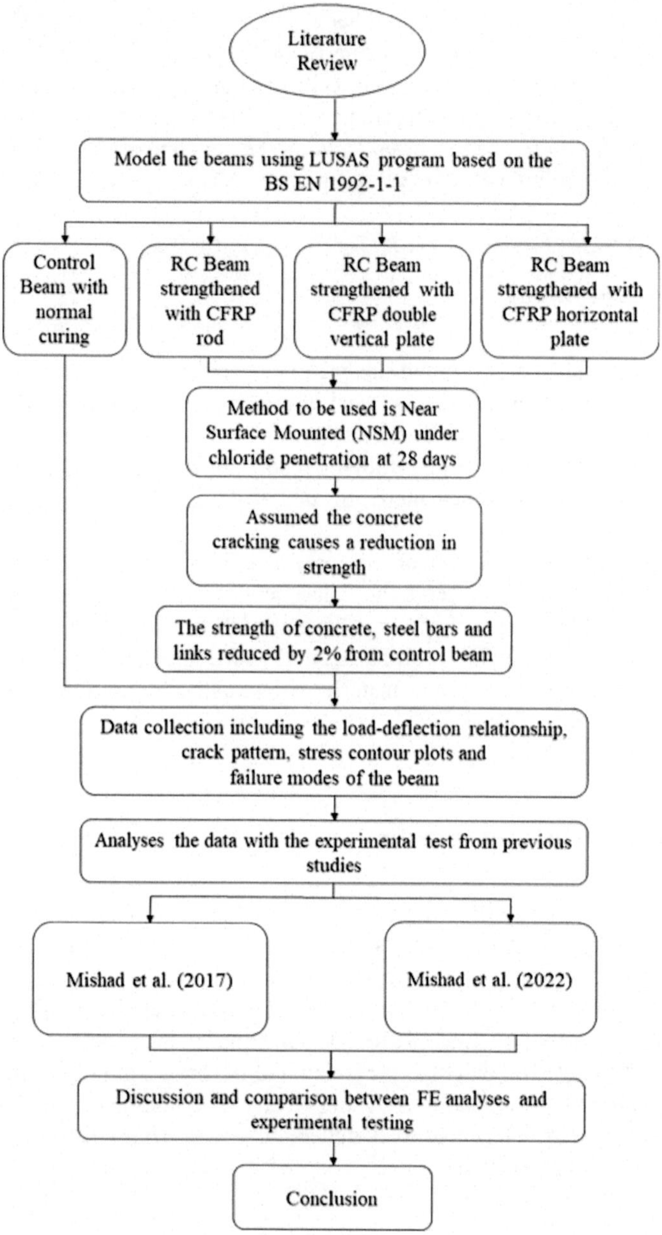

Fig. 27.1 Research methodology process

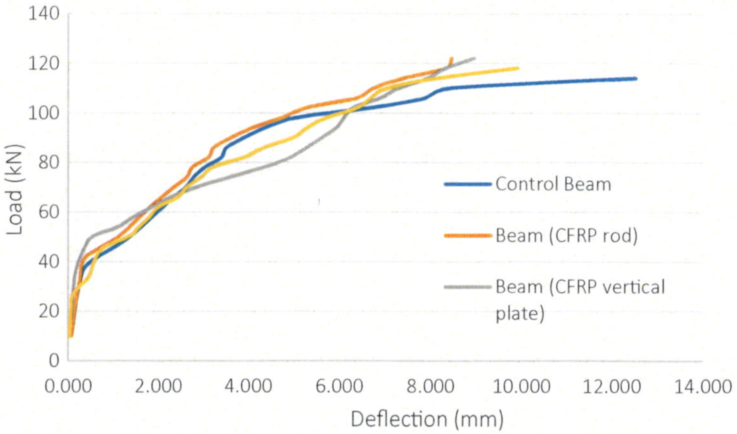

Fig. 27.2 Load versus deflection of the specimen beam under chloride penetration at 28 days

The ultimate load and deflection should be compared for all four beams. Then, the results have been identified which specimen beam has excellent flexural performance. Before that, the data were adjusted to obtain a proper result to be compared with experimental testing from previous research. The result obtained from the numerical analysis was adjusted by multiplying the load by two because the beam was subjected to a monotonically increasing four-point bending up to two-point loading to failure of the specimen. Also, the deflection was multiplied by − 1 because the deformation occurred downward. Based on Fig. 27.2, the deflection of the concrete beam was increased gradually with the increment of the load until it reaches the peak value, which is the maximum load.

Figure 27.3 and Table 27.1 show the ultimate load of all the specimen beams with displacement obtained from the numerical analysis. In this study, the highest displacement that has been observed is the control beam, which is 12.52 mm with an ultimate load of 114 kN. It is because of the absence of strengthening material in the beam. Therefore, the displacement for the control beam is slightly higher compared to the other specimen beams. The ultimate load that the beam can cater to until it fails or cracks for reinforced beams strengthened with CFRP rod, CFRP vertical plate, and CFRP horizontal plate under chloride penetration at 28 days is 122 kN, 122 kN, and 118 kN, respectively. The displacement for specimen beams strengthened with CFRP comprised 8.44 mm for a beam with a vertical plate, 8.95 mm for a beam with a vertical plate, and 9.90 mm for a beam with a horizontal plate. To be summarised, the beam strengthened with CFRP materials has the highest value of ultimate load compared to the control beam, but the displacement was smaller compared to the control beam because it was affected by the resistance of CFRP materials in the tension zone.

Plates 1, 2, 3, and 4 (Figs. 27.4, 27.5, 27.6 and 27.7) show the deformation of the specimen beams, including beams strengthened with CFRP. The deformation of the control beam as shown in Plate 14 has the greater deformation, which is 12.52 mm

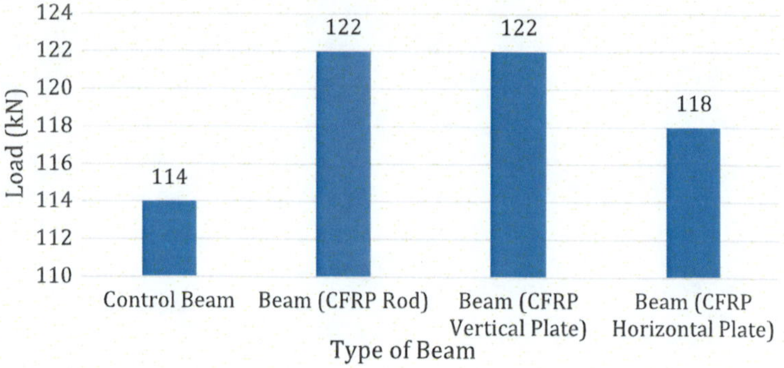

Fig. 27.3 Ultimate load for beam specimen

Table 27.1 Load and deflection of specimen beams

Beam	Ultimate load (kN)	Displacement (mm)
Control beam	114	12.52
Beam (CFRP rod)	122	8.44
Beam (CFRP vertical plate)	122	8.95
Beam (CFRP horizontal plate)	118	9.90

compared to the beam with CFRP rod, CFRP vertical plate, and CFRP horizontal plate, which have a deflection of 8.44 mm, 8.95 mm, and 9.90 mm, respectively. Based on the deformation mesh in the analytical analysis, it shows that there was significantly different deflection downward between normal concrete and a beam strengthened with NSM-CFRP.

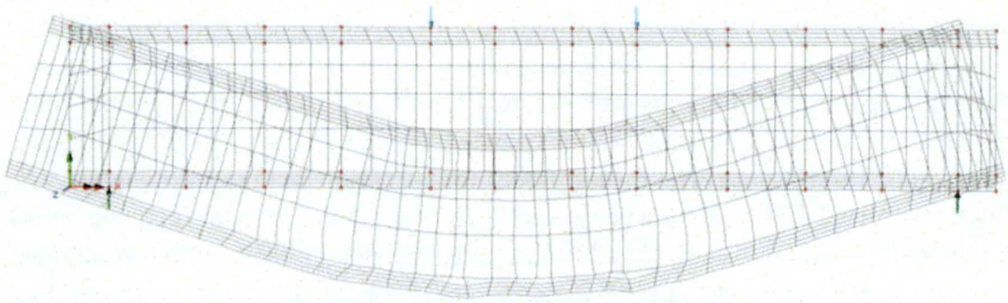

Fig. 27.4 *Plate 1* Deformation of the control beam model

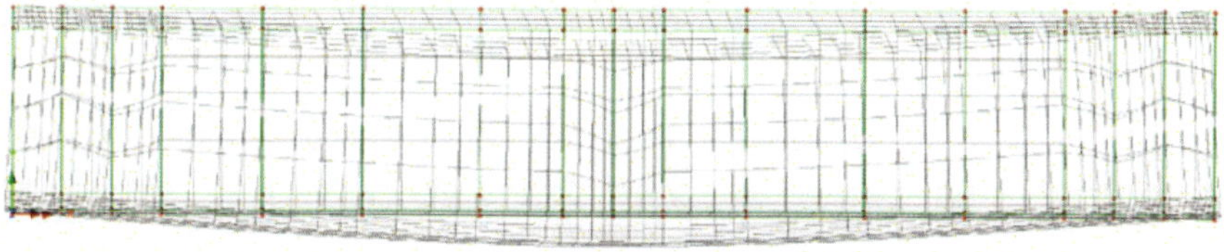

Fig. 27.5 *Plate 2* Deformation of the concrete beam strengthened with CFRP rod

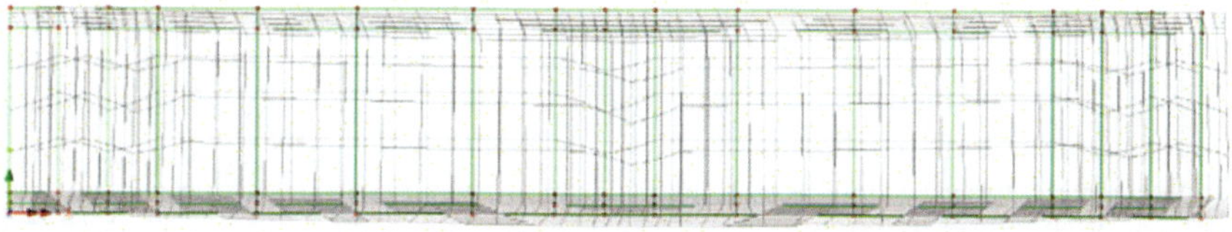

Fig. 27.6 *Plate 3* Deformation of the concrete beam strengthened with CFRP vertical plate

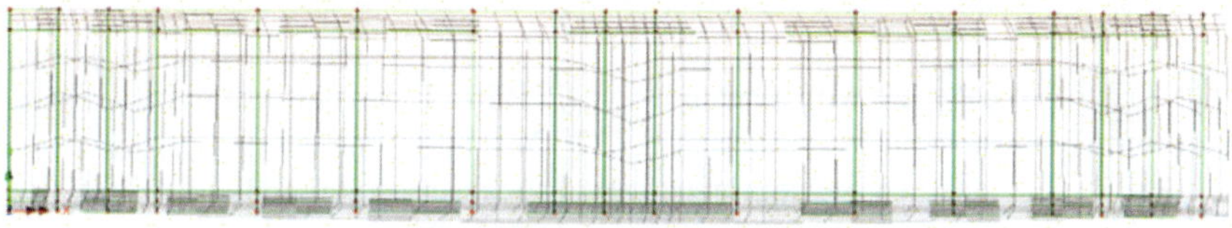

Fig. 27.7 *Plate 4* Deformation of the concrete beam strengthened with CFRP horizontal plate

27.3.2 Maximum Principal Stress Behaviour

Plate 5 (Fig. 27.8) shows the result of the stress contour for the control beam model in LUSAS software. The concrete structures are known to be very robust in compression but very weak in tension. Therefore, most concrete structures must be reinforced with steel or fibre reinforcements. Tensile areas were represented by areas that were yellow or orange in colour, whereas compressive areas were represented by areas that were green in colour. The blue colour represented the loadings and supports acting on the beam.

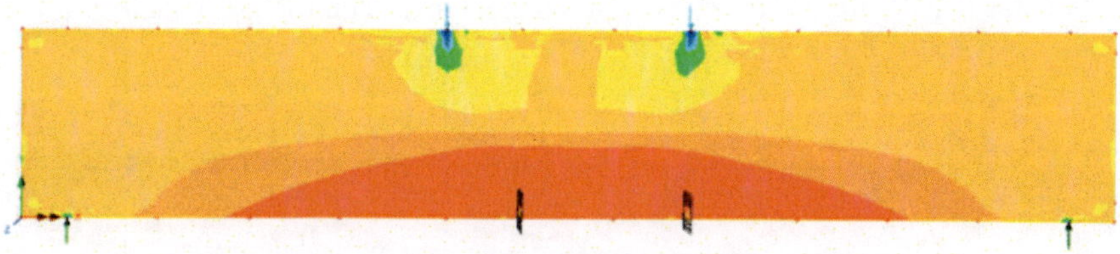

Fig. 27.8 *Plate 5* Maximum principal stress contour plots of the control beam

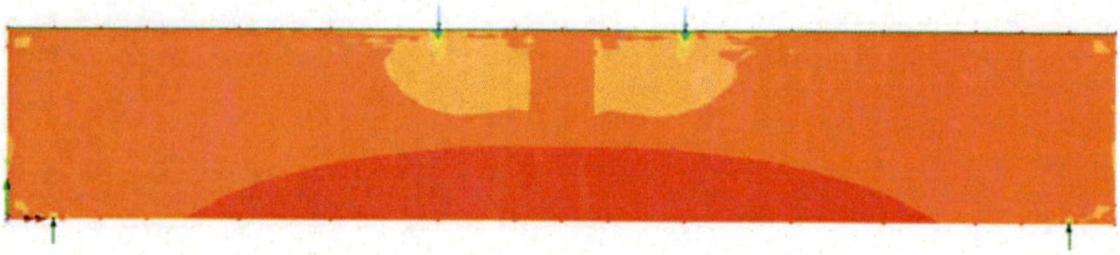

Fig. 27.9 *Plate 6* Maximum principal stress contour plots of the beam with CFRP rod

Plate 6 (Fig. 27.9) shows the stress contour for the beam strengthened with the CFRP rod, which failed at the ultimate load of 122 kN. The red colour in the contour plots shows the high tensile stress which the steel bars and CFRP rod were in the tension zone. The result can be seen in the software, and it suggests that the maximum stress in the beam is located at the bottom middle span of the beam, which can be seen to be more active. The compressive has a small presence at the beam and it can be seen in green colour. The same goes for the blue colour, representing the loadings acting on the beam.

Plate 7 (Fig. 27.10) has an identical maximum stress contour, which is a beam with CFRP vertical plate with a beam with a CFRP rod. It is because it has the same maximum load for both beams which is 122 kN. The placement of CFRP material is in the middle between the nominal cover of the beam, so the red colour indicates the high tensile stress that the steel bars and CFRP double vertical plate placed at the bottom of the beam. The maximum stress in this beam is also located at the bottom mid-span of the beam, which can be seen to be a more active area. The stress at the top of the beam has a small coverage in the beam.

Plate 8 (Fig. 27.11) shows the maximum stress of the beam with a CFRP horizontal plate, which failed at the maximum load of 118 kN. The red colour at the bottom of the beam has low coverage for this beam compared to another beam strengthened with CFRP. It is because the CFRP horizontal plate placed at the bottom face of the beam with the thickness of grooves is 5 mm, thus the tensile stress at the lowest part of the beam is the smallest. The stress contour of the beam with CFRP horizontal plate has the same as the control beam, but the red colour has small coverage at the tensile zone.

To be concluded, the steel should be provided in the concrete beam to resist the tensile stresses because it has a relatively low strength in tensile. The concrete will transfer the tensile stress to the steel reinforcement in the control beam; therefore, the

Fig. 27.10 *Plate 7* Maximum principal stress contour plots of the beam with CFRP vertical plate

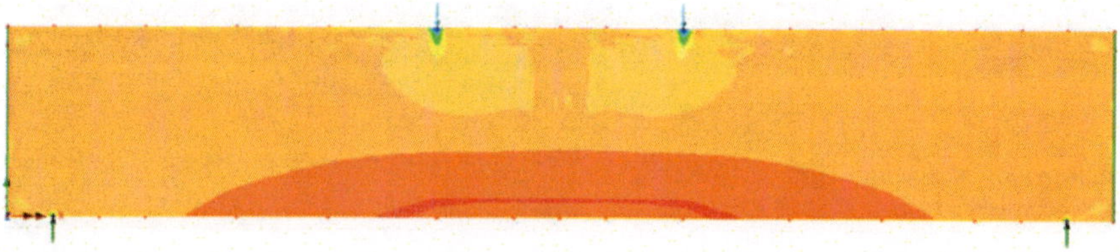

Fig. 27.11 *Plate 8* Maximum principal stress contour plots of the beam with CFRP horizontal plate

application of strengthening material to be used to resist more tensile stress. Based on the stress contour, beam with CFRP horizontal plate has the lowest maximum stress among three of the CFRP materials. Thus, in the beam specimens strengthened with CFRP material, the concrete will transfer the tensile stress to the steel bars and CFRP material.

27.3.3 Failure Modes and Cracking Pattern

The failure modes of the reinforced concrete beam specimens from analytical testing were compared with the experimental testing which the images were taken during the lab experiment. There are ten different colours in the LUSAS software to indicate the area that are most affected by cracking and displacement from range 1 to 10 scales as shown in Fig. 27.12.

In Fig. 27.13 shows the failure mode of a control beam without any strengthening material from the experimental testing. According to the experimental observation and findings, once the specimen beam has achieved its maximum compressive strain, the beam tends to fail in flexure with concrete being crushed in the compressive zone. The concrete will transfer the tensile stress to the steel reinforcement if there is no

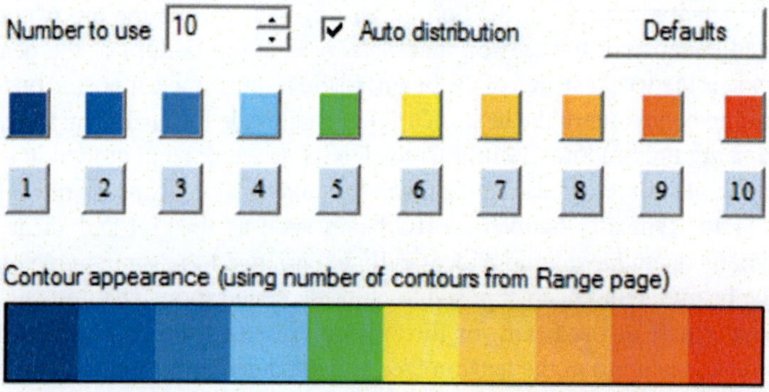

Fig. 27.12 Contour appearance

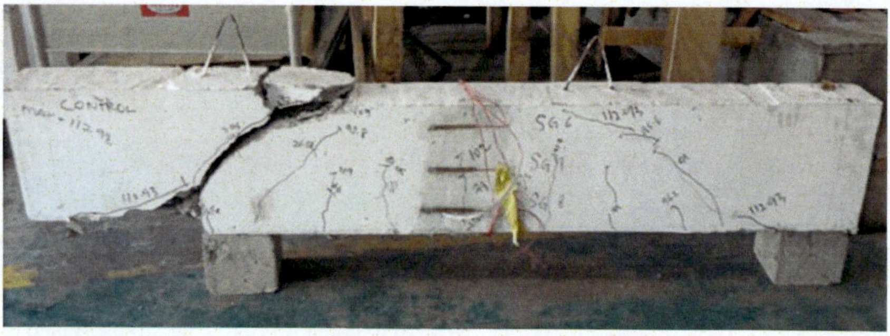

Fig. 27.13 Failure mode of the control beam from experimental testing

Fig. 27.14 *Plate 9* Failure mode of the control beam from LUSAS software

strengthening material in the control beam. The steel will yield and transmit the stress elsewhere, causing the concrete to crush as the strain rises along with the stress. However, the analytical analysis is unable to model several degradations; as a result, Plate 9 (Fig. 27.14) shows only the cracking patterns and not the behaviour of the concrete being crushed.

A failure mode and crack pattern can be observed in Plates 9 and 10 (Figs. 27.14 and 27.15) for the control beam. The crack has started to happen when the load is increased because it undergoes increasing loading. As the crack pattern extended and widened during the breakpoint, the beam began to fail. The crack pattern can be seen in the tension zone from the bottom part of the beam to the top of the beam. It happened because of the tension that developed at the bottom of the beam. The tension zone is the weakest area of the beam, so the failure crack appeared on that side of the bending behaviour of the beam. The crack pattern in the analytical analysis was evaluated using the Smoothed Multi Crack Concrete Model and plotted in contour.

The mode of failure of beam specimens for analytical analysis is shown in Plate 10 (Fig. 27.15) typically happen on RC beam such as flexural and shear failure. All specimens beam have almost identically experienced the same mode of failure despite the beam having been exposed to chloride penetration. The mid-span of the beam was the part of the beam for flexural failures tend to occur. Most of cases, flexural cracks occurred at the tension face and extended at the neutral axis. For the analytical analysis, it can be observed that the initial crack started at the mid-span

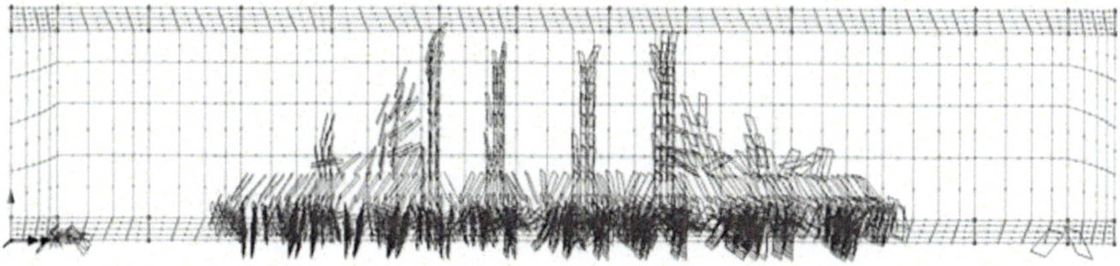

Fig. 27.15 *Plate 10* Cracking patterns of the control beam

of the beam and spread as it moved closer to the support. The tension face had the greatest widths of cracks, and it was reduced as it further towards the support.

The occurrence of shear failure in RC beams is known as diagonal tension failure. When the beam reaches the failure load, shear failure will happen. The size of the micro crack will occur and eventually develop a major crack when the load is increased. Shear cracks start at support and continue up the beam to the loading point at the top. In this analytical analysis, control beam in the Plate 10 (Fig. 27.15) shows the large crack on the mid-span of the beam upon reaching maximum load compared to the other specimen beam.

The beam specimens, including the control beam, beam strengthened with CFRP rod, CFRP vertical plate, and CFRP horizontal plate from the analytical analysis, failed at the ultimate load of 114 kN, 122 kN, 122 kN, and 118 kN, respectively. As shown in Plate 11 (Fig. 27.16), the mid-span of the beam has a small red colour indicator because there was the lowest affected of the cracking and displacement for the beam with CFRP rod. Plate 12 (Fig. 27.17) shows the crack pattern for the beam with CFRP rod has smaller than the control beam and CFRP plates. The flexural cracks that occurred on the beam strengthened with CFRP rod can be observed that the cracking spreading at a lower rate than the control beam. The cracks can be assumed to be smallest at the tension face than the control beam. The shear failure occurs when the debonding between CFRP materials and concrete occurs before the beam reaches the failure mode.

The application of CFRP vertical plate in a reinforced concrete beam helps to prevent or delay the cracks in the beam from expanding. The crack pattern obtained from the analytical analysis shows identical results for beams strengthened with CFRP vertical plate and horizontal plate. Plates 14 (Fig. 27.19) and 16 (Fig. 27.21) show that both specimen beams for CFRP plates vertically and horizontally have undergone nearly the same mode of failure and type of failure, including flexural

Fig. 27.16 *Plate 11* Failure mode of the beam strengthened with CFRP rod

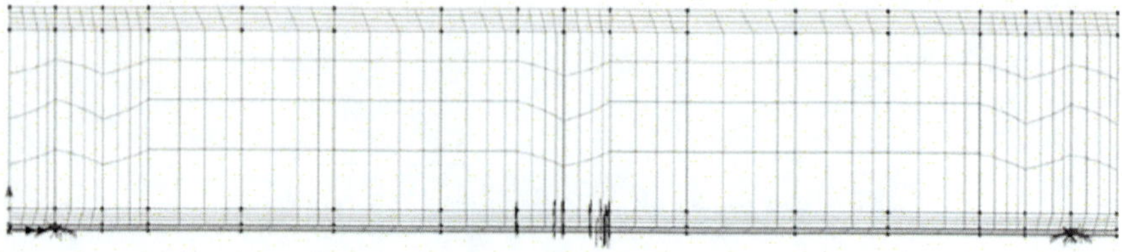

Fig. 27.17 *Plate 12* Cracking patterns of the beam with CFRP rod

and shear failure. Furthermore, shear failure will develop when the beam reaches the failure load after experiencing flexural failure first. The specimen beam with CFRP plates prevented cracks from forming on the bonded beam and delayed the process of cracking as shown in Plates 14 (Fig. 27.19) and 16 (Fig. 27.21). Plates 13 (Fig. 27.18) and 15 (Fig. 27.20) show the specimen beam with the CFRP vertical plate has the most affected crack and displacement compared to the CFRP horizontal plate, where the red colour on the beam is greater than the horizontal plate.

The analytical experiment shows that the reinforced beams with strengthening materials had the slightly highest maximum flexural reading in comparison to the

Fig. 27.18 *Plate 13* Failure mode of the beam strengthened with CFRP vertical plate

Fig. 27.19 *Plate 14* Cracking patterns of the beam with CFRP vertical plate

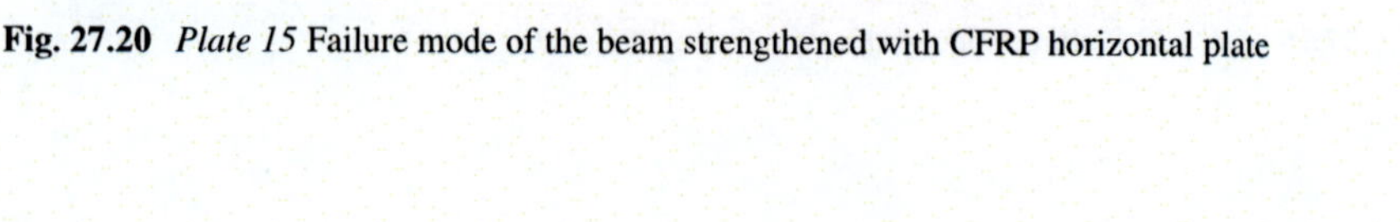

Fig. 27.20 *Plate 15* Failure mode of the beam strengthened with CFRP horizontal plate

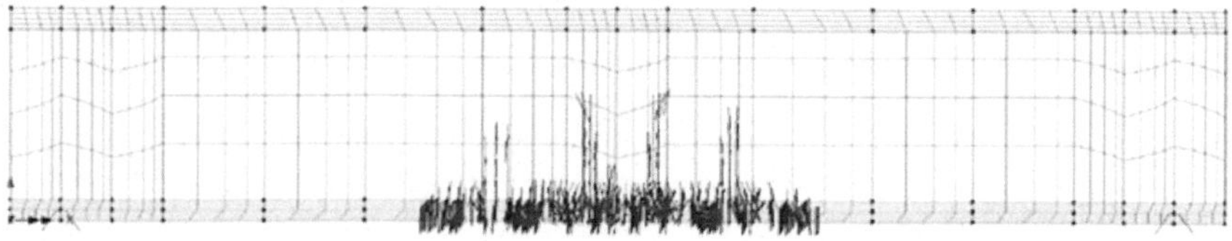

Fig. 27.21 *Plate 16* Cracking patterns of the beam with CFRP horizontal plate

control beam. In addition, the deflection data for the ultimate load for specimen beam strengthened with CFRP lower when compared to the control beam. Based on the failure mode and crack pattern as discussed above, using CFRP strengthening materials through the NSM method may increase the tension zone performance of RC beams, which can enhance the structural performance of beams reinforced with CFRP.

27.4 Conclusion

In this analytical analysis, beam specimens strengthened with CFRP using near-surface mounted (NSM) technique were modelled and analysed to determine the maximum flexural strength. The maximum flexural load obtained from FE analysis for the control beam is 114 kN, while for beam strengthened with CFRP rod is 122 kN, beam with CFRP vertical plate is 122 kN, and beam with CFRP horizontal plate is 118 kN. These results were obtained in which the beam has been exposed to the chloride penetration for a short-term duration, thus the concrete and tensile strength were reduced by 2%. Despite the specimen beams have been exposed to saltwater, it still shows the highest maximum flexural strength compared to the normal control concrete.

References

1. A. Siddika, K. Saha, M.S. Mahmud, S.C. Roy, M.A.A. Mamun, R. Alyousef, Performance and failure analysis of carbon fiber-reinforced polymer (CFRP) strengthened reinforced concrete (RC) beams. SN Applied Sciences **1**(12), 1–11 (2019)
2. A.W. Al Zand, W.H.W. Badaruzzaman, A.A. Mutalib, A.H. Qahtan, Finite element analysis of square CFST beam strengthened by CFRP composite material. Thin-Walled Structures **96**, 348–358 (2015)
3. B. Almassri, R. Francois, F. Al-Mahmoud, Behaviour of corroded reinforced concrete beams repaired with NSM CFRP rods, experimental and finite element study. Compos. B **92**, 477–488 (2016)
4. En 1992-1-1 English: Eurocode 2: design of concrete structures—part 1-1: general rules and rules for buildings (Authority: The European Union Per Regulation 305/2011, Directive 98/34/EC, Directive 2004/18/EC)

5. D.R. Haseen, R.K. Pandey, Nonlinear analysis for concrete beam strengthened in flexure with near surface mounted CFRP. International Journal of Scientific Engineering and Technology Research **02**(06), 455–461 (2013)
6. A. Mishad, M. Hisbany, M. Hashim, A. Ibrahim, S.B. Saidin, Bending performance of RC beams strenghtened with near surface mounted carbon fiber reinforced polymer (CFRP) plate or rod under long term saltwater exposure. International Journal of Civil Engineering and Technology **9**(8), 304–317 (2018)
7. A. Khene, N.-E. Chikh, H.A. Mesbah, Numerical modelling of reinforced concrete beams strengthened by NSM-CFRP technique. International Journal of Research in Chemical, Metallurgical and Civil Engineering (2016)
8. A. Mishad, M.H.M. Hashim, A. Ibrahim, M. Nafis, Flexural performance of reinforced concrete beams strengthened with double vertical carbon fiber reinforced polymer (CFRP) plate using near surface mounted (NSM). Adv. Sci. Lett. **23**(5), 4458–4462 (2017)
9. A. Mishad, M. Hisbany Mohd Hashim, A. Ibrahim, A. Newman, Double vertical carbon fiber reinforced polymer plates strengthened to reinforced concrete beams for six months saltwater exposure. IOP Conference Series: Materials Science and Engineering **513**(1) (2019)
10. N. Tankut, A.N. Tankut, M. Zor, Analiza drvnog materijala metodom konačnih elemenata. Drvna Industrija **65**(2), 159–171 (2014)
11. A. Mishad, M.H.M. Hashim, A. Ibrahim, M.H. Jamal, D.A. Baboh, RC Beams strengthened with near surface mounted carbon fiber reinforced polymer plate at short term saltwater exposure. Lecture Notes in Civil Engineering **215**, 987–998 (2022)
12. I. Olofin, R. Liu, The application of carbon fibre reinforced polymer (CFRP) cables in civil engineering structures. International Journal of Civil Engineering **2**(7), 1–5 (2015)
13. H.Y. Omran, R. El-Hacha, Nonlinear 3D finite element modelling of RC beams strengthened with prestressed NSM-CFRP strips. Constr. Build. Mater. **31**, 74–85 (2012)
14. P. Sabol, S. Priganc, Shear strengthening of concrete members using NSM method. Procedia Engineering **65**, 364–369 (2013)
15. A. Sakbana, M. Mashreib, Finite element analysis of CFRP-reinforced concrete beams. Revista Ingenieria de Construccion **35**(2), 148–169 (2020)
16. F.U.A. Shaikh, Effect of cracking on corrosion of steel in concrete. International Journal of Concrete Structures and Materials. SpringerOpen (2018)
17. A. Srirekha, K. Bashetty, Infinite to finite: an overview of finite element analysis. Indian J Dent Res **21**, 425–432 (2010)
18. K. Sumangala, C.M.F. Dani, Finite element analysis of RC beam subjected to corrosion—a review. International Journal of Research in Engineering and Science (IJRES). **9**(6), 59–63 (2021)
19. The LUSAS User's Manual, Finite Element Analysis Ltd, Nonlinear Analysis of a Concrete Beam 1–22(2016)

Chapter 28
The Application of Microbes to the Fly Ash-Based Alkali-Activated Material Performance Containing Slag

Andrie Harmaji and Januarti Jaya Ekaputri

Abstract Microbial agents are often used to enhance the performance of cementitious material. This study presents the utilization of microbes to the properties of fly ash (FA)-based alkali-activated material (AAM) paste, with ground granulated blast furnace slag (GGBFS) as partially FA substitution. Variations of GGBFS-to-FA weight ratio were between 0 and 50%. Raw material and specimens are characterized with X-ray fluorescence, X-ray diffraction, scanning electron microscope, compression test, porosity test, workability test, and setting time. Supplement of microbes to FA-based AAM enhances the strength in a limited amount up to 20% slag replacement due to its workability. AAM without GGBFS addition showed a 28-day compressive strength of 23.00 MPa. The 28-day compressive strength of 20% GGBFS substitution without and with microbes addition was 32.90 MPa and 33.50 MPa, respectively. Microbes application to 20% GGBFS substitution in AAM resulting total porosity (pf) of 19% which was lower than without microbes addition. Microbes increased the workability but fastened the setting time of fresh AAM. The X-ray diffraction (XRD) and scanning electron microscope (SEM) analysis showed the microbes image and precipitation dominated by calcite.

Keywords Alkali-activated material · Fly ash · GGBFS · Microbes · Calcite

---

A. Harmaji
Department of Metallurgical Engineering, Institut Teknologi Sains Bandung, Kabupaten Bekasi 17530, Indonesia
e-mail: harmaji@itsb.ac.id

J. J. Ekaputri (✉)
Department of Civil Engineering, Institut Teknologi Sepuluh Nopember, Kota SBY 60111, Indonesia
e-mail: januarti@ce.its.ac.id

A. Harmaji · J. J. Ekaputri
Konsorsium Riset Geopolimer Indonesia, NASDEC, Kampus ITS, Sukolilo, Surabaya 60111, Indonesia

M. A. A. Mohd Salleh et al. (eds.), *Proceedings of the Green Materials and Electronic Packaging Interconnect Technology Symposium*, Springer Proceedings in Physics 289,
https://doi.org/10.1007/978-981-19-9267-4_28

28.1 Introduction

Most industrial production lines consist of highly complex processes that create a high amount of by-product materials. Taking an example from the steelmaking industry, ground granulated blast furnace slag (GGBFS), specifically, is a waste material from the steelmaking industry made from blast furnaces cooled by water spraying method. Its crushed product was used for road base, coarse aggregate, and clinker raw material. One ton of steelmaking produces around 20% slag waste, which is quite high. Another example of industrial waste is fly ash, a by-product from coal-generated energy power plants. Both fly ash and GGBFS can be used as precursor in alkali-activated material (AAM), an inorganic polymer binder that can be made from silicon (Si) and aluminum (Al) and activated by alkali solution [1, 2]. Fly ash and GGBFS are common precursors used for geopolymer [3], and their ratio affects its binding mechanism and properties. Inclusion of 25% fly ash results in higher compressive strength than 100% GGBFS precursor. However, when fly ash content increases to 50% it decreases the compressive strength of the resulting product [4]. Both fly ash and GGBFS contain SiO2, Al_2O_3, which are the backbone of geopolymerization. Unfortunately, most of these industrial wastes are only stored as waste stockpiles when they hold an immense potential to become source materials for other production processes, such as constructions. There are economic, environmental, and social needs to increase the value of GGBFS waste resulting in reduced quantities of disposal materials. Economically, GGBFS can be used as Portland cement replacement materials due to its ability to enhance mechanical properties of concrete materials [5–7]. When GGBFS and fly ash are further processed, it is also less harmful to the environment due to less industrial waste produced. To process fly ash and GGBFS, silica and alumina in fly ash were reacted with alkali activator to generate aluminosilicate materials with CaO in GGBFS facilitated the setting of the materials. High content of GGBFS can reduce the amorphous phase in AAM paste. This also leads to autogenous shrinkage and formation of microcracks, which can decrease the strength of specimens [8]. On the other hand, microbes have been reported to be combined with AAM, which can affect the pore size in AAM. The aerobic microbes have been used by various researchers in their previous studies. One of the species, such as Bacillus *pasteurii* which is also reclassified as *Sporosarcina pasteurii* [9, 10], generates calcite in pores and tiny cavity areas [11]. Previous research has studied the ureolytic microbes addition to increase the strength of concrete and self-healing properties [12, 13]. This study focused on carbonic anhydrase-producing microbes to induce precipitation of calcium carbonate ($CaCO_3$) as micropore fillers in AAM. The addition of microbes in fly ash and GGBFS-based AAM paste forms a mechanism called biomineralization. We also discuss the ability of microbes to enhance the physical and mechanical properties of the paste. The compressive strength of the resulting AAM paste is also presented along with the morphology behavior and compound of resulting specimens using XRD and SEM characterization methods.

28.2 Methodology

28.2.1 Materials

The class-F fly ash with a fineness of 82.6% (passed from the sieve no. 325) and density of 2.5 g/cm^3 was obtained from Suralaya Coal Fired Power Plant Banten, Indonesia. The GGBFS was obtained from PT Krakatau POSCO, Cilegon, Banten, Indonesia, with a fineness and density of 98.2% and 2.9 g/cm^3, respectively. The oxides composition of these materials is shown in Table 28.1 by using X-ray fluorescence (XRF) method. Technical grades of sodium hydroxide and water glass were purchased from Bratachem, Bandung. The AAM was mixed as a combination of fly ash and GGBFS as the binder. As the main binder, the fly ash has some irregular particles that may affect the setting time even though the lime content is less than 10% [14, 15].

Binder consists of GGBFS to replace the fly ash's weight from 0–50%. A local commercially available microbes consortium for concrete was added with a dosage of 400 mL per m^3 to the mixture. The microbes were dominated with Rhizopus spp. It was obtained from Bioconc Foundation Center, Indonesia. The mass ratio of sodium silicate (WG) to sodium hydroxide (NaOH) was kept constant at 2.0. Alkali activator was prepared by mixing NaOH solution of 4 M with WG. The mix design per m^3 is listed in Table 28.2.

The microbes consortium used in this study was in the ureolytic liquid form. AAM was made by mixing class-F fly ash, GGBFS, and alkali activator to result in slurry. Then, the microbes in liquid form were added and mixed until homogenous. It was then poured into a cubical mold with a size of 50 mm × 50 mm × 50 mm. The specimens were cured under moist conditions for 7 and 28 days. Workability and setting time test were conducted to determine the fresh properties of AAM paste. Workability test was conducted by using mini slump cone (Fig. 28.1) with a size of top diameter of 25 mm, bottom diameter of 50 mm, and the height of 75 mm. The size of flow was measured at 30 min after the cone was lifted. Flow test is presented in Fig. 28.2.

The compressive strength of resulting AAM was measured using universal testing machine (UTM) confirmed to ASTM C-39. Debris from compression strength was collected for characterization purposes. The X-ray diffraction (XRD) measurement was performed on Philips Diffractometer PW1710 with Cu as anode. Resulting diffraction pattern was compared to the Joint Committee on Powder Diffraction Standards (JCPDS). Scanning electron microscope (SEM) measurement was performed

Table 28.1 Chemical composition of fly ash and GGBFS (mass%)

Material	SiO_2	Al_2O_3	CaO	Fe_2O_3	MgO	K_2O
Fly ash	52.30	26.57	6.00	7.28	2.13	0.76
GGBFS	34.20	11.70	41.20	1.43	8.81	–

Table 28.2 AAM mix design with different fly ash and GGBFS ratio

No.	Mix code	Fly Ash (kg)	GGBFS (kg)	Slag content (%)	WG (kg)	NaOH (kg)	Microbes (mL)
1	F100S0	1050.7	0.0	0	376.9	188.4	0.0
2	F90S10	945.6	105.1	10	376.9	188.4	0.0
3	F80S20	840.5	210.1	20	376.9	188.4	0.0
4	F70S30	735.5	315.2	30	376.9	188.4	0.0
5	F60S40	630.4	420.3	40	376.9	188.4	0.0
6	F50S50	525.3	525.3	50	376.9	188.4	0.0
7	F100S0M	1050.7	0.0	0	376.9	188.4	400
8	F90S10M	945.6	105.1	10	376.9	188.4	400
9	F80S20M	840.5	210.1	20	376.9	188.4	400
10	F70S30M	735.5	315.2	30	376.9	188.4	400
11	F60S40M	630.4	420.3	40	376.9	188.4	400
12	F50S50M	525.3	525.3	50	376.9	188.4	400

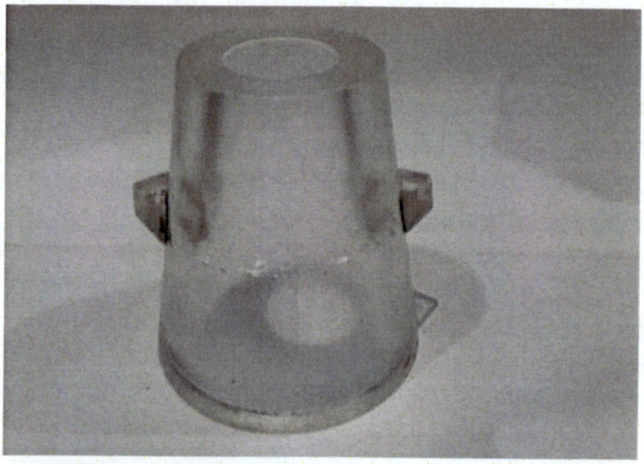

Fig. 28.1 Mini slump cone used for workability test

Fig. 28.2 Flow test of AAM paste

with HITACHI SU3500. Since the specimens are not conductive, it was coated with gold or carbon with ion sputtering method. Both of these characterizations were conducted at the Center of Advanced Sciences (CAS) Institut Teknologi Bandung, Indonesia.

28.3 Result and Discussions

This section discusses the results from workability test, setting time test using Vicat apparatus, porosity test according to ASTM C642-13 [16], compressive test, XRD, SEM, and EDS characterization.

28.3.1 Workability Test Results

Workability test conducted by testing on the slump flow of specimens to measure the diameter of paste affected by GGBFS and microbes' addition. Figure 28.3 shows that the GGBFS has less effect on the workability of AAM paste.

Addition of microbes increases the workability of specimens containing 20% GGBFS (F80S20M) resulting from the emergence of micropores. However, as the GGBFS particles are denser, finer, and more reactive than FA makes a denser paste for specimens containing 30–50% GGBS. This made the microbes more difficult to enhance the workability. In general, geopolymer paste has a good workability when the slump flow value is 110% [17]. This indicated that the materials were mixed properly without segregation. The effect of microbes' addition enhances the workability because it produces carbon dioxide gas [18]. This applies to high content of GGBFS. The microbe's effect is insignificant at F50S50M because its pores are covered with

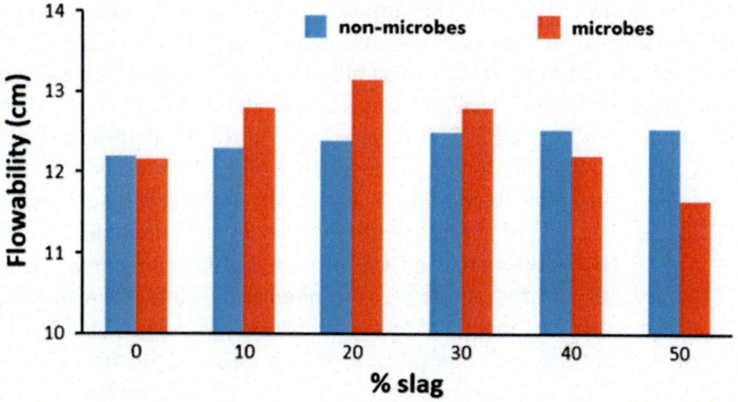

Fig. 28.3 Flow test result of AAM with and without addition of microbes

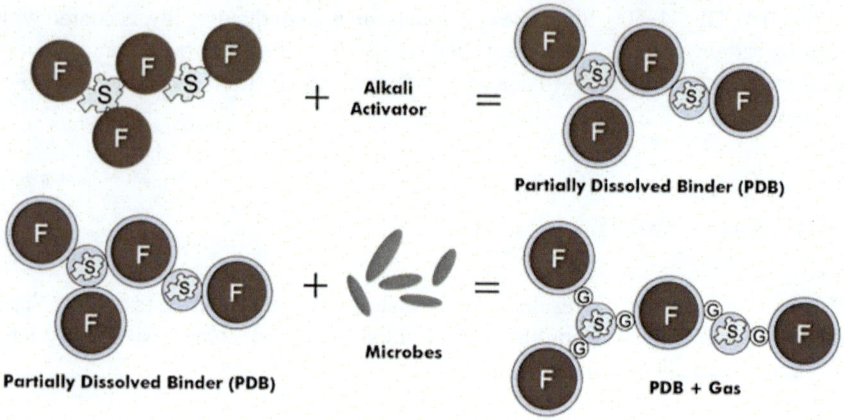

Fig. 28.4 Schematic figure of partially dissolved binder with microbes results in PDB and gas

GGBFS. In Fig. 28.3, its flowability is the lowest. Mechanism of microbes' effect to increase the flow of mixture is illustrated in Fig. 28.4 which represents the mechanism of fly ash (F) and GGBFS (S) reacting with alkali activator. Alkali activators encapsulate fly ash and GGBFS to make distance between each particle and form partially dissolved binder (PDB). When PDB reacts with microbes, it releases gas (G) that increases the distance between each fly ash and GGBFS particles. The same microbes were also applied in our previous study on concrete with high volume fly ash [19]. It was revealed that the microbes increased the workability of concrete.

28.3.2 *Setting Time Test Results*

Setting time of the geopolymer paste was conducted using Vicat apparatus and conformed to ASTM C-191 [20]. The purpose is to investigate the effect of microbes addition on setting time. Figure 28.5 shows that AAM pastes set rapidly between 15 and 30 min after pouring to Vicat apparatus bowl. This is due to the existence of waterglass (Na_2SiO_3) which accelerates the setting time of AAM.

Setting time of specimen F50S50M was slightly different at 15 min faster than specimens with less slag. Final setting is determined when the Vicat needle is unable to penetrate the slurry after three trials. These results are in accordance with some reports [21, 22]. The effect of irregular particles of fly ash is well-studied to increase its reactivity [23]. High CaO content in GGBFS decreases the setting time of resulting AAM [24]. Microbes' addition accelerated the initial setting time of AAM. However, space created by microbes made the agglomeration of geopolymerization process start rapidly. Contrasting with this process, gas formed in F50S50M delayed its final setting time. This is caused by the gas created by the microbes resisted the solidification of the paste.

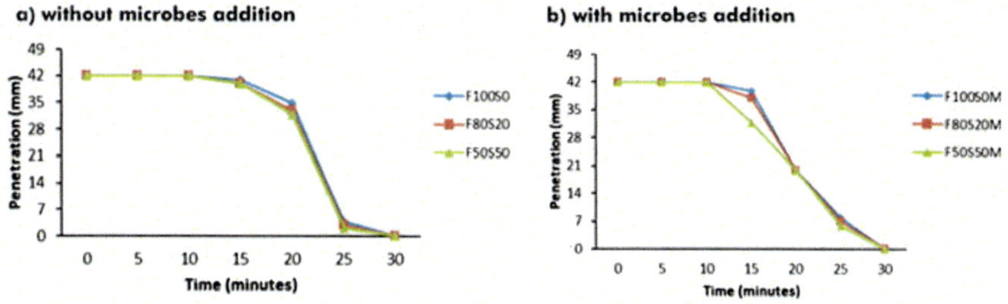

Fig. 28.5 Setting time results from Vicat apparatus testing of AAM

28.3.3 Porosity Test Results

Porosity test was conducted to calculate open porosity (p_o), closed porosity (p_f), and total porosity (p_t) conformed to ASTM C642-13. Image of 24 h paste using micro-camera is presented in Fig. 28.6. Figure 28.6 is the microstructure of AAM without microbes addition while Fig. 28.6b is the microstructure of AAM containing. Figure 28.6 shows the pores diameter of 30 μm caused by gas from microbes. The crack-like pattern is the activity of microbes to create more pores. The specimen without microbes addition revealed almost no visible pores due to the solidity of AAM. Figure 28.7 represents the porosity test result.

Both samples with or without microbes have the minimum porosity value of 20%. AAM paste is the densest of all because of the fewest total porosity (p_t) compared to concrete [25]. In general, addition of microbes in AAM resulted in denser total porosity (Fig. 28.7c). However, there is no effect of microbes on the pore in F80S20M. There is a phenomenon where microbes are still alive and make pore because it is an inorganic material that has aerobic characterization and can only metabolize if oxygen is present [26]. Spores of microbes will still be alive when oxygen is mixed into the AAM paste. Microbes release gas bubbles when removed from the container. This created greater porosity than non-microbes but did not decrease the compressive

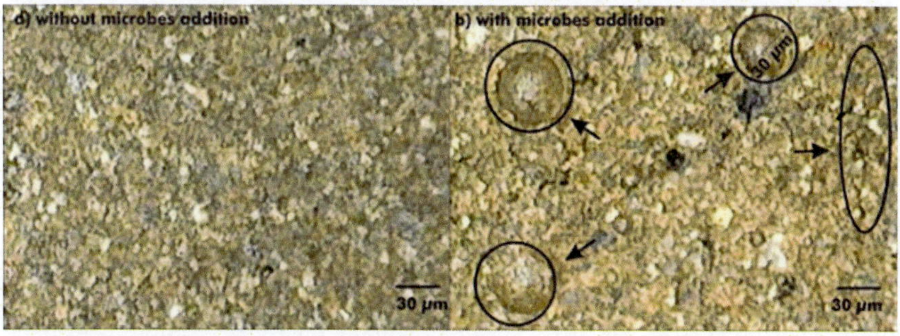

Fig. 28.6 Microstructure of AAM **a** without microbes addition **b** with microbes addition using micro-camera with a magnification of 250 ×

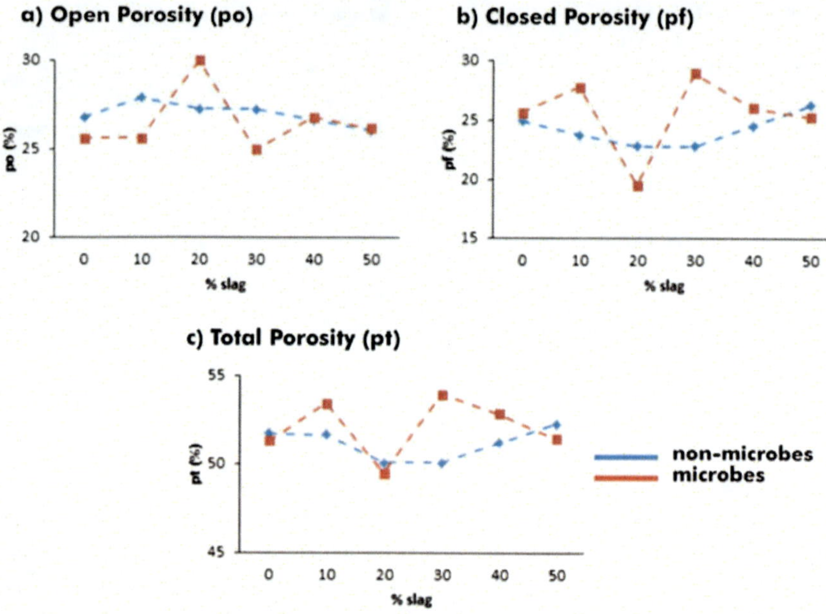

Fig. 28.7 Porosity value of resulting AAM

strength. Open porosity is higher in F80S20 while closed pore is less than that in F80S20. This showed that the microbes filled the open pores and created more closed pores. The results of porosity are related to the mechanical performance of the paste [27].

28.3.4 Compressive Strength Test Results

Compressive test is a significant indicator that expresses the effect of GGBFS and microbes addition toward fly ash-based alkali-activated materials. The test was conducted at 7 days and 28 days. Compressive strength properties of specimens are displayed in Fig. 28.8.

Figure 28.8a shows that higher resulting compressive strength at seven days of specimens without microbes (GGBFS content 10–50%). The highest achieved by F50S50 that increases its compressive strength by ± 50%. This proved that addition of GGBFS showed increasing the early compressive strength of resulting AAM [28–30]. Compressive strength results at 28 days showed that addition of more than 20% Slag will decrease the compressive strength of samples at 28 days. Hence, the optimum value of GGBFS is 20%, whether it is with or without microbes addition. The effect of microbes increased the strength without GGBFS (F100S0M). In Fig. 28.8b, the strength increases from 23 to 29 MPa. Microbes' addition increased the strength

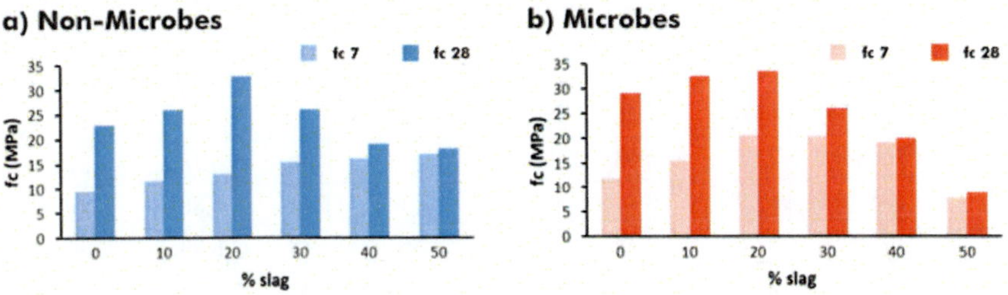

Fig. 28.8 Compressive strength result of AAM at 7 and 28 days

up to 20% slag content. This is related to the results of the porosity test where the closed porosity in F80S20M was higher than F80S20. This was influenced by biomineralization to form denser pores.

28.3.5 XRD Analysis

The X-ray diffraction (XRD) analysis was performed to identify the resulting compound from the geopolymerization process. Figure 28.9 shows that sodalite is a dominant mineral. Sodalite appeared at low molarity AAM [31, 32]. There are six major compounds appeared in resulting specimen, which are Boldine ($C_{19}H_{21}NO_4$; JCPDS #130888), Zeolite-A ($Na_{12}(AlO_2)_{12}(SiO_2)_{12}$; JCPDS #741183), Quartz (SiO_2; JCPDS #421401), Albite ($NaAlSi_3O_8$; JCPDS #030508), Sodalite ($Na_8(AlSiO_4)_6(NO_2)_2$; JCPDS #811299), and Calcite ($CaCO_3$; JCPDS #030670).

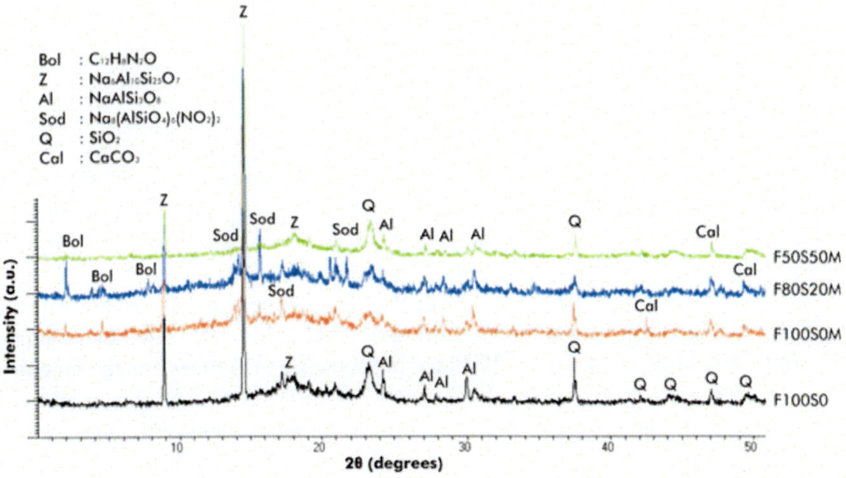

Fig. 28.9 Diffractogram of resulting samples

Table 28.3 Percentage of each compound in Fig. 28.9

No.	Code	Am (%)	Z (%)	Q (%)	Al (%)	Sod (%)	Cal (%)	Bol (%)
1	F100S0	64.90	1.26	12.81	12.81	8.25	–	–
2	F100S0M	72.58	0.63	7.62	6.35	8.88	6.18	2.74
3	F80S20M	65.98	0.42	9.93	4.91	11.44	6.88	2.63
4	F50S50M	62.83	0.39	6.18	4.56	14.78	8.88	1.90

Notes Am = Amorphous, Z = Zeolite-A, Q = Quartz, Al = Albite, Sod = Sodalite, Cal = Calcite, Bol = Boldine

Table 28.3 shows that microbes contribute significantly toward the amorphous phase in samples, but when GGBFS content increases, the amorphous phase also drops.

Formation of calcite was the result reaction of CaO (from fly ash and GGBFS) and CO_2 (from microbes). Small amount of boldine was found in specimens containing microbes. Boldine as the product of microbes formed from ammonia precipitated in the pores. This product existed in specimen F50S50M but was not formed optimally. This proves that microbes selected lower slag content as the suitable inorganic habitat. Calcite is a recurring mineral that is usually present in clam shells [33].

To link this mineral to compressive strength, this was demonstrated with the results of 20% addition of GGBFS at 7 days compressive strength. The results are 13.1 MPa (F80S20) and 20.5 MPa (F80S20M), respectively. This is due to precipitated calcite at the beginning. Minerals at F100S0 and F100S0M have different hardness values because of the absence of slag. Calcite clogs and deposition in open pores resulting in decrease of Po (open pores) and become Pf (closed pores). This increased the compressive strength. Its deposition was maximum in F80S20M and occurred at the early ages. Although the GGBFS was increased up to 50%, microbes have significant influence in seven days rather than in 28 days.

28.3.6 Morphological Analysis

The identification of resulting microstructures was studied with scanning electron microscope (SEM) characterization. Sodalite and albite appeared as dominant mineral. In general, these minerals are the products of geopolymerization, having a strong surface and building the strength of the paste.

Figure 28.10a shows microstructures of the F100S0 specimen as a control specimen. Figure 28.10b, c, and d shows the formation of calcite generated by biomineralization from microbes. Figure 28.11a and b shows the microbes and the precipitation of microbes metabolism results. In paste F50S50M, there are some unreacted slag particles and the presence of cracks because of autogenous shrinkage (water deficiency), whereas moist curing has already been conducted. It is presented in Fig. 28.12a and c. Those cracks are considered as the open pores. In some cases, these micropores are closed by the product of calcite precipitation in the presence of microbes. Furthermore, the open pores change to closed pores. This phenomenon is well-known in the mechanism of self-healing concrete [34, 35]. Porous material from SEM images can be linked with porosity test results.

Slag which is rich in calcium, absorbed water results in decrease of liquid-to-binder ratio. It induced double reactions (hydration from slag and geopolymerization

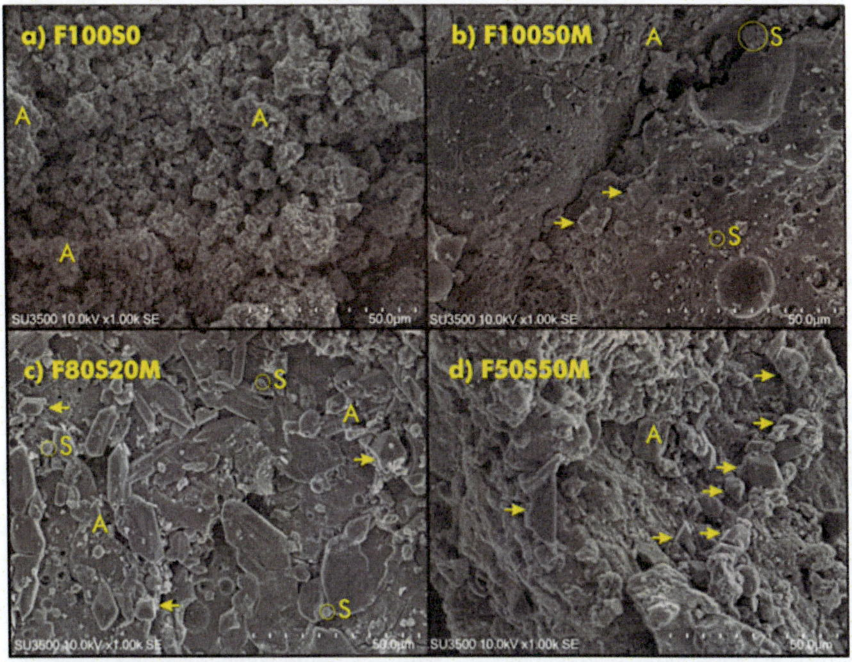

Fig. 28.10 SEM images of **a** F100S0 **b** F100S0M **c** F80S20M **d** F50S50M. All in magnification 1000×. The arrows indicate calcite formation (*Note* A = Albite, S = Sodalite)

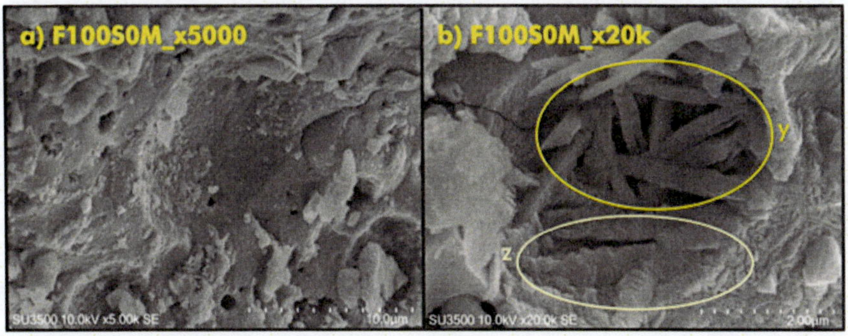

Fig. 28.11 SEM images of F100S0M (*Note*: y = microbes in the pore, z = precipitation products)

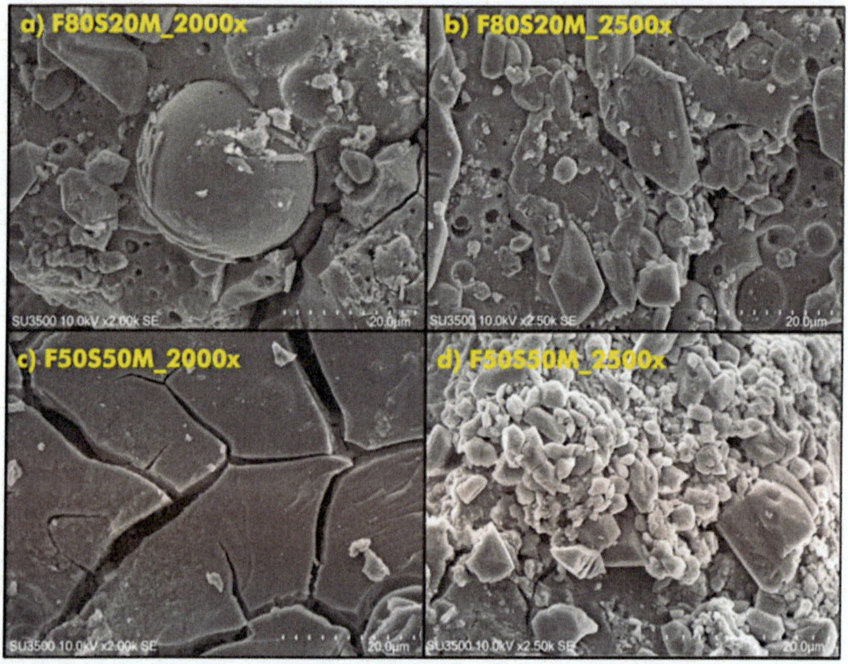

Fig. 28.12 SEM images of **a** cracks in F80S20M, **b** unreacted slag in F80S20M, **c** cracks in F50S50M, **d** unreacted slag in F50S50M

at the same time) [36]. The cracks occurred as the results of the desiccation as shown in Fig. 28.12a and c. Addition of more slag in the paste caused more cracks in the paste as represented in specimen F50S50M. GGBFS is known as the cause of self-desiccation in the concrete mixture with low liquid content causing the cracks because of the autogenous shrinkage phenomenon [37].

The unreacted slag particles as shown in Fig. 28.2b and d were not involved in the geopolymerization process. These particles are considered as the fillers only. In conjunction with many cracks in the paste, this caused the strength decreased. Hydration takes the alkali prior to the geopolymerization process that causes the reaction not to occur perfectly. It was reported that there were no significant effects of microbes and high value of unreacted slag [38, 39]. From the previous XRD characterization, it was confirmed the emergence of calcite formation. Figure 28.13 explains further about the composition in F80S20M. There are some unreacted fly ash (uFA) as well as microbes (M) present, which is similar to SEM images of original microbes magnified by 15,000×.

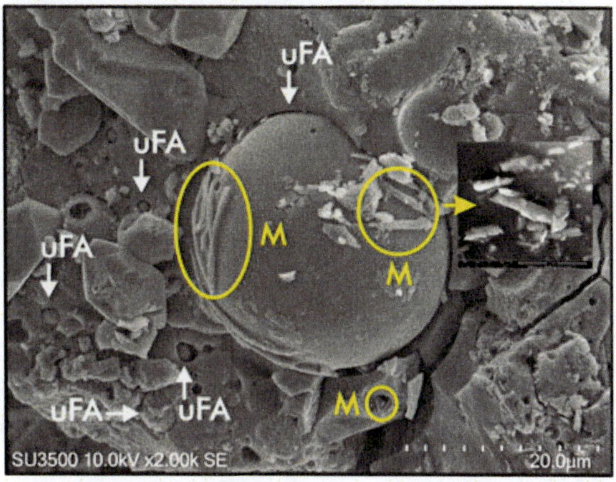

Fig. 28.13 SEM images F80S20M (*Note* M = microbes conformed to original SEM Image, uFA = unreacted fly ash)

28.3.7 EDS Analysis

Identification of resulting element was studied with energy-dispersive X-ray spectroscopy (EDS) characterization. The weight percentage of each element was presented in Table 28.4.

The Ca and C element dominantly found in F80S20M indicated the $CaCO_3$ formation was the greatest. This explains the highest compressive strength of F80S20M with microbes addition. The low percentage of Ca and C in F100S0M indicated that the slag was necessary to build up compressive strength when the microbes were applied.

Table 28.4 EDS results of resulting AAM

No.	Element	Sample			
		F100S0	F100S0M	F80S20M	F50S50M
1	C K	5.21	5.64	19.95	7.86
2	O K	49.4	53.52	49.14	51.18
3	Na K	13.12	6.05	12.19	18.62
4	Mg K	3.23	–	0.78	1.71
5	Al K	6.04	23.07	3.22	4.15
6	Si K	13.02	11.19	6.28	10.16
7	K K	0.31	–	3.16	3.69
8	Ca K	5.14	0.53	5.28	2.63

28.4 Conclusion

In summary, addition of microbes enhanced the workability of AAM paste because of gas created during mixing. This caused the pores to distribute evenly. However, higher content of GGBFS nullifies this because it clogged the pores. The microbes have an influence on the setting of alkali-activated pastes incorporating a high volume of slag. High CaO content in GGBFS reduced the initial setting time of specimens with microbes addition. This was caused by two different reactions that occurred which were hydration and geopolymerization. Microbes made the AAM denser because of calcite formation in early age to change open pores to become clogged pores. Addition of microbes to fly ash and GGBFS-based AAM paste is potential for high early strength material purposes, especially for the 20% slag by binder weight. From the compressive test result, it is recommended to apply a maximum GGBFS content of 20% in specimens with microbes addition. Its average compressive strength was ± 50% higher than the specimen without microbes. Formation of calcite was due to the carbonic anhydrase behavior of microbes. Sodalite and albite were found as the major minerals that build the strength of the paste. SEM results indicated that high GGBFS content caused cracks due to autogenous shrinkage. At 50% slag content, unreacted slag particles were also found as a filler. The cracks associated with the unreacted particles contributed to the lower strength.

Acknowledgements The authors gratefully acknowledge the support from Bioconc Center Foundation.

References

1. J. Davidovits, Sawyer JL, US Patent 1985, No. 4, 509, 985 (1985)
2. Davidovits J (1999) Chemistry of Geopolymeric Systems Terminology, in *Proceedings of Geopolymer, International Conference*, France (1999)
3. M. Chi, R. Huang, Binding mechanism and properties of alkali-activated fly ash/slag mortars. Constr. Build. Mater. **40**, 291–298 (2013). https://doi.org/10.1016/j.conbuildmat.2012.11.003
4. A.F. Abdalqader, F. Jin, A. Al-Tabbaa, Development of greener alkali-activated cement: utilisation of sodium carbonate for activating slag and fly ash mixtures. J. Clean. Prod. **113**, 66–75 (2016). https://doi.org/10.1016/j.jclepro.2015.12.010
5. M. Criado, A. Palomo, A. Fernandez-Jimenez, Alkali activation of fly ash. Part I: Effect of curing condition on The Carbonation of The reaction Products, Fuel **84**, 2048–2054 (2005). https://doi.org/10.1016/j.fuel.2005.03.030
6. O.A. Mohamed, A review of durability and strength characteristics of alkali-activated slag concrete. Materials (Basel) **12**(8):1198 (2019). https://doi.org/10.3390/ma12081198. PMID: 31013765; PMCID: PMC6515512
7. M. Teymouri, K. Behfarnia, A. Shabani, Mix design effects on the durability of alkali-activated slag concrete in a hydrochloric acid environment. Sustainability **13**(14), 8096 (2021). https://doi.org/10.3390/su13148096
8. J.J. Ekaputri, T. Ishida, K. Maekawa, Autogeneous shrinkage of mortars made with different types of slag cement. JCI Saitama **32**(1), 353–358 (2010)

9. J.H. Yoon, K.C. Lee, N. Weiss, Y.H. Kho, K.H. Kang, Y.H. Park, *Sporosarcina aquimarina* sp. nov., a bacterium isolated from seawater in Korea, and transfer of Bacillus *globisporus* (Larkin and Stokes 1967), Bacillus *psychrophilus* (Nakamura 1984) and Bacillus *pasteurii* (Chester 1898) to the genus *Sporosarcina* as *Sporosarcina globispora* comb. nov., *Sporosarcina psychrophila* comb. nov. and *Sporosarcina pasteurii* comb. nov., and emended description of th. Int. J. Syst. Evol. Microbiol. **51**(Pt 3):1079–1086 (2001). https://doi.org/10.1099/00207713-51-3-1079
10. D. Arias, L.A. Cisternas, M. Rivas, Biomineralization mediated by ureolytic bacteria applied to water treatment: a review. Crystals **7**(11), 345 (2017). https://doi.org/10.3390/cryst7110345
11. J.Z.S. Doctolero, A.B. Beltran, M.O. Uba, A.A.S. Tigue, M.A.B. Promentilla, Self-healing biogeopolymers using biochar-immobilized spores of pure- and co-cultures of bacteria. Minerals **10**(12), 1114 (2020). https://doi.org/10.3390/min10121114
12. N. De Belie, Application of bacteria in concrete: a critical evaluation of the current status. RILEM Tech Lett **1**, 56–61 (2016). https://doi.org/10.21809/rilemtechlett.2016.14
13. S.K. Khattra, M. Parmar, U.G. Phutela, Study of strength variation of concrete using ureolytic bacteria. International Journal of Engineering and Applied Sciences 3(4) (2016)
14. M. Amran, R. Fediuk, G. Murali, S. Avudaiappan, T. Ozbakkaloglu, N. Vatin, M. Karelina, S. Klyuev, A. Gholampour, Fly ash-based eco-efficient concretes: a comprehensive review of the short-term properties. Materials **14**(15), 4264 (2021). https://doi.org/10.3390/ma14154264
15. J. Ma, D. Wang, S. Zhao, P. Duan, S. Yang, Influence of particle morphology of ground fly ash on the fluidity and strength of cement paste. Materials **14**(2), 283 (2021). https://doi.org/10.3390/ma14020283
16. ASTM C642, *Density, Absorption, and Voids in Hardened Concrete* (ASTM International, 2013)
17. I.N. Guntur, M.W. Tjaronge, R. Irmawaty, J.J. Ekaputri, Compressive strength of medium calcium fly ash based geopolymer paste. IOP Conf. Ser.: Earth Environ. Sci. **419**, 012027 (2020). https://doi.org/10.1088/1755-1315/419/1/012027
18. E. Stanaszek-Tomal, Bacterial concrete as a sustainable building material? Sustainability **12**(2), 696 (2020). https://doi.org/10.3390/su12020696
19. A.R. Muhammad, J.J. Ekaputri, M. Basoeki, The effect of microbes and fly ash to improve concrete performance. Journal of Advanced Civil and Environmental Engineering **4**(2), 60–69 (2021). https://doi.org/10.30659/jacee.4.2.60-69
20. ASTM C191, *Standard Test Method for Time of Setting of Hydraulic Cement by Vicat Needle* (ASTM International, 2013)
21. T. Luukkonen, Z. Abdollahnejad, J. Yliniemi, P. Kinnunen, M. Illikainen, One-part alkali-activated materials: a review. Cem. Concr. Res. **103**, 21–34 (2018). https://doi.org/10.1016/j.cemconres.2017.10.001
22. T. Du, Y. Liu, Effect of water glass on setting time and mechanical properties of fly ash slag geopolymer. J. Phys.: Conf. Ser. **1676**, 012024 (2020). https://doi.org/10.1088/1742-6596/1676/1/012024
23. P. Risdanareni, J.J. Ekaputri, P. Puspitasari, Chemical and physical characterization of fly ash as geopolymer material. MATEC Web of Conferences **97**, 01031 (2017). https://doi.org/10.1051/matecconf/20179701031
24. G. Kumar, S.S. Mishra, Effect of GGBFS on workability and strength of alkali-activated geopolymer concrete. Civil Engineering Journal **7**, 1036–1049 (2021). https://doi.org/10.28991/cej-2021-03091708
25. P. Cong, Y. Cheng, Advances in geopolymer materials: a comprehensive review. Journal of Traffic and Transportation Engineering **8**(3), 283–314 (2021). https://doi.org/10.1016/j.jtte.2021.03.004
26. K.D. Wulandari, J.J. Ekaputri, F.C. Triwulan, D.H.E. Setiamarga, Effects of microbial agents to the properties of fly ash-based paste. MATEC Web of Conferences **195**, 01012 (2018). https://doi.org/10.1051/matecconf/201819501012
27. Z.B. Bundur, S. Bae, M.J. Kirisits, R.D. Ferron, Biomineralization in self-healing cement-based materials: investigating the temporal evolution of microbial metabolic state and material

porosity. J. Mater. Civ. Eng. **29**(8), 04017079 (2017). https://doi.org/10.1061/(ASCE)MT.1943-5533.0001838
28. Y. Tajunnisa, M. Sugimoto, T. Uchinuno, T. Sato, Y. Toda, A. Hamasaki, T. Yoshinaga, K. Shida, M. Shigeishi, Effect of GGBFS and micro-silica on mechanical properties, shrinkage and microstructure of alkali-activated fly ash mortar. Geomate Journal **13**(39), 87–94 (2017). https://doi.org/10.21660/2017.39.11341
29. M. Valente, M. Sambucci, A. Sibai, Geopolymers vs. cement matrix materials: how nanofiller can help a sustainability approach for smart construction applications-a review. Nanomaterials (Basel) **11**(8), 2007 (2021). https://doi.org/10.3390/nano11082007
30. X. Guo, X. Pan, Effects of steel slag on mechanical properties and mechanism of fly ash–based geopolymer. Journal of Materials in Civil Engineering **32**(2) (2020). https://doi.org/10.1061/(ASCE)MT.1943-5533.0003012
31. K. Juengsuwattananon, K. Pimraksa, Variable factors controlling amorphous-zeolite phase transformation in metakaolin based geopolymer. Acta Metallurgica Slovaca **23**(4), 378–386 (2017). https://doi.org/10.12776/ams.v23i4.1001
32. K.D. Wulandari, J.J. Ekaputri, Triwulan, S.B. Kurniawan, W.E. Primaningtyas, S.R.S. Abdullah, N.I. Ismail, M.F. Imron, Effect of microbes addition on the properties and surface morphology of fly ash based geopolymer paste. Journal of Building Engineering 101596 (2021). https://doi.org/10.1016/j.jobe.2020.101596
33. S. Bramhe, J.-K. Ryu, M.C. Chu, A. Balakrishnan, T.N. Kim, Synthesis of calcium phosphate minerals from biowaste clam shells using microwave heating. Korean Journal of Materials Research. The Materials Research Society of Korea (2014). https://doi.org/10.3740/mrsk.2014.24.12.700
34. S. Sangadji, Can self-healing mechanism helps concrete structure sustainable? Procedia Engineering. **171**, 238–249 (2017). https://doi.org/10.1016/j.proeng.2017.01.331
35. X. Li, S. Zhao, S. Wang, Status-of-the-art for self-healing concrete. J. Phys.: Conf. Ser. **1622**, 012011. https://doi.org/10.1088/1742-6596/1622/1/012011
36. J.J. Ekaputri, K. Maekawa, T. Ishida, Experimental study on internal RH of BFS mortars at early age. MSF **857**, 305–310 (2016). https://doi.org/10.4028/www.scientific.net/msf.857.305
37. A.L. Wijaya, J.J. Ekaputri, Triwulan (2017) Factor influencing strength and setting time of fly ash based-geopolymer paste. MATEC Web of Conferences **138**, 01010. https://doi.org/10.1051/matecconf/201713801010
38. E.D. Banks, N.M. Taylor, J. Gulley, B.R. Lubbers, J.G. Giarrizzo, H.A. Bullen, T.M. Hoehler, H.A. Barton, Bacterial calcium carbonate precipitation in cave environments: a function of calcium homeostasis. Geomicrobiol J. **27**(5), 444–454 (2010). https://doi.org/10.1080/01490450903485136
39. X. Ouyang, Y. Ma, Z. Liu, J. Liang, G. Ye, Effect of the sodium silicate modulus and slag content on fresh and hardened properties of alkali-activated fly ash/slag. Minerals. **10**(1), 15 (2020). https://doi.org/10.3390/min10010015

Chapter 29
Waste to Product: Potential of Mg-Rich Gypsum Additive for Improvement of Peat Soil

Ayah Almsedeen, Nurmunira Muhammad, and Mohd Fakhrurrazi Ishak

Abstract The stabilization of problematic soils with chemical additives has increased in demand globally. Highly developed industrial plants have made an urgent need to utilize all types of soils even the problematic soils such as organic, marine clay, lateritic or expansive clay. The use of industrial waste by-products, namely magnesium-rich gypsum, for improving the weak characteristic of peat soil has not been investigated. This paper investigated the mechanism of gypsum that contributes to the compressibility of peat soil, a typical soil in Malaysia. The optimum combination of the additives into the soil was further examined by physicochemical properties by analytical techniques such as pH, scanning electron microscopy (SEM), X-ray fluorescence (XRF), and Fourier transform infrared spectroscopy (FTIR). This technical paper is more on comparison of theoretically analyzing the characteristics of peat and gypsum that have the potential to be strongly mixed and improve the characteristics of peat soil. The significance of this result shall contribute to the potential application of industrial waste by-products by recycling methods for soil improvement techniques.

Keywords Chemical soil stabilization · Peat soil · Mg-rich gypsum · Soil stabilization · Chemical additives

29.1 Introduction

In recent years, there has been a significant increase in the intensity of land use activities, which is a direct outcome of the population and economic growth. As a

A. Almsedeen · N. Muhammad (✉)
Faculty of Civil Engineering Technology, Universiti Malaysia Pahang, Kuantan, Malaysia
e-mail: muniramuhammad@ump.edu.my

M. F. Ishak
Centre for Sustainability of Ecosystem & Earth Resources (Earth Centre), Universiti Malaysia Pahang, Kuantan, Malaysia

M. A. A. Mohd Salleh et al. (eds.), *Proceedings of the Green Materials and Electronic Packaging Interconnect Technology Symposium*, Springer Proceedings in Physics 289,
https://doi.org/10.1007/978-981-19-9267-4_29

result of population growth, the amount of land appropriate for infrastructure development has decreased, which may become a disadvantage in the near future [1]. In Malaysia, as in numerous other countries, a substantial amount of the land is covered by peat. Approximately 2.7 million hectares of Malaysia's total land area are covered with tropical peat soil [2]. Additionally, peat soils contain four separate types of organic matter, known as human, humic, fulvic, and yellow organic acids, respectively [3]. Depending on its location, the type of fiber it was created from, the relative humidity, and the temperature, peat can have various components [4]. Peat has typical characteristics, which include high natural moisture content, high compressibility, water-holding capacity, low specific gravity, low bearing capacity, and medium-to-low permeability [5]. The purpose of soil stabilization is to improve the soil stability, especially in the case of peat, as well as to reduce settlement and lateral deformation and boost bearing capacity [6].

The soil stabilization method is utilized, particularly for peat, to increase its stability by decreasing the settlement and compressibility and boosting its bearing capacity. The stabilization of a chemical admixture can be accomplished by combining additives to promote flocculation (aggregation) and the formation of chemical bonds between particles [7]. There are many chemical additives that have been studied in the stabilization of peat soil such as fly ash, lime, and cement. However, there is a paucity of studies on potential improvement in the physical and geotechnical properties of organic soil stabilized with a gypsum admixture [8]. In fact, recycled gypsum produced from gypsum wastes has a potential to be used as a stabilizer material in ground improvement since gypsum is one of the cementation materials [9]. The magnesium-rich synthetic gypsum (MRSG) has a high pH and is constituted of gypsum as well as magnesium and sulfur [6]. Due to these characteristics, the MRSG can potentially be a liming agent in the process of alleviating deficiencies produced by an acidic soil state. Due to the presence of calcium in MRSG, the pH of acidic soils will rise, leading to the formation of gibbsites or aluminum hydroxide [7]. A few studies have studied the properties of the Mg-rich gypsum wastes as a stabilization for potential improvement of peat soil. This study focused on the potential of chemical stabilization of peat using magnesium-rich gypsum waste for improving the characteristics of the peat soil.

29.2 Methodology

29.2.1 Materials

Dark brown soil was brought from a site in Gebeng, Kuantan, Pahang, Malaysia. The natural moisture content was determined immediately in the soil mechanics and geotechnical laboratory, Universiti Malaysia Pahang. The pH value was read at 3.59 following ASTM D4972 Method A (very acidic). The soil has been sieved to obtain the final soil sample for the test using sieve number (0.425 mm).

The MRSG was obtained from the waste product of the rare earth industry in Gebeng, Kuantan, Pahang, Malaysia. The Mg-rich gypsum was air-dried for 5 days. Then after drying, it was then screened through a 1.18 mm mesh sieve and stored in airtight container.

The peat soil and Mg-rich gypsum were characterized using Fourier transform infrared spectroscopy (FTIR), scanning electron microscopy (SEM), X-ray fluorescence (XRF), and pH obtain their physiochemical properties.

29.2.2 Fourier Transform Infrared Spectroscopy (FTIR)

Fourier transform infrared spectroscopy (FTIR, Nicolet iS5, the USA) was applied to provide information about the chemical structure of peat soil and Mg-rich gypsum samples. The samples are prepared with KBr salts owing to their transparency, thus offering a better resolution of the spectrum. The spectrum scopes of hydrogels the spectra were between 4000 and 500 cm^{-1}.

29.2.3 Scanning Electron Microscopy (SEM)

SEM was used to examine the morphologies and structures of the peat soil and the Mg-rich gypsum samples (SEM, FEI Quanta 450). The dry peat soil and Mg-rich gypsum were coated with gold and analyzed by SEM.

29.2.4 X-Ray Fluorescence (XRF)

An X-ray fluorescence spectrometer (XRF) (ULVAC-PHI 5000 VersaProbe II) was used to determine major and trace elements in peat soil and Mg-rich gypsum samples. Samples were introduced, via an open-ended sample plastic cell. An acid-purified sand (SiO_2) was used as the media blank for determining the detection limits of major and trace elements and heavy metals.

29.3 Result and Discussions

29.3.1 Classification of Peat Soil

The classification of peat soil was according to the results of organic content and acidity content (pH) as shown in Table 29.1.

Table 29.1 Classification of peat soil

No.	Test	Result	Peat classification based on ASTM D 4427-92 (2002)
1	pH	3.59	Very acidic
2	Organic content	57.63	High ash peat

The peat soil is very acidic having a pH of 3.59, while the Mg-rich gypsum pH value is 9.47 high alkaline following ASTM D4972 Method A. The pH of most peat soils is low due to the presence of organic acids, the exchangeable hydrogen and aluminum, iron sulfide, and other oxidizable sulfur compounds. It was found that the alkalinity of the Mg-rich gypsum was caused by the incorporation of a significant amount of calcium as well as basic oxides. The high alkalinity of the substance is assumed to be due to the high amount of calcium and basic oxides, giving it a strong capacity to neutralize acid medium increased.

29.3.2 Fourier Transform Infrared Spectroscopy (FTIR)

The spectra of the bulk peat soil and the Mg-rich gypsum samples, as well as their standard deviation spectra, are depicted in Figs. 29.1 and 29.2. The peat soil and Mg-rich gypsum spectra show that variability between samples is larger for the O–H region (3700–3600 cm^{-1}), aliphatics (2921–2850 cm^{-1}), CH_2 bending (1465.66–1435.77 cm^{-1}), aromatics (1640–1623 cm^{-1}), and polysaccharides (1032.71–1121.42 cm^{-1}), respectively [10]. Peat soil showed O–H (3000–3670 cm^{-1}) stretching vibration bands broader than the Mg-rich gypsum.

29.3.3 Scanning Electron Microscopy (SEM)

Scanning electron microscopy (SEM) was used to examine the surface morphology of peat soil and Mg-rich gypsum. The peat soil sample (Fig. 29.3) has a disjointed structure characterized by obvious voids and porosity. The gypsum crystal has an elongated shape with a few interconnected voids allowing it to fill or adhere to the porous micro-touch surfaces of the peat structure (Fig. 29.4), which causes it to fill or bond with the peat structure's porous and micro-touch surfaces [11].

29.3.4 X-Ray Fluorescence (XRF)

X-ray fluorescence (XRF) examines major and trace elements in peat soil and Mg-rich gypsum samples as instructed in Figs. 29.5 and 29.6, respectively. XRF study

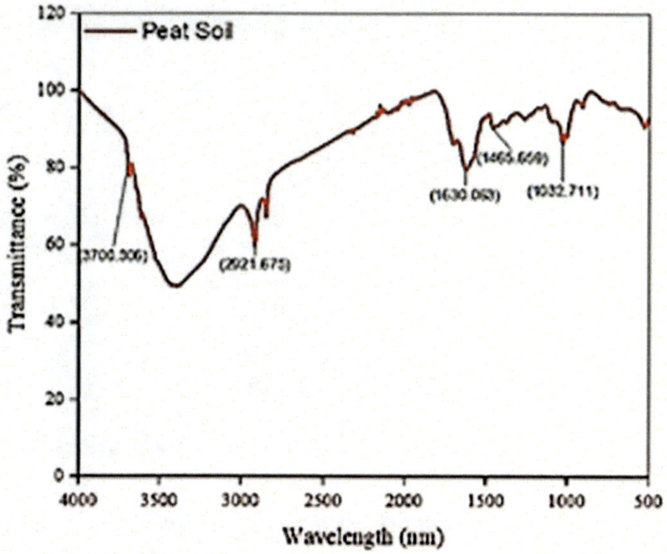

Fig. 29.1 FTIR spectra of peat soil

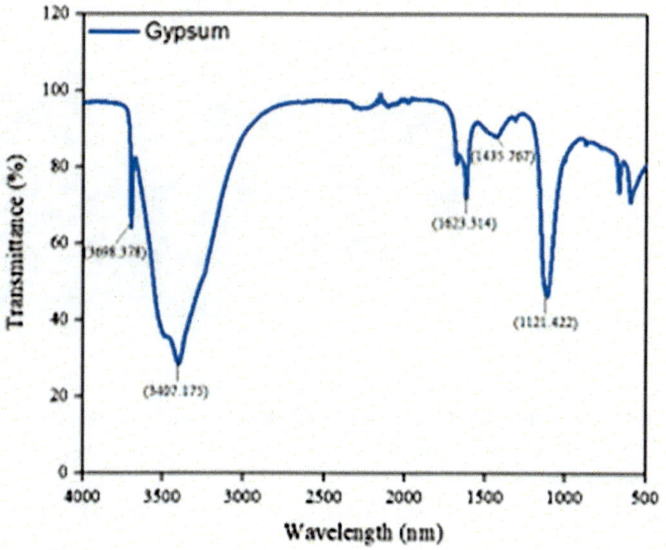

Fig. 29.2 FTIR spectra of gypsum

Fig. 29.3 Scanning electron microscopy (SEM) image for peat soil

Fig. 29.4 Scanning electron microscopy (SEM) image for gypsum

confirmed the main constituent of the studied MRSG sample as calcium and magnesium. The XRF of the peat is primarily characterized by a high weight percentage of carbon and oxygen (i.e., 88.2%), and the remaining 11.8% corresponds to the total elemental weight of aluminum (Al), silicon (Si), iron (Fe), and potassium (K). Gypsum contains a very high CaO content of 19.7%. The engineering properties of $Ca(OH)_2$-rich materials' stabilized soil are ascribed to three basic reactions, namely pozzolanic reaction, aggregation and flocculation, and cation exchange. The influence of the three reactions is basically liable for the changes in shrinkage and plasticity.

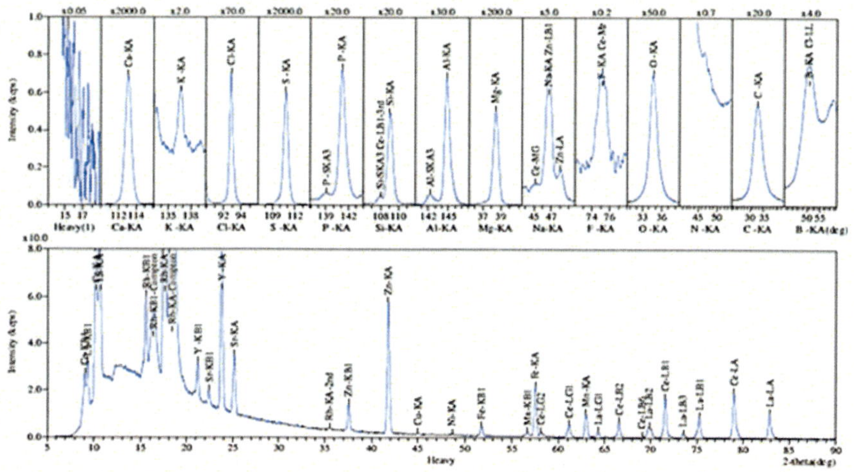

Fig. 29.5 XRF spectra of peat soil

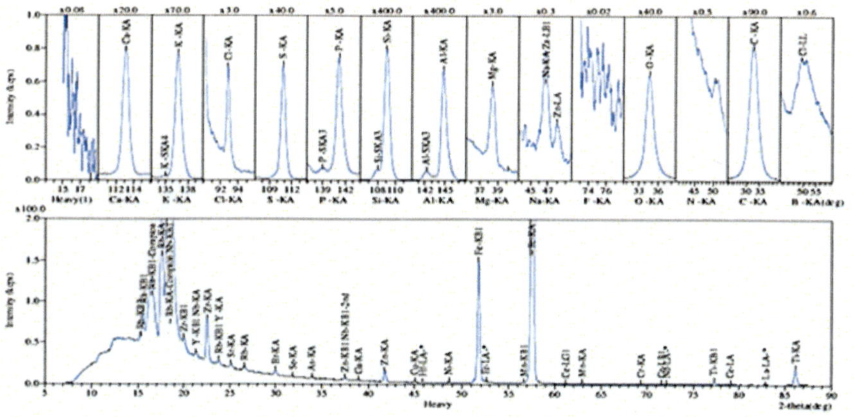

Fig. 29.6 XRF spectra of gypsum

29.4 Conclusion

In conclusion, the properties of peat soil and Mg-rich gypsum were successfully investigated using physicochemical characterizations. These results proposed that Mg-rich gypsum additive can be applied to improve peat soil.

Acknowledgements The authors would like to thank the Universiti Malaysia Pahang for laboratory facilities as well as additional financial support under Internal Research grant RDU210342 and RDU223305.

References

1. B. Surya, D.N.A. Ahmad, H.H. Sakti, H. Sahban, Land use change, spatial interaction, and sustainable development in the metropolitan urban areas, South Sulawesi Province. Indonesia. Land **9**(3), 95 (2020)
2. A. Hauashdh, R.M.S.R. Mohamed, J. Abd Rahman, J. Jailani, Analysis of leachate from soildified peat soil, in MATEC Web of Conferences, vol. 250, p. 6015 (2018)
3. M. Drobek, M. Frąc, J. Cybulska, Plant biostimulants: importance of the quality and yield of horticultural crops and the improvement of plant tolerance to abiotic stress—a review. Agronomy **9**(6), 335 (2019)
4. O.S. Misnikov, A.E. Timofeev, O.V. Pukhova, Preparation of molded sorption materials based on peat-mineral compositions. Polym. Sci. Ser. D **8**(1), 66–74 (2015)
5. P.K. Kolay, H.Y. Sii, S.N.L. Taib, Tropical peat soil stabilization using class F pond ash from coal fired power plant. Int. J. Civ. Environ. Eng. **3**(2), 79–83 (2011)
6. M.A. Rahgozar, M. Saberian, Geotechnical properties of peat soil stabilised with shredded waste tyre chips. Mires Peat **18** (2016)
7. R.C. Mamat, Engineering properties of Batu Pahat soft clay stabilized with lime, cement and bentonite for subgrade in road construction. Universiti Tun Hussein Onn Malaysia (2013)
8. J. Han, *Principles and practice of ground improvement* (Wiley, New Jersey, 2015)
9. A. Ahmed, Recycled bassanite for enhancing the stability of poor subgrades clay soil in road construction projects. Constr. Build. Mater. **48**, 151–159 (2013)
10. Z.A. Rahman, J.Y.Y. Lee, S.A. Rahim, T. Lihan, W.M.R. Idris, Application of gypsum and fly ash as additives in stabilization of tropical peat soil. J. Appl. Sci. **15**(7), 1006 (2015)
11. S. Horpibulsuk, C. Phetchuay, A. Chinkulkijniwat, A. Cholaphatsorn, Strength development in silty clay stabilized with calcium carbide residue and fly ash. Soils Found. **53**(4), 477–486 (2013)

Chapter 30
Smart Solar Safety Worker Helmet Using Accelerometer and GPS Sensor

Azlina Idris, Che Wan Nur Zulaikha Che Wan Adnan, Warid Wazien Ahmad Zailani, Suzi Seroja Sarnin, Muhammad Naufal Mansor, Miradatul Najwa Muhd Rodhi, Mohd Fadzil Arshad, and Norbaya Sidek

Abstract In construction projects, safety is a big issue. There is not an effective option to overcome the situation. The safety of the public is jeopardized during the construction process. Some incidents happen when people fall from great heights and go unreported, resulting in death due to a lack of medical assistance. However, the numbers of deaths could have been prevented. This research aims to create smart wearable gadgets, such as a safety helmet that use a variety of sensors such as solar, accelerometer, and GPS sensor to improve worker safety. The IoT-based gadgets assist in detecting worker falls and sending notifications for emergency assistance. The system starts with solar energy charging the battery and then when the accelerometer sensor detects any falls, the notification will be sent to the Blynk. Meanwhile, GPS sensor detects the location of the worker and shows in the Blynk. As a results,

A. Idris (✉) · S. S. Sarnin
School of Electrical Engineering, College of Engineering, Universiti Teknologi MARA, 40450 Shah Alam, Selangor, Malaysia
e-mail: azlina831@uitm.edu.my

S. S. Sarnin
e-mail: suzis045@uitm.edu.my

C. W. N. Z. Che Wan Adnan · W. W. Ahmad Zailani · M. F. Arshad · N. Sidek
School of Civil Engineering, College of Engineering, Universiti Teknologi MARA, 40450 Shah Alam, Selangor, Malaysia
e-mail: waridwazien@uitm.edu.my

M. F. Arshad
e-mail: mohdfadzil.arshad@uitm.edu.my

N. Sidek
e-mail: norbayasidek@uitm.edu.my

M. N. Mansor
Faculty of Electrical Engineering Techology, University Malaysia Perlis, Arau, Malaysia
e-mail: naufal@unimap.edu.my

M. N. Muhd Rodhi
School of Chemical Engineering, College of Engineering, Universiti Teknologi MARA, 40450 Shah Alam, Selangor, Malaysia
e-mail: miradatul@uitm.edu.my

M. A. A. Mohd Salleh et al. (eds.), *Proceedings of the Green Materials and Electronic Packaging Interconnect Technology Symposium*, Springer Proceedings in Physics 289,
https://doi.org/10.1007/978-981-19-9267-4_30

the prototype has been tested in a variety of situations and proved to be extremely accurate which the improvement is almost 3.3%.

Keywords Smart solar · Safety worker helmet · Accelerometer · GPS sensor

30.1 Introduction

A construction site is a location or parcel of land where construction is taking place. The terms 'building site' and 'construction site' are frequently interchanged, though 'building site' usually refers to the construction of buildings (and sometimes, more specifically, housing), whereas 'construction site' can refer to any type of work, such as road construction, sewer construction, landscaping, and so on [1]. A construction site, according to the CDM Regulations, "includes any place where construction work is being carried out or to which employees have access but does not include a workplace inside the site set aside for purposes other than construction work" [1]. In addition, construction is a high-risk industry that encompasses a wide range of construction, alteration, and/or repair activities. Construction workers are exposed to a variety of risks, including falling from rooftops, unattended machinery, being struck by heavy construction equipment, electrocutions, silica dust, and asbestos [1].

Furthermore, the number of fatalities on construction sites is increasing year after year. People's safety and health are not guaranteed on construction sites. Workers face numerous challenges and difficulties in the job as a result of an unbalanced work-life balance. They are not only physically harmed, but they are also psychologically harmed. Among all other industries, the construction industry is the greatest cause of fatalities [1]. A breakdown of construction accidents by body part revealed that the head (161 individuals, or 41.2%) and numerous body parts (123 people, or 31.5%) were the most commonly wounded, and falls were the most prevalent form of fatal construction accident (8699 people, 32.7%) [2].

Thus, because accidental falling down has such a negative impact on the worker, a falling down detection system is required to eliminate this threat. This system must be able to detect the action of falling down and alert relevant personnel, as well as position the falling location so that relevant personnel can know where to help in the shortest time [3]. Several methods for automatically detecting falls have been developed over the last decade. They are divided into three categories based on the sensor technology they use: vision-based sensors, ambient sensors, and wearable devices. Vision-based and ambient sensors, on the other hand, have a limited monitoring area and require installation, adjustment, and maintenance, which can lead to higher costs. Inertial sensors have recently become smaller and less expensive thanks to technological advances in the domains of electrical, mechanical, and computer engineering, particularly in the field of microelectromechanical systems (MEMS). An inertial sensor (such as a 3D accelerometer, 3D gyroscope, or 3D compass) can be as small as 55 mm in diameter and as inexpensive as one US dollar. As a result, it is frequently employed in the development of wearable devices that allow for the

Table 30.1 Accelerometer test on falls detection

Events	Amount of experiment tested	Number of alarms triggered	Results
Walking	50	0/50	No alarm triggered
Front fall	50	50/50	Alarm triggered for each fall
Back fall	50	50/50	Alarm triggered for each fall
Side fall	50	50/50	Alarm triggered for each fall

monitoring of physical activity in a real-world setting. Indoor and outdoor activities are included [4, 5].

Hence, accelerometer is chosen to be used on this research to detect the fall of the worker. The goal of this research is to develop wearable gadgets, such as a smart worker helmet, to monitor construction workers and provide a safer and more secure working environment. The gadgets assist in providing emergency alerts in the event of a worker's slip or accident (both in-house and out-house falls). Another goal of this study is to look at the accelerometer MPU6050 and GPS sensors to see whether they can help minimize power usage and produce a dependable system. It is also about putting in place a system that can send out early warnings to those who use IoT. On the prior research into the fall detection system, a small study was conducted. Researchers are working hard to improve the current system as well as ensure that the new system being built is more trustworthy. The comparative component utilized in the prior study is provided in Table 30.1. The major objectives of these studies are to notice a fall from the monitored individual and focusing on the monitoring the worker's condition. Furthermore, the differences between this system compared to propose system is the supply for the system is from solar energy and Wi-Fi NodeMCU is used compared to Zigbee and GSM module.

30.2 Methodology

30.2.1 Hardware Part

There are two parts in this methodology section which are hardware and software part of the smart solar safety worker helmet using accelerometer and GPS sensor system. The system's block diagram is shown in Fig. 30.1. These devices are utilized to build the hardware component. The system's inputs include an IR sensor, an accelerometer, a GPS sensor, and a push button or emergency button. In the meanwhile, Blynk smartphone apps are the system's output. The system is powered by solar energy, which is fed into the NodeMCU ESP8266. When the NodeMCU ESP8266 received enough power, it sent the instructions that were programmed into it to the sensors. The NodeMCU ESP8266 processes the data and delivers it to the Blynk apps as soon as the sensors collect it. The in-charge monitor staff will receive a notification from the Blynk apps via smartphone. When the accelerometer sensor detects a fall from the worker's helmet and from the push/emergency button on it, the Blynk applications will display a notification.

Figure 30.2 shows the prototype of the research. The sensors is located in the worker's helmet to make that the data is received by the sensor.

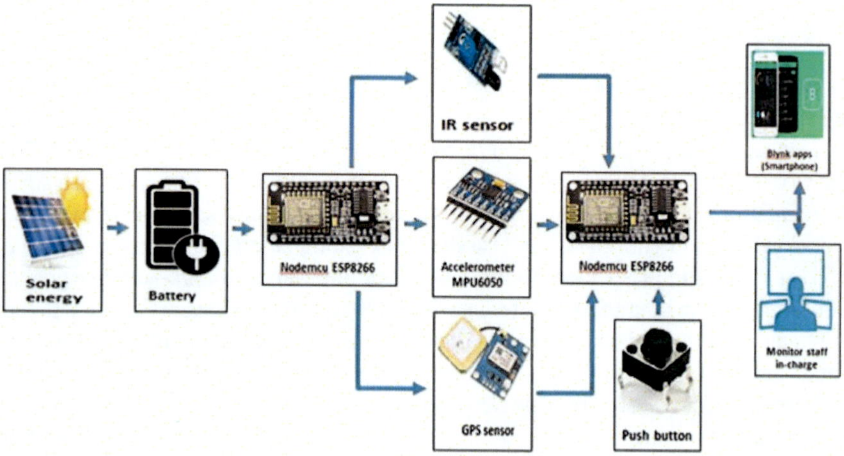

Fig. 30.1 Block diagram of the system

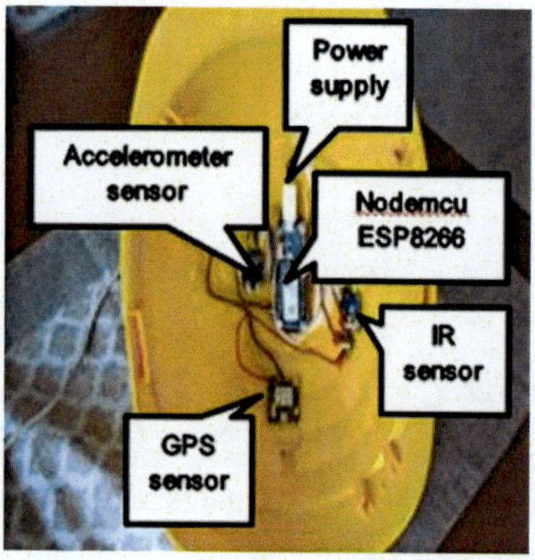

Fig. 30.2 Prototype of the research

30.2.1.1 Components Description

In this research, ESP8266 is used as Wi-Fi module compared to ZigBee which is used by existing project. This is because ESP8266 has very high speed and large network compared to Zigbee [6]. Since this is safety related project, a speed in transfer the data to the monitor system is very important. Hence, the ESP8266 is chosen. The ESP8266 chip in NodeMCU is a highly integrated chip built for the needs of the new connected world. It provides a fully self-contained Wi-Fi networking solution, allowing it to host the application or offload all Wi-Fi networking tasks to another application processor. The ESP8266 NodeMCU has robust on-board processing and storage capabilities, allowing it to be integrated with sensor-specific devices via its GPIOs with minimal development and runtime loading. ESP8266 also has a low cost and a lot of functions, making it an excellent Internet of Things module (IoT). It can be used in any application where a device has to be connected to a local network or the Internet [7].

Next, the MPU6050 accelerometer is the world's first and only 6-axis motion tracking device, developed for smartphones, tablets, and wearable sensors that require low power, low cost, and great performance. MPU6050 is a three-axis accelerometer and three-axis gyroscope micro-electro-mechanical system (MEMS). It aids in the measurement of velocity, direction, acceleration, displacement, and other motion-related characteristics [7].

Then, the Global Positioning System (GPS) provides an unrivaled breadth of services to commercial military and consumer applications regardless of time, location, or weather. The majority of these services allow aerial, land, and marine users to know their exact velocity, location, and time at any time and from any location on the planet. Other traditional positioning and well-known navigation methods and technologies, such as magnetic compasses, radio-based devices, and chronometers, have become obsolete and unworkable due to the advancement and capabilities of GPS technology [7].

30.2.2 Software Part

30.2.2.1 Blynk Applications

Blynk is an Android and iOS platform that can run a variety of hardware modules, including Raspberry Pi, Arduino, NodeMCU, and over 400 more. Furthermore, Wi-Fi, Ethernet, Cellular, USB, serial, and Bluetooth are all viable options for connecting the hardware module device to the Internet. Blynk allows you to create many apps and use them to control multiple boards linked to a device with Internet connection from anywhere in the globe using a smartphone. The Blynk app's graphic user interface (GUI) on the smartphone is very easy and user-friendly, allowing users to install widgets that they want to use and control them via connectivity options such as Wi-Fi Internet [8].

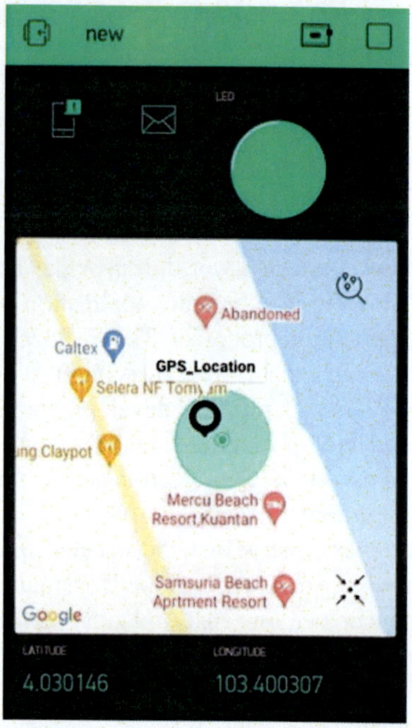

Fig. 30.3 Blynk application in smartphone

Figure 30.3 shows the data that have been sent to Blynk apps on the smartphone. Based on Fig. 30.3, the IR sensor will be on with green light once the helmet is worn. The GPS will always show the current map location of the worker with latitude and longitude values. Then, the Blynk will receive a notification if it detects fall detection or emergency button is pushed. Beside the in-charge monitor staff, other person such as family member of the worker will receive email notifications once the worker is in danger.

30.2.2.2 Flowchart of the System

Figure 30.4 shows the flowchart of the system. Solar system absorbs the sunlight to charge the battery. At the same time, the battery is used to power up the NodeMCU ESP8266. After the NodeMCU ESP8266 received power from the battery, the NodeMCU ESP8266 initialized to distribute the instructions to each component in the system. Then, the sensors which are accelerometer, GPS sensor, and IR sensor began to work as coded. Once the accelerometer detects any falls or whether emergency button is pushed, the data will be sent to the Blynk apps with notifications simultaneously with email notification. Otherwise, the Blynk will not receive any notification. In addition, GPS will always show the current location of the worker and IR sensor will always show whether the helmet is worn or not regardless of the condition emergency occurs or not.

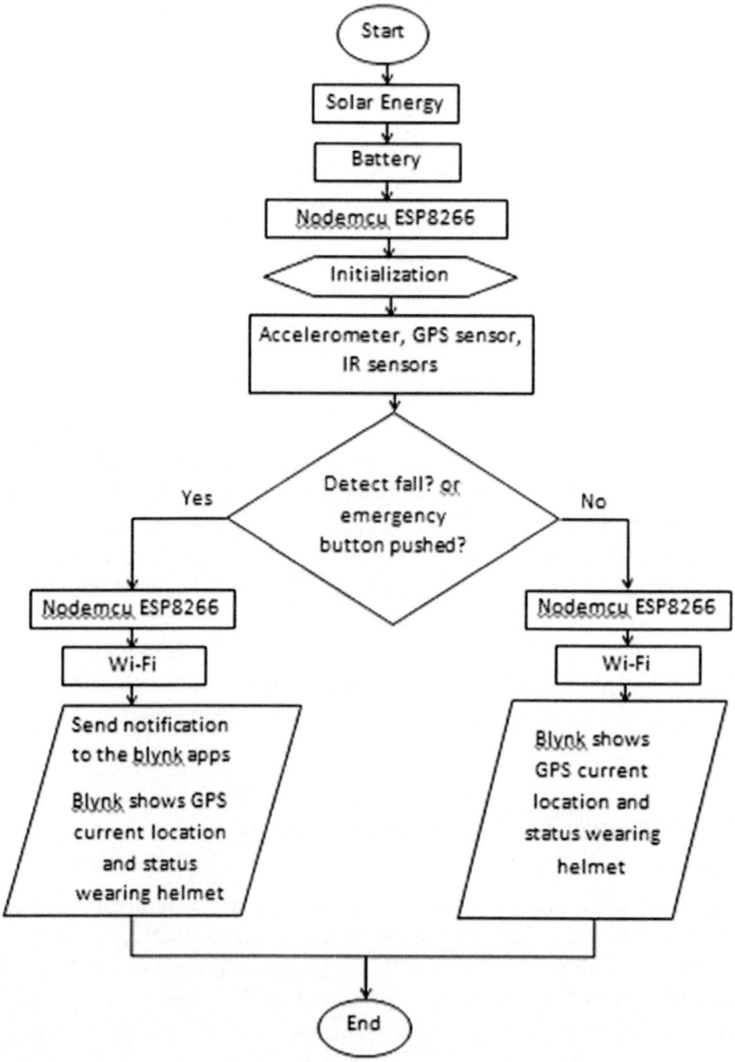

Fig. 30.4 Flowchart of the smart solar safety worker helmet using accelerometer and GPS sensor

30.3 Result and Discussions

The system is using NodeMCU, accelerometer, GPS sensor, solar, and Blynk to complete the project. The result below shows the test on accelerometer to detect fall, comparison of wireless network technology, battery lifetime, and the results of supplying power from solar energy.

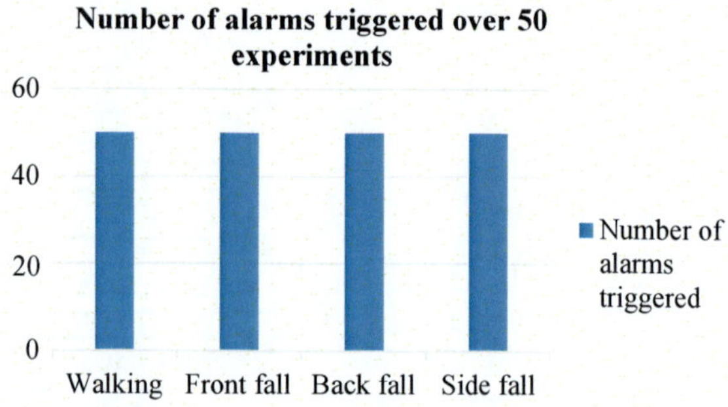

Fig. 30.5 Graph for each events vs alarm triggered over 50 times of experiments

30.3.1 Accelerometer Test on Detecting Fall

Table 30.1 shows results accelerometer test on fall detection. There are four events which are walking, front falling, back falling, and side falling that has been monitor. These entire events are tested with the same amount of experiment which is 50 times for each event. The event is test for 50 times to observe the accuracy of the sensor. Repeated trials are used to minimize the impact of errors and hence improve the dependability of experiment outcomes. As the result shown, except for the walking, the entire falls event triggered the sensor thus sending the alarm notifications to the smartphone. This shows how accurate the accelerometer sensor on detecting the fall [8] (Fig. 30.5).

30.3.2 Comparison of Wireless Technology

The range of communication is one of the system aspects. With an average range of 30–100 m, Wi-Fi is suitable for personal area network (PAN) and wireless local area network (WLAN) area networks. While Zigbee is limited to wireless personal area networks (WPANs) with a range of 10–30 m, Bluetooth is not. A Bluetooth and ultra-wideband (UWB) connection's range is between 1 and 10 m. A Bluetooth connection is said to form a personal area network due to its small range (PAN) [10]. Table 30.2 compares data rate and range from four wireless technologies.

Based on Fig. 30.6 graph, it shows that Wi-Fi and UWB are not a good selection in terms of energy usage. But since this project uses a solar system, then this problem

Table 30.2 Comparison on data rate and range from different wireless technologies

Wireless technology	Data rate (M bits/s)	Range (m)
Bluetooth	2	5
Wi-Fi	85	95
Zigbee	1	25
UWB	85	5

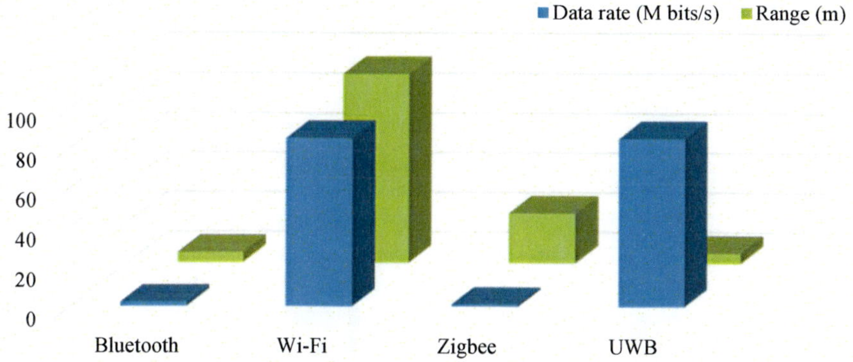

Fig. 30.6 Comparison graph on data rate and range from different wireless technology

can be solved as the solar will keep charging battery to supply power. Meanwhile, UWB is used for short range and high data rate application. In this comparison, Wi-Fi is the fastest and most flexible, with a range of 100 m compared to 10 m for ZigBee, Bluetooth, and UWB. ZigBee and Bluetooth are both low cost and power efficient. Higher data transfer rates are available with UWB and Wi-Fi with 85 (M bits/s) but have different ranges which are 5 m and 95 m, respectively. Then, in this research, Wi-Fi is used since in this research which can produce the process of transferring data faster at the wider range.

30.3.3 GPS Sensor

Figure 30.7 shows the results of GPS location in Blynk apps. This GPS will locate the current location of the helmet user with latitude and longitude values. The latitude and longitude coordinates can be used to specify any place on the earth's surface, just like every actual house has its address. As a result, it easy to specify practically any position on the globe using latitude and longitude [8].

30.3.4 Performance of Battery and Photovoltaic Voltage Output

The studies of voltage charging rate by using solar in different days have been taken. The pace of battery level charging utilizing solar panels is shown in Fig. 30.8. The test takes place at the same time and in the same location. The battery begins charging at 9 a.m. and continues until 5 p.m., with an initial voltage of 11.30 V. The test was taken as shown in Tables 30.3 and 30.4 during 9 a.m. until 5 a.m. because when the sun is shining, solar systems work very well. The strength of the sun's radiation varies

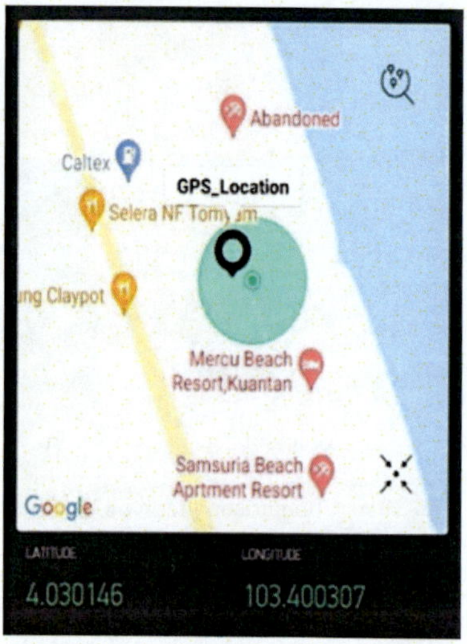

Fig. 30.7 GPS location results on Blynk apps in smartphone

with the brightness and angle of the sun throughout the day and seasons, affecting the quantity of electricity generated by a solar power system. During the day, the peak sun generation occurs between 11 a.m. and 4 pm. The battery voltage is increased to 11.40 V after eight hours of charging on day one. Day 2 and day 3 require 11.41 V and 11.42 V, respectively. Due to the weather, the varied end figures of billing for three days are different. The output of solar panels can be influenced by the weather. The initial voltage of the battery is 11.30 V, which represents a 70.83 percent voltage drop. The battery voltage is increased to 11.40 V, resulting in a voltage drop of 74.99%. The charge rate for an 8-h period is 4.16%.

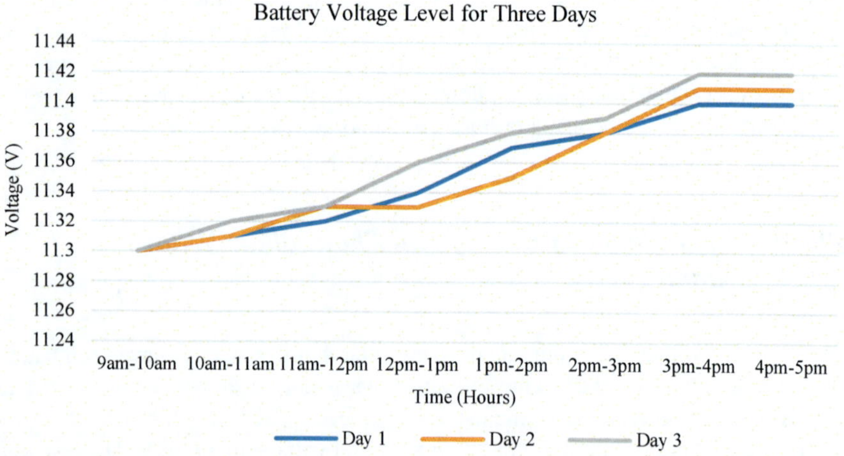

Fig. 30.8 Lead acid battery voltage level charge by solar panel

Table 30.3 Lead acid battery voltage level for three days

Time	Day 1 (v)	Day 2 (v)	Day 3 (v)
9 a.m.–10 a.m.	11.3	11.3	11.3
10 a.m.–11 a.m.	11.31	11.31	11.32
11 a.m.–12 p.m.	11.32	11.33	11.33
12 p.m.–1 p.m.	11.34	11.33	11.36
1 p.m.–2 p.m.	11.37	11.35	11.38
2 p.m.–3 p.m.	11.38	11.38	11.39
3 p.m.–4 p.m.	11.4	11.41	11.42
4 p.m.–5 p.m.	11.4	11.41	11.42

Table 30.4 Solar voltage reading for three days

Time	Day 1 (v)	Day 2 (v)	Day 3 (v)
9 a.m.–10 a.m.	11.65	11.68	11.7
10 a.m.–11 a.m.	11.66	11.69	11.75
11 a.m.–12 p.m.	11.67	11.7	11.76
12 p.m.–1 p.m.	11.69	11.75	11.8
1 p.m.–2 p.m.	11.67	11.76	11.81
2 p.m.–3 p.m.	11.66	11.75	11.85
3 p.m.–4 p.m.	11.66	11.75	11.85
4 p.m.–5 p.m.	11.65	11.74	11.86

Figure 30.9 shows the voltage of the solar panel's power input. The voltage on day 1 from 1 to 2 pm decreased. This is due to the weather conditions during that day. The weather conditions might affect the changes in peak voltage from day to day.

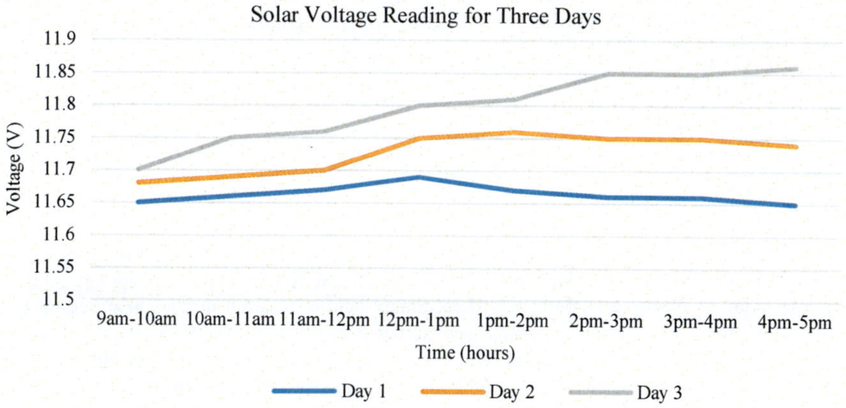

Fig. 30.9 Solar output voltage for three days

Through this experiment, it can be clearly seen that the research is successfully working using solar panel and the 12 V lead acid battery can be charge successfully. The solar charger controller circuit prevents overcharging and prevents overcharging and reverse current in the 12 V lead acid rechargeable battery at night [9].

30.4 Conclusion and Recommendation

By integrating the several sensors such as accelerometer and GPS sensors will be help much on increasing the safety of worker. This research has proposed to design a smart solar safety worker helmet that can monitor the falls or any emergency cases of the construction worker to increase the safety during working hour. Based on the result obtained, it achieved the objective which is to develop a monitoring safety system that can notify people immediately when any emergency cases happen and to locate the current location of the worker using IoT. The percentage improvement of this smart solar safety worker helmet using accelerometer and GPS sensor system is 3.3%. Since this project using solar, the amount of work and money spend on changing the battery is no longer a problem. The system can work perfectly without the need to changing or charging the battery. Furthermore, Wi-Fi is the better for this system compared to other wireless technology. For recommendation, a sensor that can detect a hazard originating from the top, in front, or back of the worker can be implemented in this project for future work related to the safety worker helmet to boost the worker's safety. In terms of wireless communication, the system can use the long-range wide area network (LoRaWAN), which allows data to be sent over long distances of up to 15 km in urban areas and 10 km in rural areas, respectively.

Acknowledgements The authors would convey their profound appreciation and gratitude to the College of Engineering, Universiti Teknologi MARA, Shah Alam, Selangor, Malaysia for permitting us to perform this study. Many thanks to the Ministry of Higher Education for the financial assistance received using Fundamental Research Grant (FRGS/1/2018/TK04/UITM/02/29) and Universiti Teknologi MARA (600-IRMI/FRGS 5/3 (016/2019). Special thanks to those who contributed to this project directly or indirectly.

References

1. S. Brown, MPH, W. Harris, R.D. Brooks, X.S. Dong, Fatal injury trends in the construction industry, CWPR the centre for construction Research and Training, pp. 1–6, February 2021
2. S.H. Kim, C. Wang, S.D. Min, S.H. Lee, Safety helmet wearing management system for construction workers using three-axis accelerometer sensor. Appl. Sci. **8**(12), 2400 (2018)
3. J. Cheng, Y. Chen, W. Bao, Y. Hu, N. Ding, X. Zeng, Positionable wearable fall detection system for elderly assisted living applications, in 2013 IEEE 10th International Conference on ASIC (ASICON 2013), The State-Key Lab of ASIC and System, Fudan University, Shanghai 201203, China, October 2013

4. A. Buke, F. Gaoli, W. Yongcai, S. Lei, Y. Zhiqi, Healthcare algorithms by wearable inertial sensors: a survey. China Communications **12**(4), 1–12 (2015); Systems, International Journal of Engineering, Science and Mathematics **7**(3), 141–149, (2018)
5. J. He, C. Hu, X. Wang, A smart device enabled system for autonomous fall detection and alert. International Journal of Distributed Sensor Networks 10 (2016)
6. Y.V. Varshney, A.K. Sharma, Design & simulation of Zigbee transceiver system using Matlab. Int. J. Eng. Trends Technol. **4**(4), 1316–1319 (2013)
7. Y.S. Parihar, Internet of things and NodeMCU a review of use of NodeMCU ESP8266 in IoT products. Journal of Emerging Technologies and Innovative Research (JETIR) **6**(6), 1085–1088 (2019)
8. N.M. Fung, J.W.S. Ann, Y.H. Tung, C.S. Kheau, A. Chekima, Elderly fall detection and location tracking system using heterogeneous wireless networks, in 2019 IEEE 9th Symposium on Computer Applications & Industrial Electronics (ISCAIE) (2019)
9. B. Chan, S.A. Jumaat, M.N. Abdullah, Solar powered paddy irrigation system using Arduino UNO microcontroller: battery performance. Journal of Physics: Conference Series (2020)
10. K. Karimi, K. Salah-ddine, A comparative study of the implementations design for smart homes/smart phone systems. International Journal of Engineering, Science and Mathematics **7**(3), 141–149 (2018)

Chapter 31
Influence of Flue Gas Desulfurization (FGD) Waste as Substitute Feldspar on the Physicomechanical Porcelain Properties

Suffi Irni Alias, Banjuraizah Johar, Syed Nuzul Fadzli Syed Adam, Mustaffar Ali Azhar Taib, and Fatin Fatini Othman

Abstract In this work, the influence of flue gas desulfurization (FGD) waste on the physical, mechanical and thermal properties of porcelain samples was investigated. The influence of the flue gas desulfurization (FGD) waste content (0–15 wt.%) which was sintered at a temperature of 1200 °C was also studied. The result showed that the substitution of feldspar by flue gas desulfurization (FGD) waste in porcelain bodies led to an increase in porosity resulted in a decrease in bulk density and mechanical strength when amount of FGD waste increased. Besides, the increasing FGD waste is also contributed to a decrease value in coefficient thermal expansion value which could reduce the thermal shock in porcelain. As a result, the sintered sample with the composition of 5 wt.% of FGD waste (S5) has the potential to be used as a porcelain tile as it meets the requirements of the standard tile (>35 MPa) despite flexural strength decreased. While the composition of 10 wt.% of FGD waste (S10) and 15 wt.% of FGD waste (S15) was found most suitable for developing porous brick because it meets the requirements of the porous brick standard (>2.5 MPa).

Keywords Flue gas desulfurization waste · Porcelain · Tile · Brick · Porosity · Bulk density · Flexural strength · Coefficient thermal expansion

---

S. I. Alias · B. Johar (✉) · F. F. Othman
Faculty of Chemical Engineering Technology, Universiti Malaysia Perlis, 01000 Kangar,, Perlis, Malaysia
e-mail: banjuraizah@unimap.edu.my

S. N. F. Syed Adam
Faculty of Mechanicall Engineering Technology, Universiti Malaysia Perlis, Sungai Chuchuh, Kampus UniCITI Alam, 02100 Padang Besar, Perlis, Malaysia

M. A. A. Taib
Division of Advanced Ceramic Materials Technology (ADTEC), Taiping, Perak, Malaysia

M. A. A. Mohd Salleh et al. (eds.), *Proceedings of the Green Materials and Electronic Packaging Interconnect Technology Symposium*, Springer Proceedings in Physics 289,
https://doi.org/10.1007/978-981-19-9267-4_31

31.1 Introduction

The flue gas desulfurization (FGD) waste is industrial waste. The FGD used in this research was provided from Nippon Electric Glass Malaysia (NEGM) Sdn Bhd. FGD is a chemical reaction process that removes sulfur dioxide (SO_2) from flue gas power plants in wet scrubbing by spraying a contaminated gas stream with limestone ($CaCO_3$) to minimize pollutant emission, resulting in FGD sludge as a waste product. The use of various wastes in the production of ceramics such as pottery, earthenware, sanitary ware, cordierite, porcelain and other materials has been the focus of several in-depth studies. Porcelain has widely used in various applications due to its excellent properties such as flexural strength of 75–41 MPa [1, 2], having low thermal coefficient of 3×10^{-6}/°C–5×10^{-6}/°C [3, 4] and having good thermal conductivity of 1.5 W/mK [5]. According to Zanelli [6], porcelain are composed of 10–20% clay, 20–30% kaolin, 35–55% feldspar, 0–3% carbonate, 2–5% bentonite and 0–5% zirconia. Thus, feldspar is the most often utilized ingredient in the formulation, comprising up to 55% of the porcelain's composition. One of the primary issues with using feldspar, an alkaline aluminosilicate, in ceramics is its high cost which is due to the long distances between feldspar extraction sites which can be thousands of kilometers away from the industries [6]. Feldspars are considered to be fluxing agents because they provide alkaline oxides, such as K_2O, calcium oxide (CaO), magnesium oxide (MgO) and sodium oxide (Na_2O), which contribute to producing a low viscosity liquid at high temperatures during firing [7]. They can fill the pores of the ceramic composition by capillary action at temperatures over 1100 °C, bringing the particles closer together and producing densification of the ceramic bodies in a process known as sintering by viscous liquid phase [7]. Additionally, waste from industries, particularly, contains alkali, alkaline earth elements and other chemical components that are usually found in the fluxing agents in ceramic production. FGD waste contains fluxing agents such as potassium oxide (K_2O), calcium oxide (CaO), magnesium oxide (MgO), sodium oxide (Na_2O), barium oxide (BaO) and zinc oxide (ZnO). Because FGD waste and feldspar have similar fluxing agents, FGD waste has the potential to be utilized as a raw material to substitute feldspar in ceramic manufacture.

31.2 Methodology

31.2.1 Sample Preparation and Characterization Process

This research focuses on the influence of using FGD waste substitute feldspar by FGD waste in the ceramic porcelain production. The samples were labelled as S0, S5, S10 and S15. A standard composition (kaolin 25 wt.%, silica 30 wt.%, ball clay 25 wt.% and feldspar 20 wt.%). Sample S0 is the reference sample which commonly used in porcelain production. In this study, the mixture powders were then pressed using a uniaxial hydraulic press at 11 MPa and sintered at a temperature of 1200 °C with a soaking time of 3 h and heating rate of 5 °C/min, respectively. The sintered samples were characterized in terms of chemical composition by X-ray fluorescence (XRF), phase transformation by X-ray diffraction (XRD), morphology property by scanning electron microscope (SEM), physical properties by density and porosity testing, mechanical property by flexural strength testing and thermal property by coefficient thermal expansion.

31.3 Result and Discussions

31.3.1 Chemical Composition

The chemical composition of various amounts of FGD waste substituted by feldspar into porcelain powder is shown in Table 31.1. Before the sinter, the mixture powder has been analyzed by XRF to determine the chemical composition. From the observation, the composition of SiO_2, SO_3, CaO, Fe_2O_3, MnO and SrO showed an increment pattern in the mixture powder. This can be explained by the addition of FGD waste to the mixture powder, and the wt.% composition of SO_3, CaO, Fe_2O_3, MnO and SrO has increased. Besides, the composition of Al_2O_3 and K_2O has decreased. This might be due to the decrement of feldspar in the mixture powder.

Table 31.1 Chemical composition of studied porcelain sample with various amounts of FGD waste powder before sintered

Compound	Value unit (wt.%)			
	S0	S5	S10	S15
SiO_2	56.000	58.600	55.600	57.400
SO_3	–	1.400	2.500	3.540
CaO	0.235	1.750	2.969	4.350
Fe_2O_3	0.447	0.452	0.462	0.498
SrO	0.004	0.005	0.005	0.006
MnO	0.006	0.006	0.007	0.009
ZnO	0.002	0.004	0.004	0.005
Al_2O_3	39.600	35.300	35.300	32.600
K_2O	2.480	1.760	1.280	0.948
Rb_2O	0.025	0.019	0.014	0.010
Others	1.100	0.704	0.859	0.634

31.3.2 Crystal Structures

The XRD pattern of sintered samples S0, S5, S10 and S15 is presented in Fig. 31.1. The XRD pattern for sintered sample demonstrates sample S0 contained two phases which were quartz (SiO_2) and mullite ($Al_4.68Si_9.66O_{1.2}$), quartz as the major phases, whereas mullite as secondary phases. The sintered sample S5 was then presented to the quartz phase and the mullite phase with the presence of a new peak assigned to anorthite ($Al_2CaSi_2O_8$). The anorthite phase's diffraction phases continued to increase as the main phase along with quartz and mullite phases which is gradually significantly decreasing, as the FGD waste continued to increase up to 5 wt.%, 10 wt.% and 15 wt.% in the sintered samples S5, S10 and S15, respectively. The formation of anorthite mineral was from the reaction of calcium oxide with decomposed product of FGD waste and clay mineral. Mainly, the formation of anorthite starts at 1100 °C [8, 9]. From the observation, the higher content of FGD waste in this sintered sample has favored the appearance of anorthite phase. This can be proven by the XRF result in which the CaO content was increased with the increasing the amount FGD waste added. Anorthite has great potential as reinforcing phase in porcelain insulator and posessess low coefficient thermal coefficient thermal expansion (4.5×10^{-6}/°C) which enable the improvement of the body strength [4, 10–12].

31.3.3 Morphology Analysis

The morphology for all sintered samples with the various amounts of FGD waste under ×500 magnification are scaled which are displayed in Fig. 31.2. Samples with white particle agglomerates exists probably due to residual quartz particle that were trapped and melted during the sintering process [13]. The sintered sample for sample S5 displayed insignificant interparticle pore structure and sample S10 displayed insignificantly growth of pores which are presented in Figs. 31.2b and c, respectively. Meanwhile, the sintered sample for sample S15 showed the formation of open and close pore structures and a glassy body significantly in Fig. 31.2d. The pore size increased rapidly with the increase of FGD waste. Smaller pores are formed first, then connected to form large pores. This is most likely associated due to more gases SO_x and CO_x released from the decomposition of FGD waste thus generated the pore structure [14, 15]. Besides, the glassy body in sintered sample S15 was related to the presence of excess alkaline earth oxide which allowed the material to melt during the sintering process and subsequently favor the formation of the glassy body.

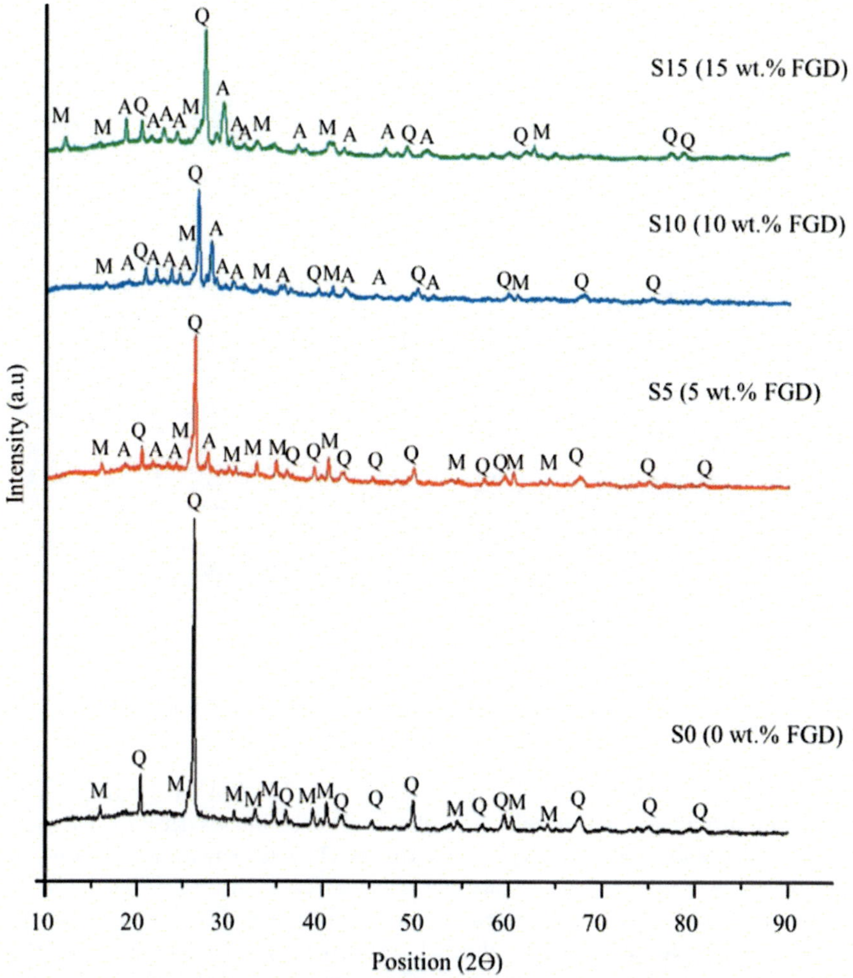

Fig. 31.1 X-ray diffraction pattern of porcelain with of various amounts of FGD waste added in the composition. (Q: quartz, M: mullite, A: anorthite)

31.3.4 Physical and Mechanical Properties

Table 31.2 shows the percentage of porosity, bulk density and flexural strength as function of FGD waste-substituted feldspar in the porcelain. The porosity of the sintered samples gradually increased as the amount of FGD waste contents increases. The more FGD waste present, the more gases such as SO_x and CO_x decomposed and released subsequently resulting in the formation of pores' structure [14, 15]. In addition, the bulk density tends to decrease with increasing the amount of FGD waste. This can be explained by the presence of phase in sintered samples S5, S10 and S15 which were anorthite with a general theoretical density of 2.74 g/cm^3 [16], while the

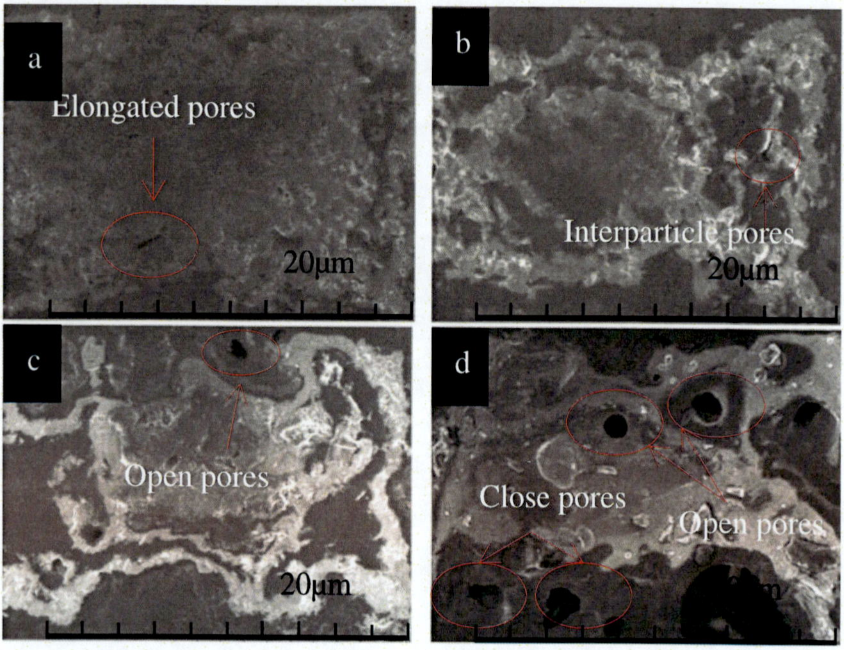

Fig. 31.2 SEM images of porcelain with various amounts of FGD waste in sintered samples for (**a**) Sample S0, (**b**) **Sample** S5, (**c**) Sample S10 and (**d**) Sample S15 under ×500 magnification

theoretical density for quartz and mullite phases was 2.65 g/cm^3 [17] and 3.2 g/cm^3 [18]. As shown in the XRD results, the mullite and quartz phases decrease and cause the phase densities also tend to decline when the FGD waste increases. As a result, the bulk density was expected to decrease due to the decreased theoretical density in the quartz and mullite phases in sintered sample. Furthermore, the increment porosity and decreased bulk density have affected the reduction in flexural strength. This circumstance can be explained, and the highest flexural strength is due to the pore formed having thicker strut and low expansion structure, whereas the lowest flexural strength is due to the thinner strut and higher expansion structure, then contributed to the collapse when forced applied. The flexural strength for sintered samples S0 and S5 were maintaining acceptable flexural strength value as minimum strength of 13,006 standard for ceramic tile which was 35 MPa [19] while the samples S10 and S15 were exceeded > 2.5 MPa which is the minimum requirement of application criteria.

Table 31.2 Elastic recovery of silver-based conductive ink at different strains with the initial length of 25 mm

Testing	Sintered samples			
	S0	S5	S10	S15
Porosity (%)	10.533	14.762	18.579	36.994
Bulk density (g/cm^3)	2.043	1.720	1.525	0.912
Flexural strength (MPa)	46.810	39.120	33.120	6.220

31.3.5 Thermal Properties

The coefficient thermal expansion (CTE) of various amounts of FGD waste at range temperature of 30–1000 °C is shown in Fig. 31.3. All plots revealed the sintered samples S0, S5, S10 and S15 began shrank at temperature range of 30–249 °C. The sintered sample for S0 shows the lowest shrunk with the lowest CTE value of 1.0×10^{-6}/°C, while the sintered sample S15 shows the highest shrunk with the highest CTE value of -2.8×10^{-6}/°C in this range temperature. The increasing of CTE value is indicating that there were more gases and pressures trapped inside the body that need to be released. When heat was applied to all sintered samples, the gases released and experience shrinkage due to increased porosity and expansion structure [20, 21]. At temperature range 249–500 °C, all sintered samples were significantly expanded with increase of CTE value. The sintered sample S0 displayed the most expand with highest CTE value of 7×10^{-6}/°C, whereas sintered sample S15 displayed the lowest expand with lowest CTE value of 3.1×10^{-6}/°C. When the temperature was increased to 500–749 °C, all of the sintered samples expanded further, with increased CTE values ranging from 3.9×10^{-6}/°C to 7.3×10^{-6}/°C. This can be explained by the sintered sample underwent the kinetic energy process in this range of temperature. All of the sintered samples were observed to shrink slightly again with decreasing CTE values at temperatures between 749 and 1000 °C. As can be seen, the sintered sample S15 shows the most shrunk with lowest CTE value of 3.6×10^{-6}/°C, whereas the sintered sample S0 shows the lowest shrunk with higher CTE value of 5.9×10^{-6}/°C. The sintered sample experienced softening temperature. The shrinkage of sintered sample and decline in CTE value occurred in this range temperature because the particle underwent viscous sintering.

At higher temperatures, the viscosity of the liquid phase decreased and it is easier for particles commence to coalescence subsequently eliminating the remaining of pores for densification and vitrification. A glassy phase with low viscosity easily flowed causing the sintered sample to tend to have a higher shrinkage rate [22]. This phenomenon can be explained by the formation of liquid phase enhanced diffusion to lower surface tension and capillary which helps consolidate the particle as well as reduce pores, resulting increasing shrinkage rate [22]. Previously, Mousumi [4] obtained the CTE value which is 4.6×10^{-6}/°C on anorthite porcelain. Thus, the sintered sample S10 with CTE value 4.6×10^{-6}/°C has a similar value of CTE which can be improved the heat stability which is similar to the porcelainized stoneware

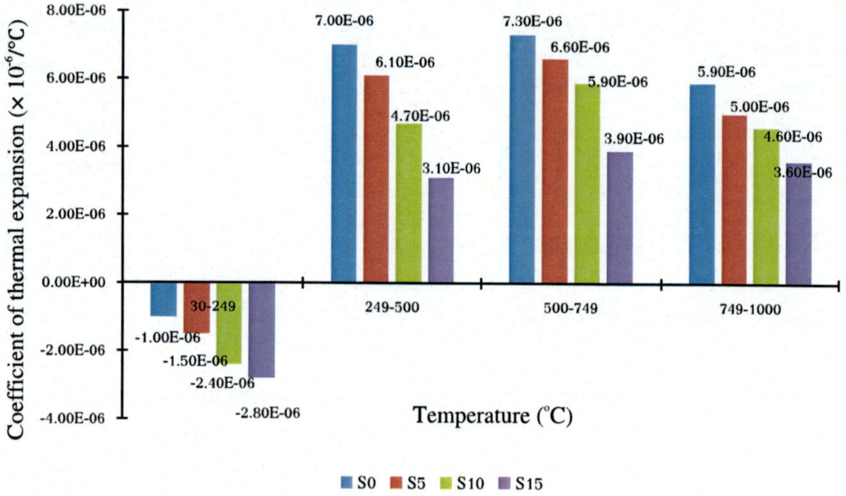

Fig. 31.3 Value of coefficient thermal expansion of porcelain with various treated FGD sludges at range temperature of 30–1000 °C

material based on anorthite suggesting its application as such as cooking wares [4], while the CTE value for the sintered sample S15 was obtained at the value of 3.5×10^{-6}/°C. Thus, the sintered sample at that range can be matched with the applicable glaze [3]. With this low CTE, the material is expected to have better thermal shock-resistant property [4].

31.4 Conclusion

The porcelain with various amounts of FGD waste substituted by feldspar was successfully fabricated. The XRF analysis has shown that SO_3 and CaO were found to increase, whereas the wt.% of Al_2O_3 was found to decrease. As a result, the amount of quartz and mullite phases tends to decrease, while the anorthite phase tends to grow increased. The morphology analysis has shown that as the amount of the FGD waste increased, the sintered samples displayed significant porous structure and the glassy-phase surface can be clearly observed due to decomposition of SO_x and CO_x gaseous and higher liquid phases forming from the FGD waste. The bulk density was tended to decrease due to the decreased theoretical density in the quartz and mullite phases in sintered sample. The flexural strength was gradually decreased due to the thinner strut and higher pores' size structure, then contributed to the collapse when forced applied. Even though the porosity has increased in the porcelain bodies and promoted the reduction in flexural strengthvalue, but the flexural strength for sample S0 and sample S5 was maintain within acceptable flexural strength for porcelain tile production. Meanwhile, samples S10 and sample S15 were fulfilled the minimum requirement on application criteria of porous brick standard [8, 9, 11].

Acknowledgements The author would like to thank the Ministry of Higher Education (MOHE) Malaysia for the financial support under Fundamental Research Grant Scheme (FRGS) (Grant No: FRGS/1/2019/STG07/UNIMAP/02/4) and Universiti Malaysia Perlis (UniMAP) for providing the research facilities. The authors also would like to thank Nippon Electric Glass Malaysia (NEGM) company for providing sludge waste. 23 24

References

1. O.I. Ece, Z.E. Nakagawa, Bending strength of porcelain. Ceram. Int. **28**(2), 131–140 (2002). https://doi.org/10.1016/S0272-8842(01)00068-2
2. R.A. Giordano II., L. Pelletier, S. Campbell, R. Pober, Flexural strength of an infused ceramic, glass ceramic and feldspar porcelain. J. Prosthet. Dent. **73**(5), 411–418 (1995). https://doi.org/10.1016/S0022-3913%2805%2980067-8
3. A. Capoglu, Elimination of discolouration in reformulated bone china bodies. J. Eur. Ceram. Soc. **25**(13), 3157–3164 (2005). https://doi.org/10.1016/j.jeurceramsoc.2004.07.008
4. P. Mousumi, D. Sukhen, D. Swapan Kumar, Anorthite porcelain: synthesis, phase and microstructural evolution. Bull. Mater. Sci. (India) **38**(2), 551–555 (2015). https://doi.org/10.1007/s12034-015-0855-6
5. Information on https://material-properties.org/porcelain-density-heat-capacity-thermal-conductivity
6. C. Zanelli, M. Raimondo, G. Guarini, M. Dondi, The vitreous phase of porcelain stoneware: Composition, evolution during sintering and physical properties. J. Non Cryst. Solids **357**(16–17), 3251–3260 (2011). https://doi.org/10.1016/j.jnoncrysol.2011.05.020
7. M. Locks, S. Arcaro, C.P. Bergmann, M.J. Ribeiro, F. Raupp-Pereira, O.R.K. Montedo, Effect of feldspar substitution by basalt on pyroplastic behaviour of porcelain tile composition. Materials (Basel) **14**(14), 3990 (2021). https://doi.org/10.3390/ma14143990
8. J. Liu et al., Feasible recycling of industrial waste coal fly ash for preparation of anorthite-cordierite based porous ceramic membrane supports with addition of dolomite. J. Eur. Ceram. Soc. **36**(4), 1059–1071 (2016). https://doi.org/10.1016/j.jeurceramsoc.2015.11.012
9. Y. Yasmin, M.N. Mazlee, A.H. Norzilah, J.B. Shamsul, R. Azmi, W.H. Chan, *The Investigation of Physical and Mechanical Properties of Porous Anorthite Ceramics Using Statistical Analysis* (2016). https://doi.org/10.1063/1.4958792
10. C.L. Chin, Z.A. Ahmad, S.S. Sow, Relationship between the thermal behaviour of the clays and their mineralogical and chemical composition: Example of Ipoh, Kuala Rompin and Mersing (Malaysia). Appl. Clay Sci. **143**, 327–335 (2017). https://doi.org/10.1016/j.clay.2017.03.037
11. M. Sutcu, S. Akkurt, A. Bayram, U. Uluca, Production of anorthite refractory insulating firebrick from mixtures of clay and recycled paper waste with sawdust addition. Ceram. Int. **38**(2), 1033–1041 (2012). https://doi.org/10.1016/j.ceramint.2011.08.027
12. N.M. Khalil, Y. Algamal, Recycling of ceramic wastes for the production of high performance mullite refractories. SILICON **12**(7), 1557–1565 (2020)
13. X. Tian, J.G. Heinrich, J. Günster, D. Li, Withdrawn: feasibility study on rapid prototyping of porcelain products. J. Eur. Ceram. Soc. (2010). https://doi.org/10.1016/j.jeurceramsoc.2010.09.015
14. D. Zheng, H. Lu, X. Sun, X. Liu, W. Han, L. Wang, Reaction mechanism of reductive decomposition of FGD gypsum with anthracite. Thermochim. Acta **559**, 23–31 (2013). https://doi.org/10.1016/j.tca.2013.02.026
15. H. Wang, Y.H. Wang, R.F. Zhang, *Bulletin of the Chinese Ceramic Society* (2013)

16. S. Zaiou et al., Sintering of anorthite based ceramics prepared from kaolin DD2 and calcite. Ceramica **62**(364), 317–322 (2016). https://doi.org/10.1590/0366-69132016623642015
17. Information on https://en.wikipedia.org/wiki/Bulk_density
18. Information on https://www.sciencedirect.com/topics/chemical-engineering/mullite
19. P. Wimuktiwan, Influence of the addition of pore foaming agent on mechanical and thermal properties of porcelain tiles. Ceram. Silik., pp. 164–171 (2020)
20. H. Boussak, H. Chemani, A. Serier, Characterization of porcelain tableware formulation containing bentonite clay. Int. J. Phys. Sci. **10**(2), 38–45 (2015)
21. A.M. Nawi, N.A. Badarulzaman, Effect of plaster of Paris waste and sintering temperatures on physical properties of pottery. Procedia CIRP **26**, 752–755 (2015). https://doi.org/10.1016/j.procir.2014.08.019
22. M. Abubakar et al., Influence of firing temperature on the physical, thermal and microstructural properties of Kankara kaolin clay: A preliminary investigation. Materials (Basel) **13**(8), 1872 (2020). https://doi.org/10.3390/ma13081872
23. Information on https://digitalfire.com/article/formulating+a+porcelain
24. Y. Chang, *Ceramic Porcelain* (Alexander Street Press, Alexandria, VA, 2018)

Chapter 32
The Mechanochemical Process and H_2SO_4 Treatment on the Rehydration of Anhydrite from FGD Sludge into Gypsum and Hemihydrate

Fatin Fatini Othman, Banjuraizah Johar, Shing Fhan Khor, Nik AKmar Rejab, and Suffi Irni Alias

Abstract The rehydration of anhydrite ($CaSO_4$) leads to the crystallization of gypsum ($CaSO_4{\cdot}2H_2O$) has been widely studied. Different process parameter had been conducted on anhydrite from the flue gas desulphurization (fgd) sludge to identify its ability to be reused as synthetic gypsum. A mechanochemical process using a high-energy planetary mill would break down the particle size and enhance the reaction of Ca^{2+} and $SO_4{}^{2-}$ with water while additives modify the number of ions present in the sample during the hydration. Difference hydrothermal temperatures (80 and 130 °C) for 30 min were conducted to dehydrate the synthetic gypsum from fgd sludge to hemihydrate ($CaSO_4{\cdot}0.5H_2O$) to identify its potential to replace natural gypsum during the plaster of Paris production. The sample without a mechanochemical process wasn't able to crystallize a high amount of gypsum while sample with 5×10^{-4} mol^{-1} of H_2SO_4 additive was able to crystallize a high amount of gypsum from the fgd sludge. After undergoing the hydrothermal process, all of the samples were able to crystallize hemihydrate at low-intensity peak due to the favourable of hemihydrate crystal's growth at plane [31-3] only, which leads to the preferred orientation of hemihydrate crystal.

Keywords Anhydrite · Fgd Sludge · Gypsum · Hemihydrate · Morphology · Transformation

F. F. Othman · B. Johar (✉) · S. I. Alias
Faculty of Chemical Engineering Technology, Universiti Malaysia Perlis, 01000 Kangar, Perlis, Malaysia
e-mail: banjuraizah@unimap.edu.my

S. F. Khor
Faculty of Electrical Engineering Technology, Universiti Malaysia Perlis, 02600 Arau, Perlis, Malaysia

N. A. Rejab
School of Materials and Mineral Resources Engineering, Universiti Sains Malaysia, 14300 Nibong Tebal, Penang, Malaysia

M. A. A. Mohd Salleh et al. (eds.), *Proceedings of the Green Materials and Electronic Packaging Interconnect Technology Symposium*, Springer Proceedings in Physics 289,
https://doi.org/10.1007/978-981-19-9267-4_32

32.1 Introduction

Flue gas desulfurization (fgd) is a waste incineration process that employs limestone slurry ($CaCO_3$) as a scrubber to remove SO_2 from exhaust gases, thereby reducing pollutant emissions. $CaCO_3$ reacted with SO_2 to form calcium sulphite ($CaSO_3$), which was then oxidized to form anhydrite ($CaSO_4$) and gypsum ($CaSO_4 \cdot 2H_2O$), also known as fgd sludge. Nippon Electric Glass Sdn Bhd (NEGM) disclosed production rates of 4.5 tonnes and monthly production of approximately 135 tonnes of fgd sludge. Massive production of fgd sludge is quite concerning, as fgd sludge disposal at landfills significantly increases the risk of chemical contamination by sulphur-rich fgd sludge waste. Thus, the reusability of fgd sludge as synthetic gypsum has been made to cope with this issues.

In addition, high demand for natural resources mined from gypsum is happening widely. Natural gypsum surface mining has significant environmental consequences, including groundwater contamination, soil erosion, and the destruction of wildlife habitats. As a result, the reusability of fgd sludge that is mainly composed of anhydrite and gypsum may help to reduce the cost of natural gypsum discoveries. The ability of fgd sludge to be transformed into gypsum and hemihydrate via different operating parameter would benefit humanity by reducing fgd sludge disposal and the exploration of natural gypsum.

Gypsum has been widely known for its use such as retarder in Portland cement and primary raw material for plaster of Paris (POP) and plaster ceiling materials [1]. Extensive research has been carried out to ascertain the rehydration process for crystallizing gypsum from various natural anhydrite. According to Sievert [2], anhydrite has very low reactivity with water, so simple anhydrite rehydration will not result in gypsum. They discovered that the maximum specific surface area obtained by the ball mill process enhanced the anhydrite-water reaction. Wet milling is a mechanochemical process that improves the reaction activity of anhydrite to rehydrate into gypsum and hemihydrate by reducing particle size and increasing the total surface area for the rehydration process. Furthermore, the use of additives such as sulphuric acid (H_2SO_4) [3] and potassium sulphate (K_2SO_4) [4] has been shown to increase the reaction energy during anhydrite rehydration, but acid treatment using H_2SO_4 has a promising ability to enhance the crystallization of gypsum and hemihydrate using fgd sludge from other manufacture [5].

Commercial POP is mainly composed of hemihydrate produced by heating raw gypsum at temperatures ranging from 80 to 160 °C [6] using a hydrothermal bomb to ensure high moisture and pressure subjected during the dehydration process. POP has been widely used in plaster mould production due to its ability to absorb water from ceramic slurry and fasten the hardening process. Thus, the ability of synthetic gypsum from fgd sludge as substitute raw material for natural gypsum has been studied. The overall reaction during the rehydration of anhydrite to gypsum and dehydration of gypsum to hemihydrate was demonstrated in Eq. 32.1.

$$CaSO_4 \rightarrow CaSO_4 \cdot 2H_2O \rightarrow CaSO_4 \cdot 0.5H_2O \tag{32.1}$$

32.2 Methodology

32.2.1 Materials and Rehydration Process

In this study, raw fgd sludge from Nippon Electric Glass Malaysia Sdn Bhd (NEGM) was used. During the mechanochemical process, 80 g of fgd sludge and 80 ml of distilled water were mixed in a tungsten jar with a glass rod. As grinding media, 30 tungsten carbide balls were used, and the milling process was carried out using a planetary mill (model Pulverisette 6 machine). The milled sample was dried at room temperature (30 °C) for several days before being examined. The same steps were repeated by adding 0.02 ml of 2.0 mol of H_2SO_4 to the sample before undergoing a mechanochemical process for identifying the effectiveness of acid treatment on the rehydration process. 30 g of the milled sample and 5 ml of distilled water were stirred and mixed with a glass rod for a minute before being heated in an oven at a temperature of 80 and 130 °C for 30 min and dried with the same process as before. A titanium high-pressure hydrothermal reactor was used to dehydrate synthetic gypsum from the mechanochemical process and additives treatment into hemihydrate phase.

32.2.2 X-Ray Diffraction (XRD)

The phase transformation of anhydrite to gypsum and hemihydrate can be identified with Bruker's X-ray Diffraction D8-Discover instrument with Cu K radiation in the 10° to 60° range at a scanning rate of 1°/min and a scan rate of 0.2 s/step. The sample's powder was crushed with a mortar and pastel and sieve through 45 μm and compacted into the sample holder. The powder must be on a smooth surface to avoid an offset peak position and randomize orientation. X'Pert HighScore Plus software was used to analyse the diffraction pattern of XRD testing. Crystal structure data for all phases present in samples were obtained from the ICSD database which was used as the initial model to generate the calculated pattern. Rietveld method is used to refine and quantify each phase present until the rwp of the sample is less than 12, indicating the refinement parameter used to be satisfactory. Parameters used during the refinement are scale factor, background, lattice parameters, profile parameters, atomic positions and displacement parameters, and scale factor.

32.2.3 Scanning Electron Microscopy (SEM)

The morphology and microstructure of the processed samples were investigated by a JSM-6460LA machine under 10 kV. Samples were glued on a sample holder using carbon tape and coated using a Rotary Pumped Coater Q150R Plus machine with a gold sputtering. Coated samples were placed into the sample stage and the SEM images of the sample were obtained under the magnification level of ×5000.

32.3 Result and Discussions

32.3.1 Phase Analysis

X-ray diffraction (XRD) patterns were used to identify the phase analysis of fgd sludge that can be observed at Fig. 32.1. The main composition of fgd sludge used in this research are gypsum ($CaSO_4 \cdot 2H_2O$) and anhydrite ($CaSO_4$), with a small amount of fluorite (CaF_2), dodecaborane ($B_{12}H_{16}$), and meyerhofferite ($Ca(B_3O_3(OH)_5)(H_2O)$).

Based on Fig. 32.1, the major phase observed in the fgd sludge are anhydrite and gypsum. Single peak of fluorite also can be found at $2\theta = 28.25°$ [111]. The presence of boron also can be observed, indicate fgd sludge from NEGM contain a high amount of glass forming agent. The presence of different boron phases in fgd sludge is due to the unstable and rapid bonding of element boron that occurs during the scrubbing process, and thus leads to the formation of dodecaborane with the presence of hydrogen while the bonding between boron, calcium, and water leads to

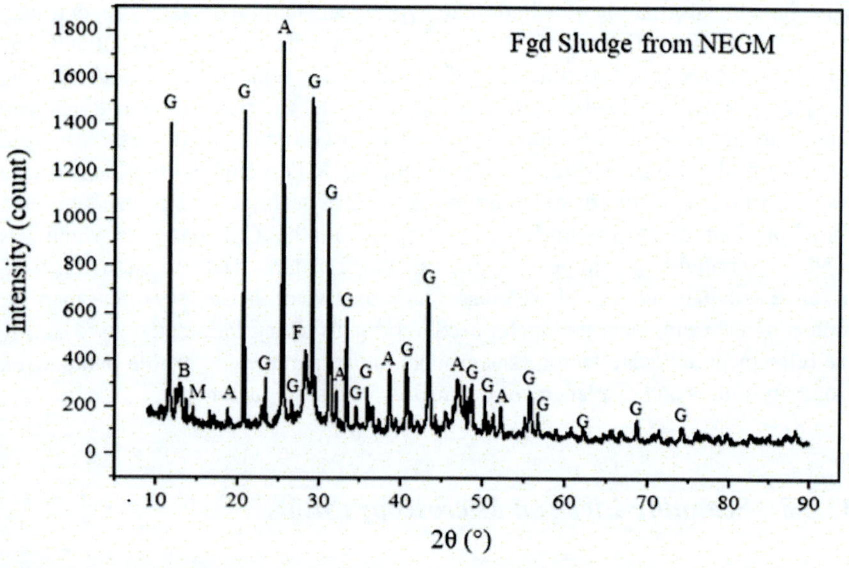

Fig. 32.1 X-ray diffraction patterns of fgd sludge, G = Gypsum, A = Anhydrite, F = Fluorite, B = Dodecaborane, M = Meyerhofferite

Table 32.1 Crystal data of refine fgd sludge

Parameter	Phase(s)				
	Anhydrite	Gypsum	Fluorite	Dodecaborane	Meyerhofferite
Reference code	98-000-5306	98-001-1992	98-002-1697	98-004-4797	98-001-1922
Chemical Composition	$CaSO_4$	$CaSO_4{\cdot}2H_2O$	CaF_2	$B_{12}H_{16}$	$Ca(B_3O_3(OH)_5)(H_2O)$
Space Group	Cmcm	C12/c1	Fm-3 m	P n a 21	P-1
Crystal Structure	Orthorhombic	Monoclinic	Cubic	Orthorhombic	Anorthic
Crystallite Size (Å)	603	694	378	226	800
Density (g/cm^3)	2.96	2.31	3.19	0.80	1.98
Percentage (%)	45.00	33.30	2.30	15.60	3.80
rwp	11.45				

the formation of meyerhofferite. The Rietveld refinement done on the experimental pattern of the fgd sludge was shown in Table 32.1.

Figure 32.2 shows the XRD pattern of sample with different operating parameter on fgd sludge. The intensity peak of gypsum is increased while the intensity peak of anhydrite diminished after undergoing the mechanochemical and acid treatment, indicating the rehydration of insoluble anhydrite to gypsum occurs.

As shown in Fig. 32.2, all of the anhydrite peaks in fgd sludge tends to diminish after undergoing the mechanochemical and hydrothermal process, indicating the rehydration of insoluble anhydrite into gypsum occurs. According to Sievert [2], the reactivity of anhydrite to react with water is very low, however, grinding the anhydrite to fine powder may enhance the responsiveness of rehydration anhydrite into gypsum, thus, explaining the high percentage amount of gypsum in the sample. During the wet milling, the particle size for fgd sludge gradually decreases and thus increases the total surface area of reaction between anhydrite and water. This statement can be supported by Zhang [7], as they were able to increase the specific surface area (SSA) up to 100 m^2/g with a planetary mill for 2 h. Thus, it helps in promoting the energy of rehydration to occur at a high surface area of anhydrite. With the addition of 5×10^{-3} mol^{-1} of sulphuric acid (H_2SO_4), the diffraction peak of the anhydrite phase at [200] shifts slightly to lower 2θ values, whereas the intensity of the diffraction peak of gypsum at [020] and [02-1] was slightly higher compared to the sample without undergoing acid treatment. This is due to the reducing d-spacing of lattice and compacted lattice volume in the fgd sludge, which has been modified by Ca^{2+} supersaturation caused by H_2SO_4 addition during the synthesis of gypsum [8]. Anhydrite is the most stable form in the calcium sulphate group, where it was hard for anhydrite to react with water. With H_2SO_4 treatment, the anhydrite bond tends to break down into Ca^{2+} and SO_4^{2-}ions, and thus with the presence of water,

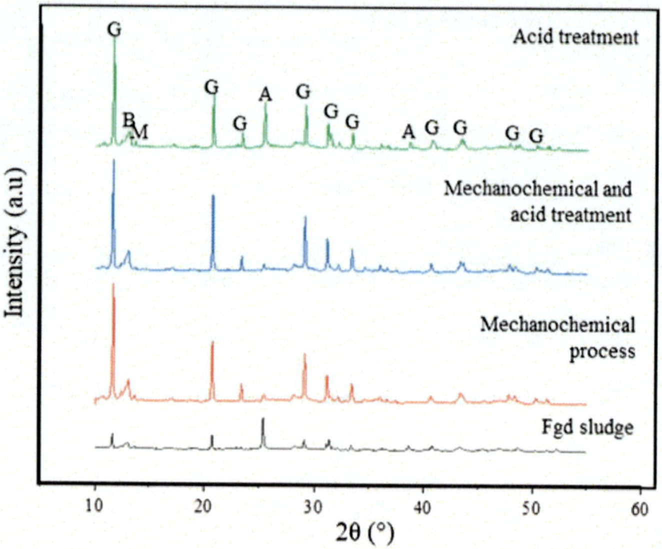

Fig. 32.2 X-ray diffraction patterns of fgd sludge after undergoing different process parameter, G = Gypsum, A = Anhydrite, F = Fluorite, B = Dodecaborane, M = Meyerhofferite

anhydrite's ion will recrystallize into gypsum. The liquid–solid equilibrium between Ca^{2+} and $SO_4{}^{2-}$ ions in the solid phase of gypsum are shown in Eq. (32.2).

$$Ca^{2+} + SO_4^{2-} + 2H_2O(\text{precursor}) \rightarrow CaSO_4 \cdot 2H_2O(\text{gypsum}) \tag{32.2}$$

Without the milling process, the sample didn't obtain a high surface area for reaction, and thus the dehydration of anhydrite to gypsum didn't occur completely even though with acid treatment. This is due to high-intensity peak of anhydrite that can be observed in the sample with acid treatment only but it able to promote the intensity peak of gypsum compared to the raw fgd sludge, indicates it ability to promote the dissolution of anhydrite into Ca^{2+} and SO_4^{2-} eventhough without the milling process. Thus, both mechanochemical process and acid treatment are required to enhance the crystallization of gypsum from fgd sludge.

To identify the ability of synthetic gypsum from fgd sludge as raw material in producing POP, different hydrothermal temperature was conducted on the sample. Figure 32.3 shows the XRD pattern of the sample with difference heating temperature.

Based on Fig. 32.3, the intensity of gypsum in each sample after undergoing different heating processes can be observed, indicating that heat treatment also helps in the rehydration process as at higher heating temperature, the intensity peak of anhydrite at 25.43° [200] slowly diminishes while the intensity peak of gypsum at 11.61° [020] shows an increment over temperature. At high temperature, the average kinetic energy between the particles increases and thus, enhances the rate of reaction with

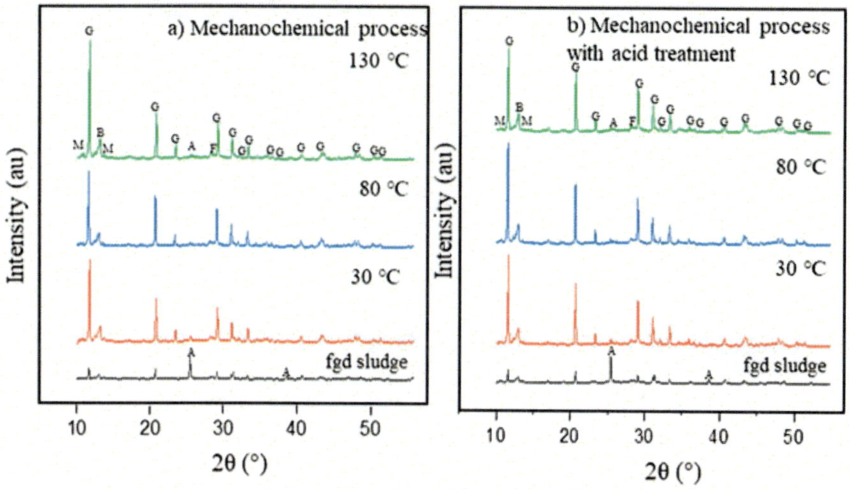

Fig. 32.3 X-ray diffraction patterns of fgd sludge after undergoing different process parameter and heated at different hydrothermal temperature, G = Gypsum, A = Anhydrite, F = Fluorite, B = Dodecaborane, M = Meyerhofferite

water to transform into gypsum. Low-intensity peak of hemihydrate was diffracted at position $2\theta = 25.75°$ belonging to the [31-3] plane, however, the highest intensity peak of hemihydrate located at 14.74° belongs to the [11-1] plane and 29.59° belong to [020] plane were not diffracted due to the preferred orientation. This phenomenon occurs when the hemihydrate crystal tends to grow faster along the *c* axis, rather than the *a* and *b* axes, and thus leads to the disappearance of the hemihydrate peak at 25.75° and 14.74° [8]. Wang [9] stated that H_2SO_4 is also able to promote gypsum dissolution and enhance hemihydrate supersaturating, which is favourable to hemihydrate crystallization to some extent but the low intensity of hemihydrate only can be observed in the close-up of XRD pattern as shown in Fig. 32.4.

All of the samples show an identical peak diffracted even though undergo different hydrothermal temperature processes. This is due to the slow drying process that took place in all samples, the dehydration of gypsum to hemihydrate didn't occur completely but the quantitative analysis from Rietveld refinement was able to reveal the significant difference between the sample with and without acid treatment against heating temperature, as shown in Fig. 32.4. From the observation, the anhydrite phase is depleted with increasing heating temperature and accelerates with H_2SO_4 addition, while the amount of gypsum is surging high, indicate anhydrite tends to rehydrate into gypsum. In addition, the hemihydrate phase is increasing with increasing temperature and surge with H_2SO_4 addition but at higher temperature (>100 °C), the formation of anhydrite took place. According to Wang [10] and Shen [11], the heating temperature required for the transformation of gypsum to hemihydrate is contradictorily lower with the addition of H_2SO_4, as the acid reduced the activity of water to bind with Ca^{2+} and SO_4^{2-}, resulting in the transformation of gypsum to anhydrite (Fig. 32.5).

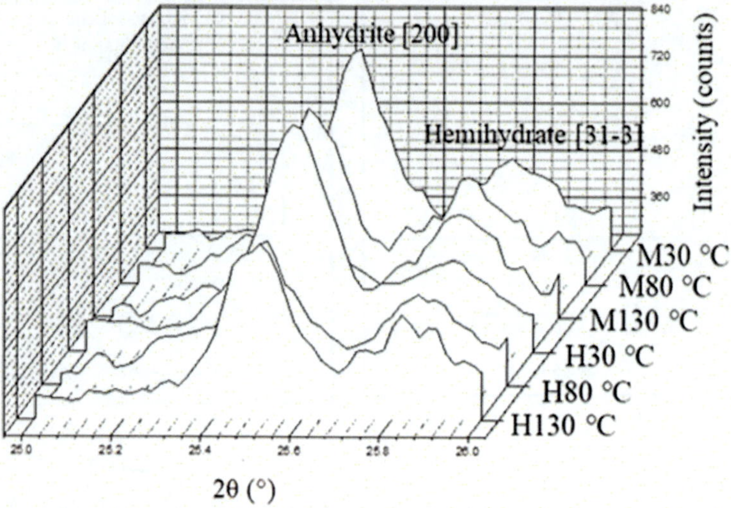

Fig. 32.4 Intensity peak of anhydrite at [200] plane and hemihydrate at [400] plane along $2\theta = 25\text{–}26°$

Fig. 32.5 Influence of H_2SO_4 treatment in phase transformation at different heating temperature **a** percentage of gypsum phase, **b** percentage of anhydrite phase, **c** percentage of hemihydrate phase

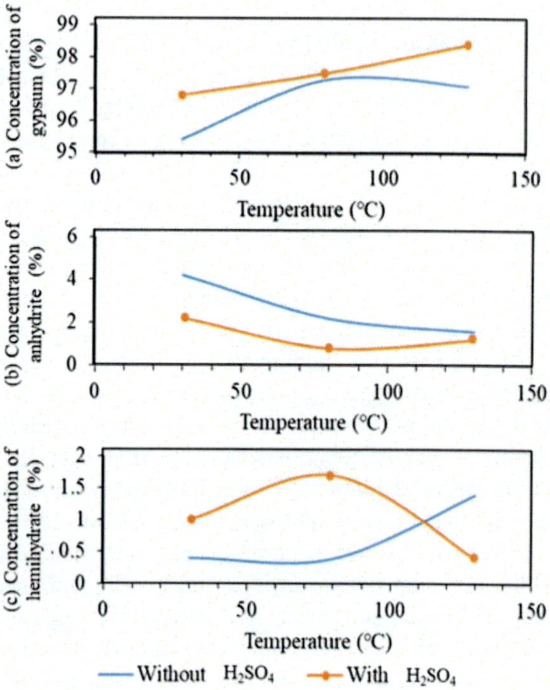

According to Ricci [12], hemihydrate material is formed by heating gypsum at 123 °C under approximately 17 psi of pressure, in steam for 5–7 h in an autoclave. Valimbe [13] were able to crystallize hemihydrate from fgd sludge at > 130 °C for 1 h while Zhang [5] required 120 °C for 30 min to crystallize hemihydrate from different fgd sludge. Thus, different raw materials used required different heating times and temperatures to promote the crystallization process. To promote the crystal growth of hemihydrate, increasing ethanol concentrations would provide high nucleation kinetics of the reaction, and thus, drive force for the incorporation of Ca^{2+} and $SO_4{}^{2-}$ ions into the hemihydrate crystal lattice. This phenomenon occurs as the adsorption of ethanol on gypsum surface would lead to the crystal growth along [111] and [110] directions where high concentrations of $SO_4{}^{2-}$ ions would promote the nucleation of hemihydrate crystal, and consequently promotes the formation of an equiaxed crystal, needle or fibre-like shapes [8].

In addition, the impurities obtained in raw material controlled the rate of crystallization of the hemihydrate phase. Engbrecht [14] stated that the presence of halite (NaCl) in natural ore gypsum exhibits a reduction in the temperature of conversion to the hemihydrate phase. However, the presence of impurities such as boron and fluorite in fgd sludge hinders the dehydration of crystalline water at 130–150 °C for 30 min as impurities may increase the boiling point of water. This is because impurities reduce the availability of water molecules to vaporize during heating, thus a great amount of heat is required to remove 1.5 of crystalline water from gypsum. In addition, Zhang [5] were able to obtain a high amount of hemihydrate from fgd sludge in China by drying the product at a temperature range of 80–100 °C for 4 h after undergoing hydrothermal treatment at 120 °C with different soaking time (20, 40, and 60 min). They observed that after extending the holding time to 60 min, the sample exhibited a long fibrous hemihydrate phase. The Ca^{2+} and SO_4^{2-} are able to grow in a certain direction at a longer holding time, resulting in an increasing in the intensity of the hemihydrate peak. Instead of drying the sample at room temperature, they dried the sample by heating in an oven at a temperature of 80 °C for 4 h. During this process, the hydroxyl bond in gypsum is weakened, and thus leads to the dehydration of gypsum into hemihydrate. Thus, the sample is required to heat in an oven during the drying process to enhance the crystallization of the hemihydrate phase. In this regard, prolonged heating time (> 30 min), ethanol addition, and drying of the wet sample in an oven are required to enhance the recrystallizing hemihydrate phase from fgd gypsum.

32.3.2 Morphology Analysis

The morphology of fgd sludge after undergoing the mechanochemical process and H_2SO_4 treatment can be observed in Fig. 32.6. From the SEM image, the powder's particle of fgd sludge tends to break into smaller particles after being subjected to higher heating temperature and H_2SO_4 treatment. At 30 °C, the sample with H_2SO_4 (H30) tends to break into small particles and leads to the formation of hemihydrate

Fig. 32.6 Scanning electron microscopy (SEM) images of the samples with and without H2SO4 treatment heated at 30 and 130 °C at magnification of 5000×

phases compared to the sample without H_2SO_4 treatment (M30). At higher temperatures, the reactivity of water with fgd sludge deteriorates and thus the particle's size of the sample with H_2SO_4 treatment heated at 130 °C (H130) didn't break into a smaller particle as the sample didn't undergo H_2SO_4 treatment.

As mentioned before, the hemihydrate phase didn't completely occur, and thus there is only a short needle-like microstructure appeared in the sample after being subjected to 130 °C. According to Wang [10] and Shen [11], the heating temperature required for the transformation of gypsum to hemihydrate is contradictorily lower with the addition of H_2SO_4, as the acid reduced the activity of water to bind with Ca^{2+} and SO_4^{2-}, resulting in the transformation of gypsum to anhydrite. This is why the formation of a fine needle-like structure in the sample with H_2SO_4 treatment is contradictorily lower than the sample with the mechanochemical process only, Thus, the SEM images and XRD results correlate with each other.

32.4 Conclusion

This study is helpful in identifying the effect of mechanochemical process and acid treatment on fgd sludge to crystallize gypsum and hemihydrate. The mechanochemical process and acid treatment able to crystallize gypsum due to the presence of high surface area of reaction. In addition, increasing heating temperature able to enhance the crystallization of the hemihydrate phase while the concentration of gypsum and hemihydrate tends to decrease at 130 °C with acid treatment due to the formation of anhydrite.

Acknowledgements The research was supported by Research Grant Scheme (FRGS) (Grant No: FRGS/1/2019/STG07/UNIMAP/02/4) and Faculty of Chemical Engineering and Technologies, Universiti Malaysia Perlis (UniMAP). The authors would also like to thanks Nippon Electric Glass Sdn. Bhd. for providing the fgd sludge.

References

1. I. Suárez-Ruiz, C.R. Ward, Coal Combustion. Appl. Coal Petrol., 85–117 (2008)
2. T. Sievert, A. Wolter, N.B. Singh, Hydration of anhydrite of gypsum (CaSO4.II) in a ball mill. Cem. Concr. Res. **35**(4), 623–630 (2005)
3. Z. Wang et al., Thermochemical behavior of three sulfates (CaSO4, K2SO4 and Na2SO4) blended with cement raw materials (CaO-SiO2-Al2O3-Fe2O3) at high temperature. J. Anal. Appl. Pyrolysis **142** (2019)
4. N.B. Singh, The activation effect of K2SO4 on the hydration of gypsum anhydrite, CaSO4(II). J. Am. Ceram. Soc. **88**(1), 196–201 (2005)
5. X. Zhang et al., Structural characteristic and formation mechanism of hemihydrate calcium sulfate whiskers prepared using FGD gypsum. Particuology **62**, 98–103 (2022)
6. W.M. Lynch, Plaster processing dynamics. Am. Ceram. Soc. Bull. **74**(1), 60–62 (1995)
7. Q. Zhang, E. Kasai, F. Saito, Mechanochemical changes in gypsum when dry ground with hydrated minerals. Powder Technol. **87**(1), 67–71 (1996)
8. Z. Pan et al., Preparation of calcium sulfate dihydrate and calcium sulfate hemihydrate with controllable crystal morphology by using ethanol additive. Ceram. Int. **39**(5), 5495–5502 (2013)
9. Y. Wang, X. Mao, C. Chen, W. Wang, W. Dang, Effect of sulfuric acid concentration on morphology of calcium sulfate hemihydrate crystals. Mater. Res. Express **7**(10) (2020)
10. W. Wang, D. Zeng, Q. Chen, X. Yin, Experimental determination and modeling of gypsum and insoluble anhydrite solubility in the system CaSO4-H2SO4-H2O. Chem. Eng. Sci. **101**, 120–129 (2013)
11. L. Shen, H. Sippola, X. Li, D. Lindberg, P. Taskinen, Thermodynamic modeling of calcium sulfate hydrates in a CaSO4-H2SO4-H2O system from 273.15 to 473.15 K up to 5 m sulfuric acid. J. Chem. Eng. Data **65**(5), 2310–2324 (2020)
12. J.L. Ricci, M.J. Weiner, S. Mamidwar, H. Alexander, Calcium Sulfate. in *Bioceramics and their Clinical Applications*, pp. 302–325 (2008)
13. P.S. Valimbe, V.M. Malhotra, Effects of water content and temperature on the crystallization behavior of FGD scrubber sludge. Fuel **81**(10), 1297–1304 (2002)
14. D.C. Engbrecht, D.A. Hirschfeld, Thermal analysis of calcium sulfate dihydrate sources used to manufacture gypsum wallboard. Thermochim. Acta **639**, 173–185 (2016)

Chapter 33
The Thermal Behavior of Cordierite-Based Ceramic with the Substitution of Treated Flue Gas Desulfurization Sludge in the Non-stoichiometric Cordierite Composition

Fatin Fatini Othman, Banjuraizah Johar, Shing Fhan Khor, Nik AKmar Rejab, and Suffi Irni Alias

Abstract The substitution of FGD sludge that is rich with a glass-forming agent in non-stoichiometric cordierite composition is able to crystallize α-cordierite at the low sintering temperature, 1250 °C via solid-state reaction methods. Even though the substitution of MgO with FGD sludge can reduce the sintering temperature, the physical and mechanical properties of cordierite-based ceramic with FGD sludge are depleted due to the decomposition of SO_3 took place during the sintering process and thus lead to the formation of voids in the ceramic. Thus, this paper is focusing on the thermal properties of cordierite-based ceramic when subjected to a high sintering temperature, 1000 °C using a dilatometric test. Different wt% of treated FGD sludge substitution gave a different type of CTE value, but 3.0 wt% of FGD sludge is able to obtain the lowest CTE value, 2.26×10^{-6}/°C, compared to other samples as this sample obtains a single phase of α-cordierite phases.

Keywords Cordierite · Dilatometer · FGD sludge · Nucleating agent · Sintering aids · Substitution

---

F. F. Othman · B. Johar (✉) · S. I. Alias
Faculty of Chemical Engineering Technology, Universiti Malaysia Perlis, 01000 Kangar, Perlis, Malaysia
e-mail: banjuraizah@unimap.edu.my

S. F. Khor
Faculty of Electrical Engineering Technology, Universiti Malaysia Perlis, 02600 Arau, Perlis, Malaysia

N. A. Rejab
School of Materials and Mineral Resources Engineering, Universiti Sains Malaysia, 14300 Nibong Tebal, Penang, Malaysia

M. A. A. Mohd Salleh et al. (eds.), *Proceedings of the Green Materials and Electronic Packaging Interconnect Technology Symposium*, Springer Proceedings in Physics 289,
https://doi.org/10.1007/978-981-19-9267-4_33

33.1 Introduction

Sludge from the flue gas desulfurization (FGD) process is the solid residue produced by the exhaust gas system that was used to limit the release of SO_2 emissions. The abundant amount of FGD sludge produced every day may contaminate groundwater resources during landfill disposal. As a result, the potential application of FGD sludge in various applications was studied such as plaster mold [1–4], sintering agent [5], and plasterboard [6]. Previous research also finds out that FGD sludge from Nippon Electric Glass Malaysia Sdn Bhd (NEGM) contains glass-forming agents such as boron trioxide (B_2O_3), silica (SiO_2), and phosphorus pentoxide (P_2O_5) that has a promising application as a nucleating agent, as well as sintering aids in crystallizing α-cordierite at low sintering temperature, 1250 °C, via solid-state reaction methods.

Cordierite is one of the useful crystalline phases in the $MgO–Al_2O_3–SiO_2$ ternary system as it is known as a new development material with numerous significant applications because of its outstanding properties, such as excellent thermal shock resistance, low thermal expansion coefficient (4.0×10^{-6}/°C) [7], low dielectric constant (4.0–5.0 at 1 MHz), good chemical durability, excellent refractoriness, and mechanical properties (55 MPa–90 MPa) [8]. The formation of α-cordierite phases with stoichiometric composition ($2MgO \cdot 2Al_2O_2 \cdot 5SiO_2$) commonly crystallizes at high sintering temperatures, approaching its incongruent melting temperature when solid-state methods are used [9]. According to Eing [10], cordierite has an incongruent melting point because the solid compound of cordierite does not melt to form the liquid of its composition, but rather dissociates to form a new solid phase. Various parameters have been conducted to crystallize a single phase of α-cordierite at low sintering temperature such as the addition of sintering aids [11], the substitution of magnesium oxide (MgO) with calcium oxide (CaO) [12–14], and the readjustment of the stoichiometric into different non-stoichiometric cordierite compositions [15, 16].

From the previous studies, we found out that FGD sludge from NEGM can too act as a sintering aid in nucleating α-cordierite phase due to the presence of boron, as well as a substitution compound in non-stoichiometric cordierite composition due to the high amount of CaO derived from anhydrite in the FGD sludge. The substitution of FGD sludge also enhances the production of voids in the sample as the decomposition of sulfur trioxide (SO_3) as shown in Eq. 33.1 took place during the sintering process and thus enhances the thermal properties of the cordierite-based ceramic. The formation of interconnected pores within the sample may be difficult for heat to be transferred within the voids and thus enhances the conductivity of porous cordierite-based with the substitution of FGD sludge. In this research, the correlation between the microstructure and the coefficient of thermal expansion (CTE) of cordierite-based ceramic was studied.

$$CaSO_4 \rightarrow CaO + SO_3. \tag{33.1}$$

Table 33.1 Four samples with different weight percentages (wt%) of treated FGD sludge used in non-stoichiometric cordierite composition

Raw material	Weight percentage (%)			
	S0	S1.5	S3.0	S4.5
Sludge	0.00	1.50	3.00	4.50
Silica	0.13	0.13	0.13	0.13
Magnesia	4.50	3.00	1.50	0.00
Alumina	4.05	4.05	4.05	4.05
Talc	29.48	29.48	29.48	29.48
Kaolin	61.82	61.82	61.82	61.82

33.2 Methodology

33.2.1 Materials and Methodology

In this study, raw FGD sludge from Nippon Electric Glass Malaysia Sdn Bhd (NEGM) was used and treated by calcination at 1100 °C for 3 h to remove any moisture, organic compound, and unstable gaseous. Calcine sludge was left overnight before being pulverized to fine and narrow particle size by using a planetary mill (Pulverisette 6 machine). The composition of treated FGD sludge in 0 wt%, 1.5 wt%, 3.0 wt%, and 4.5 wt% was chosen to substitute MgO content in non-stoichiometric cordierite formulation, as shown in Table 33.1. These ratios were selected without any consideration of impurities present in each raw material used. Each composition was weighted and milled using a planetary mill, at 300 RPM for 1 h with tungsten carbide as the grinding media to homogeneously mix the compositions. The powder mixture was then pressed into a cylindrical shape die with a 12 mm diameter under a pressure of 11 MPa to obtain a pellet structure. The samples were then sintered at 1250 °C for 3 h at a heating rate of 5 °C/min before characterization was conducted.

33.2.2 Scanning Electron Microscopy (SEM)

The morphology and microstructure of the sintered samples were investigated by Table Top SEM (HITACHI TM3000) which was used to identify the microstructure of sintered cordierite-based ceramic. The microstructure of the internal structure of sintered sample was obtained by cutting down the pellet body into two. Emscope SC500A unit machines were used to coat the solid sample using a gold sputtering and placed into the sample stage. The magnification level of ×500 and ×3000 was used in SEM analysis.

33.2.3 Dilatometer Test

A dilatometer is a high-precision system designed to measure dimensional changes in material caused by a physical or chemical process due to a thermal environment. Solid materials mostly expand in response to heating and contract cooling. This response to temperature change is expressed as a coefficient of thermal expansion (CTE). Model-Linseis DTA PT1600 was used to record the dimensional changes of the solid sample subjected to high temperatures. The CTE of sintered samples mixed with different amounts of FGD sludge was measured using a push-rod dilatometer in the range of 30–1000 °C. The applicable temperature range for CTE value is −120–600 °C, although this range may be modified based on instruments and calibration materials, and the CTE value for each sintered sample till 1250 °C is unpredictable. The end of the sintered sample was wedged between the edge of the push rod and the quartz tube. The average value of the thermal expansion coefficient was calculated using Eq. (33.2).

$$\text{CTE, } \propto = \frac{\mathrm{d}l}{l_0} \times \frac{1}{\mathrm{d}t} \times 100, \tag{33.2}$$

where $\propto$ is CTE, dl is the difference in specimen length (in unit mm) as a function of temperature, l_0 is the initial length (in unit mm) at the initial temperature, and dt is the difference of temperature (in unit °C).

33.3 Result and Discussions

33.3.1 The Properties of Cordierite-Based Ceramic

From the previous research [17], the phase identification, physical and mechanical properties of cordierite-based ceramic are tabulated in Table 33.2. The presence of 3.0 wt% of treated FGD sludge is able to help in crystallization of α-cordierite at 1250 °C using solid-state reaction methods, while the physical and mechanical performances of cordierite-based ceramic are deteriorated with the substitution of MgO with treated FGD sludge due to the formation of porosity in the sample.

33.3.2 The Microstructure of Cordierite-Based Ceramic

Based on the cross-section of the sintered sample shown in Fig. 33.1, increasing the amount of treated sludge enhances the number and size of pores. However, the presence of pores in the internal structure of the samples sintered at temperature 1250

Table 33.2 Properties of sintered cordierite-based ceramic with different weight percentages (wt%) of treated FGD sludge used in non-stoichiometric cordierite composition

Properties		Sample			
		S0	**S1.5**	**S3.0**	**S4.5**
Phase analysis (%)	α-cordierite	5.30	38.90	100.00	95.20
	β-cordierite	57.10	32.80	-	-
	Forsterite	37.50	28.20	-	4.20
	Anorthite	-	-	-	0.60
Physical	Porosity (%)	17.15	17.94	24.69	30.44
	Density (g/m^3)	2.08	2.00	1.75	1.79
	Volume of shrinkage (%)	31.00	24.08	17.62	15.64
Mechanical	Flexural strength (MPa)	54.69	45.76	41.12	34.43

°C and the accumulation of numerous irregular particles indicates that the sintered sample was not fully densified.

The increasing amount of treated FGD sludge added into the composition would lead to a higher rate of SO_3 decomposition and thus leads to the formation of pores. All of the samples with and without treated FGD sludge substitution illustrated the formation of interconnected pores and macropore size, while the pore size of all sintered samples recorded a value of more than 50 nm (0.05 μm) [18]. Although the FGD sludge had been treated at 1100 °C, the calcination temperature used was unable to completely decompose SO_3 gaseous. According to Sievert [19], anhydrite

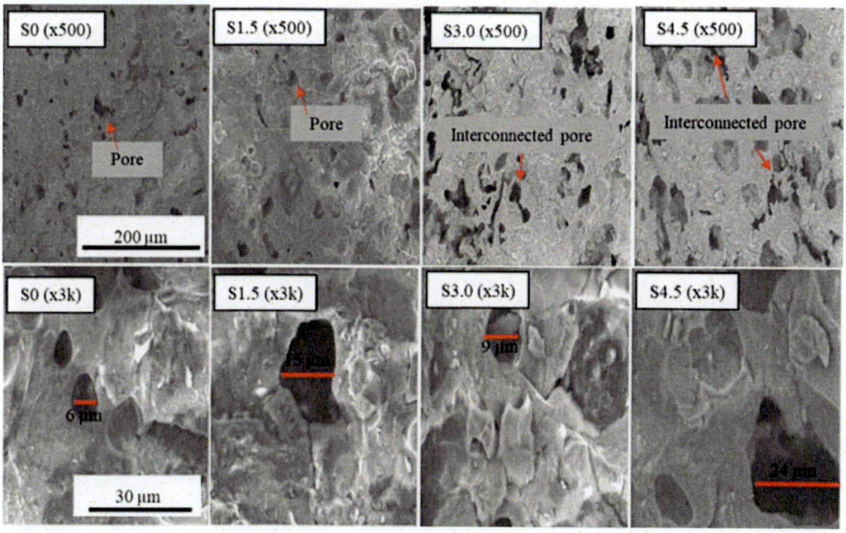

Fig. 33.1 Scanning electron microscopy (SEM) images of the samples sintered at 1250 °C with different treated FGD sludge contents at magnification of 500 × (above) and 3000 × (bottom)

is thermodynamically stable up to a temperature of 1180 °C, while Wang [20] added that the decomposition of $CaSO_4$ took place at a temperature of 1150 °C. Thus, the decomposition of SO_3 took place during the sintering of the sample and leads to the formation of pores in the internal structure of cordierite-based ceramic. The formation of the pore by the slow decomposition of gaseous during sintering is a beneficial method to produce a porous ceramic. The formation of porous in a ceramic material is beneficial for building structure material since it can reduce the thermal conductivity of the final products, especially in refractory brick.

33.3.3 The Thermal Behavior of Cordierite-Based Ceramic

The dilatometric curve of sintered cylindrical pellets with various treated FGD sludge compositions is shown in Fig. 33.2 to identify their thermal behavior. The sintering reaction of sintered materials goes through three phases: shrinkage at a temperature from room temperature to 200 °C, expansion at a temperature of 200–780 °C, and shrinkage at a temperature of 780–1000 °C. The thermal expansion coefficient of samples as a function of treated FGD sludge substitution at each stage of dimensional change is demonstrated in Fig. 33.3 and Table 33.3.

The shrinkage of all samples began at room temperature till 200 °C in the first stage, indicating that the anisotropic shrinkage of cordierite took place. All of the sintered samples also show a negative value of CTE, indicating that the shrinking process took place. The non-isotropic or anisotropic shrinkage of the cordierite sample was related to various parameters such as particle size and shape and their

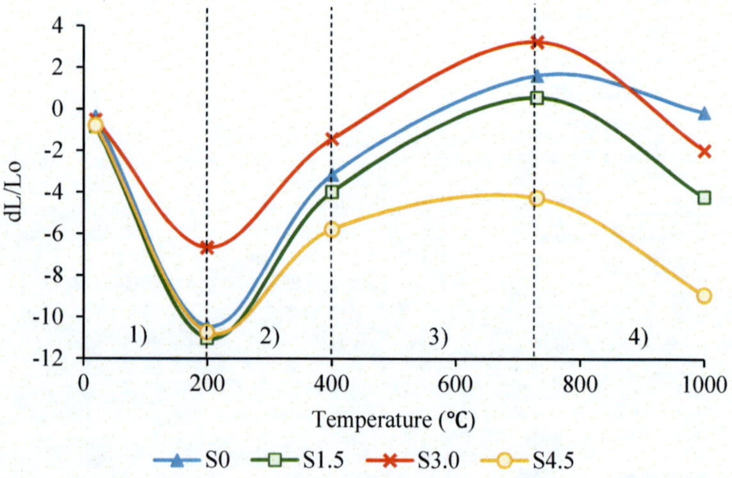

Fig. 33.2 Dilatometric curve of sintered cylindrical pellets with various treated FGD sludge compositions

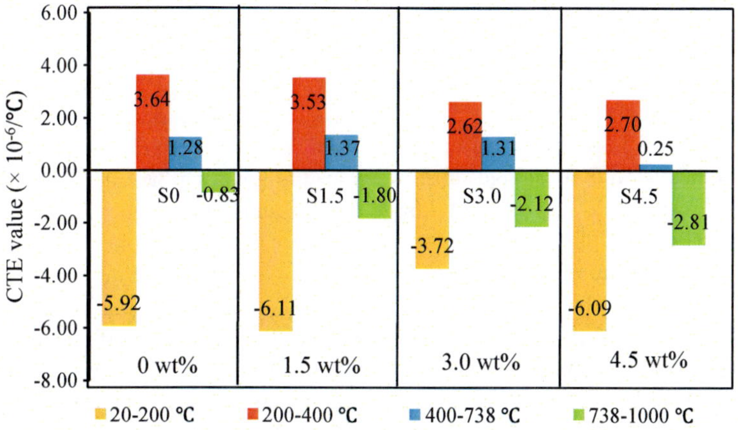

Fig. 33.3 Coefficient of thermal expansion of sample as a function of treated sludge substitution at each stage of dimensional changes

Table 33.3 Coefficient of thermal expansion of sample as a function of treated sludge substitution at each stage of dimensional changes

Stages		CTE value (× 10^{-6}/°C)			
		1	2	3	4
Temperature (°C)		30–200	200–400	400–73*	73*–1000
Sample	S0	−5.92	3.64	1.28	−0.83
	S1.5	−6.11	3.53	1.37	−1.80
	S3.0	−3.72	2.62	1.31	−2.12
	S4.4	−6.09	2.70	0.25	−2.81

* S0 = 738 °C, S1.5 = 738 °C, S3.0 = 737 °C, S4.5 = 736 °C (softening temperature)

distribution, amount, and direction of pressure applied during pressing [21]. The anisotropic shrinkage of the sintered sample took place due to the planar orientation of the pore or solid interface as all of the samples indicated a high volume of a pore. Sample S3.0 has a lower rate of shrinkage, with a CTE value of −3.72 × 10^{-6}/°C compared to other samples since it has modest volume fluctuations compared to others due to the presence of a high α-cordierite phase.

The second step of the sintering reaction began at a temperature of 200 °C, whereas all of the samples began to experience a thermal expansion. This is due to the high energy of grain's vibration which was subjected during the firing process at temperature ranges of 200–800 °C. During this stage, all of the samples expand and thus affect densification. According to Maniere [22], the expansion process would slow down the sample densification since the grain's space has a direct impact on the diffusional path of segregation. Ardebili [23] also stated that the thermal expansion rate is also subjected to the material types since different types of materials expand differently at the same heating rate due to the attachment of other elements with similar CTE values. During this stage, there are two CTE values detected, indicating that there are two different rates of expansion of the samples. At a temperature range between 200 and 400 °C, samples S0 and S1.5 show an identical trend of rate of expansion, with CTE values of 3.64×10^{-6}/°C and 3.53×10^{-6}/°C, respectively, while with high amount of treated FGD sludge substitution lead to lowering the CTE value as S3.0 and S4.5 recorded a CTE value of 2.62×10^{-6}/°C and 2.70×10^{-6}/°C, respectively. The reduction of CTE values indicated the presence of a high amount of α-cordierite crystalline in the samples. The theoretical CTE value of α-cordierite is substantially lower (approximately 1.5–4.0×10^{-6}/°C) than other phases present in the samples, which is due to the rigid tetrahedral framework [24] and anisotropic expansion of the octahedral framework of the α-cordierite structure [25]. Those criteria demonstrate α-cordierite's capacity to endure high temperature without significant volume changes, indicating that 3.0 and 4.5 wt% of treated FGD sludge substitution help in crystallizing a high quality of α-cordierite. Wool [26] stated that the CTE reflects the average distance between atoms with increasing temperature based on the atomic perspective. Generally, weaker bonds tend to have a higher CTE value due to the changes in bond length at a higher temperature. Thus, a high CTE value of S0 and S1.5 is due to the presence of low bonding between particles due to the low or lack of liquid phase that diffuses between the particles. This can be supported as forsterite recorded a high CTE value, 9.9×10^{-6}/°C, leading to a higher amount of CTE value on the sample with low treated FGD sludge substitution as XRD patterns show the high amount of forsterite formation in S0 and S1.5.

When compared to the early stages of expansion, the second rate of expansion of all samples shows a decrement even though at a high sintering temperature. According to Khattab [27], factors that may be affecting the decrease of thermal expansion are the development of micro-cracks due to the difference in thermal expansion between the glassy phase and the matrix that reversed the values of thermal expansion, the presence of lattice thermal anisotropy, and the presence of pores in the sample structure. At this stage, the rate of grain growth in all samples began to slow, indicating the activation of glassy phase formation in the sample. The presence of a significant number and size of pores in the sample with treated FGD sludge substitution easily explains this occurrence. The grain growth in the samples occurred within the pore and hence did not affect the total volume changes of the sample.

The final/third step occurs where all of the samples began to shrink with all of the samples recording a negative value of CTE, indicating the softening temperature for each sample to obtain a dense structure. This explanation was supported by Pomeroy [28], where the glass softening temperature (T_D) began at a rapid contraction of the sample length. Each sample shows different softening temperatures where increasing wt% of treated FGD sludge substitution would reduce the softening temperature of the sintered sample. Zero wt% and 1.5 wt% of treated FGD sludge substitutions recorded the same softening temperature, 738 °C, while 3.0 wt% started to shrink at 737 °C, while the shrinkage with 4.5 wt% substitutions started at 736 °C. The CTE value of the sample also decreases with an increasing amount of treated FGD sludge substitution. At this stage, the fluxing element such as CaO and B_2O_3 in the treated

FGD sludge would start to melt and produces a glassy phase. A higher amount of treated FGD sludge would produce a high amount of glassy phase that would diffuse between the atoms, helps to reduce the gaps between grains, and thus helps in crystallizing the most equilibrium compound at a lower temperature, as mentioned in SEM images. In addition, this shrinkage also can be correlated with the rate of shrinkage of the sintered sample, as the higher the amount of treated FGD sludge leads to a decreasing rate of shrinkage. Thus, the decreasing rate of shrinkage would lead to a decreasing CTE value as the sample shows the lowest ability to shrink and reduce its total volume.

33.4 Conclusion

This study helps in identifying the thermal behavior of cordierite-based ceramic with the substitution of MgO with treated FGD sludge from NEGM. Increasing treated FGD sludge in the non-stoichiometric cordierite composition would lead to high porosity due to the formation of voids from the decomposition of SO_3 during the sintering process. The CTE value of a cordierite-based ceramic sample is controlled by the major phase present in the sample and the porosity. Sample with 3.0 wt% of treated FGD sludge is able to obtain the lowest CTE value, $2.62 \times 10^{-6}/°C$, due to the presence of a single phase of α-cordierite in the sample.

Acknowledgements The research was supported by Research Grant Scheme (FRGS) (Grant No: FRGS/1/2019/STG07/UNIMAP/02/4) and Faculty of Chemical Engineering and Technologies, Universiti Malaysia Perlis (UniMAP). The authors would also like to thank Nippon Electric Glass Sdn. Bhd. for providing the FGD sludge.

References

1. O. Fatin Fatini, J. Banjuraizah, S.F. Khor, R. Nik Akmar, A. Suffi Irni, Phase transformation of FGD by-product into plaster mould via heat treatment process. Key Eng. Mater. **908** KEM, 672–677 (2022)
2. B. Guan, H. Fu, J. Yu, G. Jiang, B. Kong, Z. Wu, Direct transformation of calcium sulfite to α-calcium sulfate hemihydrate in a concentrated Ca-Mg-Mn chloride solution under atmospheric pressure. Fuel **90**(1), 36–41 (2011)
3. B. Guan, L. Yang, H. Fu, B. Kong, T. Li, L. Yang, α-calcium sulfate hemihydrate preparation from FGD gypsum in recycling mixed salt solutions. Chem. Eng. J. **174**(1), 296–303 (2011)
4. A. Zürz, I. Odler, F. Thiemann, K. Berghöfer, Autoclave-free formation of α-hemihydrate gypsum. J. Am. Ceram. Soc. **74**(5), 1117–1124 (1991)
5. S.I. Alias, B. Johar, S.N.F. Adam, M.A.A. Taib, F.F. Othman, K.S. Fhan, Phase transformation and microstructure characterization of ceramic porcelain with different ratio of treated Fgd sludge/feldspar. J. Phys. Conf. Ser. **2129**(1) (2021)
6. J.K. Lee, T.Y. Kim, Evaluation of humidity control ceramic paint using gypsum binder. J. Korean Ceram. Soc. **55**(1), 74–79 (2018)

7. M. Rundans, I. Sperberga, Porous cordierite ceramics from natural clays. Mater. Sci. Appl. Chem., pp. 33–38 (2015)
8. J.R. González-Velasco, R. Ferret, R. López-Fonseca, M.A. Gutiérrez-Ortiz, Influence of particle size distribution of precursor oxides on the synthesis of cordierite by solid-state reaction. Powder Technol. **153**(1), 34–42 (2005)
9. M.A. Camerucci, G. Urretavizcaya, A.L. Cavalieri, Sintering of cordierite based materials. Ceram. Int. **29**(2), 159–168 (2003)
10. K.K. Eing, B. Johar, L.N. Ho, Y. Zabar, Influence of sintering temperature on crystallization behavior of cordierite synthesized from non-stoichiometric formulation, in *MATEC Web Conference*, vol. 78 (2016)
11. R. Gören, A. Mergen, R. Ceylantekin, Preparation of cordierite ceramics from talc with boron oxide addition. Key Eng. Mater. **264–268**(I), pp. 301–304 (2004)
12. S. Lamara, D. Redaoui, F. Sahnoune, M. Heraiz, N. Saheb, Formation of anorthite containing cordierite materials through reaction sintering kaolin, MgO and CaO precursors. Sci. Sinter. **52**(2), 135–147 (2020)
13. G.H. Chen, Effect of replacement of MgO by CaO on sintering, crystallization and properties of MgO-Al2O3-SiO2 system glass-ceramics. J. Mater. Sci. **42**(17), 7239–7244 (2007)
14. G. hua Chen, Sintering, crystallization, and properties of CaO doped cordierite-based glass-ceramics. J. Alloys Compd. **455**(1–2), 298–302 (2008)
15. J. Banjuraizah, H. Mohamad, Z.A. Ahmad, Crystal structure of single phase and low sintering temperature of α-cordierite synthesized from talc and kaolin. J. Alloys Compd. **482**(1–2), 429–436 (2009)
16. J. Banjuraizah, H. Mohamad, Z.A. Ahmad, Thermal expansion coefficient and dielectric properties of non-stoichiometric cordierite compositions with excess MgO mole ratio synthesized from mainly kaolin and talc by the glass crystallization method. J. Alloys Compd. **494**(1–2), 256–260 (2010)
17. F.F. Othman, B. Johar, S.F. Khor, N.A. Rejab, S.I. Alias, Physical and mechanical properties of non-stoichiometric cordierite with treated FGD sludge addition sintered at 1250°C. J. Phys. Conf. Ser. **2129**(1) (2021)
18. T.J. Mays, A new classification of pore sizes. Stud. Surf. Sci. Catal. **160**, 57–62 (2007)
19. T. Sievert, A. Wolter, N.B. Singh, Hydration of anhydrite of gypsum (CaSO4.II) in a ball mill. Cem. Concr. Res. **35**(4), 623–630 (2005)
20. Z. Wang et al., Thermochemical behavior of three sulfates (CaSO4, K2SO4 and Na2SO4) blended with cement raw materials (CaO-SiO2-Al2O3-Fe2O3) at high temperature. J. Anal. Appl. Pyrolysis **142** (2019)
21. H.E. Exnet, E.A. Giess, Anisotropic shrinkage of cordierite-type glass powder cylindrical compacts. J. Mater. Res. **3**(1), 122–125 (1988)
22. C. Manière, S. Chan, G. Lee, J. McKittrick, E.A. Olevsky, Sintering dilatometry based grain growth assessment. Results Phys. **10**, 91–93 (2018)
23. H. Ardebili, J. Zhang, M.G. Pecht, Characterization of encapsulant properties. Encapsulation Technol. Electron. Appl., 221–258 (2018)
24. S.S.V.S.S. Vepa, A.M. Umarji, Effect of substitution of Ca on thermal expansion of cordierite (Mg2Al4Si5O18). J. Am. Ceram. Soc. **76**(7), 1873–1876 (1993)
25. B.P. Saha, R. Johnson, I. Ganesh, G.V.N. Rao, S. Bhattacharjee, Y.R. Mahajan, Thermal anisotropy in sintered cordierite monoliths. Mater. Chem. Phys. **67**(1–3), 140–145 (2001)
26. R.P. Wool, Bio-based composites from soybean oil and chicken feathers. Bio-Based Polym. Compos., 411–447 (2005)
27. R.M. Khattab et al., Sintering, physicomechanical, thermal expansion and microstructure properties of cordierite ceramics based on utilizing silica fume waste. Mater. Chem. Phys. **270** (2021)
28. M.J. Pomeroy, Thermal analysis techniques for technical ceramics and glasses. Encycl. Mater. Tech. Ceram. Glas. **1–3**, 676–688 (2021)

Chapter 34
Production of Porous Glass-Foam Materials from Photovoltaic Panel Waste Glass

Bui Khac Thach, Le Nhat Tan, Do Quang Minh, Ly Cam Hung, and Phan Dinh Tuan

Abstract The solar energy production is growing quickly for the global demand of renewable one, decrease the dependence on fossil fuels. However, disposing of used photovoltaic (PV) panels will be a serious environmental challenge in the future decades since the solar panels would eventually become a source of hazardous waste. The potential of waste solar panel glass to generate porous glass material with the addition of $CaCO_3$ and water glass was assessed in this study. The porous glass firing temperature range, from 830 to 910 °C, was determined using a simulation of heating microscope technique. The created samples have the smallest volumetric density of 0.25 g/cm^3 and the largest water absorption of 303.08 wt.%. This indicates that the image analysis of samples during the heating process could be used to identify the firing temperature for better foaming, which was favorably indicated by specific physicochemical parameters. The created glass-foam materials with an apparent porosity up to 81.49% could be used as a water-retaining medium in hydroponic and aquaponic systems.

Keywords Waste glass from solar panels · Porous glass · Heating microscope

34.1 Introduction

Solar power generation is expanding rapidly and providing significant benefits. The estimated lifetime of photovoltaic (PV) modules is about 25 years. Therefore, in the

B. K. Thach · L. N. Tan · D. Q. Minh (✉)
Faculty of Materials Technology, Ho Chi Minh City University of Technology, 268 Ly Thuong Kiet St., Ward 14, District 10, Ho Chi Minh City 70000, Vietnam
e-mail: mnh_doquang@hcmut.edu.vn

Vietnam National University Ho Chi Minh City, Linh Trung Ward, Thu Duc District, Ho Chi Minh City 70000, Vietnam

L. C. Hung · P. D. Tuan
Hochiminh City University of Natural Resources and Environment, 236B Le Van Sy St., Tan Binh District, Ho Chi Minh City 70000, Vietnam

M. A. A. Mohd Salleh et al. (eds.), *Proceedings of the Green Materials and Electronic Packaging Interconnect Technology Symposium*, Springer Proceedings in Physics 289,
https://doi.org/10.1007/978-981-19-9267-4_34

coming decades, solar panels may eventually become a source of hazardous waste, and disposing of PV panels will be a crucial environmental issue [1]. Furthermore, most of the solid wastes from the used solar panels would be landfilled, coursing the contamination of soil and groundwater, damaging the ecosystem in the long term [2].

After being separated from PV modules, the glass from wasted solar panels is difficult to be recycled in floating or container glass furnaces due to its impurities. The procedure of purifying the glass from waste solar panels is complicated and expensive [1]. In thermal delamination, the ethylene vinyl acetate (EVA) is eliminated, and components including glass, aluminum frames, plastics, and other components are separated [3]. Kang et al. [4] used organic solvents to recover glass from waste solar panels, after the panels were soaked in toluene for 2 days at 90 °C, the tempered glass and PV cells are separated from the swollen and dissolved EVA. Pagnanelli et al. [5] recovered approximately 91% of the glass weight using two stages treatment: a physical treatment (triple crushing and thermal treatment) and a chemical treatment. Also, melting at temperature up to 1550 °C in glass furnace consumes a large amount of energy and causes CO_2 and SO_2 emission [6].

There have been studies using waste glass flakes as a by-product in the production of ceramic materials, such as fired clay bricks and floor tiles [7, 8]. In addition, waste glass fragments have also been studied to make geopolymers [9], glass ceramics, etc. [10]. It has also been studied using solar panels as an aggregate for Portland cement [11]. Thanks to the pozzolanic activity of glass powder, a new material called "High-Performance Glass Concrete" has been created with the low water absorption, smooth surface, and high volumetric density. It also has rheological properties that enhance the workability of fresh concrete [12]. Waste glass and red mud have been mixed to make bricks by pressing the products and then sintering at the temperatures below 1000 °C [13]. Waste glass mixed with $Ca(OH)_2$ in the ratio Ca/Si = 0.83 can also form calcium silicate materials through hydrothermal treatment. As the hydrothermal time increased, the mechanical properties and density of the materials improved [14]. Wollastonite ($CaSiO_3$) material can be prepared at low temperature by using a mixture of rice husk ash, waste glass, and CaO by the hydrothermal process followed by calcination at 1000 °C [15].

Porous glass is an application of vitrification typically accomplished by adding a gas-forming decomposing agent to the molten glass to create air bubbles into the glass matrix. Porous glass has desirable properties such as high porosity, low thermal conductivity, chemical resistance, water resistance, and non-flammability [16]. In general, foaming agents are divided into two categories: foaming via decomposing agents such as calcium carbonate ($CaCO_3$) [17], silicon carbide (SiC) [18], and gaseous combustions as carbon (C), etc. The volumetric density of glass foam tends to decrease as the initial particle size decreases, and rapidly decreases when the initial size of the glass powder reaches below 125 μm. The foaming is almost non-existent if the initial size of the glass powder is larger than 0.4 mm [19, 20]. Heating rates of 5–10 °C/min are commonly used to avoid cracking due to too fast heating or premature gas generation due to too slow heating [16].

The properties of porous glass are fundamentally affected by the firing temperature, which is usually chosen between 700 and 900 °C [21]. The porous glass has been demonstrated for its filtering ability to remove undesirable minerals from brackish water [22]. It must meet high strength, lightweight, and chemical stability [23]. The material for insulation application can be fabricated with glass grain sizes of 75–150 μm and a sintering temperature of 800 °C [24]. Besides, it creates an excellent hydroponic and aquaponic growing medium. After being irrigated, the pores help to absorb water quickly, hold water well, easily re-wet and have high porosity and aeration [25]. Furthermore, porous glass has many potential applications such as moisture control, electromagnetic absorption, lightweight bricks, and used as construction materials [26].

To make the porous glass, determining the firing temperature is very important. If the firing temperature is too high, the molten glass will seal the voids, but if the firing temperature is too low, the glass particles will not form stable bonds. Moreover, the calcination temperature needs to match the decomposition temperature of the foaming agents, so that the glass sample has high porosity while still having enough mechanical strength. Glass has no fixed melting point, but changes from solid to liquid through a soft temperature range. Experimentally, the softening temperature range was determined by Heating Microscopy (HM) [27]. The firing temperature for foaming glass must be chosen within the softening temperature range. Characteristically, temperatures for the softening temperature range of glass are defined according to DIN 51,730 (1998-4)/ISO 540 (1995-03-15) as follows:

- Sintering temperature (T_{sinter}): the temperature at which the sample has had a dimensional shift of 5% in comparison to the initial image.
- Softening temperature (T_s): the temperature for which round edges was visible ($H = ¾\,H_0$).
- Sphere temperature: the temperature for which the probe appeared like a sphere ($H = D$).
- Hemisphere temperature: the temperature for which the height is half of the base ($H = 1/2D$).
- Flow temperature (T_f): the temperature for which the sample is melted down to 1/3 of its initial height ($H = 1/3H_0$).

The softening temperature range ΔT is determined by $\Delta T = T_f - T_s$. The firing temperatures of the glass samples will be selected in this temperature range.

In this study, a simulation of the HM technique was performed. Characteristic temperatures in the glass softening range from waste solar panels with $CaCO_3$ foaming agent were determined by analyzing images taken with a digital camera. Then, the firing temperatures of the glass samples were selected in the glass softening range so that the $CaCO_3$ decomposed to pores, and the sintered glass samples had high enough mechanical strength.

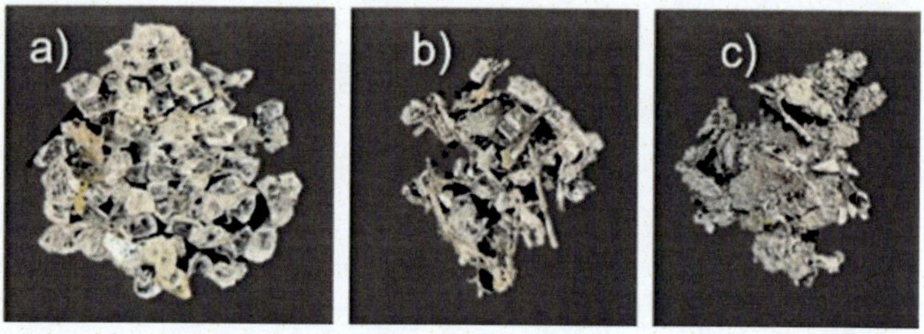

Fig. 34.1 **a** Waste glass fragments, **b** Waste glass and aluminum ribbons, **c** EVA residues

34.2 Materials and Methods

34.2.1 Raw Materials

Broken PV panels were collected from solar Vietnam JSC, Ho Chi Minh City, Vietnam. After using a heat gun to apply heat on the surface of the solar panels, the glass was manually separated from EVA layer. These separated materials contain waste glass fragments, pieces of solar cells, and aluminum ribbons (Fig. 34.1a, b).

The waste glass was sorted and collected by hand, then ground using a ball mill in 6 h. The powder then was sieved through 125 μm to remove EVA residues (Fig. 34.1c) and used to form the experimental samples. Commercially available $CaCO_3$ powders was used as high-temperature foaming agents and water glass was used as the initial binder. Their chemical compositions were analyzed using X ray fluorescence (XRF) with a loss on ignite at 950 °C in 0.5 h (ARL ADVANT'X, Thermo).

The particle size distributions of the ground waste glass and $CaCO_3$ were analyzed by Laser Diffraction Size Analysis (Mastersizer MS3000, Malvern Panalytical).

34.2.2 Experimental Methods

The Canon 700D digital camera with Lens 18–55 IS STM was used to take images of the specimen during the heating process with the heating rate of 10 °C/min. The specimen with the mixture of 10% of $CaCO_3$ and 10% water glass in the proportion of the waste glass by weight was compressed into a cylindrical block with 3 mm high and 3 mm in diameter. The characteristic temperatures were selected by analyzing the behavior of green sample under heating process.

Raw materials were mixed in ratio of 2, 4, 6, 8, and 10% of $CaCO_3$ and 10% water glass based on the weight of the waste glass. Tap water was used to control the humidity of mixture fixed at 20% (in wt.) then mixed for 5 min using an electric beater. After thorough mixing, the mixture was formed using a PVC mold about 12 mm in height and 18 mm in diameter. Then, the samples were dried in the dryer (Venticell) at

Fig. 34.2 Images of **a** PVC mold, green sample, and foam-glass sample, respectively, **b** diagram of compressive strength test of foam-glass samples

temperature of 110 °C for 4 h. Physical properties of the samples including volumetric density, water absorption, and apparent porosity were determined following the standard of ASTM C830 (2011). The total porosity was obtained from the volumetric density and the sintered powder density with the following equations:

$$\text{Total porosity} = (1 - \text{volumetric density/sintered powder density}) \times 100\%$$

To ascertain the compressive strength of the glass foam, the cylindrical specimens of 15 mm high and 30 mm diameter were tested using a loading speed of 1.5 kN/min (DTU-900MHN, Daekyung Tech, Korea). The compressive strength test was illustrated in Fig. 34.2.

34.3 Result and Discussions

34.3.1 Characteristics of Raw Materials

Chemical compositions of raw materials used to make the porous glass samples are shown in Table 34.1. The results of chemical analysis show that the glass from waste solar panels was based on soda-lime-silica glass system.

The particle size distribution of the raw materials is shown as Fig. 34.3. According to the results of this analysis, the mean size (d_{50}) of the waste glass from solar panels and $CaCO_3$ powder is 35.0 μm and 18.1 μm, respectively.

Table 34.1 The chemical composition (wt.%) of raw materials

Material	SiO_2	Na_2O	CaO	Al_2O_3	Fe_2O_3	MgO	Others	LOI
Waste glass	67.17	12.74	11.74	2.00	1.92	2.43	1.44	0.56
$CaCO_3$	–	–	56.28	–	0.01	0.32	0.01	43.38
Water glass	69.15	27.82	0.40	1.66	0.07	–	0.90	0.00

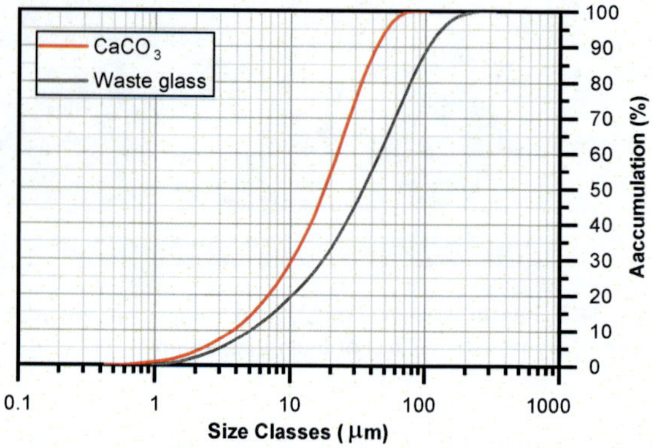

Fig. 34.3 Particle size distributions of waste glass and $CaCO_3$

34.3.2 *Investigation of Foaming Process*

The thermal behaviors of sample were shown in Fig. 34.4. The sintering temperature starts at 730 °C. With increasing temperatures, the expansion of the sample took place at 770 °C and maximum expansion 870 °C thanks to the CO_2 release from the decomposition of $CaCO_3$. Then, the shape of the sample remained unchanged until 990 °C—the softening point, and almost reached the sphere at 1000 °C when the rounding of the edges is visible. As the temperature gets higher, the hemisphere and flow point could be observed at 1110 °C and 1120 °C, respectively.

Based on the behavioral characterization of the sample, the firing temperatures were chosen at 830, 850, 870, 890, and 910 °C with the heating rate at 10 °C/min for 15 min in an electrical kiln.

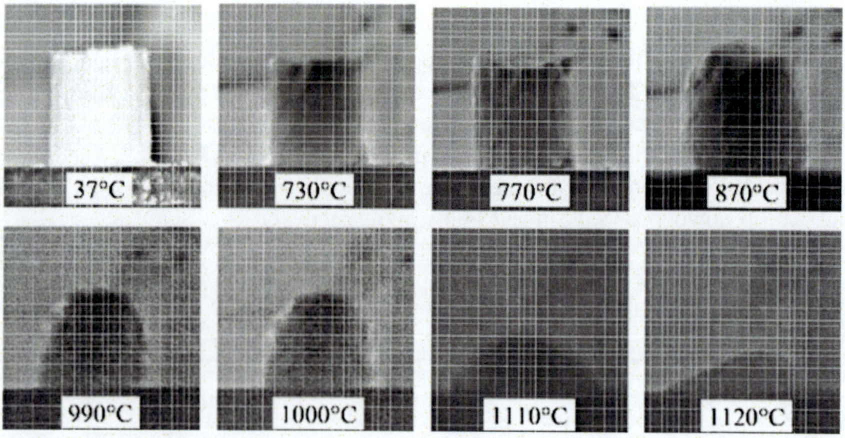

Fig. 34.4 Images of the foaming process of foam-glass specimen

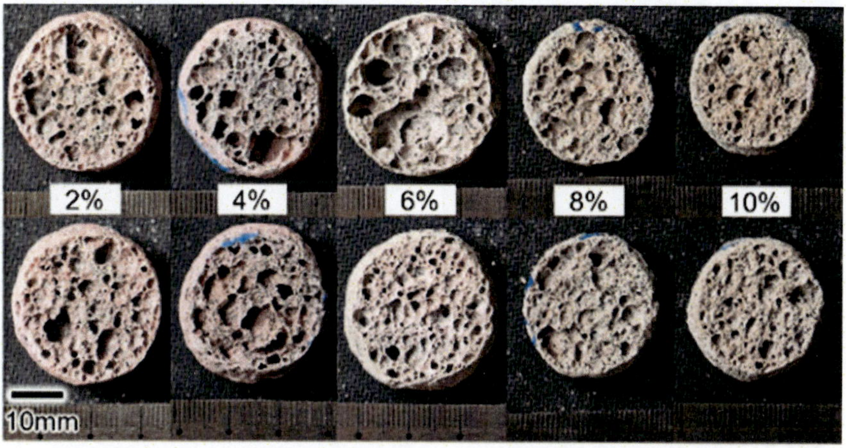

Fig. 34.5 Images of the top (first row) and bottom (second row) of the samples heating at 870 °C with different $CaCO_3$ contents from 2 to 10% (in wt.)

34.3.3 Characteristic Appearance of the Foam Samples

The pore morphology of the two surfaces of the samples sintered at 870 °C for 15 min with different $CaCO_3$ contents is shown in Fig. 34.5. It can be seen that the increase of $CaCO_3$ makes the pore size of the sintered samples more evenly. Furthermore, the pore structure of the 10% weight $CaCO_3$ sample is considered to be uniform. The porosity distribution is not uniform because the bubbles rise to the top of the samples when the vitreous viscosity is low. A maze-like network of pores in glass is created when the intense gas released during their decomposition disrupts the walls of the individual pores [28]. The pore diameter was significantly reduced in samples where the $CaCO_3$ content increased by up to 10% by weight, similar results were reported by Souza et al. [29] when eggshell was used as a foaming agent.

34.3.4 Physical and Mechanical Properties of Sintered Samples

The physical properties of the test samples are shown in Fig. 34.6. The volumetric density of the sintered samples was increased with increasing $CaCO_3$ content from 2 to 10 wt.%. Samples with high total porosity have a low volumetric density and will therefore absorb a larger amount of water. Open porosity has a significant effect on water absorption. With the increase of open porosity, water absorption increases. According to the results, 870 °C was considered as the ideal sintering temperature to maximize the water absorption for all amounts of $CaCO_3$ corresponding to the maximum volume expansion.

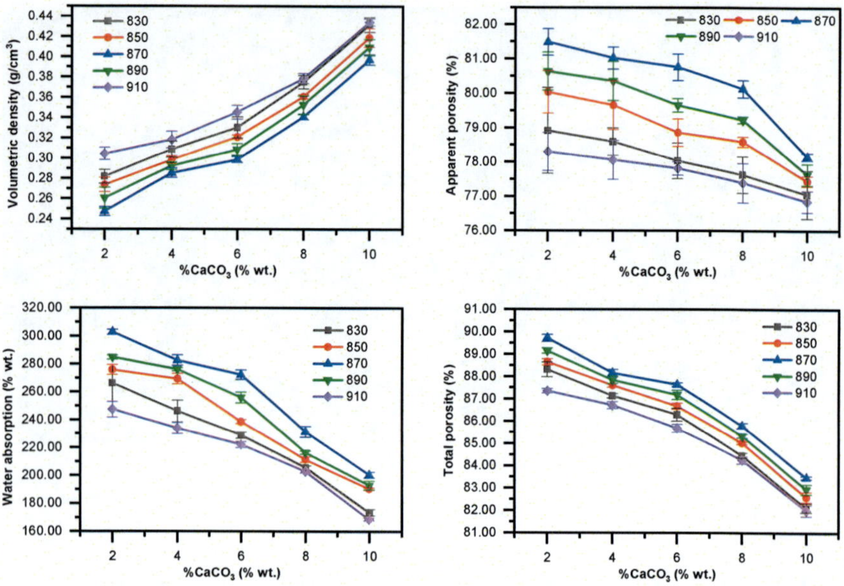

Fig. 34.6 Influence of sintering temperature and $CaCO_3$ content on the physical properties of the glass-foam samples

Comparing the sintered samples with other porous glass produced from glass bottles and shells [21], the densities of the sintered samples were significantly lower.

At the low sintering temperature, 830 °C, gas generation in the highly viscous glass matrix produces closed pores. In addition, incomplete sintering of the glass powder results in an almost porous mass. The partition walls between the pores are expected to deform at higher temperatures, 910 °C, reducing the apparent porosity. As the $CaCO_3$ content increases, the volumetric density tends to increase. The results are similar to the reported study [30], especially the samples with 10% by weight $CaCO_3$.

The compressive strength of the samples with different amounts of $CaCO_3$ is shown in Fig. 34.7. It is clear that the 10% $CaCO_3$ samples had the highest compressive strength and the lowest porosity.

34.4 Conclusion

Various samples of foam glass with different porosity have been formed from a mixture of $CaCO_3$, water glass, and glass powder from used solar panels. The method of prototyping with a 10 wt.% water glass binder and then casting it in a plastic mold has been shown suitable to create such types of materials.

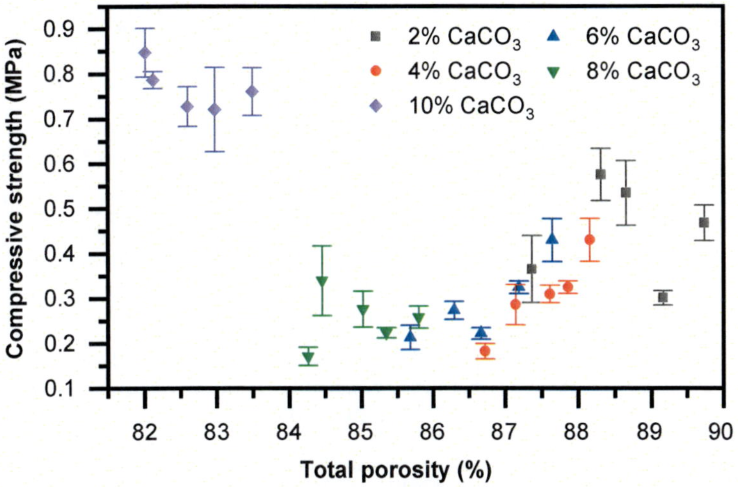

Fig. 34.7 Total porosity versus compressive strength of samples with different $CaCO_3$ content (in wt.%)

The firing temperature for foaming was selected by image analysis of the sample during the heating process, favorably tailored to specific physicochemical properties.

Under the experimental conditions, the samples had the volumetric density of 0.25–0.43 g/cm^3 and water absorption of 160.02–303.08 wt.% as the heating temperature in the range of 830–910 °C and the $CaCO_3$ contents of 2–10 wt.%.

The results indicated that as the $CaCO_3$ content increased, but the porosity decreased and the mechanical strength increased. Glass-foam-based materials with an apparent porosity of 76.85–81.49% can be used as a water-retaining material in hydroponic and aquaponic systems. It is also expected to be used as microbial media in water and wastewater treatments.

Acknowledgements The authors acknowledge financial support from the Ministry of Natural Resources and Environment of Vietnam through the Project coded 04/HĐ-BQLTDA-tm.

Conflict of Interest The authors declare no competing interests.

References

1. Md.S. Chowdhury, K.S. Rahman, T. Chowdhury, N. Nuthammachot, K. Techato, M. Akhtaruzzaman, S.K. Tiong, K. Sopian, N. Amin, An overview of solar photovoltaic panels' end-of-life material recycling. Energy Strategy Rev. **27**, 100431 (2020). https://doi.org/10.1016/j.esr.2019.100431
2. J.I. Kwak, S.-H. Nam, L. Kim, Y.-J. An, Potential environmental risk of solar cells: current knowledge and future challenges. J. Hazard. Mater. **392**, 122297 (2020). https://doi.org/10.1016/j.jhazmat.2020.122297

3. E. Klugmann-Radziemska, P. Ostrowski, K. Drabczyk, P. Panek, M. Szkodo, Experimental validation of crystalline silicon solar cells recycling by thermal and chemical methods. Sol. Energy Mater. Sol. Cells **94**(12), 2275–2282 (2010). https://doi.org/10.1016/j.solmat.2010.07.025
4. S. Kang, S. Yoo, J. Lee, B. Boo, H. Ryu, Experimental investigations for recycling of silicon and glass from waste photovoltaic modules. Renew. Energy **47**, 152–159 (2012). https://doi.org/10.1016/j.renene.2012.04.030
5. F. Pagnanelli, E. Moscardini, G. Granata, T. Abo Atia, P. Altimari, T. Havlik, L. Toro, Physical and chemical treatment of end of life panels: an integrated automatic approach viable for different photovoltaic technologies. Waste Manag. **59**, 422–431 (2017). https://doi.org/10.1016/j.wasman.2016.11.011
6. P.T. Kien, K.D.T. Kien, D.Q. Minh, Research on wasted glass as non-firing brick using hydrothermal method. J. Sci. Technol. **52**(4A), 198–204 (2014)
7. K.-L. Lin, T.-C. Lee, C.-L. Hwang, Effects of sintering temperature on the characteristics of solar panel waste glass in the production of ceramic tiles. J. Mater. Cycles Waste Manag. **17**(1), 194–200 (2014). https://doi.org/10.1007/s10163-014-0240-3
8. J. Jimenez-Millan, I. Abad, R. Jimenez-Espinosa, A. Yebra-Rodriguez, Assessment of solar panel waste glass in the manufacture of sepiolite based clay bricks. Mater. Lett. **218**, 346–348 (2018). https://doi.org/10.1016/j.matlet.2018.02.049
9. H. Hao, K.-L. Lin, D. Wang, S.-J. Chao, H.-S. Shiu, T.-W. Cheng, C.-L. Hwang, Elucidating characteristics of geopolymer with solar panel waste glass. Environ. Eng. Manag. J. **14**(1), 79–87 (2015). https://doi.org/10.30638/eemj.2015.010
10. V. Savvilotidou, A. Kritikaki, A. Stratakis, K. Komnitsas, E. Gidarakos, Energy efficient production of glass-ceramics using photovoltaic (P/V) glass and lignite fly ash. Waste Manag. **90**, 46–58 (2019). https://doi.org/10.1016/j.wasman.2019.04.022
11. K. Máčalová, V. Václavík, T. Dvorský, R. Figmig, J. Charvát, M. Lupták, The use of glass from photovoltaic panels at the end of their life cycle in cement composites. Materials **14**(21), 6655 (2021). https://doi.org/10.3390/ma14216655
12. J. Esmaeili, A. Oudah AL-Mwanes, A review: properties of eco-friendly ultra-high-performance concrete incorporated with waste glass as a partial replacement for cement. Mater. Today Proc. **42**, 1958–1965 (2021). https://doi.org/10.1016/j.matpr.2020.12.242
13. D.Q. Minh, N.H. Thang, P.T. Kien, Research on manufacturing decorative tiles from cullet and red mud, in*Green Material and Green Technology for Green "MONOZUKURI,"* Hanoi University of Science and Technology, Hanoi, Vietnam, pp. 75–81 (2013)
14. P.T. Kien, N.T.H. Phong, N.H. Thang, K.D.T. Kien, D.Q. Minh, L.H. Anh, P.T.T. Trinh, Wasted cullet in glass technology used as ecomaterials, in*Collaboration and Exchange in Advanced Science and Technology*, Can Tho University, Vietnam, pp. 73–78 (2014)
15. P. Trung Kien, T. Thi Thien Ly, P. Thi Lan Thanh, T. Pham Quang Nguyen, N. Hoc Thang, M.M.A.B. Abdullah, A novel study on using Vietnam rice hush ash and cullet as environmental materials, in *MATEC Web Conference*, vol. 97, p. 01118 (2017). https://doi.org/10.1051/matecconf/20179701118
16. G. Scarinci, G. Brusatin, and E. Bernardo, "Glass Foams," in *Cellular Ceramics (Structure, Manufacturing, Properties and Applications)*, ed. by M. Scheffler, P. Colombo (Wiley-VCH Verlag GmbH & Co. KGaA, Weinheim, FRG, 2006), pp. 158–176. https://doi.org/10.1002/3527606696.ch2g.
17. M. Zhu, R. Ji, Z. Li, H. Wang, L. Liu, Z. Zhang, Preparation of glass ceramic foams for thermal insulation applications from coal fly ash and waste glass. Constr. Build. Mater. **112**, 398–405 (2016). https://doi.org/10.1016/j.conbuildmat.2016.02.183
18. J. Bai, X. Yang, S. Xu, W. Jing, J. Yang, Preparation of foam glass from waste glass and fly ash. Mater. Lett. **136**, 52–54 (2014). https://doi.org/10.1016/j.matlet.2014.07.028
19. V. Ducman, M. Kovačević, The Foaming of Waste Glass. Key Eng. Mater. **132–136**, 2264–2267 (1997). https://doi.org/10.4028/www.scientific.net/KEM.132-136.2264
20. J. König, R.R. Petersen, Y. Yue, Influence of the glass particle size on the foaming process and physical characteristics of foam glasses. J. Non-Cryst. Solids **447**, 190–197 (2016). https://doi.org/10.1016/j.jnoncrysol.2016.05.021

21. N.A.N. Hisham, M.H.M. Zaid, S.H.A. Aziz, F.D. Muhammad, Comparison of foam glass-ceramics with different composition derived from ark clamshell (ACS) and soda lime silica (SLS) glass bottles sintered at various temperatures. Materials **14**(3), 570 (2021). https://doi.org/10.3390/ma14030570
22. Sulhadi, Susanto, A. Priyanto, A. Fuadah, and M.P. Aji, Performance of porous composite from waste glass on salt purification process. Procedia Eng. **170**, 41–46 (2017). https://doi.org/10.1016/j.proeng.2017.03.008
23. X. Fang, Q. Li, T. Yang, Z. Li, Y. Zhu, Preparation and characterization of glass foams for artificial floating island from waste glass and Li2CO3. Constr. Build. Mater. **134**, 358–363 (2017). https://doi.org/10.1016/j.conbuildmat.2016.12.048
24. S.S. Owoeye, G.O. Matthew, F.O. Ovienmhanda, S.O. Tunmilayo, Preparation and characterization of foam glass from waste container glasses and water glass for application in thermal insulations. Ceram. Int. **46**(8), 11770–11775 (2020). https://doi.org/10.1016/j.ceramint.2020.01.211
25. M. Raviv, J.H. Lieth, A. Bar-Tal (eds.), *Soilless Culture: Theory and Practice*, Second edn. (Academic Press, an imprint of Elsevier, Amsterdam, 2019)
26. M. Flood, L. Fennessy, S. Lockrey, A. Avendano, J. Glover, E. Kandare, T. Bhat, Glass fines: a review of cleaning and up-cycling possibilities. J. Clean. Prod. **267**, 121875 (2020). https://doi.org/10.1016/j.jclepro.2020.121875
27. C. Venturelli, Heating microscopy and its applications. Microsc. Today **19**(1), 20–25 (2011). https://doi.org/10.1017/S1551929510001185
28. Yu.A. Spiridonov, L.A. Orlova, Problems of foam glass production. Glass Ceram. **60**(9/10), 313–314 (2003). https://doi.org/10.1023/B:GLAC.0000008234.79970.2c
29. M.T. Souza, B.G.O. Maia, L.B. Teixeira, K.G. de Oliveira, A.H.B. Teixeira, A.P. Novaes de Oliveira, Glass foams produced from glass bottles and eggshell wastes. Process Saf. Environ. Prot. **111**, 60–64 (2017). https://doi.org/10.1016/j.psep.2017.06.011
30. J. König, R.R. Petersen, Y. Yue, Influence of the glass–calcium carbonate mixture's characteristics on the foaming process and the properties of the foam glass. J. Eur. Ceram. Soc. **34**(6), 1591–1598 (2014). https://doi.org/10.1016/j.jeurceramsoc.2013.12.020

Chapter 35
Porous Geopolymeric Materials from Diatomite

Do Quang Minh, Tran Huu Phuc, Nguyen Vu Uyen Nhi, and Kieu Do Trung Kien

Abstract A new porous geopolymer material made from diatomite is introduced in this study. The alkaline active solid precursor was diatomite from Phu Yen province. Part of the diatomite is calcined at 600 °C to form metakaolin for increased alkaline activity. Mixtures of diatomite and metakaolin in the proportions of 60, 65, 70, 75, and 80 (%wt.) were mixed with active alkaline solution (AAS) and then cast in a steel mold of size 4 × 4 × 16 (cm). The AAS was water glass (1.49 kg/L density) mixed with NaOH 10 M solution in a ratio of 20%wt. The samples were removed from the mold after two days, then cured at room temperature for seven days. Material properties were determined by volumetric density (0.6–0.7 g/cm^3) and mechanical strength (bending strength 0.5–0.8 MPa). Geopolymer bond formation was tested by FTIR and SEM.

Keywords Porous geopolymer material · Diatomite · Alkaline active · Cured at room temperatures

35.1 Introduction

Geopolymers are adhesives formed from the chemical reactions of alkaline substances (solutions of silicates, calcium hydroxide, sodium hydroxide, and/or potassium) with synthetic or natural aluminosilicate powders, usually industrial wastes [1–3]. These alkali-activated materials are currently attracting great interest in the production of unburnt porous materials for use as insulation in various applications [4–6]. Previous studies have shown the possibility of synthesizing porous

---

D. Q. Minh (✉) · T. H. Phuc · N. V. U. Nhi · K. Do Trung Kien
Faculty of Materials Technology, Ho Chi Minh City University of Technology (HCMUT), 268 Ly Thuong Kiet Street, District 10, Ho Chi Minh City, Vietnam
e-mail: mnh_doquang@hcmut.edu.vn

Vietnam National University Ho Chi Minh City, Linh Trung Ward, Thu Duc City, Ho Chi Minh City, Vietnam

M. A. A. Mohd Salleh et al. (eds.), *Proceedings of the Green Materials and Electronic Packaging Interconnect Technology Symposium*, Springer Proceedings in Physics 289,
https://doi.org/10.1007/978-981-19-9267-4_35

geopolymer materials based on alkali-active substances such as metakaolin or kaolin and potassium or sodium silicate (SS) [7, 8].

Various foaming techniques for porous materials based on the forming and growth of gases in a substrate of suitable viscosity have been extensively explored. Gases such as hydrogen peroxide (H_2O_2) for foaming in genotyping paste [9, 10] or hydrogen (H_2), produced in the redox reaction of Al, Zn, or Si in alkaline solution, have been used as porous materials additives [11, 12]. By comparing different foaming methods for the production of lightweight geopolymers, Masi et al. [13] demonstrated that homogeneous microstructures with small pores can be achieved by combining two different foaming agents, such as surfactants and hydrogen peroxide.

Diatomite is a naturally occurring porous material from a sedimentary rock composed mainly of amorphous silica (SiO_2). Diatomite exists in natural deposits with large reserves and can be extracted at a low cost. The diatom skeleton is highly porous, porous at the micro to nanometer level, very light, and chemically stable. Furthermore, diatomite can be dissolved in alkaline solutions. This is a necessary condition for creating geopolymer links.

Diatomite has been used in many applications, primarily as a filter, as a source for the production of low molecular carbon-mineral adsorbents, functional [14–17] fillers for paints and plastics, and additives pozzolanic, insulating refractory bricks. Because of its resistance to high temperatures and the effects of chemicals, it is also used in fireproof cement, as insulation, and as an absorbent in the manufacture of explosives [17].

Diatomaceous earth (DE) or fly ash (FA) was used as an activator with red mud to form geopolymer material [18–20]. Research results show that the alkaline activity of DE is not as high as that of FA. Meja et al. [21] compare the uses of rice husk ash (RHA) and DE that have been calcined from the brewing industry, using fly ash (FA) and metakaolin (MK) as geopolymer precursors. Silica fume has been used as an alkali activator to strengthen DE geopolymers [22]. Geopolymer only used DE with low strength if there is no additive to increase alkalinity [23]. In other words, the DE has low alkaline activity.

In this paper, porous materials from DE were prepared by geopolymer method. Mixing ratio DE/MK was mixed with different concentrations, from 65 to 80 (%wt.). Interestingly, to increase alkalinity, MK was made from diatomite.

Geopolymer bonding was investigated using Fourier transform infrared spectroscopy (FTIR). The mechanical properties were analyzed by bending tests and scanning electron microscopy (SEM) was used to study the morphology of the pores in the material.

35.2 Experimental Method

The alkaline active solution (AAS) was prepared by adding water glass at a density of 1.49 kg/L to the NaOH 10 M solution, and the ratio water glass/NaOH 10 M solution was 20%wt. **The alkaline active solid precursor** is DE obtained from the Phu Yen province. To increase alkaline activity, the DE was calcined at 600 °C to

Fig. 35.1 Sample preparation process

create metakaolin (MK). The alkaline active solid precursor is DE obtained from the Phu Yen mine. To increase alkalinity, the DE was calcined at 600 °C for 3 h to produce metakaolin (MK). The MK was mixed with AAS at the ratio of 60, 65, 70, 75, and 80 (%wt.), then shaped in a 4 × 4 × 16 (cm) mold by casting method.

The samples were kept at room temperature for 2 days to dry and shrink, then they were removed from the mold. Continue curing at room temperature for 7 days, then the samples are tested for density, bending strength, FTIR, and SEM characteristics (Fig. 35.1).

35.3 Result and Discussions

35.3.1 Density and Bending Strength of the Samples

The bending strengths and densities of samples are shown in Figs. 35.2 and 35.3. Accordingly, the bending strength decreased as the DE/MK ratio was increased.

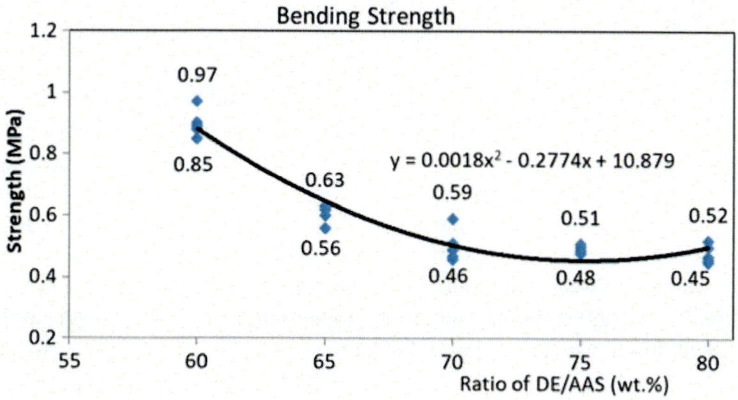

Fig. 35.2 The bending strength of the samples

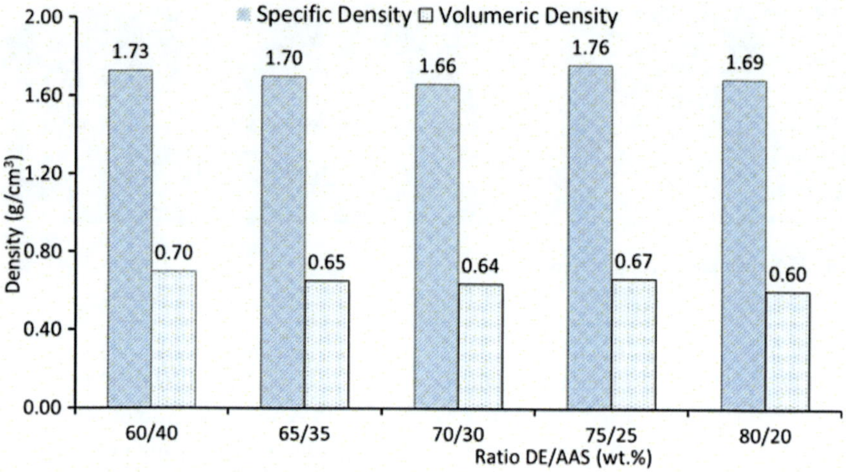

Fig. 35.3 The density of the samples

With a DE/MK ratio of 60/40, the porous material had the highest bending strength (0.9 MPa) and the lowest bending strength of 80/20 (0.48 MPa).

35.3.2 FTIR Spectra

The FTIR spectra for all samples are shown in Fig. 35.4 with peak/band designations in the 1300–400 cm^{-1} wave band. In this region, some peaks are associated with asymmetric and asymmetric-elongated geopolymer formation of the Si–O–Si and Si–O–Al bonds.

The MK spectrum prominently exhibits a wide range of oscillations between 1087 and 798 cm^{-1}, which is attributed to the overlap of the mean values of the contributions of the T–O–Si units (T ¼ Si or Al (IV)) [24] present in the amorphous aluminosilicate structure, where Al is the coordination tetrahedron. However, when MK is activated by AAS solution, geopolymerization occurs, and the peak at 1087 cm^{-1} was shifted to a other wave number of 1099, 1066, 1058, and 1060 (cm^{-1}), due to the change in the relative amount of T–O–Si (Verdolotti et al.) [25]. In the same region, absorption peaks of the geopolymer can also be observed (which MK is entirely absent), which is related to the contribution of the Q^4, Q^3, and Q^2 aggregates (Q^n – n is the –Si–O–Si-Units) obtained by the condensation reaction of AAS with diatomite. Si–O–Si(s) stretching vibration mode in the range 1000–1200 cm^{-1}, Si–O–Si(b) bending mode located around 800 cm^{-1}, -Si–O– (Q^3 groups) between 890 and 950 cm^{-1} and Si–O– (Q^2 groups) around 840 cm^{-1} [26]. In fact, the amorphous silica contained in the DE reacted with the AAS was dissolved, precipitated, and created a Si–O–Si–O polysilicate system [27].

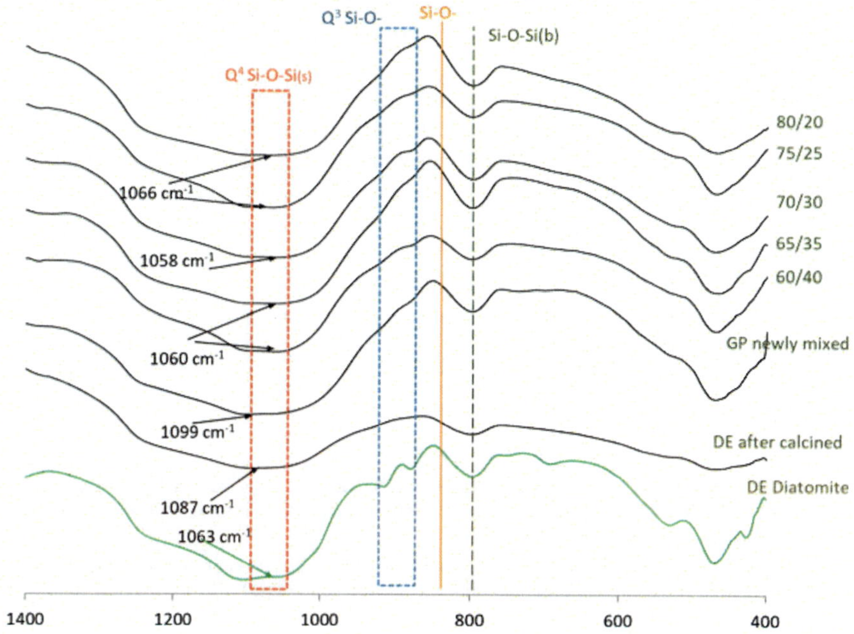

Fig. 35.4 FTIR spectra of raw material (DE), MK, and geopolymer samples

35.3.3 SEM Images

Figure 35.5 shows SEM images of DE and a geopolymer sample with a DE/MK ratio of 60/40. The structural destruction caused by the action of AAS can be clearly observed, representing the geopolymer formation.

Fig. 35.5 SEM images of DE and the geopolymer sample with a DE/MK ratio of 60/40

35.4 Conclusion

The FTIR, SEM, and bending strength result showed that porous geopolymer materials from diatomite could be formed. It is possible to use MK calcined from DE to increase the alkaline activity of this geopolymer. With a volumetric density of only 0.6–0.7 g/cm^3, the porous material can be used as heat and sound insulation.

Acknowledgements We acknowledge Ho Chi Minh City University of Technology (HCMUT), VNU—HCM for supporting this study.

References

1. J. Payá, L. Soriano, A. Font, M.V.B. Rosado, J.A. Nande, J.M.M. Balbuena, Reuse of industrial and agricultural waste in the fabrication of geopolymeric binders. Materials **14**(9), 2089 (2021). https://doi.org/10.3390/ma14092089
2. A.K. Thakur, A. Pappu, V.K. Thakur, Synthesis and characterization of new class of geopolymer hybrid composite materialsfrom industrial wastes. J. Cleaner Prod. **230**, 11–20 (2019). https://doi.org/10.1016/j.jclepro.2019.05.081
3. A. Fernández-Jiménez, N. Cristelo, T. Miranda, A. Palomo, Sustainable alkali activated materials: Precursor and activator derived from industrial wastes. J. Cleaner Prod. **162**, 1200–1209 (2017). https://doi.org/10.1016/j.jclepro.2017.06.151
4. W. Hu, Q.K. Nie, B.S. Huang, X. Shu, Q. He, Mechanical and microstructural characterization of geopolymers derived from red mud and fly ashes. J. Cleaner Prod. **186**, 799–806 (2018). https://doi.org/10.1016/j.jclepro.2018.03.086
5. K. Pimpawee, T. Chayanee, R. Silawat, T. Pajaree, P. Thammarat, F. Alexandre, L. Cristina, C. Duangrudee, Metakaolin-based porous geopolymer with aluminium powder. Key Eng. Mater. **608**, 132–138 (2014). https://doi.org/10.4028/www.scientific.net/KEM.608.132
6. K. Pimraksa, P. Chindaprasirt, A. Rungchet, K. Sagoe-Crentsil, T. Sato, Lightweight geopolymer made of highly porous siliceous materials with various Na2O/Al2O3 and SiO2 /Al2O3 ratios. Mater. Sci. Eng. A. **528**(21), 6616–6623 (2011). https://doi.org/10.1016/j.msea.2011.04.044
7. B.S.C. Kumar, K. Ramesh, Experimental study on metakaolin and GGBS based geopolymer concret. Inter. J. Civ. Eng. Technol. **9**(2), 341–349 (2017). https://doi.org/10.21817/ijet/2017/v9i1/170902311
8. I. Lancellotti, M. Catauro, C. Ponzoni, F. Bollino, C. Leonelli, Inorganic polymers from alkali activation of metakaolin: effect of setting and curing on structure. J. Solid State Chem **200**, 341–348 (2013). https://doi.org/10.1016/j.jssc.2013.02.003
9. W.V. Bonin, U. Nehen, U.V. Gizycki, Hydrogen peroxides blowing agent for foams, US Patent 3864137 A (1975)
10. V. Vaou, D. Panias, Thermal insulating foamy geopolymers from perlite. Miner. Eng. **23**(14), 1146–1151 (2010). https://doi.org/10.1016/j.mineng.2010.07.015
11. E. Prud'homme, P. Michaud, E. Joussein, C. Peyratout, A. Smith, S. Rossignol, In situ inorganic foams prepared from various clays at low temperature. Appl. Clay Sci. **51**(1–2), 5–22 (2011). https://doi.org/10.1016/j.clay.2010.10.016
12. A.R. Studart, U.T. Gonzenbach, E. Tervoort, T.J. Gauckler, Processing routes to microporous ceramics: a review. J. Am. Ceram. Soc. **89**(6), 1771–1789 (2006). https://doi.org/10.1111/j.1551-2916.2006.01044.x

13. G. Masi, W.D.A. Rickard, L. Vickers, M.C. Bignozzi, A.V. Riessen, A comparison between different foaming methods for the synthesis of light weight geopolymers. Ceram. Int. **40**(9), 13891–13902 (2014). https://doi.org/10.1016/j.ceramint.2014.05.108
14. R. Leboda, Carbon-mineral adsorbents-new type of sorbents? Part I: the methods of preparation. Mat. Chem. Phys. **31**(3), 243–255 (1992). https://doi.org/10.1016/0254-0584(92)90261-6
15. D.Q. Minh, N.H. Thang, Porous bricks from diatomite. J. Sci. Technol. **76**, 123–127 (2010)
16. A.C. Aydin, R. Gu¨l, Influence of volcanic originated natural materials as additive on the setting time and some mechanical properties of concrete. Constr. Build. Mater. **21**(6), 1277–1281 (2007). https://doi.org/10.1016/j.conbuildmat.2006.02.011
17. M. Reguerio, J.P. Calvo, E. Elizaga, V. Calderon, Spanish diatomite geology and economics. Ind. Miner. Rock. **306**, 57–67 (1993)
18. H.T. Nguyen, S.M. Gallardo, F.T. Bacani, H. Hinode, Q.M. Do, M.H. Do, M.A.B. Promentilla, Evaluating thermal properties of geopolymer produced from red mud, rice husk ash and diatomaceous earth. ASEAN Engineering Journal Part B **4**(1), 52 (2015). https://doi.org/10.11113/aej.v4.15427
19. D.Q. Minh, N.H. Thang, Characteristics of novel geopolymer composites synthesized from red mud and diatomaceous earth in autoclave conditions without using alkaline activator. J. Polym. Compos. **8**(3), 83–91 (2020). https://doi.org/10.37591/jopc.v8i3.4471
20. N.H. Thang, D.Q. Minh, H. Hinode, S.M. Gallardo, F.T. Bacani, M.A.B. Promentilla, Development of geopolymer-based materials from a ternary blend of red mud, rice husk ash and diatomaceous earth using the statistical mixture design modeling approach. J. Philip. Inst. Chem. Eng. **15**(2), 33–45 (2015)
21. J.M. Mejía, R.M. Mejía, C. Montes, Rice husk ash and diatomaceus earth as a source of silica to fabricate a geopolymeric binary binder. J. Clean. Prod. **118**, 133–139 (2016). https://doi.org/10.1016/j.jclepro.2016.01.057
22. C. Bagci, G.P. Kutyla, W.M. Kriven, Fully reacted high strength geopolymer made with diatomite as a fumed silica alternative. Ceram. Int. **43**(17), 14784–14790 (2017). https://doi.org/10.1016/j.ceramint.2017.07.222
23. A. Font, L. Soriano, L. Reig, M.M. Tashima, M.V. Borrachero, J. Monzó, J. Payá, Use of residual diatomaceus earth as a silica source in geopolymer production. Mater. Lett. **223**, 10–13 (2018). https://doi.org/10.1016/j.matlet.2018.04.010
24. L. Verdolotti, B. Liguori, I. Capasso, A. Errico, D. Caputo, M. Lavorgna, S. Iannace, Synergistic effect of vegetable protein and silicon addition on geopolymeric foams properties. J. Mater. Sci. **50**, 2459–2466 (2015). https://doi.org/10.1007/s10853-014-8801-3
25. L. Verdolotti, S. Iannace, M. Lavorgna, R. Lamanna, Geopolymerization reaction to consolidate incoherent pozzolanic soil. J. Mater. Sci. **43**, 865–873 (2008). https://doi.org/10.1007/s10853-007-2201-x
26. H. Aguiar, J. Serra, P. González, B. León, Structurak study of sol-gel silicate glasses by IR and Raman spectroscopies. J. Non-Cryst. Solids **355**(8), 475–480. https://doi.org/10.1016/j.jnoncrysol.2009.01.010
27. B. Liguori, I. Capasso, V. Romeo, M.D. Auria, M. Lavorgna, M. Caputo, S. Iannace, L. Verdolotti, Hybrid geopolymeric foams with diatomiteaddition: effect on chemico-physical properties. J. Cell. Plast. **53**(5), 1–12 (2017). https://doi.org/10.1177/0021955X1769509

Chapter 36
Effect of Reaction Temperature on Zeolite Synthesised from Oil Palm Ash

Faizul Che Pa and Muhammad Faheem Mohamad Tahir

Abstract The zeolite synthesis using silica extracted from oil palm ash (OPA) was carried out under hydrothermal conditions by activation with sodium hydroxide (NaOH) solution. The parameters involved in this study were reaction temperatures (40, 60, 80, and 100 °C), while the solid/liquid ratio, alkali concentration, and reaction time used were 0.2:1, 2 M, and 12 h, respectively. The properties of zeolites synthesised were characterised using X-ray fluorescence (XRF) and X-ray diffraction (XRD). Cation exchange capacity (CEC) values of the zeolite produced were also determined. Phase analysis by XRD shows that zeolite gismondine ($CaAl_2Si_2O_8{\cdot}4(H_2O)$) was produced after hydrothermal activation of treated OPA.

Keywords Palm ash · Zeolite · Silica · Hydrothermal

36.1 Introduction

Silica is one of the raw materials used in zeolite synthesis. Silica is the main component of palm ash. Using silica extracted from oil palm ash as the forerunner in synthesising zeolites is motivated by various elements, such as the abundance in quantity and lower production cost.

The synthesis of zeolites under hydrothermal treatment is affected by the treatment conditions, such as the hydrothermal temperature, the reaction time, and the concentration of alkaline solution during zeolite synthesis [1, 2]. This research investigates the reaction temperature's effect on zeolite synthesis using silica extracted from oil palm ash.

F. Che Pa (✉) · M. F. Mohamad Tahir
Center of Excellence Geopolymer and Green Technology (CEGeoGTech), Universiti Malaysia Perlis, Kompleks Pusat Pengajian Jejawi 2, 02600 Jejawi, Arau, Perlis, Malaysia
e-mail: faizul@unimap.edu.my

Faculty of Chemical Engineering and Technology, Universiti Malaysia Perlis, Kompleks Pusat Pengajian Jejawi 2, 02600 Jejawi, Arau, Perlis, Malaysia

M. A. A. Mohd Salleh et al. (eds.), *Proceedings of the Green Materials and Electronic Packaging Interconnect Technology Symposium*, Springer Proceedings in Physics 289,
https://doi.org/10.1007/978-981-19-9267-4_36

Table 36.1 Chemical composition of OPA

Compound	SiO_2	Al_2O_3	P_2O_5	MgO	K_2O	CaO	Fe_2O_3	Others
Composition (wt%)	45.50	5.40	5.38	3.20	23.30	12.80	3.26	1.16

Fig. 36.1 SEM photomicrographs of ground oil palm ash (Magnification 500×)

36.2 Materials and Methods

36.2.1 Oil Palm Ash (OPA)

The extraction of oil from oil palm fruitlets requires the separation of the fruitlets from fruit bunches before further processing. Palm oil mills often use empty fruit bunches as boiler fuel to generate steam for palm oil extraction and power generation [3]. This combustion process will produce oil palm ash (OPA). OPA is being disposed of as a landfill due to its limited uses, which could lead to environmental issues. OPA has about 45.5% silicon dioxide (SiO_2), as shown in Table 36.1.

The morphological investigation using a scanning electron microscope (SEM) of the ground OPA is shown in Fig. 36.1. It has a porous structure. Its main components are in an irregular and angular form, with a sizable fraction showing cellular textures.

36.2.2 Pre-Treatment Process

The OPA from the palm oil mill had undergone citric acid (3%) leaching at a temperature of 70 °C for 60 min to remove metal oxide impurities. After the leaching process, the solids (recovered by filtration) were dried at 60 °C in the oven for 60 min and then combusted at 800 °C for 30 min in the furnace. The combustion temperature of

800 °C was used to prevent the crystallisation of amorphous silica contained in the ash [4].

36.2.3 Hydrothermal Treatment

The most common method used to synthesise zeolites involves a hydrothermal process. In every hydrothermal process, 10 g of dry sample (2 g kaolin + 8 g treated OPA) was mixed and stirred with sodium hydroxide solution (alkali sources for zeolite synthesis) in a water bath. The concentration and reaction time were 2 M and 12 h, respectively. Experiments were conducted with various reaction temperatures (40, 60, 80, and 100 °C).

After the hydrothermal treatment process, the solid products obtained were recovered by filtration. Then, washed with distilled water (to remove all traces excess of alkali) and dried in the oven overnight at 60 °C. The chemical composition of the raw materials and synthesised products was determined using the X-ray fluorescence (XRF) spectrometry machine (Rigaku-R1X3000), and mineral phases were determined using the X-ray diffraction (XRD) machine (Philip PW 1729). Cation exchange capacity (CEC) values of the zeolite produced were also determined using the ammonium replacement method.

36.3 Result and Discussions

36.3.1 Hydrothermal Treatment

The precursor elements for the alkaline hydrothermal conversion into zeolites were based on the Al and Si content of the starting sample. Table 36.2 shows the XRF result of treated oil pam ash. It indicates that no Al element is absent in the treated oil palm ash.

Kaolin is added into the treated oil palm ash to add some of the Al elements so that zeolite synthesis can be performed. Kaolin with chemical composition $Al_2H_4O_9Si_2$ has been used as Si and Al sources for synthesising various zeolites, such as Linde Type A, X, Y, P, Na, chabazite, and faujasite [5].

During the hydrothermal process, the Al^{3+} from kaolin is able to occupy the position or replace Si^{4+} in the centre of the tetrahedron of four oxygen atoms due to its small size. The successful conversion of the silica extracted from oil palm ash

Table 36.2 XRF result of treated oil palm ash

Compound	SiO_2	Al_2O_3	K_2O	CaO	Fe_2O_3	Others
Concentration (wt%)	92.00	–	3.56	0.85	3.07	0.52

into zeolite under alkaline hydrothermal treatment was testimony by the detection of elements of Si and Al in the synthesised samples.

36.3.2 *Effect of Reaction Temperature*

The reaction temperature is essential in attaining optimum conditions in hydrothermal zeolite formation. A series of experiments were conducted with various reaction temperatures (40, 60, 80, and 100 °C) to determine its effect on the product.

The XRD patterns of zeolite crystals crystallised at different hydrothermal temperatures for 12 h are shown in Fig. 36.2. No zeolite peaks were detected for activation at 40 °C (Fig. 36.2a). Figure 36.2b shows a few weak synthetic zeolite peaks observed in the spectrum. It seemed that for activation of less than 60 °C, only parts of the silica extracted reacted with the sodium ions, and the reaction continued with increasing temperatures.

The crystal of zeolite gismondine ($CaAl_2Si_2O_8{\cdot}4H_2O$) grows the most after hydrothermal treatment at 80 °C. It seemed that for activation at a temperature of 60 °C and below, only part of the extracted silica reacted with the sodium ions (alkaline), and the reaction continued with increasing temperatures.

The highest level of zeolite formation achieved is from a reaction at a high temperature between NaOH and extracted silica, where species are produced that can easily dissolve in solution. Comparing Fig. 36.2b with Fig. 36.2c and Fig. 36.2d, it was demonstrated that the main peak of zeolite gismondine ($2^{\theta} = 26.71°$) is continuously increasing with increasing reaction temperature. Si^{4+} would be eluted from an amorphous phase in silica extracted while Al^{3+} from kaolin to form zeolite.

Querol et al. [6] and Murayama et al. [7] reported the successful zeolite formation of fly ash using 1–2 M NaOH solutions. This fact indicated that short-term activation could be achieved only when higher alkali concentrations and/or higher temperatures are applied. Crystallisation growth depends on the reaction temperature. The increase in reaction temperature contributed to the increased energy supply for the nucleus to achieve the activation energy for crystal growth, which increased the degree of crystallisation [8].

The growth of crystallisation depends on the reaction temperature. The increase in reaction temperature contributed to the increase in the energy supply of the nucleus to attain the activation energy for crystal growth, which increased the degree of crystallisation [8].

Table 36.3 shows the XRF result of the synthesised sample after hydrothermal treatment with different reaction temperatures. The reduction of CaO concentration is expected due to the dissolvable characteristic of the calcium oxide in an alkaline solution. During the alkaline hydrothermal treatment, the aluminium ions, Al^{3+}, can occupy the position or replace silicon ions, Si^{4+}, in the centre of the tetrahedron of four oxygen atoms due to their small size.

The zeolite synthesised was also evaluated in terms of cation exchange capacity (CEC) value. The CEC values obtained at different reaction temperatures are shown

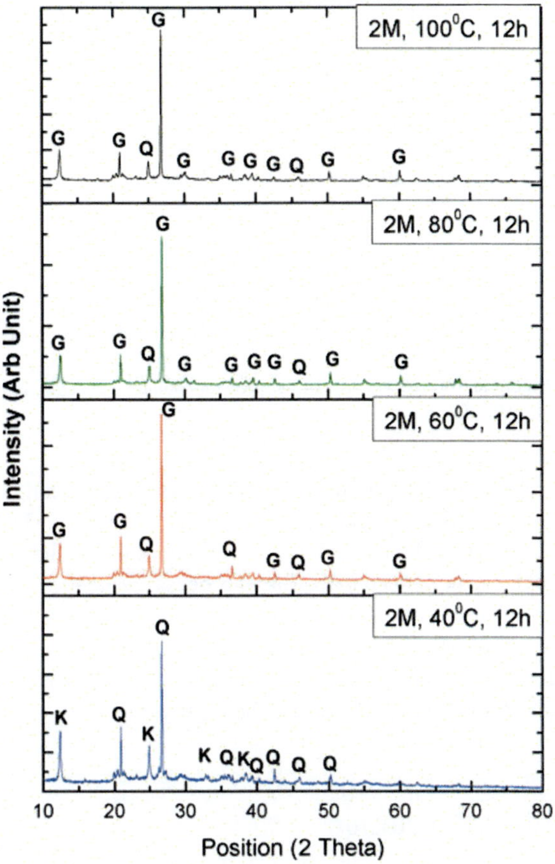

Fig. 36.2 XRD patterns obtained at different hydrothermal reaction temperatures (Q = Quartz, G = Gismondine, K = Kalsilite)

Table 36.3 XRF result of the synthesised sample after hydrothermal treatment with different reaction temperatures

Compound	40 °C	60 °C	80 °C	100 °C
SiO_2	52.70	59.00	63.50	66.80
Al_2O_3	16.00	11.10	15.70	16.75
K_2O	0.64	3.79	1.52	1.35
CaO	19.90	16.30	11.20	9.90
Fe_2O_3	4.17	3.62	3.05	2.98

in Fig. 36.3. CEC values obtained after 12 h of the reaction process with 2 M NaOH concentration increased with temperature. Even though activation at 60 °C is low enough to promote the formation of zeolite gismondine, the highest CEC value of 100 meq/100 g was reached at a temperature of 80 °C after an activation period of 12 h.

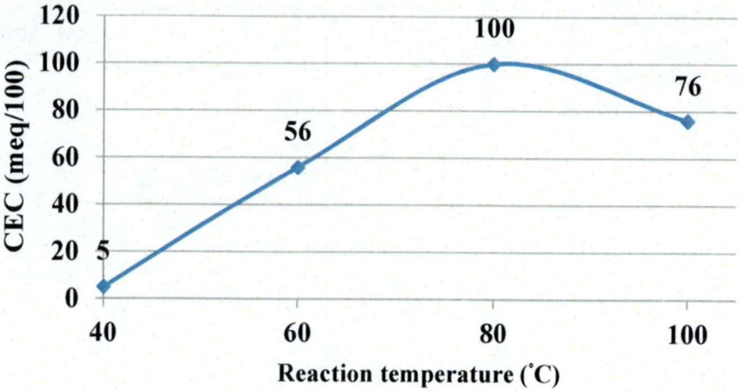

Fig. 36.3 The CEC value of the products obtained at different reaction temperatures of 40, 60, 80, and 100 °C

A higher temperature of 100 °C was also investigated, but the CEC value of the product produced at this temperature was reduced significantly. A similar result was reported by Huang et al. [9]. This reduction was due to a desilication process (similar to that experienced in the Bayer process during the digestion of bauxites in NaOH), in which the Si and Al dissolved by the alkali reaction were re-precipitated and formed a sodium-aluminium–silicate "desilication" product (DSP). DSP precipitates were formed on the surface of the silica extracted, and undisturbed interfered with the alteration process, lowering the CEC values.

36.4 Conclusion

XRD shows that zeolite gismondine ($CaAl_2Si_2O_8{\cdot}4(H_2O)$) was produced. It shows that reaction temperature affects the formation of zeolite gismondine. It seems that 80 °C is the suitable temperature to promote the formation of zeolite gismondine in terms of CEC value.

Acknowledgements The authors would like to **acknowledge** the Centre of Excellence Geopolymer and Green Technology (CeGeoGTech) and the Faculty of Chemical Engineering **and** Technology, Universiti Malaysia Perlis, for giving us the facilities to run this research work.

References

1. A. Shoumkova, V.B. Stoyanova, Zeolites formation by hydrothermal alkali activation of coal fly ash from thermal power station 'Maritsa 3', Bulgaria. Fuel **103**, 533–541 (2013). https://doi.org/10.1016/j.fuel.2012.07.076
2. Y. Liu, Q. Luo, G. Wang, X. Li, P. Na, Synthesis and characterisation of zeolite from coal fly ash. Mater. Res. Express **5**(5), 055507 (2018). https://doi.org/10.1088/2053-1591/aac3ae

3. C.P. Faizul, C. Abdullah, B. Fazlul, Review of extraction of silica from agricultural wastes using acid leaching treatment. Adv. Mat. Res. **626**, 997–1000 (2013). https://doi.org/10.4028/www.scientific.net/AMR.626.997
4. C.P. Faizul, C. Abdullah, B. Fazlul, Palm ash as an alternative source for silica production. Key Eng. Mater. **673**, 3–20 (2016). https://doi.org/10.1051/matecconf/20167801062
5. M. Gougazeh, J.-C. Buhl, Synthesis and characterisation of zeolite A by the hydrothermal transformation of natural Jordanian kaolin. J. Assoc. Arab Univ. Basic Appl. Sci. **15**(1), 35–42 (2014). https://doi.org/10.1016/j.jaubas.2013.03.007
6. X. Querol, N. Moreno, J.C. Umaña, A. Alastuey, E. Hernández, A. López-Soler, F. Plana, Synthesis of zeolites from coal fly ash: an overview. Int. J. Coal Geol. **50**(3), 413–423 (2002). https://doi.org/10.1016/S0166-5162(02)00124-6
7. N. Murayama, H. Yamamoto, J. Shibata, Mechanism of zeolite synthesis from coal fly ash by alkali hydrothermal reaction. Int. J. Miner. Process. **64**, 1–17 (2002). https://doi.org/10.1016/S0301-7516(01)00046-1
8. A.R. Siti Haslina, Y. Leny, R. Zainab, Rapid synthesis and characterization of nano sodalite synthesized using rice husk ash. Malays. J. Anal. Sci. **16**(3), 247–255 (2012)
9. G.U. Ryu, G.M. Kim, H.R. Khalid, H.K. Lee, The effects of temperature on the hydrothermal synthesis of hydroxyapatite-zeolite using blast furnace slag. Materials **12**(13), 2131 (2019). https://doi.org/10.3390/ma12132131

Chapter 37
Enhanced Mechanical and Thermal Properties of Acrylonitrile Butadiene Rubber Compounds (NBR) by Using High-Density Polyethylene (HDPE)

Huynh Khanh Tuong and Cao Xuan Viet

Abstract Nowadays, acrylonitrile butadiene rubber (NBR) has been applied to various products and equipment to serve industrial production and human life. These include products that are resistant to oil and chemical environments, especially with high wear resistance. In this study, composites of acrylonitrile butadiene rubber/high-density polyethylene (NBR/HDPE) reinforced with silica (60 phr) were prepared. The effect of HDPE content on the mechanical and thermal properties of silica-filled NBR/HDPE composites was investigated. The results show that by increasing the HDPE content up to 30 phr, the mechanical properties of compounds such as tensile strength, elongation, hardness, and abrasion resistance increased significantly. Scanning electron microscope (SEM) shows good dispersion of HDPE into the rubber matrix at low HDPE content. DSC, TGA, and DMTA results suggest the thermal enhancement by the addition of HDPE.

Keywords Nitrile butadiene rubber · HDPE · Composites

37.1 Introduction

Nitrile rubber, known as acrylonitrile butadiene rubber (NBR), is a synthetic rubber from acrylonitrile and butadiene [1], created by emulsion copolymerization, with basic stages including polymerization of NBR in the form of NBR latex, coagulate the latex, and dry the final product [2]. Acrylonitrile participates in polymerization reaction with butadiene in the presence of a redox catalyzed system of potassium persulfate and triethanolamine to produce two different products: Nitrile rubber and 4-ciano cyclohexene [2, 3]. The properties of nitrile rubber depend on the acrylonitrile

H. K. Tuong · C. X. Viet (✉)
Department of Polymer Materials, Faculty of Materials Technology, Ho Chi Minh City University of Technology (HCMUT), 268 Ly Thuong Kiet Street, District 10, Ho Chi Minh City, Vietnam
e-mail: caoxuanviet@hcmut.edu.vn

Vietnam National University Ho Chi Minh City, Linh Trung Ward, Ho Chi Minh City, Vietnam

M. A. A. Mohd Salleh et al. (eds.), *Proceedings of the Green Materials and Electronic Packaging Interconnect Technology Symposium*, Springer Proceedings in Physics 289,
https://doi.org/10.1007/978-981-19-9267-4_37

content in the macromolecular chain as well as on the molecular weight of the material. Currently, there have been many studies to improve the mechanical properties of nitrile rubber, especially high abrasion resistance, in many different ways [4–8]. But the most prominent method is the surface modification of silica, mixed with thermoplastics to create thermoplastic rubber materials. High-density polyethylene (HDPE) is a semi-crystalline material with outstanding chemical resistance, durability, and low cost. Xueshen Liu et al. [9], studied the influence of particle size and dispersion of SiO_2 on nitrile rubber. By FTIR, SEM, abrasion measurement, the results showed that the grain size of SiO_2 has a great influence on its dispersion in nitrile rubber, and the surface modification of SiO_2 by TESPT (Silane 69) has can greatly improve its dispersion. However, the abrasion resistance is still not high because silica is easy to agglomerate. Mohammad Barghamadi et al. [10], have shown that a blend of NBR and polyvinyl chloride (PVC) is used due to its unique oil resistance and high strength in combination. The two fillers used, graphene nanosheet (GNP) and organoclay montmorillonite (OMMT) on vulcanization, recovery, and mechanical properties of NBR/polyvinyl chloride (NBR/PVC) nanocomposite rubber were investigated.

Based on the combination of rubber and thermoplastic to create a material with the advantages of each component—elasticity like rubber and machinability like thermoplastic—this material is called thermoplastic rubber (TPE). A great number of research papers and books were published on this issue [11–15]. Fabrication of thermoplastic rubber (TPE) materials from a mixture of NBR and HDPE rubber contributes to the development of manufacturing rubber rice rollers products. Currently, the rice rollers used in the rice milling machine are available on the market with many different brands, sizes, and prices [16, 17]. However, there are also some drawbacks such as a short time of use due to limited wear resistance, high cost, easy surface cracking during use, and rice grain. Therefore, the aim of this research is to investigate the enhancement of NBR compound with HDPE addition for rice roller application to overcome aforementioned limitations. The mechanical and thermal properties of these composites were evaluated.

37.2 Methodology

37.2.1 Materials

Acrylonitrile–butadiene rubber (NBR 35L, *acrylonitrile, 35 wt%*) provided by Kumho–Korea, high-density polyethylene (HDPE) provided by Dow Plastic (USA) (density is 0.95 g/cm^3), Silica Tokusil 185G with 93% SiO_2 and 1% max Al_2O_3 + Fe_2O_3 (Thailand), and silane 69 with 22.5% sulfur (China).

Stearic acid, PEG 4000, SPP antioxidant, Tetramethylthiuram disulfide (TMTD), N-Cyclohexyl-2-benzothiazolesulfenamide (CBS) as accelerators, and sulfur were

Table 37.1 Formulation of silica-filled NBR/HDPE composites

Ingredients	phr
NBR	100
HDPE	0/20/30/40
Silica	60
Silane 69	6
Stearic Acid	2
SPP Antioxidant	1
PEG	8
TMTD	0.5
CBS	1
S	2.2

obtained from the commercial suppliers. The formulation of NBR compounds is shown in Table 37.1.

37.2.2 Preparation and Curing of NBR/HDPE/Silica Composites

Ingredients were first blended in internal mixer and followed by further mixing with curing agents (accelerators and sulfur) on the two-roll mill. After that, the sample was stabilized for 24 h prior to rheometer measurement to determine the curing parameters. Compounds were pressed on a hydraulic press with temperature of 160 °C, pressure of 50kgf/cm^2, and curing time determined from the vulcanization curve. Samples were then cut into shapes for various testing and characterization.

37.2.3 Mechanical Properties

- Abrasion resistance: measured by AKRON method according to JIS K 6264-1: 2005 standards.
- Hardness: measured by using Shore hardness scale (Shore A).
- Aging resistance: the aging test was conducted at 70 °C for 24 h according to TCVN 2229-77 standards. Samples after the aging test were subjected to tensile measurement.
- Tensile measurement: tested under tension force at a crosshead speed of 500 mm/min using M350-5 testometric according to ASTM D412 standards.

37.2.4 Morphology and Thermal Properties

The morphology characteristic of the composite was studied by scanning electron microscope (SEM, HITACHI-S-4800). Differential scanning calorimetry (DSC) and Thermogravimetric analysis (TGA) were performed on the integrated TGA/DSC Meter—Toledo in a nitrogen atmosphere, room temperature to 600 °C, heating rate 10 °C/min. Dynamic mechanical thermal analysis (DMTA) was performed on MCR302 instrument, Anton Paar from −100 to 200 °C at heating rate: of 3 °C/min, and frequency of 1 Hz.

37.3 Result and Discussions

37.3.1 Mechanical Properties

The variation in mechanical properties with the addition of HDPE to NBR in the silica filled NBR/HDPE composites is shown in Table 37.2, with σ_u: tensile strength (MPa), ε: elongation at break (%).

The Shore A hardness value increases by increasing HDPE content in the NBR/HDPE composites. Specifically, when increasing the HDPE content, the hardness also increased linearly from 75 to 96 Shore A, increasing to 28%. The sample without HDPE has the lowest hardness (75 Shore A), and the sample with 40 HDPE phr has the highest hardness (96 Shore A). The increase in stiffness is due to an increase in the proportion of HDPE; HDPE is served as the hard segment, while rubber is the soft matrix; therefore, the hardness of the mixture gradually increases.

It can be observed that the tensile strength of the composites increases with increasing HDPE content. When increasing the HDPE content from 0 to 30 phr, the tensile strength at break also increases from 17.68 to 20.09 MPa, but at HDPE 40 phr content, the tensile strength goes down to 8.83 MPa. The sample with HDPE content of 30 phr has the highest tensile strength at break (20.09 MPa). It can be seen that the tensile strength of the composites increases with HDPE addition to some extent. At high HDPE content, it is not well dispersed, remained in large size, and prevents filler encapsulation leading the tensile strength to decrease drastically

Table 37.2 The mechanical properties of silica-filled NBR/HDPE composites

Compound	Hardness (Shore A)	Akron Abrasion (cm^3/1.61km)	σ_u (MPa)	ε (%)
NBR/HDPE (100/0)	75±0.2	0.42±0.01	17.68±0.46	572.43±15.33
NBR/HDPE (80/20)	85±0.25	0.24±0.01	19.45±0.64	301.63±15.02
NBR/HDPE (70/30)	90±0.19	0.19±0.02	20.09±0.45	162.77±7.94
NBR/HDPE (60/40)	96±0.25	0.17±0.02	8.83±0.38	20.5±4.72

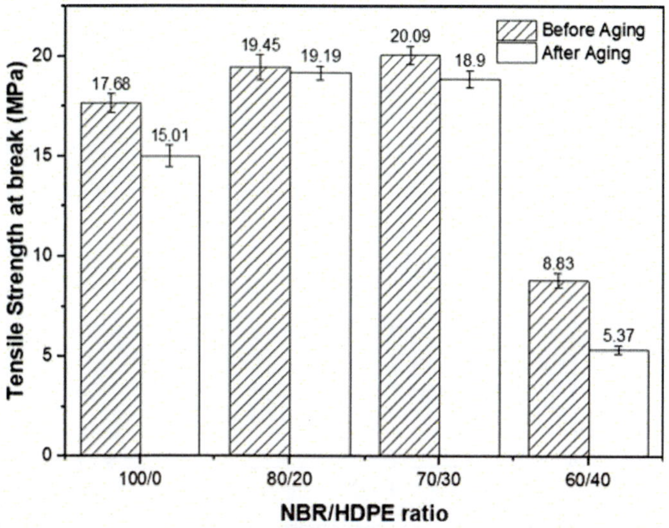

Fig. 37.1 Dependence of tensile strength at break of the composites on HDPE contents after thermal aging

[4]. Figures 37.1 and 37.2 display the tensile results before and after the aging test. Tensile properties of the composite remain quite well at the NBR/HPDE (70/30) ratio. However, at 40 phr HDPE, the tensile strength and elongation at break reduce due to the low compatibility of the two components NBR and HDPE can cause larger phase separation under high-temperature conditions.

Increasing the HDPE content from 0 to 40 phr, the Akron abrasion decreased linearly from 0.42 to 0.17 (cm^3/1.61 km) (147.1%). The higher HDPE contents is, the lower values of Akron abrasion of the composites shows. This result indicates that HDPE can significantly enhance the wear resistance of NBR [1]. The improvement in wear resistance of NBR compound by incorporation of HDPE is due to the excellent wear resistance of HDPE.

37.3.2 Morphology of the Composites

The SEM images of silica-filled NBR/HDPE composites are shown in Fig. 37.3. NBR sample (a) has a smooth surface. Silica particles can be observed on the NBR matrix. At 30 phr, HDPE (Fig. 37.3b) plastic phase is well dispersed into the rubber phase [2]. When HDPE content is 40 phr, the phase separation surface appears clearer leading to a drastic reduction in tensile strength, and the elongation of the composite material.

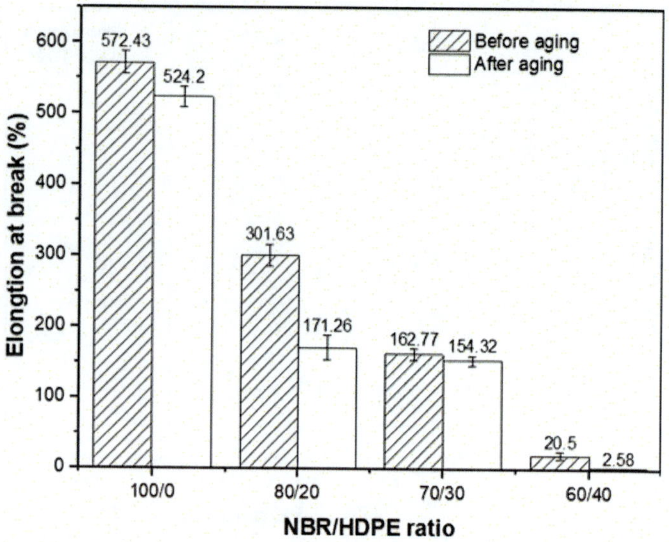

Fig. 37.2 Dependence of elongation at break of the composites on HDPE contents after thermal aging

37.3.3 Thermal Analysis

Based on mechanical results, the NBR/HDPE (70/30) composites with the best overall performance in terms of high hardness, Akron abrasion resistance, and aging resistance were selected to compare with NBR compound (100/0) for thermal properties.

The DSC and TGA results of the NBR/HDPE (100/0) and the NBR/HDPE (70/30) composites are shown in Figs. 37.4 and 37.5. For NBR/HDPE (70/30) composite, a melting temperature of HDPE at 138.81 °C is clearly observed. The starting temperature of decomposition is about 280 °C, which is earlier than the NBR/HDPE (100/0). It can be found that the decomposition of the HDPE takes place in the range of 280–390 °C, followed by the decomposition of NBR, plasticizers, and other substances [4]. DSC and TGA results reveal that the composite had a lower onset temperature, but the maximum decomposition rate is lower than that of the NBR sample.

The DMTA curves of the NBR/HDPE (100/0) and the NBR/HDPE (70/30) composites are shown in the Figs. 37.6, 37.7 and 37.8.

In Fig. 37.6, the storage modulus curve of the NBR/HDPE (100/0) has three distinct regions. The high modulus region at low temperature, the transition region where the modulus decreases sharply with temperature (about −25 °C), and the strong temperature transition of E′ indicate a transition state from the glass region to the highly soft region. The temperature at which the modulus drops the most is considered the glass transition temperature of (about −14.6 °C). The molecular chains begin to shift in region 3 (high-soft region). In Fig. 37.7, the storage modulus

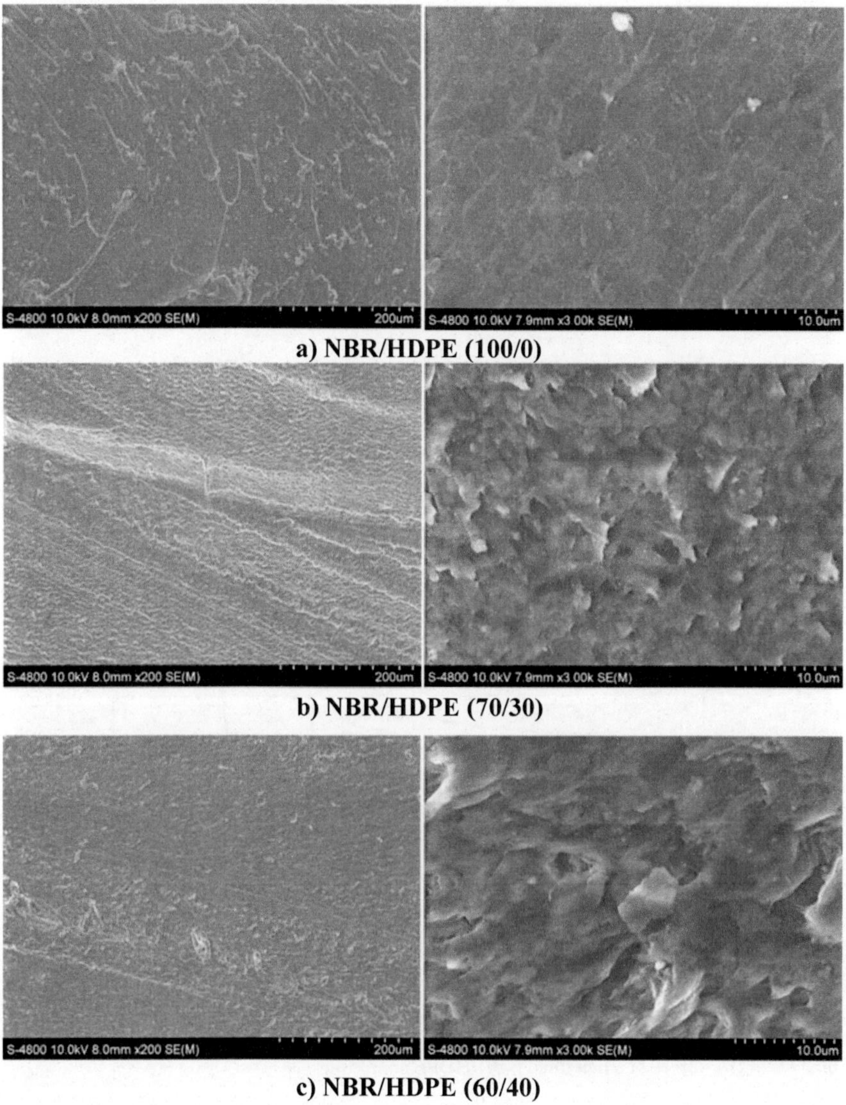

Fig. 37.3 SEM images of silica-filled NBR/HDPE composites at 200X and 3000X: **a** NBR/HDPE (100/0); **b** NBR/HDPE (70/30) composite; **c** NBR/HDPE (60/40) composite

curve of the NBR/HDPE (70/30) composites has four distinct regions due to the presence of HDPE (region 4). The glass transition temperature of NBR is slightly increased (−11.48 °C) by the addition of HDPE.

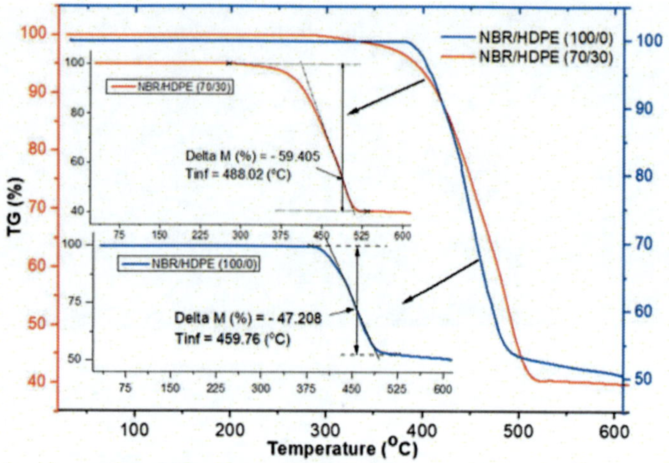

Fig. 37.4 DSC curves of silica-filled NBR/HDPE composites

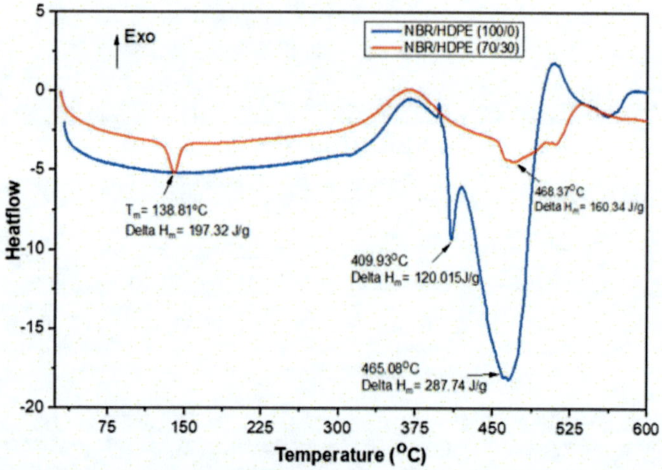

Fig. 37.5 TGA curves of silica filled NBR/HDPE composites

Figure 37.8 shows the storage modulus curve (E′) of the NBR compound and composite. In the NBR compound, the storage modulus drops drastically at glass transition temperature of NBR, then remained almost unchanged. In the temperature from 25 to 100 °C, the storage modulus of the NBR/HDPE (70/30) composites has a higher value than that of the NBR compounds which indicates good mechanical properties of NBR/HDPE (70/30) composite can be obtained.

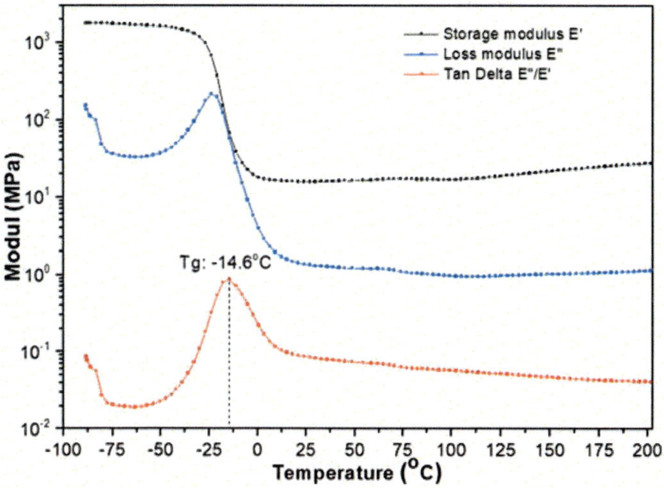

Fig. 37.6 DMTA curve of NBR compound (100/0)

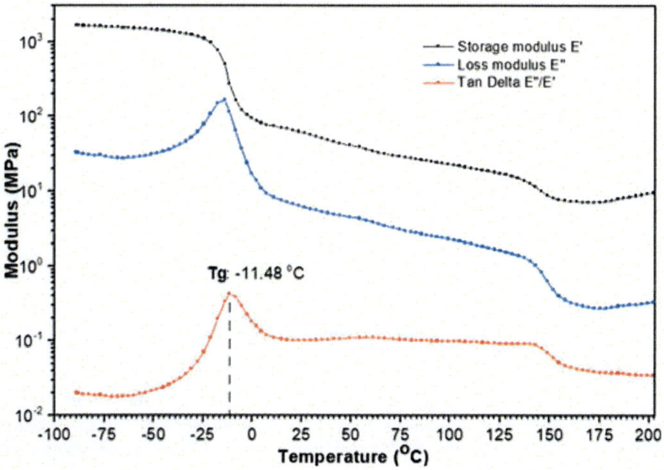

Fig. 37.7 DMTA curve of NBR/HDPE (70/30) composite

37.4 Conclusion

Based on the results, the mechanical properties of NBR compounds significantly improved with HDPE addition. Some conclusions could be drawn as follows:

- Silica-filled NBR/HDPE composite with 30 phr HDPE has the best overall mechanical properties with tensile strength of 20.09 MPa, a hardness of 90 Shore A, and an abrasion resistance of 0.19 cm^3/1.61 km.

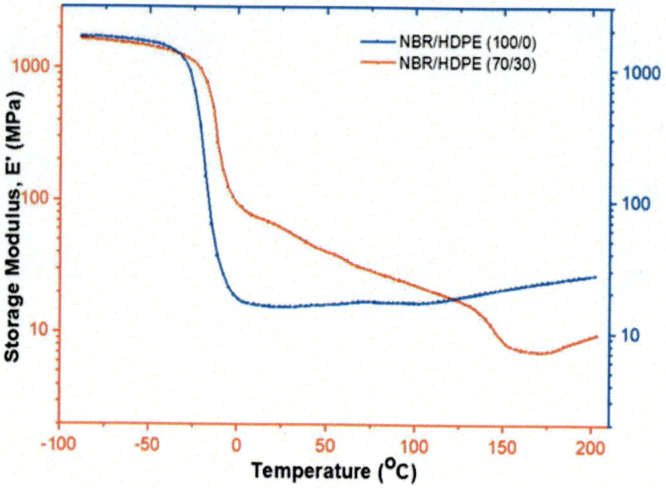

Fig. 37.8 Storage modulus E′ of the NBR/HDPE (100/0) and the NBR/HDPE (70/30) composite

- The maximum decomposition rate in NBR/HDPE (70/30) composite is lower than that of the NBR sample.
- DMTA shows that the storage modulus of NBR/HDPE (70/30) composite is higher than that of the NBR compound.

Acknowledgements We acknowledge the support of time and facilities from Ho Chi Minh City University of Technology (HCMUT), VNU-HCM for this study.

References

1. J. George, K.T. Varughese, S. Thomas, Dynamically vulcanised thermoplastic elastomer blends of polyethylene and nitrile rubber. Polymer **41**(4), 1507–1517 (2000). https://doi.org/10.1016/S0032-3861(99)00302-X
2. J.F. Rabek, *Experimental Methods in Polymer Chemistry: Physical Principles and Application* (A Wiley—Interscience Publication, 1983), pp. 22–23
3. P.R. Sruthi, S. Anas, An overview of synthetic modification of nitrile group in polymers and applications. J. Polym. Sci. **58**(8), 1039–1061 (2020). https://doi.org/10.1002/pol.20190190
4. K. Ahmed, An investigation on chloroprene-compatibilized acrylonitrile butadiene rubber/high density polyethylene blends. J. Adv. Res. **6**, 811–817 (2015). https://doi.org/10.1016/j.jare.2014.06.003
5. F.K. El-Nemr, Effect of different curing systems on the mechanical and physical–chemical properties of acrylonitrile butadiene rubber vulcanisates. Mater. Des. **32**(6), 3361–3369 (2011). https://doi.org/10.1016/j.matdes.2011.02.010
6. N. Murugan, P. Amrishkumar, G.B. Nando, N.K. Singha, Thermoplastic elastomer blend based on EMA and NBR; optimization of process parameters. J. Appl. Polym. Sci. **137**(27), 48900 (2020). https://doi.org/10.1002/app.48900

7. Y. Guo, Z. Zhang, Z. Cao, D. Wang, Wear behavior of hollow glass beads (HGB) reinforced nitrile butadiene rubber: effects of silane coupling agent and filler content. Mater. Today Commun. **19**, 366–373 (2019). https://doi.org/10.1016/j.mtcomm.2019.03.003
8. M.M. Salehi, T. Khalkhali, A.A. Davoodi, The physical and mechanical properties and cure characteristics of NBR/silica/MWCNT hybrid composites. Polym. Sci. Ser. A **58**(4), 567–577 (2016). https://doi.org/10.1134/S0965545X16040131
9. X. Liu, X. Zhou, C. Yang, J. Huang, F. Kuang, H. Wang, Study on the effect of particle size and dispersion of SiO_2 on tribological properties of nitrile rubber. Wear **6**, 460–461 (2020). https://doi.org/10.1016/j.wear.2020.203428
10. M. Barghamadi et al., Effects of two types of nanoparticles on the cure, rheological, and mechanical properties of rubber nanocomposites based on the NBR/PVC blends. J. Appl. Polym. Sci. **136**(25), 47550 (2019). https://doi.org/10.1002/app.47550
11. S.M.R. Paran, G. Naderi, M.H.R. Ghoreishy, Microstructure and mechanical properties of thermoplastic elastomer nanocomposites based on PA6/NBR/HNT. Polym. Compos. **38**, 451–461 (2017). https://doi.org/10.1002/pc.23936
12. P. Mahallati, A. Arefazar, G. Naderi, Thermoplastic elastomer nanocomposites based on PA6/NBR. Int. Polym. Proc. **25**(2), 132–138 (2010). https://doi.org/10.3139/217.2311
13. Z. Asgarzadeh, G. Naderi, Morphology and properties of unvulcanized and dynamically vulcanized PVC/NBR blend reinforced by graphene nanoplatelets. Int. Polym. Proc. **33**(4), 497–505 (2018). https://doi.org/10.3139/217.3515
14. R. Syafri, I. Ahmad, I. Abdullah, Effect of rice husk surface modification by LENR the on mechanical properties of NR/HDPE reinforced rice husk composite. Sains Malaysiana **40**(7), 749–756 (2011)
15. T. Sharika et al., Excellent electromagnetic shield derived from MWCNT reinforced NR/PP blend nanocomposites with tailored microstructural properties. Compos. B Eng. **173**, 106798 (2019). https://doi.org/10.1016/j.compositesb.2019.05.009
16. A. Baker, R.S. Dwyer-Joyce, et al., Effect of different rubber materials on husking dynamics of paddy rice. J. Eng. Tribol. **226**(6), 516–528 (2012). https://doi.org/10.1177/1350650111435601
17. A.F. Adisa, K.C. Mamah, A.A. Aderinlewo, S.O. Ismaila, Effectiveness of industrial rubber as roller material for rice processing machine. Agric. Nat. Resour. **54**, 98–104 (2020)

Chapter 38
Evaluation of Petrographical Characteristics of Deteriorated Cement Concrete Containing Potential ASR

I. Ibrahim, A. Rahim, K. Ramanathan, R. A. Abdullah, and W. M. W. Ibrahim

Abstract Understanding the alkali–silica reaction (ASR) is a very important step in order to measure the structural integrity of a certain building and establishment. ASR can affect the life expectancy of a structure, making it deteriorate much faster than how it is supposed to be. The purpose of this research is to focus on the characteristics of deteriorated cement concrete that is suspected to be containing ASR and its effect of its chemical composition and relative strength. By using the petrographic analysis method, ASR can be found in the suspected samples and it can be further researched by running the X-ray fluorescent spectrometry (XRF) test to examine the mineral content of the respective samples. Furthermore, the relative strength of the samples can be determined by using the Schmidt Rebound Hammer test. Overall, several minerals in the samples such as 〖SO〗 _3 and CaO were found to be affecting the Estimated Concrete Strength (f_cu) of the samples the most. This study definitively answers the question regarding the correlation between the chemical composition of concrete and its Estimated Concrete Strength.

Keywords ASR · XRF · Schmidt rebound hammer · Estimated concrete strength

38.1 Introduction

Alkali–silica reaction (ASR) had been found in the USA in the early 1940s. Since then, a lot of concrete structures from all around the world have been verified to be deteriorated due to ASR, a very discreet internal–chemical reaction. ASR has been identified to be the chemical–physical expansive reaction between mineral phases in

I. Ibrahim · A. Rahim (✉) · K. Ramanathan · R. A. Abdullah
School of Civil Engineering, Faculty of Engineering, Universiti Teknologi Malaysia, 81310 Johor Bahru, Malaysia
e-mail: afikahrahim@utm.my

W. M. W. Ibrahim
Faculty of Mechanical Engineering Technology, Universiti Malaysia Perlis, 026000 Arau, Perlis, Malaysia

M. A. A. Mohd Salleh et al. (eds.), *Proceedings of the Green Materials and Electronic Packaging Interconnect Technology Symposium*, Springer Proceedings in Physics 289,
https://doi.org/10.1007/978-981-19-9267-4_38

aggregates and alkali hydroxides ($Na+$ $K+$ $OH-$) in mortar or concrete [1, 3, 4]. Moreover, a reactive form of silica is needed to be present in order for it to occur. After the alkali–silica have reacted with each other, a gel will be formed as it will draw water from the cement paste while swelling. Additionally, the products from this reaction have a great affinity with moisture causing it to swell and expand to create pressure and cracking of the surrounding paste and aggregate.

Seawater is an aqueous solution which contains aggressive constituents like sulfate, chloride, and magnesium ions. Exposing concrete to this solution for an extended amount of time will increase the possibility of the concrete to react with these constituents leading to its deterioration [5]. This mechanism is called chemical attack. In 1900s, concrete deterioration had been determined to be the natural phase of the aging of the concrete [6]. After that, a few effects of the deterioration of concrete had been found such as low performance and reduced service life. It had also been assessed that there may be some anomalies in some areas such as chloride content, compressive and tensile strength, water infiltration depth, and permeability that would lead to such effects.

38.2 Characteristics of Deteriorated Cement Concrete

Rapid worldwide urbanization is taking place in developing countries. This is the cause of the massive growth of population that demands buildings and infrastructures in the cities. In order to be able to fulfill those demands, the production of concrete must be parallel with the advancement of technology to make sure that the capability of the concrete to sustain load is on par with the rapid growth of the population [2]. Concrete is a composite material usually composed of cement, aggregates, and water. Aggregates in many mixtures of concrete have the highest percentage of the cement and considered as its major component. It should be neutral without any harmful substance that can change the desired properties of the concrete. The negligence to detect ASR from the cement concrete will result in the failure in structure without any prior preparations which will may proceed to casualties. Thus, the determination of the characteristics of deteriorated cement concrete containing potential ASR is very important in order to prevent any unfortunate accidents. This can be done with petrographical analysis in order to identify ASR and microcracks in the concrete. Moreover, other tests such as Schmidt Rebound Hammer test and XRF analysis also could be carried out to further compliment the petrographical analysis. In this study, potential ASR and microcracks in deteriorated cement concrete were identified by using petrographical analysis. Next, the relationship of salt water intrusion on potential ASR and microcracks has been found, and the relationship of chemical composition of the concrete and its relative strength was established. Petrographical analysis will be carried out in order to determine the characteristics of the ASR in the samples obtained. From the observation of the samples, the materials used for the concrete are assumed to be ordinary Portland cement.

38.3 Characteristics of Alkali–Silica Reaction (ASR) in Concrete

Contemporary concrete hugely includes fine and coarse aggregates mixed with ordinary Portland cement (OPC) with the presence of water. The OPC will be treated as an organic binder [1]. Aggregates are defined as particles that are over 4 mm as coarse aggregates, and a fraction below the value will be classified as fine aggregates. Moreover, the cement used will most probably be modified with the addition of some fly ash or many other pozzolanic materials. Very few cement mixes would be added some chemical admixtures in order to fix the settling characteristics of the concrete. Furthermore, the hydrated cement will also have its own ratio between water and cement (W/C ratio) to maximize the strength and workability of the concrete. As time goes on, the concrete will be exposed to many conditions such as aggressive chemicals such as sodium chloride; an alkali solution can be detrimental to the health and service life of the designed concrete.

The most popular form of alkali-aggregate reaction is the alkali–silica reaction (ASR). This is a generic term mainly used for reactions between certain mineral phases inside the aggregates and the alkaline concrete pore solution. ASR manifests itself at the aggregate and cement paste scales as local silica dissolution, the growth of microcracks, their filling with ASR materials, and the overall expansion of aggregates and eventually paste. Inside the meso-structure, microcracks are uniformly distributed. Their orientation is determined by the level of stress [7]. Crack patterns can range from randomly oriented to line up with the stress of the structure. This may cause a decrease in stiffness and tensile strength. The expansion will increase exponentially when the relative moisture becomes more than 80%. That is when the significant expansions are only observed [5]. A further hypothesis on this topic was also done about the swelling of the ASR after absorbing moisture [3]. Furthermore, ASR product can also be classified into two that are amorphous and crystalline. However, it was proven very recently that the crystalline ASR product cannot absorb water efficiently, causing swelling to be negligible for the crystalline ASR [9]. On the other hand, amorphous crystalline can absorb water very well causing serious swelling which will cause harm to the structure.

Petrographic analysis is the first step of examining the presence of ASR in a cement concrete. This process generally strives to classify aggregates, in the form their content of possible reactive forms of silica as follows:

(1) Class 1—very unlikely to be alkali-reactive,
(2) Class 2—alkali reactivity is not certain,
(3) Class 3—very likely to be alkali-reactive.

In most cases, the common practice includes two techniques that will work on the observation procedures. First is the macroscopic petrography—a process to find out the modal contents of rock types in coarse aggregate by visual inspection of each sample and separating the particle manually. Next is the thin-section petrography by using an optical microscope and point counting to analyze the mineral and compositions [3].

As stated before, there are many types of rocks that were used as the aggregate for concrete. Thus, this petrographic analysis is very significant in order to trace any reactive silica present in the rock used. In order to reach the conclusion, the equipment for the mineral tracing must be used properly. In this case, it is the polarizing microscope. First, what is the definition of polarization? As an example, in a loosely hanging rope, if vibrations are generated, it will most probably propagate in a certain plane. The phenomena are identified as the plane of vibration or vibration direction. In the same case, electromagnetic waves also will vibrate in certain planes [6]. Lights that consist of just one particular number of vibration direction are called linearly polarized light. On the other hand, if it contains an infinite number of different variations of vibration, it is identified as white light or also known as natural light. Polarizers will help in the production of light waves of uniform vibration direction.

X-ray fluorescent spectrometry (XRF) method is commonly used in examining the contents of a mineral in order to detect any changes due to certain effects. It measures the fluorescent X-ray that is emitted by a sample after it is excited by a source of a primary X-ray [8]. The elements that are detected will produce a certain characteristic of "fingerprints" that can only be found on a specific element. The process includes a sample being exposed to a primary X-ray source from a controlled tube. Then, after the ray collides with the atom inside the sample (provided that the energy of the X-ray is higher than that of the atom's K or L shell binding energy), one of the atom's inner orbital shells which will be dislodged will have its electron ejected. However, this atom will regain its stability from the empty space left by the ejected electron by filling it with an electron from its outer shell. The movement if these electrons will induce rays based on its behavior. Ejected electron will produce K X-ray, while the electron that filled the vacancy will produce L X-ray. The measurement of both of these types of energy is what XRF analysis is based on. Figure 38.1 shows the mechanism of XRF on a sample.

The energy produced by this reaction will vary with different intensities and will create a graph with a few energy peaks. These energy peaks will determine what element is contained in the sample and also will indicate its concentration.

Next, referred to as Swiss Hammers, Rebound Hammer or Schmidt Hammer is one if an effective method to estimate the compressive strength of a concrete. This method is also being applied broadly and widespread due to it being easy to use and high performance [9]. It is invented in 1954 by Ernst O. Schmidt; hence, the name Schmidt Hammer was distributed by a Swiss company, Proceq. This hammer is mostly being used in order to test a hardened concrete, in order to determine its structural integrity. A simple examination of test hammer data enables for quick, cost-effective, and well-informed structural appropriateness judgments. It highlights the lower strength area. Thus, it can isolate area that was damaged by fire or freezing. The damage exerted on the concrete is very minimal. Usually, it will only leave small dent on the concrete's surface.

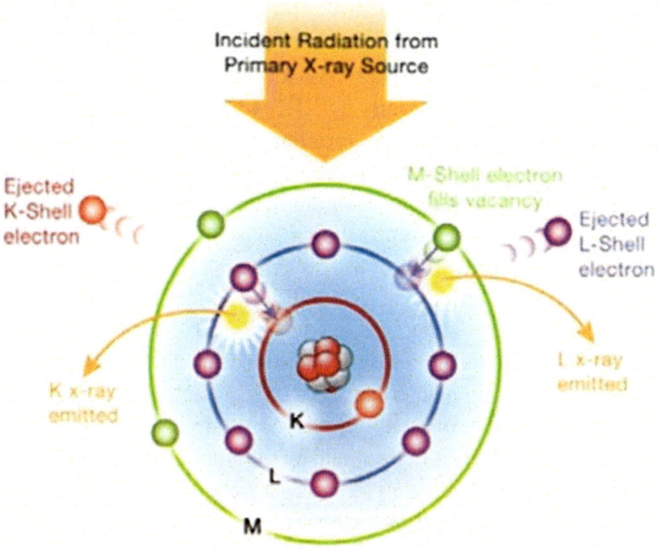

Fig. 38.1 Mechanism of XRF analysis by determining the level of L X-ray and K X-ray energy

38.4 Sample Investigation

This research was carried out with four main activities including petrographical analysis using a microscope, XRF analysis, and Schmidt Rebound Hammer test. The textures and mineral compositions of samples are determined by petrographic analysis. Petrographic research entails examining a rock or concrete sample under a microscope to assess its mineralogical and chemical properties. This research is extremely useful for identifying potentially harmful minerals in aggregates. Block samples or cores may be used to obtain petrological inspection samples. They have been resin-filled. Sawing, grinding, and polishing yield suitable surfaces (polished or thin sections). Examine the sample with a reflected (transmitted) light rock (geologically polarized) microscope. Monochromatic, ultraviolet, or polarized light may be used. A petrographic analysis can include describing core samples from an exploration well, looking at thin parts of the reservoir or reservoir boundary, using a scanning electron microscope (SEM) to describe fracture surfaces from microstructures, using X-ray diffraction (XRD) to determine the exact mineral assemblages of a sample, or any other useful petrographic technique.

For this analysis by XRF, the test materials were prepared by the mixing of one part of the sample's powder by weight (0.7 g) that is dried at 110°C alongside five parts by weight of dried lithium metaborate flux. Then, it was left to fuse inside a muffle furnace for 15 min at 1100°C. To ensure that the solution is homogenous, it was then swirled repeatedly and directly poured on to a pre-heated hotplate to be pressed forming a flat surface disk (1.5 mm thick with 35 mm in diameter). To avoid the loss-on-ignition (LOI) due to the loss of constituents such as water and carbon dioxide, the sample was weighed before and after heating on a pre-ignited silica-crucible to 1000°C for 1 h. These constituents must be recorded in this way because it cannot be measured directly by XRF. Moreover, it will be lost during the glass formation. Next, to trace the major elements, the rock powder (~10 g) was mixed with a few drops of polyvinylpyrrolidone–methylcellulose binder (~0.7 ml). To produce the powder pellets (35 mm in diameter), it will then be pressed to form a flat-surfaced pellets. The minimum thickness of these pellets must be not less than 5 mm. It was the dried overnight at 110°C. The XRF analysis was then carried out at the Department of Earth Sciences and Environment at Universiti Kebangsaan Malaya.

The plunger of a rebound hammer is slammed on the concrete surface in the procedure of a rebound hammer test, causing a spring-controlled mass with a constant energy to strike the concrete surface. A spring-controlled mass rebounds at the same moment. The surface hardness is determined by the amount of rebound on a graduated scale. This measured number is referred to as a rebound number or a rebound index. The rebound number or rebound index of reinforced concrete structures with low compressive strength and stiffness is lower. For the rebound number, an average of six readings should be taken for each point of testing.

Several precautions need to be taken while the test is taking place.

1. The sample surface should be smooth, clean, and dry.
2. Any loose surface should be removed by a grinder before testing.
3. The point of impact between the rebound hammer and the sample must be 20 mm away from the edge.
4. The test should not be performed on a rough surface caused by incomplete concrete compaction, grout loss, or a damaged or tooled surface.

As a precaution step, before performing the test on hardened concrete, the rebound hammer must be calibrated. Testing a concrete cube of size (150 * 150 * 150 mm) with a rebound hammer and subsequently with a compression testing machine is a good way to examine the calibration of the rebound hammer (CTM). It is ready to use if the rebound hammer test strength findings exactly match the CTM results. However, if there is a significant variation in the results, we must calibrate the rebound hammer before testing.

38.5 Result and Discussion

The result from the petrographical analysis, XRF analysis, and Schmidt Rebound Hammer Test is discussed here. For petrographical analysis, by utilizing the microscope, the position of ASR had been concluded as shown in Fig. 38.2.

Aggregates are uniformly dispersed throughout the concrete. The aggregate is fairly well graded. The coarse aggregate fraction in this concrete shows relatively potential for alkali reaction as denoted by the presence of boundary reaction between

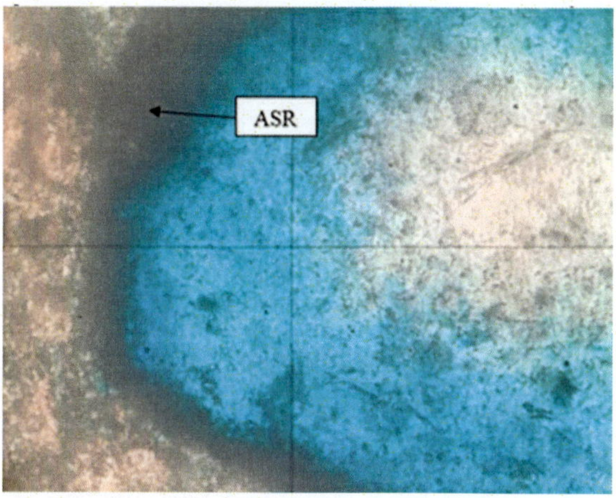

Fig. 38.2 Thin-section photomicrograph of an alkali–silica reaction section from sample SWJ-1-20

the cement and the coarse aggregates. The coarse aggregate is abundant with angular–subangular quartzo-feldspathic mineral. Reaction rims or gel was observed around the aggregates. The fine aggregates are made up of quartz sand, minor amount of plagioclase, and rock fragments. The fine aggregates are presented as angular–subangular fractions and show uniform distribution and grading. The grain size varies between 0.5 and 3 mm.

Figure 38.3 shows the petrographic textures of the sample SWJ-1-20 from computerized method.

Microcracks are found pass through an aggregate and very minor through the cement paste. Aggregates are uniformly dispersed throughout the concrete. The aggregate is fairly well graded. The coarse aggregate is abundant with angular feldspar mineral. Reaction rims or gel is absent. The fine aggregates are made up of quartz sand, minor amount of plagioclase, and rock fragments. The fine aggregates are presented as angular–subangular fractions and show uniform distribution and grading. The grain size varies between 0.5 and 3 mm. The concrete is fairly in good condition with absence of silica gel and alkali–silica reaction. Photomicrograph A shows presence of biotite flakes formed in fine-grained quartz aggregates. Note, quartz is free from any alteration effect due to its simple chemical framework (SiO_2). Next, Photomicrograph B shows subangular-shaped and fine-grained aggregate of plagioclase feldspar with very minor alteration to sericite (dusty like) at lower half of its grain. Note, quartz is free from any form of alteration effect. Moreover, Photomicrograph C shows subangular large fine-grained feldspar with distinctive perthitic texture. Note that, the feldspar grain is well bounded by very fine-grained crystal of mainly quartzo-feldspathic and cement as denoted by its smooth boundary.

After obtaining the XRF results from the Department of Earth Sciences and Environment at Universiti Kebangsaan Malaya, the results from Table 38.1 can be concluded.

Based on Table 38.1, CaO and SO_3 both have a pattern where it is higher for four samples that are SWJ-1-20, SWJ-3-55, SWJ-7-120, and SWJ-9-240. Because

Fig. 38.3 Petrographic textures of sample SWJ-1-20 from computerized method, A B C

Table 38.1 Composition of minerals determined by using XRF analysis

Samples	Composition (%)								
	Al_2O_3	SiO_2	SO_3	Fe_2O_3	CaO	MgO	Na_2O	K_2O	TiO_2
SWJ-1-20	7.77	38.6	4.33	5.04	53.6	1.24	1.81	1.69	0.531
SWJ-2-30	5.50	37.6	1.30	3.01	39.8	1.41	1.09	0.84	0.488
SWJ-3-55	7.36	40.2	3.84	3.32	49.3	1.16	1.52	1.17	0.471
SWJ-4-65	5.12	46.9	2.58	1.97	34.6	1.64	1.42	0.63	0.473
SWJ-5-75	6.22	40.3	2.30	3.23	33.5	1.56	1.31	0.32	0.455
SWJ-6-115	7.23	39.4	1.00	2.33	32.3	1.12	0.21	0.24	0.358
SWJ-7-120	4.30	36.8	4.07	2.89	51.9	0.99	0.20	0.37	0.480
SWJ-8-145	4.41	29.1	1.35	2.55	40.2	0.87	0.95	0.17	0.462
SWJ-9-240	4.27	32.9	3.75	2.60	51.3	0.99	1.31	0.35	0.420

Notes Al_2O_3 = Aluminum oxide; SiO_2 = Silicon oxide; SO_3 = Sulfur trioxide; Fe_2O_3 = Ferric oxide; CaO = Calcium oxide; MgO = Magnesium oxide; Na_2O = Sodium oxide; K_2O = Potassium oxide; TiO_2 = Titanium dioxide of the direct interaction with sea water, the samples have a higher SO_3 level. The high levels of calcium oxide CaO in the samples indicate that the cement content used for constructions has a lot of calcium hydroxide ($Ca(OH)_2$), which causes deterioration due to ASR. Meanwhile, for other elements, there is no pattern shown, so no conclusion can be derived from it.

The Schmidt Rebound Hammer was designed to test the concrete's strength and homogeneity. An elastic rebound of the hammer on the front surface of the beam, column, or slab can accomplish this. The surface was cleaned prior to the test to ensure that it was free of any debris or dust. Next, a 4 × 3 grid was drawn to take the average from different sides of the concrete sample as shown in Fig. 38.4.

The Rebound Hammer test results are summarized in Fig. 38.5. From the figure, the test results show that the estimated strength for the pile jacket is mostly above 50 N/mm^2. However, it can be concluded that SWJ-1-20, SWJ-3-55, SWJ-7-120, and SWJ-9-240 have lower values of Estimated Concrete Strength fcu (N/mm^2).

Fig. 38.4 4 × 4 gridlines drawn on the beam surface

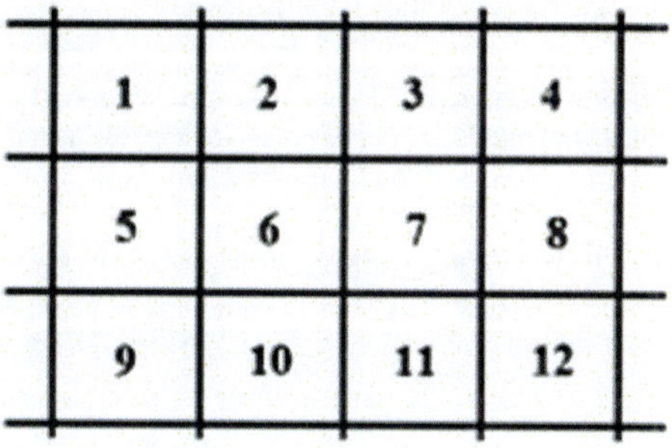

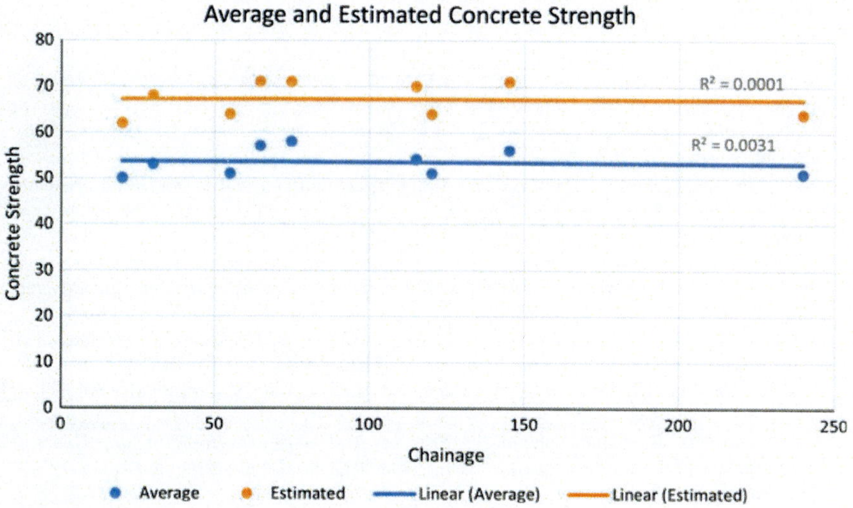

Fig. 38.5 Results from Schmidt Rebound Hammer test

38.6 Conclusion

Overall, petrographical analysis can be useful in determining the characteristics of ASR contained inside the reinforced concrete. The findings of this research can be summarized as follows. Potential ASR and microcracks inside a deteriorated cement concrete can be determined by using the petrographical analysis method, as it is aligned with the results from XRF analysis and Schmidt Rebound Hammer test. Samples SWJ-1-20, SWJ-3-55, SWJ-7-120, and SWJ-9-240 that are expected to have ASR from the petrographical analysis clearly show higher composition of CaO and SO_3 during the XRF analysis. Moreover, during the Schmidt Rebound Hammer test, these four samples also showed a lower Estimated Concrete Strength compared to the other five samples. Salt water intrusion, which will increase the amount of minerals such as CaO and SO_3 inside the concrete, will increase the probability of the minerals to react with the aggregates and deteriorate the concrete, thus decreasing its Estimated Concrete Strength.

Acknowledgements This work was supported by the Fundamental Research Grant Scheme (FRGS) awarded by the Ministry of Education of Malaysia entitle "Enhanced Coalbed Methane Recovery Processes by Coupling Fluid Flow of Mukah-Balingian Coal Deposits" with the referral no. R.J130000.7851.5F371.

This work was supported by the Universiti Teknologi Malaysia Encouragement Research Grant (UTMER) awarded by Universiti Teknologi Malaysia entitle "Diagnosis of Alkali-Aggregate Reaction–Polarizing Microscopy and SEMEDS Analysis" with the referral no. Q.J130000.3851.19J79.

References

1. Z.S. Andreas Leemanna, *Characterization of Amorphous and Crystalline ASR Products Formed in Concrete Aggregates* (Elsevier, Trondheim, 2020)
2. D.J.-N. Aneta Antolik, *Assessment of the Alkali-Silica Reactivity Potential in Granitic Rocks* (Elsevier, Warsaw, 2021)
3. E. Boehm-Courjaulta, S. Barbotin, *Microstructure, Crystallinity and Composition of Alkali-Silica Reaction Products in Concrete Determined by Transmission Electron Microscopy* (Elsevier, Dübendorf, 2020)
4. E.R. Gallyamov, A.C. Ramos, *Multi-Scale Modelling of Concrete Structures Affected by Alkali-Silica Reaction: Coupling the Mesoscopic Damage Evolution and the Macroscopic Concrete Deterioration* (Elsevier, Durham, 2020)
5. H. Sun, S. Pashoutani, *Nondestructive Evaluation of Concrete Bridge Decks with Automated Acoustic Scanning System and Ground Penetrating Radar* (sensors, Omaha, 2018)
6. D. Jana, *DEF and ASR in Concrete—A systematic Approach from Petrography* (ResearchGate, 2008)
7. M. Nematzadeh, S. Fallah-Valukolaee, *Erosion Resistance of High-Strength Concrete Containing Forta-Ferro Fibers Against Sulfuric Acid Attack With an Optimum Design* (ELSEVIER, Babolsar, 2017)
8. M.H. Ramsey, P.J. Potts, *An Objective Assessment of Analytical Method Precision: Comparison of ICP-AES and XRF* (Elsevier, London, 2009)
9. M. Kazemi, R. Madandoust, Compressive strength assessment of recycled aggregate concrete using Schmidt rebound hammer and core testing. (Elsevier, Rasht, 2019)

Chapter 39
Mechanical Analysis of Golf Ball Retriever Prototype

Iszmir Nazmi bin Ismail, Nur Najwa Umirah Binti Nor Azman, Nursyadzatul Tasnim Roslin, M. H. Zawawi, N. M. Zahari, S. Z. Abidin, Ahmad Wafi Mahmood Zuhdi, M. R. Aridi, Hassan Mohamed, M. Z. Ramli, M. H. Mansor, Fevi Syaifoelida, A. A. Zakaria, M. R. Isa, Daud Mohamad, M. F. Jaafar, N. A. Rahmat, and M. S. Abd Rahman

Abstract The paper highlighted the mechanical analysis on ball retriever prototype that can be used on the driving range. This analysis emphasizes in optimizing the size, strength and materials that will be used in building the prototype. The prototype has been designed after doing the market survey analysis, patent search and concept generation. One of the final parts of the design process is to run mechanical analysis on the prototype model by using CAD software to validate the design before fabrication.

Keywords Golf · Golf ball · Mechanical analysis · CAD · FEA · FEM

39.1 Introduction

A golf driving range is a facility that involves golf players, golf attendants, golf balls and retriever equipment. Developed countries such as Japan, South Korea, the UK and the USA had already implemented high-technology ball retriever equipment including an intelligent robot ball picker [1]. In Malaysia, the majority of driving ranges that are using heavy-duty driving range ball pickers are normally owned by big organizations. However, the stand-alone golf driving range is still using conventional methods like balls' scooping to retrieve the golf balls in the field due to the factors such as cost and demand. The highlighted problems are the golf attendant is not able to collect the balls effectively and they will spend more time collecting the balls and the golfers had to pause hitting the balls to ensure the safety of the attendant in the field which consumes both golfers and attendant's time. The current conventional

I. N. bin Ismail (✉) · N. N. U. B. N. Azman · N. T. Roslin · M. H. Zawawi · N. M. Zahari · S. Z. Abidin · A. W. M. Zuhdi · M. R. Aridi · H. Mohamed · M. Z. Ramli · M. H. Mansor · F. Syaifoelida · A. A. Zakaria · M. R. Isa · D. Mohamad · M. F. Jaafar · N. A. Rahmat · M. S. A. Rahman
College of Engineering, Universiti Tenaga Nasional, Jalan IKRAM-UNITEN, 43000 Kajang, Selangor, Malaysia
e-mail: iszmir@uniten.edu.my

M. A. A. Mohd Salleh et al. (eds.), *Proceedings of the Green Materials and Electronic Packaging Interconnect Technology Symposium*, Springer Proceedings in Physics 289,
https://doi.org/10.1007/978-981-19-9267-4_39

way to retrieve balls will cause strain on the collector's wrist, and it requires extra force to transfer the balls into the wheelbarrow. Lastly, the number of balls collected is relatively low where the current method only manages to collect up to 30 balls at a time and the collector had to repeat the same process which is energy-draining; therefore, it will affect the efficiency and productivity. The outcome of the study is the final 3D design of the device with engineering analysis such as stress and strain analysis on the proposed prototype [2–6]. It is designed to overcome the weaknesses of existing design without jeopardizing the main function and the economic value.

39.2 Methodology

39.2.1 Modifications and Finalized Design Selection

This paper discusses the result and analysis of the proposed design. The drawing of finalized concept with detailed dimension, bill of materials (BOM) and Finite Element Analysis (FEA) by using CAD software will be analyzed and discussed in this section. Hence, the analysis is crucial to validate the reliability of the prototype. The design integrates the mechanism of wheel and axles, pushing or pulling force as well as trajectory motion. It consists of a frame with detachable storage compartment supported by a set of disks that act as wheel and axle engaging the ground as well as castor wheel in order to assist the maneuverability of the cart. This simple mechanism helps the user to retrieve more than 50 balls at a time with the storage capacity of 300 balls. When the cart is pushed, the set of disk and castor wheel will rotate and retrieve the balls on the ground before the balls are transferred into the storage unit by trajectory motion as shown in Fig. 39.1.

Figure 39.2 shows the whole prototype 3D model, while Fig. 39.3 shows that the dimension of each of the component has been identified based on the engineering requirements and comparison of existing designs.

Dimension of the main frame is 1050 mm in length and width of 720 mm. The storage unit or basket has the length of 613 mm, 598 mm in width and 184 mm in height, respectively. Besides, the design is equipped with telescopic handles which is aimed to let the operator to work ergonomically, whereby the handle can be adjusted to optimize the comfort of utilizing the cart [7]. The telescopic handle has the length of 1493.86 mm and 100 mm in width, where the tube diameter is designed to 30 mm to connect the telescopic shaft and two units of telescopic bearing. To avoid the operator from colliding with the replacement disks, the maximum allowable height for the handle is 147 cm. On top of that, the user can adjust the handle to the desired height; however, the optimum height advised by Das et al. is always 5 cm above the elbow of the operators [8]. Figure 39.3 shows that the handle can be adjusted according to the height of the operator. A swivel castor wheel is placed at the front part of the frame to assist the movement of the cart where it was first introduced to model and minimize wheel shimmy as well as to navigate the direction of a device, cart or trolley [9]. The castor wheel has a diameter of 152.4 mm which is equipped with brake, and the top

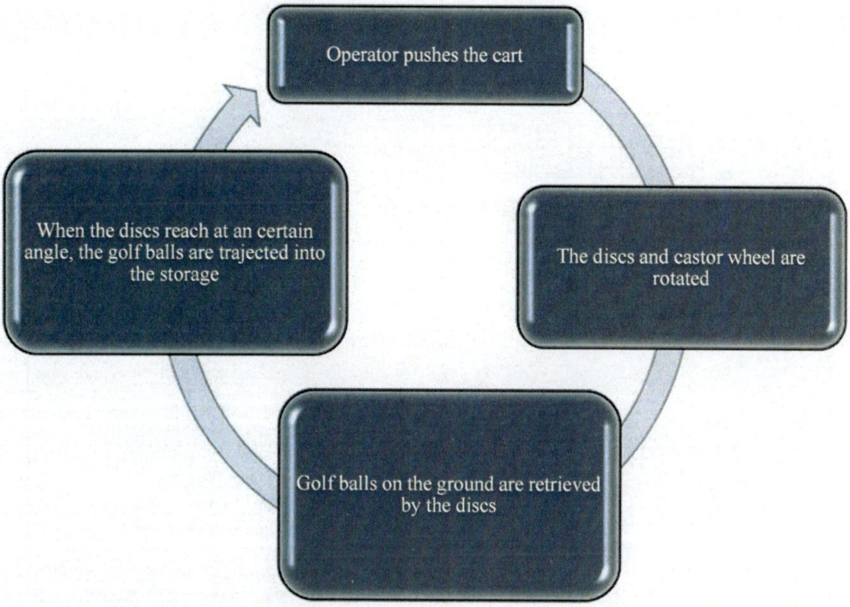

Fig. 39.1 Operating system of the design

Fig. 39.2 Three-dimensional model of final ball retriever prototype

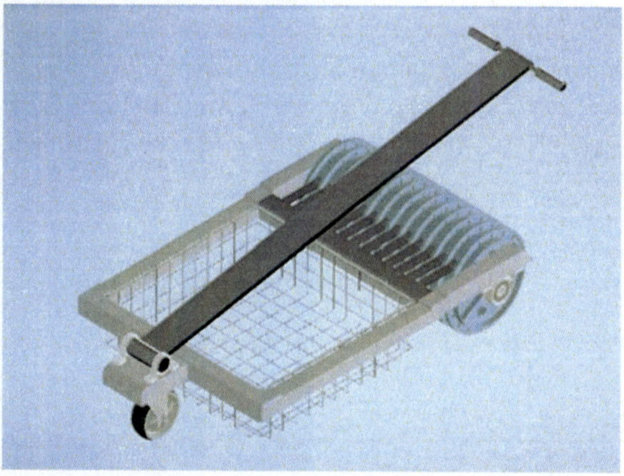

plate is fitted with four units of AS 1427-M10 × 16(6) ISO metric machine screws. It is inspired by the heavy-duty caster wheel that is commonly used for industrial applications due to its durability and load capacity. The technical specifications of the caster wheel are determined by comparing the existing product which are available in the current market [10]. Moreover, 14 pieces of disks are joined together with a shaft (axle), two pieces of mounting bearing as well as retriever finger as a support. The set of disks act as a retriever unit to retrieve the balls on the ground. It is designed to

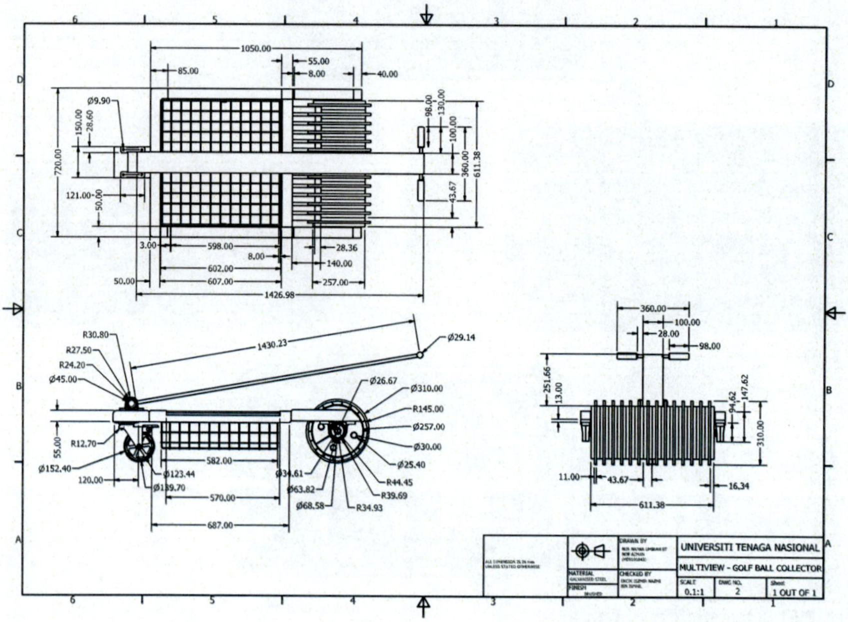

Fig. 39.3 Dimension for the design in millimeters (mm)

endure uneven or bumpy areas without destroying the grass. When the cart is pushed, the castor wheel and the disks will rotate to retrieve the balls on the ground before it is trajected into the storage unit. The retriever finger will aid the movement of the balls into the wire basket (storage unit). Table 39.1 shows the specifications of the product.

39.3 Result and Discussions

39.3.1 von Mises Stress Analysis

Design optimization is a method of engineering design that employs a mathematical formulation of a design problem to help in the selection of the best design among several choices. It comprises the following stages such as variables, objectives, constrains and feasibility [11]. The Finite Element Analysis (FEA) is the numerical simulation of any physical phenomenon using the Finite Element Method (FEM). Engineers use FEA software to minimize the number of physical prototypes and tests in their design process and optimize components to produce better quality of a product as well as reduce the time and cost-effective [12]. In this project, Finite Element Analysis (FEA) is applied to analyze the finalized design and to identify

Table 39.1 Detailed design specifications

Product design specifications		
Components and material	Replacement discs	PVC
	Body frame	Galvanized steel
	Golf retriever finger	
	Wire basket	
	Telescopic shaft	
	Telescopic handle	
	Basket clipper	
	Shaft	
	Telescopic bearing	Chrome steel—SAE 52,100
	Mounting bearing	
	Castor wheel	Rubber tire, galvanized steel for top plate and brakes
	AS 1427—M10 × 16(6)	ISO standards
Dimension (mm)	Overall dimension (L × W × H)	1050 × 720 × 310
	Wire basket (Storage unit) (L × W × H)	598 × 613 × 184
	Length of telescopic handle	1493.86
	Length of handle	360
	Diameter of castor wheel	152.4
	Distance between discs	42
Special features	Telescopic handle	Height can be adjusted
	Castor wheel	Comes with braking system
	Replacement discs	Can be replaced if broken
Life span	Maximum	15 years
Mechanism	Wheels and axles	
Method of retrieving balls	Balls are rolling into the replacement discs	
Method of storing balls	Trajectory motion	
Storage capacity	300 balls	
No. of balls retrieved at a time	50–100	
Weight (kg)	32.23	
Operating system	Push and pull forces (Manual)	

the minimum and maximum allowable stress. Besides, the factor of safety can be determined to ensure that the design is safe and reliable to operate.

The von Mises stress is a measurement that can be used to evaluate if a material will yield or fracture. It is commonly used on ductile materials like metals. The von Mises yield criterion asserts that a material will yield if its von Mises stress under load is equal to or greater than the yield limit of the same material under simple

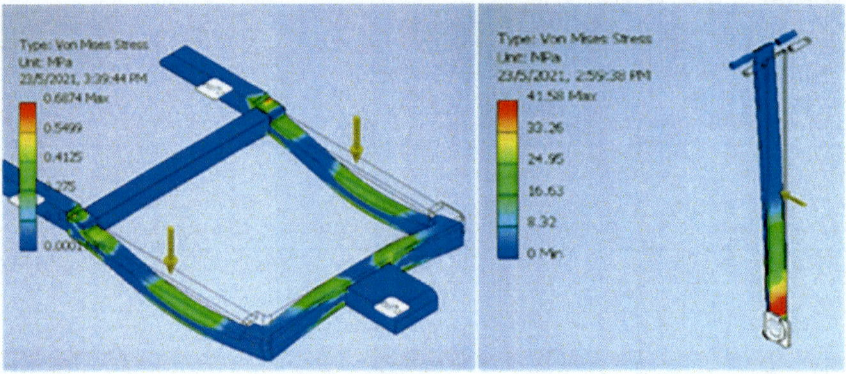

Fig. 39.4 von Mises stress analysis on the body frame (left) and telescopic handle (right)

tension. A von Mises stress analysis has been conducted on different parts such as main frame and the handle. Based on Fig. 39.4, a 250 N magnitude force was applied on the body frame with a fixed constraint by using the material of galvanized steel. The minimum von Mises stress for this part is 0.000115649 MPa and the maximum value is 0.687358 MPa. It can be concluded that the critical part is on the left, right and front part of the frame where the wire basket is located as it holds the load of the wire basket. There are two constraints applied in this analysis which are pin constraint and fixed constraint that have magnitude force of 1812.94 N and 1562.91 N, respectively. According to the analysis that was performed as shown in Fig. 39.4, the minimum von Mises stress is 0.0007585 MPa and the maximum is to be 41.5784 MPa. The fixed constraint is applied at the bottom part of the handle which makes that it has the highest value of von Mises stress.

Figure 39.5 shows a von Mises stress analysis when all the components are assembled. A fixed constraint with magnitude force of 132 N is applied in which the maximum value of von Mises stress is 1.64759 MPa. The mass of a golf ball is said to be 0.04593 kg [13], so if 300 balls are stored in the wire basket, the total mass will be 13.78 kg which is equivalent to the amount of magnitude force applied. Hence, this golf ball retriever is safe to store up to 300 balls and the components will not yield.

39.3.2 Displacement (Strain) Analysis

Strain is defined as the amount of deformation experienced by the body in the direction of force applied, divided by the body's initial dimensions. It is a useful method for assessing a material's changes as a result of different loading circumstances. The strain analysis has been conducted on the parts such as body frame, telescopic handle and the assembled golf ball retriever. The material selected for this body frame is galvanized steel and the maximum strain value on the body frame is 0.006049 mm

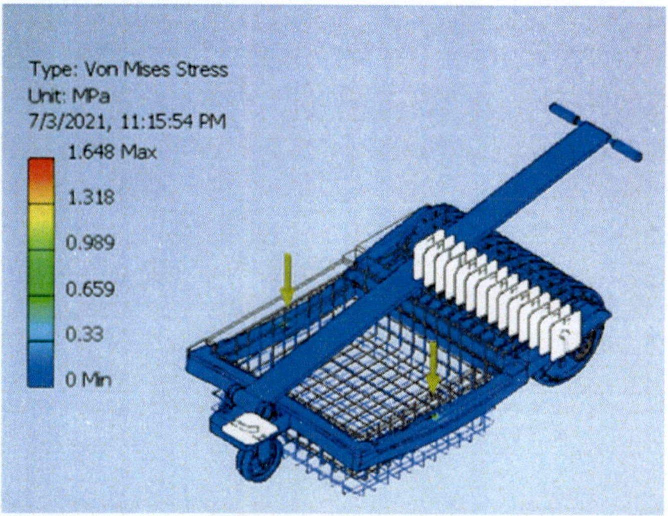

Fig. 39.5 von Mises stress analysis on the golf ball retriever

which is significantly small, and it is proven that the material has high durability and the deformation will not occur as shown in Fig. 39.6. Based on Fig. 39.6, it can be stated that the highest displacement occurs at the handle grip where the highest force is exerted that gives a value of 16.32 mm. The rest of the parts have lower value of displacement because the amount of force exerted is smaller than the handle grip. It shows that galvanized steel is suitable to be used for the telescopic handle as it has high strength and is able to reduce the stiffness.

According to Fig. 39.7, a displacement analysis has been performed on the golf ball retriever. It depicts that the displacement is at minimal point except for the wire

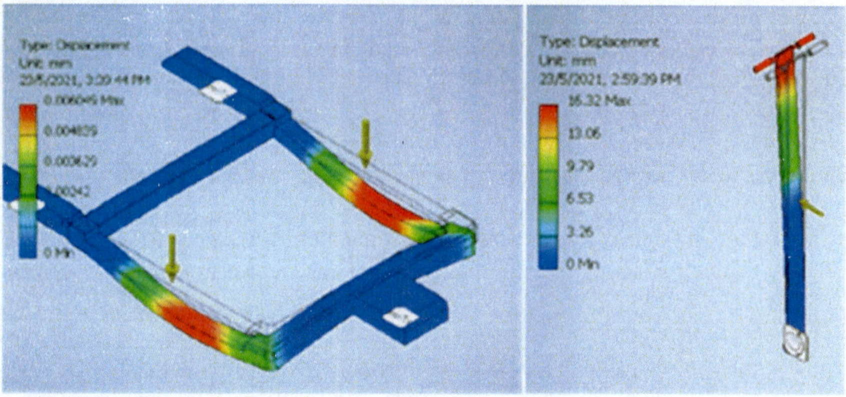

Fig. 39.6 Strain analysis on the body frame (left) and telescopic handle (right)

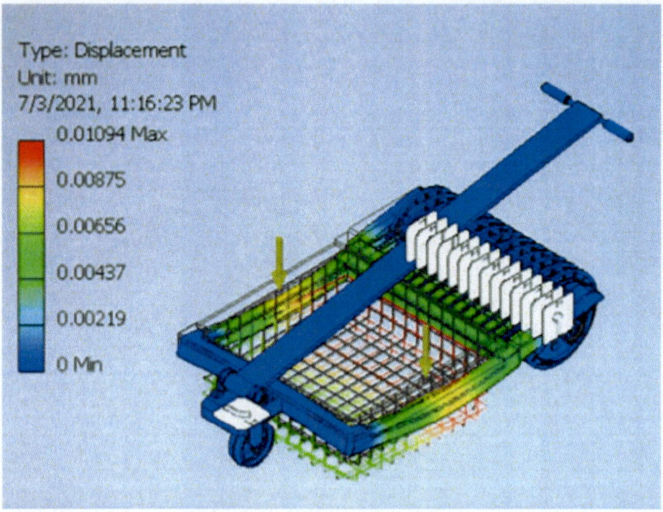

Fig. 39.7 Strain analysis on the golf ball retriever

basket where the highest force is applied. However, the maximum displacement on this analysis is only 0.0109359 mm, in which the possibility for a deformation and stiffness to occur is relatively low.

39.3.3 Factor of Safety

The factor of safety is primarily utilized to ensure that no unexpected failures, deformations or defects will occur during structural design. It also improves people's safety while lowering the danger of a product failing. When it comes to fall prevention and safety equipment, the aspect of safety is crucial. There is a risk of harm and death, as well as financial loss, if a structure fails. The materials used in every structure are determined by the necessary factor of safety. The amount of material employed is determined by the realized factor of safety [14]. A factor of safety is tested on the handle part which has the material of galvanized steel as in Fig. 39.8. The minimum factor of safety is 4.9785, and the maximum factor of safety is 15. As the contours shown in Fig. 39.8, the factor of safety is considered reliable and safe to be applied because the factor of safety specifies under the red zone area is not safe and reliable.

Both body frame part and the prototype have the same value factor of safety that is 15 as shown in Fig. 39.9. The material applied for the body frame is galvanized steel. Any industrial design that has the factor of safety value higher than 3 is acceptable according to the OSHA standards such as 1910 section 1C Design for System Components [14]. It can be concluded that the factor of safety is high and it is suitable to be used for structural design and it is able to withstand the force of 125 N.

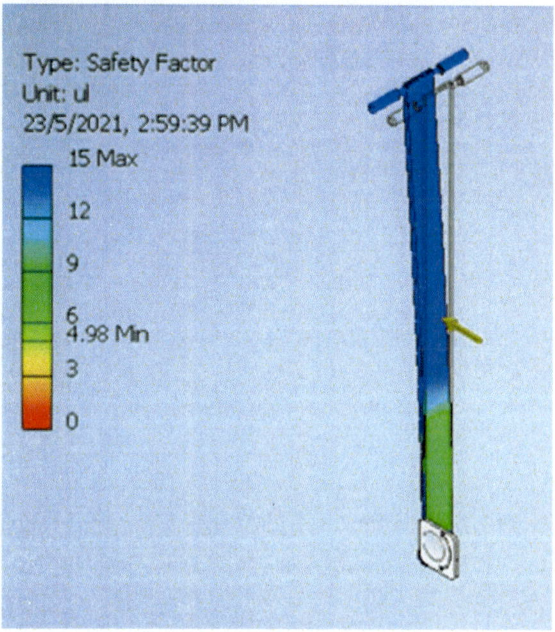

Fig. 39.8 Factor of safety of the telescopic handle

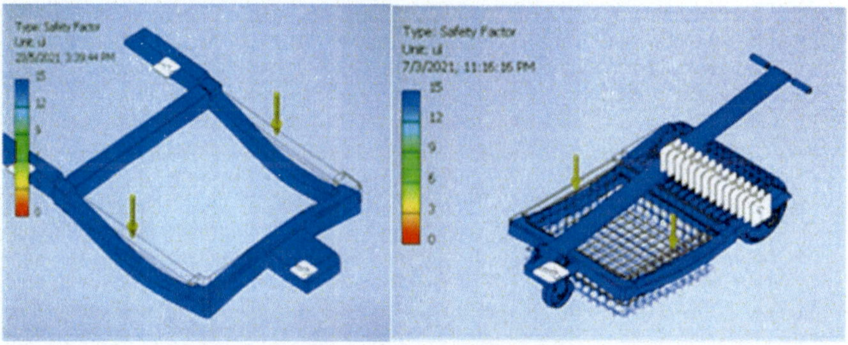

Fig. 39.9 Factor of safety of the frame (left) and the whole golf ball retriever prototype (right)

39.4 Conclusion

Stress and strain simulation of the final prototype design is very crucial before the product is made. CAD software can be used to perform Finite Element Analysis (FEA). Further research is conducted to modify and improve the conceptual design as a finalized design. Material selection is another crucial process so that the lifespan of the product is longer and deformation and yielding will not occur in the structural design. This design also meets the factor of safety which will prevent it from fail due to high load when carrying balls. Overall, this prototype reliability is proven from the FEA data.

Acknowledgements The authors would like to acknowledge Universiti Tenaga Nasional (UNITEN) for giving the YCU Grant 202210050YCU to publish this paper.

References

1. B. The Grass Masters, BallPicker—Belrobotics. *Belrobotics* (2020)
2. I.N. Ismail, K.A. Halim, K.S.M. Sahari, A. Anuar, M.F.A. Jalal, F. Syaifoelida, M.R. Eqwan, Design and development of platform deployment arm (PDA) for boiler header inspection at thermal power plant by using the house of quality (HOQ) approach, in *Procedia Computer Science*, vol. 105, p. 296–303 (8 p.) (2017)
3. I.N. bin Ismail, A.H. bin Iskandar, M.R. Eqwan, A.W.M. Zuhdi, D. Mohamad, M.R. Isa, N.M. Zahari, M.H. Zawawi, H. Mohamed, M.Z. Ramli, M.H. Mansor, Design and development an automatic plant pot prototype, in *Green Design and Manufacture: Advanced and Emerging Applications. AIP Conference Proceeding*, vol. 2030, pp. 020201-1–020201-7 (2018)
4. I.N. bin Ismail, P. Jayakumar, M.R. Eqwan, A.W.M. Zuhdi, D. Mohamad, M.R. Isa, N.M. Zahari, M.H. Zawawi, H. Mohamed, M.Z. Ramli, M.H. Mansor, Design and development of smart sorting recycle bin prototype, in *Green Design and Manufacture: Advanced and Emerging Applications. AIP Conference Proceedings*, vol. 2030, pp. 020202-1–020202-8 (2018)
5. I.N. Ismail, A. Anuar, K.S.M. Sahari, M.Z. Baharuddin, M. Fairuz, A. Jalal, J.M. Saad, Development of in-pipe inspection robot: a review, in *2012 IEEE Conference on Sustainable Utilization and Development in Engineering and Technology, STUDENT 2012—Conference Booklet*, pp. 310–315 (6 p.) (2012)
6. K.S.M. Sahari, A. Anuar, S.S.K. Mohideen, M.Z. Baharuddin, I.N. Ismail, N.M.H. Basri, N.S. Roslin, M.A. Aziz, B. Ahmad, Development of robotic boiler header inspection device, in *6th International Conference on Soft Computing and Intelligent Systems, and 13th International Symposium on Advanced Intelligence Systems, SCIS/ISIS 2012*, p. 769–773 (5 p) (2012)
7. R. Ady, W. Bachta, P. Bidaud, Development and control of a one-wheel telescopic active cane, in *5th IEEE RAS/EMBS International Conference on Biomedical Robotics and Biomechatronics*, Sao Paulo, Brazil, pp. 461–466 (2014). https://doi.org/10.1109/BIOROB.2014.6913820
8. B. Das, J. Wimpee, B. Das, Ergonomics evaluation and redesign of a hospital meal cart. Appl. Ergon. **33**(4), 309–318 (2002). https://doi.org/10.1016/s0003-6870(02)00018-2. ((PMID: 12160335))
9. D. de Falco, G. Di Massa, S. Pagano, On the castor dynamic behavior. J. Franklin Inst. **347**(1), 116–129 (2010)
10. *Industrial Wheels Castors For Sale-Dashi Factory Wholesale Exporter*. DASHICASTER (2019)
11. P. Papalambros, D. Wilde, *Principles of Optimal Design*, 3rd edn. (Cambridge University Press, Cambridge, 2017), pp.27–35
12. J. Fish, T. Belytschko, *A First Course in Finite Elements by Jacob Fish and Ted Belytschko* (Wiley, 2007)
13. About IGFGolf (2020) [Online]. https://www.igfgolf.org/about-igf
14. L. Wilhite, *Factor of Safety: What is it and Why is it Important?* Onsite Safety (2018)

Chapter 40
Customer Survey Analysis for Design and Development of Golf Ball Retriever Prototype

Iszmir Nazmi bin Ismail, Nur Najwa Umirah Binti Nor Azman, Nursyadzatul Tasnim Roslin, M. R. Isa, N. M. Zahari, S. Z. Abidin, Ahmad Wafi Mahmood Zuhdi, M. R. Eqwan, Hassan Mohamed, M. Z. Ramli, M. H. Mansor, Fevi Syaifoelida, A. A. Zakaria, M. H. Zawawi, Daud Mohamad, M. F. Jaafar, Kamarulzaman Kamarudin, and Mohamed Saiful Firdaus Hussin

Abstract This paper focuses on using customer analysis in the design and development of golf ball retriever that can be operated in the driving range. The identified problems with conventional method of retrieving the balls are the golf attendant is not able to collect the balls effectively, the number of balls collected is relatively low and the operator will experience strain in their wrists while collecting the balls with the method of scooping. Hence, the methodology and results are presented according to the customer and engineering requirements. Quality Function Deployment (QFD) method by using House of Quality (HOQ) is used to align both the customers' and engineering requirements.

Keywords Golf · Golf ball · Customer · Ball retriever · Design · QFD · HOQ · House of quality

---

I. N. bin Ismail (✉) · N. N. U. B. N. Azman · N. T. Roslin · M. R. Isa · N. M. Zahari · S. Z. Abidin · A. W. M. Zuhdi · M. R. Eqwan · H. Mohamed · M. Z. Ramli · M. H. Mansor · F. Syaifoelida · A. A. Zakaria · M. H. Zawawi · D. Mohamad · M. F. Jaafar
College of Engineering, Universiti Tenaga Nasional, Jalan IKRAM-UNITEN, 43000 Kajang, Selangor, Malaysia
e-mail: iszmir@uniten.edu.my

K. Kamarudin
School of Mechatronic Engineering, Universiti Malaysia Perlis, Kampus Pauh Putra, 02600 Arau, Perlis, Malaysia

M. S. F. Hussin
Fakulti Teknologi Kejuruteraan Mekanikal dan Pembuatan, Universiti Teknikal Malaysia Melaka, Hang Tuah Jaya, 76100 Durian Tunggal, Melaka, Malaysia

M. A. A. Mohd Salleh et al. (eds.), *Proceedings of the Green Materials and Electronic Packaging Interconnect Technology Symposium*, Springer Proceedings in Physics 289,
https://doi.org/10.1007/978-981-19-9267-4_40

40.1 Introduction

Golf is one of the primogenital systemized sports, and the golf rules were introduced in 1754 in the city of St Andrews in Scotland [1]. In 1958, the International Golf Federation (IGF) was established and recognized by International Olympic Committee (IOC) to inspire the development of the game internationally and as the engagement to promote sportsmanship [2]. Golf has become one of the trendy recreation activities that has obtained developing attentions among Malaysians. As a result, numerous numbers of golf courses and golf driving range have been built and upgraded to meet the increasing demand. The amount of golf balls required to practice the swing is vast; therefore, the golf attendants of the driving range need to drive a cart or use golf ball picker to retrieve all the balls that have been hit in the golf range. The method and technology of retrieving balls that has been implemented in the developed countries are cutting-edge compared to the methods that are being practiced in Malaysia [3]. The key factors that are often preferred by golf players are the golf range design and condition as well as services and esthetics of the facility. These are the attributes that the golfers would focus on their practice range in which the golf authority and management play vital roles to attract the players to their facilities [4]. The most important criterion in golf practice fields is that players always have access to balls. This following necessity is to keep the ball dispensing machine from being empty [5]. As a result, some technique, human or mechanical must be devised to constantly restock the machine by picking up golf balls and returning them to the dispensing machine [6]. In many circumstances, a worker with only a basket performs the golf ball plucking operation [7]. Such work is thought to be tiresome and can result in a variety of health issues, including back pain [8].

40.2 Methodology

40.2.1 Product Definition and Market Analysis

Surveys are commonly used by SMEs or manufacturers to gain better understanding of what users need or required. It comes with various forms such as customer satisfaction, employee questionnaires, exit interviews and other types of market research. One of the main reasons to conduct the survey is to identify the current issues that they are experiencing as a golf attendant. Besides, the survey is to determine if the issues addressed are aligned with the objectives and goals of this study. Survey is the initial step to start on Quality Function Deployment (QFD) or House of Quality (HOQ). It is not only to measure users' satisfaction, but it is aimed to improve credibility of a research. Google Forms is used as the online survey platform as it prioritizes the privacy of respondents with minimal restrictions on the number of questions created as well as the number of respondents.

40.2.2 Customer and Benefits

According to the informal interview in, Respondent 1 or Customer 1 stated that he has been working in one of the Selangor driving range for almost five years as a golf attendant. Besides, his schedule to retrieve the golf balls in the field is more than twice a day depending on the availability of the golf balls in the storage room. The number of balls collected daily is depending on the players' visit to the range where he and the team would normally collect 10,000 to 20,000 balls per day. In addition, the driving range is still practicing the conventional method of collecting the golf balls in the field. This is the most economical alternative to collect the golf balls; however, he would regularly feel the strain on his wrist during the collection session. He also mentioned that the cost to purchase a retriever or picker equipment is high and the owner requires to allocate the budget for maintenance cost. The conventional method of collecting balls by scooping is the most chosen compared to the other existing methods; however, it is not suitable for a stand-alone driving range due to the financial budget. Based on this informal interview, it can be concluded that the operator is having difficulty to collect the golf balls on the ground and there is a need to overcome these issues.

40.2.3 Surveying Phase

An online survey was distributed after the information was obtained through an onsite interview with the targeted user. This is to ensure that the design and development of the golf ball retriever will fulfill the customers' demand. Qualitative research has been conducted by using Google Forms and distributed particularly to the golf attendants in three different golf driving ranges which are in Petaling Jaya, Subang and Dungun, Terengganu. There are 12 respondents who managed to respond to the questionnaires. The daily routines of the golf attendants are to maintain the grass of the range and to retrieve the golf balls in the field depending on their schedule. Interviews with the operators depict an incredible measure of what they are looking into in a golf ball retriever equipment and the key elements that need to be taken into consideration to improvise the current method that they are using. This survey has drawn a clear vision of what this study must emphasize in designing and developing the golf ball retriever. It could be a first step where most of the companies would made to obtain the feedback from customers or users. Additionally, the benchmarking that is set by these respondents must be taken into consideration before executing a product design. The preferences priorities of the users should be aligned with goals of this product design and development.

Table 40.1 Customer's requirement

Customers' requirement
• Easy to set up and maneuver
• Improve the outcome of collecting the balls
• High ergonomics value
• Cost saving
• Efficiency by reducing the time of retrieving the balls
• Materials that can withstand hot and cold weather
• Non-corrosive materials
• Lightweight
• Low maintenance

40.2.4 Customers' Requirements

Customers are the main priority to an organization's sustainability and success in which it would require the organization to produce a good quality product to strive in meeting the customer requirements. Any organization that able to identify the customer requirement and needs as well as know how to manage the issues quickly are going to succeed and increase their revenue [9]. The main objective of this survey is to identify the problems with the current method of retrieving the golf balls. From the users or customers feedback, it helps to address the key requirement and need that can be improvised in future design. Table 40.1 shows the customers' requirement based on the justification made through an online survey.

40.2.5 Engineering Requirements

Engineering characteristic and requirements must be taken into consideration as well as customer requirements to innovate a good design. Based on evaluation made, several engineering requirements have been selected to meet the customer needs as shown in Table 40.2 and considered for the final design.

40.3 Result and Discussions

40.3.1 Summary of the Survey

According to the survey conducted, several questions were asked in relation to the information of retrieving golf balls. The data and information were successfully

Table 40.2 Engineering requirements

Engineering requirements
• Weight of equipment
• Cost saving
• Load capacity
• Size of the handle
• Size of storage unit
• Size of retriever unit
• Dimension of tires
• Dimension of the equipment
• Types of material used
• Number of systems applied

Table 40.3 Survey summary

No	Question	Feedback
1	As a golf attendant, how many times a day do you collect the balls in the field?	66.7% of the respondents collecting balls more than three times a day followed by three times at 25% and 8.3% is twice a day
2	The overall number of balls retrieved per day	91.7% of the respondent collecting 6000–15,000 balls and 8.3% is more than 15,000 balls collected
3	How many balls can be collected at instance with the current method?	41.7% of the respondents can collect more than 30 balls at a time, followed by 33.3% between 20 and 30 balls and 25% collect less than 20 balls

gathered from 12 respondents from different background and locations. The questionnaire's content was meant to focus on the problem statements and objectives of the study. Hence, the result is to be analyzed and justified to solve the requirement of the issues. Table 40.3 shows the summary from the survey.

40.3.2 Quality Function Deployment (QFD)

Quality Function Deployment (QFD) or generally known as House of Quality is the basic design tool of the management approach which was introduced by Mitsubishi at Kobe shipyard site in the year of 1972. It identifies the desire and requirement of a customer, determines its importance as well as discovers the engineering requirement [10]. In this concept, quality is measured by voice of customers (VOC) with a product or service that is provided by the manufacturer. QFD is a structured method that applies several management and planning tools to recognize and highlight the

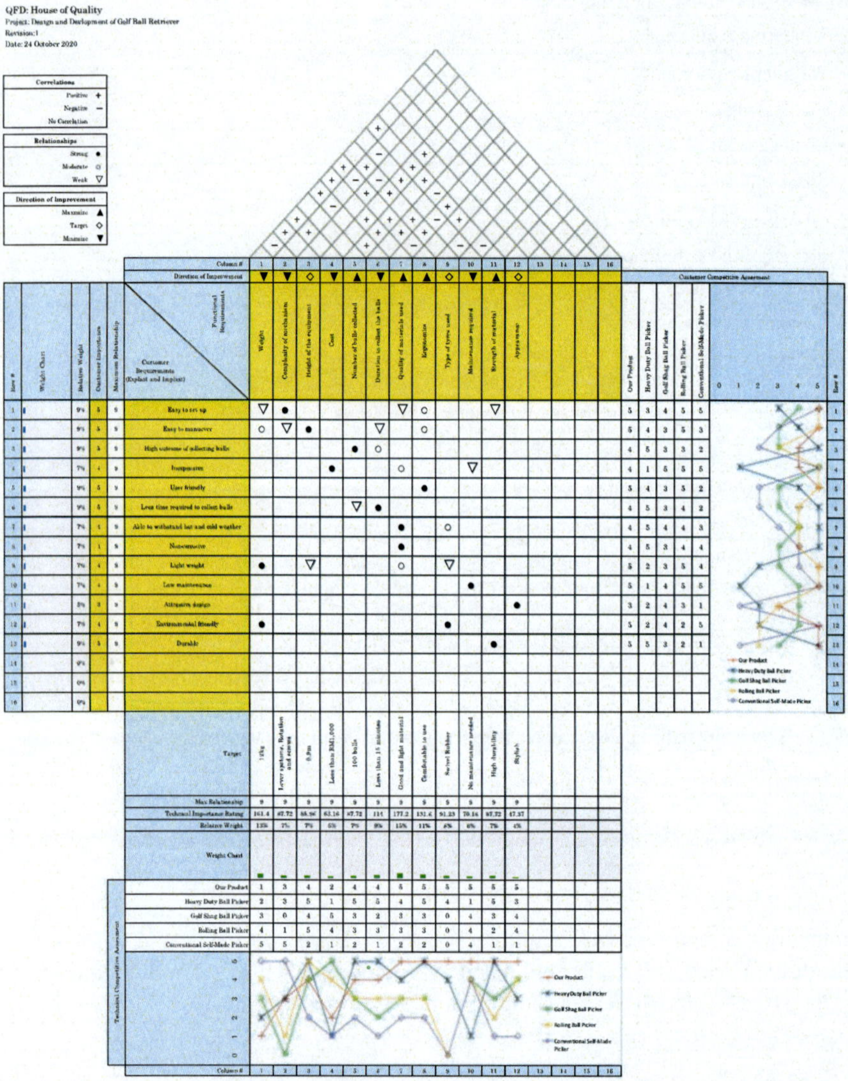

Fig. 40.1 Level 1 house of quality

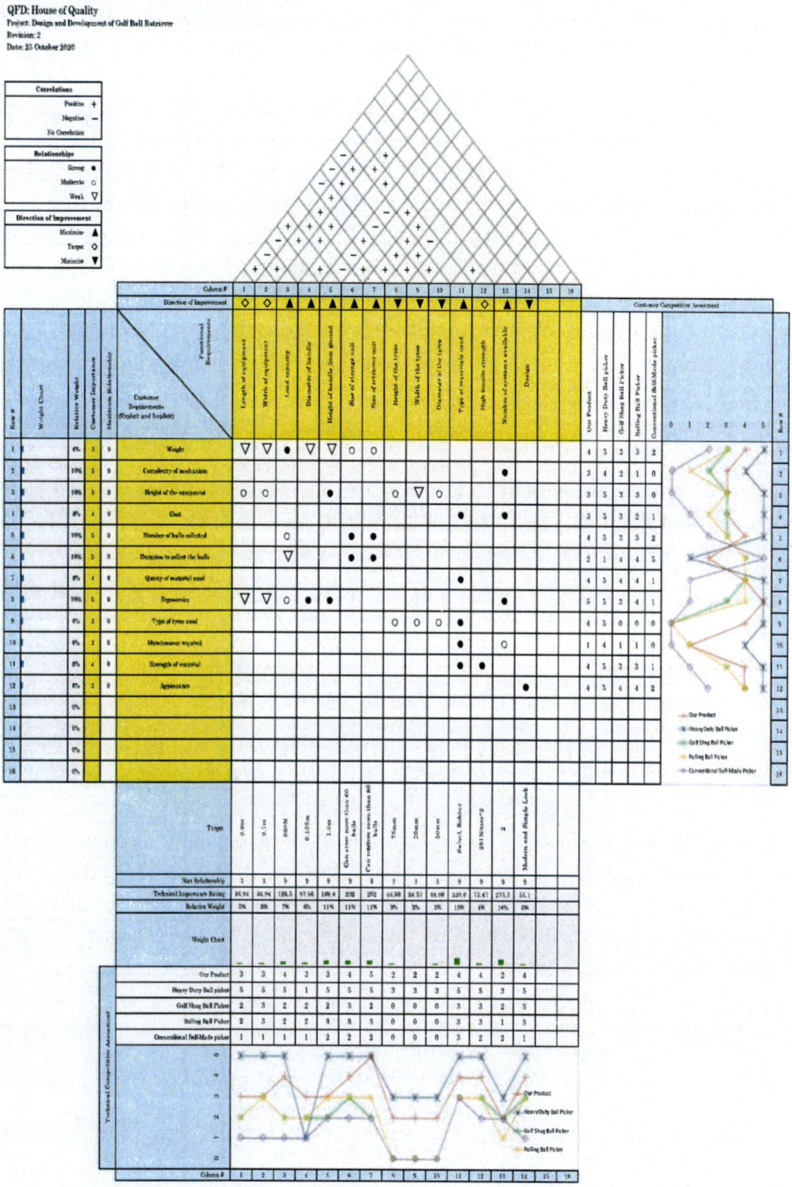

Fig. 40.2 Level 2 house of quality

issues to meet the customers' anticipations quickly and effectively [11–15]. The first level of QFD is the design requirements or system process where the voice of customers (VOC) or "WHATS" are collected and translated into product specifications. It is also a competitive analysis where the competitors are evaluated based on customer requirements. This initial design concept is based on product performance and product specifications. Apart from that, this phase on QFD will be a reference for future phase of development. Besides that, the importance factor is rated based on customers' priorities. The "HOWS" or design features and technical requirements are aligned with the VOC. After that, the "HOWS" are ranked according to the correlation of each of the "WHATS" fulfilling the customers' requirement. The roof ranking system is used to indicate how the design requirement interacts with each other. The Fig. 40.1 is the first level of QFD.

The second phase of QFD is the system or parts' requirements or product development process which includes the identification of critical parts and assemblies. Besides, the critical product characteristics are evaluated and the functional specifications are then defined. The steps are similar to level 1 of QFD but the "HOWS" and "WHATS" are different. Figure 40.2 is the structure of level 2 QFD.

40.4 Conclusion

This survey has drawn a clear vision of what this study must emphasize in designing and developing the golf ball retriever. It could be a first step where most of the companies would make to obtain the feedback from customers or users. Additionally, the benchmarking that is set by these respondents must be taken into consideration before executing a product design. The preferences and priorities of the users should be aligned with goals of this product design and development. Implementing QFD is very important to verify both customers' and engineering requirements before going to the next design process which is concepts generation and CAD analysis.

Acknowledgements The authors would like to acknowledge Universiti Tenaga Nasional (UNITEN) for giving the YCU Grant 202210050YCU to publish this paper. The authors would also like to thank fellow authors from Universiti Malaysia Perlis (UniMAP) and Universiti Teknikal Malaysia Melaka (UTeM) for contributions in this paper.

References

1. C. McGrath, D. McCormick, J. Garrity, *The Ultimate Golf Book* (Houghton Mifflin, Boston, 2006)
2. About IGFGolf (2020) [Online]. https://www.igfgolf.org/about-igf
3. H. Bryan, Leisure value systems and recreation specialization: the case of trout fishermen. J. Leis. Res. **9**, 174–187 (1977)

4. D. Tassiopoulos, N. Haydam, Golf tourists in South Africa: a demand-side study of a niche market in sports tourism. Tour. Manage. **29**, 870–882 (2007)
5. B. Ayneendra, M. Salman, J. Ojyok Attari, E. Marvan, Trolley turning and lifting mechanism. Int. J. Latest Eng. Res. Appl. (IJLERA) **02**(05), 67–74 (2017).
6. S.-L. Wu, M.-Y. Cheng, W.-C. Hsu, Design and implementation of a prototype vision-guided golf-ball collecting mobile robot, in *IEEE International Conference on Mechatronics*, 2005. ICM'05, Taipei, pp. 611–615 (2005). https://doi.org/10.1109/ICMECH.2005.1529329
7. I. Elamvazuthi, J. Law, V. Singh, M.K.A. Ahamed Khan, S. Parasuraman, M. Balaji, M. Chandrasekaran, Development of an autonomous tennis ball retriever robot as an educational tool. ELSEVIER Proc. Comput. Sci. **76**, 21–26 (2015). ISSN 1877-0509. https://doi.org/10.1016/j.procs.2015.12.270
8. A. Roihan, P.A. Sunarya, C. Wijaya, Auto tee prototype as tee golf automation in golf simulator studio, in *2018 6th International Conference on Cyber and IT Service Management (CITSM)*, Parapat, Indonesia (2018), pp. 1–5. https://doi.org/10.1109/CITSM.2018.8674249
9. D.S. Pottruck, T. Pearce, *Listening to Customers in the Electronic Age*. Fortune, 1 May 2000
10. J. Hauser, D. Clausing, *The House of Quality*. Harvard Business Review (1988)
11. I.N. Ismail, K.A. Halim, K.S.M. Sahari, A. Anuar, M. F. A. Jalal, F. Syaifoelida, M.R. Eqwan, Design and development of platform deployment arm (PDA) for boiler header inspection at thermal power plant by using the house of quality (HOQ) approach, in *Procedia Computer Science*, vol. 105, pp. 296–303 (8 p) (2017)
12. I.N. bin Ismail, A.H. bin Iskandar, M.R. Eqwan, A.W.M. Zuhdi, D. Mohamad, M.R. Isa, N.M. Zahari, M.H. Zawawi, H. Mohamed, M.Z. Ramli, M.H. Mansor, Design and development an automatic plant pot prototype, in *Green Design and Manufacture: Advanced and Emerging Applications. AIP Conference Proceedings*, vol. 2030, pp. 020201-1–020201-7 (2018)
13. I.N. bin Ismail, P. Jayakumar, M.R. Eqwan, A.W.M. Zuhdi, D. Mohamad, M.R. Isa, N.M. Zahari, M.H. Zawawi, H. Mohamed, M.Z. Ramli, M.H. Mansor, Design and development of smart sorting recycle bin prototype, in *Green Design and Manufacture: Advanced and Emerging Applications. AIP Conference Proceedings*, vol. 2030, pp. 020202-1–020202-8 (2018)
14. I.N. Ismail, A. Anuar, K.S.M. Sahari, M.Z. Baharuddin, M. Fairuz, A. Jalal, J.M. Saad, Development of in-pipe inspection robot: a review, in *2012 IEEE Conference on Sustainable Utilization and Development in Engineering and Technology, STUDENT 2012—Conference Booklet*, pp. 310–315 (6 p.) (2012)
15. K.S.M. Sahari, A. Anuar, S.S.K. Mohideen, M.Z. Baharuddin, I.N. Ismail, N.M.H. Basri, N.S. Roslin, M.A. Aziz, B. Ahmad, Development of robotic boiler header inspection device, in *6th International Conference on Soft Computing and Intelligent Systems, and 13th International Symposium on Advanced Intelligence Systems, SCIS/ISIS 2012*, pp. 769–773 (5 p.) (2012)

Chapter 41
The Effect of GGBFS and Additional Cement, Water, and Superplasticizer on the Mechanical Properties of Workable Geopolymer Concrete

Iqlima Nuril Amini and Januarti Jaya Ekaputri

Abstract Obtaining a consistent performance of geopolymer concrete for structural application is not simple. It is influenced by the main material, alkali activator, chemical additive, etc. In this paper, the mechanical properties of fresh and hardened fly ash-based-geopolymer concrete (FAGC) with 10% GGBFS and the addition of cement, water, and superplasticizer are investigated. The specimens were made using 10 M solutions from two types of alkaline: KOH and NaOH. Each solution was mixed with Na_2SiO_3 with a mass ratio of 1. The cylindrical specimens with a diameter of 50 mm and a height of 100 mm were tested for compressive strength, splitting tensile strength, and porosity. Setting time and slump test were also conducted for fresh concrete. It was found that GGBFS accelerated the setting time, reduced the workability, and increased the strength. The type of alkaline activator also influenced both of fresh and hardened properties. In the presence of GGBFS, specimens with KOH produced better workability, higher compressive strength, and lower porosity than those with NaOH. Furthermore, the addition of cement, water, and superplasticizer improved workability and mechanical properties.

Keywords Geopolymer · GGBFS · Superplasticizer · Alkaline activator

41.1 Introduction

Geopolymer is one of the green concrete technologies. Its non-Portland binder is rich in aluminosilicate. The binder is activated with the alkaline activators, such as sodium and potassium to form geopolymerization [1]. Geopolymer concrete has excellent mechanical properties, high bond strength, and high durability such as chemical attack and fire resistance [2–6] which is influenced by its matrix permeability or the porosity. Geopolymer concrete is denser and has lower porosity and permeability

I. N. Amini · J. J. Ekaputri (✉)
Department of Civil Engineering, Institut Teknologi Sepuluh Nopember, Surabaya 60111, Indonesia
e-mail: januarti@ce.its.ac.id

M. A. A. Mohd Salleh et al. (eds.), *Proceedings of the Green Materials and Electronic Packaging Interconnect Technology Symposium*, Springer Proceedings in Physics 289,
https://doi.org/10.1007/978-981-19-9267-4_41

than Portland concrete [7–9]. This is one of the reasons for its greater durability to against aggressive environment. Geopolymer can also improve soil properties. It has excellent ability as soil stabilizing material and repair cracked in soil [10]. It also has a good impact for repair the soil contaminated waste because of its high strength and low permeability [11]. This proves that geopolymer is a promising material for green construction and concrete industry.

One of the factors influencing the mechanical properties of geopolymer concrete is alkaline activator. In general, soluble silicate is combined with potassium hydroxide (KOH) or sodium hydroxide (NaOH) [12, 13]. The type, concentration, and ratio of the alkali activator affect both of fresh and hardened concrete [14–17]. Saxena and Kumar [17] studied the effect of different types of alkaline activator. It resulted that using KOH provides higher compressive strength than NaOH in 80 °C curing temperature because of increasing polymerization. Rocha et al. [18] also showed the combination of KOH and Na_2SiO_3 produces the same compressive strength, with a combination of NaOH and Na_2SiO_3 up to 80 MPa at room temperature. The two types of alkaline activators produce different activation properties because of the different size of Na^+ and K^+ ion [19].

Industrial waste, including fly ash and ground granulated blast furnace slag (GGBFS), is often used as an aluminosilicate source [20, 21]. Fly ash production in Indonesia has been predicted by approximately 6% of the total coal in power plants [22]. It has been studied that it contributes to excellent mechanical properties and durability for geopolymers [23]. However, curing temperature is one factor that influences the mechanical properties of fly ash-based-geopolymer concrete (FAGC) [24]. Sometimes, alkaline activator cannot increase fly ash reactivity because of the slow rate of chemical reaction at low temperatures [25, 26]. Using GGBFS has been indicated to improve the reactivity of fly ash at room temperature and contributed to producing C–A–S–H formation because of high of CaO content [27].

One problem of fresh geopolymer concrete is poor workability because of its rapid setting. This makes the application of geopolymer concrete difficult, especially as a thin-walled concrete material such as pipes that require a good workability. The workability was contributed by water, which is the geopolymerization product [28, 29]. The addition of water to the fresh geopolymer concrete increases the workability, but it decreased the compressive strength and increased the porosity [29]. In contrast to Portland concrete, which uses a superplasticizer to increase the workability, superplasticizer cannot work optimally on geopolymer mixtures because of high alkaline media [30]. However, another study stated that superplasticizer can improve workability, it used low volume Portland cement to trigger superplasticizer [31, 32]. Hence, a geopolymer concrete mixture proportion with good workability is required. This paper will investigate the effect of GGBFS and additional cement, water, and superplasticizer on the mechanical properties of fresh and hardened geopolymer concrete.

Table 41.1 Chemical composition of fly ash, GGBFS, and cement

Chemical composition (%)	Fly ash	GGBFS	PCC
SiO_2	54.31	39.91	19.19
Al_2O_3	17.62	13.09	5.42
Fe_2O_3	10.14	0.51	3.11
CaO	5.62	42.44	63.08
MgO	2.98	1.17	1.77
Na_2O	3.57	0.34	0.22
K_2O	1.60	0.31	0.57
SO_3	0.86	1.15	1.54
C_3S	–	–	59.23
C_3A	–	–	9.13
C_4AF	–	–	9.44
C_2S	–	–	10.43
LOI	3.3	1.08	N/A

41.2 Methodology

41.2.1 Materials

The main material in this mixture is class F fly ash obtained from Tanjung Jati power plant. Ground Granulated Blast Furnace slag (GGBFS), Portland composite cement (PCC), water, and superplasticizer (SP) were also used as the additional materials. Superplasticizer used is Sika viscocrete 1003 with the chemical composition is modified polycarboxylate copolymers. The chemical composition obtained from X-ray Fluorescence (XRF) analysis is presented in Table 41.1. The activator's solution comprises two types of alkaline: sodium hydroxide (NaOH) and potassium hydroxide (KOH) with a concentration 10 M. Sodium silicate (Na_2SiO_3) was used combined with NaOH or KOH in the ratio of 1 by mass. Both of molarity and alkali ratio were changed due to the additional water. The molarity decreased from 10 M to 7.06 M for KOH and 7.11 M for NaOH. Meanwhile, the alkali ratio changed to 1.03. Natural sand and crushed stone were used as aggregate in saturated surface dried (SSD) condition.

41.2.2 Method

The mixture proportion contains 6 variations are presented in Table 41.2. The mixing procedure is divided into four steps. Step 1 is dry mixing of fly ash and cement. Step 2 continued by mixing the liquid components (alkaline hydroxide, water, and SP) first and continued with sodium silicate separately. This step aimed to activate the fly

Table 41.2 Mix proportion

Code Material (kg/m^3)	N.G (1)	N.G.SP (2)	N.G.A (3)	N.A (4)	K.G.A (5)	K.A (6)
Fly ash	518.4	518.4	518.4	518.4	518.4	518.4
GGBFS	51.8	51.8	51.8	–	51.8	–
Coarse aggregate	985.0	985.0	985.0	1008.0	985.0	1008.0
Fine aggregate	648.6	648.6	648.6	672.0	648.6	672.0
NaOH	100.8	100.8	100.8	100.8	–	–
KOH	–	–	–	–	100.8	100.8
Na_2SiO_3	100.8	100.8	100.8	100.8	100.8	100.8
Cement	–	–	15.6	15.6	15.6	15.6
Superplasticizer	–	10.4	10.4	10.4	10.4	10.4
Water	–	–	31.1	31.1	31.1	31.1

ash with the hydroxide solution. Step 3 was mixing the natural sand and continuing with the crushed stone. Step 4 was mixing the GGBFS. GGBFS was added at the last step to avoid the rapid setting of the mixture because of its high reactivity. The fresh geopolymer concrete was placed into a cylindrical mold with a diameter of 50 mm and a height of 100 mm. Specimens were demolded after 1 day of casting and cured at room temperature in a moist condition until the test time. All hardened specimens were tested for compressive strength, tensile strength, and porosity. Fresh geopolymer concrete and fresh paste were tested for slump test. A standard Vicat test was conducted for fresh paste The mini slump was used to test geopolymer paste workability with the size 1:4 by the ratio of the Abram cone's original size. All the tests were carried out according to American Standard Testing and Material (ASTM).

41.3 Result and Discussions

41.3.1 Physical Properties of Fresh Geopolymer Mixture

The geopolymer paste slump test was carried out using a 7.5 cm height-mini-slump cone with a top diameter of 25 mm and the bottom diameter of 50 mm (Fig. 41.1a). The slump measurement was determined based on the flow diameter (d) (Fig. 41.1b).

Six proportions were used to identify the paste slump by using GGBFS and the addition of cement, water, and SP (Table 41.2). The mixtures 3 (N.G.A) and 5 (K.G.A) were stiffer than the mixtures 4 (N.A) and 6 (K.A). This indicated that adding GGBFS produced less workable mixture. Using a superplasticizer (N.G.SP) was expected to increase the workability of geopolymer paste with GGBFS, but the workability remained the same. Using the combination of additional cement, water, and superplasticizer (N.G.A and K.G.A) contributes to more workable mixture.

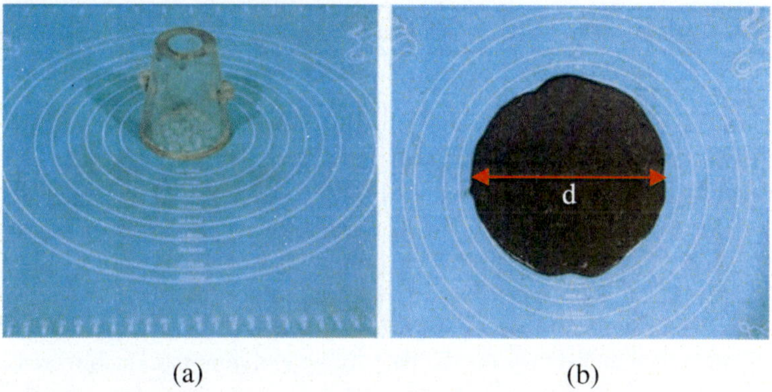

Fig. 41.1 Slump test of geopolymer paste using a mini slump **a** mini slump **b** paste slump

The fresh geopolymer concrete workability was tested using Abram's cone according to ASTM C-143. All proportions showed great flow (Fig. 41.2). The measurement was conducted by counting the time of the mixture to flow until ± 56 cm (the diameter of concrete to full flow). The difference between the whole mixture is the flow time (Table 41.3).

Like geopolymer paste, the use of GGBFS in geopolymer concrete makes the mixture stiffer. Adding superplasticizer showed less contribution to the flowability. However, combining cement, water, and superplasticizer increased the workability. The mixture flow faster and produce excellent workability.

Fig. 41.2 Slump test of geopolymer concrete using Abrams cone

Table 41.3 Paste slump

Mix variation	N.G	N.G.SP	N.G.A	N.A	K.G.A	K.A
Concrete slump flow time (s)	55	83	45	23	16	13

41.3.2 *Effect of GGBFS and Alkaline Activator on Geopolymer Concrete Mechanical Properties*

The proportion of mix no. 3 (N.G.A), 4 (N.A), 5 (K.G.A) and 6 (K.A) were selected to determine the effect of alkali activator and GGBFS. The slump obtained in geopolymer paste is proportional to the time for geopolymer concrete to flow. Figure 41.3 shows the slump results on geopolymer concrete and paste activated with different alkaline. Using KOH (K.A) resulted in a higher diameter of paste slump and the fastest flow time. GGBFS also affected the workability of both alkaline activators. It reduced the flow diameter of geopolymer paste and increased the flow time.

Figure 41.4 shows the effect of alkali activator type and GGBFS on the setting time of geopolymer paste. The fastest of both initial and final setting was obtained from K.G.A. It is shown that using KOH as an alkaline activator resulted in the fastest setting time than NaOH. However, this is contrary to the slump test results for different activators. Using KOH resulted in excellent workability but generate the fast setting compared to NaOH. Leong et al. [33] also showed the same result. It was caused by higher solubility of KOH. In addition, GGBFS in both alkaline activators reduced the initial and final set. This proves that adding GGBFS to the proportion was effective in accelerating the setting at room temperature. This result is comparable to the slump test results in which GGBFS decreases the workability because of the fast setting. According to Ekaputri et al. [34], BFS accelerated the solidification process and caused shrinkage. This was also caused by the hydration reaction which accelerated the setting time.

The compressive strength and tensile strength are shown in Figs. 41.5 and 41.6. K.G.A provides higher compressive strength than other proportion. However, it is different from non-slag concrete. N.A. resulted higher compression test than K.A. The difference between both alkaline is the ionic size. K^+ is higher than Na^+, which contributes to the formation of larger silicate oligomers with which $Al(OH)_4$ prefers to bind; it makes more geopolymer precursors exist and contributes to better setting and higher compressive strength [13]. In addition, adding GGBFS to the proportion

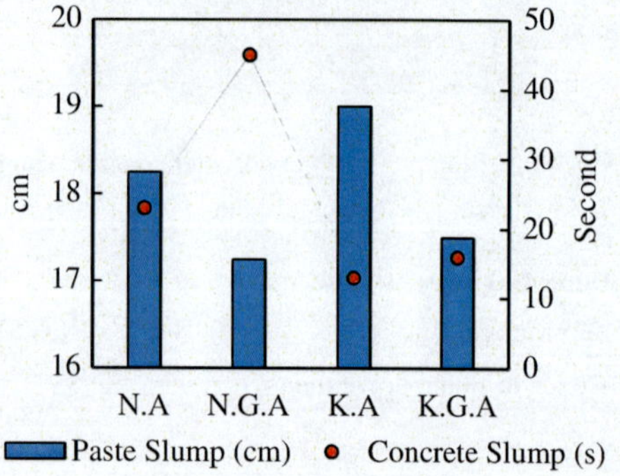

Fig. 41.3 Effect of GGBFS and alkaline activator on the paste and concrete slump

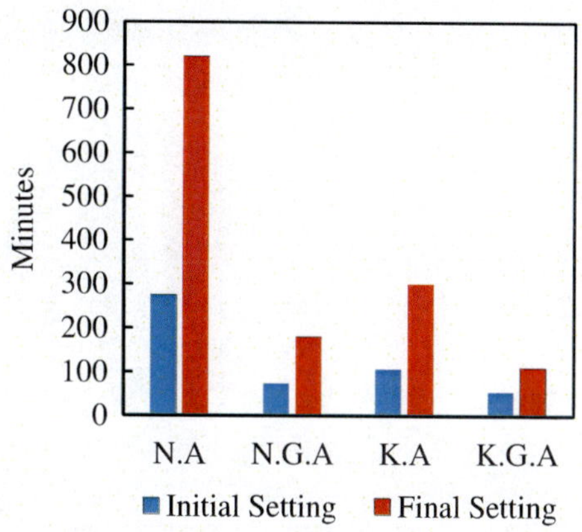

Fig. 41.4 Effect of GGBFS and alkaline activator on the setting time

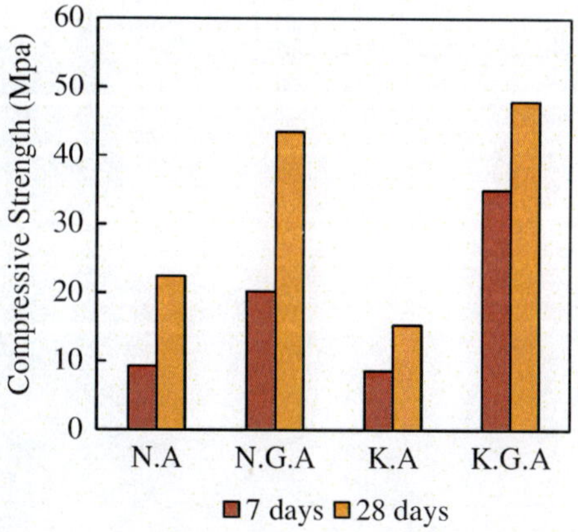

Fig. 41.5 Effect of GGBFS and alkaline activator on the compressive strength of concrete

increased compressive strength at 7 and 28 days. Puligilla and Mondal [27] found that this was because of the contribution of the GGBFS reaction, which produces an additional binding product containing calcium aluminosilicate hydrate (C–A–S–H) gel.

Similar with compressive strength, the highest non-slag concrete tensile strength was obtained in the geopolymer concrete activated with NaOH. In addition, it is contrary when GGBFS is added. The highest split tensile strength of FAGC with GGBFS was obtained on geopolymer concrete activated with KOH. Figures 41.7 and 41.8 represent the physical conditions of splitting tensile test. The presence of GGBFS resulted in a smoother split surface than non-slag specimen. Split aggregates showed that the slag matrix performed better than the non-slag concrete indicating the slag contributes good binding at the interfacial transition zone. This also revealed that GGBFS increased the hardness of the matrix. Gao et al. [35] also found that GGBFS strengthens the bond of the paste with the aggregate and made the microstructure denser.

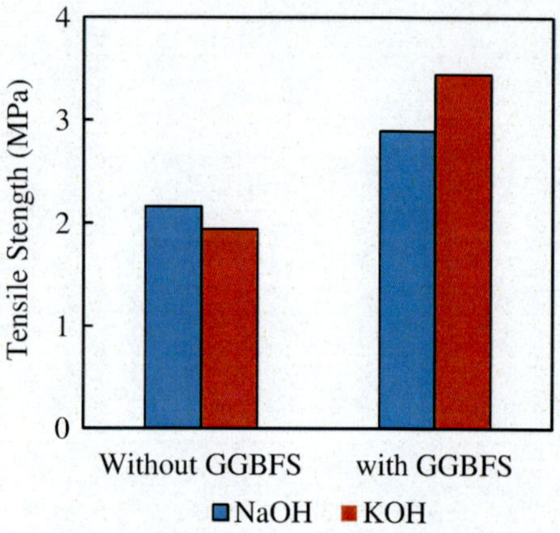

Fig. 41.6 Effect of GGBFS and alkaline activator on the tensile strength

Fig. 41.7 Splitting tensile test for **a** non-slag concrete and **b** slag concrete activated with KOH

Fig. 41.8 Splitting tensile test for **a** non-slag concrete and **b** slag concrete activated with NaOH

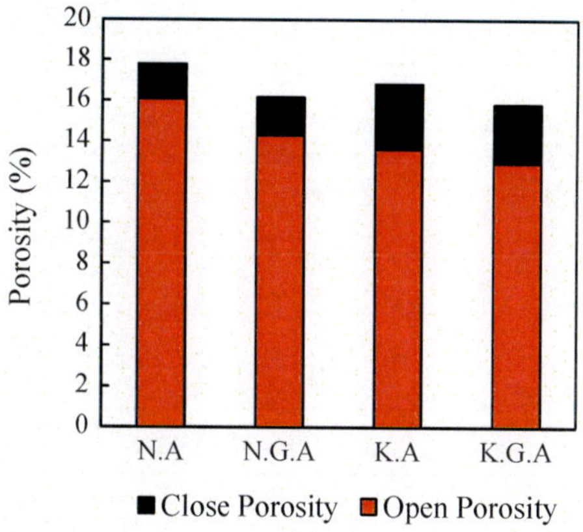

Fig. 41.9 Effect of GGBFS and alkaline activator on the concrete porosity

Workability and compaction of the mixtures change the porosity. Poor workability produces macropores and decreases the mechanical properties [36]. Figure 41.9 shows the results of porosity. Total porosity represents the number of closed and open porosities. Using NaOH produces higher porosity than KOH does. However, the use of GGBFS in both alkaline conditions was indicated to decrease the porosity. These results are related to the compressive strength in Fig. 41.5. Lower porosity indicates a denser matrix structure and results in high compressive strength. It also contributed from GGBFS to produces a dense structure. According to Bernal et al. [37], higher porosity of FAGC due to unreacted particles from fly ash and more porous N–A–S–H gel product.

41.3.3 Effect of Additional Cement, Superplasticizer, and Water on Geopolymer Concrete Mechanical Properties Containing GGBFS

The proportion of Mix no. 1 (N.G), 2 (N.G.SP), and 3 (N.G.A) used to determine the effect of additional cement, water, and superplasticizer. The aim of adding these additional was to improve the workability of the mixture. Figure 41.10 shows the result of a geopolymer paste and concrete slump. Using SP (N.G.SP) showed an increase in the slump paste diameter and the slump flow time. It differs from the proportion of N.G.A. This proportion obtains a higher slump paste diameter and the fastest time to flow. Cement was added to improve superplasticizer performance and additional water was required for cement hydration. These additional worked optimally and increased the paste slump diameter to 23% and accelerate the flow time to 18%. This proportion is highly recommended to produce self-compacting geopolymer concrete (SCGC) with some modifications to meet the requirements of self-compacting concrete (SCC).

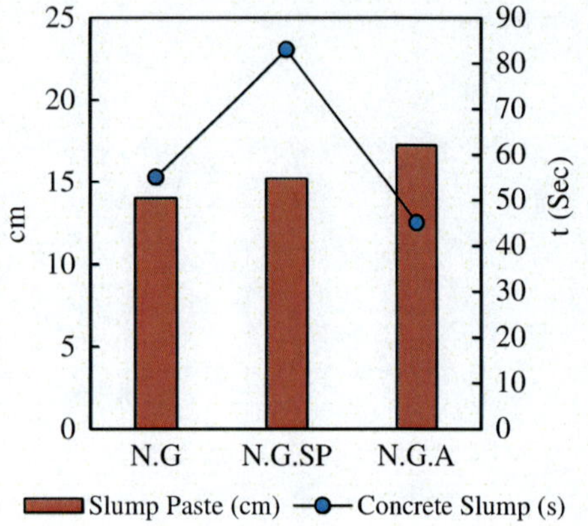

Fig. 41.10 Effect of additional cement, superplasticizer, and water on the paste and concrete slump

Figure 41.11 shows the result of setting time of paste. Using SP on the mixture N.G.SP gives a retarding effect. A previous study also showed the same result that polycarboxylate-based superplasticizer provides a retarding effect in high alkaline media [38]. In addition, additional cement, water, and SP (N.G.A) accelerated the setting time compared to the control (N.G). This is because of the calcium content in cement forming a hydration reaction. It contributed to acceleration of the setting time at room temperature [39].

The compressive strength results at 7 and 28 days are shown in Fig. 41.12. N.G.SP provides higher compressive strength in 7 days. Jang et al. [38] also got the same result by using 2% SP positively affects early age strength. At 28 days, N.G.A provides higher compressive strength. It increases the compressive strength to 7% compared to N.G. It is also a contribution from cement addition. Cement utilizes the released water from geopolymerization to form a hydration reaction. Hydration reaction provides heat and forms $Ca(OH)_2$, which causes increasing the alkalinity, enhancing fly ash reactivity [40]. The different behavior is shown in Fig. 41.13. The

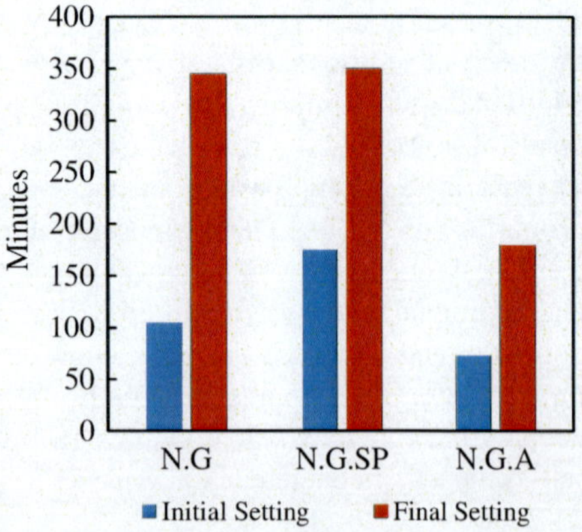

Fig. 41.11 Effect of additional cement, superplasticizer, and water on the setting time

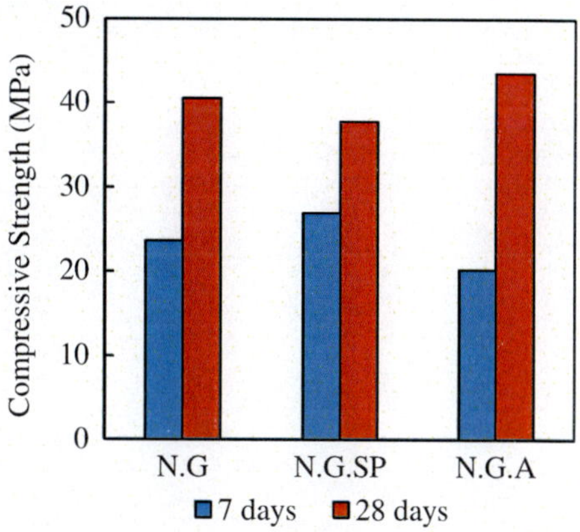

Fig. 41.12 Effect of additional cement, superplasticizer, and water on compressive strength

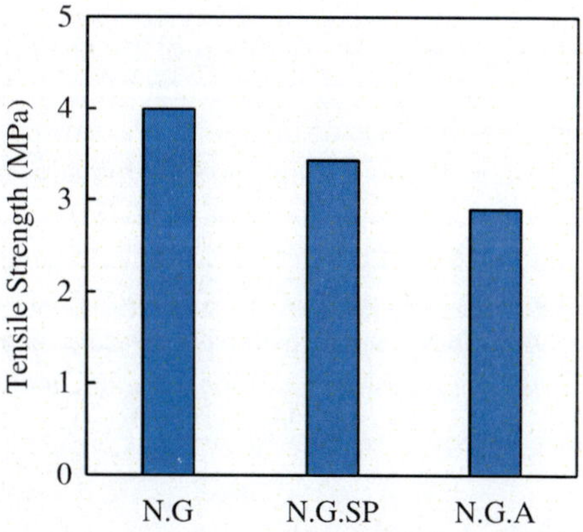

Fig. 41.13 Effect of additional cement, superplasticizer, and water on tensile strength

tensile strength decreased when the additional water, cement, and superplasticizer were added. According to Aliabdo et al. [29, 41], the addition of water reduces the tensile strength, but the addition of cement increases the tensile strength. This indicates that in this proportion, the tensile strength was decreased because of the addition of water.

Figure 41.14 represents the result of porosity affected by additional cement, water, and superplasticizer. N.G obtained higher total porosity. Higher porosity of N.G occurred because of a viscous mixture, resulting in air. Porosity was decreased when SP was added. The lowest total porosity was obtained by geopolymer concrete with additional cement, water, and superplasticizer (N.G.A). This result is related to the good compressive strength and workability of N.G.A. Another study explained that added water only on geopolymer mixture obtained good workability because in geopolymerization water showed less contribution to the chemical reaction, but it obtained high porosity and decreased concrete strength [29]. This statement relates to the high open porosity in N.G.A due to additional water and decreasing of the

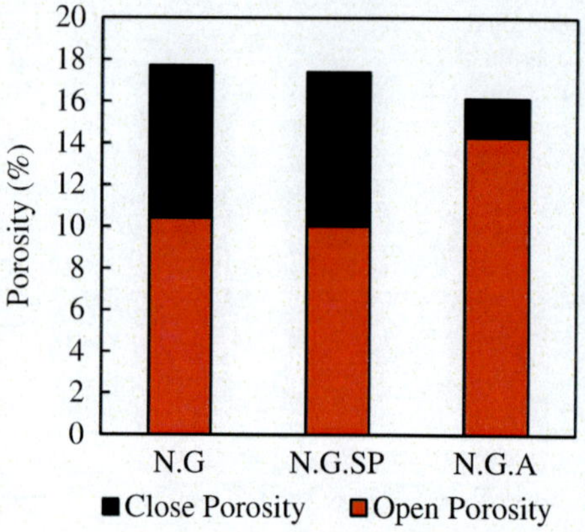

Fig. 41.14 Effect of additional cement, superplasticizer, and water on the concrete porosity

close porosity. The addition of water decreased molarity and increased the alkali ratio. This made the matrix more porous but still produce an excellent mechanical property because of the mixture is self-compacted. This also guaranty the solidification and reduce total porosity. In addition, total porosity reduced by utilizing the hydration reaction of cement with free water in the geopolymer [41]. Consequently, combination of GGBFS, SP, and cement improved the mechanical properties of fresh and hardened concrete. This mixture also has a potential to create denser matrix as it showed less total porosity. Therefore, it can be applied as a self-compacting mixture to against aggressive ion for durable concrete.

41.4 Conclusion

The effect of GGBFS and additional cement, water, and superplasticizer on fly ash-based-geopolymer concrete were investigated. The results of the study are summarized as follows:

1. GGBFS improved the compressive strength and tensile strength. It increases to 48 and 25% when activated with NaOH and 68 and 44% when activated with KOH. This indicates that reactivity of fly ash-based-geopolymer increased with the additional of slag. GGBFS also contributes a positive effect total porosity reduction. The slag may induce hydration reaction as an additional process aside to geopolymerization. However, GGBFS has a disadvantage of reducing workability. The flow time getting slower to 49% when GGBFS was added. This is related to the faster setting time.
2. The addition of cement, water, and superplasticizer increased the workability to 18%. It has a positive effect on increasing the mechanical properties. The mechanical strength increased to 7%, but the tensile strength decreased to 27%. The total porosity also decreased to 9%, but the open porosity increased. This is

a contribution of the double reaction from cement hydration aside from geopolymerization. This composition is recommended for geopolymer concrete mixtures that require excellent workability, such as self-compacting geopolymer concrete (SCGC).

Acknowledgements The authors gratefully acknowledge financial support from the financial support of the World Class funding by Universitas Diponegoro through the decreed number 11-20/UN7.6.1/2021.

References

1. P. Zhang, Z. Gao, J. Wang, J. Guo, S. Hu, Y. Ling, Properties of fresh and hardened fly ash/slag based geopolymer concrete: a review. J. Clean. Prod. **270**, 122389 (2020)
2. M. Albitar, M.S. Mohamed Ali, P. Visintin, M. Drechsler, Durability evaluation of geopolymer and conventional concretes. Constr. Build. Mater. **136**(April 2019), 374–385 (2017)
3. F. Qu, W. Li, K. Wang, S. Zhang, D. Sheng, Performance deterioration of fly ash/slag-based geopolymer composites subjected to coupled cyclic preloading and sulfuric acid attack. J. Clean. Prod. **321**, 128942 (2021)
4. L.N. Halim, J.J. Ekaputri, Triwulan, The influence of salt water on chloride penetration in geopolymer concrete. MATEC Web Conf. **97** (2017)
5. A. Rahmadina, J.J. Ekaputri, Mechanical properties of geopolymer concrete exposed to combustion. MATEC Web Conf. **138** (2017)
6. E.S. Dewi, J.J. Ekaputri, The influence of plain bar on bond strength of geopolymer concrete. AIP Conf. Proc. **1855** (2017)
7. M.C.M. Nasvi, P.G. Ranjith, J. Sanjayan, The permeability of geopolymer at down-hole stress conditions: application for carbon dioxide sequestration wells. Appl. Energy **102**, 1391–1398 (2013)
8. A.A. Aliabdo, A.E.M.A. Elmoaty, M.F. Mohamed, Permeability indices and corrosion resistance of geopolymer and Portland cement concretes. Mag. Concr. Res. 1988 (2017)
9. J. Kwasny, T.A. Aiken, M.N. Soutsos, J.A. McIntosh, D.J. Cleland, Sulfate and acid resistance of lithomarge-based geopolymer mortars. Constr. Build. Mater. **166**, 537–553 (2018)
10. Y.K. Firmansyah, R.A.A. Soemitro, D.D. Warnana, J.J. Ekaputri, Cracked soil behaviour treated with geopolymer for river embankment using a fuzzy C-means method. J. Eng. Sci. Technol. **17**(1), 508–522 (2022)
11. N. Sutra, R.A.A. Soemitro, D.D. Warnana, J.J. Ekaputri, T. Mukunoki, Ardiansah, Influence of fly ash-based geopolymers on treated-contaminated soil properties. Int. J. Adv. Sci. Eng. Inf. Technol. **11**(6), 2240–2246 (2021)
12. J. Davidovits, Properties of geopolymer cements, in *First International Conference on Alkaline Cements and Concretes* (1994), pp. 131–149
13. H. Xu, J.S.J. Van Deventer, The geopolymerisation of alumino-silicate minerals. Int. J. Miner. Process. **59**(3), 247–266 (2000)
14. P. Nath, P.K. Sarker, Effect of GGBFS on setting, workability and early strength properties of fly ash geopolymer concrete cured in ambient condition. Constr. Build. Mater. **66**, 163–171 (2014)
15. M. Verma, N. Dev, Sodium hydroxide effect on the mechanical properties of flyash-slag based geopolymer concrete. Struct. Concr. **22**(S1), E368–E379 (2020)
16. T. Rahmiati, K.A. Azizli, Z. Man, L. Ismail, M.F. Nuruddin, The effect of KOH concentration on setting time and compressive strength of fly ash-based geopolymer. Appl. Mech. Mater. **625**, 94–97 (2014)

17. S.K. Saxena, M. Kumar, Influence of alkali solutions on properties of pond fly ash-based geopolymer mortar cured under different conditions. Adv. Cem. Res. **30**(1), 1–7 (2018)
18. T. da Silva Rocha, D.P. Dias, F.C.C. França, R.R. de Salles Guerra, L.R.C.O. Marques, Metakaolin-based geopolymer mortars with different alkaline activators (Na+ and K+). Constr. Build. Mater. **178**, 453–461 (2018)
19. A.D. Hounsi et al., How does Na, K alkali metal concentration change the early age structural characteristic of kaolin-based geopolymers. Ceram. Int. **40**(7 PART A), 8953–8962 (2014)
20. G.S. Ryu, Y.B. Lee, K.T. Koh, Y.S. Chung, The mechanical properties of fly ash-based geopolymer concrete with alkaline activators. Constr. Build. Mater. **47**(2013), 409–418 (2013)
21. M.N.S. Hadi, N.A. Farhan, M.N. Sheikh, Design of geopolymer concrete with GGBFS at ambient curing condition using Taguchi method. Constr. Build. Mater. **140**, 424–431 (2017)
22. A. Umah, *Bukan Limbah, Potensi FABA di Indonesia Capai 11 Juta Ton* (CNBC Indonesia, 2021). [Online]. Available: https://www.cnbcindonesia.com/market/20210421154337-17-239670/bukan-limbah-potensi-faba-di-indonesia-capai-11-juta-ton.
23. M. Amran, S. Debbarma, T. Ozbakkaloglu, Fly ash-based eco-friendly geopolymer concrete: a critical review of the long-term durability properties. Constr. Build. Mater. **270**, 121857 (2021)
24. Triwulan, J.J. Ekaputri, N.F. Priyanka, The effect of temperature curing on geopolymer concrete. MATEC Web Conf. **97**, 0–5 (2017)
25. M.T. Junaid, A. Khennane, O. Kayali, A. Sadaoui, D. Picard, M. Fafard, Aspects of the deformational behaviour of alkali activated fly ash concrete at elevated temperatures. Cem. Concr. Res. **60**, 24–29 (2014)
26. A. Fernández-jiménez, Engineering properties of alkali-activated fly ash. ACI Mater. J. (2006)
27. S. Puligilla, P. Mondal, Role of slag in microstructural development and hardening of fly ash-slag geopolymer. Cem. Concr. Res. **43**(1), 70–80 (2013)
28. J. Davidovits, Geopolymers. J. Therm. Anal. **37**(8), 1633–1656 (1991)
29. A.A. Aliabdo, A.E.M. Abd Elmoaty, H.A. Salem, Effect of water addition, plasticizer and alkaline solution constitution on fly ash based geopolymer concrete performance. Constr. Build. Mater. **121**, 694–703 (2016)
30. M.S. Laskar, S. Talukdar, Influence of superplasticizer and alkali activator concentration on slag-flyash based geopolymer, in *Urbanization Challenges in Emerging Economies* (2018), pp. 330–337
31. Purwanto, J.J. Ekaputri, Nuroji, B.R. Indriyantho, A. Han, B.S. Gan, Shear-bond behavior of self-compacting geopolymer concrete to conventional concrete. Constr. Build. Mater. **321**(2021) (2022)
32. Purwanto, A. Han, J.J. Ekaputri, Nuroji, B.H. Prasetya, Self-compacting-geopolymer-concrete (Scgc) retrofitted haunch. Int. J. Eng. Appl. **9**(4), 180–189 (2021)
33. H.Y. Leong, D.E.L. Ong, J.G. Sanjayan, A. Nazari, The effect of different Na_2O and K_2O ratios of alkali activator on compressive strength of fly ash based-geopolymer. Constr. Build. Mater. **106**, 500–511 (2016)
34. J.J. Ekaputri, K. Maekawa, T. Ishida, Experimental study on internal RH of BFS mortars at early age. Mater. Sci. Forum **857**, 305–310 (2016)
35. J.M. Gao, C.X. Qian, H.F. Liu, B. Wang, L. Li, ITZ microstructure of concrete containing GGBS. Cem. Concr. Res. **35**(7), 1299–1304 (2005)
36. J.J. Ekaputri, C. Fujiyama, N. Chijiwa, T.D. Ho, H.T. Nguyen, Improving geopolymer characteristics with addition of poly-vinyl alcohol (PVA) fibers. Civ. Eng. Dimens. **23**(1), 28–34 (2021)
37. S.A. Bernal et al., Gel nanostructure in alkali-activated binders based on slag and fly ash, and effects of accelerated carbonation. Cem. Concr. Res. **53**, 127–144 (2013)
38. J.G. Jang, N.K. Lee, H.K. Lee, Fresh and hardened properties of alkali-activated fly ash/slag pastes with superplasticizers. Constr. Build. Mater. **50**, 169–176 (2014)
39. P. Nath, P.K. Sarker, Use of OPC to improve setting and early strength properties of low calcium fly ash geopolymer concrete cured at room temperature. Cem. Concr. Compos. **55**, 205–214 (2015)

40. L. Assi, S.A. Ghahari, E.E. Deaver, D. Leaphart, P. Ziehl, Improvement of the early and final compressive strength of fly ash-based geopolymer concrete at ambient conditions. Constr. Build. Mater. **123**, 806–813 (2016)
41. A.A. Aliabdo, A.E.M. Abd Elmoaty, H.A. Salem, Effect of cement addition, solution resting time and curing characteristics on fly ash based geopolymer concrete performance. Constr. Build. Mater. **123**, 581–593 (2016)

Chapter 42
Status and Challenges of Determining Sustainable Technology of Landfill Leachate Treatment on Municipal Solid Waste (MSW): An Update

Lim Jia Mei, Mohamad Anuar Kamaruddin, Rasyidah Alrozi, and Mohd Mustafa Al Bakri Abdullah

Abstract Municipal solid waste (MSW) generation is highly correlated with population growth and remains in increasing trend each year. Landfilling method is prevalent in developing country especially in Malaysia for MSW final disposal due to economical aspect and involved a simpler operational mechanism. However, leachate containing high strength of wastewater characteristic was produced when water percolates through solid waste during bio-decomposition processes. Sustainable and holistic approach for leachate treatment had sparked huge concern continuously. Particularly, advanced technologies and implementation on existing treatment techniques had received significant interest for environmental sustainability. Innovation on the physicochemical and biological treatment methods was reviewed in present study to enhance treatment ability and alleviate on the environmental impact associated with approach employed. Natural-based coagulant and adsorbent were received increasing interest as a sustainable treatment approach. The capability of physicochemical treatment in improving the leachate biodegradability plays an important role for a substantial performance in biological treatment subsequently. Therefore, various implementation methods for both physicochemical and biological treatment were summarized in the present study. The review is of utmost important in revealing the effective technologies applied to safeguard our environment as well as meeting

L. J. Mei · M. A. Kamaruddin (✉)
Environmental Technology Division, School of Industrial Technology, Universiti Sains Malaysia, 11800 Gelugor, Penang, Malaysia
e-mail: anuarkamaruddin@usm.my

M. A. Kamaruddin · M. M. Al Bakri Abdullah
Center of Excellence Geopolymer and Green Technology, Universiti Malaysia Perlis, 01000 Arau, Perlis, Malaysia

R. Alrozi
School of Chemical Engineering, College of Engineering, Universiti Teknologi MARA (UiTM), Cawangan Pulau Pinang Kampus Permatang Pauh, 13500 Permatang Pauh, Pulau Pinang, Malaysia

M. A. A. Mohd Salleh et al. (eds.), *Proceedings of the Green Materials and Electronic Packaging Interconnect Technology Symposium*, Springer Proceedings in Physics 289,
https://doi.org/10.1007/978-981-19-9267-4_42

government stringent regulation. A more sustainable, new technologies and cost-effective treatment methods are, therefore, the mutual goals of landfill manager in dire to achieve.

Keywords Solid waste · Landfill leachate · Treatment methods · Advanced treatment

42.1 Introduction

Malaysia is reportedly generating an immense amount of municipal solid waste (MSW) of around 40,566 tonnes daily in 2016, surpassing the projected waste generation of 30,000 tonnes per day projected in 2020 [1]. The population growth coupled with the change in consumption behavior and lifestyle habit increased the complexity of the waste composition along with the exponential surge of undesirable waste generation every year. In Malaysia, up to 94.5% of the municipal solid waste were being sent to disposal landfill sites while the remaining sent to waste recovery facilities for recycling or composting [2]. Massive generation of solid waste and the end-of-life disposal method in landfill had stressed on existing landfill lifespan. Most of the landfill facilities in Malaysia are non-sanitary had exacerbated on the environmental degradation. This practice has caused excessive generation of leachate when water percolates through the waste and carries high strength of pollutants and toxic substances, such as organic contaminants, ammonia, suspended solids, inorganic substances, and heavy metals could be detrimental to surrounding environment without proper control [3].

Leachate could potentially seep into soil layers causing pollution to groundwater and risking aquatic ecosystem as well as impact on social and economy. Tremendous efforts by government include action to close and rehabilitate most of existing unsanitary landfills and upgrade to sanitary landfills to protect human health and safeguard our environment from leachate exposure [4]. Various approaches have been employed in landfill leachate treatments and basically can be categorized into biological processes and physicochemical processes, offering different performance in alleviating the adverse effect of leachate discharge into water bodies. However, by utilizing biological or physicochemical treatment approach alone is inadequate to meet the regulation discharge limit. Multiple of treatments or combination of various treatment approaches was adopted to improve the treated quality. Different technology and innovations made on conventional treatment have continuously explored in enhancing treatment performances as well as toward sustainability. Therefore, present work aims to review on improvement made on conventional treatment system by evaluating on their removal performance reported in recent 10 years based on selective works to provide a better insight on the aspects in establishing effective treatment technology and help decision-makers on appropriate selection of treatment methods.

Generation of municipal waste is inevitable whenever there is development and population growth. Owing to variability of the leachate characteristic, leachate treatment system can be complicated to ensure good quality of the leachate effluent and compliance with discharge standard. Applying a suitable and appropriate treatment system can be a challenge due to their high organic loading and complex chemical composition that contain a large amount of toxic [5]. Development of new technologies which is viable and economy feasible in leachate treatment is, therefore, crucial in meeting stringent regulation act as well as safeguard our natural resources. A more sustainable, new technologies, and cost-effective treatment methods are, therefore, the mutual goals of landfill manager in dire to achieve.

The purpose of the study mainly to review and discuss on the existing and new mechanism of treatment developed for leachate treatment in recent years, which include:

(i) To evaluate on the technological innovations reported for landfill leachate treatment.
(ii) To compare the different techniques developed based on the pollutant removal efficiency of leachate treatment.
(iii) To establish the gap of knowledge for recent landfill leachate treatment in the past 10 years.

42.2 Methodology

The study was carried out based on the analysis of the published works searched from renowned databases, such as ScienceDirect, Scopus, Taylor & Francis, SpringerLink, Web of Science, IEEE Explore, and others, by applying keywords "landfill leachate treatment technology." Various documents in form of original articles, review papers, and conference papers were selected based on systematic reviews approach. The scope of study includes the treatment technologies being reported in current 10 years, basically from 2010 to 2020. In first round of review, the title, abstract, highlights, and conclusion were being screened. All related articles on leachate treatment are being categorized as depicted in Fig. 42.1. There were a total 112 documents being finalized at the end of process and will be adopted in the present study. The number of published articles selected by year for present review is increasing each year. The increasing numbers of publications suggested on the awareness of landfill leachate issue are increased at rapid pace around the world. Hence, simplification in the leachate treatment technologies evaluating on their removal performances as well as limitations is crucial as a cornerstone supporting for adequate decision-making.

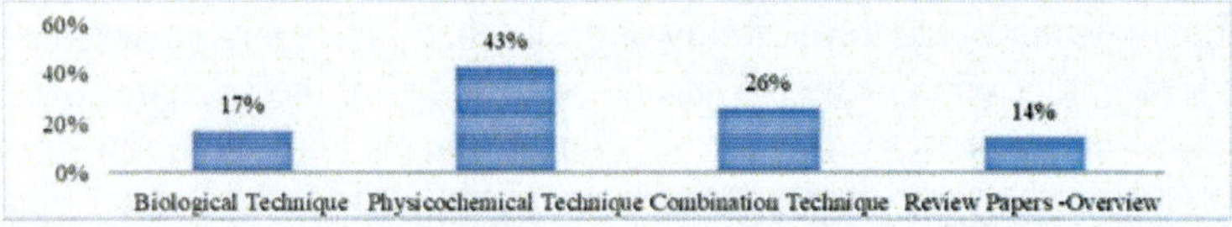

Fig. 42.1 Categories of various methods and techniques adopted in reviewed articles

42.3 Results and Discussion

The choice of the best treatment technology of landfill leachate usually considering on the few aspects, flow generation and physicochemical composition of leachate, footprint available for treatment plant, investment capacity, and operation of the landfill as well as compliance with the laws and regulations bounded by local environmental agency or authority. Economy and sustainability are mutually exclusive, and a balance in between is what stakeholders aspire to. Selection of appropriate technology that is more sustainable without compromise on the effectiveness of the respective approach adopted was mostly advocated. The improvement in conventional methods of leachate treatment in achieving satisfactory removal efficiency of treated quality is summarized in the present study.

42.3.1 Innovative Technology of Physicochemical Treatment

Physicochemical treatment methods are widely used as first stage but also key functioning of the treatment due to their discernible advantage in removing stabilized leachate with low biodegradability and high toxicity that impact on the effectiveness of biological processes. Innovative approaches were constantly gaining researchers attention in contributing toward advance technology in leachate treatment without causing secondary pollutant to environment. The wide application of alum-based coagulant caused higher environmental impacts, such as risking aquatic life and health impact associated with the prolonged exposure to aluminum. The feasibility of biodegradable and natural coagulant has been extensively investigated to replace inorganic coagulant to avoid secondary pollution as well as steering toward sustainable treatment solution. The performance of different natural coagulant was summarized as in Table 42.1.

Adsorption using activated carbon is well known with its high efficiency in removing organic and inorganic contaminant from stabilized landfill leachate as well as on its simplicity in design and ease of operation. However, the absorbent cost is crucial criteria in evaluating the feasibility of the treatment process as the frequent process of regeneration of columns or an equivalent high consumption of powdered activated carbon. The food waste derivative-based activated carbon is expected to contribute toward a sustainable environment by satisfying both environmental and

Table 42.1 Performance of landfill leachate treatment using natural coagulant

Coagulation	Leachate characteristic	Treatment condition	Performance removal %
Guar gum seed [6]	COD 1020 mg/L; pH 8.5; SS 720 mg/L; NH_4–N 2571 mg/L	Dosage 44.39 mg/L; pH 8.56 mixing speed 79.27 rpm	COD 22.57%
Tannin-based natural coagulant [3]	COD 893 mg/L, NH_3–N 531 mg/L, Fe 0.8 mg/L, Zn 280 µg/L, Cu 42 µg/L, Cr 45 µg/L, Cd 0.6 µg/L, Pb 4 µg/L, As 17 µg/L, Co 11 µg/L	Tannin dosage 0.73 g, pH 6, 45 min	COD 52.8%, NH_3–N 66.3%, Fe 89%, Zn 94%, Cu 94%, Cr 90%, Cd 17%, Pb 94%, As 86%, Co 84%
Dimocarpus longan seed powder (LSP) [7]	COD 3036.82 mg/L SS 745 mg/L Color 5517.5 PtCO	PACl as coagulant at dosage 5 g/L, pH 6 LSP as main coagulant at dosage 2 g/L, pH 4 LSP as coagulant aid with PACl LSP dosage 2 g/L PACl dosage 2.75 g/L; pH 6	COD 67.44%, SS 99.47%, color 98% COD 39.4%, SS 22.2%, color 28.3% COD 69.19%, SS 99.5%, color 98.8%
Oil palm trunk starch (OPTS) and cross-link oil palm trunk starch (C-OPTS) [8]	COD 3036.82 mg/L Turbidity 306.25 NTU SS 745 mg/L Color 5517.5 PtCO Mn 5.83 mg/L PO_4^{3-} 53.25 mg/L NH_3–N 794 mg/L	PAC dosage 8000 mg/L, pH 7 OPTS as main coagulant: Dosage: 500 mg/L, pH 7 C-OPTS as main coagulant: Dosage: 1000 mg/L, pH 8.3	COD 29.5%, turbidity 90%, SS 88.8%, color 98.9%, Mn 71.43%, PO_4^{3-} 87.9%, NH_3–N 5% COD 45.6%, turbidity 38%, SS 36.3%, color 24.7%, Mn 57.1%, PO_4^{3-} 100%, NH_3–N 12.6% COD 57.5%, turbidity 43.2%, SS 25%, color 29.8%, Mn 100%, PO_4^{3-} 100%, NH_3–N 3.34%

economic aspects. The optimum performance among low-cost adsorbent derived from food waste was evaluated and summarized in Table 42.2.

Magnetic adsorbent is a fairly new technology recently attracted much attention due to capability in shortening the separation time and increasing the capacity of treatment plants [19, 20]. Magnetic adsorbent also attended to the problem of recovery of conventional powdery adsorbents, frequently used in targeting pollutants, such as humic acid (HA) and fulvic acid (FA). The efficiency of magnetic adsorbent is usually bound to material structural and target pollutants [21].

Although magnetic adsorbent is much costly and eventually higher environmental impact in current study compared to conventional activated carbon, a lot more research is required before accurate conclusion can be made on the feasibility of magnetic adsorbent. The performance of magnetic adsorbents in selective pollutant removal was summarized in Table 42.3.

Fenton process has been commonly used in conventional leachate treatment due to their ability to remove organic load and micropollutants decomposition in wastewater and landfill leachate. The efficiency of the treatment process is largely dependent on the pH of leachate, contamination strength, reagents dosage, and the reaction time. However, removal of ammoniacal nitrogen is remained poor in Fenton process. Various methods of advance oxidation processes pertinent to improvement of dissolved organic matters and ammoniacal nitrogen were being summarized in Table 42.4.

Membrane process is currently leading technology adopted in water reuse, offering significant purification potential. However, membrane fouling and limited recovery flowrate remain the biggest challenges faced in membrane technology. Membrane with high permeability, high rejection, and suitable antifouling characteristic is favorable in offering high separation treatment process. Thus, modified membrane with superior performance in improving the quality was included in the findings as in Table 42.5.

42.3.2 Innovative Technology in Biological Treatment

Biological treatment method is still being one of the favorable means in treating leachate owing to its capability in destruction of organic compounds, sulfides, and toxicity [40, 41]. Conventional biological treatments are still reliable and frequently adopted for young leachate with high BOD content. Thus, maximizing on the potential of biological treatment process, in achieving nitrogen removal would be the object of current review. State of the art of biological treatment in recent years enhancing on nitrogen removal was being elucidated as Table 42.6.

Table 42.2 Performance of adsorbent derived from food waste

Raw material	Leachate characteristics	Adsorption capacity	Removal efficiency %
Banana frond [9]	COD 2336 mg/L, NH_3–N 2550 mg/L, boron 7.50 mg/L, iron 9.31 mg/L	Boron 11.09 mg/g, Iron 26.15 mg/g	Boron 97.45%, iron 95.14%
Coffee ground [10]	COD 1478 mg/L, color 2510 PtCO, NH_3–N 3796.75 mg/L, iron 4.57 mg/L, PO_4–P 260 mg/L	NA	Iron 77% PO_4–P 84%
Durian peel [11]	COD 3100 mg/L, NH_3–N 2900 mg/L, color 3286 PtCO	COD 61.72 mg/g, color 100 PtCO/g	COD 41.98%, Color 39.68%
Tamarind fruit seed [12]	COD 2236 mg/L, color 5095 PtCO, NH_3–N 2550 mg/L	Color 168.57 PtCO/g, COD 64.93 mg/g	Color 91.23%, COD 79.93%
Sugarcane bagasse [13]	COD 2700 mg/L, NH_3–N 2550 mg/L, PO_4–P 285 mg/L	NH_3–N 138.46 mg/g PO_4–P 12.81 mg/g	NH_3–N 79.63% PO_4–P 85.06%
Sugarcane bagasse [14]	COD 1490–1570 mg/L, color 3300–3500 PtCO, NH_3–N 1860–1950 mg/L	Color 555.56 PtCO/g, COD 126.58 mg/g, NH_3–N 14.62 mg/g	Color 94.74%, COD 83.61%, NH_3–N 46.65%
Orange peel [15]	TOC 430 mg/L	NA	TOC 59.7%
Rice husk carbon composite [16]	COD 7330–9530 mg/L, NH_3–N 685–735 mg/L	COD 3.11 mg/g, NH_3–N 12.9 mg/g	COD 27.61%, NH_3–N 51%
Rice husk residue [17]	TN 62 mg/L, COD 1000 mg/L, NH_4^+–N 18 mg/L	NA	TN 84%, COD 82%, NH_4^+–N 100%
Biochar [18]	COD 3098 mg/L, BOD 185 mg/L, TOC 550 mg/L, Ni 0.23 mg/L, Zn 1.5 mg/L, Cu 0.32 mg/L, Cd 0.00303 mg/L, Pb 0.01396 mg/L	COD 147.92 mg/g	COD 53%, Ni 84%, Zn 98%, Cu 80%, Cd 87%, Pb 95%

Table 42.3 Performance of magnetic adsorbent in selective pollutants removal

Adsorbent	Characteristic	Removal performance/adsorption capacity
Magnetic chitosan nanoparticles [19]	Ability to desorb HA at pH > 9 Regeneration with performance above 86% after 5th cycles	HA 32.6 mg/g
Magnetic graphene oxide [22]	Regeneration of HA/Pb(II) MGO exhibit 80.7% for HA and 73% for Pb(II) after 3rd cycles Regeneration of FA/Pb(II) MGO exhibit 70.5% for FA and 68.7% for Pb(II) after 3rd cycles	HA 98.82 mg/g FA 75.38 mg/g Pb 58.43 mg/g
Silica-coated Fe_3O_4 nanoparticles [23]	Regeneration of HA exhibit in range 37.97–43.17 mg/g for 5 consecutive cycles	HA 181.82 mg/g
Magnetic multi-walled carbon nanotubes decorated with calcium [24]	Regeneration of HA exhibit almost same removal efficiency 89.2% after 5th consecutive cycles	HA 90.45%
Magnetic titanium dioxide [25]	UV illumination enhanced in regeneration performance with performance up to 67% after 5th cycles	NOM 65%

42.4 Conclusion

In summary, the improvement of conventional landfill leachate treatment is important toward development of more sustainable landfill leachate management. The improvement made was toward the goals of enhancing the treatment ability and alleviating on the environmental impact. Extensive findings on the replacement material for inorganic coagulant and adsorbent derived from natural substances with acceptable efficiency was considered cornerstones toward achieving sustainable treatment method. The review is of utmost important in revealing the effective technologies applied to safeguard our environment, giving particular attention to feasible and economic solution for leachate treatment management.

Table 42.4 Summary of improvement on advance oxidation process methods

Treatment system	Operating condition	Initial concentration	Removal efficiency
AOPs ferrous ion-activated persulphate process [26]	24 h oxidation period with dosage $S_2O_8^{2-}$/COD = 0.25/1 ($[S_2O_8^{2-}]_0$ = 330 mM) $S_2O_8^{2-}/Fe^{2+}$ = 5/1 ($[Fe^{2+}]_0$ = 330 mM)	COD 21,153 mg/L, BOD_7 9428 mg/L, DOC 6185 mg/L, DN 1920 mg/L, TPh 87 mg/L	COD = 30%, BOD_7 = 30%, DOC = 24%, TPh = 95%, DN = 29%
AOPs sulfate radical-based process [27]	pH 8.2 DOD 0.5–2.0	NH_3–N 560 mg/L, COD 1096 mg/L, Cl^- 980 mg/L	NH_3–N 93% COD 99%
Iron-carbon microelectrolysis-Fenton process [28]	Fe–C dosage 55.72 g/L, H_2O_2 12.32 mL/L and pH 3.12	COD 4980 mg/L, BOD_5 548 mg/L, NH_3-N 1850 mg/L	COD 74.59%
Photo-Fenton using UV/ Cu^{2+} /H_2O_2 [29]	Photo-Fenton dosage rate: H_2O_2/Fe^{2+} = 1.75 ml/1.6875 ml at pH 4 UV/Cu^{2+}/H_2O_2 dosage rate: H_2O_2/Cu^{2+} = 1.75 ml/3.0 ml at pH 6	COD 3125.2 mg/L, TOC 486.2 mg/L, NH_4^+-N 1232 mg/L	Photo-Fenton: COD 74.21%, TOC 74%, UV/Cu^{2+}/H_2O_2: COD 65.23%, TOC 64.67%
UV-TiO_2 [30]	TiO_2 dosage of 5 gm/L, reaction time 60 min	pH 8.3, COD 8900 mg/L, TN 1050 mg/L, TDS 18,320 mg/L BOD 2200 mg/L	COD 60%
AOPs with g-C_3N_4 [31, 32]	A combination of Phanerochaete chrysosporium and photocatalysis with g-C_3N_4 Ti^{3+}–TiO_2/g-C_3N_4	TOC 100 mg/L	TOC 74.99%
	Pd/g-C_3N_4 nanoparticles in 60 min treatment time	BPA 20 mg/L	BPA 91%
Ozone-based AOPs [33]	Photoreactor (FluHelik) coupled bubble column, O_3/UVC/H_2O_2 = 500 mg/L UVC = 6W	pH 3.7, DOC 395 mg/L, COD 1073 mg/L, TSS 170 mg/L	DOC 57.5%, COD 67.1%
ScWO [34]	400–500 °C, 30 min and 100% OE	TOC 6797 mg/L, TN 2187.5 mg/L	TOC 98.2%, TN 92.2%

(continued)

Table 42.4 (continued)

Treatment system	Operating condition	Initial concentration	Removal efficiency
ScWO/Zeolite [35]	Under pressure 23 MPa at 600 and 700 °C with zeolite	pH 8.3, color 1836 uH, Turbidity 141 NTU, BOD 219 mg/L, COD 2572 mg/L, TOC 1787 mg/L, NH_3–N 505 mg/L, NO_2–N 30 mg/L, NO_3–N 8.4 mg/L	NH_3–N 90%, NO_2–N 100%, NO_3–N 98%, color 98%, turbidity 98%, TOC 81%, COD 74%

Table 42.5 Performance of different modified membrane in membrane filtration process

Material	Operating condition	Feed concentration	Rejection performance
GO/MoS_2-PVA composite membrane [36]	Operating pressure 5 bar with water flux 3.96 L/m^2h	NaCl 2000 ppm K 415.67 ppm, Ca 73.45 ppm, Mg 60.73 ppm	NaCl 89% Metal ions 86.5–99.8%
PES-GO-ZnO-PVP membrane [37]	Operating pressure of 4.8 bar with water flux of 10.46 LMH	TOrCs 50 mg/L	TOcCs 78%
Cu-BTC metal–organic framework modified membrane [38]	Operating pressure 3 bar with water flux of 194 LMH	COD 4000 mg/L	COD 40%
TiO_2/carbon membrane [39]	A three-stage fixed bed electrochemical reactor using flat sheet membrane improved by nano TiO_2 particles	Hg 0.236 mg/L, Cd 0.125 mg/L, Cr 0.227 mg/L, Cr^{6+} 0.103 mg/L, As 0.058 mg/L, Pb 0.076 mg/L	Hg > 99%, Cd > 99%, Cr 96.5%, Cr^{6+} 99%, As > 99%, Pb 34.7%

Table 42.6 Summarize of biological methods pertinent on nitrogen removal efficiency

Biological treatment	Focused pollutants	Removal performance
GSBR [42]	TAN 498 mg/L, TP 2–6 mg/L, COD 810 mg/L	TAN > 99%, TP 49%, COD 67–87%
Two stage UASB-SBR [43]	COD 11,950 mg/L, NH_4^+–N 982.6 mg/L, TN 1176 mg/L	COD 96.7%, NH_4^+–N 99.7%, TN 98.3%
SBBR [44]	NH_4^+–N 1000 mg/L, TN 1100 mg/L, BOD_5 4500 mg/L, COD 6500 mg/L	COD 85% TN 98%
MBBR inoculated with biocarrier [45]	NH_4^+–N 443.1 mg/L	NH_4^+–N 60–68%
PAC-SBR [46]	COD 1655 mg/L, color 3627 PtCO, TDS 4951 mg/L, NH_3–N 600 mg/L	COD 64.1%, color 71.2%, NH_3–N 81.4%, TDS 1.33%
AS + BF [47]	COD 2185 mg/L, BOD_5 247 mg/L, TKN 1651.8 mg/L, NH_4–N 2024.1 mg/L, nitrite 41.5 mg/L, nitrate 2.1 mg/L, phosphate 50.7 mg/L, color 3796.8 PtCO	COD 82.6%, BOD_5 90.7%, TKN 36.4%, NH_4–N 53.2%, nitrite 52.4%, nitrate 100%, phosphate 63.2%, color 21.8%
Microaerobic bioreactor [48]	COD 3623–5357 mg/L, NH_4–N 371–2495 mg/L, NO_3–N 14–666 mg/L, NO_2–N 0.1–552 mg/L, TN 2907–3042 mg/L	COD 2.6%, TN 7.5%
MFCs [49]	COD 3480 mg/L, BOD 93 mg/L, TKN 6033 mg/L, NH_4–N 5449 mg/L	COD 37%, NH_4–N 20%
Constructed wetlands (CW) [50, 51]	COD 20.1 g/m^{2}d, NH_4–N 5.0 g/m^{2}d, TN 8.3 g/m^{2}d	COD 47.8–86.6%, TN 68.9–98.5%

Acknowledgements The authors would like to acknowledge Universiti Sains Malaysia (USM) for the opportunity of research work to be conducted. Special thanks to those who contributed to this project directly or indirectly.

References

1. Green Technology and Water (KeTTHA) Ministry of Energy, *Ministry of Energy, Green Technology and Water (KeTTHA)* (2017)
2. S.T. Tan, C.T. Lee, H. Hashim, W.S. Ho, J.S. Lim, Optimal process network for municipal solid waste management in Iskandar Malaysia. J. Clean. Prod. **71**(2014), 48–58 (2014)
3. T. Banch, M. Hanafiah, A. Alkarkhi, S. Abu, Factorial design and optimization of landfill. Polym. (Basel) (2019)
4. E.P.U. Epu, Tenth Malaysia plan. J. Chem. Inf. Model. **53**(9), 1689–1699 (2013)

5. L. Bu, K. Wang, Q.L. Zhao, L.L. Wei, J. Zhang, J.C. Yang, Characterization of dissolved organic matter during landfill leachate treatment by sequencing batch reactor, aeration corrosive cell-Fenton, and granular activated carbon in series. J. Hazard. Mater. **179**(1–3), 1096–1105 (2010)
6. S.Y. Cheng et al., Sustainable landfill leachate treatment: optimize use of guar gum as natural coagulant and floc characterization. Environ. Res. **188** (2020)
7. H.A. Aziz, N.A. Rahim, S.F. Ramli, M.Y.D. Alazaiza, F.M. Omar, Y.T. Hung, Potential use of Dimocarpus longan seeds as a flocculant in landfill leachate treatment. Water (Switzerland) **10**(11), 1–15 (2018)
8. M.S. Yusoff, H.A. Aziz, M.Y.D. Alazaiza, L.M. Rui, Potential use of oil palm trunk starch as coagulant and coagulant aid in semi-aerobic landfill leachate treatment. Water Qual. Res. J. Canada **54**(3), 203–219 (2019)
9. K.Y. Foo, L.K. Lee, B.H. Hameed, Preparation of banana frond activated carbon by microwave induced activation for the removal of boron and total iron from landfill leachate. Chem. Eng. J. **223**, 604–610 (2013)
10. S.L. Ching, M.S. Yusoff, H.A. Aziz, M. Umar, Influence of impregnation ratio on coffee ground activated carbon as landfill leachate adsorbent for removal of total iron and orthophosphate. Desalination **279**(1–3), 225–234 (2011)
11. M.A. Kamaruddin, Treatment of semi-aerobic landfill leachate using durian peel-based activated carbon adsorption—optimization of preparation conditions. Int. J. Energy Environ. **3**(2), 223–236 (2012)
12. K.Y. Foo, L.K. Lee, B.H. Hameed, Preparation of tamarind fruit seed activated carbon by microwave heating for the adsorptive treatment of landfill leachate: a laboratory column evaluation. Bioresour. Technol. **133**, 599–605 (2013)
13. K.Y. Foo, L.K. Lee, B.H. Hameed, Preparation of activated carbon from sugarcane bagasse by microwave assisted activation for the remediation of semi-aerobic landfill leachate. Bioresour. Technol. **134**, 166–172 (2013)
14. N.B. Azmi, M.J.K. Bashir, S. Sethupathi, C.A. Ng, Anaerobic stabilized landfill leachate treatment using chemically activated sugarcane bagasse activated carbon: kinetic and equilibrium study. Desalination (2015)
15. Z. Xie, W. Guan, F. Ji, Z. Song, Y. Zhao, Production of biologically activated carbon from orange peel and landfill leachate subsequent treatment technology. J. Chem. **6**, 2014 (2014)
16. A.A. Halim, N.N.Z. Abidin, N. Awang, A. Ithnin, M.S. Othman, M.I. Wahab, Ammonia and COD removal from synthetic leachate using rice husk composite absorbent. J. Urban Env. 24–31 (2011)
17. Y. Luo, D. Li, Y. Chen, X. Sun, Q. Cao, X. Liu, The performance of phosphoric acid in the preparation of activated carbon-containing phosphorus species from rice husk residue. J. Mater. Sci. **54**(6), 5008–5021 (2019)
18. A. Kwarciak-Kozłowska, R. Włodarczyk, K. Wystalska, Biochar compared with activated granular carbon for landfill leachate treatment. E3S Web Conf. **100** (2019)
19. C. Dong, W. Chen, C. Liu, Preparation of novel magnetic chitosan nanoparticle and its application for removal of humic acid from aqueous solution. Appl. Surf. Sci. **292**, 1067–1076 (2014)
20. M.A. Kamaruddin, M.S. Yusoff, L.M. Rui, A.M. Isa, M.H. Zawawi, R. Alrozi, An overview of municipal solid waste management and landfill leachate treatment: Malaysia and Asian perspectives. Environ. Sci. Pollut. Res. **24**(35), 26988–27020 (2017)
21. Q. Yang, Y. Jiang, X. Li, Y. Yang, L. Hu, Magnetic-supported cucurbituril: a recyclable adsorbent for the removal of humid acid from simulated water. 1167–1174 (2014)
22. J. Zhang et al., Simultaneous removal of humic acid/fulvic acid and lead from landfill leachate using magnetic graphene oxide. Appl. Surf. Sci. **370**, 335–350 (2016)
23. J. Wang, L. Bi, Y. Ji, H. Ma, X. Yin, Removal of humic acid from aqueous solution by magnetically separable polyaniline: adsorption behavior and mechanism. J. Colloid Interface Sci. **430**, 140–146 (2014)

24. S. Li, M. He, Z. Li, D. Li, Z. Pan, Removal of humic acid from aqueous solution by magnetic multi-walled carbon nanotubes decorated with calcium. J. Mol. Liq. **230**, 520–528 (2017)
25. M. Ng, E.T. Kho, S. Liu, M. Lim, R. Amal, Highly adsorptive and regenerative magnetic TiO_2 for natural organic matter (NOM) removal in water. Chem. Eng. J. **246**, 196–203 (2014)
26. E. Kattel, N. Dulova, Ferrous ion-activated persulphate process for landfill leachate treatment: removal of organic load, phenolic micropollutants and nitrogen. Environ. Technol. (United Kingdom) **38**(10), 1223–1231 (2017)
27. C. Chen, H. Feng, Y. Deng, Re-evaluation of sulfate radical based–advanced oxidation processes (SR-AOPs) for treatment of raw municipal landfill leachate. Water Res. **153**, 100–107 (2019)
28. L. Wang et al., Advanced landfill leachate treatment using iron-carbon microelectrolysis-Fenton process: process optimization and column experiments. J. Hazard. Mater. **318**, 460–467 (2016)
29. L. Li et al., Process parameters study and organic evolution of old landfill leachate treatment using photo-Fenton-like systems: Cu^{2+} versus Fe^{2+} as catalysts. Sep Purif Technol **211**(October 2018), 972–982 (2019)
30. M.S. Mahtab, I.H. Farooqi, UV-TiO_2 process for landfill leachate treatment: optimization by response surface methodology. Int. J. Res. Eng. Appl. Manag. **05**(12), 14–18 (2020)
31. L. Hu et al., Organic matters removal from landfill leachate by immobilized Phanerochaete chrysosporium loaded with graphitic carbon nitride under visible light irradiation. Chemosphere **184**, 1071–1079 (2017)
32. Y. Wang, D. Cao, M. Liu, X. Zhao, Insights into heterogeneous catalytic activation of peroxymonosulfate by Pd/g-C_3N_4: the role of superoxide radical and singlet oxygen. Catal. Commun. **102**(May), 85–88 (2017)
33. A.I. Gomes, T.F. Soares, T.F.C.V. Silva, R.A.R. Boaventura, V.J.P. Vilar, Ozone-driven processes for mature urban landfill leachate treatment: organic matter degradation, biodegradability enhancement and treatment costs for different reactors configuration. Sci. Total Environ. **724** (2020)
34. V.F. Marulanda Cardona, P.A. Marulanda Buitrago, D.H. Alvarado Acosta, Landfill leachate treatment by batch supercritical water oxidation. Cienc. Ing. Neogranadina **27**(2), 5–26 (2017)
35. A.P.J. Scandelai, J.P. Zotesso, V. Jegatheesan, L. Cardozo-Filho, C.R.G. Tavares, Intensification of supercritical water oxidation (ScWO) process for landfill leachate treatment through ion exchange with zeolite. Waste Manag. **101**, 259–267 (2020)
36. S. Yadav, I. Ibrar, A. Altaee, A.K. Samal, R. Ghobadi, J. Zhou, Feasibility of brackish water and landfill leachate treatment by GO/MoS_2-PVA composite membranes. Sci. Total Environ. **745** (2020)
37. O.T. Mahlangu, R. Nackaerts, J.M. Thwala, B.B. Mamba, A.R.D. Verliefde, Hydrophilic fouling-resistant GO-ZnO/PES membranes for wastewater reclamation. J. Memb. Sci. **524**(November 2016), 43–55 (2017)
38. M. Mazani, S.A. Aktij, A. Rahimpour, N.T. Hassan Kiadeh, Cu-BTC metal-organic framework modified membranes for landfill leachate treatment. Water (Switzerland) **12**(1), 1–14 (2020)
39. H. Hui et al., A three-stage fixed-bed electrochemical reactor for biologically treated landfill leachate treatment. Chem. Eng. J. **376**(February) (2019)
40. M.A. Kamaruddin, M. Yusoff, H.A. Aziz, Y.T. Hung, Sustainable treatment of landfill leachate. Appl. Water Sci. **5**(2), 113–126 (2015)
41. M.A. Kamaruddin, M.S. Yusoff, H.A. Aziz, N.K. Basri, Removal of COD, ammoniacal nitrogen and colour from stabilized landfill leachate by anaerobic organism. Appl. Water Sci. **3**(2), 359–366 (2013)
42. Y. Ren, F. Ferraz, M. Lashkarizadeh, Q. Yuan, Comparing young landfill leachate treatment efficiency and process stability using aerobic granular sludge and suspended growth activated sludge. J. Water Process Eng. **17**(February), 161–167 (2017)
43. H. Sun, Q. Yang, Y. Peng, X. Shi, S. Wang, S. Zhang, Advanced landfill leachate treatment using a two-stage UASB-SBR system at low temperature. J. Environ. Sci. **22**(4), 481–485 (2010)

44. W. Yin, K. Wang, J. Xu, D. Wu, C. Zhao, The performance and associated mechanisms of carbon transformation (PHAs, polyhydroxyalkanoates) and nitrogen removal for landfill leachate treatment in a sequencing batch biofilm reactor (SBBR). RSC Adv. **8**(74), 42329–42336 (2018)
45. J. Xiong, Z. Zheng, X. Yang, J. He, X. Luo, B. Gao, Mature landfill leachate treatment by the MBBR inoculated with biocarriers from a municipal wastewater treatment plant. Process Saf. Environ. Prot. **119**, 304–310 (2018)
46. S.Q. Aziz, H.A. Aziz, M.S. Yusoff, M.J.K. Bashir, Landfill leachate treatment using powdered activated carbon augmented sequencing batch reactor (SBR) process: optimization by response surface methodology. J. Hazard. Mater. **189**(1–2), 404–413 (2011)
47. H.A. Aziz, T.J. Ling, A.A.M. Haque, M. Umar, M.N. Adlan, Leachate treatment by swim-bed bio fringe technology. Desalination **276**(1–3), 278–286 (2011)
48. R. He, X.M. Wei, B.H. Tian, Y. Su, Y.L. Lu, Characterization of a joint recirculation of concentrated leachate and leachate to landfills with a microaerobic bioreactor for leachate treatment. Waste Manag. **46**, 380–388 (2015)
49. S. Puig, M. Serra, M. Coma, M. Cabré, M. Dolors Balaguer, J. Colprim, Microbial fuel cell application in landfill leachate treatment. J. Hazard. Mater. **185**(2–3), 763–767 (2011)
50. E. Wojciechowska, Potential and limits of landfill leachate treatment in a multi-stage subsurface flow constructed wetland—evaluation of organics and nitrogen removal. Bioresour. Technol. **236**, 146–154 (2017)
51. V. Cano, D.V. Vich, H.H.B. Andrade, D.T.P. Salinas, M.A. Nolasco, Nitrification in multistage horizontal flow treatment wetlands for landfill leachate treatment. Sci. Total Environ. **704**, 135376 (2020)

Chapter 43
Feasibility of Multilayer Perceptron (MLP) Network to Correlate Air Quality Index (AQI) and COVID-19 Daily Cases

M. I. F. Abd Maruzuki, T. S. A. Tengku Zahidi, K. Khairudin, M. S. Osman, N. F. Jasmy, B. Abdul Hadi, M. S. Akbar, A. Z. U. Saufie, M. Fathullah, D. S. Nor Syamsudin, and N. B. Mohd Nazeri

Abstract A movement control order (MCO) was implemented in Malaysia starting from March 18th, 2020, as a pandemic control strategy that restricted all movement and daily outdoor activities to curb the transmission of COVID-19 pandemic. The most populated area in Malaysia, Petaling Jaya, Selangor, was selected to investigate the relationship between the COVID-19 outbreak and air pollution. Multilayer perceptron (MLP) model was used in this study to correlate air quality index

---

M. I. F. Abd Maruzuki
Centre for Electrical Engineering Studies, Universiti Teknologi MARA, Cawangan Pulau Pinang, Permatang Pauh Campus, 13500 Pulau Pinang, Malaysia

T. S. A. Tengku Zahidi · K. Khairudin · M. S. Osman (✉) · N. F. Jasmy · D. S. Nor Syamsudin · N. B. Mohd Nazeri
EMZI-UiTM Nanoparticles Colloids and Interface Industrial Research Laboratory (NANO-CORE), Centre for Chemical Engineering Studies, Universiti Teknologi MARA, Cawangan Pulau Pinang, Permatang Pauh Campus, 13500 Pulau Pinang, Malaysia
e-mail: syazwan.osman@uitm.edu.my

M. Fathullah
School of Manufacturing Engineering, Universiti Malaysia Perlis (UniMAP), Pauh Putra Campus, 02600 Arau, Perlis, Malaysia

M. S. Osman · M. Fathullah
Center of Excellence Geopolymer and Green Technology (CEGeoGTech), Universiti Malaysia Perlis (UniMAP), 01000 Kangar, Perlis, Malaysia

B. Abdul Hadi
Centre for Civil Engineering Studies, Universiti Teknologi MARA, Cawangan Pulau Pinang, Permatang Pauh Campus, 13500 Pulau Pinang, Malaysia

M. S. Akbar
Faculty Science, Universiti Teknologi Malaysia, 81310 UTM Skudai, Johor, Malaysia

A. Z. U. Saufie
Faculty of Computer and Mathematical Sciences, Universiti Teknologi MARA, 40450 Shah Alam, Selangor, Malaysia

M. A. A. Mohd Salleh et al. (eds.), *Proceedings of the Green Materials and Electronic Packaging Interconnect Technology Symposium*, Springer Proceedings in Physics 289,
https://doi.org/10.1007/978-981-19-9267-4_43

(AQI) with COVID-19-related cases/deaths. The underlying hypothesis is that a pre-determined particulate concentration can encourage COVID-19 airborne transmission and make the respiratory system more susceptible to this infection. The in-silico strategy employed an innovative machine learning (ML) methodology, specifically MLP network using AQI data from the Department of Environment (DOE), Malaysia as input data and number of COVID-19 cases from the Ministry of Health, Malaysia as target data. The MLP model was trained for 10,000 times. Based on the results obtained, the model starts to converge after 1000 epochs with a small mean absolute error (MAE) (173.0–118.9) from day 1 to day 14. However, there is no definitive correlation between predicted COVID-19 patients and the AQI with respect to day configuration.

Keywords COVID-19 · Air pollution · Artificial neural network · Correlation · Multilayer perceptron (MLP)

43.1 Introduction

The recent global pandemic caused by COVID-19 is some pathogens that specifically attack the human respiratory system. COVID-19 is caused by the severe acute respiratory syndrome coronavirus 2 (SARS-CoV-2), which was also called the 2019 novel coronavirus (2019-nCoV). In late December 2019, pinpointed Wuhan, Hubei Province, People's Republic of China lays the foundation of the beginning of pandemic that eventually goes global in time. Due to the alarming spread growth, the World Health Organization (WHO) declared Public Health Emergency of International Concern (PHEIC). The COVID-19 is rapidly spreading throughout Asia (Thailand, Japan, Singapore, and Malaysia), Australia, and Europe. Countless studies unveil that older age ranging from 80 years or older face high mortality susceptibility and a 21.9% probability of death after being infected with COVID-19 [1]. Researchers advocated for the spread of COVID-19-based on socioeconomic and health status factors such as migration scale, occupation, and living conditions.

Furthermore, air pollution has a significant impact on the aerosol transmission of SARS-CoV-2. This is due to the particular virus able to remain in aerosols condition for a few hours resulting airborne transmission to be plausible. Fattorini and Regoli researched the interrelation of COVID-19 spread and the prolonged exposure of air pollution and had presumption that pollutants in the air present is favorable to the diffusion of SARS-CoV-2 [2]. It has been speculated that there are more than enough reasons why ambient air pollutants affect the transmission of SARS-CoV-2.

Air quality throughout Malaysia that has continuously worsen ought to be managed and controlled by the Malaysian government for a better enhancement of air quality. In accordance with the World Health Organization (WHO), of about 5.5 million of human population died due to the results of air pollution and poor air quality [3]. The issue of air pollution is widely recognized as a serious one, particularly in developing countries such as Malaysia. The emission of these types of air pollution, namely, sulfur dioxide (SO_2), nitrogen dioxide (NO_2), carbon monoxide (CO), ozone (O_3), and particulate matter (PM10 and PM2.5) should be fully monitored as it will increase the possibility of respiratory infections and inflammations. Pollutants gained access through respiratory tract and able to damage other organs resulting too many health problems and deaths in children and adults. Additionally, pollutants from the surrounding atmosphere results in inflammatory response which leads to higher risk of any sort of viruses' infection against the respiratory system. Prolonged exposure of air pollutants inside the respiratory tract generates the vulnerability toward COVID-19 infection. Even though it has not been clearly proven, certain evidence confirms that SARS-CoV-2 might have attached onto the particulate matter of the atmospheric air [4]. In respect to it, air pollution may contribute to the rate of COVID-19 pandemic spread as it could extensively travel through the respiratory tract on the surface of particulate matter. Initially understood the COVID-19 transmission is originated by the droplets of the infected victim towards the next susceptible host [5]. Indirect transmission of COVID-19 through aerosols was also suggested as there is direct involvement of particulate matter distinctly PM2.5 that acts as virus transportation agent [6]. Although there is no correlation between the concentration of the virus and the diameter of particulate matter, there is a positive involvement between PM2.5 and few of other respiratory viruses such as influenza reported [7]. PM2.5 is a mixture of organic and inorganic particles with a diameter of 2.5 μm and less suspended in the atmospheric air. The majority of the pollution came from outdoor resources used by humans in their daily activities, such as industrial emissions and transportation vehicles. Long-term exposure of PM2.5 would deteriorate human health conditions as this fine matter can effortlessly penetrate deep into the lungs. Apparently, PM2.5 has a longer lifetime in the air than liquid droplets, which pose a significant viral exposure risk if SARS-CoV-19 attaches to them [8, 9]. Typically, AQI is calculated based on five main pollutants including SO_2, NO_2, CO, O_3, and particulate matter. In conjunction, it is hypothesized that there is a relationship between the AQI and the number of cases of COVID-19 infection.

This study uses artificial neural network (ANN) method to determine the interrelations of AQI and COVID-19 distributions. This method is developed in the form of mathematical models that generalize biological nervous systems. The foundation elements present in neural network are artificial neurons, simply neurons, or even nodes. In a simpler mathematical model, synapse is displayed with the connection weights that design the effects of correlated input signals, as well as nonlinear element shown by neurons from the transfer function. The impulses of neurons are computed being sum of the input signals that have already been weighted, which are then revamp by transfer function. The ability to learn the information given for an artificial neuron is achieved by adapting different weights from the chosen algorithm of the study [10]. The ANN generally work by stimulating the way biological should behave by combining inputs of AQI data as well as COVID-19 daily cases for target data of the chosen date and time, which can then give out own signals and output that helps to predict future results.

While ANN has been shown to outperform other methods in estimating various correlation problems [11], to the best of our knowledge, no ANN model is currently being developed to estimate the correlation between AQI and daily COVID-19 cases. Thus, the main aim of this project is to correlate AQI and daily COVID-19 cases with the ANN model.

43.2 Methodology

43.2.1 Materials

ANN is a form of artificial intelligence that possesses the ability to learn from features or data. The use of ANN is currently wide spreading through the applications of pattern recognition, control, estimation, and classification [12–15]. The model was chosen based on MLP network, which is a class of feedforward ANN. The typical structure of an MLP network consists of three categories of layers: input layer, hidden layer, output layer as shown in Fig. 43.1. The structure of the network is often designed with nonlinear activation function to provide nonlinear behavior, which is crucial in performing aforementioned applications [11]. Figure 43.1 shows a typical model of an MLP network.

The ability to learn in MLP is partly attributed to back-propagation (BP) method [11]. This training algorithm is based on error minimization at output layer. One factor that can improve learning ability of MLP is the number of hidden nodes in hidden layer.

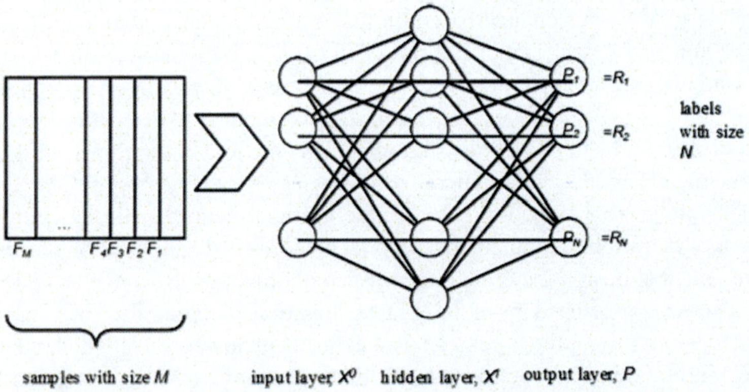

Fig. 43.1 Hypothetical example of multilayer perceptron network [10]

43.2.2 Study Area

The location chosen for this study targets the area of Petaling Jaya where this area is a part of the state of Selangor, Malaysia. Owing to the fact that this location is one of the core districts that incorporate of the highest COVID-19 cases which in turn could help this research for a more accurate result.

The district of Petaling Jaya is under Petaling Jaya City Council (MBPJ), and segment of this district could also be under Subang Jaya City Council (MPSJ) as well as Shah Alam City Council (MBSA). The location of Petaling Jaya is directed right next to the Federal Territory of Kuala Lumpur that is acclaimed as the focus and attraction in the public eye that has gained the advantages for further growth of the Petaling Jaya land concerning urbanization, industrial development, institutions, in returns creates the employment opportunities. Until now, Petaling Jaya continuously to grow rapidly as a more developed city that targets investment, industry, education, and living quarters [16]. The overall area of Petaling Jaya district is about 97.2 km^2 (Fig. 43.2).

43.2.3 Data

The data was obtained from the Department of Environment (DOE), Malaysia to design the output of the ANN. Out of all six main types of air pollutants, this experimentation exclusively requiring air quality index (AQI) data of the selected duration in terms of its time and date. It is particularly in need of an hourly data of the chosen date which are from March 1, 2020 to May 31, 2020. The related data is in the form of International System of Units (SI units). As the data are in hourly, it must first uncover the average numerals to substantiate the compatibility with the COVID-19 data.

This study is also in need of the data of number of COVID-19 daily cases from the March 1, 2020 to May 31, 2020. The COVID-19 daily cases data was derived from the official Ministry of Health Malaysia (MOH) as it is recognized as the most accurate number of cases provided. All in all, both data are directed at Petaling Jaya.

There are total of 80 days of AQI and COVID-19 data. Then, the COVID-19 data was extracted and arranged into a dedicated number of delay days. Day-01 means a particular AQI reading will be used to correlate with a number of COVID-19 patients within same date, while Day-14 means an AQI reading will be used to correlate with a number COVID-19 in the next 14 days later. Table 43.1 shows example of data collected for both AQI and number of COVID-19 patients.

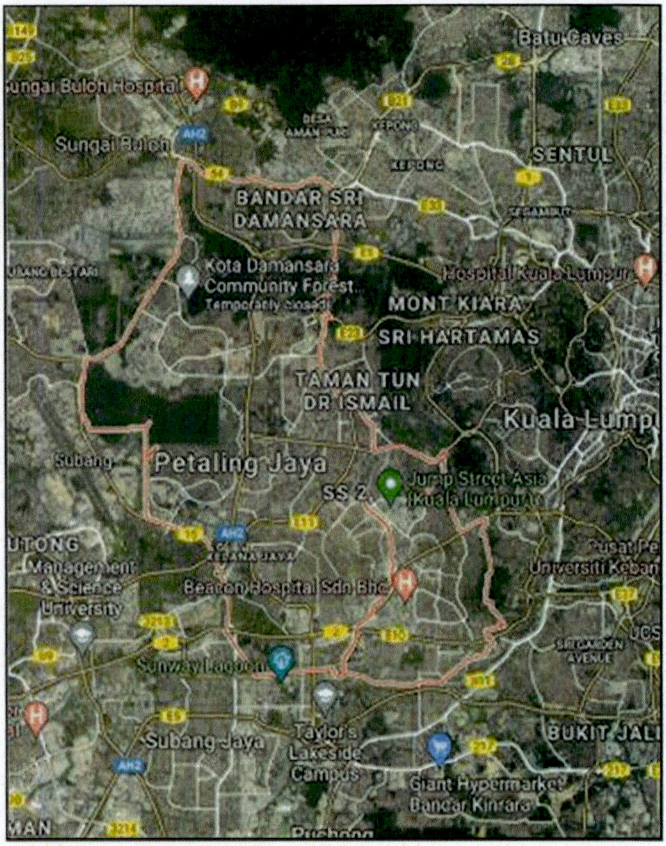

Fig. 43.2 Location of Petaling Jaya, Selangor [16]

43.2.4 MLP Configuration

14 MLP models were developed which correspondent to the days in Table 43.1. All MLP models used similar configurations as detailed in Table 43.2, while Fig. 43.3 shows the actual architecture of one of our models. Exponential linear unit (*elu*) activation function from Eq. (43.1) is applied on the outputs of all hidden layers due to its superior performance as suggested by many previous works [18]:

$$X^L = \begin{cases} X^{L-1} & \text{if } X^{L-1} > 0 \\ a \times \exp\left(X^{L-1} - 1\right) & \text{else} \end{cases} \tag{43.1}$$

where L is current layer, $L - 1$ is previous layer and X^{L-1} is the input from previous layer and a is a trainable hyperparameter.

Table 43.1 Sample data of AQI and number of COVID-19 patients in Petaling Jaya [17]

AQI	Day-00	Day-01	Day-02	Day-03	Day-04	Day-05	Day-06	Day-07	Day-08	Day-09	Day-010	Day-011	Day-012	Day-013
80	193													
124	263	193												
91	340	263	193											
83	249	340	263	193										
76	319	249	340	263	193									
75	261	319	249	340	263	193								
58	283	261	319	249	340	263	193							
69	271	283	261	319	249	340	263	193						
78	142	271	283	261	319	249	340	263	193					
82	137	142	271	283	261	319	249	340	263	193				
79	128	137	142	271	283	261	319	249	340	263	193			
66	124	128	137	142	271	283	261	319	249	340	263	193		
69	67	124	128	137	142	271	283	261	319	249	340	263	193	
68	191	67	124	128	137	142	271	283	261	319	249	340	263	193
74	33	191	67	124	128	137	142	271	283	261	319	249	340	263
70	136	33	191	67	124	128	137	142	271	283	261	319	249	340
68	246	136	33	191	67	124	128	137	142	271	283	261	319	249

Table 43.2 Architecture configuration for the linear regression model

Model	Layer	Type	Total neutrons	Activation function
MLP	X0	Normalization	1	*elu*
	X1	Fully connected	64	*elu*
	X2	Fully connected	64	*elu*
	P	Output	1	*linear*

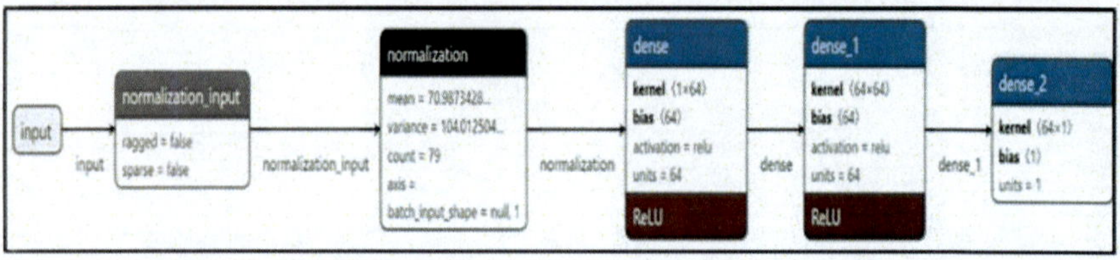

Fig. 43.3 Architecture visualization of our proposed model

Finally, a linear activation function at the output layer is responsible for returning the values from the previous layers into predefined output using Eq. (43.2):

$$P_n = X^{L-1} \tag{43.2}$$

where n is number of outputs at P [19].

43.2.5 *Training Methodology*

Table 43.3 shows the default hyperparameter for training configuration of all 14 MLP models. The dataset was divided to 80% training, 10% validation, and 10% in test. In this work, all training procedures and each MLP model are developed from scratch. The models are trained for 10,000 epochs. At the beginning of every training epoch, the training samples are shuffled to ensure that the model can learn robustly and produce better learning performance.

Glorot normalization is performed in the weight initialization process at the start of every new training repetition [20]. As for the learning algorithm, all models use Adam as optimizer [21].

Mean absolute error (MAE) was used to evaluate fitness of the MLP model. MAE is computed using Eq. (43.3):

$$\text{MAE} = \left(\frac{1}{M} \sum^{M} D_m \right) \tag{43.3}$$

Table 43.3 Training configuration

CPU	Intel(R) Core(TM) i5-8300H CPU @ 2.30 GHz
GPU	GeForce GTX 1050 4 GB
RAM	16 GB
Framework	TensorFlow, Keras
OS	Windows 10
Dataset and model	AQI and no. of Covid-19 patients
Training epoch	10,000
Optimizer	Adam: $\beta 1 = 0.9$, $\beta 2 = 0.99$
Initial learning rate	0.001

$$D_m = |\boldsymbol{P}_m - \boldsymbol{R}_m| \tag{43.4}$$

where m are total samples in the training dataset. Low MAE means higher ability to correlate between input and targeted data.

43.3 Result and Discussions

The learning performance of the proposed models was examined by studying their estimation accuracy on validation dataset. Figures 43.4 and 43.5 show results for the proposed MLP models, with hidden nodes setting in Table 43.2 within 10,000 training epochs. As shown in Fig. 43.4, the proposed MLP model for estimating Day-14th COVID-19 patient using current AQI shows the lowest MAE, at 118.9. The validation performance was also examined in every epoch for each estimated day. Figure 43.5 illustrates the learning curves of our proposed model during validation. Overall, the propose MLP model converges after 1000 epochs for all days configurations. However, the steep MAE curves for Day-14 signifies a good generalization performance of the proposed model.

However, even though the proposed model shows promising results in terms of MAE, the correlation between AQI and number of COVID-19 is not definitive. As shown from the estimation accuracy in Fig. 43.6, none of the models is able to correctly predict or estimate the number of patients based solely on the AQI data. This demonstrates the inferiority of AQI-based solutions in solving the correlation problem. Among the factors that affect the inaccuracy performance are the low number of data and the single type of data involve in building or correlation. Thus, in order to improve our correlation, for future work, we suggest including other factors as data in the proposed model as well. Examples of data that can be taken as consideration are CO level, O_3 level, PM2.5 level, SO_2 level, NO_2 level, as

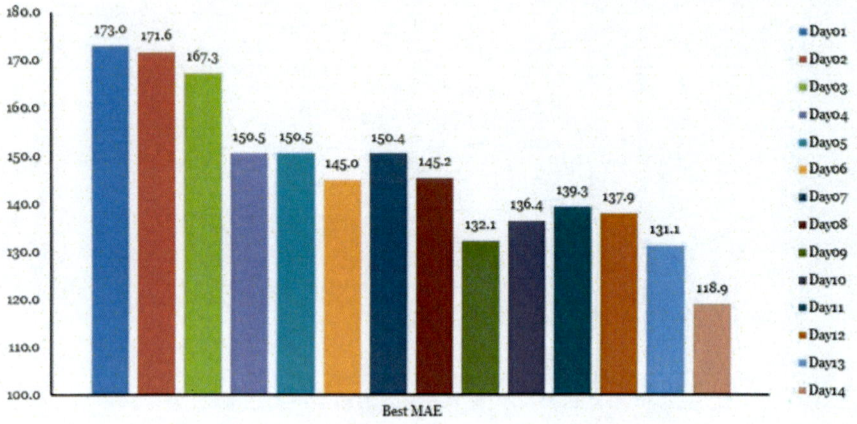

Fig. 43.4 Laser the best performance in terms of MAE after 10,000 epochs

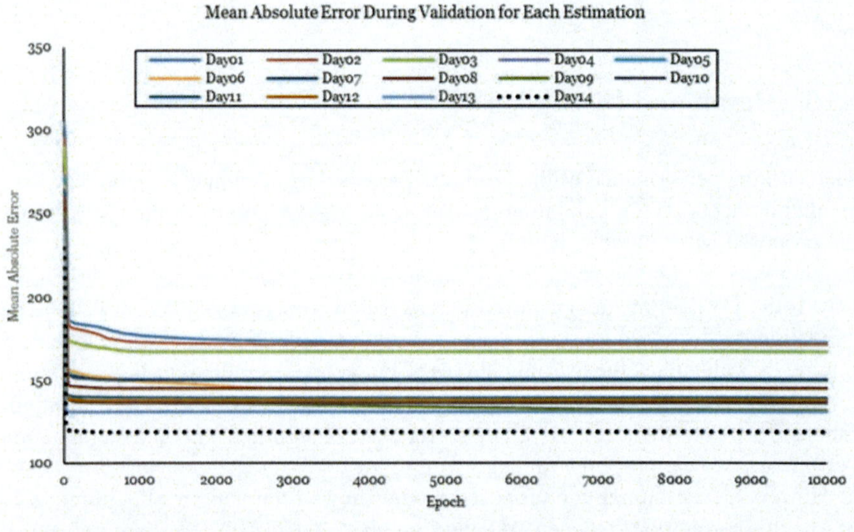

Fig. 43.5 MAE performance in validation set during 10,000 epochs

well as humidity and temperature. While currently there is no other reference that attempts similar work, for future work an in-depth analysis is required for objective benchmarking of the proposed solution against other work by other researchers.

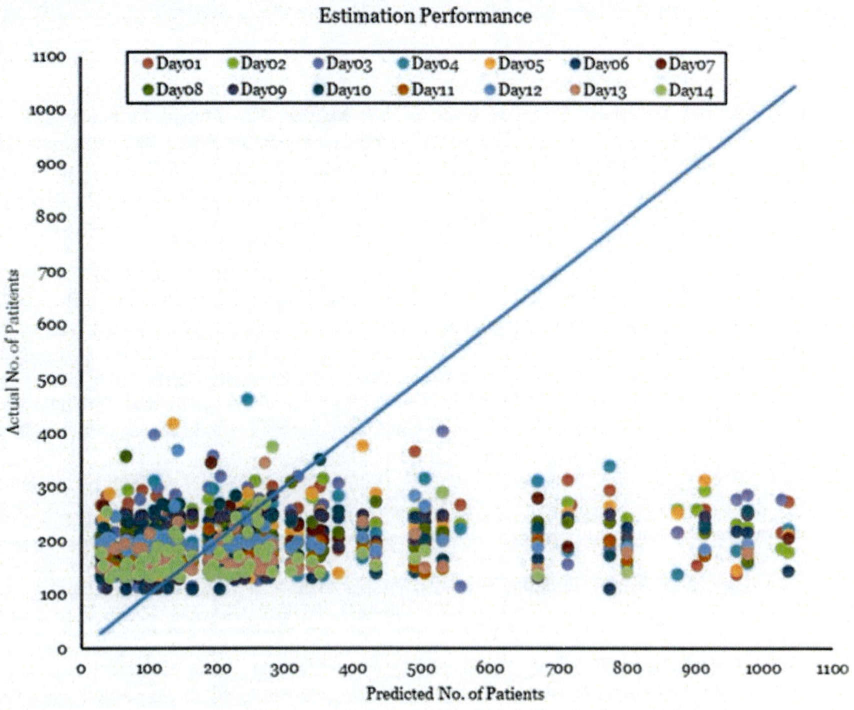

Fig. 43.6 Estimation performance of our proposed model in testing set after training

43.4 Conclusion

As a conclusion, we have described our recent work on building correlation between air pollution and COVID-19 patients. In this paper, we developed 14 MLP models that take AQI of an air pollution index, as its input and correlate the number of COVID-19 patients. As a result, the best MLP model, namely, Day14 is able to correlate the number of COVID-19 cases based on the level of air pollution. However, the developed model has very low performance in term of accurately determined the number of patients due to the lack of data from DOE and MOH, which hinders salient features learning in MLP.

Acknowledgements The authors would like to gratitude Kasih Kirana Resources Sdn Bhd for supporting this research financially under the research grant (100-TNCPI/PRI 16/6/2 (029/2020)) and Universiti Teknologi MARA, Cawangan Pulau Pinang, 13500 Permatang Pauh, Penang, Malaysia.

References

1. S. Abdullah, A.A. Mansor, N.N.L.M. Napi, W.N.W. Mansor, A.N. Ahmed, M. Ismail, Z.T.A. Ramly, Air quality status during 2020 Malaysia movement control order (MCO) due to 2019 novel coronavirus (2019-nCoV) pandemic. Sci. Total Environ. **729**, 139022 (2020)
2. X. Zhang, M. Tang, F. Guo, F. Wei, Z. Yu, K. Gao, M. Jin, J. Wang, K. Chen, Associations between air pollution and COVID-19 epidemic during quarantine period in China. Environ. Pollut. **268**, 115897 (2021). https://doi.org/10.1016/j.envpol.2020.115897
3. M.S.M. Nadzir, M.C.G. Ooi, K.M. Alhasa, M.A.A. Bakar, A.A.A. Mohtar, M.F.F.M. Nor, M.T. Latif, H.H.A. Hamid, S.H.M. Ali, N.M. Ariff, J. Anuar, F. Ahamad, A. Azhari, N.M. Hanif, M.A. Subhi, M. Othman, M.Z.M. Nor, The impact of movement control order (MCO) during pandemic COVID-19 on local air quality in an urban area of Klang valley, Malaysia. Aerosol Air Qual. Res. **20**(6), 1237–1248 (2020). https://doi.org/10.4209/aaqr.2020.04.0163
4. A. López-Feldman, D. Heres, F. Marquez-Padilla, Air pollution exposure and COVID-19: a look at mortality in Mexico City using individual-level data. Sci. Total Environ. **756**, 143929 (2021). https://doi.org/10.1016/j.scitotenv.2020.143929
5. Y. Liu, Z. Ning, Y. Chen, M. Guo, Y. Liu, N.K. Gali, L. Sun, Y. Duan, J. Cai, D. Westerdahl, X. Liu, K. Xu, K.F. Ho, H. Kan, Q. Fu, K. Lan, Aerodynamic analysis of SARS-CoV-2 in two Wuhan hospitals. Nature **582**(7813), 557–560 (2020).https://doi.org/10.1038/s41586-020-2271-3
6. N.S.M. Nor, C.W. Yip, N. Ibrahim, M.H. Jaafar, Z.Z. Rashid, N. Mustafa, H.H.A. Hamid, K. Chandru, M.T. Latif, P.E. Saw, C.Y. Lin, K.M. Alhasa, J.H. Hashim, M.S.M. Nadzir, Particulate matter (PM2.5) as a potential SARS-CoV-2 carrier. Sci. Rep. **11**(1), 1–6 (2021).https://doi.org/10.1038/s41598-021-81935-9
7. W. Su, X. Wu, X. Geng, X. Zhao, Q. Liu, T. Liu, The short-term effects of air pollutants on influenza-like illness in Jinan, China. BMC Public Health **19**(1), 1–13 (2019). https://doi.org/10.1186/s12889-019-7607-2
8. Y.F. Xing, Y.H. Xu, M.H. Shi, Y.X. Lian, The impact of PM2.5 on the human respiratory system. J. Thorac. Dis. **8**(1), 69–74 (2016). https://doi.org/10.3978/j.issn.2072-1439.2016.01.19
9. A. Zwoździak, I. Sówka, A. Worobiec, J. Zwoździak, J. Nych, The contribution of outdoor particulate matter (PM1, PM2.5, PM10) to school indoor environment. Indoor Environ. **1**, 320–321 (2015). https://doi.org/10.1177/1420326X9500400602
10. S.J. Kwon, Artificial neural networks. 1–426 (2011). https://doi.org/10.15864/jmscm.1104
11. M.W. Gardner, S.R. Dorling, Artificial neural networks (the multilayer per- ceptron)—a review of applications in the atmospheric sciences. Atmos. Environ. **32**(14–15), 2627–2636 (1998)
12. H. Huang, R. He, Z. Sun, T. Tan, Wavelet-SRnet: a wavelet-based CNN for multi-scale face super resolution, in *Proceedings of the IEEE International Conference on Computer Vision, France, 13–16 Oct. 2003, Nice, France* (2017)
13. K. Zhang, W. Zuo, Y. Chen, D. Meng, L. Zhang, Beyond a gaussian denoiser: Residual learning of deep CNN for image denoising. IEEE Trans. Image Process. **26**(7), 3142–3155 (2017). https://doi.org/10.1109/TIP.2017.2662206
14. K. Zhang, W. Zuo, S. Gu, L. Zhang, Learning deep CNN denoiser prior for image restoration, in *Proceedings of the IEEE Conference on Computer Vision and Pattern Recognition, USA, 21–26 July 2017, Honolulu, HI, USA* (2017). https://doi.org/10.1109/CVPR.2017.300
15. Y. Wei, W. Xia, M. Lin, J. Huang, B. Ni, J. Dong, Y. Zhao, S. Yan, HCP: a flexible CNN framework for multi-label image classification. IEEE Trans. Pattern Anal. Mach. Intell. **38**(9), 1901–1907 (2015). https://doi.org/10.1109/TPAMI.2015.2491929
16. Y. Marinah, *Proses Pembangunan di Petaling Jaya, Selangor* (2021). https://www.scribd.com/doc/118598737/PROSES-PEMBANGUNAN-DI-PETALING-JAYA-SELANGOR
17. Ministry of Health (2021). https://www.moh.gov.my/
18. D.A. Clevert, U. Unterthiner, S. Hochreiter, Fast and accurate deep network learning by exponential linear units (elus) (2015). arXiv preprint arXiv:1511.07289
19. I. Goodfellow, Y. Bengio, A. Courville, *6.2.2.3 Softmax Units for Multinoulli Output Distributions* (Deep Learning, MiT Press, 2016)

20. X. Glorot, Y. Bengio, Understanding the difficulty of training deep feed-forward neural networks, in *Proceedings of the Thirteenth International Conference on Artificial Intelligence and Statistics, Italy, 13–15 May 2010, Chia Laguna Resort, Sardinia, Italy* (2010)
21. D.P. Kingma, J. Ba, Adam: a method for stochastic optimization (2014). arXiv preprint arXiv: 1412.6980

Chapter 44
Petrographical Analysis on Microcracks and Delayed Ettringite Formation (DEF) of Saltwater Intruded Concrete

M. N. Jusoh, A. Rahim, K. Ramanathan, R. A. Abdullah, T. L. Goh, and W. M. W. Ibrahim

Abstract The purpose of this study is to analyze deteriorated concrete samples from a port that located at southwest of Johor. The structure is exposed to aggressive environment which is seawater. Seawater contains chloride and sulfate which can promote to the concrete and structure deterioration due to chemical and mechanical reactions. Microstructural study which is petrographic analysis is aimed to examine the presence of microcrack and delayed ettringite formation (DEF) in micro-sized scale. Microcrack and DEF images are identified under polarizing microscope. The other method is used X-ray diffraction method as to determine the mineral content while rebound hammer test validates the mechanical strength of the port's structure. Microcracks and zonal boundaries are obviously observed in relation to the DEF content to samples with low mechanical strength. DEF is proven its growth with the 5.7–10.4% of mineral composition, especially ettringite caused the material strength reduced.

Keywords Deteriorated concrete · Delayed ettringite formation · Petrographic analysis · X-ray diffraction · Microcrack

---

M. N. Jusoh · A. Rahim (✉) · K. Ramanathan · R. A. Abdullah
School of Civil Engineering, Faculty of Engineering, Universiti Teknologi Malaysia, 81310 Johor Bahru, Malaysia
e-mail: afikahrahim@utm.my

T. L. Goh
Department of Earth Sciences and Environment, Faculty of Science and Technology, Universiti Kebangsaan Malaysia, 43600 Bangi, Selangor Darul Ehsan, Malaysia

W. M. W. Ibrahim
Faculty of Mechanical Engineering Technology, Universiti Malaysia Perlis, 026000 Arau, Perlis, Malaysia

M. A. A. Mohd Salleh et al. (eds.), *Proceedings of the Green Materials and Electronic Packaging Interconnect Technology Symposium*, Springer Proceedings in Physics 289,
https://doi.org/10.1007/978-981-19-9267-4_44

44.1 Introduction

Concrete is the most material applied in construction, and there are 9.5 billion m^3 or about 22 billion tons of concrete produced for global industry. The production of modern concrete includes Ordinary Portland Cement (OPC), binder, course and fine aggregates, water for hydration, and also additives in order to control their workability and consistency [1]. The American Society for Testing and Materials (ASTM) has standardized five Portland cements in the USA: ordinary (Type I), modified (Type II), high early strength (Type III), low heat (Type IV), and resistant sulfate (Type V). In other countries/regions, type II is omitted, and type III is called rapid hardening. Type V is called Ferrari cement in some European countries. There are numerous other special types of Portland cement. Colored cement is made by grinding 5–10% of suitable pigments of white or ordinary gray Portland cement. Air-entrained cement is made by adding a small amount of about 0.05% of organic reagents during grinding, which can cause very fine bubbles to be entrained in the concrete. This increases the ability of concrete to resist freeze–thaw damage in cold climates. When making concrete, air-entraining agents can be added to the mixture as a separate component. The outstanding durability of Portland cement concrete is the main reason why it has become the most widely used building material in the world. However, concrete deterioration can cause by the limitations of materials, design and construction practices, and harsh exposure conditions. May cause aesthetic, functional, or structural problems. So that, it is very important to know the relationship between chemical reactions in the concrete with the concrete strength.

44.2 Mechanisms of Delayed Ettringite Formation (DEF)

The prevalent concrete will include coarse and fine aggregate with a combination of inorganic binder which is hydrated ordinary Portland cement (OPC). At the same time, additives such as super-plasticizer and retarding are used. The presence of mineral materials like pozzolanas or fillers can improve concrete properties. During the process of casting the concrete, the imperfections leading to local inhomogeneity and anisotropy are virtually inevitable [1]. The structure will expose to the various meteorological conditions with different climate changes such as wet and dry seasons. The mechanisms of delayed ettringite formation (DEF) and alkali–silica reaction (ASR) can both cause severe degradation and consequent reduction in commission lifetime of concrete structures [2]. These situations can influence the properties of the concrete and building service life of the structure.

The durability of concrete is shown by the capability of climate change especially weather and exposure to the weather, the reaction of chemical substances like seawater and acid-based solution. Nowadays, the use of minerals in concrete can cause the deterioration of concrete. Alkali–silica reaction (ASR) is a reaction when there is the presence of hydroxyl ions which is from the pore of the concrete and silica.

Silica can be found in the aggregate. The expansion of the concrete was definitely caused by ASR. ASR can also cause cracking of concrete and will result in major effects for the concrete structure. The mechanisms of delayed ettringite formation (DEF) and alkali–silica reaction (ASR) can both cause severe degradation and consequent reduction in commission lifetime of concrete structures. ASR could be a widely known deterioration process with long-running scientific conferences dedicated to its research and investigations. DEF, however, may be a relatively new deterioration process which became noticed within the late 1980s and research on that has been intense since the 1990s [2]. Severe map cracking of concrete within the massive pile caps of bridge piers in state was observed about four years after construction. In Thailand, random cracks are found on the surface of massive concrete structures within several years of construction and have extended even after repair [3].

The X-ray diffraction method (XRD) is the most appropriate test for detecting the presence of DEF. X-ray powder diffraction (XRD) is used as to measure the chemical reaction and mineral growth, especially DEF which developed due to the aggregate existence. XRD is a technique that involves irradiating a material with incoming X-rays and then measuring the intensities and scattering angles of the X-rays that depart the sample. The development of ettringite will reduce the concrete strength. Rebound hammer test will show the concrete strength on the selective sample. Therefore, this study to examine does chemical and mechanical factors which can influence the deteriorated sample. Other than that, the microstructural condition and delayed ettringite formation (DEF) were investigated. The relationship between chemical reactions due to mineral growth and material strength was identified. Finally, relationship between microstructural analysis and chemical reaction with the material strength was established. There are several tests that have been executed based on the guidelines of practice. X-ray powder diffraction (XRD) is used as to measure the chemical reaction and mineral growth especially DEF which developed due to the aggregate existence, called as delayed ettringite formation (DEF); XRD will determine the composition of mineral and the density of ettringite that is corporate with the degree of weathering which influence the strength of deteriorated sample. Therefore, rebound hammer test (secondary data) is used to determine the strength of the structure which affected by the saltwater intrusion.

44.3 Concrete Microcracks

Ettringite is that the mineral name for calcium sulfoaluminate ($3CaO \cdot Al_2O_3 \cdot 3CaSO_4 \cdot 32H_2O$), which is generally found in Portland cement concretes. Calcium sulfate sources, such as gypsum, are intentionally added to Portland cement to control early hydration reactions to forestall flash setting, improve strength development, and reduce drying shrinkage. The term 'delayed ettringite formation' (DEF) is usually wont to talk to the doubtless deleterious reformation of ettringite in moist concrete, mortar, or paste after the destruction of primary ettringite by heat. The sulfate present in cement under normal conditions will bring about to the formation of hydrated sulphoaluminate compounds, and these form while the concrete is in an exceedingly plastic state, or simply after the concrete has begun its initial set, and don't cause any significant localized disruptive action [4]. It has been claimed that DEF can cause damage in concrete that has not experienced an elevated temperature. This was attributed to excessive contents of SO_3 within the clinker or, more precisely, to the presence of that SO_3 in phases from which it had been only slowly released [5].

A microcrack could be a variety of material damage consisting of cracks sufficiently small to want magnification to watch. A microcrack is a sign of fabric failure which will ultimately cause complete failure. It should occur on a coating during the appliance or drying process, or during load strain of a coating or material. Conventionally, microcracks are defined as cracks whose width is about few micrometers (<10 μm). Microcrack measurements can vary significantly with magnification and are markedly stricken by the degree of resolution of the observation technique. Such a phenomenon originates from the heterogeneous and multi-scale structure of cement-based materials. Manual counting of microcracks is tedious and constitutes a limitation for a scientific analysis. Test results can also be biased by the subjectivity of the operator who often needs to convert a posh picture into a binary (yes/no) result [6].

A salt solution, mostly sodium chloride and magnesium sulfate, wets concrete exposed to saltwater. Damage to concrete, if it happens, is typically the consequence of a failure to apply excellent practices in concrete construction, and is frequently the result of freezing and thawing or wetting and drying, as much as or more than the impacts of seawater. Most, if not all, of the components of hardened Portland cement paste, particularly the aluminate ingredient, may be attacked by magnesium sulfate; chlorides may accelerate steel corrosion; and alkalis may participate in the alkali-aggregate reaction. Thus, concrete exposed to saltwater should be constructed with regulated aluminate content cement and nonreactive aggregate, embedded steel should be properly covered by low permeability concrete, and appropriate building methods should be followed. Because corrosive substances such as chloride ions will infiltrate the concrete matrix through the porous network channels, mitigating the alkaline environment and encouraging rebar corrosion. When designing and optimizing the types of concrete component materials and the mix ratio for RC structures in coastal, atmospheric salt spray, ocean, and other conditions, special attention should be devoted to increasing the density of concrete and slowing the influence of chloride penetration [7].

Sulfate products will form in narrow pores during the dormant period of ESA, causing supersaturation of the pore solution, and the concrete expansion can cause crystallization pressure [8]. At this stage, microcracks are initiated in the cement slurry near the pores, but they do not penetrate each other. Secondly, the continuous increase of crystallization pressure causes microcracks in the cement slurry to penetrate each other, which will cause the concrete to expand significantly. Therefore, within a harmful duration, the interpenetration of microcracks will cause the disappearance of closed pores, which will lead to the crystallization of sulfate products and the rapid reduction of supersaturation of the pore solution [8]. These failures eventually led to high reconstruction costs and a heavy financial burden on the municipality [2].

For multiple mechanisms, petrography can effectively determine the cause of degradation at a relatively low cost, including alkali-aggregate reaction (AAR), sulfate attack (DEF, TSA), leaching, another fluid interaction, fire, normal carbonation and poor workmanship, [1]. Alkali–silica reaction (ASR) is the regular form of alkali-aggregate reaction (AAR). This is the hydroxide ion in the pore water and the reactive silica in the aggregate is mainly quartz, but there are also amorphous or poor crystallinity polymorphs [1]. The use of X-ray diffraction to characterize crystalline materials is an effective non-destructive technique. It contains details about structures, phases, desired crystal orientations (texture), and other structural parameters including average grain size, crystallinity, strain, and crystal defects. XRD peaks are generated by the positive interference of a monochromatic beam of X-rays dispersed at specific angles from each group of lattice planes in a sample. The atomic locations within the lattice planes decide the peak intensities. A fast online search of a common database for X-ray powder diffraction patterns allows for rapid phase detection of a wide range of crystalline samples. The XRD study was performed in a Bragg–Brentano scheme using a Rigaku diffractometer with an angular range of 2 = 10–80 degrees, a radiation of = 154,182, a voltage of 25 kV, and an amplitude of 25 mA.

In order to achieve the third objective which is to study the relationship between microstructural analysis, chemical reaction with the material strength, the strength of sample was tested as collected at site by using rebound hammer test. The rebound hammer is a non-destructive testing device that measures the rebound of a spring-driven mass after contact with a concrete surface. The rebound hammer's output is known as the rebound number, and it is connected to the surface hardness of concrete. By creating a proper association between the rebound index and the compressive strength of concrete, the rebound hammer technique gives a practical and quick indicator of the compressive strength of concrete.

44.4 Concrete Sample Analysis

This research consisted of three key activities which are preparation of the sample, analyzing sample, and secondary data. Firstly, in thin section sample, there are seven steps for thin section procedure. A thin section is a 30-μm-thick rock slice that is epoxy-attached to a glass slide. Step 1 is cutting a slab. A suitable size slab for mounting on a slide is cut from a piece of rock or drill core with a diamond saw. Step 2 is initial lapping of the slab. The slab is labeled on one side, and the other side is lapped flat and smooth first on a cast iron lap with 400 grit carborundum, then finished on a glass plate with 600 grit carborundum. Step 3 which is glass slide is added. After drying on a hot plate, a glass slide is glued to the lapped face of the slab with epoxy. Step 4 slab is sectioned. Using a thin section saw, the slab is cutoff close to the slide. The thickness is further reduced on a thin section grinder. Step 5 is final lapping. A finished thickness of 30 μ is achieved by lapping the section by hand on a glass plate with 600 grit carborundum. A fine grinding with 1000 grit prior to polishing is optional. Step 6 is polishing. The section is placed in a holder and spun on a polishing machine using nylon cloth and diamond paste until a suitable polish is achieved for microscope or SEM studies. Step 7 is final inspection.

Next, Microstructural Study/Petrographic Analysis. Petrographic analysis functions as determined by the textures and mineral compositions of samples. Petrographic testing by using a microscope to examine a rock or concrete sample to determine its mineralogical and chemical properties. This analysis is very useful for the identification of potentially deleterious minerals in the aggregates. The petrological inspection samples can be obtained from block samples or cores. They are impregnated with resin. Suitable surfaces (polished or thin sections) produced by sawing, grinding, and polishing. Examine the sample through a rock (geologically polarized) microscope that reflects (transmitted) light. The lighting can be ordinary, ultraviolet, or polarized light. A petrographic analysis can be describing core samples from an exploration well, looking at thin sections of the reservoir or reservoir boundary, using a scanning electron microscope (SEM) to characterize the fracture surfaces from microstructures, doing X-ray diffraction (XRD) to identify exact mineral assemblages of a sample, or some other useful petrographic technique that can be used to identify the chemical or physical features of a specific sample.

After that, the use of X-ray diffraction to characterize crystalline materials is an effective non-destructive technique. It contains details about structures, phases, desired crystal orientations (texture), and other structural parameters including average grain size, crystallinity, strain, and crystal defects. XRD peaks are generated by the positive interference of a monochromatic beam of X-rays dispersed at specific angles from each group of lattice planes in a sample. The atomic locations within the lattice planes decide the peak intensities. A fast online search of a common database for X-ray powder diffraction patterns allows for rapid phase detection of a wide range of crystalline samples. The XRD study was performed in a Bragg–Brentano scheme using a Rigaku diffractometer with an angular range of 2 = 10–80°, a radiation of = 154,182, a voltage of 25 kV, and an amplitude of 25 mA. For secondary data, the data was obtained from the technical report. The strength of sample was collected at site by using rebound hammer test. The rebound hammer is a non-destructive testing device that measures the rebound of a spring-driven mass after contact with a concrete surface. The rebound hammer's output is known as the rebound number, and it is connected to the surface hardness of concrete by creating a proper association between the rebound index and the compressive strength of concrete and the rebound hammer technique.

44.5 Result and Discussion

The results on the petrographic analysis and X-ray powder diffraction are discussed herein. The relationship between petrographically analysis, chemical mineral content with material strength and relationship between chemical and minerals strength, as presented later.

Fig. 44.1 Sample of concrete at SWJ 1 (medium)

For the sample descriptions using the microstructural analysis, Sample SWJ 1 with medium in shape and has strong strength. In the microscopic view, the cement-coarse aggregate bond and cement-fine aggregate bond are good, no major cracks and carbonation of the cement paste during the sample collection occurred (Fig. 44.1).

By using the petrographical analysis, microcracks are found passing through an aggregate and through the cement paste. The aggregate is fairly well-graded. The coarse aggregate fraction in this concrete shows relatively potential for alkali reaction as denoted by the presence of boundary reaction between the cement and the coarse aggregates. The coarse aggregate is abundant with angular–subangular quartzo-feldspathic mineral. Reaction rims or gel were observed around the aggregates. The fine aggregates are made up of quartz sand, minor amount of plagioclase and rock fragments. The fine aggregates are presented as angular–subangular fractions and uniformly distributed with varies in their weathering. The grain size varies between 0.5 and 3 mm (Fig. 44.2).

The cement paste is light to medium grayish and showed a vitreous luster, microgranular texture, and hackly fracture. The binder is slightly loose, especially at the

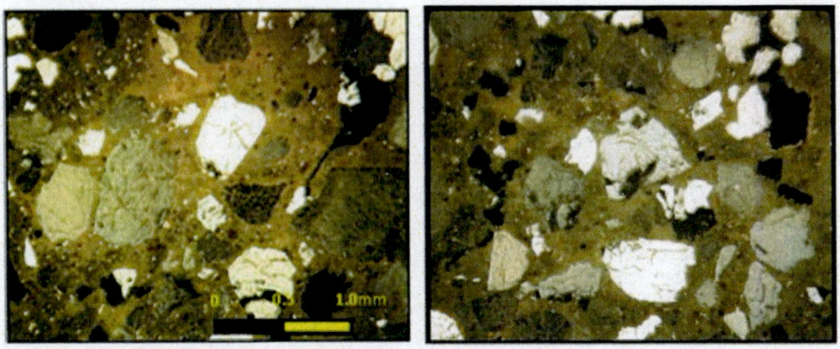

Fig. 44.2 (Left) Photomicrograph shows the presence of microcracks or shrinkage in some of the fine aggregates which composed of predominant quartz. (Right) Photomicrograph shows significant presence of void spaces mark by dark color

aggregate's boundary which indicated the weathering grades have the minerals itself and separated from the other minerals with the chemical reaction. The amount of entrained air voids is generally low and normally clustered in ranges from 3 to 5%. The entrapped is observed but again it is very small and scattered. Small percent of air void can result in low deteriorated process. The concrete indicates evidence of slightly alkali–silica reaction and small clustered of coarse aggregates has shown it incipient reaction (zonal degradation) (Figs. 44.3 and 44.4).

Based on Table 44.1, the value of ettringite per others mineral in the sample is 0.104 Which is, 10.4% of ettringite in the sample compare to the others mineral. The observation is true because the proportion of ettringite under microscopic analysis is low (Fig. 44.5; Table 44.2).

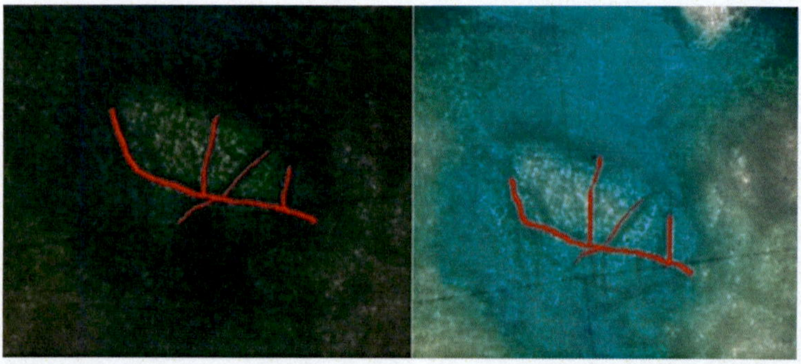

Fig. 44.3 Ettringite found embedded between the minerals under plane polarized light (left). Under cross-plane polarized light, the ettringite form through aggregate, the weathering looks matured (right)

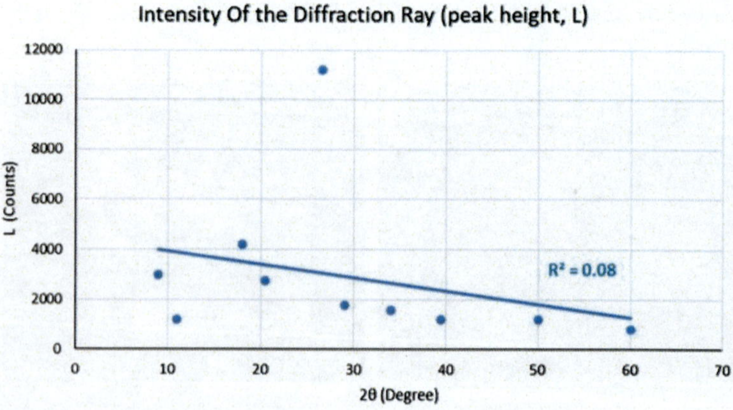

Fig. 44.4 XRD pattern for SWJ 1 (medium)

Table 44.1 Crystal lattice spacing and intensity of the diffraction ray for sample SWJ 1

Mineral	Crystal lattice spacing (diffraction angle, 2θ)	Intensity of the diffraction ray (peak height, *L*)
Ettringite (E)	9	3000
Quartz (Q)	26.5	11,200
Gypsum (G)	11	1200
	20.5	2760
C3S/C2S (H)	29	1800
Calcium silicate hydrate (C)	39.5	1200
	60	800
Calcium aluminate silicate (S)	50	1200
Calcium hydroxide (CH)	18	4200
	34	1600

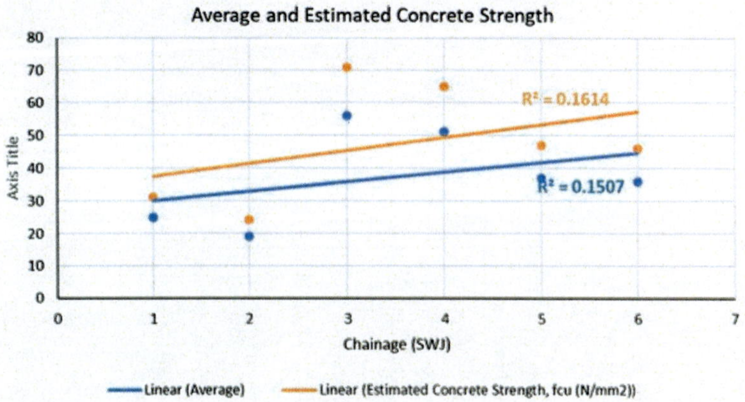

Fig. 44.5 Average and estimated concrete strength based on rebound hammer test results

Relationship is between petrographical analyses and chemical mineral content with the material strength. Based on secondary data for concrete strength, the estimated concrete strength is between 24 and 71 N/mm^2 for every chainage. Estimated concrete strength for SWJ 5 the concrete strength is 24 N/mm^2 that means the condition for SWJ 5 is severe.

Table 44.2 Estimated concrete strength based on rebound hammer test results

	Chainage	Condition	Position	Rebound number												Average	Estimated concrete strength, f_{cu} (N/mm^2)
				1	2	3	4	5	6	7	8	9	10	11	12		
Component	SWJ 6	Severe	A	24	25	31	30	26	22	26	23	22	21	20	30	25	31
	SWJ 5	Severe	A	15	15	19	21	20	21	19	18	20	22	18	20	19	24
	SWJ 4	Good	A	54	55	54	54	51	58	56	55	57	57	55	58	56	71
	SWJ 3	Good	A	45	52	52	44	48	48	54	48	50	49	48	50	51	65
	SWJ 2	Medium	A	38	36	39	35	36	38	34	36	32	34	38	36	37	47
	SWJ 1	Medium	A	33	37	35	31	33	37	38	36	40	38	37	37	36	46

44.6 Conclusion

Throughout the research, we are able to discuss and explain all the objective:

1. The petrographical analysis examine by using polarizing microscope, showed that microcrack (minor to severe) and for severely deteriorated ettringite are seen in six samples proportionately on its physical condition of weathering states. The highly weathered samples (samples SWJ 5 and SWJ 6) show the abundant of zonal boundaries between minerals, microcracks, and premature to ettringite occurrences. The amount of entrained air voids is generally low and normally clustered in ranges from 3 to 5%. The entrapped is observed but again it is very small and scattered. Small percent of air void can result low deteriorated process.
2. The chemical reaction occurs in the sample which is DEF resulted in lower strength of concrete. The ettringite found embedded between the minerals under cross-plane polarized light, the ettringite form through aggregate, the weathering looks matured thus making the concrete strength decreases.
3. The mineral composition in the sample examine by using X-ray powder diffraction showed that the intensity of the diffraction ray and crystal lattice spacing. The highly ettringite content ratio per other mineral and the lower strength of the concrete samples is seen in severely deteriorated samples.

Acknowledgements This work was supported by the Fundamental Research Grant Scheme (FRGS) awarded by the Ministry of Education of Malaysia entitle 'Enhanced Coalbed Methane Recovery Processes by Coupling Fluid Flow of Mukah-Balingian Coal Deposits' with the referral no. R.J130000.7851.5F371.

This work was supported by the Universiti Teknologi Malaysia Encouragement Research Grant (UTMER) awarded by Universiti Teknologi Malaysia entitle 'Diagnosis of Alkali-Aggregate Reaction—Polarizing Microscopy and SEMEDS Analysis' with the referral no. Q.J130000.3851.19J79.

References

1. I. Fernandes, M.A.T.M. Broekmans, Alkali–silica reactions: an overview. Part I (2013)
2. S. Ekolua, G. Rakgosia, D. Hooton, Long-term mitigating effect of lithium nitrate on delayed ettringite formation and ASR in concrete—microscopic analysis. Mater. Charact. (2017)
3. S. Asamoto, K. Murano, I. Kurashige, A. Nanayakkarad, Effect of carbonate ions on delayed ettringite formation. Constr. Build. Mater. (2017)
4. A.B. Poole, I. Sims, *Concrete Petrography* (2015)
5. H.F.W. Taylor, C. Famy, K.L. Scrivener, Delayed ettringite formation (2001)
6. A. Ammouche, J. Riss, D. Breysse, J. Marchandc, Image analysis for the automated study of microcracks in concrete. Cement Concr. Compos. 267–278 (2001)
7. J. Liu, W. Zhang, Z. Li, H. Jin, L. Tang, Influence of deicing salt on the surface properties of concrete specimens after 20 years. Constr. Build. Mater. (2021)
8. G.-J. Yin, X.-B. Zuo, X.-N. Li, Y.-X. Zou, An integrated macro-microscopic model for concrete deterioration under external sulfate attack. Eng. Fract. Mech. (2020)

Chapter 45
Development of Low-Cost, High-Efficiency Powder Transfer System for Food Industry

M. S. Ismail, A. I. M. Shaiful, M. I. Hussain, and C. W. Chai

Abstract Powder Transfer System (PTS) has been widely used in various industries to transport powder from one workstation to another. It speeds up the loading process while decreasing operator fatigue and manual handling. Majority of food industry in Small and Medium Enterprise (SME), Perlis, is still using the manual handling methods to transfer the powder from mixer to filling funnel station which requires higher needs of manpower. Hence, PTS is introduced to overcome the problems in the industry with the aim of development of low cost (affordable by SME industries) yet high efficiency system. Development of this system started from selecting the most suitable transfer system for food grade powder with the consideration of the working principle and safety for food and environment. Pneumatic Vacuum Conveying System (PVCS) is selected as the most suitable system in this project through the system selection process. This project also includes the system design which involves the designation of system function and design based on requirements by using the functional decomposition and engineering CAD drawing software—SolidWorks, material selections for the system components which suitable for food grade conditions, and cost estimation. Lastly, validation of the system design structure stability, transfer rate analysis, and development of the prototype are carried out at company M, one of the SME companies in Perlis, Malaysia.

Keywords Powder transfer system · Pneumatic vacuum conveying system

45.1 Introduction

Majority of the food industries in Small and Medium Enterprise (SME), Perlis, is still using the manual handling methods to transfer the powder from mixer to filling station funnel which requires higher needs of manpower. The methods used by the

M. S. Ismail (✉) · A. I. M. Shaiful · M. I. Hussain · C. W. Chai
Faculty of Mechanical Engineering Technology, Universiti Malaysia Perlis, Kampus Tetap Pauh Putra, 02600 Arau, Perlis, Malaysia
e-mail: mohdshihabudin@unimap.edu.my

M. A. A. Mohd Salleh et al. (eds.), *Proceedings of the Green Materials and Electronic Packaging Interconnect Technology Symposium*, Springer Proceedings in Physics 289,
https://doi.org/10.1007/978-981-19-9267-4_45

food industry to transfer powder is not effective, is of poor hygiene, and lack of dust control. From the market survey nowadays, most of the Powder Transfer Systems (PTSs) for transferring food grade powder are high in cost due to the material used, system principle, etc. As such, a better performance PTS can be supplied to the industry to increase their productivity. PTS is introduced to solve the problems met in this SME industries. Operators are no longer to physically load the process machines or handle heavy equipment like buckets, bags, and other containers to transfer the powders with the used of PTS. It speeds up the loading process while decreasing operator fatigue and manual handling [1].

This project aims to develop a low-cost, high-efficiency PTS for food industries that is affordable by SME industries. The suitable PTS for food grade powder is determined in terms of working principle of the system such as Pneumatic Vacuum Conveying System (PVCS), food grade material used for food contacting components, and desired system characteristics such as safety requirement, efficiency, cost, etc. Next, the PTS is designed by using SolidWorks based on the requirements stated by the food industry and the material selection and cost estimation are included. Finally, the system design and transfer rate are validated by using formula calculation or simulation of Ansys Workbench to make sure it met with the requirements set by the industry.

Pneumatic conveying system is one of the PTS which is suitable in this project, and it is using both negative and positive pressures without relying on gravity to transfer powders as if they were liquid eliminating the need for multi-floor processes [2]. Pneumatic conveying is a method of moving bulk solid materials such as powders from one processing station to the next one using air or pressurized gas. Material is moved through an enclosed pipeline using a combination of pressure differential of airflow from a blower or vacuum pump.

45.2 Methodology

The methodology is distributed into 3 phases and matched with the 3 objectives accordingly. The phases include system selection on the PTS, system design, validation of the system design, and prototype development of the system. In a first phase, the project starts with study and review on the PTS in the market. Then, system selection is implemented for the most suitable PTS for the food industry in Perlis to overcome the problems they are facing in terms of cost, efficiency, hygiene etc. Next, the selected PTS is designed to meet with the requirements of the industry and which can work at the existed workplace by drawing the system using SolidWorks software. The material selection and cost estimation are included in this section. Finally, the system design is validated by analyzing the stability of the PTS by using calculation or Ansys Workbench. The transfer rate analysis is also done to validate the industry's requirement and complete the project. The validation is carried out at company M, one of the SME food industry in Perlis.

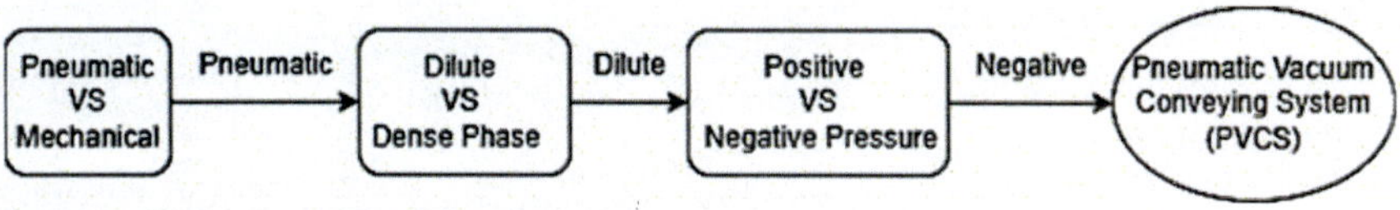

Fig. 45.1 Flowchart of system selection

45.2.1 System Selection and Design

The method of comparison is used to choose the most suitable PTS. A total of three stages are included in the system selection, which are from the comparison between pneumatic and mechanical conveying system, where pneumatic conveying system is selected. Followed by the comparison between dilute and dense phase system, dilute phase system is selected. Lastly, comparing the negative or positive pressure, negative pressure is selected. So, Pneumatic Vacuum Conveying System (PVCS) is selected as the most suitable PTS. Figure 45.1 shows the flowchart of the system selection.

System design includes industry requirements and discussion, the functional decomposition of the PTS, details of design drawing of PTS, material selection, and cost estimation (bill of materials) for the PTS developed. System design is one of the important sections in this project due to one of the objectives related with it. Through this section, the product's details are finalized and visualized.

45.2.2 Stability Analysis of Design Structure

The stability analysis of this system is mainly considered as overturning due to the variable factor of the lateral loads. The lateral load in this system is the weight of the feeding materials, which is the powder vacuum from the buffer tank to the conveyor body. Calculation is selected due to the lack of data and information of the design such as the real weight, condition, position of center of gravity, etc. Based on the research of the equilibrium of rigid body [3], the detailed procedure to analyze the stability of the design structure of PVCS is shown as below:

1. Prepare for free-body diagram: Construct the x–y coordinate in the suitable orientation. Remove the fix and support on the system and draw the outline of the system. Label all the forces and couple of moments acting on the body. List all the data including the unknown to be found and directions relative to x–y directions. Indicate the dimensions necessary to calculate the unknown.
2. Application of equation of equilibrium: Apply the moment equation of equilibrium about the point that lies at the middle of 2 unknown forces. By this setting, the moments of these unknowns are 0 about the point, and third unknown can be determined. The direction of the moment movement is assumed to the problem, if the solution of the equilibrium equations yields a negative scalar for force or

couple moment magnitude, and this indicates that it is the opposite direction of the assumption.

45.2.3 Transfer Rate Analysis

The aim of the transfer rate analysis is to validate the transfer rate of the PTS designed in this project as whether it has met with the requirements stated or not. The aim of the transfer rate analysis is to determine the required vacuum capacity for the PVCS which must met the industry's requirements. The system's transfer rate value must be higher than the flow rate of the filling funnel system to ensure that there is no break off gap or pause between the process of PTS and filling funnel system. After the aim and requirements are clear, the calculation of conveyor body volume analysis is carried out to find the volume and mass of feeding materials and the required vacuum capacity. Next, the suitable vacuum pump based on the vacuum capacity is determined. Lastly, the transfer rate of the PTS is validated with the requirements of increase the transfer rate from 1200 to 1600 kg/day.

45.3 Result and Discussions

45.3.1 System and Product Design

The results of the system selection and system design involve the development of PTS, simulation, and transfer rate which will be presented and discussed. For system selection, Pneumatic Vacuum Conveying System (PVCS) is selected. PVCS reduces the material loss and is comparatively more hygiene. Environment cleanliness and risk for explosion are more secure. It is more flexible and requires less components, which leads to easier assemblage and work in an existing workplace. The usage of less components also leads to the cheaper cost in manufacturing the system. There are few requirements from industry which include movement (portable and adjustable height), fast transfer rate, food grade materials, easy to use and clean, durable, and low-cost, high-efficiency system. The results of the 3D drawing by using SolidWorks are presented in this section. Figure 45.2 shows the isometric view of the PVCS drawing.

The material for the conveyor body, vacuum filter, pipeline, and parts expose to environment is selected as shown in Table 45.1.

The costing for the materials and components is done by market review and estimation. The total estimation cost for the materials and components to build a PVCS is RM5340, which is cheaper than the market price of the fully assembly system at around RM20000. From the cost estimation, future analysis of cost with the other types of PVCS can be done. Functional decomposition represents a strategy used in the early stages of concept design generation. This section includes some

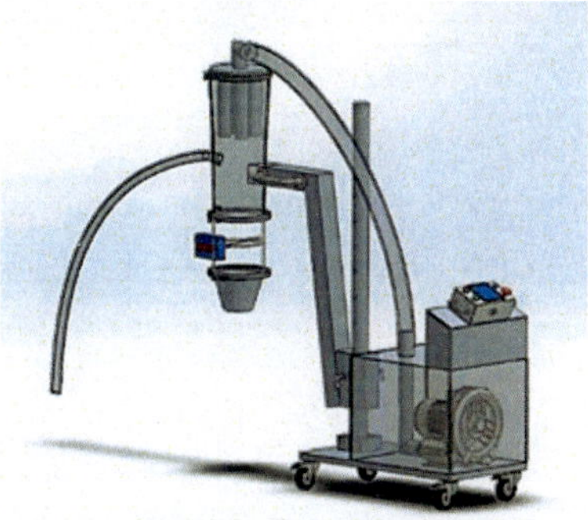

Fig. 45.2 Isometric view of PVCS

Table 45.1 Summary of material selection

Part	Material
Conveyor body	Stainless steel 304
Vacuum filter	Stainless steel 316 powder sintered filter
Pipeline	Food grade silicone braided hose
Parts expose to environment	Stainless steel 304

parts, in summation functional structure block diagram has shown all the details as in Fig. 45.3. The function block in the diagram represents the transformation done by the system, assembly of components, and labeling of each block by selecting function names. Flowlines are drawn with arrows to indicate the direction and are labeled to define the flow connected to the function block. The function of the automation process is included by adding the capacitance probe level sensor and electric control valve.

45.3.2 Product Stability Analysis

When the designation of the PVCS is done, the next stage of the project is to analyze the design of the system. From the designation of the PVCS, stand rod can be rotated 180° to meet the requirements of aligning the conveying body with the filling funnel during the discharging process. There are two optimum positions where the conveyor body is in a straight line of the portable body and perpendicular of the portable body when discharging, as shown in Fig. 45.4.

From Fig. 45.4a, the maximum weight, Wmax, of the lateral load is 73.45 kg. Weight of lateral load is the sum of weight of conveyor body and powder. From the analysis of volume of powder capacity, the weight of the powder at maximum level is 8.5 kg, while the conveyor body is approximately 10 kg. So, the total weight, W,

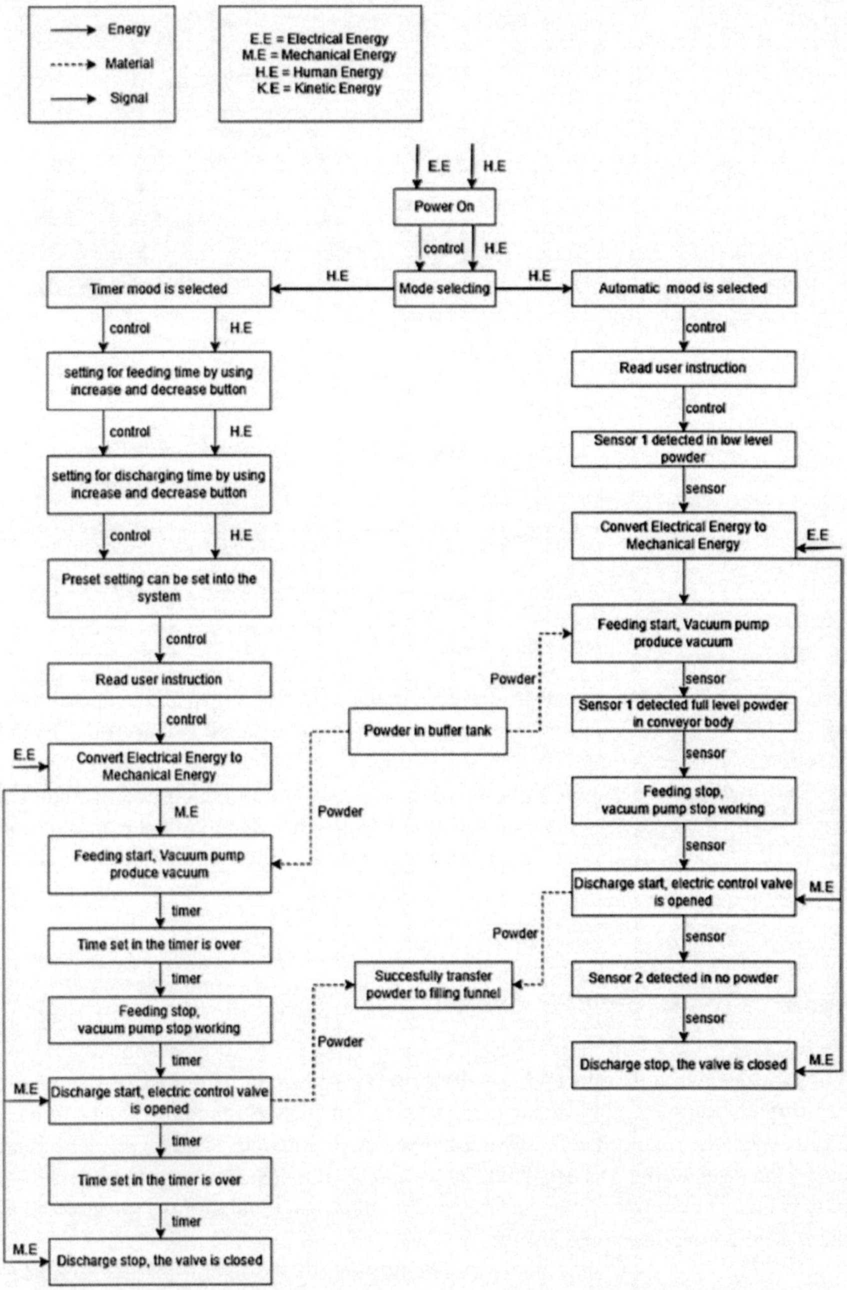

Fig. 45.3 Functional structure block diagram of pneumatic vacuum conveying system

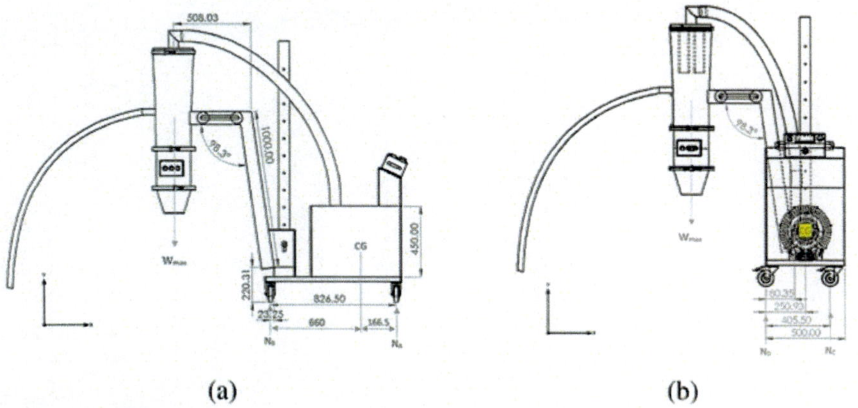

Fig. 45.4 Free-body diagram at **a** position 1 and **b** position 2

of the lateral loads is 18.5 kg. $W < W_{\max}$, the total weight is less than the maximum weight. Therefore, there is no overturn happened to the system. From Fig. 45.4b, the maximum weight, $W_{\max}$, of the lateral load is 37.20 kg. Weight of lateral load is the sum of weight of conveyor body and powder. From the analysis of volume of powder capacity, the weight of the powder at maximum level is 8.5 kg, while the conveyor body is approximately 10 kg. So, the total weight, W, of the lateral loads is 18.5 kg. $W < W_{\max}$, the total weight is less than the maximum weight. Therefore, there is also no overturn happened to the system.

45.3.3 Transfer Rate Analysis

As mentioned in the methodology, the aim of the transfer rate analysis is to determine the required vacuum capacity for the PVCS which must met with the industry's requirement of the system's transfer rate higher than the flow rate of the filling funnel system and increment of the transfer rate from 1200 to 1600 kg/day. Volume and mass capacity of the feeding material in the conveyor body must be determined before proceeding to the further calculation. Figure 45.5 shows the dimensions and the capacity level of the feeding materials in the conveyor body.

There are two different shapes which are occupied by the powder (feeding material): V_1 and V_2 are the shapes of frustum and cylinder, respectively [4]. The total volume of V_1 and V_2 is 0.0131 m^3. Next, the mass of the feeding material is calculated by using the formula of density. The density of the feeding material (cocoa powder) is given by the food industry, $\rho = 648.065$ kg/m^3. By calculation, the mass of the feeding material is 8.5 kg.

Fig. 45.5 Dimension and capacity level of feeding material in conveyor body

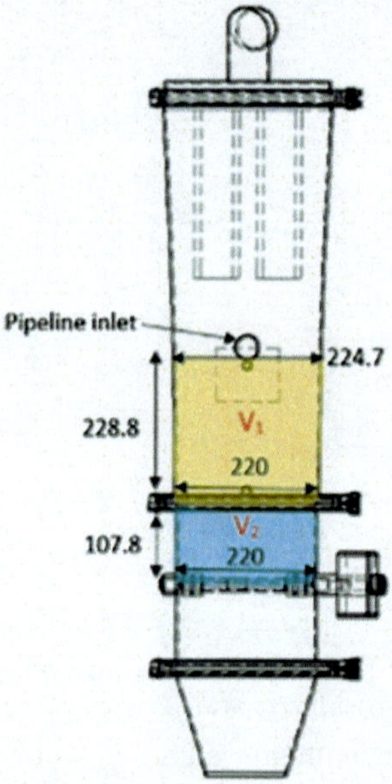

45.3.4 Vacuum Capacity

From the information of food industry, the top filling funnel system capacity is 20 kg and the filling funnel flow rate is 1 kg/10 s. Therefore, for 20 kg capacity, the time required to fully fill the funnel is 200 s. PVCS needs 2.4 cycles to fully fill the filling funnel. So, one cycle of PVCS process needs 83.33 s. This is the maximum allowable time for one cycle to ensure that no break off gap or pause between PVCS and filling funnel system. Therefore, the time taken for 1 cycle of PTS must less than 83.33 s. A cycle process of PVCS included the time for feeding and discharge. The maximum allowable time taken for a cycle process, $T_{\max}$, is 83.33 s and the time taken for discharge is approximately 8 s for 8.5 kg of cocoa powders. Therefore, the maximum allowable time for feeding (vacuum) to the conveyor body is 75.33 s. With the maximum feeding time, the minimum vacuum capacity for the vacuum pump is 408 kg/h.

Next, the suitable vacuum pump is selected based on the required vacuum capacity which is larger than 408 kg per hour to ensure the smooth process between the system. After comparison of the models of vacuum pump, side channel blower vacuum pump with motor power 2.2 kW and 650 kg/h highest vacuum capacity is selected. This model's vacuum capacity is larger than the required vacuum capacity with the lowest motor power. This model can deal with the lower component cost and reduction in the electric consumption. With the minimum vacuum capacity of 408 kg/h, the total transfer rate of the cocoa powder in an 8-h, non-stoppable one day work is 3264 kg/day. Therefore, the requirement of increment in transfer rate to 1600 kg/day is successfully validated.

45.4 Conclusion

PVCS is selected as the most suitable PTS for the SME food industry from the procedure of system selection. From the validation carried out at company M, the PVCS could be assembled and could work well in an existing workplace. It was achieved by the procedural steps of the system design, which includes the requirements of the design; clarified and visualized system function by functional decomposition of the system; design of products based on the requirements using the 3D CAD drawing software—SolidWorks; material selection on the design parts based on the requirements; and estimation of the cost of the product's components and materials. Finally, the analysis of system design stability and transfer rate is validated with the requirements. In conclusion, PVCS has achieved the aim of development of the low-cost, high-efficiency PTS. From the perspective of costing and efficiency, the cost of materials used by PVCS for product manufacturing is comparatively lower than the full assemblage of the system in the market. Optimization of moving parts in PVCS also leads to reduction in maintenance fees. PVCS can also be adjusted to seamlessly connect or run through facilities or structures in industry. This makes it more adaptable, cost-effective in the long run, and cuts off the production time. High efficiency in transfer rate is achieved by using the side channel blower and automation of the system function, which most importantly results in reduction in the manpower and decrease in the labor cost.

Acknowledgements The authors would like to acknowledge UniPRIMA 2021 internal grant awarded by Universiti Malaysia Perlis (UniPRIMA: 9002-00137 (Industrial Fund)) to carry out this project. Special thanks to those who contributed to this project directly or indirectly.

References

1. W. Michaeli, G. Potsch, *Injection Molding: An Introduction* (Hanser, Fachbuchverlag, Munchen, 1995)
2. D. Annicchiarico, J.R. Alcock, Mater. Manuf. Proces. **29**, 662–682 (2014)
3. Y.C. Lam, L.Y. Zhai, K. Tai, S.C. Fok, Int. J. Prod. Res. **42**, 2047–2061 (2004)
4. D. Montgomery, *Design and Analysis of Experiments*, 6th edn. (Wiley, New York, 2005)

Chapter 46
Morphological Characterization of Silver Nanowires (AgNWs)-Embedded Polymeric Film for Flexible Wearable Antenna

M. S. Osman, N. H. Abd Halim, K. Khairudin, A. Abu Bakar, A. R. Razali, M. Fathullah, A. K. Abdul Razak, and M. S. Akbar

Abstract Silver nanowires (AgNWs) transparent electrodes have recently been selected as an alternative conductor for use in various electronic devices. The synthesis of AgNWs for a flexible wearable antenna is successfully demonstrated on PDMS/PVDF elastomeric substrate. In order to consider its suitability for use in flexible antenna applications, the structural properties of AgNWs-polymeric film are discussed. This type of flexible antenna is prepared using the casting method by embedding the AgNWs on the surface of the PDMS/PVDF elastomeric substrate. Analysis of FTIR and SEM showed that AgNWs are reliably transferred to the elastomeric substratum surface, which then served as a conductive radiating factor for the antenna system. Uniform average porosity suggested sponge formation as a porous structure on the film surface that is beneficial for the AgNWs catchment area. The as-made AgNWs-polymeric substrate revealed antennas shaped − 18 dB return loss with a frequency of 2.725 GHz. These results have shown the feasibility

M. S. Osman (✉) · N. H. Abd Halim · K. Khairudin · A. K. Abdul Razak
EMZI-UiTM Nanoparticles Colloids and Interface Industrial Research Laboratory (NANO-CORE), Centre for Chemical Engineering Studies, Universiti Teknologi MARA, Cawangan Pulau Pinang, Permatang Pauh Campus, 13500 Permatang Pauh, Pulau Pinang, Malaysia
e-mail: syazwan.osman@uitm.edu.my

M. Fathullah
School of Manufacturing Engineering, Universiti Malaysia Perlis (UniMAP), Pauh Putra Campus, 02600 Arau, Perlis, Malaysia

M. S. Osman · M. Fathullah
Center of Excellence Geopolymer and Green Technology (CEGeoGTech), Universiti Malaysia Perlis (UniMAP), 01000 Kangar, Perlis, Malaysia

A. Abu Bakar · A. R. Razali
Centre for Electrical Engineering Studies, Cawangan Pulau Pinang, Universiti Teknologi MARA, Permatang Pauh Campus, 13500 Permatang Pauh, Pulau Pinang, Malaysia

M. S. Akbar
Faculty Science, Universiti Teknologi Malaysia, 81310 Skudai, Johor, Malaysia

M. A. A. Mohd Salleh et al. (eds.), *Proceedings of the Green Materials and Electronic Packaging Interconnect Technology Symposium*, Springer Proceedings in Physics 289,
https://doi.org/10.1007/978-981-19-9267-4_46

of this AgNWs' polymer film in wireless communications' applications for wearable systems.

Keywords Silver nanowire · Flexible wearable antenna · PDMS · Casting · Synthesis · Characterization

46.1 Introduction

Wearable antenna for body-centric wireless communication area is expected to be soft, flexible and even stretchable to be more human-friendly in the future. As the result of human body movement, an antenna that is attached on the human body will be stretched, folded or twisted. To enable conformable contact with moving body, it requires the device to endure with the operations without reducing the antenna performance. In other words, a wearable antenna should meet the requirement of being bent, stretch, compressed, twisted and deformed while maintaining its excellent electrical performance. Combination of nanotechnology with electronic and electrical engineering has the potential to contribute to the positive development of this type of human-friendly smart devices. By exploiting nanotechnology in this system, it is a promising method to provide deformable characteristic in electronic devices in virtue of maintaining electrical conductivity and flexibility properties in them [1]. Silver nanowires (AgNWs) have been chosen as an alternative transparent conductor as they have excellent electrical and optical properties and also mechanical strength comparing with other conductive materials such as metal nanoparticles, carbon nanotube and conducting polymer [1].

Therefore, due to higher electrical conductivity and excellent flexibility, AgNWs offer unique advantages in the fabrication of a wearable antenna for the wearable wireless communication industry. AgNWs can be considered as a viable option for stretchable conductor since silver not only possesses the highest electrical conductivity among metals, but they also are known to have superior yield strength and Young's modulus over bulk silver [2]. Polydimethylsiloxane (PDMS) polymer is believed to be an excellent candidate to be incorporated with AgNWs in this wearable antenna. One of the main benefits of PDMS as it is not only due to enhanced stretchability and biocompatibility, it is also superhydrophobic to maximize the wettability of wearable antenna [2, 3]. Since antenna should be in small size and must be easy to manufacture, PDMS shows an advantage to be applied as a substrate in this industry as it can be molded to form thin films at low cost compared with other types of polymers [4]. Physical and chemical attributes of PDMS [5] including (1) low glass transition temperature ($T_g = -125$ °C), (2) its low Young's modulus value which is 1 Mpa, (3) its low shear modulus which varies between 100 kPa and 3 Mpa, (4) being able to use over a wide temperature range at least from − 100 °C up to + 100 °C, (5) low chemical reactivity and non-toxic nature have enabled PDMS to be applied as a flexible encapsulation material for wearable antenna without influencing the sensitivity of the device.

Herein, in this work, a facile method to fabricate wearable antenna made up of AgNWs embedded onto PDMS elastomer substrate via casting method has been described. Structural properties of AgNWs and PDMS elastomeric substrate were characterized to correlate the relationship between structure and electrical properties in enhancing the performance of wearable antenna as highly conductive flexible wireless communication device. Fabricated AgNWs is anticipated to be able to demonstrate use in an application whereby too high electrical conductivity and mechanical flexibility under a considerable stretching strain while meanwhile owes a strong adhesion between these two are required.

46.2 Methodology

Antenna mold for thin-film fabrication is obtained from the Faculty of Electrical Engineering UiTM Cawangan Pulau Pinang. As shown in Fig. 46.1, the antenna mold is fabricated through rapid prototyping method according to the following dimension in Table 46.1.

46.2.1 Synthesis of AgNWs

For AgNWs' synthesis, 0.334 g of polyvinylpyrrolidone (PVP) and 20 mL of ethylene glycol (EG) are being heated and thermally stabilized at a temperature of 170 °C. About 0.025 g of silver chloride (AgCl) is being grounded finely and added to the mixture once the temperature is already stable. This step is the initial step nucleation of the silver seeds. After three minutes, 0.11 g silver nitrate ($AgNO_3$) is being slow titrated into the mixture for 10 min to avoid the formation of the various shapes of nanowires. Then, the mixture is heated for an additional 30 min to ensure complete growth of nanowires. Subsequently, allow the solution to cool down before centrifuged the solution three times at 6000 ppm for 30 min each time. This step is to remove EG, PVP and impurities present in the supernatant. After the final centrifuge, the precipitate is being dispersed in 5 mL of ethanol and being centrifuged again for

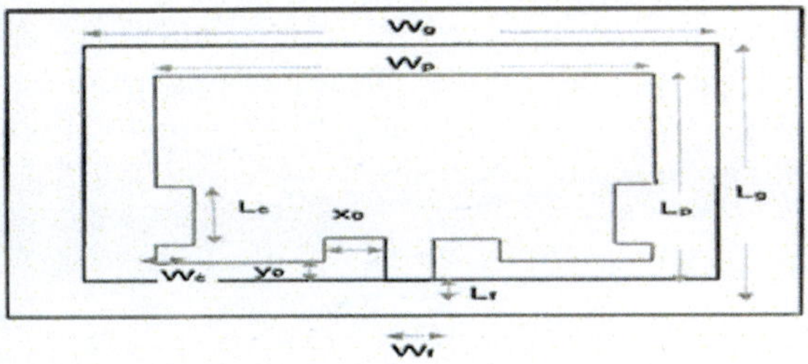

Fig. 46.1 Two-patch antenna geometry

Table 46.1 Dimension for patch antenna

Parameters	Optimized dimensions (mm)
Patch length, L_p	35
Patch width, W_p	43
Ground length, L_g	47
Ground width, W_g	56
Substrate length, L_{ss}	47
Substrate width, W_{ss}	56
Inset feed length, y_0	3
Inset feed width, x_0	3
Feed length, L_f	3
Feed width, W_f	1.9
Patch cutout width, W_c	2
Patch cutout length, L_c	8
Substrate encasing length, L_{es}	53
Substrate encasing width, W_{es}	62

20 min at 6000 ppm. The final precipitate is then dispersed in 10 mg/mL of ethanol for characterization.

46.2.2 *Synthesis of PDMS Layer*

Polydimethylsiloxane (PDMS) thin film was prepared by incorporated polyvinylidene fluoride (PVDF) with PDMS substrate. Initially, 1 wt% PVDF is dissolved in *N*-dimethylformamide (DMF) acetone solution with the 1:1 by volume at 60 °C by stirring for 2 h. Then, the desired amount of PDMS (1, 2, 3 and 4 g) was added into the PVDF solution and being magnetically stirred up until 60 °C for two hours. The homogenous gel solution formed then was poured into antenna mold by using a facile drop-casting method to form a thin film. The thin film together with the mold was immersed in the 0.5 wt% sodium dodecyl sulfate. Subsequently, the film was peeled off from the mold and was soaked in deionized water to remove residue of the solvent. PVDF was blended in together with PDMS during preparation in order to improve material structure integrity. Since PVDF is an excellent piezoelectric polymer, thus adding PVDF is excellent in converting mechanical force into electrical energy of the flexible wearable antenna.

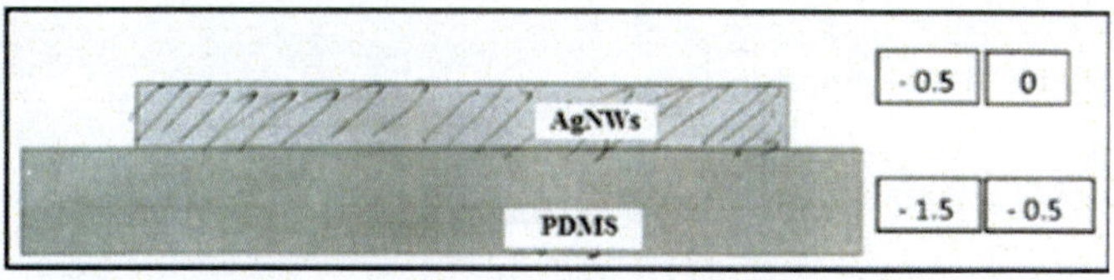

Fig. 46.2 Layer of AgNWs-polymeric film

Table 46.2 Thickness for each layer of polymeric film

Layer	Thickness (mm)
Patch (AgNWs)	0.5
Substrate (PDMS)	1.0

46.2.3 Fabrication of AgNWs-Polymeric Film

PDMS thin film was prepared by incorporated PVDF within PDMS substrate solution with weight ratio of 1:1, 2:1, 3:1 and 4:1. The casting solution was magnetically stirred up until 60 °C for two hours. The homogenous gel solution formed then was poured into antenna mold (Fig. 46.1) by using a simple drop-casting method to form a thin film. The thin film together with the mold was immersed in the 0.5 wt% sodium dodecyl sulfate. Subsequently, the film was peeled off from the mold and was soaked in deionized water to remove residue of the solvent. AgNWs with an average diameter of 20 nm × 12 m that dispersed in ethanol with a concentration of 2–8 mg/L were embedded into the PDMS thin film forming a polymeric surface film. Before being embedded on the substrate film, AgNWs were sonicated for 30 min in the water bath. The homogeneous solution of AgNWs was drop cast on the PDMS film; the layer of AgNWs-PDMS formed is being illustrated in Fig. 46.2 and their thickness is being tabulated in Table 46.2.

46.2.4 Structural Characterization

The polymeric film was characterized through few characterization methods to evaluate its structural and electrical properties. The Fourier Transform Infrared Spectroscopy (FTIR) was employed to determine the presence of a functional group on the resultant film. The sample was analyzed using Thermo Fisher FTIR from 400 to 4000 cm^{-1} at room temperature. Scanning Electron Microscopy (SEM) analysis determines morphology, spatial distribution and also structural arrangement of the AgNWs-polymeric film. The porosity of film was analyzed using pat and weight method. *S* parameter was collected using an Agilent Technologies N5230A PNA.L Network Analyzer to measure resonant frequency and reflection coefficient.

46.3 Result and Discussions

46.3.1 The FTIR Spectra of Polymeric Film and AgNWs-Polymeric Film

The FTIR spectra of the composite PDMS/PVDF film with AgNWs (a) and pure PDMS/PVDF film (b) are depicted in Fig. 46.3. FTIR analysis was applied to confirm the molecular structure of PDMS/PVDF and AgNWs-PDMS/PVDF composite film. These IR spectra provide information about the molecular environment of the organic molecules that exist on the surface of polymeric films. Typically, IR spectra are utilized to investigate the surface functional groups of the polymeric film and possible interaction between nanowire with PDMS and also PVDF. Based on all the IR spectra shown in the figure, it proved the existence of PDMS in both polymeric films. According to Fig. 46.3, sharp peaks at 762, 1178 and about 1066 cm^{-1} represent the functional group of (Si-CH$_3$)$_2$, Si-(CH$_3$)$_2$ and O–Si–O of PDMS molecule. The transmittance peak at 2162 cm^{-1} in both spectra corresponds to the Si–H stretching vibration. The comparison between both spectra indicated that the peak intensity of Si–H groups got reduced after attachment of AgNWs on the film surface, while the peaks at 3022 and 1403 cm^{-1} that were related to asymmetric vibration and CH2 wagging vibration of PVDF proved that PVDF successfully blended in together with PDMS during preparation method. In overall, even though the two films have slightly different intensities, the transmittance profiling for PDMS/PDMS film was similar to AgNWs-PDMS/PVDF film. With this observation, relatively implied that synthesis of both polymeric films inherits almost similar properties.

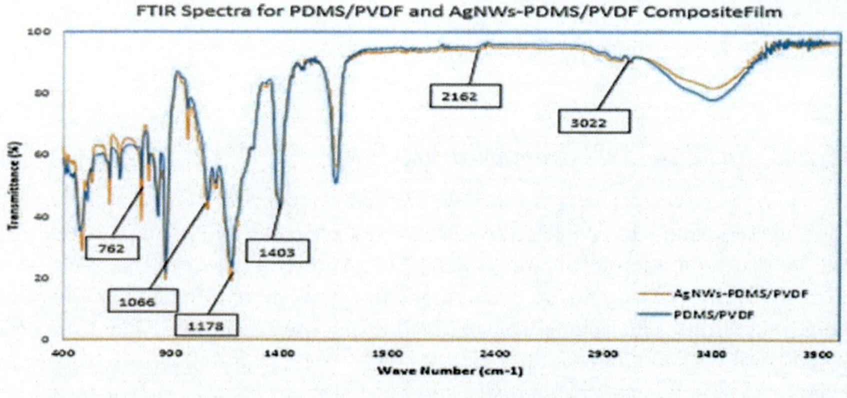

Fig. 46.3 FTIR spectra for AgNWs-PDMS/PVDF and PDMS/PVDF film

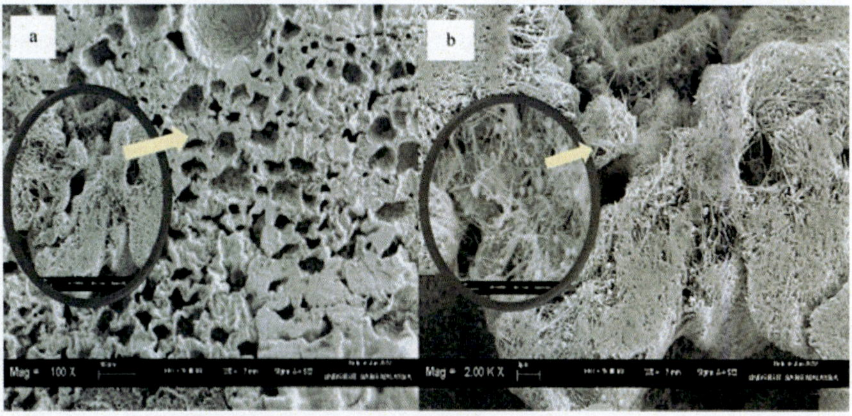

Fig. 46.4 **a** Top view of SEM of porous PDMS/PVDF film at 100X magnification, **b** cross-sectional view of SEM of AgNWs attached on the surface of PDMS/PVDF film at 2000X magnification

46.3.2 Morphologies of the AgNWs-Polymeric Film

Figures 46.4a and b show the surface morphologies of SEM analysis from a top view and cross-sectional view. As revealed in Fig. 46.4a, the surface of the AgNWs-PVDF/PDMS film was composed of macroporous structure with an average diameter of about 69 m $\pm$ 10. The average diameter of the sponge-like macroporous was obtained through a mean calculation for $n = 30$ porous structure formed on the surface film surface using ImageJ software. Formation of porous structure on the film might be due to the exchange between solvent and non-solvent across the film interface during film preparation [6]. Right after casting the PDMS/PVDF solution onto the antenna mold, the film is immediately being immersed in a non-solvent bath containing 5% SDS and a large amount of water. This phenomenon is called phase inversion, where thermodynamic and kinetics interaction between polymer–solvent and non-solvent plays a significant role.

On the other hand, Fig. 46.4b illustrates that the AgNWs are well attached to the surface of the polymeric substrate. The variation of macropores' size formed on the film is yielding large surface area for the AgNWs attachment. AgNWs act as a radiating element in the wearable antenna application; thus, effective networking of AgNWs formed on the pores of the substrate is essential in enhancing both electrical conductivity and mechanical deformability of the film.

46.3.3 Porosity of AgNWs-Polymeric Film

The influences of PDMS/PVDF mass ratio on porosity of the film were investigated. PDMS/PVDF blend film was prepared at a different mass ratio between PDMS and

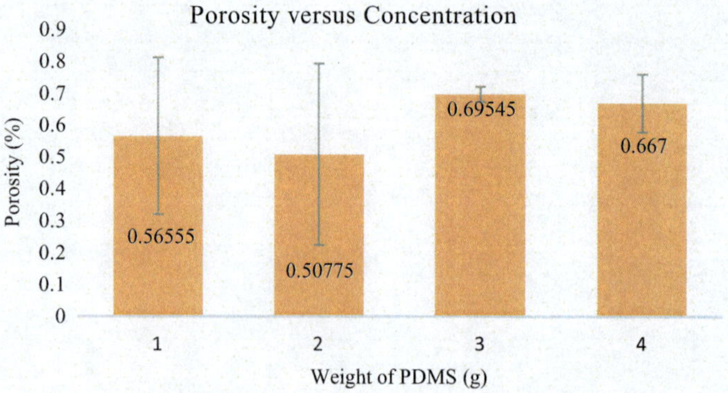

Fig. 46.5 Porosity of polymeric film versus concentration of PDMS in PDMS/PVDF casting solution

PVDF. Figure 46.5 illustrates the effect of mass ratio with the average porosity of the polymeric film formed. It was observed that the average porosity of the PDMS/PVDF film increased as the PDMS/PVDF mass ratio increased, and it achieved its maximum value when the value was 4:1. PDMS concentration determined the viscosity of the casting solution formed. Thus, the addition of PDMS increased the stability of the casting solution and hindered the solvent and non-solvent diffusion rates during the phase inversion process. This trend will reduce the precipitation rate and favor the formation of a sponge-like porous structure and is indicating a more uniform pores' structure formed [7]. High precipitation rate of casting polymer would result in finger-like macro voids, whereas a low precipitation rate would result in sponge-like structure [8]. The increase in the average porosity of films was expected due to the low precipitation rate. Uniform porous structure in the PDMS/PVDF polymeric film formed is illustrated through the average film porosity obtained, which is almost consistent at the range between 0.5 and 0.7. The sponge-like porous structure of thin film is favorable in the antenna design as it will increase the collecting area for AgNWs that mainly function as radiating element in this kind of device. Besides being a favor for the improvement of film porosity, sponge-like porous structure also helps to improve the mechanical properties of the polymeric film which are important factors for the performance of flexible wearable antenna.

46.3.4 Feasibility of the AgNWs-Polymeric Film Antenna

The feasibility of the composite film as a flexible antenna further was validated through *S* parameter analysis. The antenna was experimentally characterized by measuring the scattering parameter. From Fig. 46.6, the fabricated AgNWs flexible antenna reflected power (S11) propagating over the air at 2.725 GHz was − 37.881 dB. Since S11 is the ratio between incident power and the reflected power, thus − 37.881 dB indicated that about 37% of power would reflect meanwhile, about 63% of power transmitted with an operational frequency of 2.725 GHz, which is suitable for wearable WLAN application [9]. There was some deviation between the simulation and experimental results. This deviation expected caused by a different types of polymer is employed as the substrate during the antenna analysis. Experimentally, the flexible wearable antenna is fabricated by using PDMS/PVDF as the substrate due to enhanced mechanical integrity, whereas during simulation analysis, the substrate applied for the antenna was PDMS only as the input for the software does not allow composite material. The substrate provided mechanical strength to the antenna design and allowed the propagation of a wave through it. Thus, changes in substrate material will give a massive effect to the system performance as the dielectric constant will be changed [9, 10].

The flexibility of the fabricated antenna using AgNWs and PDMS/PVDF substrate is demonstrated by subjecting it to two deformation modes which are bending and twisting as illustrated in Fig. 46.7. The ability of the antenna to return to its original state after undergoes deformation proved that AgNWs-PDMS polymeric film flexible antenna is reversibly deformable and robust [11].

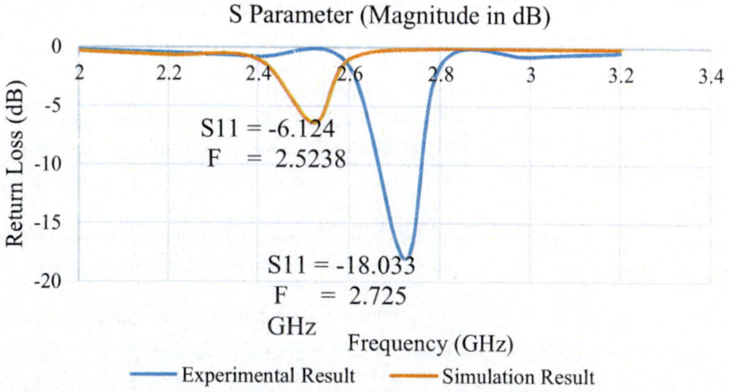

Fig. 46.6 Difference of *S* parameter in dB for experimental and simulation results

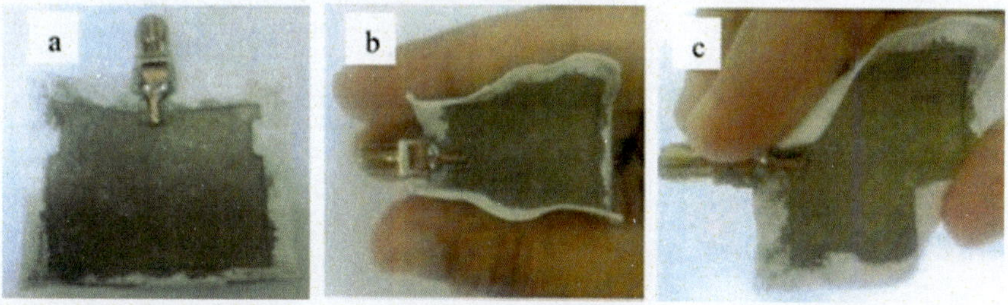

Fig. 46.7 Photograph of a stretchable patch antenna composed of AgNWs/PDMS film flexible conductor: **a** relaxed, **b** bent, **c** twisted

46.4 Conclusion

In conclusion, a novel method to demonstrate a flexible wearable antenna by combining AgNWs with elastomeric PDMS/PVDF substrate was successfully employed. The structural advantages of AgNWs and PDMS/PVDF enabled this network to exhibit low resistance and good flexibility. These properties resulted in reliable performance of antenna including − 18 dB return loss and 2.725 GHz radiation frequency.

References

1. Y. Ahn, H. Lee, D. Lee, Y. Lee, Highly conductive and flexible silver nanowire-based microelectrodes on biocompatible hydrogel. ACS Appl. Mater. Interfaces **6**, 18401–18407 (2014)
2. N.G. Shimpi, D.P. Hansora, R. Yadav, S. Mishra, Performance of hybrid nanostructured conductive cotton threads as LPG sensor at ambient temperature: preparation and analysis. RSC Adv. **5**, 99253–99269 (2015)
3. D. Ebert, B. Bhushan, Transparent, superhydrophobic, and wear-resistant surfaces using deep reactive ion etching on PDMS substrates. J. Colloid Interface Sci. **481**, 82–90 (2016)
4. Y. Qin, M.M.R. Howlader, M.J. Deen, Y.M. Haddara, P.R. Selvaganapathy, Chemical polymer integration for packaging of implantable sensors. Sens. Actuators B Chem. **202**, 758–778 (2014)
5. J.C. Lotters, W. Olthuis, P.H. Veltink, P. Bergveld, The mechanical properties of the rubber elastic polymer polydimethylsiloxane for sensor applications. J. Micromech. Microeng. **7** (1997)
6. A. Rahimpour, S.S. Madaeni, S. Zereshki, Y. Mansourpanah, Preparation and characterization of modified nano-porous PVDF membrane with high antifouling property using UV photo grafting. Appl. Surf. Sci. **255**, 7455–7461 (2009)
7. D. Sun, M. Liu, J. Guo, J. Zhang, B. Li, D. Li, Preparation and characterization of PDMS PVDF hydrophobic microporous membrane for membrane distillation. Desalination **370**, 63–71 (2015)
8. S. Simone, A. Figoli, A. Criscuoli, M.C. Carnevale, A. Rosselli, E. Drioli, Preparation of hollow fibre membranes from PVDF/PVP blends and their application in VMD. J. Memb. Sci. **364**, 219–232 (2010)
9. A. Kumar, T. Ugc, Compact tri-band microstrip patch antenna system (2013) (2017)
10. G. Kaur, E.S. Goyal, To study the effect of substrate material for microstrip patch antenna. Int. J. Eng. Trends Technol. **36**, 490–493 (2016)
11. L. Song, A.C. Myers, J.J. Adams, Y. Zhu, Stretchable and reversibly deformable radio frequency antennas based on silver nanowires. ACS Appl. Mater. Interfaces **6**, 4248–4253 (2014)

Chapter 47
Prediction of Selected Water Quality and Macronutrients Parameters in an Aquaponic System Using Artificial Neural Network (ANN)

M. S. Osman, Q. K. Abdul Rahman, S. Setumin, M. I. F. Maruzuki, S. F. Senin, M. I. Nizam, M. Fathullah, N. B. M. Nazeri, and M. S. Akbar

Abstract Aquaponics, essentially incorporates multiple aquaculture and hydroponics into a single closed-loop system, is an excellent alternative method for generating reduced environmental waste by recycling nutrients (fish waste) for crop development. The pH, dissolved oxygen (DO), total ammoniacal nitrogen (TAN), and total percentage sludge of phosphorus (P) and nitrogen (N) of empirical experiments obtained from the aquaponic system have been investigated (N). To obtain the predicted value, an artificial neural network (ANN) modelling function employing the Levenberg–Marquardt (LM) and scaled conjugate gradient (SCG) training functions was utilized in this research. The outputs of the training function model suggested the optimal number of neurons for each pH parameter, DO, and TAN for neurons 6, respectively. In terms of total sludge (nitrogen and phosphorus), the optimal number of neurons is three. On the other hand, the most optimum neuron number at neurons

M. S. Osman (✉) · Q. K. Abdul Rahman · M. I. Nizam · N. B. M. Nazeri
EMZI-UiTM Nanoparticles Colloids and Interface Industrial Research Laboratory (NANO-CORE), Centre for Chemical Engineering Studies, Universiti Teknologi MARA, Cawangan Pulau Pinang, Permatang Pauh Campus, 13500 Pulau Pinang, Malaysia
e-mail: syazwan.osman@uitm.edu.my

M. Fathullah
School of Manufacturing Engineering, Universiti Malaysia Perlis (UniMAP), Pauh Putra Campus, 02600 Arau, Perlis, Malaysia

M. S. Osman · M. Fathullah
Centre of Excellence Geopolymer and Green Technology (CEGeoGTech), Universiti Malaysia Perlis (UniMAP), 01000 Kangar, Perlis, Malaysia

S. Setumin · M. I. F. Maruzuki
Centre for Electrical Engineering Studies, Cawangan Pulau Pinang, Universiti Teknologi MARA, Permatang Pauh Campus, 13500 Pulau Pinang, Malaysia

S. F. Senin
Centre for Civil Engineering Studies, Cawangan Pulau Pinang, Universiti Teknologi MARA, Permatang Pauh Campus, 13500 Pulau Pinang, Malaysia

M. S. Akbar
Faculty Science, Universiti Teknologi Malaysia, 81310 Skudai, Johor, Malaysia

M. A. A. Mohd Salleh et al. (eds.), *Proceedings of the Green Materials and Electronic Packaging Interconnect Technology Symposium*, Springer Proceedings in Physics 289,
https://doi.org/10.1007/978-981-19-9267-4_47

4 for parameters pH, DO, and TAN, while for the SCG training function was the most optimal neuron number is 4 for the total sludge parameter (N and P), respectively. Meanwhile, in trained neural network's evaluated prediction parameters accuracies, LM outperformed SCG in terms of correlation coefficient (R) and Mean Square Error (MSE).

Keywords Aquaponics · Artificial neural network · Water quality parameters · Macronutrients

47.1 Introduction

Recently, climate change [1], population growth [2], and water and land scarcity [3] have lately emerged as serious and important concerns. These issues require immediate attention and an urge to find solutions. According to the World Population Data Sheet 2020 reported by Population Reference Bureau [4], it is projected that the world's population will increase by more than 9 billion people. As a result, the demand for foods such as fish, vegetables, and medicinal plants will grow exponentially [5, 6], necessitating greater work and space to supply the need. Therefore, a method known as the aquaponics system was developed and practised. Due to its unique characteristics and benefits, the aquaponics system is seen to be a viable choice for resolving these difficulties. In fact, many researchers have recently claimed aquaponics technology as a sustainable approach [7, 8] that promotes the minimization of by-product flow from economic operations. The system plays a significant role as an alternative way of enhancing the food production of aquatic proteins and vegetables to meet the high demands of human consumption due to the growing population on an annual basis. Aquaponics is the integration of aquaculture and hydroponics [9]. It can be described as a closed-loop system where water from fish waste enters the hydroponic system and is recycled back to the fish tank after the process of bio filtering [10, 11]. In addition, aquaponics can recycle nutrients for plant growth with very little effluent discharge compared to conventional aquaculture techniques [12].

Typically, in the aquaponic system, nitrogen compounds originate mainly from fish faeces and leftover fish food. Nitrogen, which is typically found in a nitrate form, is used to supply plant nutrients, Nitrosomonas sp. and with the Nitrobacter sp. These two species are the common species of nitrifying bacteria that convert ammonia (NH_3) to nitrite (NO_2). Nitrites will be used as their energy source to produce nitrate (NO_3) [13]. According to Rakocy [14], in monitoring water quality, pH is one of the critical parameters that need to be maintained. This is to ensure optimum water quality, which has a direct influence on the health and requirements of fish and plants [15]. Research showed essential parameters that should be taken into account in the water quality of aquaponics units such as pH, ammonia, nitrite, and dissolved oxygen (DO) [2]. To increase the fish productivity, the DO should be greater than 4 mg/L with a minimal ammonia concentration (< 0.2 mg/L) and pH between 6 and 9. The mineral contents such as nitrate and phosphate should be presented for tracking chemical parameters [16, 17]. Diver [18] argued that other factors that influence the quality of fish tank water include the stocking density of fish, their growth rate, the rate at which they are fed, and the volume of water and environmental conditions of the system. Nelson [19] noted that the optimum pH for fish is 6.5–7.0 at pH and favours tilapia's survival at a range of 3–10 mg/L DO levels. On the other hand, Graber and Junge [20] argued that the concentration of NO_3-N, NH_4-N, and NO_2-N tolerance limits must be less than < 150 mg/L, < 1.0 mg/L, and < 0.2 mg/L, respectively. They also added that the DO level must be higher than six mg/L using the Light-expanded clay aggregate (LECA) as a trickling filter. To add, plant disposal capacity also positively impacts water quality and the growth and production of fish [19]. In other words, optimum water quality in a fish tank is vital to ensure the healthy growth of fish and vegetables.

Distinct advantages have been applied to the artificial neural network (ANN) in many areas, such as pattern recognition, identification, classification, vision, speech, and control systems. ANN training can be categorized into three different learning situations: supervised learning, unsupervised learning, and enhanced learning. Feed-forward neural networks are the most widely used because of the method used for training, and the network is also known as the (error) back-spreading network. It is wise to note that the experimental data from the aquaponic system is affected by a variety of factors, based on the operation or environmental circumstances that occur or are being built throughout the system's process. As a result, simulating the system, particularly to maintain the aquaponic system's optimum state, is challenging. However, in recent years, ANN has been used in many science and engineering areas, particularly in forecasting water quality in wastewater due to its simplicity rather than the highly complex conventional mathematical modelling [21]. For example, Negoita et al. [22] used ANN for the estimation of computational nuclear physics and macronutrients such as total river nitrogen concentrations. Nabeel [23] modelled the water quality index (WQI) by differentiating the importance of turbidity rather than suspended solid (SS) using ANN. To forecast the ANN predicted outcomes between the experimental findings, a sophisticated tool such as ANN is necessary. In terms of pH, DO, TAN, and total sludge (N and P) value, such a prediction is necessary to identify the sensitivity analysis of the examined variables to evaluate whether the input variables have substantial influences on the water quality and accumulated nutrients in the aquaponic system. Therefore, this paper aims to propose and evaluate the feasibility of ANN with various training algorithms to predict the water quality and macronutrients parameters of the aquaponic system. The capacity of the ANN model to predict satisfactory output data in the aquaponic system will be determined by deploying it with various training functions. This research examines empirical aquaponics activity data and compares it to data from a simulated artificial neural network (ANN).

47.2 Methodology

47.2.1 Input and Target Data for pH, DO, and TAN

The targeted data such as pH, DO, and total ammonical nitrogen (TAN) used in this study was estimated based on the range of 3 different loading rates which is low flow rate (LFR), medium flow rate (MFR), and high flow rate (HFR). The flow rate treatment consisting flow rate (L/day), Hydraulic Loading Rate (HLR) (m^3/m^2/day), and Hydraulic Retention Time (HRT) (h) as inputs' data. The study conducted a water regulation flow rate to examine the water quality deep water culture (DWC) recirculation in the aquaponic system. For this purpose, 13 samples for each input, HLR (3.3–1.1 m^3/m^2/day), flow rate (138–46 L/day), and HRT (5.8–17.3 h) were estimated based on the calculated value obtained by Refs. [24, 25].

47.2.2 Input and Target Data for Total Sludge (Nitrogen and Phosphorus)

This study developed a dynamic design approach of decoupled aquaponics system (DAPS) model which involves an integrated recirculating aquaculture system (RAS) and a hydroponic unit based on the selected input data found in the literature. The literature cover aquaculture, sludge treatment, and hydroponics field to predict nutrient (N and P) levels to develop a theoretical concept of DAPS. The stocking density of fish was estimated by running the experiment with multiple parameters before macronutrients values, specifically N and P flow, were simulated. The input parameters consisting of feed, fish retention, and water (dissolved excretion) in per cent values were observed from the tilapia feeding in RAS cage production. The target data were based on the parameters of per cent total sludge accumulated. 12 samples were estimated based on the range value of the study conducted for input and target data which obtained from input simulation data of feed [100% for both nitrogen (N) and phosphorus (P)] flow, fish retention (30–50% of N and 60–70% of P) as well as water-soluble excretion (20–30% N and 5–10% P) and then simulated data output of total sludge (15–25% N and 35–45% P) [26].

47.2.3 Artificial Neural Network Modelling

ANN was simulated using MATLAB software v9.2.0 (R2017a) which used a feed-forward back-propagation (BP) that consists of input nodes, hidden nodes, and output nodes. In this study, Levenberg–Marquardt (LM) and scaled conjugate gradient (SCG) training function was used by varying the number of neurons in the hidden layer to measure those parameters to obtain the predicted value. In ANN, LM was used to optimize the system's efficiency due to efficient of LM although needs more capable computer memory [27]. Concepcion II et al. [28] stated that the scaled conjugate gradient is more memory efficient, which uses gradient calculations. He et al. [29] noted that the number of synaptic weights should be less than the number of samples in the training set which is a rule of thumb when selecting the number of hidden nodes. The suggested number of neurons in a hidden layer ranges from $[((n_i + n_o)/2) + 1]$ to $[(n_i + n_o) + 10]$ where n_i is the number of input nodes and no is the number of output nodes. For parameter such as pH, DO and TAN, one hidden layer with 4, 6, 8, 10, 12, and 13 neurons were studied. Meanwhile, one hidden layer with 3, 4, 6, 9, and 12 neurons were studied for total sludge (% N and % P). Different range neurons value was used for pH, DO, TAN, and total sludge (N and P) due to different input data found in the literature.

47.3 Result and Discussions

47.3.1 Cyclic Loading Test Results

Forecast data of water quality and macronutrients dynamics applied feed-forward back-propagation (BP) network with the LM and SCG learning algorithms. In this network, a single hidden layer with 4, 6, 8, 10, 12, and 13 neurons in each layer was used for parameter pH, DO, and TAN, and single hidden layer with 3, 4, 6, 9, and 12 neurons in each layer were used for total sludge (% N and % P). The developed ANN block diagram is composed of a two-layer which are tan sigmoid (TANSIG) for the hidden layer and linear transfer function (PURELIN) for the output layer. The best Mean Squared Error (MSE) and regression coefficient (R) from each trained neuron number from 3 to 13 will be determined, as shown in Fig. 47.1. The result for the most optimum neuron number can be explained by the value of Sum Squared Error (SSE) and Mean Absolute Percentage Error (MAPE%) with the lowest value. In ANN, MSE is the average of the squared error that is used as the loss function for least squares regression, and it is the sum for over all the data points, of the square of the difference between the predicted and actual target variables, divided by the number of data points. The figures show comparisons of the results between feed-forward ANN and existing data at most optimum and least optimum neuron number for LM and SCG algorithms from Figs. 47.2, 47.3, 47.4 and 47.5.

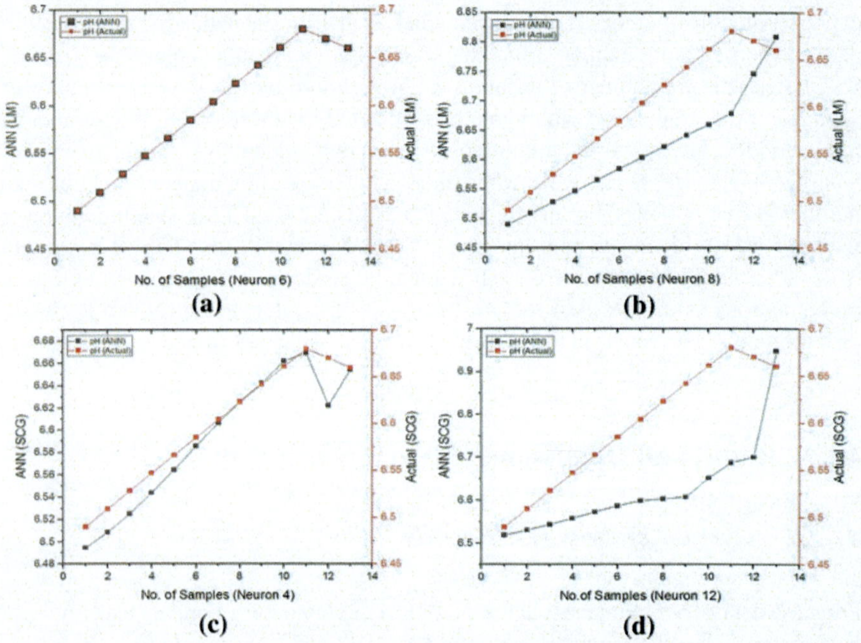

Fig. 47.1 Comparison between predicted and measured values of the pH for LM algorithm with **a** the most optimum, **b** the least optimum, and SCG algorithm with **c** the most optimum, **d** the least optimum

The results for pH parameters are shown in Fig. 47.2 of LM and RP algorithms. These figures show that the best number of neurons for the LM algorithm is at neuron 8. On the other hand, the best number of neurons was observed in neuron 4 for the SCG algorithm. Figure 47.3 indicates dissolved oxygen, DO with neuron 8 and neuron 4 are the best for LM and SCG algorithms, respectively. As shown in Fig. 47.4, neuron 8 is the least optimum neuron for total ammonium nitrate (TAN), as well as with neuron 13 was selected as the least optimum neurons for TAN in the SCG algorithm. Furthermore, Fig. 47.4 relates to the LM algorithm in Figs. 47.2 and 47.3 that neuron 6 are the most optimum neurons for pH, DO, and TAN, respectively. Figure 47.5a and c displays that at neuron 3 and neuron 9 are the best neurons for total sludge (% N), respectively, for LM and SCG algorithm. Besides, Fig. 47.5b and d represents that the least optimum neuron number for the LM and SCG algorithm at neuron 9 and neuron 6, respectively. Next, Fig. 6a and b shows the best and the less optimum neuron number for parameter total sludge (% P), which at neuron 3 and neuron 12 were selected for LM algorithms, respectively. As shown in Fig. 6c and d with neuron 4 is the most optimum neuron for parameter total sludge (% P), as well as with neuron 6 was selected as the least optimum neurons for total sludge (% P) in SCG algorithm. Tables 47.1, 47.2, and 47.3 demonstrate the ANN model's performance results with LM and SCG training function of, DO, TAN, pH, and total sludge (% N and % P). The tables display each trained neuron number such as epoch

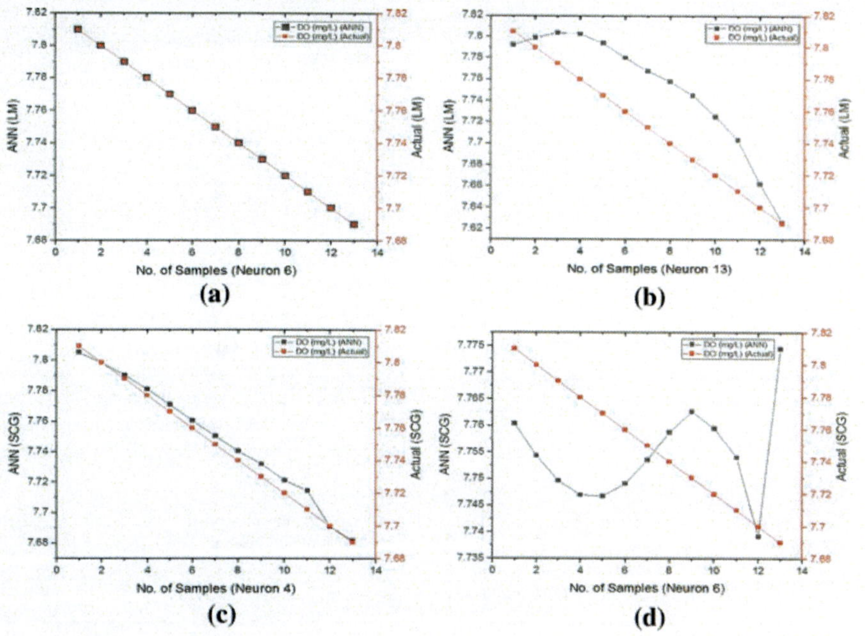

Fig. 47.2 Comparison between predicted and measured values of the DO for LM algorithm with **a** the most optimum, **b** the least optimum, and SCG algorithm with **c** the most optimum, **d** the least optimum

number, MSE and R for the LM, and SCG algorithms. Mishra et al. [30] stated that a random relationship or a close relationship displays R-value, which equals zero or one. It highlights the relationship between the targets and the output data. The MSE value less than the target data in predicting DO levels in seawater indicates the ANN model was successful [31]. The learning algorithm has good predictive potential when MSE and R^2 value is closed to one [32]. Then, R^2, RMSE, SSE, and MAPE% are calculated based on the ANN results data. The MAPE% with the highest value indicates that the model performed poorly [33]. It is observed that SSE and MAPE% close to one approve that the LM algorithm has better predictive potential.

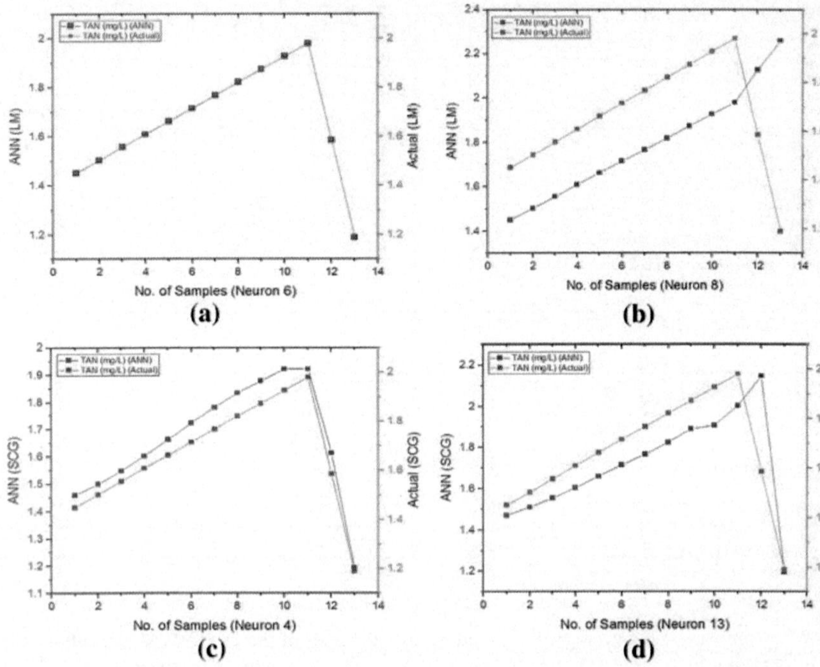

Fig. 47.3 Comparison between predicted and measured values of the TAN for LM algorithm with **a** the most optimum, **b** the least optimum, and SCG algorithm with **c** the most optimum, **d** the least optimum

47.4 Conclusion

This paper predicts water quality and macronutrients in an aquaponic system is vital for the system balance and optimum condition and profitable growth and healthy food products (fish and plants). This study aims to predict critical parameters in an ANN model's aquaponics system and select the best learning algorithms. Therefore, pH, DO, TAN, and total sludge (% N and % P) is preferred due to the critical parameter taken into account in the aquaponic system. ANN method is that it does not need any mathematical relationship between input and output data. It is a promising tool for simulation, modelling, and prediction of water quality and macronutrient performance in an aquaponic system. The feed-forward back-propagation neural network was used to predict the output of modelling, which can be used for sensitivity analysis and studying the water quality and macronutrient's dynamic behaviour. LM and SCG algorithms of this network were compared to each other. By these measures, the best neurons and layers were indicated based on lower SSE and MAPE% values. The neural network has shown the right prediction.

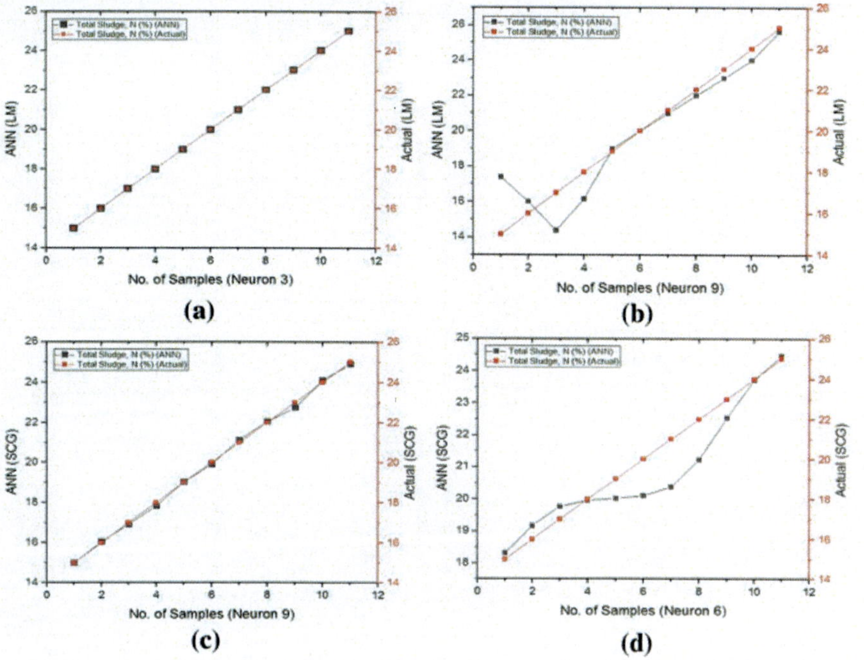

Fig. 47.4 Comparison between predicted and measured values of the total sludge (% N) for LM algorithm with **a** the most optimum, **b** the least optimum, and SCG algorithm with **c** the most optimum, **d** the least optimum

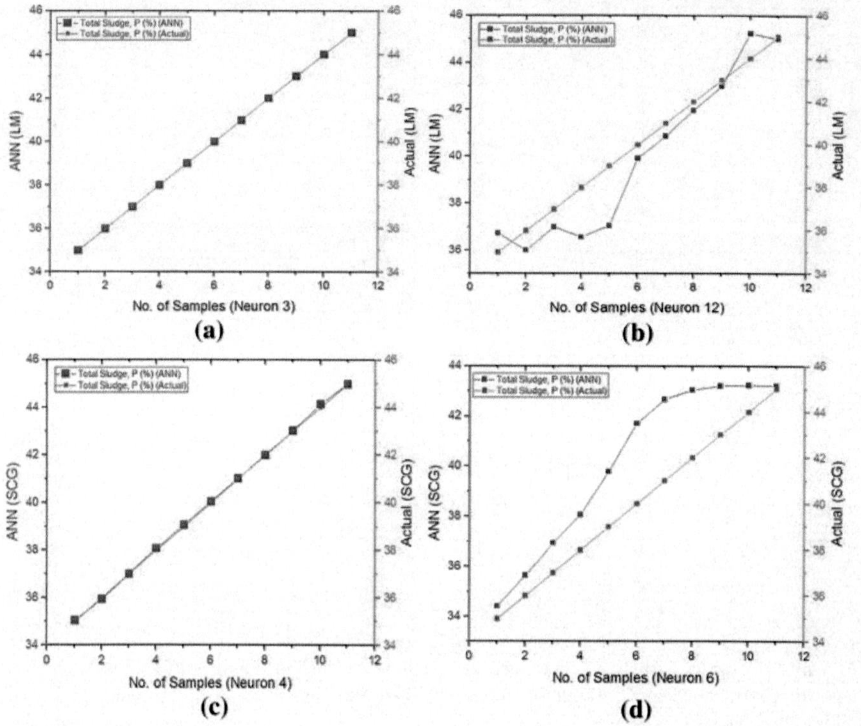

Fig. 47.5 Comparison between predicted and measured values of the total sludge (% P) for LM algorithm with **a** the most optimum, **b** the least optimum, and SCG algorithm with **c** the most optimum, **d** the least optimum

Table 47.1 Performance results of the ANN model with LM and SCG training function of DO, TAN, and pH

Training function	Neuron	Epoch	SSE			MSE	RMSE	R	R^2	MAPE%		
			DO	TAN	pH					DO	TAN	pH
LM	4	16	0.0000587	0.000586	0.00012	8.40E−07	9.17E−04	1	1	0.018115	0.207406	0.0302562
	6	278	1.76E−09	1.10E−08	5.29E−09	9.66E−11	9.83E−06	1	1	1.02E−04	1.26E−03	1.70E−04
	8	12	0.005693	1.435197	0.028128	2.01E−08	1.42E−04	0.99762	0.995246	0.102798	9.561905	0.263908
	10	232	2.41E−03	2.57E−01	1.02E−02	6.12E−11	7.82E−06	0.99954	0.99908	4.92E−02	2.46E+00	1.16E−01
	12	74	1.42E−04	4.83E−02	5.60E−04	7.25E−11	8.51E−06	0.99991	0.99982	1.73E−02	1.56E+00	3.57E−02
	13	7	0.0084	0.0248	0.0078	7.98E−07	8.93E−04	0.99993	0.99986	2.61E−01	1.56E+00	2.18E−01
SCG	4	50	0.0002	0.0048	0.0024	1.86E−05	4.31E−03	0.99999	0.99998	0.035514	0.704735	0.09654
	6	13	0.0215	0.1	0.0152	7.42E−04	2.72E−02	0.99977	0.99954	4.62E−01	4.61E+00	4.12E−01
	8	18	0.0084	0.0279	0.0082	5.69E−04	2.39E−02	0.99992	0.99984	0.241023	2.202791	0.325858
	10	18	0.0115	0.0506	0.0192	2.41E−04	1.55E−02	0.99986	0.99972	3.23E-01	1.96E+00	3.03E−01
	12	16	0.0073	0.1649	0.0876	2.86E−04	1.69E−02	0.99957	0.99914	2.46E−01	3.22E+00	5.59E−01
	13	30	0.0009	0.319	0.0027	2.45E−05	4.95E−03	0.99945	0.9989	5.46E−02	3.23E+00	1.50E−01

Table 47.2 Performance results of the ANN model with LM and SCG training function of nitrogen

Training function	Neuron	Epoch	SSE	MSE	RMSE	R	R^2	MAPE%
LM	3	1000	1.98E−05	4.92E−09	7.01E−05	1	1	0.003494
	4	23	0.017494	5.27E−06	2.30E−03	0.99996	0.99992	0.18256
	6	6	2.136575	1.08E−29	3.29E−15	0.99025	0.980595	1.747711
	9	10	16.28837	1.09E−21	3.30E−11	0.93713	0.878213	4.001259
	12	4	4.518046	1.04E−19	3.22E−10	0.98925	0.978616	1.322292
SCG	3	14	0.365589	1.84E−03	4.29E−02	0.99838	0.996763	0.72509
	4	15	0.330777	4.90E−03	7.00E−02	0.99869	0.997382	0.668899
	6	8	35.24835	3.99E−01	6.32E−01	0.94335	0.889909	7.897725
	9	32	0.157783	1.15E−03	3.39E−02	0.99935	0.9987	0.467015
	12	10	15.62896	2.04E−01	4.52E−01	0.95145	0.905257	5.014022

Table 47.3 Performance results of the ANN model with LM and SCG training function of phosphorus

Training function	Neuron	Epoch	SSE	MSE	RMSE	R	R^2	MAPE%
LM	3	1000	1.98E−05	4.92E−09	7.01E−05	1	1	0.001526
	4	10	0.040615	9.63E−06	3.10E−03	0.99993	0.99986	0.116866
	6	5	0.06799	4.72E−20	2.17E−10	0.99988	0.99976	0.17165
	9	5	5.551292	3.93E−27	6.27E−14	0.97503	0.950684	0.995134
	12	5	10.33458	9.05E−26	3.01E−13	0.95698	0.915811	1.546788
SCG	3	8	11.22837	1.63E−02	1.28E−01	0.95707	0.915983	2.142392
	4	33	0.041812	1.01E−03	3.18E−02	0.99985	0.9997	0.114528
	6	11	11.7606	3.78E−01	6.15E−01	0.94906	0.900715	2.007497
	9	20	1.532032	1.29E−02	1.14E−01	0.99508	0.990184	0.696239
	12	12	2.261406	2.27E−02	1.51E−01	0.9911	0.982279	0.912882

References

1. Y.F.G. Hutton, G. Feulner, P.D. Lund, S. Henson, J.-A. Røttingen, S.J. Hoffman, D. Butler, Global challenges—an innovative journal for tackling humanity's major challenges. Glob. Challenges **1**, 3–4 (2017). https://doi.org/10.1002/gch2.1004
2. T. Shafeena, Smart aquaponics system: challenges and opportunities (2016)
3. T.Y. Kyaw, A.K. Ng, Smart aquaponics system for urban farming. Energy Procedia **143**, 342–347 (2017). https://doi.org/10.1016/j.egypro.2017.12.694
4. Population Reference Bureau, *World Population to Reach 9.9 Billion By 2050|News|SDG Knowledge Hub|IISD* (International Institute for Sustainable Development, 2020). https://sdg.iisd.org/news/world-population-to-reach-9-9-billion-by-2050/. Accessed 29 Dec 2021
5. A.B. Dauda, I. Natrah, M. Karim, M.S. Kamarudin, A.H. Bichi, African catfish aquaculture in Malaysia and Nigeria: status, trends and prospects. Fish Aquac. J. **09** (2018). https://doi.org/10.4172/2150-3508.1000237

6. K.N. Azad, M.A. Salam, M.Z. Hossain, Performance evaluation of different bedding media in aquaponic system for growth and production of okra and tilapia. Int. J. Eng. Res. Sci. **4** (2018). https://doi.org/10.5281/zenodo.1441224
7. A. Endut, F. Lananan, S.H. Abdul Hamid, A. Jusoh, W.N. Wan Nik, Balancing of nutrient uptake by water spinach (Ipomoea aquatica) and mustard green (Brassica juncea) with nutrient production by African catfish (Clarias gariepinus) in scaling aquaponic recirculation system. Desalin. Water Treat. **57**, 29531–29540 (2016). https://doi.org/10.1080/19443994.2016.1184593
8. E.A. Espinosa Moya, C.A. Angel Sahagún, J.M. Mendoza Carrillo, P.J. Albertos Alpuche, C.A. Álvarez-González, R. Martínez-Yáñez, Herbaceous plants as part of biological filter for aquaponics system. Aquac. Res. **47**, 1716–1726 (2016). https://doi.org/10.1111/are.12626
9. C. Li, B. Zhang, P. Luo, H. Shi, L. Li, Y. Gao, C.T. Lee, Z. Zhang, W.-M. Wu, Performance of a pilot-scale aquaponics system using hydroponics and immobilized biofilm treatment for water quality control. J. Clean. Prod. **208**, 274–284 (2019). https://doi.org/10.1016/j.jclepro.2018.10.170
10. L. Silva, D. Valdés-Lozano, E. Escalante, E. Gasca-Leyva, Dynamic root floating technique: an option to reduce electric power consumption in aquaponic systems. J. Clean. Prod. **183**, 132–142 (2018). https://doi.org/10.1016/j.jclepro.2018.02.086
11. R. Calone, G. Pennisi, R. Morgenstern, E. Sanyé-Mengual, W. Lorleberg, P. Dapprich, P. Winkler, F. Orsini, G. Gianquinto, Improving water management in European catfish recirculating aquaculture systems through catfish-lettuce aquaponics. Sci. Total Environ. **687**, 759–767 (2019). https://doi.org/10.1016/j.scitotenv.2019.06.167
12. B. Zhang, P. Luo, H. Pang, Y. Gao, Z. Zhang, C. Li, Nitrogen pollutants removal characteristics in aquaponic system. Chin. J. Environ. Eng. **12**, 1501–1509 (2018). https://doi.org/10.12030/j.cjee.201710116
13. N.Z. Mamat, M.I. Shaari, N.A.A. Abdul Wahab, The production of catfish and vegetables in an aquaponic system. Fish Aquac. J. **07** (2016). https://doi.org/10.4172/2150-3508.1000181
14. J. Rakocy, Ten guidelines for aquaponic systems (2007). https://cdn.aquaponicsassociation.org/uploads/2019/10/Aquaponics-Journal-10-Guidelines.pdf. Accessed 30 Dec 2021
15. H.A. Yousif, S. Köhler, A Fortran code for the scattering of EM plane waves by two cylinders at normal incidence. Comput. Phys. Commun. **59**, 371–385 (1990). https://doi.org/10.1016/0010-4655(90)90185-4
16. L. Anyanwu, J.R. Solomon, Physio-chemical parameters of fishpond used for Dutch clarias hybrid fed coconut chaff and bambara nuts. Afr. J. Environ. Pollut. Health. (2015)
17. G. Bhatnagar, Singh, culture fisheries in village ponds: a multi-location study in Haryana, India. ABJNA **1**, 961–968 (2010). https://doi.org/10.5251/abjna.2010.1.5.961.968
18. S. Diver, *Aquaponics—Integration of Hydroponics with Aquaculture* (2006)
19. J.-Y. Liang, Y.-H. Chien, Effects of feeding frequency and photoperiod on water quality and crop production in a tilapia–water spinach raft aquaponics system. Int. Biodeterior. Biodegradation **85**, 693–700 (2013). https://doi.org/10.1016/j.ibiod.2013.03.029
20. A. Graber, R. Junge, Aquaponic systems: nutrient recycling from fish wastewater by vegetable production. Desalination **246**, 147–156 (2009). https://doi.org/10.1016/j.desal.2008.03.048
21. E.S. Elmolla, M. Chaudhuri, The use of artificial neural network (ANN) for modelling, in *Simulation and Prediction of Advanced Oxidation Process Performance in Recalcitrant Wastewater Treatment* (2011).https://doi.org/10.5772/14920
22. G.A. Negoita, G.R. Luecke, J.P. Vary, P. Maris, A.M. Shirokov, I.J. Shin, Y. Kim, E.G. Ng, C. Yang, Deep learning: a tool for computational nuclear physics (2018)
23. N.M. Gazzaz, M.K. Yusoff, A.Z. Aris, H. Juahir, M.F. Ramli, Artificial neural network modeling of the water quality index for Kinta River (Malaysia) using water quality variables as predictors. Mar. Pollut. Bull. **64**, 2409–2420 (2012). https://doi.org/10.1016/j.marpolbul.2012.08.005
24. H.M. Resh, *Hydroponic Food Production* (CRC Press, Boca Raton, FL, 2012)
25. C. Somerville, M. Cohen, E. Pantanella, A. Stankus, A. Lovatelli, Small-scale aquaponic food production integrated fish and plant farming (2014)

26. S. Goddek, C. Espinal, B. Delaide, M. Jijakli, Z. Schmautz, S. Wuertz, K. Keesman, Navigating towards decoupled aquaponic systems: a system dynamics design approach. Water **8**, 303 (2016). https://doi.org/10.3390/W8070303
27. H. Al-Saadi, R. Al-Sayed, M.A.-S. Hasan, H. Al-Bahadili, Simulation of maximum power point tracking for photovoltaic systems. J. Energy Power Eng. **8** (2014)
28. R.S. Concepcion II, P.J.M. Loresco, R.A.R. Bedruz, E.P. Dadios, S.C. Lauguico, E. Sybingco, Trophic state assessment using hybrid classification tree-artificial neural network. Int. J. Adv. Intell. Informatics **6**, 46 (2020). https://doi.org/10.26555/ijain.v6i1.408
29. B. He, T. Oki, F. Sun, D. Komori, S. Kanae, Y. Wang, H. Kim, D. Yamazaki, Estimating monthly total nitrogen concentration in streams by using artificial neural network. J. Environ. Manage. **92**, 172–177 (2011). https://doi.org/10.1016/j.jenvman.2010.09.014
30. S. Mishra, R. Prusty, P.K. Hota, Analysis of levenberg-marquardt and scaled conjugate gradient training algorithms for artificial neural network-based LS and MMSE estimated channel equalizers (2015).https://doi.org/10.1109/MAMI.2015.7456617
31. S. Palani, S.-Y. Liong, P. Tkalich, An ANN application for water quality forecasting. Mar. Pollut. Bull. **56**, 1586–1597 (2008). https://doi.org/10.1016/j.marpolbul.2008.05.021
32. N. Esmaile, S. Shabaneh, S. Mofavvaz, M.R. Sohrabi, B. Torabi, Spectrophotometric determination of trace amounts of benzotriazole in aqueous solutions using gold nanoparticles: artificial neural network modeling. ChemistrySelect **5**, 5712–5719 (2020). https://doi.org/10.1002/slct.202000922
33. A.M. Melesse, S. Ahmad, M.E. McClain, X. Wang, Y.H. Lim, Suspended sediment load prediction of river systems: an artificial neural network approach. Agric. Water Manag. **98**, 855–866 (2011). https://doi.org/10.1016/j.agwat.2010.12.012

Chapter 48
A Comparative Experimental Investigation Between the Mineral Oil and Vegetable Oil-Based Mono Nanofluids for Transformer Application

M. Syarafi Shuhaimi and V. Vicki Wanatasanappan

Abstract Throughout the process of transmitting the electrical energy, losses continuously tend to occur in the transformer in the form of heat. Therefore, insulating and cooling oil is required to increase to maintain the temperature of power transformers. Coolants are introduced into the transformer to carry out the cooling process. Usually, conventional mineral oil is commonly applied as a cooling liquid for the transformer. However, it is believed that traditional mineral oil is a highly flammable liquid and lacks biodegradability. Thus, palm oil and palm oil/ZnO nanofluid are introduced to overcome the limitations of conventional mineral oil. This research aims to evaluate the effect of nanoparticle volume concentration on the thermal conductivity, dynamic viscosity and dielectric strength of palm oil/ZnO-based nanofluid by an experimental comparison with mineral oil at different temperatures. The samples prepared are conventional mineral oil, pure palm oil and palm oil nanofluid with ZnO at concentrations of 0.05, 0.15 and 0.25%. The breakdown voltage and thermal conductivity results show that pure palm oil and palm oil/ZnO nanofluid give enhanced results compared to conventional mineral oil. However, the viscosity properties, which represent the resistance for the coolant to circulate around the transformer, prove that conventional mineral oil has the lowest viscosity of other samples. Palm oil suspended with ZnO nanoparticles has good potential to be used as insulating oil in transformers.

Keywords Mono nanofluid · Vegetable oil · Transformer cooling · Mineral oil

M. Syarafi Shuhaimi
Department of Mechanical Engineering, Universiti Tenaga Nasional, 43000 Kajang, Selangor, Malaysia

V. Vicki Wanatasanappan (✉)
Institute of Power Engineering, Universiti Tenaga Nasional, 43000 Kajang, Malaysia
e-mail: vignesh@uniten.edu.my

M. A. A. Mohd Salleh et al. (eds.), *Proceedings of the Green Materials and Electronic Packaging Interconnect Technology Symposium*, Springer Proceedings in Physics 289,
https://doi.org/10.1007/978-981-19-9267-4_48

48.1 Introduction

The transformer can be known as a device that is used in transmitting electrical energy to another circuit through electromagnetic induction, is usually used to either increase or decrease the voltage level and is used in various fields such as distribution sector and electric transmission energy; this device, which is the first idea came from Michael Faraday, can be classified into several types based on different factors. These include the medium used in the core, winding arrangement, installation location and working voltage range which consists of a step-up transformer and step-down transformer [1]. Essentially, a transformer is divided into three parts. One of them is primary winding which, when connected to an electrical source, produces magnetic flux. Other than that, the magnetic core where the path for magnetic flux to pass through is generated by the primary winding and, lastly, connected to the third part, a secondary winding, to provide desirable output from the transformer [2].

Throughout the process of transmitting the electrical energy, losses continuously tend to occur in the transformer in the form of heat. For this reason, the transformer's cooling needs to be done to reduce the loss. The reliability of the transformer depends on the insulation system within it. According to literature, the average life span of a transformer is about 18 years, and one of the reasons for the failure of a power transformer is due to insulation failure [3]. Conventional mineral oil produced from petroleum is mostly used as insulation oil in oil-filled power transformers for the past decades [4]. This is because conventional mineral oil has good properties such as dielectric strength, dynamic viscosity and moderate thermal performance [5, 6]. Besides, traditional mineral oil is said to be suitable for a transformer that operates at normal temperature and outdoor installations [7]. Plus, conventional mineral oil also offers a lesser price than other oil cooling liquid, which may be the biggest bargain. However, it is believed that conventional mineral oil is a highly flammable liquid which is why the oil is only suitable for average temperature operating transformers [8]. This mineral oil is an insulating and cooling agent for the transformer.

The two main properties of transformer oil are thermal and dielectric properties. Many researchers explored the potential way to enhance these properties further. One breakthrough in the enhancement of mineral oil for transformers is using nanotechnology. The nano-sized (< 100 nm) particle suspension in any base fluid is known as nanofluid. The nanofluid technology was used in mineral oil for transformer by Rafiq et al. [9] when they evaluated the dielectric property of Fe_3O_4-based mineral oil by experimental work. The authors dispersed the Fe_3O_4 nanoparticles in mineral oil at various concentrations' weight to volume of 5–80%. Their results show that the AC and positive impulse breakdown strength of nanofluids increase by adding Fe_3O_4 nanoparticle concentration up to 40%. Hessien et al. [10] synthesized nanocrystalline $Mn0.2Ni0.8Fe_2O_4$ powder and dispersed it in transformer oil by sonication technique. Their findings reported a higher breakdown voltage for transformer oil with a manganese nickel ferrite system-based concentration of 0.04 g/l.

Even though the use of nanotechnology can enhance the property of transformer oil, the dependency on mineral oil continues. Mineral oil is not biodegradable and is produced from a nonrenewable source. Therefore, it has environmental and sustainability issues. One of the possible alternatives to replace mineral oil is vegetable oil. Vegetable oils extracted from oil seeds have a dark color and contain solid constituents such as proteins, fibers, fats and oil. Both fats and oil are triglyceride esters of fatty acids [9]. In recent years, many researchers explored the use of inedible oil-based insulating fluids for transformers. For example, Yu et al. [11] researched renewable low-viscosity dielectrics based on vegetable oil methyl esters. Their study indicates that the methyl ester products had good dielectric strengths, high water saturation and enough fire resistance for transformer application. Besides, they also concluded that canola oil methyl ester has excellent potential as low-viscosity vegetable oil-based dielectric fluid. However, vegetable oil still has several limitations that affect its applicability as transformer oil.

Therefore, further enhancement in vegetable oil's thermal and electrical properties is crucial to enable insulation and cooling oil use in power transformers. This study aims to formulate a palm oil-based ZnO hybrid nanofluid and experiment with the thermophysical and electrical properties. The experimental results of palm oil nanofluid will be compared with the properties of mineral oil. The effect of ZnO nanoparticles on the thermal conductivity, viscosity and AC breakdown voltage of palm oil was evaluated experimentally. The research work is crucial and beneficial to explore the alternative for transformer oil.

48.2 Methodology

48.2.1 Materials Selection

The oil samples that were prepared are conventional transformer coolant which is mineral-based oil, where a product from MASW high-performance transformer oil was used throughout this research, and vegetable-based oil, whereas in this research, food grade palm oil has been chosen. At the same time, zinc oxide nanoparticles and CTAB surfactant that were selected were supplied by Sigma-Aldrich, which are crucial in preparing the nanofluid. Table 48.1 provides the property for both ZnO nanoparticles and CTAB surfactant [12, 13].

Table 48.1 Properties for ZnO nanoparticles and CTAB surfactant

Product	Color	Molecular weight (g/mol)	Mean particle size (nm)	Density (g/cm^3)
ZnO	White	81.39	≤ 100	5.68
CTAB	White	364.45	–	2.30

Table 48.2 Mass needed for 250 g nanofluid preparation

Parameter	0.05% ZnO + palm oil	0.15% ZnO + palm oil	0.25% ZnO + palm oil
Mass of base oil (g)	249.875	249.625	249.375
Mass of ZnO (g)	0.125	0.375	0.625
Mass solution (g)	250.000	250.000	250.000
Mass CTAB (g)	0.013	0.038	0.063

48.2.2 Nanofluid Preparation

Equation 48.1 is applied to calculate the mass of zinc oxide (ZnO) nanoparticles needed to prepare 0.05, 0.15 and 0.25% mass percent of palm oil with ZnO nanoparticles. Using the formula, 250 g of nanofluid was prepared at a time with total of 500 g needed per mass percentage of nanofluid. As for surfactant, the mass is 1/10 from ZnO nanoparticles. All masses including CTAB surfactant needed to be weighed are shown in Table 48.2.

$$\text{Mass Percent}, m_{\text{per}} = \frac{\text{Mass of Solute}, m_{\text{sol}}}{\text{Mass of Solution}, m_{\text{soln}}} \times 100\,\% \tag{48.1}$$

After that, the samples were placed on a magnetic stirrer with all the substances that were weight and put into a beaker and were left to stir for 2 h. Then, the samples were placed under an ultrasonication probe for 1 h and 30 min while controlling the temperature by putting ice around the sample's beaker. Lastly, the samples then were placed in an ultrasonication bath for 2 h while controlling the temperature between 20 and 40 °C.

48.2.3 Nanofluid Sedimentation Test

After a nanofluid sample was completely prepared, the sample undergone sedimentation test. This test was carried out to observe changes in prepared nanofluid for sedimentation in the solution by visual observation every 24 h for a period of three days.

48.2.4 Breakdown Voltage Test

This test was carried out at the University of Malaya, Selangor, Malaysia. The equipment used for this test is a 60 kV Portable Oil Test Set by Megger. Basically, this experiment was carried out to determine the minimum voltage required to produce a spark between two electrodes for each sample. The samples used for this experiment are conventional mineral oil, pure palm oil, 0.05% ZnO + palm oil nanofluid, 0.15%

ZnO + palm oil nanofluid and 0.25% ZnO + palm oil nanofluid. As for the gap between the two electrodes in the test vessel, the gap is set to 2.5 mm, following the IEC 60156 standard.

48.2.5 Thermal Conductivity Measurement

This experiment was conducted at Sunway University, Petaling Jaya, Selangor, Malaysia, for the thermal conductivity test. Generally, this experiment aims to study the potential for all samples in conducting heat away from the transformer. The samples used for this experiment are conventional mineral oil, pure palm oil, 0.05% ZnO + palm oil nanofluid, 0.15% ZnO + palm oil nanofluid and 0.25% ZnO + palm oil nanofluid. In measuring the thermal conductivity for all samples, a KD2 Pro thermal analyzer equipment for displaying the data was used. Other than that, a sensor was connected to the thermal analyzer and will be assisting in measuring the property. Besides, a water bath was also used to control the temperature for oil samples and a thermocouple. In addition, the thermal conductivity measurement for all samples was performed at three different temperatures, which are 25, 40 and 60 °C.

48.2.6 Viscosity Measurement

The viscosity of prepared oil samples was also measured at Sunway University, Petaling Jaya, Selangor, Malaysia. Primarily, the main objective of this test to be conducted is to determine the resistance of all oil samples to flow. Same with the other two experiments, the samples needed for this experiment are conventional mineral oil, pure palm oil, 0.05% ZnO + palm oil nanofluid, 0.15% ZnO + palm oil nanofluid and 0.25% ZnO + palm oil nanofluid. For this experiment, 10 mL amount of each sample was required. The viscosity for each sample is determined by equipment named Rheometer Model MCR92 manufactured by Anton Paar. As for the shear rate for this test, it was set to 50/s. Lastly, similar to the thermal conductivity test, this viscosity measurement was also assessed at three different temperatures, which include 25, 40 and 60 °C.

48.3 Result and Discussions

48.3.1 Sedimentation Analysis

The samples were observed daily for three days for sedimentation tests. According to Figs. 48.1a and b, for the first sample, 0.05% ZnO nanofluid, from the pictures taken, it can be seen that there was no sedimentation from the first day until the third day. This proves that the ZnO nanoparticles are completely dispersed into the palm-based oil and that the sample is stable. Next, for the second sample, 0.15% ZnO nanofluid, from visual observation (Fig. 48.1c), it can be observed that there were no changes from the first day and after the second day. However, on the third day, there were two shades in the sample, which can be seen in Fig. 48.1d. Anyhow, it can be said that the ZnO nanoparticles were quite fully dispersed into the palm-based oil and still stable. Based on visual observation for the last sample as illustrated in Fig. 48.1e, 0.25% ZnO, the sample showed an early change which started to show sedimentation after only one day. After three days of observation, the sedimentation in the 0.25% ZnO sample becomes more obvious and can be seen clearly in Fig. 48.1f. Therefore, it can be said that the ZnO nanoparticles for 0.25% ZnO nanofluid are not fully dispersed into the palm oil. Perhaps, the reason why the ZnO nanoparticles are not fully dispersed into the palm oil is that the time for ultrasonic probe and ultrasonication bath is not enough since it has the highest ZnO mass percentage, 0.25%, which is about 66% higher than 0.15% ZnO mass percentage. Since the nanoparticles for 0.25% are not fully dispersed, it slightly affects the thermophysical and electrical properties. This is because higher mass concentration nanoparticles will enhance the properties more theoretically. However, it will be not achieved if the nanoparticles are not fully dispersed.

48.3.2 Breakdown Voltage Analysis

Figure 48.2 represents a bar graph that shows the breakdown voltage of mineral oil, pure palm oil and palm oil/ZnO nanofluid at all concentrations investigated. According to the results obtained, pure palm oil without any suspension has a higher breakdown voltage value than mineral oil. The breakdown voltage measured for palm oil is about 26.3 kV. Meanwhile, the breakdown voltage of MASW mineral oil is 20.3 kV. This indicates that palm oil has better electrical properties than conventional mineral oil used in transformers. Furthermore, with the presence of ZnO, the BDV value was enhanced, and it can be seen from the graph that all samples with ZnO nanoparticles produced higher BDV values compared to pure palm oil. Other than that, high ZnO nanoparticle mass concentration gives a higher BDV value. This clearly can be seen from the bar graph in Fig. 48.2, where palm oil with 0.05% ZnO has a higher BDV value of 30.1 kV than palm oil without ZnO, which recorded 26.3 kV. The highest BDV value of 35.7 kV was measured for the palm oil/ZnO

Fig. 48.1 Nanofluid sedimentation observation **a** first day (ZnO 0.05%), **b** third day (ZnO 0.05%), **c** first day (ZnO 0.15%), **d** third day (ZnO 0.15%), **e** first day (ZnO 0.25%), **f** third day (ZnO 0.25%)

nanofluid with 0.15% concentration with 42.3% enhancement relative to the base oil. However, 0.25% ZnO nanofluid produced the lowest BDV value of 27.5 kV when compared to 0.05 and 0.15% despite containing the largest ZnO mass concentration. The reason behind this is that as mentioned earlier, the ZnO nanoparticles for 0.25% are not dispersed completely into the palm oil. Thus, it proves that unstable palm/ZnO nanofluid affects the BDV significantly. In this case, the 0.25% has a lower BDV value compared to 0.05 and 0.15%. Nevertheless, the value acquired for 0.25% ZnO nanofluid is still higher than pure palm oil, thus proving that the presence of nanoparticles enhances the breakdown voltage compared to pure palm oil.

48.3.3 Thermal Conductivity Analysis

A line graph of thermal conductivity value for each sample for three different temperatures was plotted and can be observed in Fig. 48.3. Based on Fig. 48.3, the thermal conductivity increasing trend for all sample tested is the same, whereas the temperature increases, the thermal conductivity also increases. It occurs because as temperature increases, more kinetic energy is presented to the particles in each sample, and this is similar to the finding reported in the literature for water as a base fluid [14–17]. Consequently, vibration motion will also rise, which causes the oil molecules to collide with each other more frequent, thus increasing the thermal conductivity.

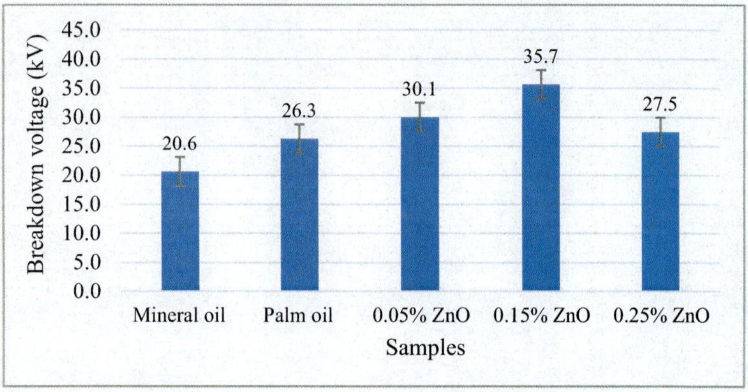

Fig. 48.2 Average breakdown voltage for all samples

Other than that, between palm oil and conventional mineral oil, the results show that pure palm without any nanoparticle suspension exhibits a higher thermal conductivity value than conventional mineral oil. The MASW mineral oil was found to have the lowest thermal conductivity value for temperatures 25, 40 and 60 °C. Next, when comparing palm oil with the presence of ZnO nanoparticles and without nanoparticles, it seems that palm oil with ZnO nanoparticles offers higher thermal conductivity than the one without nanoparticles. In fact, as illustrated in the graph from Fig. 48.3, higher ZnO nanoparticles provide a greater value of thermal conductivity, which can be observed that 0.15% ZnO nanofluid has the highest thermal conductivity at every temperature except for 0.25% ZnO nanofluid since it was proved that ZnO nanoparticles were not fully dispersed from the result of nanofluid sedimentation test. The maximum thermal conductivity enhancement of palm oil due to the ZnO nanoparticle mass concentration is about 14.7% relative to the base oil. The reason for the increase in thermal conductivity of base fluid with increasing ZnO nanoparticle concentration is due to the influence of larger solid volume fraction of nanoparticles, thus influencing the frequency of collision between particles to be increased.

48.3.4 Dynamic Viscosity Analysis

The result for the viscosity test can be seen from the line graph provided in Fig. 48.4. The results have shown that the dynamic viscosity of mineral oil, palm oil and palm oil/ZnO nanofluids decreases with an increase in temperature. All samples represent the same trend because the average speed of molecules in a sample rises when temperature increases. This will cause the period for each molecule to collide with each other decreases and lead the sample resistance to flow to decrease or, in other words, reduces the viscosity. Consequently, decreasing viscosity at high temperatures will facilitate the circulation of coolant when applied to the transformer [2]. When

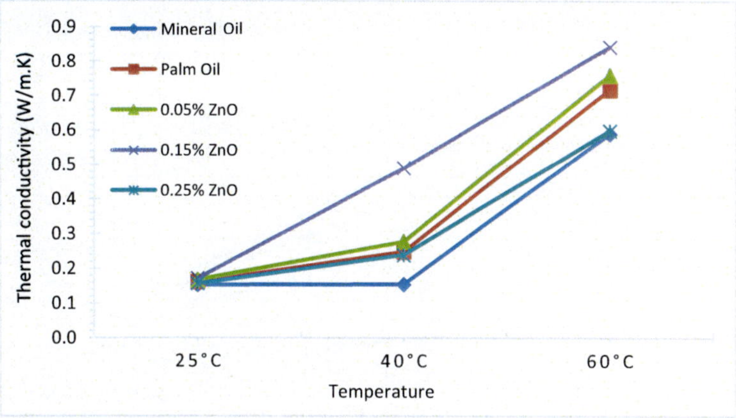

Fig. 48.3 Thermal conductivity of nanofluids and mineral oil

comparing between samples, mineral oil has the lowest viscosity in all three different temperatures, 25, 40 and 60 °C, compared to the other samples, and from the data obtained, mineral oil has around three times lower viscosity value than palm oil at 25 °C which is 13.716 mPa·s. When compared to mineral oil, the viscosity of palm oil/ZnO (0.25%) is around 5 times more than mineral. In addition, from the graph plotted, it can be noticed that the more weight concentration of ZnO nanoparticles provides higher viscosity. This can be proved clearly from the graph where 0.25% ZnO nanofluid has the greatest viscosity value of 67.998 mPa·s at 25 °C compared to 55.021 mPa·s and 58.469 mPa·s for 0.05% ZnO nanofluid and 0.1% ZnO nanofluid, respectively. The reason behind these results is that a larger weight nanoparticle concentration will lead to a larger nanoparticle size present in the sample [3]. This will influence the viscosity to become higher as nanoparticles will be organized into a structure and cause the fluid unlikely to flow and circulate [4]. Moreover, since the viscosity of palm oil with ZnO nanoparticles is much higher than mineral oil, especially at the starting temperature of 25 °C, as can be seen from Fig. 48.4, the only concern for this type of coolant is the suitability for application at low temperature.

48.4 Conclusion

In a nutshell, this research was done to evaluate the effect of ZnO concentration on the thermal conductivity, dynamic viscosity and dielectric strength of palm oil/ZnO-based nanofluid. From the first analysis on dielectric strength, acquired results show that 0.05% palm oil/ZnO nanofluid has higher average breakdown voltage of 30.1 than 26.3 kV for palm oil without any ZnO nanoparticles. As for higher ZnO weight concentration, 0.15% palm oil/ZnO nanofluid, average breakdown voltage obtained, 35.1 kV, is higher than 0.05% weight concentration. However, the highest weight

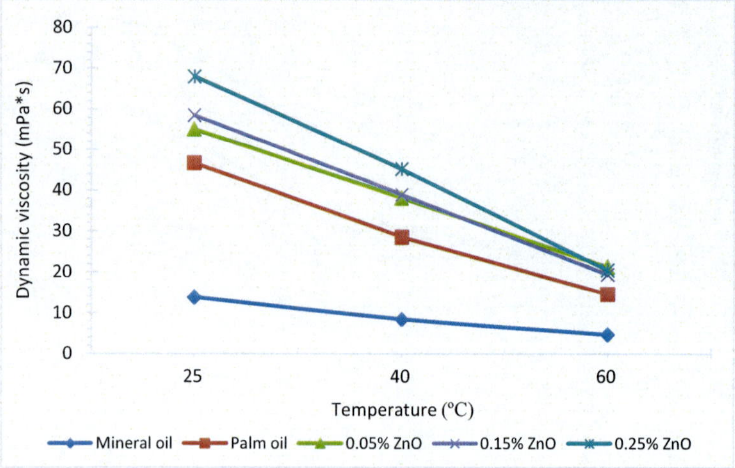

Fig. 48.4 Dynamic viscosity of nanofluid and mineral oil

concentration palm oil/ZnO nanofluid, 0.25%, does not offer the largest breakdown voltage value since the nanoparticles are not fully dispersed. In the meantime, for second experiment which is thermal conductivity test, the results prove the same argument as first experiment where higher weight concentration of ZnO nanoparticles enhanced the properties of a coolant. Thereupon, along with dielectric strength, larger weight concentration of palm oil/ZnO also leveled up the thermal conductivity for a coolant. Next, despite higher concentration of palm oil/ZnO nanofluid will enhance the properties of dielectric strength and thermal conductivity, it is quite contrast when it comes to viscosity. Besides, it can be concluded that palm oil/ZnO nanofluid offered better enhancement for thermal conductivity and dielectric strength compared to conventional mineral oil. Ultimately from this study, palm oil-based ZnO nanofluid is obviously a good choice replacement for conventional mineral oil.

Acknowledgements The authors acknowledge Universiti Tenaga Nasional for the financial support through YCU research grant (202210032YCU) and BOLD research grant (J510050002/2021075). Special thanks to those who contributed to this project directly or indirectly.

References

1. Aakash, *"Transformer—Definition, Types, Working Principle, Diagram." Electrical4U, "What is a Transformer (And How Does it Work)?"* (2021)
2. C.G. Núñez, W.T. Navaraj, E.O. Polat, R. Dahiya, Energy-autonomous, flexible, and transparent tactile skin. Adv. Funct. Mater. **27**(18) (2017). https://doi.org/10.1002/adfm.201606287
3. B.S. Sarkar, J K. John, V T. Corp, *Different Fluids for Transformer Cooling* (Virginia Transformer Corp)

4. M. Rafiq, Y.Z. Lv, Y. Zhou, K.B. Ma, W. Wang, C.R. Li, Q. Wang, Use of vegetable oils as transformer oils—a review. Renew. Sustain. Energy Rev. **52**, 308–324 (2015). https://doi.org/10.1016/J.RSER.2015.07.032
5. R. Walvekar, D.A. Zairin, M. Khalid, P. Jagadish, N.M. Mubarak, G. TCSM, Stability, thermophysical and electrical properties of naphthenic/POME blended transformer oil nanofluids. Therm. Sci. Eng. Prog. **23**, 100878 (2021). https://doi.org/10.1016/j.tsep.2021.100878
6. E.T. Nkouetcha, G.M. Mengounou, A.M. Imano, Elaboration and performance analysis of a bio-based insulating liquid from castor oil for power transformers. **6**, 1–16 (2019). https://doi.org/10.4236/oalib.1105404
7. S.S. Sinan, S.N. Shawaludin, J. Jasni, N. Azis, M.Z.A.A. Kadir, M.N. Mohtar, Investigation on the AC breakdown voltage of RBDPO Olein, in *IEEE Innovative Smart Grid Technologies—Asia (ISGT ASIA)* (2014), pp. 760–763. https://doi.org/10.1109/ISGT-Asia.2014.6873888
8. M. Rafiq, Y. Lv, C. Li, A review on properties, opportunities, and challenges of transformer oil-based nanofluids. J. Nanomater. **2016**, 8371560 (2016). https://doi.org/10.1155/2016/8371560
9. M. Rafiq, C. Li, Y. Ge, Y. Lv, K. Yi, Effect of Fe_3O_4 nanoparticle concentrations on dielectric property of transformer oil, in IEEE International Conference on High Voltage Engineering and Application (ICHVE) (2016), pp. 1–4
10. M.M. Hessien, N.A. Sabiha, S.S.M. Ghoneim, A.A. Alahmadi, Enhancement of dielectric characteristics of transformer oils with nanoparticles. Int. J. Appl. Eng. Res. **12**(24), 15668–15673 (2017). http://www.ripublication.com/
11. H. Yu, P. Yu, Y. Luo, Renewable low-viscosity dielectrics based on vegetable oil methyl esters. J. Electr. Eng. Technol. **12**(2), 820–829 (2017). https://doi.org/10.5370/JEET.2017.12.2.820
12. S. Aldrich, *Safety Data Sheet ZnO Nanoparticles* (2022)
13. S. Aldrich, *Safety Data Sheet CTAB* (2022)
14. W.V. Vicki, M.Z. Abdullah, P. Gunnasegaran, Thermophysical properties of Al_2O_3-CuO hybrid nanofluid at different nanoparticle mixture ratio: an experimental approach. J. Mol. Liq. **313**, 113458 (2020). https://doi.org/10.1016/j.molliq.2020.113458
15. V. Wanatasanapan, M.Z. Abdullah, P. Gunnasegaran, Effect of TiO_2-Al_2O_3 nanoparticle mixing ratio on the thermal conductivity, rheological properties, and dynamic viscosity of water-based hybrid nanofluid. J. Mater. Res. Technol. **9**(6), 13781–13792 (2020). https://doi.org/10.1016/j.jmrt.2020.09.127
16. N. Asokan, P. Gunnasegaran, V. Vicki Wanatasanappan, Experimental investigation on the thermal performance of compact heat exchanger and the rheological properties of low concentration mono and hybrid nanofluids containing Al_2O_3 and CuO nanoparticles. Therm. Sci. Eng. Progress. **20** (2020). https://doi.org/10.1016/j.tsep.2020.100727
17. W.V. Vicki, M.Z. Abdullah, P. Gunnasegaran, Effect of volume concentration and nanofluid temperature on the thermal conductivity of mono and hybrid Al_2O_3-TiO_2 nanofluid, in *AIP Conference Proceedings (ICONGDM 2020)* (AIP Publishing, 2020)

Chapter 49
Experimental Analysis of Geological Structure to the Water Intrusion into Tunnel

M. I. Zalrusli, A. Rahim, K. Ramanathan, R. A. Abdullah, T. L. Goh, and W. M. W. Ibrahim

Abstract This paper presents a geological characteristic in Karak for water transfer tunnel. A dense population that exists year by year in the urban area created a problem which is high demand water consumption. In order to avoid running out of water in the future, the government has planned to make a water transfer tunnel that connects from catchment area from Karak to Langat. The objectives of this study are to study the joint pattern and overburden at NATM-2 and to investigate the relationship between joint pattern, overburden, and water inrush. Fieldwork has been carried out to collect the data from the site. The data collected is tunnel face mapping. From the data, we can accomplish the objective needed. Analysis for this study involved the analysis of the structural map, geological map, tunnel face mapping along 1927 m of total length at NATM-2. The geological structures especially in the strength and water inrush that presents within meta-sedimentary rock have affected the excavation process and decreased the tunnel performance. Three types of joints orientations observed are parallel with the main lineament orientation, perpendicular to the tunnel, and crossing random voids. Most water inrush occurrences are seen as in line with the joint sets orientation which are parallel orientations, perpendicular to the tunnel drive direction and 1 most least is cross voids that crossing with random pocket water zones.

Keywords Joint pattern · Overburden · Water penetration

M. I. Zalrusli · A. Rahim (✉) · K. Ramanathan · R. A. Abdullah
School of Civil Engineering, Faculty of Engineering, Universiti Teknologi Malaysia, 81310 Johor Bahru, Malaysia
e-mail: afikahrahim@utm.my

T. L. Goh
Department of Earth Sciences and Environment, Faculty of Science and Technology, Universiti Kebangsaan Malaysia, 43600 Bangi, Selangor Darul Ehsan, Malaysia

W. M. W. Ibrahim
Faculty of Mechanical Engineering Technology, Universiti Malaysia Perlis, 026000 Arau, Perlis, Malaysia

M. A. A. Mohd Salleh et al. (eds.), *Proceedings of the Green Materials and Electronic Packaging Interconnect Technology Symposium*, Springer Proceedings in Physics 289,
https://doi.org/10.1007/978-981-19-9267-4_49

49.1 Introduction

Due to high population densities, the need for subterranean urban infrastructure rises. During these modern days, a traditional water piping was not enough to sustain the high demand in urban area. Even in this most advanced day to predict hazard in rock tunneling during underground excavations is almost impossible to be done. As stated by Moon and Jeong [1], the best record from highly previous structural features in predicting hazard to the tunnel only cover at a distance greater than two tunnel diameters from the excavation. Other than that, it comes with an extreme consequence resulting in flooding and abandonment of the tunnel excavation. Example can be found related with these geological features such as the long Zagros Tunnel and Alborz Tunnel in Iran.

To acknowledge with accuracy, the water inrush into tunnels is the hardest part knowing that the problem can come from rock mass permeability and environmental conditions. The geological structures like fault and open fractures were the main paths for groundwater to flow in low permeability or hard rock formation. Or in different cases like construction tunnel and underground cavern, these were the characteristics for groundwater to flow. According to Zimmerman and Bodvarrson, these characteristics have its own behaviors related with hydraulic conductivity and depend on the properties' experience. Between the two characteristics of fracture, which are frequency and aperture, the aperture fracture shows a solid evidence with permeability compared to frequency. According to Huang et al., an example can be seen from coal mine in Shandong Province, Eastern China, where hydraulic conductivity of groundwater was strongly connected to rock structure, as the values of hydraulic conductivity were approximately 50 and 1500 greater in cataclastic structure than the thinly layered structure and a layered structure.

A lot of analyses emphasizing analytical, empirical, and numerical have been studied in the last decades regarding the groundwater inrush underground. The study of hydrogeological characteristics using analytical and numerical tools has often failed because of generalization and simplification of important parameters and can never be simple especially at terrain relating to crystalline hard rock such as granite or metamorphic rock. So the accuracy of the data and the sources of the data on location, and the amount of groundwater inrush are important.

49.2 Ground Water Inrush

Excavation of tunnel boring machine (TBM-1) at Karak is 2000 m that full of groundwater inrush and joint orientations. The geology that explains along the tunnel is mostly the Main Range Granite Batholith with a lesser extent of metasedimentary rocks of the Karak Formation [2]. To predict water inrush in tunnels, all the properties have been studied to the core. Fractures properties like aperture, frequency, length, orientation interconnection, angle, filling materials, and characteristics of the fracture plan have strong relation to the hydraulic conductivity to enable better transmissivity of groundwater within the permeable of rock mass. Studies have mentioned that a lot of factors come up to measure the flow pattern, for example fracture geometric parameters, in-situ stresses and excavation disturbance. As stated by Holmøy and Nilsen, the most accepted hypothesis on the significance of geological parameters is on the predicting groundwater inrush related to the quality of rock mass using Q-system, the behavior of joints towards the major principal stress and contact boundary zone of rock formations. Having accurate geological data and construction of a model for each geological characteristics is difficult to support with complicated geological characteristics. Hence, the objective of this research is to study the joint pattern and overburden at NATM-2 and to investigate the relationship between joint pattern, overburden, and water inrush. The research was conducted by creating the joint pattern modeling of meta-sedimentary rock in NATM-2 from TD1024.08 until TD1919.46, Karak, Pahang, during the excavation process. The main source information for this study is the geological document which infers to tunnel face mapping. The data is classified and study of the joint pattern utilizes the available image processing and computer vision system in RocScience Software. Geological Map and Topographical Map that have been used as a site investigation tool for this study is used to study the relationship between the joint pattern, overburden, and the water inrush occurrences.

49.3 Geology of Study Area

NATM-2 is mostly self-possessed of medium to very coarse-grained, porphyritic biotite granite, which is cut by minor porphyritic differentiates and fine-grained variety of granite that represents a later phase of granite intrusion. Some quartz dikes (quartzite) have developed within and parallel to the linear zone of deformation such as the stretch of lineament lines L-B and L-C, faults or shear zones. Near such faults zones, proto-mylonite, mylonite, cataclasite, and breccia are produced as a result of shearing within the granite. For some points, aplite is also presented in granite rock mass. Other foundation rocks that present within the main range granite are microgranite, monzonite, aplite, granite porphyry, quartz porphyry, felsite, granodiorite and diorite and quartz vein. The area involved in this study is within Karak Formation which is composed mostly meta-sedimentary rock and also pass through under Sungai Karak. The rock type that involved in Karak Formation is the Devonian age of metamorphic rock. In geology, it is when transforming in mineral assemblage and texture when a rock experiences pressures and temperatures different to the conditions under which the rock was originally formed. The magmatic rocks and exaggerated by tectonic movements reflecting post-depositional events in the development of the sector are intruded into metasedimentary rocks. According to Padget, they normally represent the first recognizable event though little is said or known about the basement on which they accumulated. However, the meta-sedimentary rock and igneous rock such as granite were compared metasedimentary rock is more harder.

Padget stated the regular types present are quartzite, micaceous quartzites and politic (argillaceous) gneisses. Regnier 2003 informed that a remarkable crisis in polyorogenic settings is that some rocks maintain tectonometamorphic fabrics of an older orogeny awfully well, and that these fabrics are not distorted during a severe orogenic overprint. Olson stated that an essential question in structural geology is what influences the geometric feature of joint pattern in rock. Contribution of geologist in the evaluation of usual risks and stability of a site for engineering works has a long record that developed from the lore of our forefathers. The rationale for structural geologist attempts of analyzing rock fractures is to allow evaluation of each fracture set or geological setting to be made on a wider context. The existence of variation loading histories, material properties or fracture mechanisms were affected the differences in joint patterns. The effects of displacement and stress boundary conditions on resultant fracture length were greatly reviewed. It has been proven where the areal distribution of joints and their lengths were controlled by the nature of the velocity law essentially controlled. According to Olsen, special emphasis is placed on the clustering of parallel joints into narrow, intensely fractured zones and on the relationship stress shielding around the zones and interzone spacing.

Rock masses are normally discontinuous in nature as a result of various geological processes. The most intense in the great mountain belts of the world is deformation, where sedimentary rocks that originally horizontal and below sea level are now folded contorted, fractured and in some places completely overturned. According to Humblin and Christiansen, these folded and warped rock layers testify to the continuing motion of the lithosphere and the deformation it produces. The failure of rock mass during construction usually occurs because of weathering action, construction extensive blasting, hydrostatic pressure and minor earthquake. During the fieldwork, more than 1242 joint measurements were mapped during the construction of the tunnel consisting of 895.38 m length from TD1024.08 to TD1919.46. In order to obtain pattern of relationship between the jointing systems (joint orientation) with the water amount, the characteristic of joints that manipulate the rate of water inrush along the tunnel is separated into groups of classification. Dry joint, wet joint, and joint with high water inrush are the three groups that represent the joint conditions. These groups are analyzed by using quantity of inrush in unit L/min.

It is well known that the groundwater inrush will flow through any porous media. Based on Darcy's law, it is stated that the combination of rock permeability and hydraulic gradient within the rock mass control the groundwater inrush. Example as lineament, fault area and joints are deformation processes that are compulsory to determine the relationship between water inrush in jointed rock mass. According to Karasaki et al., hydrogeological processes control the hydraulic gradient around the fault zone such as rates of groundwater recharge forcing topography-driven flow, anthropogenic influences such as extraction of groundwater and deeper processes such as fluid flow driven by sediment compaction.

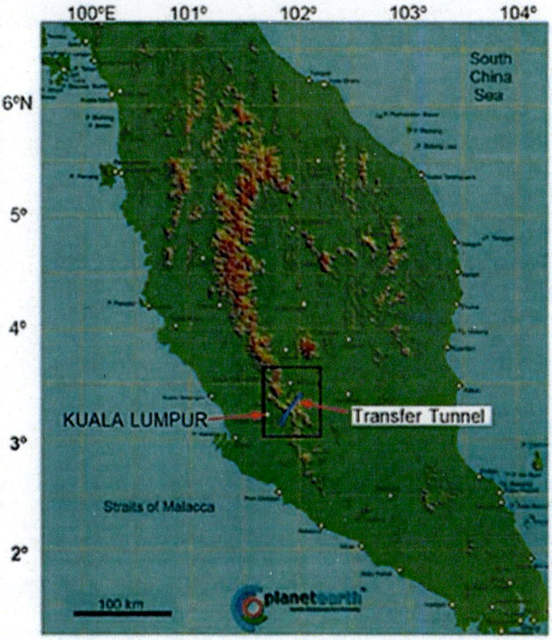

Fig. 49.1 Map of West Malaysia

49.4 Sample Analysis

This research consisted of several key activities which are desk study, data collection, data acquisition, and data analysis. The location of study is located under the Titiwangsa Mountains range, from Pahang River Basin (Kelau and Telemong River), into upstream of Langat River Basin. To be accurate, at NATM-2, based on geological map of West Malaysia (Fig. 49.1) and Pahang (Fig. 49.2), the study area is under Titiwangsa Mountains range, respectively.

The data collection was based on secondary data. In this research, it used Tunnel Mapping Visit. The purpose is to collect the geological data, mapping of discontinuities and overburden data. After that, analysis of graph overburden, joint and water inrush was done. In this part, all the joints, overburden, and water inrush were distinguished (Figs. 49.3 and 49.4).

The tunnel mapping data being convert into pole diagram, and rosette diagram by using RocScience Software (Figs. 49.5 and 49.6).

49.5 Discussion

Every joints appear at the tunnel face that has its own characteristic. There are 1242 joints noted from overall of tunnel face. From the total joints, there are 357 faults, 228

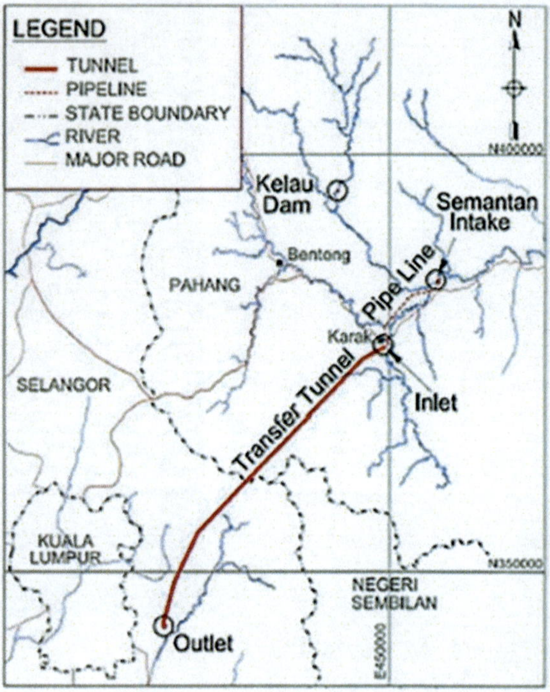

Fig. 49.2 Map of Pahang

plane, 109 foliation, and 548 random joints. Most of the joints focused at 10°–40°. Most of the directional of joint pattern followed the regional fault pattern except at TD1497.36 to TD1637.55, TD1644 to TD1788.43, TD1692.95 to TD1794.61, and TD1678.84 to TD1687.14.

At the tunnel inlet, the overburden of the tunnel is low compared to the other part. This will cause the structure of rock in that part undergoes weathering process. This process will alter the joint pattern and cause irregularity in the rock structure. Because of that, the directional dominant joint slightly strayed from the regional fault pattern. It shows that the weathering process is one of the causes of joint development.

At the tunnel inlet, the overburden of the tunnel is low compared to the other part. This will cause the structure of rock in that part undergoes weathering process. This process will alter the joint pattern and cause irregularity in the rock structure. Because of that, the directional dominant joint slightly strayed from the regional fault pattern. It shows that the weathering process is one of the causes of joint development.

By evaluating engineering geological properties of the rock mass along the tunnel, the correlation of joint pattern and water inrush can be further analyzed. Rosette diagram was used in analyzing all joint sets, as shown in Fig. 49.8. The joint pattern along the study area shows dissected figure. The orientation and dip angle of the two or three cross joints sets increase the probability for high water inrush to occur as the area is oriented in the northeast/southwest direction with a dip toward the south. The three cross-cutting trends follow up with the highest joint densities. Table 49.1

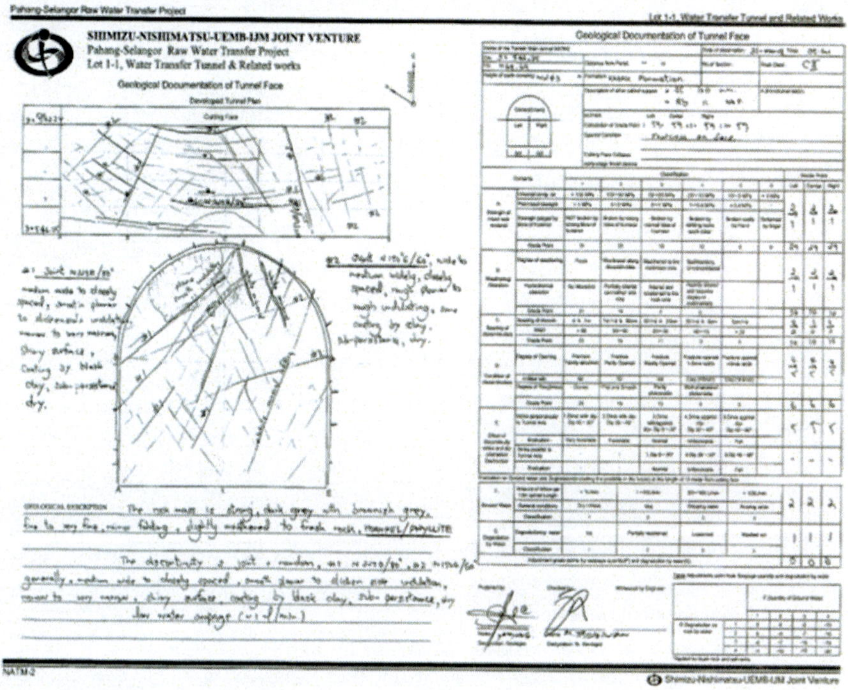
Pahang-Selangor Raw Water Transfer Project
Lot 1-1, Water Transfer Tunnel and Related Works
SHIMIZU-NISHIMATSU-UEMB-IJM JOINT VENTURE
Pahang-Selangor Raw Water Transfer Project
Lot 1-1, Water Transfer Tunnel & Related works
Geological Documentation of Tunnel Face
Developed Tunnel Plan
Geological Documentation of Tunnel Face
NATM-2
Shimizu-Nishimatsu-UEMB-IJM Joint Venture

Fig. 49.3 Sample tunnel mapping

Fig. 49.4 Example of tunnel face

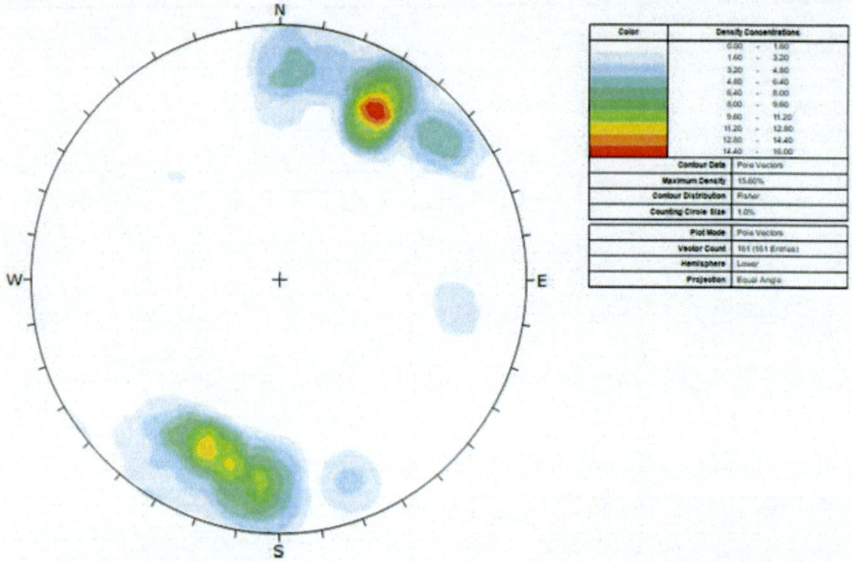

Fig. 49.5 Example of pole diagram

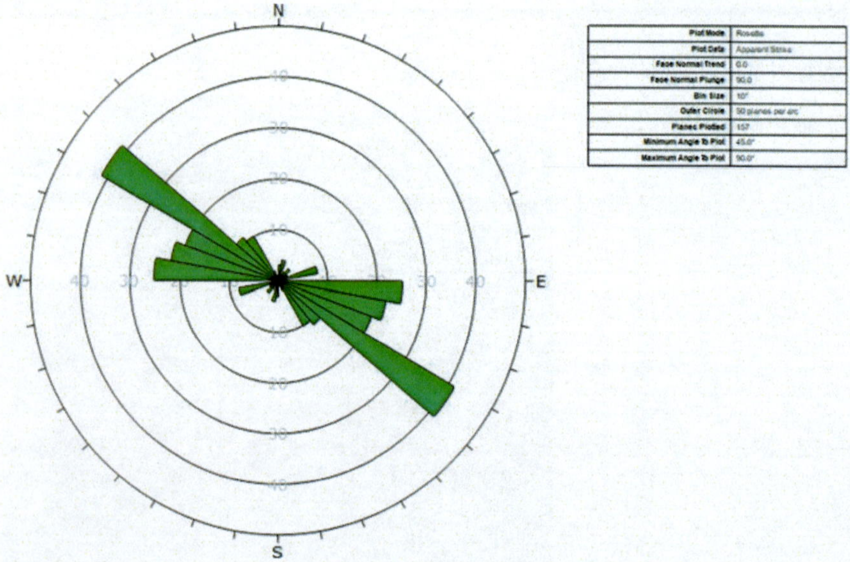

Fig. 49.6 Example of rosette diagram

Table 49.1 Elastic recovery of silver-based conductive ink at different strains with the initial length of 25 mm

No	Tunnel distance, TD (m)	Joint orientation (Dominant)	Water inrush (L/min)
1	1604.85–1611.85	N140E/70, N060-090E/70–90	5
2	1755.16–1791.67	N330-340E/70–90, N270E/40–50, N80E/40–80	3–5
3	1867.09–1870.51	N070-100E/60–70, N140E/60–90	5–10

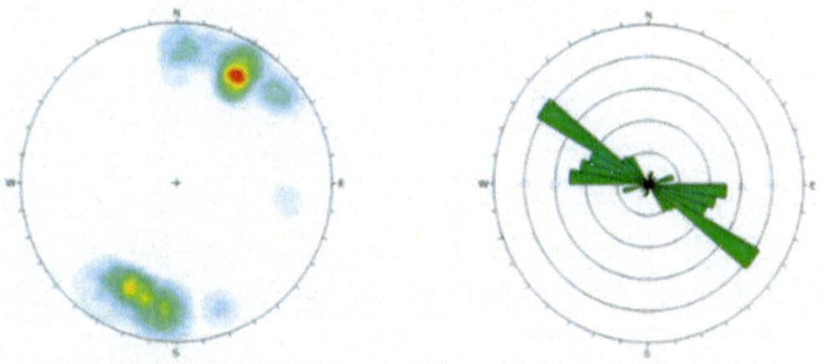

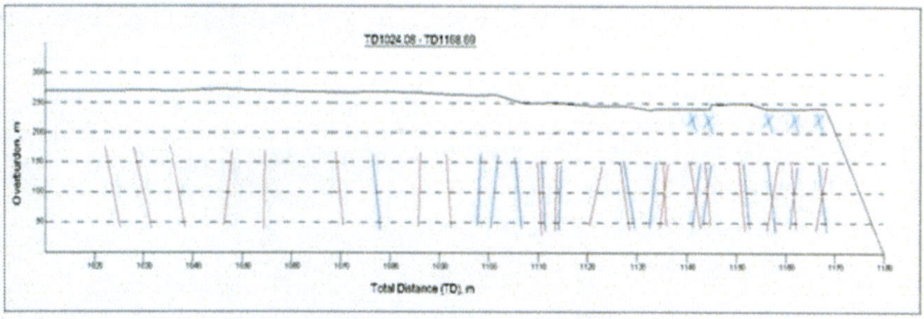

Fig. 49.7 Graph overburden, joint and water inrush at TD1024.08–TD1168.69

shows the summary of overall geological map from TD1024.08 until TD1919.46 (Figs. 49.7 and 49.9).

49.6 Conclusion

By understanding the methods of discontinuities zone and hydrogeology in this research, the true objective can be found. The specific effect on joint pattern produces deformation in small or large scale, especially at the fault zone. This circumstance interpolates heterogeneity and anisotropy in permeability which contributes to significant effect on various processes such as regional groundwater inrush and hydrothermal fluid circulation. Existing of the fault zones introduces hydraulic

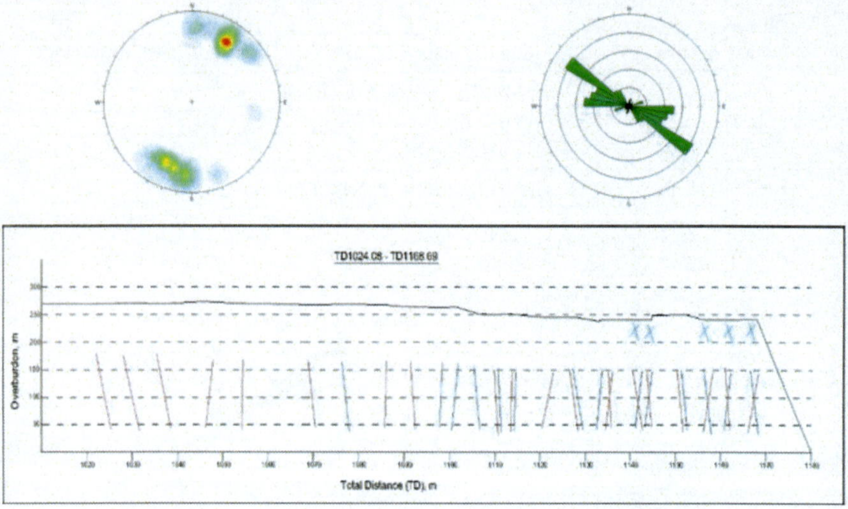

Fig. 49.8 Graph overburden, joint pattern and water inrush at TD1024.08–TD1168.69

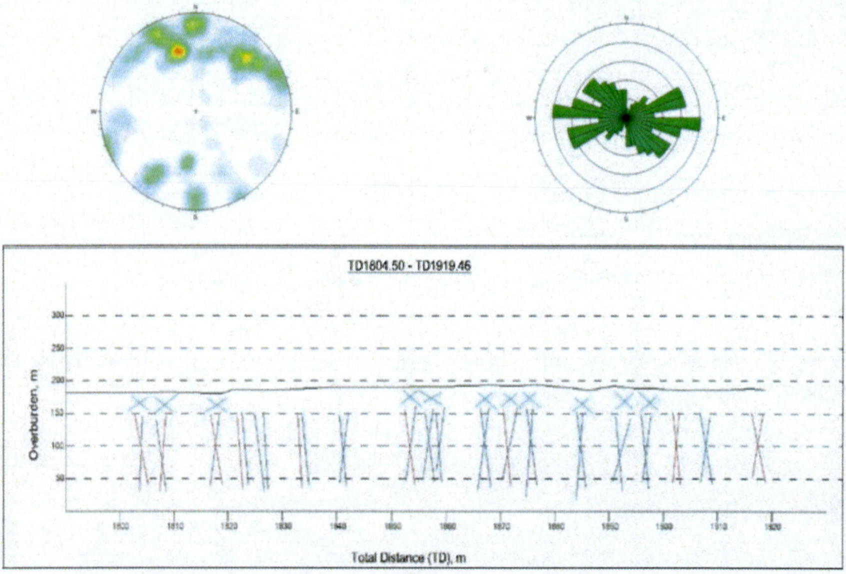

Fig. 49.9 Graph overburden, joint pattern and water inrush at TD1804.50–TD1919.46

channel connecting the underground, but at the same time the formation of effective barriers to groundwater flow occurs from a group of fault cores. Trend of the joint sets can be seen from the highest rate of water inrush. The severe area can be found at TD1604.85–TD1611.85, TD1755.16–TD1761.54, TD1761.54–TD1775.90, TD1775.90–TD1778.34, TD1788.21–1791.67, and TD1867.09–TD1870.51.

Acknowledgements This work was supported by the Fundamental Research Grant Scheme (FRGS) awarded by the Ministry of Education of Malaysia entitle 'Enhanced Coalbed Methane Recovery Processes by Coupling Fluid Flow of Mukah-Balingian Coal Deposits' with the referral no. R.J130000.7851.5F371.

This work was supported by the Universiti Teknologi Malaysia Encouragement Research Grant (UTMER) awarded by Universiti Teknologi Malaysia entitle 'Diagnosis of Alkali-Aggregate Reaction—Polarizing Microscopy and SEMEDS Analysis' with the referral no. Q.J130000.3851.19J79.

References

1. J. Moon, S. Jeong, Effect of highly pervious geological features on ground-water flow into a tunnel. Eng. Geol. **117**(3–4), 207–216 (2011)
2. H. Zabidi, A. Rahim, M. Trisugiwo, Structural controls on groundwater inflow analysis of hardrock TBM (2019)

Chapter 50
Effect of Aging Between Untreated Bamboo, Treated Bamboo, and Salvaged Bamboo from Bamboocrete Panel

Muhamad Syaqir Sha'rani, Noor Aina Misnon, Mohamad Ikhwan Syafiq Zainan, Norazman Mohamad Nor, Norhasliya Mohd Daud, Azrul Affandhi Musthaffa, and Nur Liza Rahim

Abstract The use of bamboo as a substitute for steel promotes "green technology" or "green steel" and encourages the use of sustainable resources in construction. Bamboo becomes one of the most high-quality and environmentally friendly building materials due to its good qualities such as high tensile strength and is available at a low cost. In the previous study, bamboo was used as reinforcement in concrete panels namely bamboocrete panels, and the strength and durability of bamboocrete panels cast over three years period were investigated. This study was carried out to investigate the effect of aging between untreated bamboo, treated bamboo, and salvaged bamboo from bamboocrete panels. The tension and compression tests were carried out to determine the mechanical properties of the bamboo. The compressive strength of salvaged bamboo was about 12% lower than treated bamboo; however, it was 11% higher than untreated bamboo of the same age samples. Similar pattern on tensile strength was also observed.

Keywords Bamboocrete panel · Green steel · Green technology · Sustainability

50.1 Introduction

Bamboo is one of the fast-maturing plants as a natural raw material. Bamboo is much stronger when compared to other types of wood and bricks [1]. It is a sustainable raw

M. S. Sha'rani · N. A. Misnon (✉) · M. I. S. Zainan · N. M. Nor · N. M. Daud
Faculty of Engineering, Universiti Pertahanan Nasional Malaysia, Kem Sungai Besi, Sungai Besi, 57000 Kuala Lumpur, Malaysia
e-mail: nooraina@upnm.edu.my

A. A. Musthaffa
Stare Resources Sdn. Bhd., 53300 Kuala Lumpur, Malaysia

N. L. Rahim
Faculty of Civil Engineering and Technology, Universiti Malaysia Perlis, 02600 Jejawi, Arau, Perlis, Malaysia

M. A. A. Mohd Salleh et al. (eds.), *Proceedings of the Green Materials and Electronic Packaging Interconnect Technology Symposium*, Springer Proceedings in Physics 289,
https://doi.org/10.1007/978-981-19-9267-4_50

material and could be a cheap substitute for steel reinforcement. While timber takes a longer period to reach maturity which is 30–50 years, bamboo only takes around 3–5 years [2]. The bamboo plantation also can help mitigate global warming issues and at the same time offset greenhouse gases. Bamboo reaches its full growth in just a few months when it reaches its maximum mechanical resistance [3]. Commonly, as the bamboo culms mature the layer of the sheath that protects the culm falls off. Bamboo culms are partitioned by a diaphragm which can be seen by the ring around the culm of bamboo. Internode is what the in-between node of the bamboo culm is called. The diameter and thickness of bamboo culm decrease as it grows upward. The internodal length, however, is the largest in the middle of the culm [3].

Modern constructions around the world utilized conventional materials like concrete and steel, and the demand for these materials increase exponentially over the decades [4]. Steel rebars are the most common choice for reinforcing concrete as it increases the tensile strength of the concrete as commonly known that concretes are high in compression strength and durability but low in tensile strength. Steel is undoubtedly a non-renewable material, and its production cost is very high and pollutes the environment. Therefore, the study of alternative materials is in dire need to lower the cost of production thus at the same time preserving the environment against pollution.

Bamboo's fast growth rate and high tensile strength make it a good candidate for reinforcing concrete. Bamboo can sustain tension and compression loading, unlike any other natural material that can only sustain tension loading in major, due to its dense fibrous structure along the grains of the bamboo [3]. The mechanical properties of bamboo and its availability in developing regions have led to its empirical use as reinforcement in concrete structures. The possibility of the utilization of bamboo as a sustainable substitute for steel poses a concern in the industry due to its structural capacity and compatibility [1].

Bamboo has the potential to be a more affordable alternative to traditional steel rebar for building construction due to its benefits in self-weight reduction [5]. The purpose of this study was to compare the effects of aging on bamboo that had been treated, untreated bamboo, and salvaged bamboo from bamboocrete panels.

50.2 Methodology

The type of bamboo used in this study is Buluh Semantan. The average diameters of the bamboo were ranging from 45 to 60 mm, with the height of the culm is 3 m approximately. The samples were prepared with two types of bamboo conditions, i.e., (i) untreated and (ii) treated. The untreated bamboo samples were left to be dried under the shed after harvest, and the treated samples were soaked in borax and boric acid and then dried under a shed as per shown in Figs. 50.1 and 50.2. This study is continuation from study conducted in December 2016 [1] to investigate the effect of aging of the bamboo over the time. At the time when the study was conducted, the age of the bamboo samples was 3-year-old. The bamboo samples were classified into four categories of samples which labeled as new treated bamboo (A), 3-year-old treated bamboo (B), 3-year-old untreated bamboo (C), and salvaged bamboo 3-year-old treated (D). The salvaged bamboo samples were taken from the tested panels which were treated and cast in October 2017. The procedure and methods carried out in this study were accordance with the International Standards ISO 22157-1 [6] and ISO 22157-2 [7].

Fig. 50.1 Drying process of the treated bamboo

Fig. 50.2 Dried and treated bamboo samples

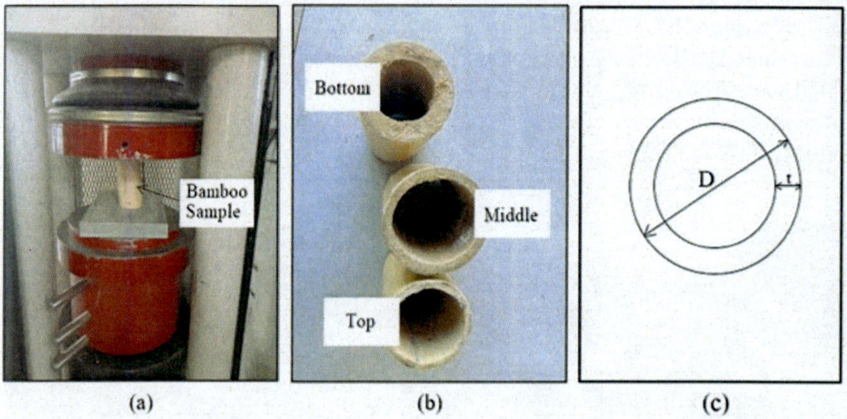

Fig. 50.3 Test setup **a** compression machine setup, **b** top, middle, and bottom samples of bamboo, and **c** illustration of external diameter and thickness of the bamboo sample

50.2.1 Compression Test of Samples

Prior to the compression test, the bamboo samples were cut into three sections of each culm from the top, middle, and bottom (see Fig. 50.3b). The height of each sample was cut into two times of the external diameter of the bamboo samples. The compression test setup is shown in Fig. 50.3a. The ultimate compressive strength was then calculated using Eq. (50.1).

$$\sigma_{\text{ult}} = F_{\text{ult}}/A \tag{50.1}$$

where σ_{ult}: compressive stress (MPa), F_{ult}: maximum load (N), and A is the bamboo cross-sectional area (mm^2) and was calculated using Eq. (50.2).

$$\text{Cross sectional area, } A = \left(\pi \times \left[D^2 - (D - 2t)^2\right]\right)/4 \tag{50.2}$$

where D is the external diameter of the bamboo sample (mm) and t is the thickness of the bamboo sample (mm) (see Fig. 50.3c).

50.2.2 Tensile Test of Samples

Samples were taken from each of the three-section (top, middle, and bottom) and cut into the strip with the dimension of as shown in Fig. 50.4. According to the International Standards ISO 22157-1 [6], the gauge length shall be between 50 to

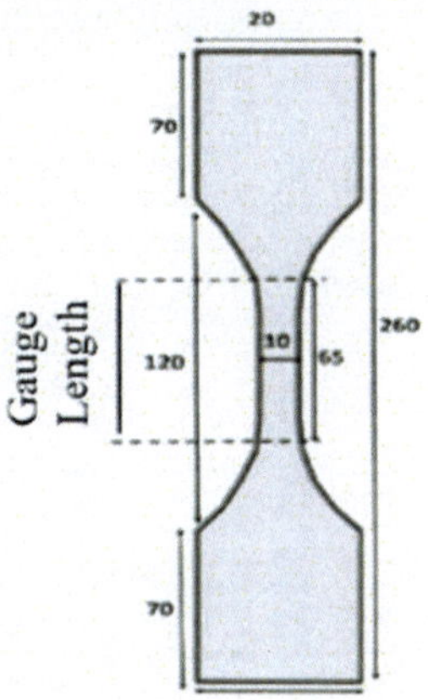

Fig. 50.4 Illustration of bamboo strip sample

100 mm. The ultimate tensile strength of the bamboo strip was then calculated using Eq. 50.3.

$$\sigma_{t.\mathrm{ult}} = F_{t.\mathrm{ult}}/A \tag{50.3}$$

where $\sigma_{t\ \mathrm{ult}}$: ultimate tensile strength (MPa), $F_{t\ \mathrm{ult}}$: maximum tensile load (N), and A: bamboo strip cross-sectional area (mm^2). The cross-sectional area, A, was calculated using Eq. 50.4.

$$\text{Cross sectional area, } A = w \times t \quad \text{(at gauge length)} \tag{50.4}$$

where w: width at gauge length (mm) and t: thickness at gauge length (mm).

50.3 Result and Discussions

50.3.1 Compressive Strength

The compressive strength of the bamboo samples is shown in Table 50.1 and Fig. 50.5.

Table 50.1 Compressive strength of the bamboo samples

Samples	Compressive strength (MPa)
New bamboo treated, A	44.98
3-year-old treated, B	50.43
3-year-old untreated, C	40.11
Salvaged bamboo 3-years-old, treated, D	44.47

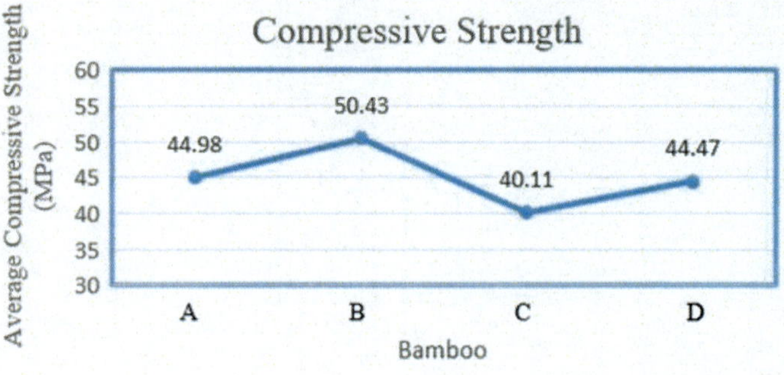

Fig. 50.5 Compression strength of the bamboo samples

From the results given in Table 50.1, the highest average ultimate compressive strength of the bamboo samples is 50.43 MPa which is from the treated bamboo harvested in December 2016 (sample B). The samples showed 25% higher than the untreated bamboo kept for the same duration and under the shed (sample C). This indicates the effectiveness of the bamboo treatment with borax and boric acid. The treatment successfully preserves the bamboo samples from early biodegradation and maintains the strength capacity in compression. The compressive strength of the salvage bamboo (sample D) in the tested panel is 44.47 MPa which is about 12% lower when compared with sample B. The reduction of the strength could be caused by the chemical effect from the concrete during the hydration process. The untreated and aged bamboo samples (sample C) showed the lowest compressive strength with value of 40.11 MPa resulting from biodegradation effect of the bamboo.

50.3.2 Tensile Strength

The tensile strength test result of the bamboo samples is shown in Table 50.2 and Fig. 50.6.

From Fig. 50.6, sample B showed the highest average ultimate tensile strength with 154.29 MPa. The tensile strength is about 4% and 17% higher than the sample A and sample C, respectively. The tensile strength of the sample D is 139.74 MPa which is approximately 6% and 9% lower than sample A and sample B, respectively. However,

Table 50.2 Tensile strength of bamboo samples

Categories	New bamboo, treated (A)	3 years old, treated (B)	3 years old, treated (C)	Salvage bamboo, 3 years old, treated (D)
Tensile strength (MPa)	147.72	154.29	128.25	139.74

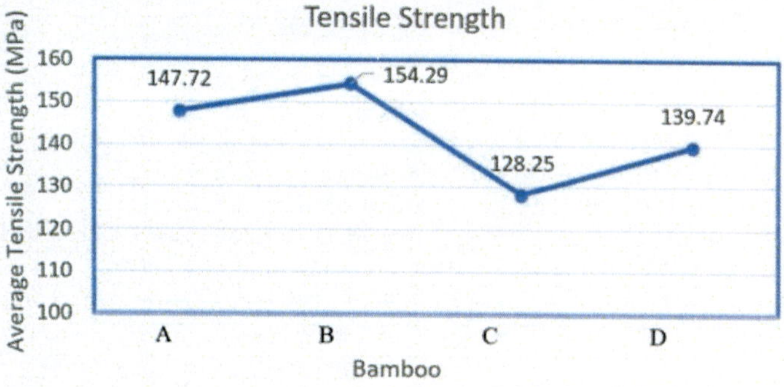

Fig. 50.6 Tensile strength of the bamboo samples

sample B showed 9% higher tensile strength than untreated sample C. Similar pattern to compressive strength was observed for the tensile strength of the bamboo samples. This implies that the compressive strength has significant relationship to the tensile strength of the bamboo samples.

50.4 Conclusion

The study on the effect of aging between untreated bamboo, treated bamboo, and salvaged bamboo from bamboocrete panels discovered aging affects the overall strength of the bamboo. From the study, it can be inferred that:

(i) Bamboo treatment with borax and boric acid showed significant improvement to the aged samples when compared to untreated bamboo,
(ii) Treated bamboo samples deteriorates over time, but the strength still satisfies the capacity as reinforcement materials,
(iii) Sample A (new treated bamboo) showed slightly lower strength than sample B (aged, treated bamboo), due to the non-homogeneity nature characteristics since these two samples were harvested from different batches.

Further studies on the weathered samples of bamboo should be carried out to ensure these materials are promising as good materials in building construction.

Acknowledgements This project was funded by Universiti Pertahanan Malaysia under grant number PS0018-UPNM/2021/GPPP/TK/10. The authors gratefully acknowledge the assistance of the technical staff of the structural and heavy material laboratory.

References

1. N.M. Daud, N.M. Nor, M.A. Yusof, A.A.M. Al Bakhri, A.A. Shaari, The physical and mechanical properties of treated and untreated Gigantochloa Scortechinii bamboo, in *AIP Conference Proceedings*, vol. 1930 (2018). https://doi.org/10.1063/1.5022910
2. K. Ghavami, Bamboo as reinforcement in structural concrete elements. Cem. Concr. Compos. **27**(6), 637–649 (2005). https://doi.org/10.1016/j.cemconcomp.2004.06.002
3. K. Yathushan, S. Kishok, B.E. Thevarajah, M. Nithurshan, Bamboo cane as an alternative reinforcement in reinforced concrete beam, in *MERCon 2021—7th International Multidisciplinary Moratuwa Engineering Research Conference* (2021), pp. 154–159. https://doi.org/10.1109/MERCon52712.2021.9525805
4. N.M. Daud et al., Economic comparison: bamboocrete multi-purpose panel and typical reinforced concrete panel (Perbandingan Ekonomi: Panel Serba Guna Bamboocrete dan Panel Konkrit Bertetulang). J. Kejuruter. SI **4**(1), 43–47 (2021) [Online]. Available: https://doi.org/10.17576/jkukm-2021-si4
5. N.M. Daud, N.M. Nor, M.A. Yusof, M.A. Yahya, V. Munikanan, Axial and flexural load test on untreated bamboocrete multi-purpose panel. Int. J. Integr. Eng. **10**(2), 28–31 (2018). https://doi.org/10.30880/ijie.2018.10.02.006
6. ISO 22157-1, *Bamboo-Determination of Physical and Mechanical Properties—Part 1: Requirements* (2004)
7. J.J.A. Janssen, *Bamboo-Determination of Physical and Mechanical Properties—Part 2: Laboratory Manual* (2004)

Chapter 51
Resistance Spot Welding and Laser Welding Effect on Nickel Tab for Electric Vehicle Battery Development

M. Syafiq, N. H. Jamadon, A. Syahmi, S. Janasekaran, T. Zaharinie, and R. Rangappa

Abstract In this study, the joint ability of two different joining techniques, namely resistance spot welding and laser welding of nickel weld interfaces and joint microstructure were addressed. Similar materials of nickel joint were used to investigate the parallel connection of tab cells for electric vehicle (EV) battery development. It was found that using the parameters discussed, the joining of two similar nickel was successfully achieved via these welding methods. The results show that the material joints fabricated using laser welding had the highest average stress value of 155.86 MPa, whereas the average for nickel joints of resistance spot welding was 122.25 MPa. In addition, the results of FESEM analysis on the cross-section of the weld material also found that the length of the surface for the material being connected and affects the stress value. For resistance spot welding process, 0.09 s of welding time promoted the formation of sound weld beads with better control of penetration depth and excellent joint interface. Compared to the resistance spot

M. Syafiq · N. H. Jamadon (✉) · A. Syahmi
Department of Mechanical and Manufacturing Engineering, Faculty of Engineering and Built Environment, Universiti Kebangsaan Malaysia, 43600 Bangi, Selangor, Malaysia
e-mail: nashrahhani@ukm.edu.my

M. Syafiq
Jabatan Kejuruteraan Mekanikal, Politeknik Banting Selangor, Persiaran Ilmu, Jalan Sultan Abdul Samad, 42700 Banting, Selangor, Malaysia

S. Janasekaran
Faculty of Engineering, Built Environment & Information Technology, Center for Advanced Materials and Intelligent Manufacturing, SEGi University, 47810 Petaling Jaya, Selangor, Malaysia
e-mail: shaminijanasekaran@segi.edu.my

T. Zaharinie
Department of Mechanical Engineering, Faculty of Engineering, University of Malaya, 50603 Kuala Lumpur, Malaysia
e-mail: tzaharinie@um.edu.my

R. Rangappa
Department of Mechanical Engineering, , BMS Institute of Technology and Management, Bengaluru, Visvesvaraya Technological University, Belagavi 590018, India
e-mail: ravichandra@bmsit.in

M. A. A. Mohd Salleh et al. (eds.), *Proceedings of the Green Materials and Electronic Packaging Interconnect Technology Symposium*, Springer Proceedings in Physics 289,
https://doi.org/10.1007/978-981-19-9267-4_51

welding process, laser welding produced a better weld interface, which contributes to a higher strength value.

Keywords Electric vehicle · Joining techniques · Resistance spot welding · Laser welding

51.1 Introduction

EV batteries consist of battery cells arranged in large numbers in series or parallel. It is important to ensure that optimal power and energy capacity can be supplied to electric vehicles [1]. In addition, vehicle batteries also need to travel through harsh environments. A variety of severe driving conditions such as vibration, bad weather, and humidity, the connections in the battery pack will experience possible mechanical fatigue, high-temperature condition and corrosion [2, 3]. The selection of a good joining method is important from the perspective of mechanical strength and microstructure of the welded joint [4].

Connecting battery cells in large quantities is difficult to be implemented due to multi-layers of thin iron which need to be welded together with high conductivity [5]. A very crucial point in an EV is the design and the performance of battery system (traction battery) as part of the electric powertrain [6, 7]. The battery system needs to have both high energy density and power with optimal range and sufficient driving dynamics.

The lithium-ion batteries (Li-ion) were preferred for energy storage due to its characteristic of high-power density. The voltage and current density of a single battery cell will be inadequate for operation in a battery pack. As a result, a battery pack requires the assembly and connection of numerous battery cells, either in series or parallel. The most common way to connect battery cells will be either in series or parallel form, but this is also the main cause of the degradation of consistency between cells [8, 9].

In general, spot welding technique was employed extensively compared with laser welding due to its lower operating cost. However, the quality of the spot welding technique was poorer than the laser welding technique, which it requires some control in several parameter such as the welding force from the electrode welder, electric voltage, electric current, welding time, and electrode geometry as well as the electrode material [10–12]. In addition, many battery packs will be manufactured each year for production due to increase in demand. Therefore, batteries must be assembled by using a robust joining process and the development of effective joining technology for battery manufacturing as an important requirement for car manufacturers.

The current research work was focused on the analysis of welded Ni and Ni joints similar materials by using the spot welding process. Three main parameters in spot welding will be used: electrode force, welding current, and welding duration. Based on the findings from literature review, the welding duration parameters will be modified, meanwhile the electrode force and welding current will be maintained throughout the experiment. The effects of different welding parameters such as weld pressure and weld time on the mechanical testing and microstructure of joint area will be studied. In comparison with the laser welding which has very much shorter weld cycle and able to create stronger joints on the basis of the same weld region will be observed.

51.2 Methodology

This study focuses on laser welding and resistance spot welding against two similar materials, which was the Ni. Resistance and laser welding were employed, as well as assessing the surface of the welding material. The methodology commenced with specimen preparation, continued with the welding procedure, mechanical and microstructural observation, and concluded with the interpretation of the obtained results. Figure 51.1 exhibits the flowchart of the overall procedure for this experiment. The experimental plan was developed based on the joining requirements for various time of welding parameters.

The specimens to be used were precisely cut into 50 × 8 mm dimensions, with the thickness of 0.1 mm, as shown in Fig. 51.1. A 50 mm of nickel was superimposed with a nickel sample of equal size at 10 mm, which portraits in Fig. 51.2. The size of the nickel sample which had been joined was 90 mm after two nickel samples were welded. The sample was analyzed by using Scanning Electron Microscope (SEM) and energy dispersive X-ray (EDX) analysis approach to identify the purity of the specimen used. The resistance spot welding will then be performed using three parameters: welding time, welding current, and electrode force. This experiment will be carried out in three sets, with the only difference being the welding time, which will range between 0.03 and 0.09 s with a step size of 0.03 s, as given in Table 51.1.

For laser welding, only one parameter was used in this study to compare between the two welding techniques. The parameters used are given in Table 51.2. Laser welding is carried out using Ytterbium Fiber Laser Marking Machine.

Mechanical testing will be performed on the specimen which was completed the spot welding and laser welding processes. Tensile shear tests were carried out on a universal testing equipment in accordance with the ASTM D1002 standard. The

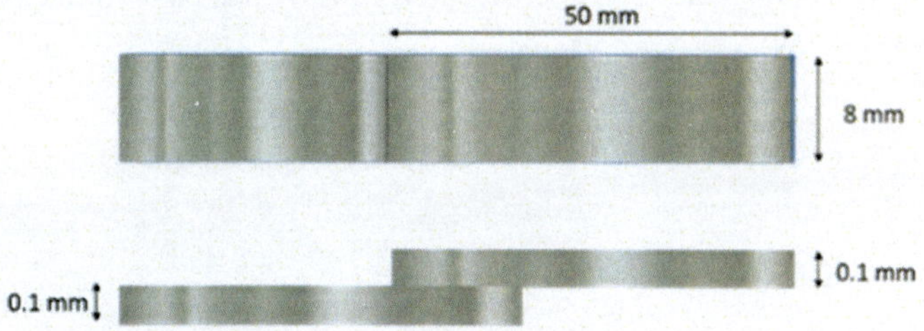

Fig. 51.1 Ni–Ni welded sample

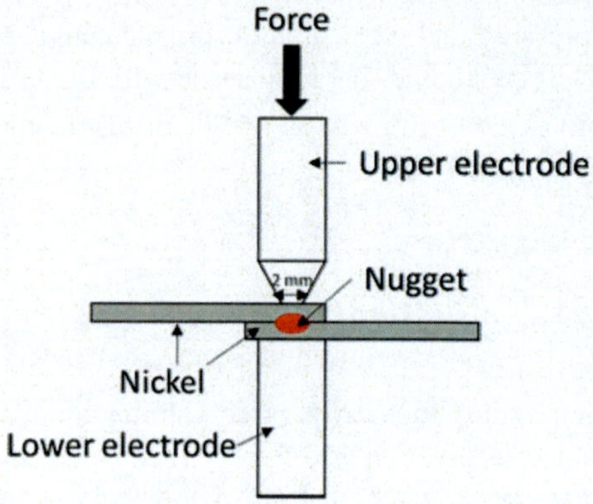

Fig. 51.2 Schematic of resistance spot welding

Table 51.1 Resistance spot welding parameters were used in the study

Sample	Set 1	Set 2	Set 3
Time	0.03 s	0.06 s	0.09 s
Current	5 kA		
Electrode force	0.3 MPa		
Electrode material	Copper		
Spot welding arrangement	1 spot		

Table 51.2 Laser welding parameters were used in the study

Parameter	Value
Pulse width	2 ms
Frequency	250 Hz
Power	200 W

tensile shear test was conducted at 0.5 and 1 mm/min. The mechanical performance of the joint results was described by the failure mode. The welded sample will be cut with an EDM wire cutting machine to examine the microstructure, where this sample will then go through with mounting process. The previously analyzed samples will be ground with a grinding machine. Following that, an optical microscope and field emission scanning electron microscopy (FESEM) will be used to examine the microstructure.

51.3 Result and Discussions

51.3.1 Joining Samples and Optical Microscopy of Welded Surface

Resistance spot welding was used because it is the most suitable method for small scale and limited production. This is because spot welding requires low initial costs and low maintenance costs. It is easy to use and produce good welding quality. Spot welding requires simple setup and can be done semi-automatically or fully automatically. Figure 51.3 shows spot welding samples with various timings and laser welded samples. It can be seen that spot welding performed at 0.09 s produces (Fig. 51.3c) highly noticeable weld on the surface of the joint material compared to spot welding at 0.03 and 0.06 s (Fig. 51.3a, b, respectively). In addition, the welding results became more significant in parallel with the increase in time. Besides, the sample that was welded using laser welding was observed to result in three layers of circles, as shown in Fig. 51.3d. Appearances of laser weld specimen are almost similar with the results obtained by Long et al. [13]. The behavior of laser welded sample will be discussed in detail.

Microscope images of welded specimens are shown in Fig. 51.4. Figure 51.4a–c show microscope images of resistance spot welding samples with welded time of 0.03 s, 0.06 s, and 0.09 s, respectively. It was found that the size of the resulting weld on the nickel surface was the same as the size of the copper electrode used (2 mm). In addition, the welding results show it became more significant in parallel with the increase in time. This is because spot welding performed at 0.09 s produces a more noticeable weld on the surface of the joint material compared to spot welding at 0.03 and 0.06 s.

Laser welding was used to join nickel sheets because it is a non-contact process, which can produce high precision welds and has shorter cycle time. The orientation of the nickel sheet during laser welding was the same as in resistance spot welding. It was found that the sample welded using laser welding has three layers of circles as resulted in Fig. 51.3d. Spot 1 is the part of the joint which was exposed to the laser light directly, while spot 2 is the part of the heat-affected zone (HAZ). Next, spot 3

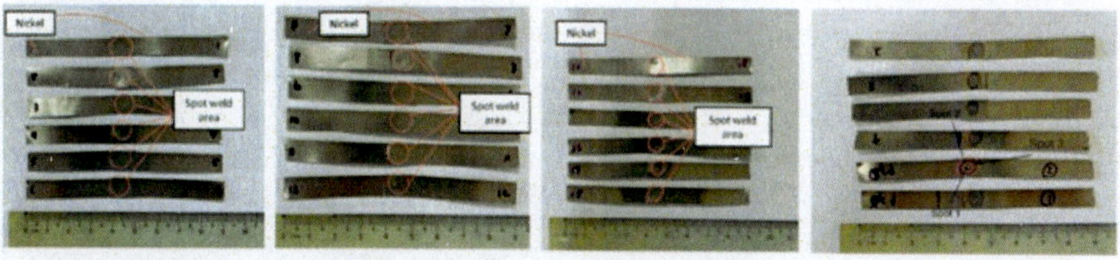

(a) Spot weld 0.03 s (b) Spot weld 0.06 s (c) Spot weld 0.09 s (d) Laser weld 2 ms

Fig. 51.3 Ni–Ni welded samples at various weld times of **a** 0.03, **b** 0.06 s, **c** 0.09 s, and **d** laser welded sample

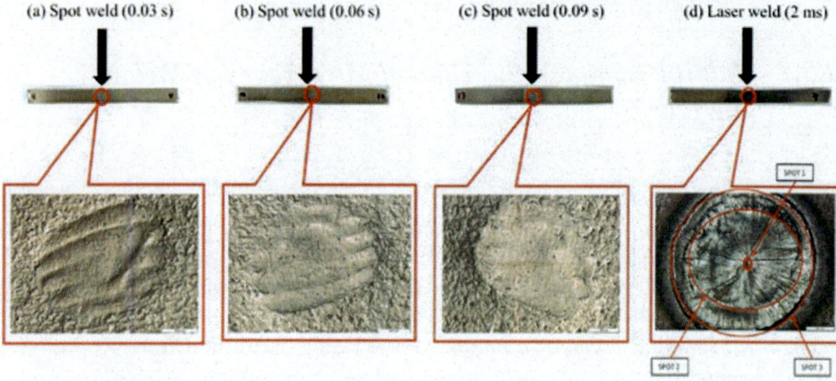

Fig. 51.4 Surface microstructure of the joining parts

is the part resulting from oxidation on the outside of the connection. This may occur due to the reaction that occurs between the laser and oxygen as it was performed in normal atmosphere conditions [11]. In contrast, laser welding produces three spots on the surface of the joint, with spot 1 having the same size as the groove's radius (beam radius) created during welding.

51.3.2 *Joint Strength*

Figure 51.5 exhibits the shear test of the joint strength for spot welding and laser welding. From the shear test result, the highest stress strength resulted in spot welding at the 0.09 s parameter which was 132.83 MPa while the minimum stress resulted in spot welding using the 0.03 s parameter at 125.48 MPa, whereas the joint strength of spot welding sample welded at 0.06 s was 127.37 MPa. The results show that increasing of welding time will eventually rises the heat generated in the welding zone and causes the weld area to expand, resulting in an increase in joint strength. As such, more in-depth testing can be done using different parameters.

Shear tests were also performed on welded joints using laser welding. The average stress value resulting from laser welding was 155.86 MPa. It was found that laser welding can produce a stronger Ni–Ni connection than resistance spot welding at parameters as due to the optimal welding geometry that can be achieved and has a high-power density. The heat source produced by the laser beam was highly concentrated. This is in line with the study by Das et al. [14] which states that various types of metals and conductors with thickness of several millimeters can be welded due to the high-power density of the laser beam. Hence, it may be assumed that a smaller spot makes it possible to weld with less power, minimizing the amount of heat produced by the process. Additionally, faster welding speeds can be used, which makes production more efficient and enhanced the joint reliability.

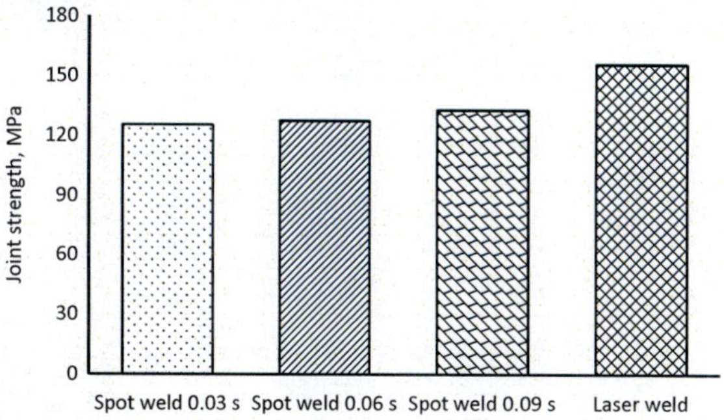

Fig. 51.5 Shear strength of weld sample

51.3.3 Fracture Surface Morphology

The morphology of fracture surface conditions was seen in Fig. 51.6. According to the shear test findings, all of the samples for spot welding, Fig. 51.6a–c had tearing failure on the base material, which is on the Ni section, but no failure on the welding joint part. This occurs because the strength of the resulting weld joint is greater than the strength of the material. It may be presumed that all of the samples welded using both techniques were completely joined at the spot weld area. There are several possibilities for failure. An adhesive failure happens when the bond between the adhesive layer and one of the adhesives fails. Joint failure frequently comprises many failure mechanisms and was measured as a proportion of adhesive failure. From this study, it can be assumed that the tearing that occurs in the base material for resistance spot welding may be caused by the orientation of the sample during the tension test, as depicted in Fig. 51.7, which permits tearing to occur despite the lower stress value produced on the resistance spot welding joint sample.

In laser welding, however, the connection is not severed at base material, but rather at spots 1, 2 and 3, as shown in Fig. 51.6d. This could be the result of the Ni undergoing metallurgical fusion. It was also discovered that the laser weld has a heat-affected zone (HAZ), which increases the joint area and thus influences the joint ability. It was revealed that fracture occurred in the fused zone on the laser weld region when welding occurred in a larger spot. This result is comparable to that of a previous study by Dimatteo et al. [15]. This demonstrates that the link between the size of the welded surface and the measured stress value was directly proportional.

51.4 Conclusion

Based on the results obtained, it was found that the joining of Ni materials using laser welding has the highest average stress value of 155.86 MPa while the average for connection of materials using resistance spot welding was 122.25 MPa. In addition, for the joining of materials using resistance spot welding, three parameters have

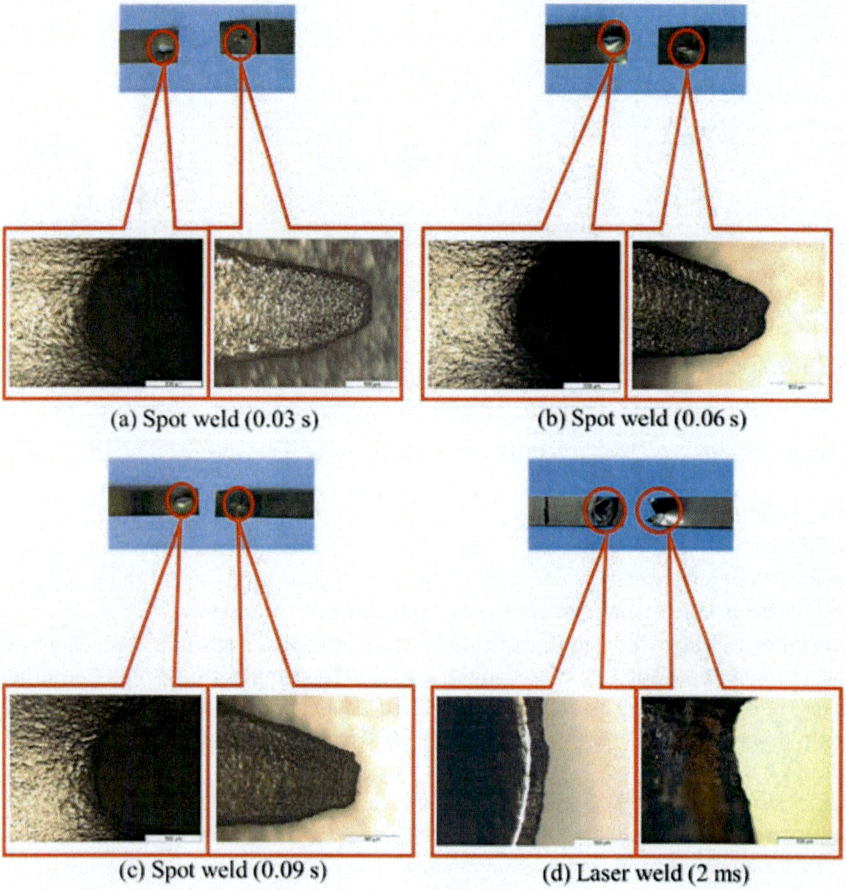

Fig. 51.6 Morphology of fractured surfaces of Ni–Ni samples **a** spot weld at 0.03 s, **b** spot weld at 0.06 s, **c** spot weld at 0.09 s, and **d** laser weld 2 ms

been used which were at 0.03, 0.06, and 0.09 s, and based on the data obtained, the joining of materials using resistance spot welding at 0.09 s was able to produce higher stress values compared to the other parameter at 132.83 MPa. From the observation through optical microscope, it was found that tearing has occurred on all resistance spot welding material samples. The laser weld joining occurs in the heat-affected zone (HAZ), where Ni undergoes metallurgical fusion, contributing to the greatest joint strength value of 155.86 MPa. Based on observations, it was found that both types of welding were successful in joining two similar Ni materials using the parameters that have been discussed.

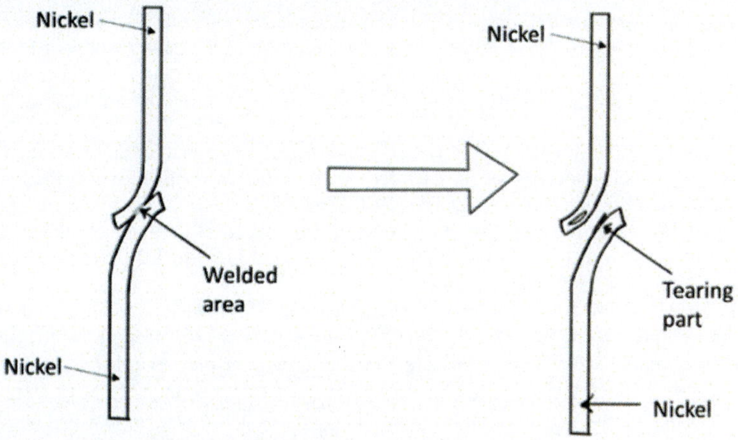

Fig. 51.7 Orientation of the sample during the tension test

Acknowledgements The authors gratefully acknowledge the financial support by the Universiti Kebangsaan Malaysia (GGPM-2019-064). Special thanks to the contributors from University of Malaya, Segi University, and BMS Institute of Technology and Management, India, for this project.

References

1. A.K.M.A. Habib, M.K. Hasan, M. Mahmud, S.M.A. Motakabber, M.I. Ibrahimya, S. Islam, A review: Energy storage system and balancing circuits for electric vehicle application. IET Power Electron. 1–13 (2020). https://doi.org/10.1049/pel2.12013
2. M.Y. al Shdaifat, R. Zulkifli, K. Sopian, A.A. Salih, Basics, properties, and thermal issues of EV battery and battery thermal management systems: Comprehensive review. Proc. Inst. Mech. Eng. Part D: J. Automob. Eng. **0**(0), 09544070221079195. https://doi.org/10.1177/09544070221079195
3. P.M.W. Salehen, M.S. Su'Ait, H. Razali, K. Sopian, in *Battery Management Systems (BMS) Optimization for Electric Vehicles (EVs) in Malaysia*. AIP Conference Proceedings, Apr 2017, vol. 1831. https://doi.org/10.1063/1.4981173
4. A. Das, D. Li, D. Williams, D. Greenwood, Joining technologies for automotive battery systems manufacturing. World Electr. Veh. J. **9**(2) (2018). https://doi.org/10.3390/wevj9020022
5. M.J. Brand, P.A. Schmidt, M.F. Zaeh, A. Jossen, Welding techniques for battery cells and resulting electrical contact resistances. J. Energy Storage **1**, 7–14 (2015). https://doi.org/10.1016/j.est.2015.04.001
6. M. Burridge, S. Alahakoon, in *The Design and Construction of a Battery Electric Vehicle Propulsion System—High Performance Electric Kart Application*. IOP Conference Series: Earth and Environmental Science, Jul 2017, vol. 73, no. 1. https://doi.org/10.1088/1755-1315/73/1/012016
7. K. Sayed, A. Kassem, H. Saleeb, A.S. Alghamdi, A.G. Abo-Khalil, Energy-saving of battery electric vehicle powertrain and efficiency improvement during different standard driving cycles. Sustainability (Switzerland) **12**(24), 1–26 (2020). https://doi.org/10.3390/su122410466

8. J. Jiang, Y. Zhang, W. Shi, J. Xu, W. Diao, H. Guo, in *An Analysis of Optimized Series and Parallel Method for Traction Lithium-Ion Batteries*. 2014 International Conference on Intelligent Green Building and Smart Grid (IGBSG), 2014, pp. 1–7. https://doi.org/10.1109/IGBSG.2014.6835264
9. M. Ye, X. Song, R. Xiong, F. Sun, A novel dynamic performance analysis and evaluation model of series-parallel connected battery pack for electric vehicles. IEEE Access **7**, 14256–14265 (2019). https://doi.org/10.1109/ACCESS.2019.2892394
10. P.R. Pati, S. Das, M.P. Satpathy, B.C. Routara, S.K. Sahoo, S.K. Bhuyan, Ultrasonic spot welding of Al-Cu sheets: a comprehensive study. Mater. Today: Proc. **33**, 5168–5173 (2020). https://doi.org/10.1016/j.matpr.2020.02.873
11. E. Haddad, J. Helm, A. Olowinsky, A. Gillner, Nanosecond pulsed fiber laser as a tool for laser micro welding. Proc. CIRP **94**, 571–576 (2020). https://doi.org/10.1016/j.procir.2020.09.077
12. S. Janasekaran, N.H. Jamadon, F. Yusof, in *Feasibility Study on Autogenously Lap Joint Between Non-Ferrous Metal Alloys Using Low Power Fiber Laser*. Proceedings of SPIE—The International Society for Optical Engineering, 2019, vol. 11205. https://doi.org/10.1117/12.2542500
13. J. Long et al., Fiber laser spot welding of molybdenum alloy in a hyperbaric environment. Opt Express **28**(6), 7843 (2020). https://doi.org/10.1364/oe.385079
14. A. Das, R. Beaumont, I. Masters, P. Haney, Macro-modelling of laser micro-joints for understanding joint strength in electric vehicle battery interconnects. Materials **14**(13) (2021). https://doi.org/10.3390/ma14133552
15. V. Dimatteo, A. Ascari, E. Liverani, A. Fortunato, Experimental investigation on the effect of spot diameter on continuous-wave laser welding of copper and aluminum thin sheets for battery manufacturing. Opt. Laser Technol. **145** (2022). https://doi.org/10.1016/j.optlastec.2021.107495

Chapter 52
A Comparative Study on the Use of Fine and Ultra-Fine-Crushed Lime Foliar Fertilizer for Rice Growth

Nguyen Ngoc Tri Huynh, Tran Ho Van Anh, Nguyen Vinh Phuoc, and Nguyen Khanh Son

Abstract In recent years, the development of nanotechnology enhanced the delivery of agrochemicals and agricultural productivity. From raw lime, ultra-fine-crushing processes were carried out to prepare the nano calcite particles. This study aimed to investigate the effects of foliar application of nano calcite on the growth of rice plants. Particle size distribution by dynamic light scattering (DLS), specific surface area (BET), and microstructure (SEM) of nano calcite as necessary parameters were established before applying to rice plants as foliar fertilizer in laboratory conditions and small-area field-scale. In addition, experiments on the characteristics of leaves (leaf length, leaf width) and the rate of photosynthesis and chlorophyll content were carried out and demonstrated the considerable effect of using nano calcite spraying fertilizer.

Keywords Foliar · Fertilizer · Ultra-fine-crushing · Nano calcite

52.1 Introduction

Generally, every plant needs macro- and micro-nutrients for healthy growth and reproduction. Nitrogen, phosphorus, and potassium are mainly macronutrients; they are soluble and easy to absorb by plant roots. Due to their tiny quantities, micro-nutrients such as calcium, magnesium, and sulfur are necessary for plant development. Usually, some common symptoms appear on leaves, roots, and fruits related

N. N. T. Huynh · T. H. Van Anh · N. K. Son (✉)
Department of Silicate Materials, Faculty of Materials Technology, Ho Chi Minh City University of Technology (HCMUT), 268 Ly Thuong Kiet Street, District 10, Ho Chi Minh City, Vietnam
e-mail: ksnguyen@hcmut.edu.vn

Vietnam National University Ho Chi Minh City (VNU HCM), Linh Trung Ward, Thu Duc City, Ho Chi Minh City, Vietnam

N. V. Phuoc
SANOMAT Co., Ltd, No.9B, 1A Road, Bien Hoa Industrial Zone 2, Bien Hoa City, Dong Nai Province, Vietnam

M. A. A. Mohd Salleh et al. (eds.), *Proceedings of the Green Materials and Electronic Packaging Interconnect Technology Symposium*, Springer Proceedings in Physics 289,
https://doi.org/10.1007/978-981-19-9267-4_52

to the phenomenon of calcium deficiency, for example, stunted growth, brown roots, fruit abnormalities, leaves deformation, and yellow chlorosis. Some primary roles of calcium in plant growth are nutrient transportation into and within the plant, cell wall deposition, and, more critically, facilitating metabolic activities and increasing plant resistance to outside attacks. In soil, abundant calcium resources can be found in calcite minerals, and the calcium absorption of plant roots is in the form of Ca^{2+} solutions. Calcium carbonate ($CaCO_3$) of natural calcite minerals is relatively insoluble. Low soil pH in the environment around the root system can also reduce the ability of calcium uptake by plants. Those two factors may contribute significantly to the calcium disturbances commonly seen in crop plants.

Nowadays, foliar fertilization is widely adopted in agriculture to complement soil fertilization. Supplemental doses of macro- and micro-nutrients, plant hormones, stimulants, and other beneficial substances could be supplied via the foliar route. Such application methods are often adjusted over time to meet the nutrient requirement at each stage of growth or fruiting of a plant. The formulation of fertilizer is also adjusted accordingly time after time. Another significant benefit of this approach is the simultaneous use of sprinkler irrigation systems that are now very common in large-scale farming. Furthermore, the integration of nanotechnology in fertilizer industries has recently given birth to nano fertilizers. Since nano fertilizers or nanoparticle size fertilizers gain much attention due to different benefits relating to both nutrient values and environmental impact, such nano fertilizers could easily dissolve nutrients in water considering their high surface area. So, they could be perfectly adapted to the foliar application.

Literature study reveals that some commercialized products of nano calcite fertilizer can be converted from natural calcite through a mechanical activation process. Some authors also tried to interpret the acting mechanism of nano calcite with the plant cells. Nano calcite allows CO_2 to be generated quickly with high concentration, helping promote rubisco activation reaction, and CO_2 fixation photosynthesis [1–3]. Such availability of CO_2 makes photosynthesis increase significantly. These nanoparticles enter plant cells through the opening of the stomata because the particles of calcite nanoparticles are much smaller than the stomata. Inside the leaf, this seed breaks down into CO_2 and CaO, which are immediately available to plants. In addition, $Ca(OH)_2$ can be generated because CaO combines with water to have an appropriate pH, balance cytoplasm, and enhance metabolism.

In this study, we aim to investigate the comparative effects of calcite fertilizers (fine and ultra-fine) on the growth of plants, especially in the case of Vietnamese rice. The quality of nano calcite after the fine-crushed process was first analyzed. Next, experimentally evaluate the effect on rice growth in the laboratory and the field, monitor and determine the appropriate level of using foliar spray on crops and the soil substrate and draw conclusions about the advantages and limitations of using ultra-fine-crushed lime (Fig. 52.1).

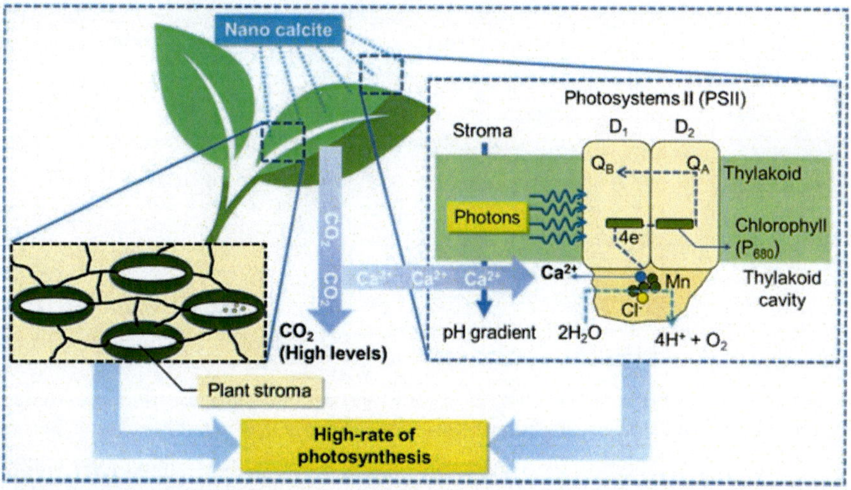

Fig. 52.1 Mechanism of action when using ultra-fine-crushed lime as a high-rate photosynthesis foliar fertilizer [4, 5]

52.2 Methodology

52.2.1 Materials

By ultra-fine-crush technology, nano calcite was supplied from Saigon Nanomat Co., Ltd in Vietnam. Nano calcite level of the product according to the purity is 98%. A high-speed beads mill was employed to prepare nano-size calcite particles from raw lime, and polycarboxylate superplasticizers PCE was used to disperse the ground particles. As the main result of the fine-crush process or mechanical activation, the particle size distribution of calcite was measured by photon correlation spectroscopy from dynamic light scattering (DLS) of the aqueous dispersions (using The HORIBA SZ-100 Nanoparticle Size and Zeta Potential Analyzer) and by scanning electron microscopy (SEM) of the dried particles (using Hitachi S-4800 FE-SEM). Besides, Brunauer–Emmett–Teller (BET) surface area analysis (Quantachrome NOVA 2200E BET Surface Area Analyzer) was used to check and control the stability across fine-crushed batches.

After evaluating the quality of nano calcite after fine-crushed process, experiments using nano calcite were designed to compare with using raw lime without crushing and without using foliar fertilizer. This study selected Vietnam long grain white rice 5451 (OM 5451) to apply the nano calcite fertilizer. The rice seeds were cultivated in trays (65 × 45 × 21 cm) containing suitable nutrient sources for their growth. Also, rice seeds were sowed on 100% clean sand to control the supplied nutrient for evaluating the effect of fertilizer.

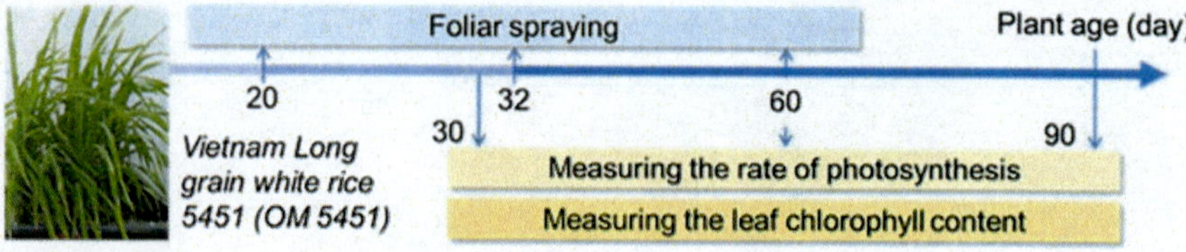

Fig. 52.2 Schematic diagram to describe different stages of foliar spraying

52.2.2 Design of Experiment Setup

Figure 52.2 summarizes the timeline of the experiment process that provides a schematic diagram to describe the different stages of foliar spraying. Note that all the plants were exposed to the same lighting and temperature in laboratory conditions. The exact concentration (100 ppm) of nano calcite and raw lime was applied to compare the reference group, which used water only. Foliar spraying with nano calcite fertilizers was applied for the first time at 20-day-old and continued at 32-day-old and 60-day-old plants. According to this strategy, the experiment measured different traits, including the number of leaves, leaf length, and leaf width at different periods to compare. Also, the rate of photosynthesis and the contents of chlorophyll were measured. Dual-wavelength (660 and 940 nm) was used for optical absorbance measurements from leaf samples to calculate the relative chlorophyll content value in the sample (values in the range of 0–2000 units). Generally, photosynthesis is the formation of carbohydrates from CO_2 and a source of hydrogen in the chlorophyll-containing tissues of plants exposed to light. Therefore, in this study, the photosynthetic rate was determined by the "CI-340 Handheld Photosynthesis System" by measuring CO_2 before and after it enters the leaf chamber to calculate the rate of CO_2 assimilation by a known leaf area.

52.3 Result and Discussions

52.3.1 Characteristic of Ultra-Fine-Crushed Lime

Figure 52.3a shows the mean diameter of fine-crushed lime particles after 45 min grinding with the rotor speed of 1200 rpm. The average particle size of ground calcium carbonate was around 52 nm. It can be seen that particles about 50 nm in size account for a large proportion of the size distribution. Figure 52.3b shows SEM micrographs taken from representative samples of fine-crushed calcite, which were image-processed to evaluate the length-average particle size. Similar result as using DLS, the size of the particles from SEM image-analysis ranged from 106 to 12 nm, with a mean value of 50 nm and focused on the range of 30–60 nm. In addition, although deposition and agglomeration still occur through the not very high

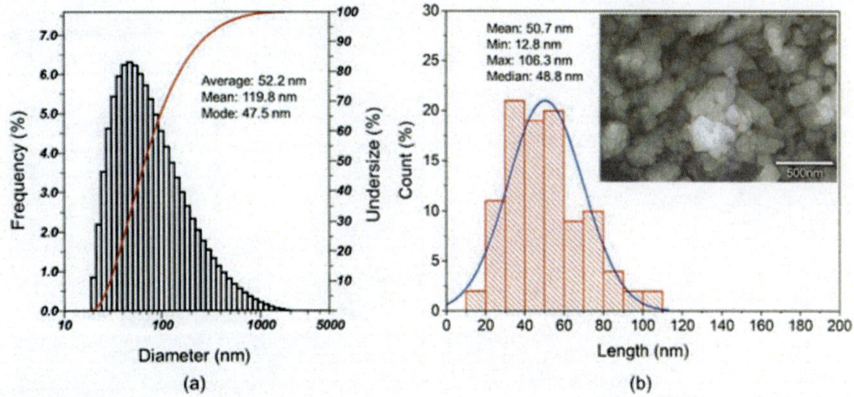

Fig. 52.3 Laser scattering particle size distribution (**a**) and length distribution by SEM image-analysis (**b**)

specific surface area (about 16 m^2/g), the influence level is almost insignificant when most of the particles are still relatively discrete when observed on SEM images (Fig. 52.4). Although remaining polymer films covered the particles, the border between particles was sharp and could still be observed. As a consequence, some "cluster-like" submicrometric aggregates can be formed. Therefore, nano calcite was separated from the bottom sediment of the container before use, using only the liquid above to overcome the deposition and agglomeration. Simultaneously, the stirring process was also applied to re-separate the agglomerated particles.

52.3.2 *The Use of Ultra-Fine-Crushed Lime Foliar Fertilizer for Rice Growth*

Figure 52.5a shows that increase in plant height is observed in each treatment. Figure 52.5b–d show the physiology of leaves. After 90 days, the height of plants sprayed with nano calcite reached the highest level. This group also shows a stable and steady growth rate in each period. Leaf characteristics such as the number of leaves, leaf length, and width gradually increased in all three groups. The group of plants using nano calcite showed the highest results with a relatively stable trend. Foliar application of nano calcite had an increasing effect on leaves number. However, the difference between using nano calcite and the other two groups is insignificant. This result can partly be explained by good nutritional conditions maintained throughout, resulting in a lack of harshness to highlight the effects of calcite nano foliar fertilizers. However, the collected experimental values also show the positive influence of providing CO_2 and the calcium-rich environment from nano calcite on the growth and development of plants, especially leaves.

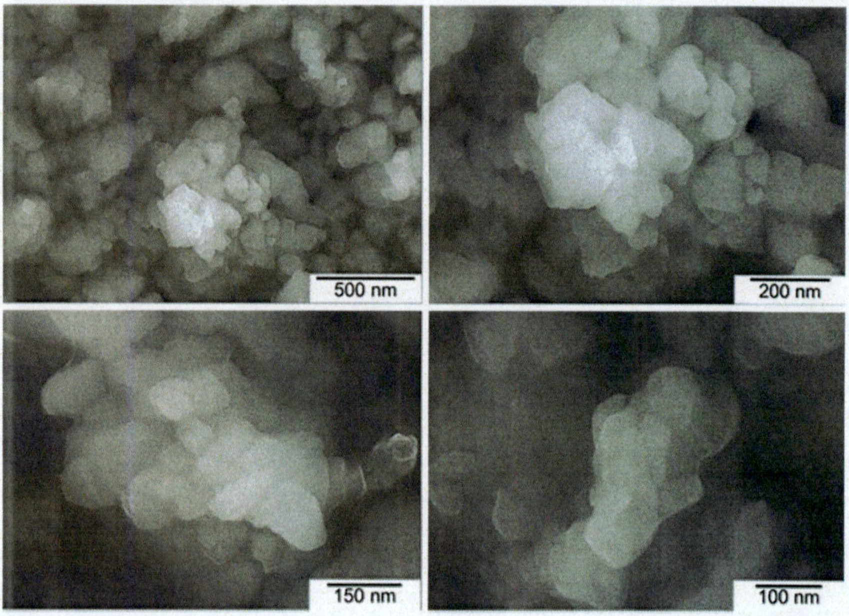

Fig. 52.4 SEM images of fine-crushed calcite products with well-dispersed nano-size particles and some "cluster-like" submicrometric aggregates

It can be seen that the photosynthetic rate in the group using nano calcite is significantly higher than that of the other two groups. Although there was a decrease at the 60-day mark, the photosynthetic rate rapidly increased significantly to reach the peak during the period of flowering and preparation for harvest (Fig. 52.6a). This positive signal shows the effectiveness of using nano calcite as foliar fertilizer. Two reasons can explain the recorded period of decline at the 60-day mark. The first reason is that photoinhibition interferes with photosynthesis. The high degree of photosynthetic stimulation after the 30-day mark (the first nano calcite spray) strongly promotes light absorption, metabolism, and leaf cell activation but also produces the most inhibition later. Compared with the group using nano calcite, the decrease in photosynthetic intensity at the 60-day mark of the regular lime group and the reference group was insignificant. The second cause is the change in the chlorophyll content in the leaves, the most critical factor determining the rate of photosynthesis. Several factors govern chlorophyll contents and the rate of photosynthesis. The increase in chlorophyll content is directly proportional to the rate of photosynthesis (Fig. 52.6b). However, at a certain point, the increase in chlorophyll content no more contributes to photosynthesis due to other environmental limitations. In general, the photosynthetic tendency of plants in the absence of foliar application is stable and grows steadily according to the natural growth cycle. In contrast, the sudden increase and decrease with mutations in nutrient composition and CO_2 concentration on the leaf surface when lime spraying, especially nano calcite, clearly show a clear impact of

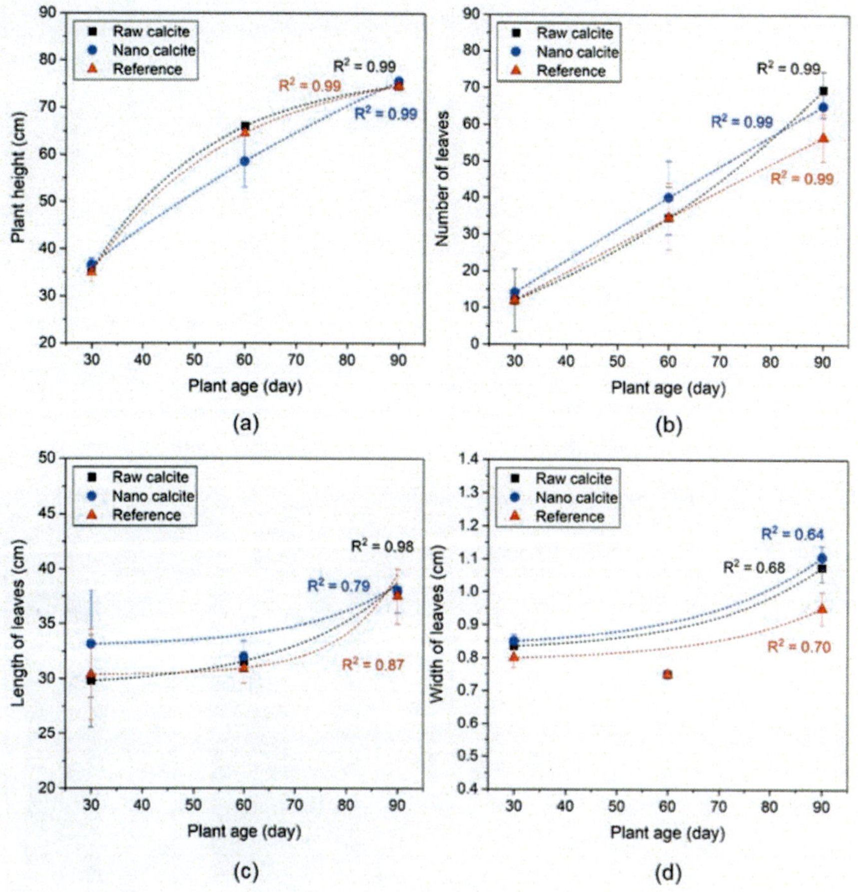

Fig. 52.5 Age-dependent plant physiology compared among different groups of specimens

this fertilizer. Besides, an increase in the photosynthesis process due to the use of nano calcite fertilizers is associated with higher levels of CO_2. Therefore, photosynthetic materials are introduced into different plant sites, including the stems, and the plant height increases.

Nano calcite is related to CO_2 storage and photosynthesis as an accelerating factor for increasing the rate of photosynthesis. Also, it can shorten the vegetative growth period and help the plant quickly enter the reproductive or flowering stages. After harvesting, the rice yield is depicted in Fig. 52.7. Although only tested on a small area, the increase in rice yield when using nano calcite foliar spray also affected plant growth and development. Furthermore, the experiment needs to be expanded to a larger area and to survey plants of different densities in the field to have a more comprehensive view of the effect of the spraying process and activation mechanism.

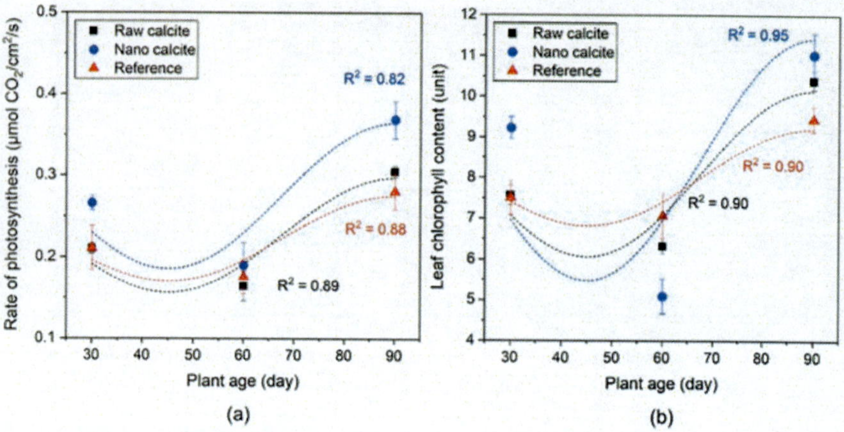

Fig. 52.6 Rate of photosynthesis and leaf chlorophyll content

Fig. 52.7 Comparing results of rice yield after harvesting

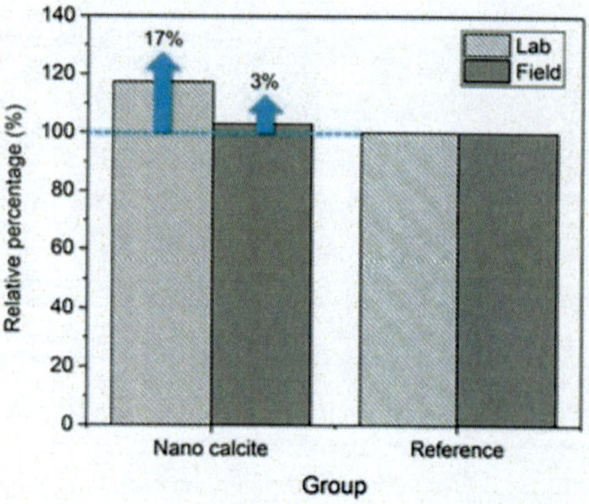

52.4 Conclusion

After a theoretical overview of plant nutrition and experimental studies on the effects of using fine-crush calcite in the super-speed mill as a foliar spray on the growth and development of the plant, using a simple, cost-effective, and industrially scalable process for preparing nano calcite, ultra-fine-crush technology was demonstrated to quickly reduce raw lime material into nano-size (around 50 nm). For the finely ground lime material prepared in the super-speed ball mill at SANOMAT Co. Ltd. , the analysis results also show that the surface area is equivalent to the fine particles of nanometer and micrometer size, and the pH is approximately neutral. These nano calcites were used as leaf fertilizers for photosynthesis enhancement by cycle-spraying in a specific case of rice plants. Experimental results on the characteristics of leaves and the rate of photosynthesis and chlorophyll content demonstrated the considerable effect of using nano calcite spraying fertilizer. More studies are required to investigate the role of nano calcite fertilizer in plant physiological processes and their effect on the activity of enzymes during the growth period. Soon, large-scale testing with environmental factors and controlled conditions will be necessary before applying this promising technology.

Acknowledgements We acknowledge Ho Chi Minh City University of Technology (HCMUT), VNU-HCM, for supporting this study.

References

1. G. Schneider, Y. Lindqvist, C.-I. Branden, RUBISCO: structure and mechanism. Annu. Rev. Biophys. Biomol. Struct. **21**, 119–143 (1992)
2. R.E. Blankenship, *Molecular Mechanisms of Photosynthesis* (Wiley, 2021)
3. J.F. Allen, Photosynthesis of ATP—electrons, proton pumps, rotors, and poise. Cell **110**, 273–276 (2002)
4. C.W. Hoganson, G.T. Babcock, A metalloradical mechanism for the generation of oxygen from water in photosynthesis. Science **277**, 1953–1956 (1997)
5. H. Lodish, A. Berk, C.A. Kaiser, C. Kaiser, M. Krieger, M.P. Scott, A. Bretscher, H. Ploegh, P. Matsudaira, *Molecular Cell Biology* (Macmillan, 2008)

Chapter 53
Effect of Adding Waste Shredded Rubber Tyre on the Shear Strength Properties of Clay Soil

Norbaya Sidek, A. S. A. Rahman, W. W. A. Zailani, Mohd Fadzil Arshad, and N. Abdul Latif

Abstract Soil stabilization is an effective strategy for improving soil mechanical properties such as shear strength. This can be accomplished by incorporating suitable admixtures into the soil for instance waste products like fly ash. The cost of these additions has grown dramatically in recent years and is now prohibitively expensive. These have led to the development of substitutes for additives such as polymers as soil stabilizers. The difficulties of society, such as reducing waste and creating usable resources from non-useful waste materials, may be met by utilizing this novel approach of soil stabilization and innovative technology. Waste shredded rubber tyre is expanding daily, causing several environmental difficulties. The disposal of rubber tyre trash without producing environmental issues has become a significant concern. As a result, because there is a scarcity of high-quality soil for building projects, using shredded rubber tyre waste as a soil stabilizer is a cost-effective solution. Thus, this study looks at the viability of employing shredded rubber tyre waste for soil stabilization. To assess the physical qualities of soil, many laboratory tests were performed, including sieve analysis, moisture content, particle density, hydrometer analysis, cone penetration test, Atterberg limit, and Standard Proctor Compaction Test. The Unconfined Compression Test (UCT) will be performed to establish whether the usage of shredded rubber tyre waste as a soil stabilizer can replace the pricey additives employed by industry. Various prior study findings were compared to ascertain engineering features. The optimal proportion of shredded rubber tyre will also be established by comparing engineering parameter comparison findings from prior experiments conducted by previous researchers.

Keywords Waste shredded rubber tyre · Unconfined Compression Test (UCT) · Soil stabilization

---

N. Sidek · A. S. A. Rahman (✉) · W. W. A. Zailani · M. F. Arshad · N. Abdul Latif
School of Civil Engineering, Universiti Teknologi MARA, 40450 Shah Alam, Selangor, Malaysia
e-mail: abdulsamad@uitm.edu.my

N. Abdul Latif
e-mail: abdulsamad@uitm.edu.my

M. A. A. Mohd Salleh et al. (eds.), *Proceedings of the Green Materials and Electronic Packaging Interconnect Technology Symposium*, Springer Proceedings in Physics 289,
https://doi.org/10.1007/978-981-19-9267-4_53

53.1 Introduction

Soil is an important component of all sorts of buildings. These infrastructures transfer their weights to the foundation and then transport them to the land and underground soil. Problematic soils result from structural soils or a change in environmental conditions. Over 75% of Malaysia's land area is dominated by residual soil aged from granite, sedimentary, and meta-sedimentary formation rocks. The rest portions are made up of alluvium and clay discovered at river valleys and coasts [1]. The clay dispersion in Malaysia is primarily concentrated in North Peninsular Malaysia [2]. Clay soils have a higher water retention rate, are more cohesive, and contain relatively little air. Malaysia's soils are categorized as tropic soils. Clay is easily shaped into something like a shape that they keep when dried, but when heated, they harden and lose their flexibility [3]. Geotechnical engineers differentiate silts and clay depending on the plasticity characteristics of the soil as evaluated by the soil.

Malaysians generate an average of 30,000 tonnes of rubbish daily, with just 5% reusing it [4]. One of them is scrap or discarded tyres. One of the greatest ways to reduce waste tyres is to redesign them into a variety of sizes of elastic particles and then use them in various initiatives. Rubber tyres are not biodegradable and take up a significant landfill area. Furthermore, improperly disposed of rubber tyres contribute greatly to fire dangers. It is advantageous to treat poor soil's physical and chemical characteristics using additional materials in the stabilization process. The stabilized soft soils can be enhanced by a variety of chemical additions. Portland cement reacts with water to generate a hard mass because it is a hydraulic binding agent. Therefore, in this research, more focus will be on soil modification by using cement. Soil–cement reactions such as hydration and pozzolanic reactions improved the engineering qualities of soil by creating primary and secondary cementitious materials [5].

Tyre chips have become more prevalent, but environmental concerns remain, especially the possible impact on ground and surface waters and aquatic life due to leaching. However, strategies for dealing with this issue have been revised. Reference [6] said tyres deposited in groundwater produce harmful leachate to aquatic life; however, if there is a sufficient separation zone from the water bodies, the harmful effects might indeed lessen through the time the leachate takes to reach the surface water. Reference [7] found that tyre shreds did not influence the concentration of metals in drinking water that met basic (health-based) requirements.

According to the findings of many studies, the use of solid waste has a largely beneficial influence on the environment owing to the decrease of pollutants and residues in landfills, as well as an improvement in the lifetime and durability of structures [8]. The use of shredded rubber tyres in geotechnical engineering to improve soil qualities has recently garnered much attention. The study's purpose was to use waste material for soil stabilization to prevent environmental pollution.

Based on the research on enhancing clay shear strength, adding 2%, rubber fibres were advised for maximum strength [9]. It was discovered that including shredded rubber tyres enhanced the material's apparent cohesiveness while decreasing the friction angle [10]. The advantages of utilizing shredded tyres as a soil stabilizer are that they are cost-effective, lessen environmental concerns caused by tyre waste disposal, and reduce waste amounts by creating usable materials from non-useful waste materials. Since high-quality clay soil is scarce for construction projects, using shredded tyres as a soil stabilizer is a cost-effective solution.

53.2 Materials and Methods

53.2.1 Soil and Additive

This research was carried out at Kampung Jawa, Klang. Klang has long been renowned as an area with abundance amount of clay. A total of 25 kg of clay soil was gathered from the research site. Since each test necessitated a specific instrument, the soil was sent to the laboratory for testing. The containers holding the disturbed soil are sealed to prevent moisture loss or gain. For physical testing, 2 kg of clay soil was weighed and separated. The balance of clay soil was oven dried for 24 h at 105 °C before being broken into smaller pieces for engineering testing.

While, for rubber shreds as shown in Fig. 53.1, are formed from discarded tyres using tyre cutting equipment. These cutting devices may slice the tyre into two halves and remove the tyre's sidewalls from the tread. Smaller tyre shreds and tyre chips typically dimension 13 mm up to 76 mm, requiring primary and secondary shredding as the manufacturing stage to accomplish a reduction in size.

Fig. 53.1 Shredded rubber tyre waste

53.3 Methodology

The research for this study was conducted utilizing quantitative methodologies. This research is based on the findings of laboratory testing and earlier investigations. The studies reported in this research aim to establish the physical parameters of clay soil. Aside from that, this study intends to compare past research results to evaluate engineering qualities such as soil shear strength and the usefulness of waste shredded rubber tyres as a soil stabilizer. Finally, this study seeks to determine the optimal amount of waste shredded rubber tyre as a soil stabilizer for enhancing soil qualities.

Many laboratory tests were performed on the clay soil, including sieve, hydrometer, moisture content, particle density, Atterberg limit, and Standard Proctor Compaction Test to determine the clay soil's physical characteristics, shear strength, optimum moisture content (OMC), and maximum dry density (MDD) were all measured. The results of the Unconfined Compression Test (UCT) were contrasted to the original soil condition and condition of the soil that contained shredded rubber tyre to recognize the efficiency of waste shredded rubber tyre as a soil stabilizer. The optimum amount of waste shredded rubber tyre was observed by differentiating engineering properties and results data from prior research.

The shredded tyre and control sample of clay soil were prepared which use a 1.2 kg of every three average samples. The 0, 5, and 10% of shredded tyre were added, and 2 and 4% of cement were included as the binding agent. The test samples had optimum moisture content and their unit dry weight during preparation in the mould was considered as the maximum dry density obtained by compaction was 18% by weight to the soil. Three identical cylindrical mould of diameter 50 mm with height 100 mm were used to place the soil and compacted. After an instantaneous weight recording, samples were stored and cured for 7 days before being subjected to testing in a room with controlled humidity and a temperature of 25 °C. This was done to ensure that the diffusion of water was complete. A strain was then applied at the rate of 1 mm/min by using compressive strength measurement machine for testing by gradually increasing the loads until the samples ruptured. The methods were repeated for two additional results, and the sample's average maximum corrected deviator stress and shear strength value was generated.

53.4 Result and Discussions

The results that were obtained for the physical properties of the clay soil in Kampung Jawa indicate that the soil is classified as a well-graded sandy-clay soil. According to the findings of the wet sieving process, 65.99% of the residual soil was found to be retained on the 0.063 mm sieve, while 34.10% was found to be fine soil that passed through the 0.063 mm sieve. The average moisture content in the clay soil of Kampung Jawa is 50.25%. For this study, the liquid limit (LL), the plastic limit (PL),

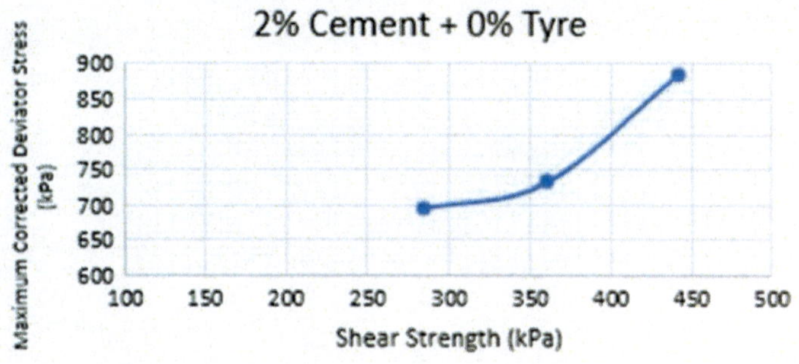

Fig. 53.2 Unconfined compression strength of 2% cement and 0% tyre

and the plasticity index (PI) of the clay soil in Kampung Jawa were calculated to be 51.58%, 21.11%, and 30.47%, respectively. The clay soil is categorized as clay of high plasticity based on the plasticity chart. The average particle density of the clay soil obtained for this research is 2.46.

The shear strength is the highest resistance of a soil to shear forces that was prior to failure. When the soil is compressed, shear forces are generated. When the applied compressive loads cause shear stresses that exceed the soil's shear strength, a soil mass will fail in shear. Figure 53.2 presents a magnified image of failure of fractures and their granular rubber particles. It was shown that the presence of rubber near the fracture increased the failure strain and minimized the loss of strength after the peak [9].

It can be observed from Table 53.1 that stress of the soil is the lowest as it is unreinforced specimen which is the maximum UCS is 388 kPa and shear strength is 194 kPa. Cemented clay specimens have a higher peak axial stress than control samples. Cementation reaction may be the cause for the substantial rise in peak axial stress. All the data collected based on the Table 53.1 is extracted and plotted. The findings are explained through each of the plotted graph.

According to Fig. 53.3, the highest corrected deviator stress, which indicates the Unconfined Compressive Strength (UCS), is 884 kPa among three samples, whereas the shear strength is 442 kPa. The greatest UCS is 1084 kPa, while the shear strength is 542 kPa, as shown in Fig. 53.4. Based on Fig. 53.5, the highest value of UCS is 966 kPa, and the shear strength is 483 kPa. The average UCS for 2% cement clay soil with shredded tyre contents of 0%, 5%, and 10% is 770.67 kPa, 903.67 kPa, and 905.67 kPa, respectively. These contents correspond to shear strengths of 362.00 kPa, 451.67 kPa, and 454.33 kPa, respectively.

At 4% cement based on graph in Fig. 53.6, the maximum Unconfined Compressive Strength (UCS) is 1556 kPa, while the shear strength is 778 kPa. Then, Fig. 53.7 displays that the highest UCS is 1646 kPa and the shear strength is 823 kPa. As for Fig. 53.8 represents the greatest value of UCS is 1231 kPa and the shear strength is 615 kPa. The average UCS for 4% cement clay soil with shredded tyre contents of 0%, 5%, and 10% are 950.00 kPa, 1075.67 kPa, and 1128.00 kPa, respectively. The

Table 53.1 Result of unconfined compressive strength of clay soil

Sample	Maximum shear strength (kPa)	Maximum corrected deviator stress (kPa)
Control sample	194	388
2% cement + 0% tyre	442	884
2% cement + 5% tyre	542	1084
2% cement + 10% tyre	483	966
4% cement + 0% tyre	778	1556
4% cement + 5% tyre	823	1646
4% cement + 10% tyre	615	1231

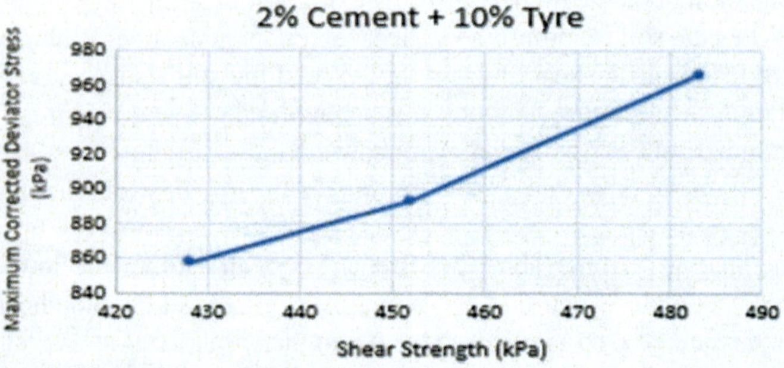

Fig. 53.3 Unconfined compression strength of 2% cement and 5% tyre

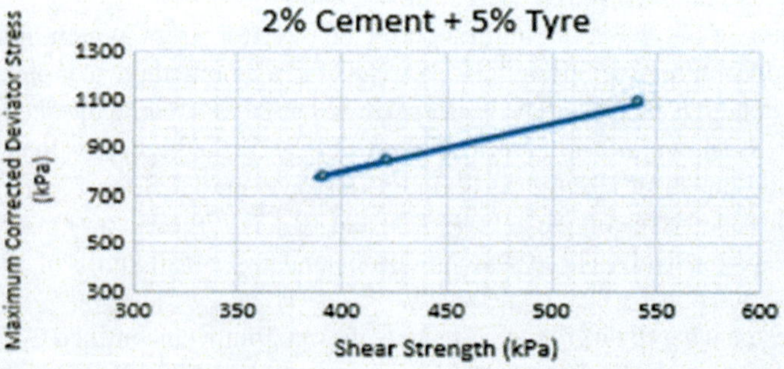

Fig. 53.4 Unconfined compression strength of 2% cement and 10% tyre

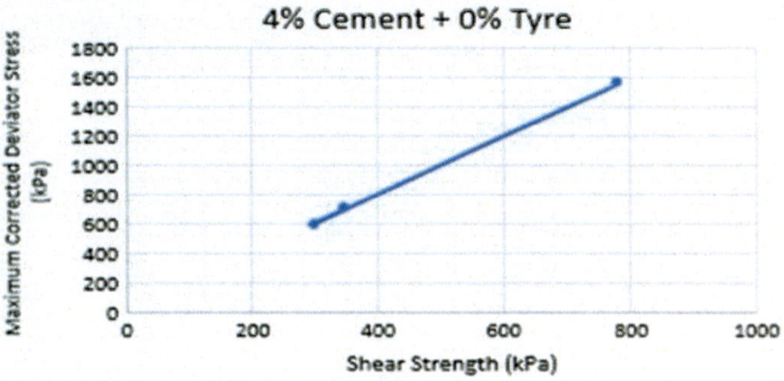

Fig. 53.5 Unconfined compression strength of 4% cement and 0% tyre

corresponding shear strength with these contents are 474.67 kPa, 538.00 kPa, and 564.00 kPa, respectively.

In Fig. 53.8, it can be observed that the UCS of stabilized clayey soil by 2% cement increases with increasing shredded tyre content. However, the strength only slightly increased after 5% of the shredded tyre. The unconfined compressive strength

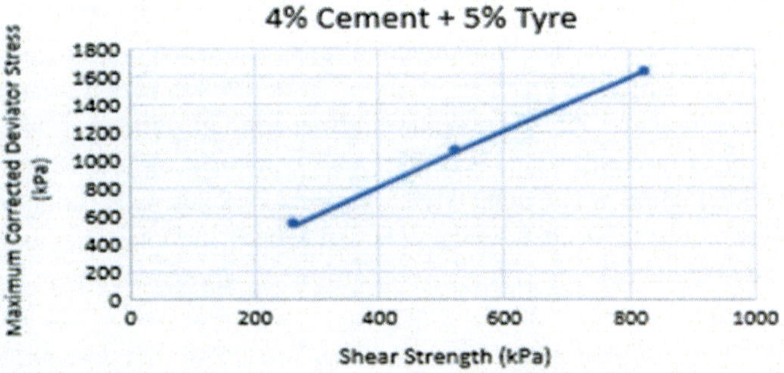

Fig. 53.6 Unconfined compression strength of 4% cement and 5% tyre

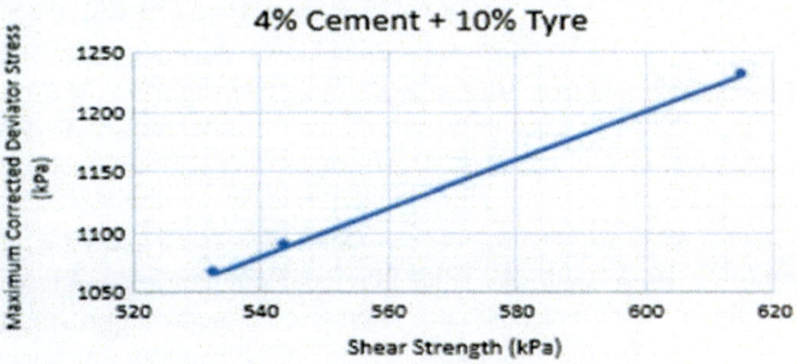

Fig. 53.7 Unconfined compression strength of 4% cement and 10% tyre

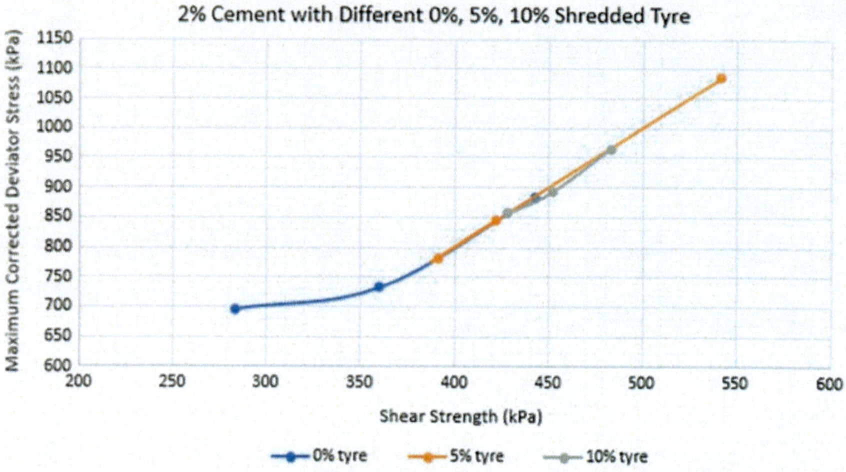

Fig. 53.8 Unconfined compression strength of 2% cement with 0, 5, and 10% tyre

for 2% cement has risen from 782 to 1084 kPa at 5% tyre. The frictional resistance produced by the shredded tyre was primarily responsible for the increase in strength. When clay particles and shredded tyres interacted, significant frictional resistance was mobilized. However, when the content of shredded tyres increased, this interaction lessened, hence preventing further strength gains.

The amount of shredded tyre was high over 5% content, such as at 10% content. As a result, the interaction of shredded tyre maximized while the interaction of clay and shredded tyre reduced. The increased interaction between clay particles and shredded tyre resulted in an increase in clay stiffness. However, at greater content levels, such an effect was minimized because of the presence of more tyre, which reduced the interaction between soil and tyre crumbles [11]. Reference [12] discovered that the optimal tyre chips proportion is 5% by weight.

The UCS-shear strength of clay containing 4% cement with different percentage of shredded tyre is shown in Fig. 53.9. It can be inferred from Fig. 53.9 that UCS of clay specimen increases marginally from1556 kPa to 1646 kPa with the addition of 5% shredded tyre. When 4% cement was added to the soil, it stopped being soft and showed its maximum strength, while adding 5% tyre chips made the shear strength parameters much better. According to Fig. 53.9, the difference reduction of UCS between 5% tyre and 10% tyre content may be due to increase in interaction between the shredded tyres.

From previous research, the drop in peak axial stress at 7.5 and 10% rubber fibre concentration may be a result of rubber fibre accumulation [13]. The addition of rubber fibres slows the rate of post-peak strength loss, hence improving the clay's ductility [14]. A review of the literature on the effects of increasing rubber content on the strength of materials indicates a rise in some situations and a decrease in others. The lower friction and weaker adhesion between clay and rubber contribute to the decreased strength of the mixture [9].

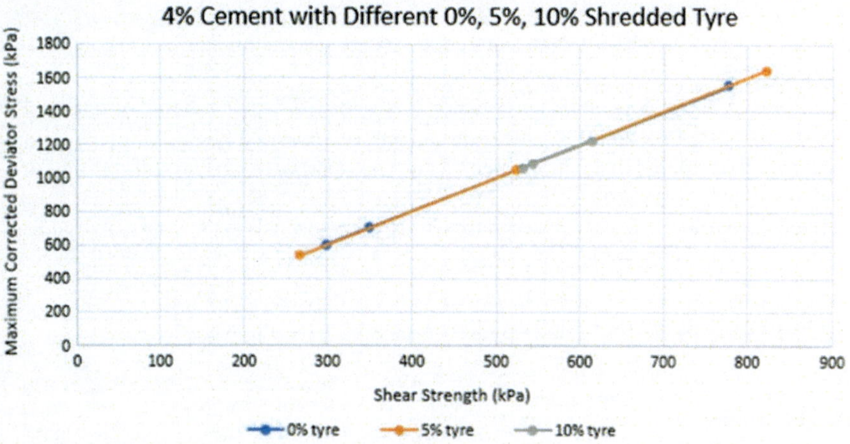

Fig. 53.9 Unconfined compression strength of 4% cement with 0, 5, and 10% tyre

According to prior research by [15], Unconfined Compression Test results of 2% cement shows more lower value than the Unconfined Compression Test results of 4% cement. Despite that, the value of UCS of 5% tyre added remains as the maximum compared to 0 and 10% tyre. Nevertheless, the results of research by [16] also stated that inclusion of the optimum tyre is at 5% which has been proved to increase the UCS of the soil. Rubber's flexibility and elastic reactivity during compaction inhibit crack development [14], which accounts for the rise in mixture shear strain [17]. After the peak strain, the elastic compression of rubber particles causes strain hardening [18].

53.5 Conclusion and Recommendations

For this research, the analysis of maximum shear strength of clay soil was evaluated at the optimum percentage of 5% shredded tyre. Even though a higher shredded tyre amount which is 10% was carried on, it does not improve the shear strength of the clay soil but decreasing the strength. Shredded tyre and cement help to promote strength and can effectively increase the properties and behaviour of clay soil. The optimal proportion of shredded tyres is 5%, according to the results of prior research. The outcomes also demonstrate that clay soil's engineering properties were enhanced by adding 5% shredded tyres and cement to the soil. The outcomes also demonstrate that shredded tyres work well as soil stabilizers and could take the place of costly additives currently used by industry. The findings also prove that shredded tyres work well as soil stabilizers and could alternate with costly additives currently used by industry. The recommendation for this research is to advance the investigation into clay soil stabilization in Kampung Jawa. Further laboratory data must be collected to bring a precise conclusion. There should be more research done on the use of shredded tyres to stabilize soil. As a result, using shredded tyres as a soil stabilizer is a cost-effective use that could eventually replace the industry's use of pricey additives. The number of laboratory tests that can be performed should vary to create a range of results that can be compared based on the ideal proportion of shredded tyres. All these criteria need to be considered to get more precise and trustworthy results on soil stabilization.

Acknowledgements The authors would like to extend our gratitude to the School of Civil Engineering, College of Engineering, Universiti Teknologi MARA for providing necessary facilities for this research.

References

1. A.S.A. Rahman, M.J.M. Noor, I.B.M. Jais, N. Sidek, J. Ahmad, in *Shear Strength of Granitic Residual Soil in Saturated and Unsaturated Conditions*. AIP Conference Proceedings, 2020. https://doi.org/10.1063/1.5062629
2. M.R. Umor, A Hussin, N. Muda, The physical properties and geochemical of clay from the Bestari Jaya, Kuala Selangor, Selangor, Malaysia for potential usage. Eur. J. Eng. Res. Sci. **5**(10), 1231–1236. https://doi.org/10.24018/ejers.2020.5.10.2182
3. J.A. Odubiyi, *The Properties of Clay* (2010)
4. V. Munikanan, M. Yahya, M.A. Yusof, M.H.F. Radzi, in *Fine Granular of Shredded Waste Tyre for Road Kerb Application as Improvised Road Furniture*. AIP Conference Proceedings, 1930. https://doi.org/10.1063/1.5022955
5. W.J. Al-Jabban, *Soil Modification by Adding Small Amounts of Binders: A Laboratory Study* (2019)
6. H. Hazarika, K. Yasuhara, in *Scrap Tire Derived Geomaterials: Opportunities and Challenges*. Proceedings of the International Workshop on Scrap Tire Derived Geomaterials—Opportunities and Challenges (IW-TDGM 2007), 23–24 Mar 2007, Yokosuka, Japan (Taylor & Francis, 2008)
7. D.N. Humphrey, L.E. Katz, in *Field Study of the Water Quality Effects of Tire Shreds Placed Below the Water Table*. Proceedings of the International Conference on Beneficial Use of Recycled Materials in Transportation Applications, Arlington, VA, pp. 699–708
8. D.S. Vijayan, D. Parthiban, in *Effect of Solid Waste Based Stabilizing Material for Strengthening of Expansive Soil—A Review*. Environmental Technology and Innovation, vol. 20 (Elsevier B.V., 2020). https://doi.org/10.1016/j.eti.2020.101108
9. D. Akbarimehr, A. Eslami, E. Aflaki, Geotechnical behaviour of clay soil mixed with rubber waste. J. Clean. Prod. 271. https://doi.org/10.1016/j.jclepro.2020.122632
10. B. Tiwari, M. Asce, B. Ajmera, S.M. Asce, S. Moubayed, A. Lemmon, K. Styler, J. Guerrero Martinez, *Improving Geotechnical Behavior of Clayey Soils with Shredded Rubber Tires—Preliminary Study*
11. A. Priyadarshee, A. Kumar, D. Gupta, P. Pushkarna, Compaction and strength behavior of tire crumbles-fly ash mixed with clay. J. Mater. Civ. Eng. **30**(4), 04018033 (2018). https://doi.org/10.1061/(asce)mt.1943-5533.0002171
12. S.M. Anvari, I. Shooshpasha, S.S. Kutanaei, Effect of granulated rubber on shear strength of fine-grained sand. J. Rock Mech. Geotech. Eng. **9**(5), 936–944 (2017). https://doi.org/10.1016/j.jrmge.2017.03.008
13. M. Tajdini, A. Nabizadeh, H. Taherkhani, H. Zartaj, Effect of added waste rubber on the properties and failure mode of kaolinite clay. Int. J. Civ. Eng. **15**(6), 949–958 (2016). https://doi.org/10.1007/s40999-016-0057-7
14. J.S. Yadav, S.K. Tiwari, Influence of crumb rubber on the geotechnical properties of clayey soil. Environ. Dev. Sustain. **20**(6), 2565–2586 (2018). https://doi.org/10.1007/s10668-017-0005-y

15. M.V.R. Reddy, I.S. Reddy, D.S.v. Prasad, Improvement of soil characteristics using shredded rubber. IOSR J. Mech. Civ. Eng. (IOSR-JMCE) **13**(2), 44–48 (2016). https://doi.org/10.9790/1684-1302014448
16. H.K. Gupta, M. Choubey (2021) Soil stabilization using waste shredded rubber tyre chips with addition of cement. www.ijcrt.org
17. Y. Mitarai, Y. Kikuchi, K. Yasuhara, J. Otani, T. Nagatome, The ductility of cement treated clay with added scrape tire chips and the change of its permeability under shear deformation. Doboku Gakkai Ronbunshuu C (2008). https://doi.org/10.2208/jscejc.64.181
18. Y. Kim, H. Kang, Engineering characteristics of rubber-added lightweight soil as a flowable backfill material. J. Mater. Civ. Eng. (2011). https://doi.org/10.1061/(ASCE)MT.1943-5533.0000307

Chapter 54
A Study of Physical and Mechanical Properties of Josephine and Yankee Pineapple Leaf Fibres for Potential Yarn Production

Nur Hanis Badaruddin, Nasaie Zainuddin, Najua Tulos, Noriza Arzain, Muhammad Hisyam Zakaria, and Asliza Aris

Abstract The purpose of this study is to compare the physical and mechanical properties of different species of pineapple leaf fibres using different testing methods including linear density, single yarn strength, colour comparison and colour analysis of the yarn. This study also examines the surface morphological structure using Scanning Electronic Machine (SEM) for different species. Josephine pineapple fibre located in Johor and Yankee pineapple leaf fibre located in Selangor have been used for the study of pineapple sepsis. The fibres have gone through several processes such as extraction, combing and spinning before testing and evaluation were conducted. The fibre extraction method uses Pineapple Leaf Fibre Extraction Machine PALF M1. The research is done after all findings of physical and mechanical properties of Josephine and Yankee have been collected to see the potential for the yarn production.

Keywords Stretchable electronics · Conductive ink · Cyclic loading · Morphology

54.1 Introduction

Recently, the concern of sustainable development has encouraged efforts towards increasing utilisation of raw resources and reducing waste production. In this context, reusing of the waste will lead to significant reduction of generated waste into textile material and product [1]. Apart from that, reused waste products in the development of textile manufacturing will lead to significant reduction of generated waste and hence decrease in environmental impact [2]. These outcomes and by-products manufactured will become one of the additional product sources for farmers and will help develop the diversification of the agricultural industry [3]. A large classification of plant fibre,

---

N. H. Badaruddin · N. Zainuddin (✉) · N. Tulos
Faculty of Applied Sciences, Universiti Teknologi Mara, 40100 Shah Alam, Malaysia
e-mail: nasaiezainuddin@uitm.edu.my

N. Arzain · M. H. Zakaria · A. Aris
College of Creative Arts, Universiti Teknologi Mara, 40100 Shah Alam, Malaysia

M. A. A. Mohd Salleh et al. (eds.), *Proceedings of the Green Materials and Electronic Packaging Interconnect Technology Symposium*, Springer Proceedings in Physics 289,
https://doi.org/10.1007/978-981-19-9267-4_54

pineapple leaf fibre, is one of the most popular natural fibres today but still lacks deeper studies. In most cases, natural fibre is cheaper than synthetic fibres and causes less hazard and health problems for workers compared to latex and glass fibre-based composites [4].

Josephine pineapple is one of the choice fruits all over the world because of its pleasant taste and flavour. Malaysia was ranked 18th in the list of pineapple world producers and twenty second as the world's pineapple exporter [5]. Pineapple leaves contain long fibre which can be used for different textile and non-textile applications. In one hectare of land, Josephine pineapples are cultivating around 30,000 pineapple plants [6]. The average weight of a single matured leaf may vary about 65–70 g which is of 30–60 cm long and 4 cm wide, consisting of fibre yield to a extent of 2–2.5% of each leaf, with Malaysia having potential of producing around 22,000 metric tonnes of pineapple leaves fibre (PALF) per year, which is worth around 114.2 million in the international market [7].

This looks for the opportunity to introduce Josephine pineapple leaf fibre compared with Yankee pineapple leaf fibre. Hence, the purpose of this study is to compare the properties of Josephine and Yankee pineapple leaf fibres by investigating the properties suitable for yarn production. The method of extraction of pineapple fibre is divided into two methods, which is conventionally or using more sophisticated machine methods. There are various methods of raking pineapple leaves all over the world [8]. In the Philippines, there are still some communities involved with the industry of raking pottery dishes that have been broken for the purpose of extraction [9]. The extraction of fibre using machine was done using a shape blade that located in the PALF M1 machine. The extraction needed to run through the machine is an one-time-only process. This is to prevent the outer layer of leaf being removed without harming the lignin layer of pineapple fibre [10].

Conventional machines (Fig. 54.1) have been used in the process of extracting pineapple leaf fibre and it is growing very rapidly around Malaysia, especially in Pontian Johor and the area around the interior of Pahang. Synthetic fibre made from organic source is attractive because it is cheaper compared to chemical fibre, and it can produce a large amount of fibres in a time. Moreover, nowadays synthetic organic materials attract the attention of many consumers due to their environment and cheaper prices.

Scanning Electron Microscope (SEM) photos were taken from two different methods of PALF extraction. Figure 54.2 shows the extraction using hand scraping and PALM M1. The result on hand scrapping shows unparallel fibre arrangement compared to parallel and finer delicate structure extracted using PALF M1. As conclusion, the fibres extracted from PALF M1 are better and parallel compared to hand scrapping, and this might be due to the uneven force that is being distributed during scrapping process which leads to fibre damage and breakage at certain point.

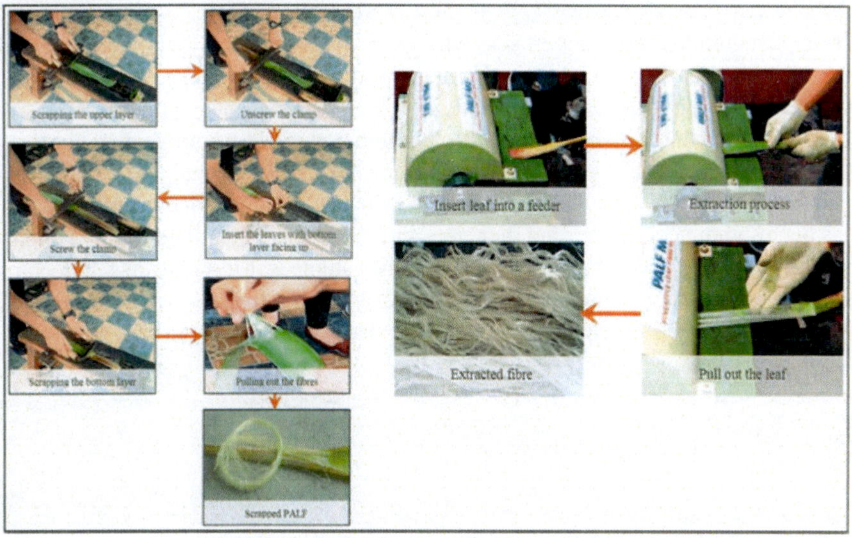

Fig. 54.1 The extraction process of pineapple leaf fibre using traditional and conventional methods

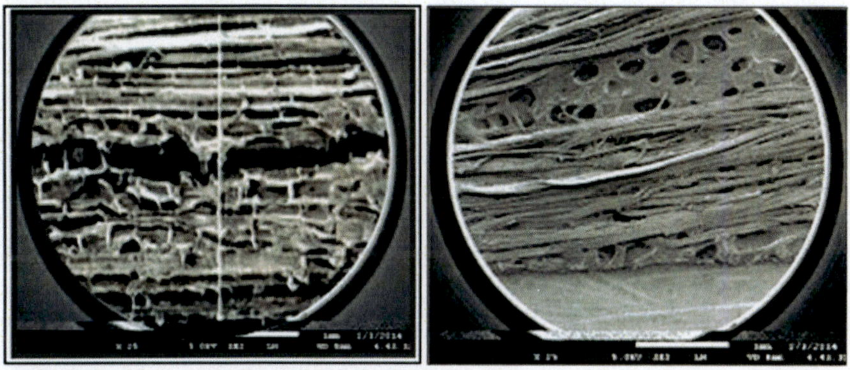

Fig. 54.2 SEM microscopic photomicrograph on PALF

54.2 Methodology

54.2.1 Materials and Method

The sample used in this research is the freshly cut Josephine and Yankee leaf fibres that have the maturity of about 9 months with approximately 60 cm long. The leaf was going through "*ketam*" and "*bangku*" conventional methods to extract the outer layer of fibre. The fibre when pulled out has got a fine hairy fibre strand and has then went through the drying process. After 24 h, the fibre was completely dry. The dried fibre must go into a spinning process where it combines several long fine yarns into spinning by twisting method. Then, a fine long thread was produced and was ready to be tested by a textile testing machine.

The method that has been used in this research is extraction of both pineapple leaf fibres from different species which are Josephine and Yankee pineapple leaves fibres. The fibres have gone through spinning process which is 2 ply yarn of every yarn that has been spinning. The fine yarn then goes into testing which initiates every sample. The estimation of total fibres used is 100 g. The sample was tested using single yarn strength, appearance, colour comparison and surface morphology. All the samples were recorded for data analysis. The research flowchart is shown in Fig. 54.3.

54.3 Result and Discussions

54.3.1 Results

After spinning process, yarn is measured and weighed using electronic balance weight to calculate the tex of both yarn (1 m/weight). Linear density and colour comparison testing methods are carried out on Josephine and Yankee pineapple leaf fibres to identify the yarn samples. The tex for Josephine pineapple leaf fibre is 0.3446, while Yankee pineapple leaf fibre is 0.3108. The Denier for Josephine pineapple leaf fibre is 3101.4, while Yankee pineapple leaf fibre is 2.8. The Josephine yarn's tex is higher compared to Yankee yarn. This shows that Josephine yarn was thicker and bulkier compared to Josephine yarn.

For Josephine pineapple leaf fibre, the motorised testing average force is 4942.6 lb at average 4.32% inches elongation. For Yankee pineapple leaf fibre, the motorised testing average force is 1337.6 lb on average 4.32% inches elongation.

The data collected show that Josephine has better strength compared to Yankee, and this is also supported because the tex of Josephine is higher, so that the yarn is thicker and can support more weight and have a better strength. If Yankee yarn is thinner, it cannot hold more strength. Colour comparison for Josephine and Yankee pineapple yarns is carried out under lighting box. Josephine yarn is light yellow in colour when compared to Yankee yarn that is browner. The appearance of Josephine yarn is smoother compared to Yankee yarn because Josephine is bulkier in size and appearance.

54.3.2 Yarn Preparation from Fibre (Fibre Selection)

The fibre was cleaned and then selected before combing process started. This is to make sure that the fibre was clean and easy for combing. Removing parts of fibre that

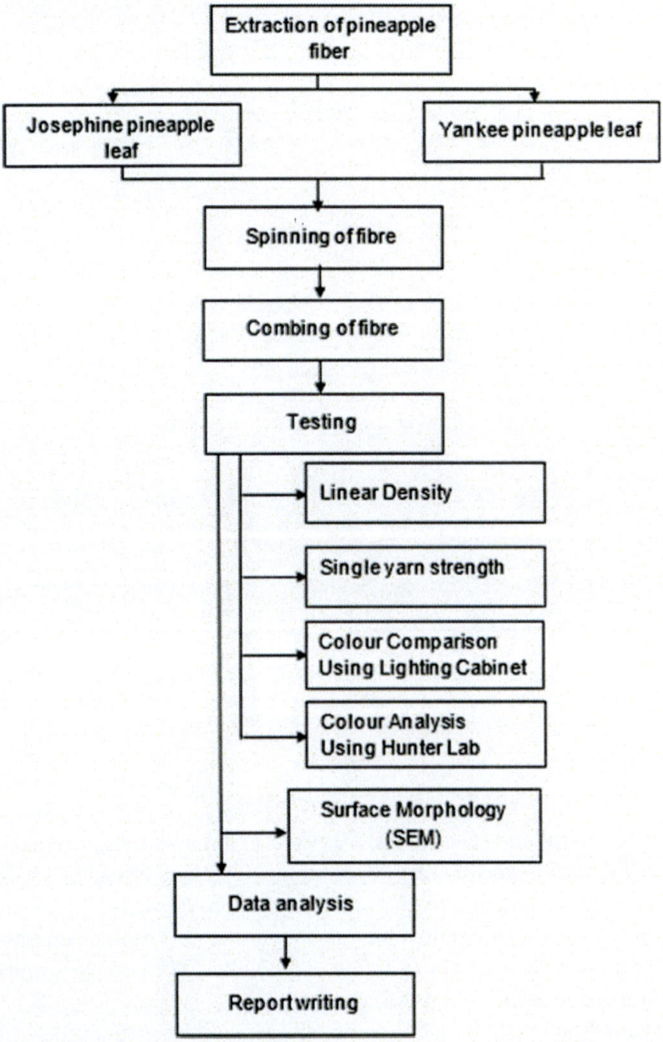

Fig. 54.3 Research flowchart

have sharp scales (Fig. 54.4) also helps to improve the surface of yarn production when spinning.

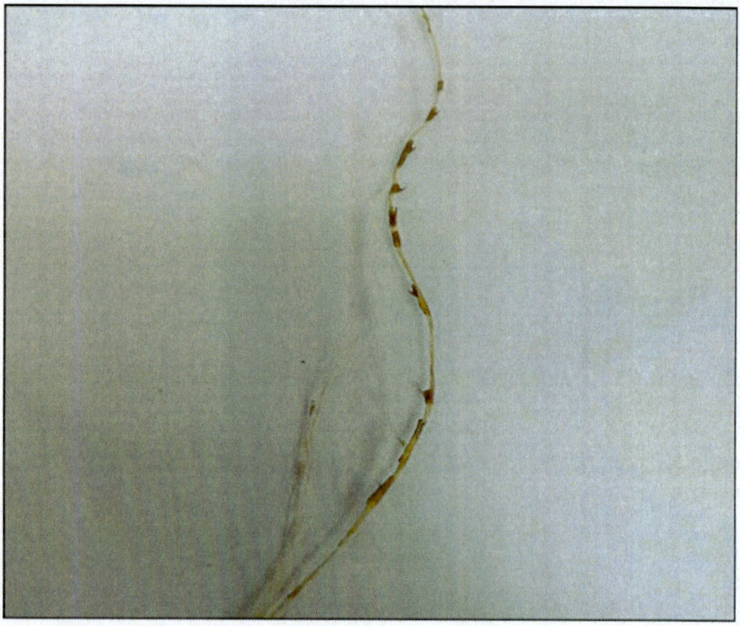

Fig. 54.4 Scale of pineapple

54.3.3 Combing

The fibre was cleaned and selected before combing process started. This is to make sure that the fibre was clean and easy for combing. Combing is a process which eliminates the short fibres, impurities, naps, immature fibres from the sliver, straightens the fibres and makes them parallel along the sliver length. On the combing process, the tools that have been used are G-clamp and comber. The comparison before and after combing can be seen in Figs. 54.5 and 54.6.

After the combing process, the fibre is straighter, and the visual appearance of the fibre was improved. Combing process also helps to increase the strength as only the long fibre is selected. Combing also helps to increase the degree of evenness and low hairiness, and remove immature and entangled fibre that can achieve high quality of yarn.

Fig. 54.5 Fibre before combing process

Fig. 54.6 Fibre after combing process

54.3.4 *Spinning of Fibre*

Spinning is a process of adapting twisting technique to form a fibre from a yarn. The fibre was drawn out from a mass and was then twisted together to form a continuous thread of yarn. The machine used in this spinning process is hand spinning machine (Fig. 54.7). The fibre is finally twisted in Z direction, and the speed was consistent throughout the process.

54.3.5 *Yarn Preparation*

After spinning process, the yarn is measured and weighed using electronic balance weight to calculate the tex of both yarn (1 m/weight).

54.3.6 *Linear Density, Count, Tex and Danier*

Table 54.1 shows the result of linear density, count, tex and danier of Josaphine and Yankee pineapples. The tex for Josephine pineapple leaf fibre is 0.3446, while Yankee pineapple leaf fibre is 0.3108. The denier for Josephine pineapple leaf fibre is 3101.4 while Yankee pineapple leaf fibre is 2.8. The Josephine yarn's tex is higher compared to Yankee yarn. This shows that Josephine yarn is thicker and bulkier compared to Josephine yarn.

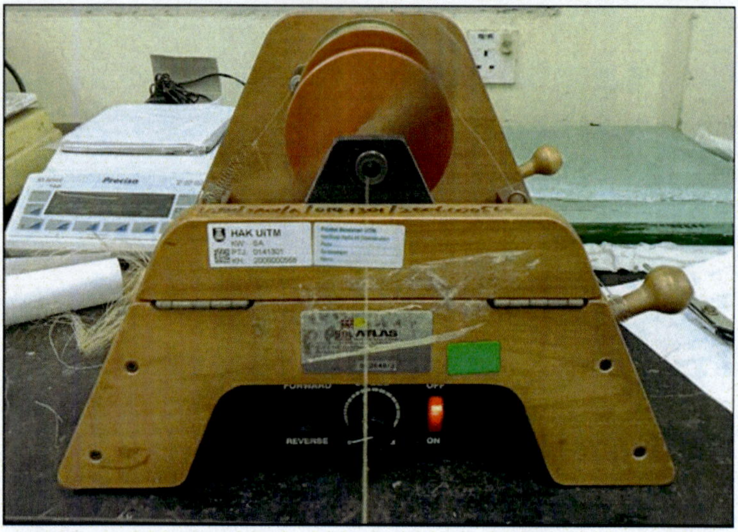

Fig. 54.7 Yarn hand spinning machine

Table 54.1 Linear density, count, tex and Danier

No.	Sample—Josephine
1	0.350
2	0.370
3	0.310
4	0.331
5	0.340
Average	0.3446 g
No.	Sample—Yankee
1	0.301
2	0.33
3	0.31
4	0.321
5	0.296
Average	0.3108 g

54.3.7 *Single Yarn Strength*

Tables 54.2, 54.3 and Fig. 54.8 shows the result of single yarn strength test for both pineapple leaves fibre.

Table 54.2 Single yarn strength for Josaphine Pineapple yarns

Sample (Josephine)	Motorised	
	Force (lb)	Elongation (inches)
1	4992	4.50
2	4187	3.90
3	5364	3.30
4	5943	4.50
5	4227	5.40
Average	4942.6	4.32

Table 54.3 Single yarn strength for Yankee Pineapple yarn

Sample (Yankee)	Motorised	
	Force (lb)	Elongation (inches)
1	2138	2.30
2	2294	3.20
3	2187	3.80
4	2559	2.70
5	2510	3.60
Average	2337.6	3.12

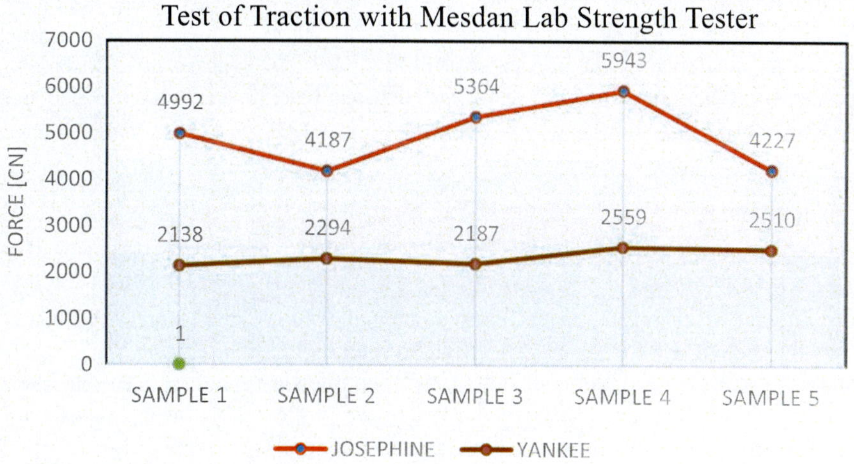

Fig. 54.8 Single yarn strength for Josaphine and Yankee Pineapple yarns

Fig. 54.9 Pineapple yarn under lighting box

54.3.8 Colour Comparison Using Lighting Cabinet

Colour comparison for Josephine and Yankee Pineapple yarns was carried out under lighting box. The yellow light colour indicates the colour of Josephine yarn, while brown colour indicates that of Yankee as in Fig. 54.9. The appearance of Josephine yarn is smoother compared to Yankee yarn because Josephine is bulkier in size and appearance.

54.3.9 Colour Analysis Using Hunter Lab

Colour analysis of Josephine and Yankee Pineapple yarns using Hunter LAB test is shown in Fig. 54.10. Josephine yarn is light in colour when compared to Yankee yarn that is more accurate. The result shows that Josephine yarn is lighter in colour when compared to Yankee yarn. This explains that different species of pineapple fibre may result in different colour shades of fibre after extraction.

The result of colour comparison test is tabulated in Fig. 54.11. The result shows that the colour spectrum of both yarns is different.

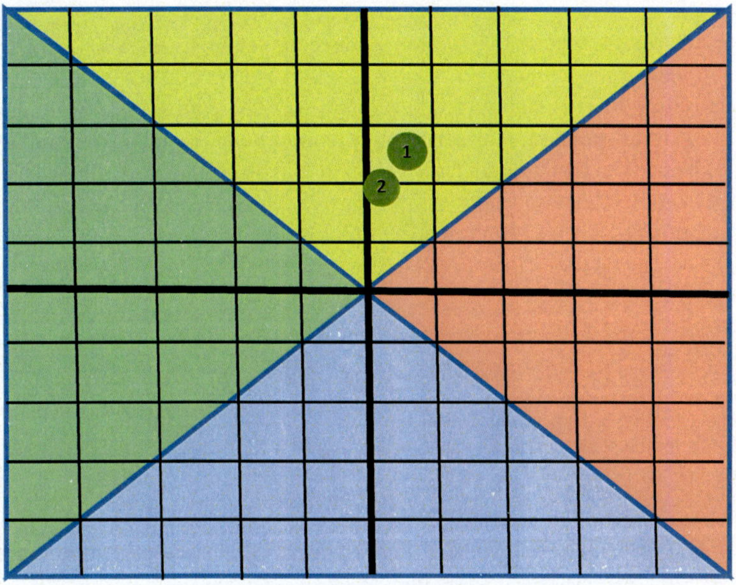

Fig. 54.10 2D colour plot-1

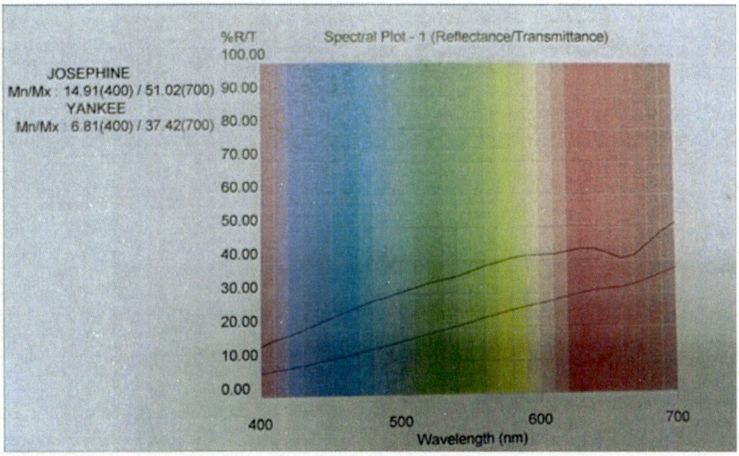

Fig. 54.11 Visible light spectrum

54.4 Conclusion

This study is helpful in enhancing the quality of moulded parts produced where the objective is to optimise warpage of the front panel housing moulded part that has been achieved. Based on the results, the warpage has been optimised by using

Response Surface Methodology (RSM) approach. The results also show that by using RSM, the significant mathematical model function can be obtained in order to predict warpage value with reasonable accuracy. From the ANOVA results, melt temperature is the most significant factor influencing the warpage condition on the moulded part, followed by coolant temperature and cooling time. The optimal processing parameter obtained from RSM has optimised warpage by 47.1% which is from 0.2650 mm from the simulation result to 0.1403 mm after optimisation.

References

1. J. Kang, C. Liu, S. Kim, H. Int. J. Consum. Stud. **37**(4), 442–452 (2013)
2. Information on Malaysia explores the market potential for Josapine Pineapple in the Middle East—TFNet—International Tropical Fruits Network (itfnet.org)
3. A.S. Rao, Asian J. Multidisp. Stud. **5**(9), 107–108 (2019)
4. S. Balakrishnan, G. Wickramasinghe, U.S. Wijayapala, Investigation on improving banana fiber fineness for textile application. Text. Res. J. **89**(21–22), 4398–4409 (2019)
5. N.A. Sukri, M.P. Saiman, K. Ishak, M.S. Wahab, N.F.A. Rashad, J. Adv. Manuf. Technol. **14**(2), 77–88 (2020)
6. M. Asim, K. Abdan, M. Jawaid, M. Nasir, Z. Dashtizadeh, M.R. Ishak, M.E. Hoque, Int. J. Polym. Sci. **2015**, 1–16 (2015)
7. R. Mahmud, A. Momin, R. Islaim, A.B. Siddique, A.N. Khan, Compos. Adv. Mater. **31**, 1–10 (2022)
8. R.S. de Queiroz, A.P.V. da Silva, A.C.L. Broega et al., Springer Nat. Appl. Sci. J. **72**(2), 1–12 (2020)
9. Y. Yusof, S.S. Yahya, A. Adam, Proc. CIRP. 756–760 (2015)
10. J. Jain, S. Jain, S. Sinha, J. Elastomers Plast **51**(3), 224–243 (2019)

Chapter 55
Performance of Energy Encryption for Medium Field Wireless Power Transfer System by Optimization Switching Frequency

Nur Hazwani Hussin, Muhammad Mokhzaini Azizan, Azuwa Ali, Norhidayu Rameli, Nur Hazirah Zaini, and Shahnurriman Abdul Rahman

Abstract Encryption is a very important technique used to protect energy transmission channels from unauthorized receiver. It can be utilized to transmit data securely over the wireless medium. Encryption techniques of wireless power transfer (WPT) are important in the research on the effects of security key for security of energy transfer to the authorized receiver. The energy encryption scheme of WPT is proposed to chaos theory. Chaos theory is applicable to the logistic map to propose as a security key to chaotically regulate the switching frequency. Furthermore, chaos theory characteristic affects power and distance performance. Therefore, this paper investigates mainly effective power based on mobile charging application. This research is focusing on the performance of energy encryption in medium field for wireless power transfer system by optimization switching frequency. This research is dedicated to the performance of energy encryption for medium field wireless power transfer system based on mobile charging application. The optimization to transport the power in this research based on power is 10 W. The research utilized MATLAB simulation to compare the performance.

Keywords Energy encryption · Optimization switching frequency · Wireless power transfer system · Medium field

N. H. Hussin (✉) · M. M. Azizan · N. Rameli · N. H. Zaini · S. A. Rahman
Faculty of Engineering and Built Environment, Universiti Sains Islam Malaysia, 71800 Nilai, Negeri Sembilan, Malaysia
e-mail: nurhazwanihussin@yahoo.com

A. Ali
Faculty of Electrical Engineering Technology, Universiti Malaysia Perlis, Kampus Tetap Pauh Putra, 02600 Arau, Perlis, Malaysia

M. A. A. Mohd Salleh et al. (eds.), *Proceedings of the Green Materials and Electronic Packaging Interconnect Technology Symposium*, Springer Proceedings in Physics 289,
https://doi.org/10.1007/978-981-19-9267-4_55

55.1 Introduction

WPT is one of the techniques for transferring electrical power from source transmitter to load a receiver without interconnected wire. WPT technology is holding good potential to switch the way people lead their lives by contributing new levels of convenience, mobility, and safety. Basically, there are three types of WPT which are near-field, medium field, and far field. WPT can be classified into inductive coupling, capacitive coupling, magnetic resonance, and electromagnetic radiation [1]. The advantage of the near field and medium field is safe and not being absorbed by the receiver. In near field, the principal method of distance is selected for short-range application and no radiation on the process to transfer power. Then, distance for medium field is selected in mid-range application and no radiation. Thus, for far field are being in extensive range of distance and it has radiation during process transfer power. So, it is not good for the health of the environment and surrounding people. WPT optimization criteria are needed for two uses, namely continuous uninterrupted power delivery and periodic charging. For continuous charging, whether under the stationary (EVs) or moving states (mobile charging), wireless communication should be fast, reliable, and energy efficiency [2]. The WPT system is increasingly attracting attention in various fields, such as charging portable electronic devices and implanting medical devices [3]. WPT is also suitable for electric vehicles (EVs) application, such as battery charging for normal vehicular operation and energy exchange [4–7].

Encryption in wireless communication channels is more vital and necessary in the present than in the past. Data need to be well-encrypted during transmission, so the data securely transfer over the wireless medium. Energy is expected to transfer to a specific receiver and switch off other unauthorized energy transmission channels. Thus, the security of energy transmission is really one of important issues in WPT system [8]. Few types of techniques are used in encrypting the energy such as a password system, radio frequency signal, and the new one is chaos theory. The newest technology and currently being explored is chaos theory technique. In order to encrypt energy of WPT, chaos theory unpredictable behavior manipulated to generate unique security key by reducing the complexity of equation. The presence of chaos is supported by calculation of the Lyapunov exponents [9]. Therefore, this paper will focus more on chaos theory technique for encrypting the energy mainly concentrate on medium field WPT system. In 1901, Nikola Tesla started constructing his famous Wharncliffe Tower near Long Island [10]. The tower was used to broadcast sound by employing wireless communication and transmitting power without utilizing conducting wires [10]. Tesla's work was impressive when it's given the lack of radio wave technologies at that time, the experiment was unsuccessful even though it attempted to demonstrate its feasibility. In the late twentieth century, the near-field inductive power transfer would become a concern when cordless charging of consumer devices gained popularity [11]. The aim is to seek ways for effectively transmitting power from a source to a device using the principle of electromagnetic induction, such as the operation of a transformer on the inductive power transfer. The system is non-radiative as it does not rely on propagation of electromagnetic waves [12]. In the inductive power transfer applications of a few kilowatts (kW), such as charging of electric vehicles (EVs) and mobile phone, 90% of transmission efficiency can be achieved by increasing the operating frequency and more than 70% of efficiency can be reached for low power, such as the maximum 5 W mobile phone charging [13]. Moreover, the operating frequency range of the inductively coupled technique is generally from 20 kHz to several MHz [13]. However, when efficiency is achieved, coil distances further enlarge or have more freedom in positioning the source and load relative to each other. Then, to solve the problem, the Massachusetts Institute of Technology (MIT) research group examined many techniques for transmitting power over medium-range distances and developed a non-radiative resonance coupled scheme to enhance transmission efficiency [12, 13]. Compared with electromagnetic radiation, resonant coupling has advantages, such as higher efficiency in omnidirectional transmission and insensitivity to the surrounding environment [14]. The operating frequency range from 10 kHz to approximately 200 MHz has been used in several studies on resonant coupled WPT [14]. However, electromagnetic radiation can be classified into unidirectional and omnidirectional radiation based on energy transmitting direction [15]. The far-field approaches balance between directionality and transmission efficiency [15]. Radio frequency (RF) and microwave systems are examples that can benefit from the transfer power over several kilometers at 90% efficiency by using high-gain antennas [15]. In summary, the WPT system can be divided into three concepts according to technology, transmission, and applications. These three concepts are near-field WPT through electrical and magnetic induction, medium field WPT through coupled resonant circuits, and far-field WPT through microwave on RF rectifier circuits.

55.2 Methodology

55.2.1 Energy Encryption for WPT System

According to design the energy encryption scheme of WPT system, Fig. 55.1 shows the block diagram for the system configuration. The system is divided into several part which is transmitter and receiver part. By the specific value of power source, the system is power up. In conventional WPT system, the operation is as depicted previously. Basically, in magnetic resonant categorizes has two coil system are added with resonant coil for medium field application. In long-distance range, the resonant coil used to transfer power on the WPT system. Thus, in order of chaos theory technique is implemented to add security key in WPT system. Generally, mathematically algorithm will be embedded in the WPT system by using chaos theory technique.

Based on Fig. 55.1, the main part is transmitter part and receiver part. In transmitter part, it is embedded from power source to primary coil. The discrete time chaotic values are generated by using logistic map algorithm in chaos theory technique. Otherwise, in logistic map it has the Lyapunov exponent to generate security key

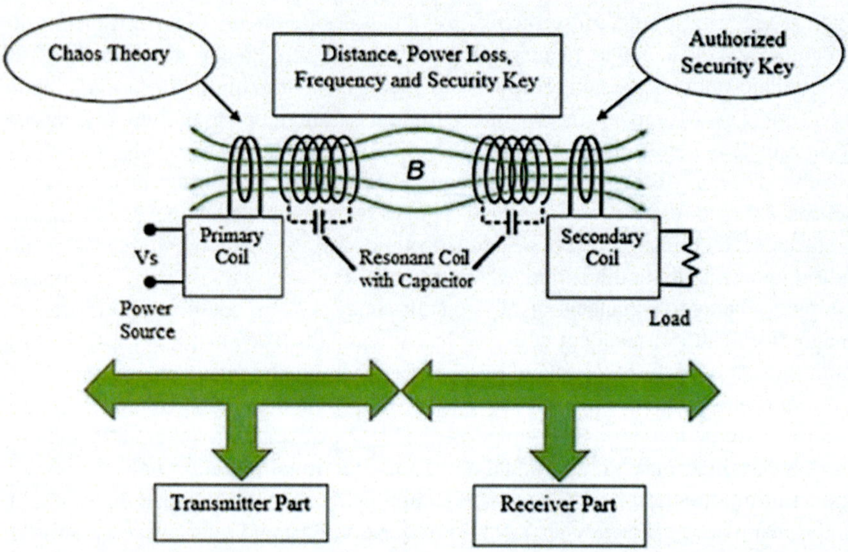

Fig. 55.1 Block diagram for energy encryption of medium field WPT systems

which is 0 and 1 for secure the system. For initialize logistic map by using Eq. 55.1:

$$\xi i + 1 = (1 - \xi i), \quad A \in [0, 4], \tag{55.1}$$

where ξi indicates the sequence and A means the bifurcation parameter. The phase portraits of ξi and $\xi i + 1$ display different topological structures alongside with the expansion in A. In addition, ξi goes about as a consistent value for $A \in [0, 1)$, a period 1 oscillation for $A \in [1, 3)$, a period n oscillation for $A \in [3, 3.57)$, and a chaotic oscillation for $A \in [3.57, 4]$. Likewise, inward the logistic map it has the largest Lyapunov exponent as a mathematical expression of the chaotic behavior. Hence, the biggest Lyapunov exponent gets to be distinctly positive when $A > 3.57$ it is a result of in chaotic oscillation period, and at same time, the chaotic behavior happens if $A = 3.9$ [11]. In this way, $A = 3.9$ is chosen to create the irregular limited security key $\xi i \in (0, 1)$ for the energy encryption scheme.

Notwithstanding, the resonant circuit is included at both sides which is resonant for primary and secondary. The resonant circuit is interconnected with primary and secondary coils for transfer power in long transmission distance with no affecting on the radiation although for changing the frequency resonant in magnetic resonant coupling is troublesome. Besides, the working frequency can control the transfer power exhibitions. The distance exhibitions are incorporate the viable power, switching frequency and security key [16–18]. Firstly, the distance to transfer power relies on the application that can be utilized.

In this project, mobile charging application is chosen. Subsequently, all the detail of specification is as indicated by mobile charging application. Along these lines, the

Table 55.1 Parameters for energy encryption in medium field of WPT systems by optimization switching frequency

Parameters	Values
Optimal switching frequency	100 kHz
Optimization switching frequency	102 kHz
Power	10 W
Transmission distance level	4 cm

power chosen is 10 W with ideal optimal switching frequency of 100 kHz. At that point, for distance level it varies from 3 to 5 cm whereby to perceive which estimation of distance has higher execution. The parameter depends on Qi-standard [13–15]. Table 55.1 demonstrates that general parameters utilized for energy encryption in medium field WPT system.

All things considered, for the switching frequency can be directed by utilizing algorithm for using the greatest power transfer is made. So that, the transmitter coil, resonant coil, and receiver coil resonant at a similar frequency. The tolerant different frequency around ± 5%. On the off chance that the frequency resonant at different frequency, the system neglect to work. Fundamentally, this process will work synchronize with security key.

$$\omega = \delta i \omega o \tag{55.2}$$

where

ω Switching frequency
ωo Optimal switching frequency

$$\delta i = a + (b - a), \quad 0 < a < b \tag{55.3}$$

where

δi Chaotic security key
a Transmission distance
b Power level

Moreover, the security key likewise has a calculation in Eq. 55.3, which have related to each other with switching frequency in Eq. 55.2. In Eq. 55.3, the values a and b are the parameter to be utilized for power transmission and distance to transfer power of WPT system. These parameters will be controlled to recognize and at same time to locate the best performance with the mix parameter. For the security key likewise, it must have matching security key from transmitter to receiver. It is to avoid from the stolen in power transfer process. The matching process of switching frequency and security key is the point at which it holds up a request from the receiver. As that process, if switching frequency and security key is matching synchronize at both parts, it is demonstrated that the power transfer to authorized receiver. As it were, the power is productively transferred to right receiver when the ideal switching frequency happens. Aside from same resonance frequency, security key at both transmitter and receiver should have the same value all together for the best performance. The value of security is either 0 or 1 are purpose in energy encryption for WPT system.

55.2.2 Optimization Procedure of Energy Encryption for WPT System

Figure 55.2 shows that flowchart of the proposed optimization procedure. Optimization switching frequency is determined by design constraints based on mobile charging application. Design constraints to optimize by size receiver, distance, and diameter of transmitter coil.

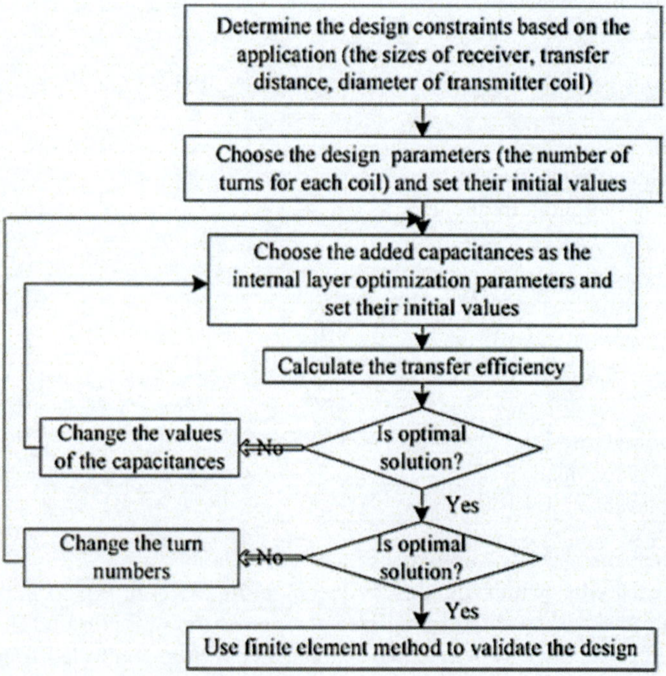

Fig. 55.2 Flowchart of the proposed optimization procedure

55.3 Result and Discussions

This section discusses the findings from the simulation. The computer simulation is carried out by using MATLAB programming to evaluate the security and performance of the resonant magnetic suggested coupling WPT system. Table 55.2 presents the results of energy encryption in medium field WPT system on the mobile charging application system. This paper varies the possible taking power based on mobile charging application as reference for value of power. So, it will conclude that, the best power level that can be transmitted are used and identify at which power level has optimal performance. Therefore, through simulation it proposed a comparison of the result. The minimum and maximum power for the simulation are taken 10 W to realize the best execution in the system. Table 55.2 demonstrates the information of energy encryption of medium field WPT system for fixed value of power with optimization switching frequency. In 10 W power, it can be seen that, from section selector 0 until 7, the security key value at transmitter increase, while the logistic map still inconsistently.

In addition, switching frequency data raise from the section selector 0 until 7. From the usual statistics in Table 55.2, the best execution synchronizes switching frequency and security key is section selector 0. The best security key at the transmitter to the receiver is 1.02 (transmitter) and 0.86 (receiver) with 102 kHz optimization switching frequency. Table 55.2 demonstrates the results for optimal switching frequency with optimization switching frequency. It can be realized that the matching switching frequency with optimal switching frequency is accurate power 10 W when optimization switching frequency at 102 kHz of 100 kHz. Thus, the matching process succeeds when the switching frequency and securely key occur simultaneously. In addition, at the receiver part of the security key value is 0.86 and maintain constant for 0 until 7 section selectors. Then, at the transmitter part of the security key value is almost closed to the receiver which at 1.02 for 10 W. Based on the result from

Table 55.2 Energy encryption of medium field WPT system result by optimization switching frequency

Section selector, *yi*	Power (W)	Distance (cm)	Optimal switching frequency, *f* (kHz)	Optimization switching frequency, *f* (kHz)	Security key, Di transmitter receiver		Logistic map, Xn1
0	10	4	100	102	1.02	0.86	0.3539
1	10	4	100	102	2.45	0.86	0.7333
2	10	4	100	102	3.71	0.86	0.9311
3	10	4	100	102	4.64	0.86	0.9942
4	10	4	100	102	5.18	0.86	0.9990
5	10	4	100	102	6.66	0.86	0.8910
6	10	4	100	102	7.53	0.86	0.7454
7	10	4	100	102	9.90	0.86	0.0388

simulation that was conducted, it can be concluded that when increasing the value of section selector, the matching process for switching frequency and security key is automatically fluctuated based on the complexity in the system. The effective power is 10 W and distance 4 cm in energy encryption for medium field of wireless power transfer system.

Figure 55.3 tabulates the comparison load voltage of authorized and unauthorized receiver. The load voltage is 5 V for authorized receiver while for unauthorized receiver is around 0.15 V. Furthermore, the authorized receiver differentiates depending on value of load voltage for transferring the power. Based on the simulation finding above, the best performance in powering the mobile charging application system is 10 W. It is because the overall specification of 10 W is fulfilled. The specification that has been investigated and proved earlier is made based on the three characteristics of chaos theory which are switching frequency, security key, and the logistic map. All the characteristics are important to realize the energy encryption of medium field for wireless power transfer system and in order to achieve the main objective of this research which is to protect the system from the unauthorized receiver.

Acknowledgements This research work was funded by the Ministry of Higher Education (MOHE) Malaysia under grant Fundamental Research Grant Scheme (Grant No.: FRGS/1/2020/ICT09/UNIMAP/02/3) titled "Chaotic behavior based Secured Encryption Algorithm for Near Field Wireless Power Transfer."

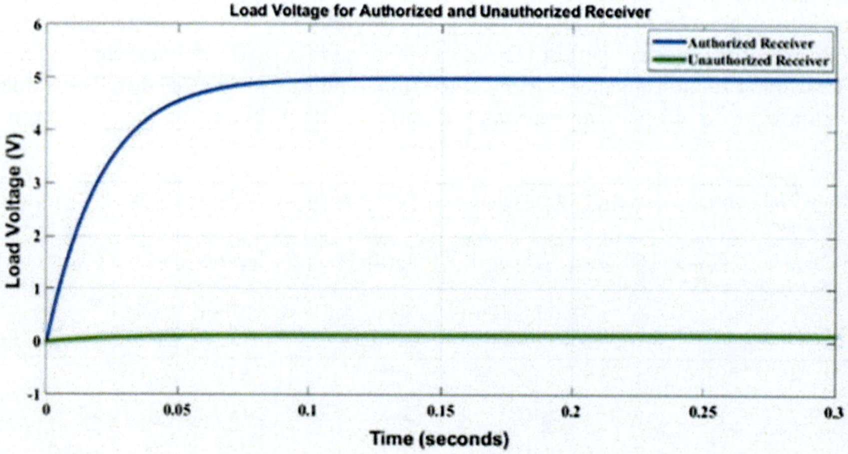

Fig. 55.3 Comparison load voltage of authorized and unauthorized receiver

References

1. X. Mou, H. Sun, in *Wireless Power Transfer: Survey and Roadmap.* IEEE Vehicular Technolology Conference, vol. 2015, pp. 1–13, 2015
2. J. Hirai, K. Tae-Woong, A. Kawamura, Wireless transmission of power and information and information for cableless linear motor drive. IEEE Trans. Power Electr. **15**, 21–27 (2000)
3. O.H. Stielau, G.a Covic, in *Design of Loosely Coupled Inductive Power Transfer Systems.* International Conference on Power System Technology Proceedings, vol. 1, pp. 85–90, 2000
4. H.F. Liew, A.R. Rosemizi, M.Z. Aihsan, I. Muzamir, I. Baharuddin, Wind characterization by three blade savonius wind turbine using IoT. IOP Conf. Ser. Mater. Sci. Eng. **932** (2020). https://doi.org/10.1088/1757-899X/932/1/012080
5. W. Chwei-Sen, O.H. Stielau, G.A. Covic, Design considerations for a contactless electric vehicle battery charger. IEEE Trans. Ind. Electron. **52**, 1308–1314 (2005)
6. A. Woojin, J. Sungkwan, L. Wonkyum, K. Sangsik, P. Junseok, S. Jaegue, K. Hongseok, K. Kyoungchoul, in *Design of Coupled Resonators for Wireless Power Transfer to Mobile Devices Using Magnetic Field Shaping.* IEEE International Symposium on Electromagnatic Compatibility, pp. 772–776, 2012
7. J.I. Agbinya, *Wireless Power Transfer*, vol. 45 (River Publishers, 2015)
8. R.A. Bercich, D.R. Duffy, P.P. Irazoqui, Far-field RF powering of implantable devices: safety considerations. IEEE Trans. Biomed. Eng. **60**, 2107–2112 (2013)
9. T.P. Duong, J.-W. Lee, Experimental results of high-efficiency resonant coupling wireless power transfer using a variable coupling method. IEEE Microwave Wirel. Compon. Lett. **21**, 442–444 (2011)
10. A. Waser, *Nikola TESLA's Wireless Systems* (English Publishers, 2000), pp. 1–14
11. J.C. Schuder, Powering an artificial heart: birth of the inductively coupled-radio frequency system in 1960. Artif. Organs Electr. Power 909–915 (2002)
12. A. Karalis, J.D. Joannopoulos, M. Soljacic, Efficient wireless non-radiative mid-range energy transfer. Ann. Phys. **323**, 34–48 (2008)
13. S.L. Ho, J. Wang, W.N. Fu, M. Sun, A comparative study between novel witricity and traditional inductive magnetic coupling in wireless charging. IEEE Trans. Magn. **47**, 1522–1525 (2011)
14. W.C. Brown, The history of power transmission by radio waves. IEEE Trans. Microw. Theory Tech. **32**, 1230–1242 (1984)
15. J.O. McSpadden, J.C. Mankins, Space solar power programs and microwave wireless power transmission technology. IEEE Microwave Mag. **3**, 46–57 (2002)
16. M.Z. Aihsan, R.B. Ali, J.H. Leong, in ***Design and Implementation of Single-Phase Modified*** *SHEPWM Unipolar Inverter*. 2015 IEEE Conference on Energy Conversion, CENCON 2015, pp. 337–342, 2015
17. M.A.S.B. Bimazlim et al., in *Comparative Study of Optimization Algorithms for SHEPWM Five-Phase Multilevel Inverter*. 2020 IEEE International Conference on Power and Energy (PECon), Penang, Malaysia, 2020, pp. 95–100. https://doi.org/10.1109/PECon48942.2020.9314491
18. M.S.M.A. Walter et al., Selective harmonic elimination pulse width modulation for five-phase cascaded multilevel inverter using non-notch and notch switching technique. J. Adv. Res. Dyn. Control Syst. **12**(7), 1196–1207 (2020). https://doi.org/10.5373/JARDCS/V12SP7/20202220.f

Chapter 56
Setting Time of Treated Sludge Containing Blended Binder

Nurshamimie Muhammad Fauzi, Mohd Fadzil Arshad, Ramadhansyah Putra Jaya, Mazidah Mukri, Sajjad Ali Mangi, and Warid Wazien Ahmad Zailani

Abstract Disposal of sludge from water treatment facilities poses a significant difficulty for the environment and landfills management. Leachate generated by landfill disposal leads to soil and groundwater contamination. Numerous research was published to treat the sludge; however, limited study on the setting time characteristics of sludge after being treated with a blended binder was found. This paper presents the setting time of water treatment sludge in a semi-solid form treated with different types of a blended binder. This paper also examines the effect of various binder types on the consistency index to determine the amount of water needed in the water binder content. Industrial waste materials used in this study are fly ash (FA), waste paper sludge ash (WPSA), and palm oil fuel ash (POFA). Industrial waste was used as replacement materials for ordinary portland cement (OPC) at a ratio of 50:50, 60:40, and 60:40 of OPC: FA, OPC: WPSA, and OPC: POFA blended binders with and without sludge as a comparison. The water/binder ratio of mix proportion without sludge was based on a standard consistency test. In comparison, the sample with sludge is blended binders mixed with WTS only at a mixed proportion of binder: WTS of 1.1. All treated sludge samples were then used to determine their set time. The results show that WTS treated with OPC: FA and OPC: POFA has a longer setting time than OPC. Meanwhile, WTS treated with OPC: WPSA provides a 65% faster setting time than WTS treated with OPC.

N. Muhammad Fauzi · M. F. Arshad (✉) · M. Mukri · W. W. A. Zailani
School of Civil Engineering, College of Engineering, Engineering Complex, Tunku Abdul Halim Muadzam Shah, Universiti Teknologi MARA, 40450 Shah Alam, Selangor, Malaysia
e-mail: mohdfadzil.arshad@uitm.edu.my

N. Muhammad Fauzi
e-mail: 2015185749@student.uitm.edu.my

R. P. Jaya
Department of Geotechnics and Infrastructure, Faculty of Civil Engineering and Earth Resources, Universiti Malaysia Pahang, 23600 Pahang, Gambang, Malaysia

S. A. Mangi
Department of Civil Engineering, Mehran University of Engineering Technology, SZAB Campus, Khaipur Mir's, Sindh, Pakistan

M. A. A. Mohd Salleh et al. (eds.), *Proceedings of the Green Materials and Electronic Packaging Interconnect Technology Symposium*, Springer Proceedings in Physics 289,
https://doi.org/10.1007/978-981-19-9267-4_56

Keywords Blended cement · Treated sludge · Setting time · Water treatment sludge

56.1 Introduction

Water treatment sludge (WTS) is a semi-solid slurry produced through the water treatment process in the water treatment plant. The sludge from the water treatment plant is presented in solid and liquid form and is considered a waste. Landfilling is the most common WTS disposal method [1]. Lohani [2] reports that landfilling is the easiest and the cheapest technology available. Unfortunately, the disposal of WTS creates a tremendous effect on the environment due to the limitation of disposal area and environmental reduction measures, such as leachate collection systems and lining materials. Moreover, WTS contains contaminants such as bacteria, aluminium, nitrates, metals, trace quantities of toxic materials, and salts. In addition, sludge disposal produces leachate, a hazardous liquid through surface runoff that contributes to soil and groundwater pollution [1]. Hence, improving the disposal method of WTS is needed.

Solidification is one technique used to remediate sludge before its disposal in a landfill. A common solidification approach is mixing the sludge with a specified binder to decrease the leaching of pollutants from the sludge, either by physical or chemical processes [3]. Consequently, the sludge can be disposed of in a safe and environmentally friendly form. Zhang [4] concluded that solidification/stabilisation (S/S) had been extensively utilised to dispose of different types of hazardous waste and to reclaim contaminated disposal sites.

Ordinary portland cement (OPC) is the most widely used material in the S/S technique. Nevertheless, the OPC is a costly binder due to its high energy consumption, expense, and environmental degradation. In addition, earlier investigations concurred that the cement sector is one of the leading contributors to greenhouse gas (GHG) emissions, particularly CO_2 emissions [5–7]. Therefore, it has been discovered that industrial waste materials can be used as a cement substitute material to solidify and stabilise sludge, hence reducing the use of OPC in solidifying WTS [8–12].

Industrial activities generate industrial waste during manufacturing factories, mills, and mining operations. The FA, POFA, and WPSA are the industrial wastes utilised for this study. The specified industrial wastes are added to the OPC mixture as a binder. These industrial waste materials are selected because they are reasonably inexpensive and stable over time [10]. As indisputable evidence, the WTS has been solidified and stabilised with industrial waste material and is capable of mitigating adverse environmental impacts [8–10]. The S/S technology often entails combining WTS with a binder to avoid the environmental leaching of contaminants. Previously, the efficacy of this approach was assessed using compressive strength and leachability tests. Nonetheless, the setting time of WTS solidified by the blended binder was still unknown. Therefore, this paper focused on the effect of various binders on the initial and final setting times of treated WTS.

56.2 Methodology

56.2.1 Materials and Testing

The ordinary portland cement (OPC), fly ash (FA), waste paper Sludge ash (WPSA), palm oil fuel ash (POFA), and water treatment sludge (WTS) are the materials utilised in this study. The OPC was supplied by local manufacturers and met MS522: Part 1:1989 requirements. The FA was supplied by Stesen Janakuasa Elektrik, Sultan Salahuddin Abdul Aziz Shah, Kapar, while WPSA was supplied by Malaysian Newsprint Industries (MNI), Mentakab, Pahang. In addition, POFA is received directly from Kilang Sawit Jengka 21, Bandar Jengka, Pahang. The WTS was obtained at the Gunung Semanggol Water Treatment Plant in Perak.

For the setting time and standard consistency tests, four (4) mix proportions were prepared for samples with and without sludge. In this study, a consistency test is conducted following BS EN 196-3-1995 to determine the amount of water in the binder paste. This test reveals the water concentration required to obtain the standard consistency index. The consistency will allow a Vicat plunger with a 10 mm diameter and a 50 mm length to penetrate the paste to a depth of 5 $\pm$ 1 mm, measured from the bottom of the mould. The needle is made of polished brass with a diameter of 10 $\pm$ 0.05 mm, and its lower end is flat. The consistency index is calculated by dividing the amount of water added by the amount of cement applied and multiplying the result by 100%.

Setting time constitutes a transition from a fluid to a rigid state. Initial setting and final setting times define the duration of the setting. In addition, setting time is required for a paste to harden to a particular consistency. The initial setting time occurs when the paste loses its plasticity, whereas the final setting time occurs when it loses all its plasticity. The testing procedure follows BS EN 196-3-1995 for setting time. The initial set is the time from the addition of water until the paste ceases to be fluid and pliable, whereas the final set is the time from the addition of water until the paste attains a certain degree of hardness. The initial setting occurs when water is applied to the cement, and a 1 mm square section needle positioned 5–7 mm from the bottom. The needle's surface area for the initial setting time is 1 mm^2, and the end needle is flat. The final setting is the time after adding water to the cement, during which a 1 mm needle can penetrate the paste in a mould, but a 5 mm needle cannot. The final setting test uses a circular needle with a metal attachment and 1 mm^2 surface area.

56.2.2 Chemical Properties

The chemical compositions of each substance were determined using X-ray fluorescence (XRF) analysis. The chemical composition of the materials required in this study is shown in Table 56.1.

Table 56.1 Chemical composition of materials

Compound	OPC (%)	WTS (%)	FA (%)	WPSA (%)	POFA (%)
CaO	61.43	0.17	6.9	30.52	4.92
SiO_2	18.62	26.74	59	28.15	59.62
Al_2O_3	4.75	20.72	21	15.77	2.54
MgO	3.21	–	1.4	1.94	4.52
Fe_2O_3	3.02	–	3.7	1.05	5.02
SO_3	2.29	–	1	0.57	1.28
Na_2O	1.51	–	–	0.67	0.76
K_2O	1.42	0.74	0.9	0.45	7.52
LOI	3.55	–	4.62	17.23	8.25
Others	–	9	–	–	–

56.2.3 Mix Proportion

The blended binder was produced by mixing OPC, FA, WPSA, and POFA. The selection of mixes proportion 50:50 for OPC: FA and 60:40 for OPC: WPSA, and OPC: POFA was chosen based on optimum results obtained on compressive strength of binder paste during preliminaries studies. Four mix proportions for the sample without sludge and sample with sludge were prepared for setting time and consistency test. For each proportion of sludge-containing mixture, 200 g of WTS and 200 g of the blended binder were utilised. The sample without sludge is depicted in Table 56.2, while the sample including sludge is depicted in Table 56.3.

Table 56.2 Mix proportions of a sample without sludge

Mix	OPC (%)	FA (%)	WPSA (%)	POFA (%)	Weight of binder (g)	Percentage of water (%)
M1	100	–	–	–	400	32
M2	50	50	–	–	200 + 200	37
M3	60	–	40	–	240 + 160	54
M4	60	–	–	40	280 + 120	31

Table 56.3 Mix proportions of a sample with sludge

Mix	OPC (%)	FA (%)	WPSA (%)	POFA (%)	Weight of binder (g)	WTS (g)
M5	100	–	–	–	200	200
M6	50	50	–	–	100 + 100	200
M7	60	–	40	–	120 + 80	200
M8	60	–	–	40	120 + 80	200

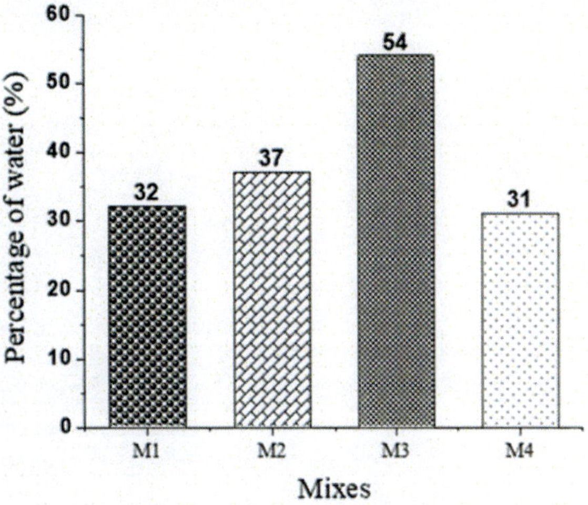

Fig. 56.1 Standard consistency index for sample without sludge

56.3 Result and Discussions

56.3.1 Standard Consistency Test Results

Figure 56.1 depicts the outcome of the standard consistency test for the sample without sludge. Standard consistency of Mix 1 (OPC) is attained due to the assumption of a 5 mm height of penetration with 32% amount of water used as presented in Fig. 56.1. Consequently, for Mix 2 (OPC + FA), 37% of water is utilised to achieve the standard height of 5 mm. Concurrently, Mix 4 (OPC + WPSA) and Mix 3 (OPC + POFA) suggest an approximate water content of 54% and 31%, respectively. The OPC + WPSA sample requires 10% more water than the other samples, which require the most significant proportion of water. Mix 2 (OPC + FA) showed a higher percentage of consistency than coarser pastes when the material had a high fineness and porosity [13]. The light particle of the material formed for Mix 3 (OPC + POFA) necessitates more water than the other mixtures. In conclusion, the cementitious material's surface area increases as its particle size decrease [13].

56.3.2 Setting Time Test Results

Figure 56.2 illustrates the initial and final setting times for samples without sludge.

The results indicated Mix 1 (OPC) required 180 min to set. This is followed by the 270-min Mix 2 (OPC + FA). Mix 3 (OPC + WPSA) required 150 min, whereas Mix 4 (OPC + POFA) required 300 min to harden. Based on the results illustrated in Fig. 56.2, Mix 4 (OPC + POFA) takes the longest time to harden. Mix 3 (OPC + WPSA) has the quickest setting time among the other samples. Mix 2 (OPC + FA) has

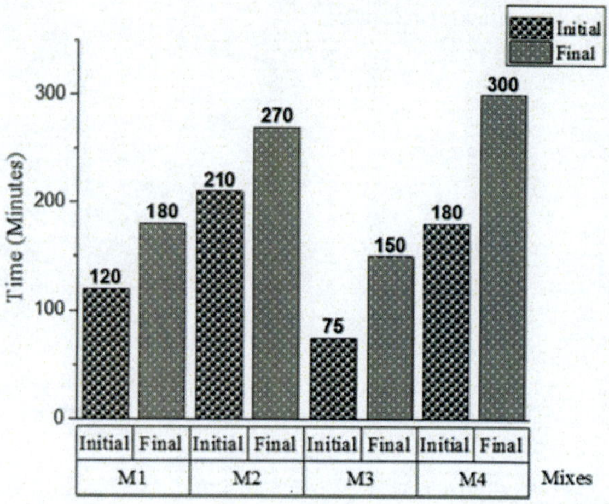

Fig. 56.2 Setting time index for sample without sludge

a longer setting time than Mix 1 (OPC) and Mix 2 (OPC + WPSA). Referring to Table 56.1's chemical composition of POFA, POFA has less calcium oxide (Cao), delaying the setting time of concrete emissions [14, 15]. Therefore, the setting durations for mixtures to harden have been raised to 300 min. It is often slower than cement hydration. The prolonged setting times of POFA concrete are likely attributable to the pozzolanic interaction between POFA and calcium hydroxide [16, 17].

Previously, based on Table 56.2's chemical composition, FA had a lower calcium concentration of 6.90%. Moreover, the total amount of SiO_2 + Al_2O_3 + Fe_2O_3 exceeds 70%, which is designated as FA in class type F [18]. The FA falls under class type F since it contains less than 8% calcium oxide, which might reduce the setting time of mixes [18]. Hence, using FA replacement cement class F may lengthen the setting time of mixtures.

Furthermore, the OPC mixture with WPSA exhibited a low setting time. The more significant CaO component in WPSA could boost the hardening of concrete. Besides, the similar properties of OPC and WPSA contributed to the double setting time impact. Compared to other materials, the high CaO and alumino-silicate concentration of OPC and WPSA accelerate the setting time of concrete. Although Mix 3 had the most significant proportion of water based on the previous consistency test, it hardened in concrete quickly. Therefore, the WPSA contains alumino-siliceous material that has been consolidated with calcium, improving its quality and the quickest setting time among the mixes [19].

The outcome of setting time for a sample including sludge is depicted in Fig. 56.3. First, Mix 5 (OPC + WTS) took 1920 min, as indicated by the results. Next, the setting time for Mix 6 (OPC + FA + WTS) was 4725 min. Then, the 660-min Mix 7 (OPC + WPSA + WTS) followed. Referring to Table 56.1, the fly ash specification deficient in CaO could decrease the hardening of concrete and lengthen its setting time [18]. In contrast, POFA, which contains less than 5% of CaO, belongs to class F. This result also demonstrates that the CaO concentration of POFA is less than 5%

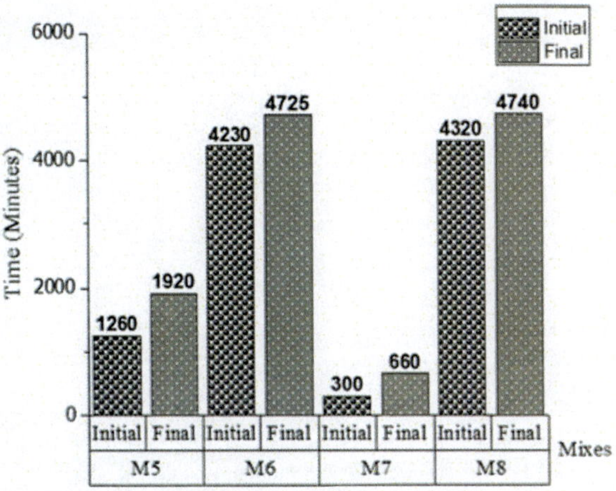

Fig. 56.3 Setting time index for sample with sludge

[10, 16]. Due to the chemical composition of WTS, the sample required the longest time to harden. The more significant amount of SiO_2 in the sludge, as stated in Table 56.1, causes a sample's setting time to be slightly slower. In addition, the outcome is slightly different in a sample including sludge since the hardening of the binder depends on its properties: the less CaO present, the slower the process of hardening [10]. Therefore, the samples with the longest setting time, Mix 6 and Mix 8, have a low CaO content and a high SiO_2 content, contributing to the sluggish setting time.

Mix 7's setting time is the one that has the quickest time compared to the others. It only took 660 min for Mix 7 (OPC + WPSA + WTS) to reach its final state. This condition occurs since OPC and WPSA have a chemical composition that is quite similar; both have a high proportion of the minerals CaO, SiO_2, and Al_2O_3, which are responsible for the rapid hardening of the concrete [8, 9]. Consequently, a higher concentration of WPSA results in a reduced setting time. In addition, OPC and WPSA are comparable and capable of bonding the solidifying sludge material. Therefore, Mix 7 (OPC + WPSA) was preferable to the others.

56.4 Conclusion

The results from this study have been obtained to determine the influence of different blended binders on the initial and final setting times of treated sludge. The present study supports that Mix 7 (OPC + WPSA) with sludge exhibited the lowest initial and final setting times, showing that it expedites the hardening process and increases the strength, demonstrating more potential in lowering leachability than the other mixes. Besides, Mix 7 (OPC + WPSA) with sludge has demonstrated that a blended binder is viable for solidification and stabilising of WTS.

Acknowledgements The authors would convey their profound appreciation and gratitude to the College of Engineering, Universiti Teknologi MARA of Malaysia for permitting us to perform this study. Many thanks to the Ministry of Education for the financial assistance received (600-IRMI/FRGS 5/3 (0120/2016)). Moreover, we gladly recognise Malaysian Newspaper Industry Sdn. Bhd. (MNI) for supplying waste paper sludge ash (WPSA). Special thanks to those who contributed to this project directly or indirectly.

References

1. S. Nanda, F. Berruti, Municipal solid waste management and landfilling technologies: a review. Environ. Chem. Lett. **19**, 1433–1456 (2021). https://doi.org/10.1007/s10311-020-01100-y
2. S.P. Lohani, M. Keitsch, S. Shakya, D. Fulford, Waste to energy in Kathmandu Nepal-A way toward achieving sustainable development goals. J. Sustain. Dev. **29**, 906–914 (2021). https://doi.org/10.1002/sd.2183
3. M.U. Kankia, L. Baloo, B.S. Mohammed, S.B. Hassan, E.A. Ishak, Z.U. Zango, Review of petroleum sludge thermal treatment and utilization of ash as a construction material, a way to environmental sustainability. Int. J. Adv. Appl. Sci. **7**, 68–81 (2020). https://doi.org/10.21833/ijaas.2020.12.008
4. Y. Zhang, L. Wang, L. Chen, B. Ma, Y. Zhang, W. Ni, D.C. Tsang, Treatment of Municipal solid waste incineration fly ash: state of the art technologies and future perspectives. J. Hazard. Mater. **411** (2021). https://doi.org/10.1016/j.jhazmat.2021.125132
5. W.F. Lamb et al., A review of trends and drivers of greenhouse gas emissions by sector from 1990 to 2018. Environ. Chem. Lett. **16**(7) (2021). https://doi.org/10.1088/17489326/abee4e
6. A. Talaei, D. Pier, A.V. Iyer, M. Ahiduzzaman, A. Kumar, Assessment of long-term energy efficiency improvement and greenhouse gas emissions mitigation options for the cement industry. Energy. J. **170**, 1051–1066 (2019). https://doi.org/10.1016/j.energy.2018.12.088
7. E. Benhelal, E. Shamsaei, M.I. Rashid, Challenges against CO_2 abatement strategies in cement industry: a review. J. Environ. Sci. **104**, 84–101 (2021). https://doi.org/10.1016/j.jes.2020.11.020
8. I. Nurliyana, M.A. Fadzil, H.M. Saman, W.K. Choong, in *Waste Paper Sludge Ash (WPSA) as Binder in Solidifying Water Treatment Plant Sludge (WTPS)*. Proceeding of the International Civil and Infrastructure Engineering Conference, pp. 497–504, 2016
9. I. Nurliyana, M.A. Fadzil, H.M. Saman, W.K. Choong, Water treatment sludge stabilizer binder by waste paper sludge ash for solidification/stabilization technique. Int. J. Integr. Eng. **11**(1), 113–119 (2019) [Online]. Available: https://penerbit.uthm.edu.my/ojs/index.php/ijie/article/view/2460
10. I. Nurliyana, M.A. Fadzil, H.M. Saman, W.K. Choong, in *Palm Oil Fuel Ash and Ceramic Sludge as Partial Cement Replacement Materials in Cement Paste*. InCIEC 2014 (Springer, Singapore, 2015), pp. 1087–1092. https://doi.org/10.1007/978-981-287-290-6_96d
11. N. Khalid, M. Mukri, F. Kamarudin, M.A. Fadzil, Clay soil stabilized using waste paper sludge ash (WPSA) mixtures. Electron. J. Geotech. Eng. **17**, 1215–2122 (2012)
12. N. Khalid, M.A. Fadzil, M. Mukri, F. Kamarudin, H. A. Ghani, F. Baharudin, in *Soft Soil Subgrade Stabilization Using Waste Paper Sludge Ash (WPSA) Mixtures*. InCIEC 2014 (Springer, Singapore, 2015). https://doi.org/10.1007/978-981-287-290-6_38
13. P. Byoungsun, C.C. Young, Investigating a new method to assess the self-healing performance of hardened cement pastes containing supplementary cementitious materials and crystalline admixtures. J. Mater. Res. Technol. (2019). https://doi.org/10.1016/j.jmrt.2019.09.080
14. H.M. Hamada, B.S. Thomas, F.M. Yahaya, K. Muthusamy, J. Yang, J.A. Abdalla, R.A. Hawileh, Sustainable use of palm oil fuel ash as a supplementary cementitious material: a comprehensive review. J. Build. Eng. **40**, 102286 (2021). https://doi.org/10.1016/j.jobe.2021.102286

15. A.A. Jhatial, W.I. Goh, N. Mohamad, S. Sohu, M.T. Lakhiar, Utilization of palm oil fuel ash and eggshell powder as partial cement replacement: a review. Civ. Eng. J. **4**(8), 1977–1984 (2018). https://doi.org/10.28991/cej-03091131
16. M. Amran, Y.H. Lee, R. Fediuk, G. Murali, M.A. Mosaberpanah, T. Ozbakkaloglu, Y.Y. Lee, N. Vatin, S. Klyuev, M. Karelia, Palm oil fuel ash-based eco-friendly concrete composite: a critical review of the long-term properties. J. Mater. **14**, 7074 (2021). https://doi.org/10.3390/ma14227074
17. B. Alsubari, P. Shafigh, Z. Ibrahim, M.F. Alnahhal, M.Z. Jumaat, Properties of eco-friendly self-compacting concrete containing modified treated palm oil fuel ash. Constr. Build Mater. **158**, 742–754 (2018). https://doi.org/10.1016/j.conbuildmat.2017.09.174
18. K. Korniejenko, N.P. Halyag, G. Mucsi, Fly ash as a raw material for geopolymerisation-chemical composition and physical properties. IOP Conf. Ser. Mater. Sci. Eng. **706**, 012002 (2019). https://doi.org/10.1088/1757-899X/706/1/012002
19. N.S. Aini, M.H.M. Haniff, M.A. Fadzil, A.R. Ridzuan, in *Effect of WPSA Particle Size to the Morphology and Compressive Strength Properties of Hydrated Cement Paste Contain WPSA as SCM*. 2013 IEEE Business Engineering and Industrial Applications Colloqium (BEIAC), pp. 301–305, 2013. https://doi.org/10.1109/BEIAC.2013.6560136

Chapter 57
Characterization of Cassava/Sugar Bagasse-Derived Biochar: The Effect of Batch Mixing

Pham Trung Kien, Tran Ngo Quan, Nguyen Cong Tuan Anh, Nguyen Minh Phong, Le Thi Kim Phung, Ho Jin Sung, Chae-Eun Yeo, Se-yoon Hong, and Hwansoo Jung

Abstract In this research, we report the fabrication of biochar through torrefication process at 300 °C for 3 h. The starting materials used were cassava powder (CP) and sugar baggase powder (BP) which were considered as products wasted in agriculture in Tay Ninh province, Vietnam. These materials were batch mixed with different weight ratios such as CP/BP of 100/0; 75/25; 50/50; 25/75; and 0/100, followed by pellet forming in the presence of KOH which uptakes up to 20 wt. batch mixtures. The green pellet was torrefication at 300 °C for 3 h to find the condition that can uptake the highest C% in biochar. The result shows that the ratio of CP/BP as 75/25 can uptake up to 89%C in biochar, leading to reduction in the CO_2 emission.

Keywords Biochar · Torrefication · Cassava · Baggase · CO_2 emission

P. T. Kien (✉) · T. N. Quan · N. C. T. Anh · N. M. Phong
Faculty of Materials Technology, Ho Chi Minh City University of Technology (HCMUT), 268 Ly Thuong Kiet Street, District 10, Ho Chi Minh City, Vietnam
e-mail: phamtrungkien@hcmut.edu.vn

P. T. Kien · T. N. Quan · N. C. T. Anh · N. M. Phong · L. T. K. Phung
Vietnam National University Ho Chi Minh City, Linh Trung Ward, Thu Duc District, Ho Chi Minh City, Vietnam

P. T. Kien
Polymer Research Center, Ho Chi Minh City University of Technology (HCMUT), 268 Ly Thuong Kiet Street, District 10, Ho Chi Minh City, Vietnam

L. T. K. Phung
Faculty of Chemical Engineering, Ho Chi Minh City University of Technology (HCMUT), 268 Ly Thuong Kiet Street, District 10, Ho Chi Minh City, Vietnam

H. J. Sung · C.-E. Yeo · S. Hong
Plant Engineering Center, Institute for Advanced Engineering, 175-28, Goan-Ro, 51 Beon-Gil, Baegam-Myeon, Cheoin-Gu, Yongin-Si 17180, Gyeonggi-Do, Korea

H. Jung
Industry-Academic Convergence Campus, Hanbat National University, 75 Techno 1-Ro, Yuseong-Gu, Daejeon 34104, Korea

M. A. A. Mohd Salleh et al. (eds.), *Proceedings of the Green Materials and Electronic Packaging Interconnect Technology Symposium*, Springer Proceedings in Physics 289,
https://doi.org/10.1007/978-981-19-9267-4_57

57.1 Introduction

Nowadays, Vietnam is the leading country in producing biomass resources, especially sugar cane. Through the process of producing sugar cane, the sugar factory can obtain sugar bagasse. To increase the adding value of sugar industry, research on the use of sugar bagasse focuses recently in Vietnam as well as in the global scale [1–8]. Mostly, the sugar bassage biomass in Vietnam was burnt to recycle the heat waste for sugar industry as well as to generate electricity use. This waste-to-energy process releases CO_2 during the O_2 gas atmosphere burning environment leading to negative impact on the environment. However, the stock of cassava and sugar bagasse vary throughout the season. Thus for the torrefication, the factory needs to mix between cassava and baggase with different mixing weight ratios to ensure the continuous working of torrefication equipment. Thus, this research focuses on the characterization of cassava-/sugar bagasse-dervied biochar: the effect of batch mixing to find the condition for process working.

57.2 Methodology

57.2.1 Materials and Sample Preparation

Preparation the sugarcane bagasse and cassava materials: these materials were supplied by Thanh Thanh Cong Sugar (TTCS) Co., Tay Ninh province, Vietnam. The sugarcane bagasse materials and cassava were ground by fast-grinding machine with the power of 3 kW (3A model, Tuan Tu Agriculture manufacturer, Vietnam), and was then passed through the 0.45 mm sieve to collect the fined sugarcane bagasse powder (BP) as well as the fine cassava powder (CP), respectively.

Preparation of CP-/BP-based green sample: the fine CP and BP were mixed together with different mixing weight ratios between CP:BP such as 25/75; 50/50; and 75/25, respectively, to investigate the effect of different mixing ratio of raw materials. The batch after homogenize were second mixed with KOH with weight ratio of batch:KOH is 80:20 as previous report [1] for easy disk forming. The KOH-content batch mixing were form to disk (1 cm Diameter × 1 cm Height) as CP/BP-based green sample for further torrefication treatment.

Torrefication process: The CP-/BP-based green pellet was torrefication in inert atmosphere at 300 °C for 3 h based on the thermal analysis data of BP materials. The flowchart of synthesize CP-/BP-derived biochar is shown in Fig. 57.1.

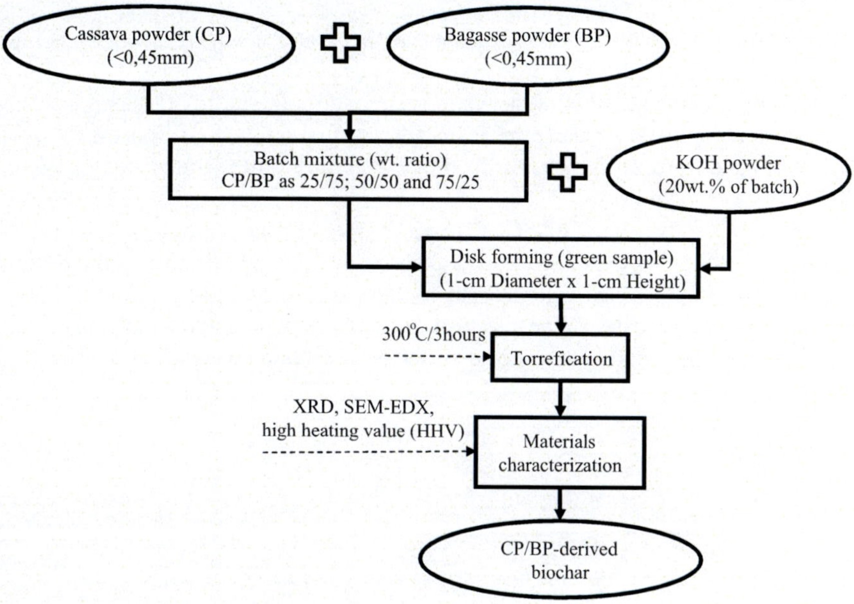

Fig. 57.1 The flowchart of synthesized CP-BP-based biochar

57.2.2 *Materials Characterization Method*

Phase investigation: raw materials and production were characterized using X-ray diffraction (XRD). The sample was ground and put in XRD machine (Bruker D8 Advance, Germany) with the 2theta scanning from 5° to 40°, operated at 40 kV and 40 mA.

The thermal analysis using DSC/TG: the sample was heated up to 600 °C using DSC/TGA technique (STA 449, Netzsch, Germany).

The chemical composition analysis using Xray Fluorescing (XRF): the sample was characterized by X-ray at 40 kV (ARL Advant, Thermo Scientific).

The morphology analysis using SEM/EDS: the sample was scattered on carbon tape stick to cupper substrate, and the SEM analysis (Hitachi S-4800) was carried at 10 kV. The amount of C and O contents was identified by EDS mapping technique.

57.3 Result and Discussions

Figure 57.2 shows the XRD pattern of (a) CP raw materials and (b) BP raw materials. The CP was composed of Laumonite mineral ($CaAl_2Si_4O_{12}.4H_2O$) PDF#26-1047 with the strongest peak at 2theta of 20° and Quartz (SiO_2) PDF#46-1045 with the strongest peak at 2theta of 26.4°. In addition, the BP (shown at Fig. 57.2b) indicate the same peak of Laumonite and Quartz. However, in case of BP, the peak of Laumonite can be observed at 2theta of 36°, indicating that amount of Ca-Al-Si containing Laumonite mineral in BP is higher than Laumonite mineral in CP materials. The origin of Laumonite and Quartz in CP and BP raw materials can be from the Ca–Al–Si in earth and fertilizer which absorbed by CP and BP during the growing process of cassava and sugar tree.

Figure 57.3 shows the TG/DSC of BP-KOH-dervied green sample before torrefication. The BP-KOH exposed the endothermic peak at 100 °C due to the drying process of cellulose-based materials. In contrast, the exothermic peaks of 220 and 285 °C were observed due to the torrefication of cellulose-based materials. Thus, the torrefication temperature was selected as 300 °C for 3 h in this research.

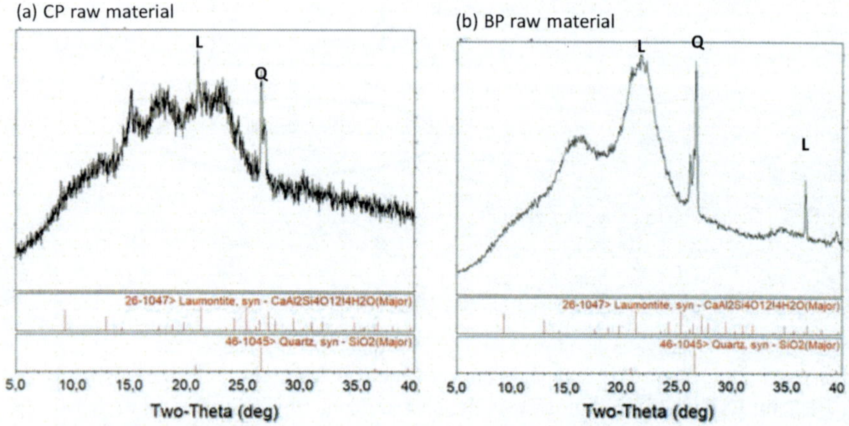

Fig. 57.2 XRD patterns of starting materials: **a** CP raw materials and **b** BP raw materials

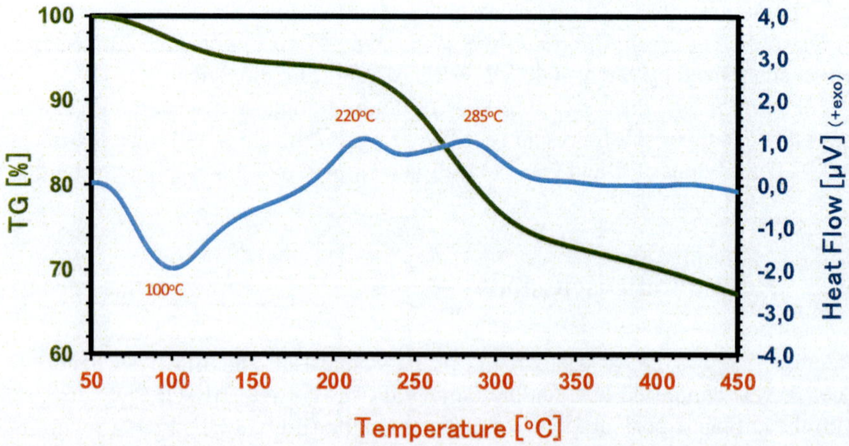

Fig. 57.3 DSC/TGA patterns of BP-KOH green sample before torrefication

Figure 57.4 shows the over view of XRD patterns of biochar sample with different CP/BP weight ratio such as: (c) CP-KOH-derived biochar; (d) BP-KOH-derived biochar; (e) CP/BP-KOH-25/75-dervied biochar; (f) CP/BP-KOH-50/50-dervied biochar; and (g) CP/BP-KOH-75/25dervied biochar. All the green samples were mixed with KOH with the weight ratio of KOH at 20 wt%, and then followed by torrefication at 300 °C for 3 h to obtain the biochar. The CP-KOH-dervied biochar indicated that after torefication at 300 °C for 3 h, the biochar was composed of Laumonite mineral (PDF#26-1047) and Quartz (PDF#46-1045) at 2theta of 21° and 26° as shown in Fig. 57.4c. Similar findings were observed upon torrefication of the BP-KOH-dervied biochar as shown in Fig. 57.4d. In contrast, the mixture of CP/BP with different weight ratios of CP/BP such as 25/75 50/50 and 75/25 shows the XRD peak of Potassium Carbonate Hydrate (PDF#46-1045) at 2theta of 32° as shown in Fig. 57.4e–g, respectively. The formation of Potassium Carbonate Hydrate due to the chemical reaction of CO_2 and H_2O in cellulose structure of CP and BP reacted with alkaline activator KOH during the torrefication process as shown in Eq. (57.1):

$$4KOH + 2CO_2 + H_2O \rightarrow 2K_2CO_3.1.5H_2O \quad (57.1)$$

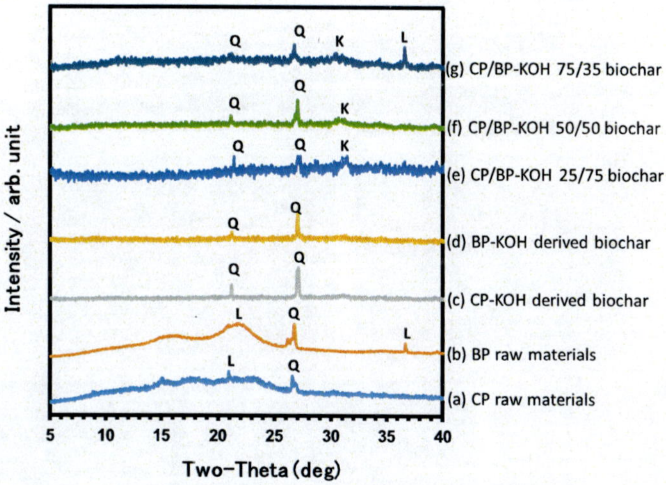

Fig. 57.4 XRD patterns of starting materials: **a** CP raw materials and **b** BP raw materials. From (**c**) to (**g**) were XRD patterns of biochar synthesize at different CP/BP mixing ratio: **c** CP-KOH-derived biochar; **d** BP-KOH-derived biochar; **e** CP/BP-KOH-25/75-derived biochar; **f** CP/BP-KOH-50/50-derived biochar; and **g** CP/BP-KOH-75/25-derived biochar. All the green sample were mixed with KOH with the weight ratio of KOH is 20 wt%, and then followed by torrefication at 300 °C for 3 h to obtain the biochar. Code: Q-Quartz (SiO_2) PDF#46-1045; L-Laumonite ($CaAl_2Si_4O_{12}.4H_2O$) PDF#26-1047; K-Potassium Carbonate Hydrate ($K_2CO_3.1.5H_2O$) PDF#11-0655

Figure 57.5 shows the SEM images of starting materials: (a) CP raw materials and (b) BP raw materials. From (c) to (g) were SEM images of biochar synthesize at different CP/BP mixing ratio: (c) CP-KOH-derived biochar; (d) BP-KOH-derived biochar; (e) CP/BP-KOH-25/75-derived biochar; (f) CP/BP-KOH-50/50-derived biochar; and (g) CP/BP-KOH-75/25-derived biochar. All the green samples were mixed with KOH with the weight ratios of KOH at 20 wt%, and then followed by torrefication at 300 °C for 3 h to obtain the biochar. The CP raw materials show the spherical shape while the BP exposed the plate shape crystal (Fig. 57.5a, b). Upon torrefication, all the samples shown plate-like crystal with the glassy morphology block the insider pore (Fig. 57.5c–g).

Figure 57.6 shown the EDS spectra of starting materials: (a) CP raw materials and (b) BP raw materials. From (c) to (g) were SEM images of biochar synthesize at different CP/BP mixing ratios: (c) CP-KOH-derived biochar; (d) BP-KOH-derived biochar; (e) CP/BP-KOH-25/75-derived biochar; (f) CP/BP-KOH-50/50-derived biochar; and (g) CP/BP-KOH-75/25-derived biochar. Before torrefication,

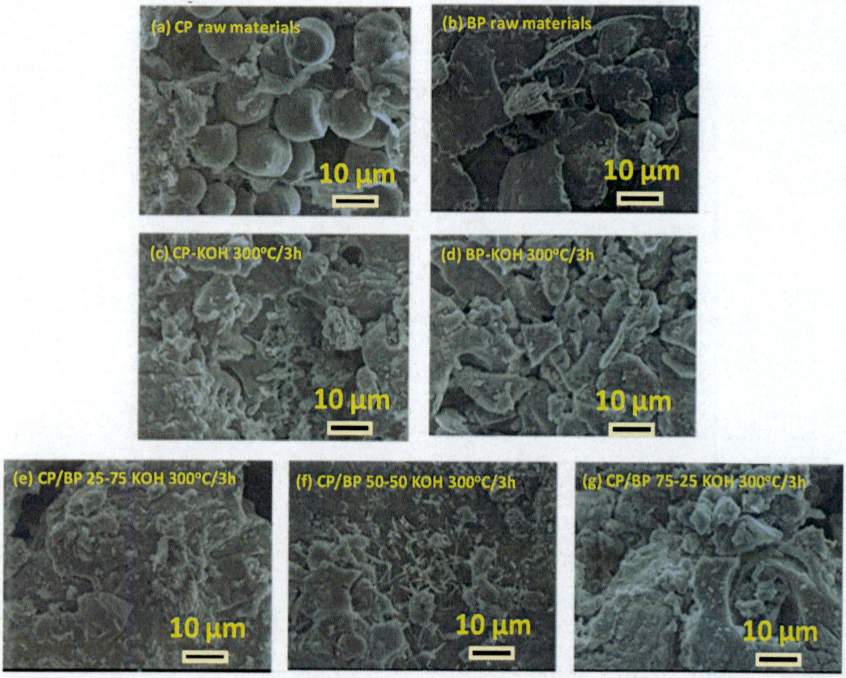

Fig. 57.5 SEM images of starting materials: **a** CP raw materials and **b** BP raw materials. From (**c**) to (**g**) were SEM images of biochar synthesize at different CP/BP mixing ratio: **c** CP-KOH-derived biochar; **d** BP-KOH-derived biochar; **e** CP/BP-KOH-25/75-derived biochar; **f** CP/BP-KOH-50/50-derived biochar; and **g** CP/BP-KOH-75/25-derived biochar. All the green samples were mixed with KOH with the weight ratio of KOH at 20 wt%, and then followed by torrefication at 300 °C for 3 h to obtain the biochar

the peak of Si and K could not be observed as given in Fig. 57.6a, b. After torrefication, the peaks of Si and K could be obtained at energies of 2 and 3.3 keV, respectively, as indicating in Fig. 57.6c–g. Based on the intensity of C, O, Si and K peaks, the amount of these elements is given in Table 57.1.

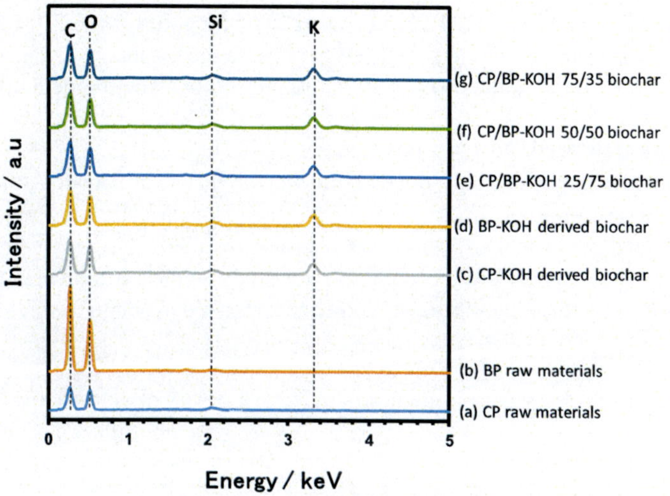

Fig. 57.6 The EDS spectra of starting materials: **a** CP raw materials and **b** BP raw materials. From (**c**) to (**g**) were EDS spectra of biochar synthesize at different CP/BP mixing ratio: **c** CP-KOH-derived biochar; **d** BP-KOH-derived biochar; **e** CP/BP-KOH-25/75-derived biochar; **f** CP/BP-KOH-50/50-derived biochar; and **g** CP/BP-KOH-75/25-derived biochar

Table 57.1 Amount of C, Si and K in biochar at different CP/BP ratio

Materials		Elemental analysis				% C uptake into biochar after torrefication	High heating value (HHV)* (MJ/kg)
		C	O	Si	K		
Before torrefication	CP	58.28	40.37	0.27	1.52		14.68
	BP	59.20	39.27	0.51	–		17.58
After torrefication (300 °C for 3 h)	CP-KOH	46.35	41.04	0.27	13.82	79.53	13.56
	BP-KOH	50.05	34.21	5.05	9.84	85.88	14.75
	CP/BP 25/75 KOH	44.63	41.04	0.27	13.82	76.58	13.87
	CP/BP 50/50 KOH	40.62	33.55	0.61	24.55	69.70	14.68
	CP/BP 75/25 KOH	52.29	30.97	0.55	15.65	89.72	14.05

*HHV were measured according to ISO 18125:2017

Apparently, the amount of C in solid bichar can be up taken into the biochar up to 50 wt% after torrefication due to the incorporation of C into alkaline activator KOH. As mentioned in the introduction, the present of KOH accelerates the binding of cellulose materials, resulting in easy pellet forming. In addition, the %C uptake into biochar after torrefication may vary from 69 to 89%. It needs to emphasized that the residual KOH and $K_2CO_3.1.5H_2O$ could block the pore size of CP-/BP-derived biochar; thus, our research group suggests to activate the biochar by acidic treatment such as HCl acidic treatment to obtain the highly porous environmental treatment materials.

This research needs to put into the context of experiment conditions. At the end of the research, the KOH-treating biochar could be obtained. Although the porosity and adsorption capacity of KOH-treating biochar could not be measured, the KOH-treating biochar still uptook the C inside up to 89% upon using the mixture of CP/BP 75/25-dervied biochar. Ongoing research on cleaning process of KOH-treating biochar with acidic treatment will be investigated to obtain the highly porous materials.

57.4 Conclusion

This study is helpful in enhancing the mixture parameter between cassava and bagasse as resource biomass in Vietnam toward the incorporation C through biochar process. The aim of the study is to find the process for easy pellet forming biochar, as well as reduce the CO_2 emission. The result show that:

- Different mixing ratio between CP/BP leading to different amount of fixed C in CP-/BP-dervied biochar, thus varying the HHV from 13 to 14 MJ/kg.
- The residua KOH as well as the byproduct potassium carbonate hydrate could block the inside porous CP/BP biochar; thus, we suggest to do activation treatment by acidic treatment such as HCl acidic treatment to obtain the highly porous environmental treatment materials.

Acknowledgements This work is supported by the Korea Agency for Infrastructure Technology Advancement (KAIA) grant funded by the Ministry of Land, Infrastructure and Transport (Grant 22DPRB-C161884-02). We would like to thank Ho Chi Minh City University of Technology (HCMUT), VNU-HCM, for the support of time and facilities for this study.

References

1. P.T. Kien, L.H. Tai, L.H.T. Anh, Effects of KOH on some biochar components from pyrolyzed fine sugar bagasse (FSB). Vietnam J. Sci. Technol. **59**(6A), 43–50 (2021)

2. Z. Amir, K. Sardar, Popular wood and sugarcane bagasse biochars reduced uptake of chromium and lead by lettuce from mine-contaminated soil. Environ. Pollut. **263**(Part A), 114446 (2020). https://doi.org/10.1016/j.envpol.2020.114446
3. A. Paulo, M. Cerqueira, Upgrading from batch to continuous flow process for the pyrolysis of sugarcane bagasse: structural characterization of the biochars produced. J. Environ. Manage. **285**, 112145 (2021). https://doi.org/10.1016/j.jenvman.2021.112145
4. M. Yongfei, Q. Yong, Adsorptive removal of imidacloprid by potassium hydroxide activated magnetic sugarcane bagasse biochar: adsorption efficiency, mechanism and regeneration. J. Clean. Prod. **292**, 126005 (2021). https://doi.org/10.1016/j.jclepro.2021.126005
5. A. Nahid, R. Fayez, Sugarcane bagasse biochar modulates metal and salinity stresses on microbial functions and enzyme activities in saline co-contaminated soils. Appl. Soil. Ecol. **167**, 104043 (2021). https://doi.org/10.1016/j.apsoil.2021.104043
6. N. Chengrong, Y. Xing, Impact of sugarcane bagasse-derived biochar on heavy metal availability and microbial activity: a field study. Chemosphere **200**, 274–282 (2018). https://doi.org/10.1016/j.chemosphere.2018.02.134
7. C.D. Sheng, Estimating the higher heating value of biomass fuels from basic analysis data. Biomass Bioenerg. **28**, 499–507 (2005). https://doi.org/10.1016/j.biombioe.2004.11.008
8. N.H. Thang, N.H.L. Ngoc, V.T.N. Uyen, P.T. Kien, Novel materials synthesized from red mud, bagasse, and bentonite for gas treatment by CO_2 absorption. MATEC **207**, 1–4 (2018)

Chapter 58
Research on the Content of KOH on Microstructure Property of Rice Hull-Based Biochar Using Pyrolyzed Process

Pham Trung Kien, Bui Van Tien, Huynh Dai Phu, Le Huynh Tuyet Anh, Tran Ngo Quan, Nguyen Cong Tuan Anh, Nguyen Minh Phong, and Tran Van Khai

Abstract In this research, we report the fabrication of biochar through the mixing of rice hull with different amount of KOH content such as 0, 20, and 40 wt% follow by pyrolysis at 550 °C for 3 h. The result shows that adding KOH can make the forming process easily, also varies the microstructure of RH/KOH-based biochar. The mechanism to uptake the fixed C into RH/KOH-based biochar is due to the chemical reaction between KOH and C-based cellulose in the rice hull. This research aims to produce the C-content biochar, thus reducing the CO_2 emission in Vietnam as well as global scale.

Keywords Biochar · Pyrolysis · Rice hull · CO_2 emission

58.1 Introduction

Currently, Vietnam is the top leading country in producing the biomass resources, especially the rice hull. Through the process of producing the rice, the rice farmer can obtain the rice hull. Mostly, the rice hull biomass in Vietnam was burnt in the open air to recycle the heat waste for rice drying. This waste-to-energy process releases the CO_2 during the O_2-gas atmosphere burning environment leading to negative

P. T. Kien (✉) · B. Van Tien · H. D. Phu · L. H. T. Anh · T. N. Quan · N. C. T. Anh · N. M. Phong · T. Van Khai
Faculty of Materials Technology, Ho Chi Minh City University of Technology (HCMUT), 268 Ly Thuong Kiet Street, District 10, Ho Chi Minh City, Vietnam
e-mail: phamtrungkien@hcmut.edu.vn

Vietnam National University Ho Chi Minh City, Linh Trung Ward, Thu Duc District, Ho Chi Minh City, Vietnam

P. T. Kien · B. Van Tien · H. D. Phu
Polymer Research Center, Ho Chi Minh City University of Technology (HCMUT), 268 Ly Thuong Kiet Street, District 10, Ho Chi Minh City, Vietnam

M. A. A. Mohd Salleh et al. (eds.), *Proceedings of the Green Materials and Electronic Packaging Interconnect Technology Symposium*, Springer Proceedings in Physics 289,
https://doi.org/10.1007/978-981-19-9267-4_58

impact on environment. To increase the adding value of rice industry for sustainable goal, research on the use of rice hull is focus recently in Vietnam as well as global scale [1–8]. In order to reduce the C emission to the atmosphere for carbon neutral production process, my research group suggest use the rice hull as source of biomass pyrolysis (without oxide atmosphere) instead of normal burning process (with oxide atmosphere). The treatment of biomass product can be used for rice hush, rice hush ash, and also sugarcane bagasse from our previous study [9–13]. During the biomass pyrolysis, the main product with be heat and carbon production, and the carbon were keep inside the biochar instead of CO_2 release to the open air. This biochar can be used as potential source for biochar-based power plan materials or removal the toxicity substance from wastewater. However, the pellet forming of rice hull-based biochar was hard due to the present of Silica inside the rice hull. Thus this research aims to add KOH with rice hull for easy shaping to improve the microstructure property of rice hull-based biochar.

58.2 Methodology

58.2.1 Materials and Sample Preparation

Preparation the rice hull materials: These materials were supplied by Loc Troi Group (LTG), An Giang province, Vietnam.. The rice hull materials were ground by fast-grinding machine with the power of 3 kW (3A model, Tuan Tu Agriculture manufacturer, Vietnam), then passed through the 0.45 mm sieve to collect the fined rice hull powder (RH).

Preparation of RH-based green sample: The fine RH were mixed together with KOH different KOH content such as 0, 20, and 40 wt%, respectively, to investigate the effect of different amount of KOH on the microstructure property of RH-based biochar. The KOH-content batch mixing were form to disk (10 mm Diameter × 10 mm Height) as RH/KOH-based green sample for further pyrolysis treatment.

Pyrolysis treatment: The RH/KOH-based green pellet was pyrolysis in inert atmosphere at 550 °C for 3 h based on the thermal analysis data of RH materials. The flowchart of synthesize RH/KOH-based biochar was shown in Fig. 58.1.

58.2.2 Materials Characterization Method

Phase investigation: Raw materials and production were characterized using X-ray Diffraction (XRD). The sample was ground and put in XRD machine (Bruker D8 Advance, Germany) with the 2 theta scanning from 5° to 40°, operation at 40 kV and 40 mA.

Fig. 58.1 The flowchart of synthesized RH/KOH-based biochar

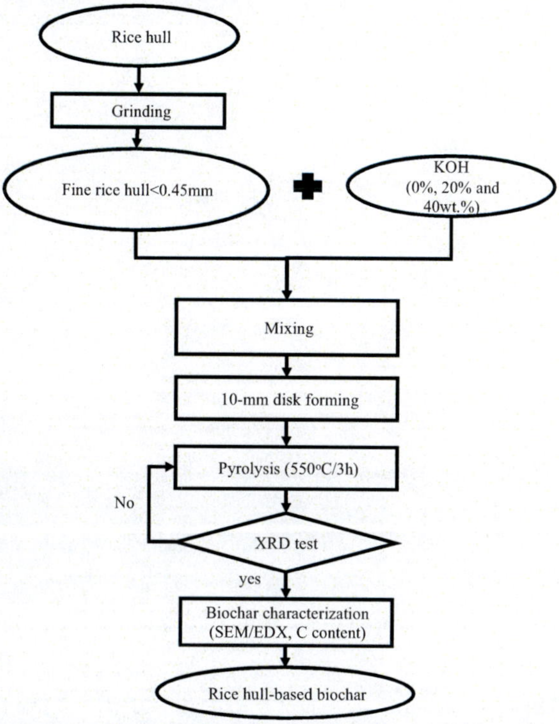

The thermal analysis using DSC/TG: The sample was heated up to 600 °C using DSC/TGA technique (STA 449, Netzsch, Germany).

The chemical composition analysis using Xray Fluorescing (XRF): The sample was characterized by Xray at 40 kV (ARL Advant, Thermo Scientific).

The morphology analysis using SEM/EDS: The sample was scattered on C tape stick to copper substrate, and carried the SEM analysis (Hitachi S-4800) at 10 kV. The amount of C, O, Si, K content was identified by EDS mapping technique.

58.3 Result and Discussions

The XRD of the fine RH was given in Fig. 58.2. Basically, the RH show the low crystal degree of Laumontite $CaAl_2Si_4O_{12}.4H_2O$ (PDF#26-1047) at 2theta = 21.5°, and the strong peak of Cristobalite SiO_2 (PDF#39-1425) at 2theta = 22°. These data indicate that the RH were composed of amorphous carbon, Ca, Al, and Si element, thus providing the good materials to keep the C content for C-zero-emission goal.

The chemical composition of RH was given in Table 58.1, with the loss on ignition (LOI) around 93 wt% indicating the cellulose layer of RH. The high LOI indicates the high energy calorific value upon burn in the inner atmosphere. Among the

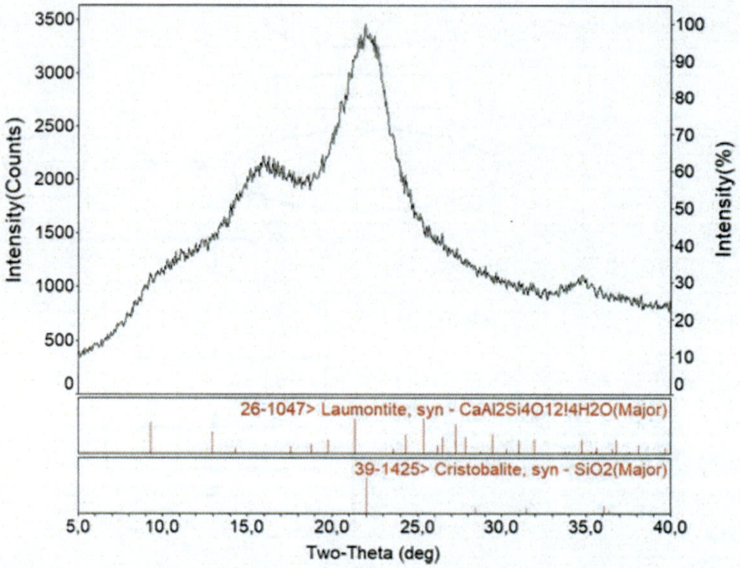

Fig. 58.2 XRD patterns of RH starting materials

Table 58.1 Element analysis of rice hull by XRF

Oxide	SiO_2	K_2O	Fe_2O_3	CaO	LOI[a]	Other	Total
Weight %	6.08	0.37	0.12	0.11	93	0.32	100

[a] *LOI* Loss on ignition

other remaining element, silica is the strongest element, around 6.08 wt% due to the adsorbent of SiO_2-containing material from farmer fertilizer.

Figure 58.3 shows the TG/DSC of RH sample. The RH exposed the endothermic peak at 100 °C due to the drying process of cellulose-based materials. In contrast, the exothermic peak of 292 and 358 °C was observed due to the pyrolysis of cellulose-based materials. The TG curve (the green line) reduces from 90 to 40 wt% at 400 °C, and shows almost no change after 500 °C. Thus, the pyrolysis temperature was selected as 550 °C for 3 h in this research.

Figure 58.4 shows the XRD patterns of biochar sample with different KOH content such as 0, 20, and 40% KOH. The 0% KOH-derived biochar indicated that after pyrolysis at 550 °C for 3 h, the biochar were composed of Quartz (PDF#46-1045) at 2 theta of 16 and 26 degree as shown in Fig. 58.3b. In contrast, the 20% KOH and 40% KOH/RH-based biochar were included the Quartz and Potassium Carbonate Hydrate (PDF#46-1045) at 2 theta of 30 and 32 degree as shown in Fig. 58.4c, d, respectively. It is obviously to produce the Potassium Carbonate Hydrate by chemical reaction of CO_2 and H_2O in cellulose structure of rice hull reacted with alkaline activator KOH during the pyrolysis process as shown in Eq. (58.1):

$$4KOH + 2CO_2 + H_2O \rightarrow 2K_2CO_3.1.5H_2O \quad (58.1)$$

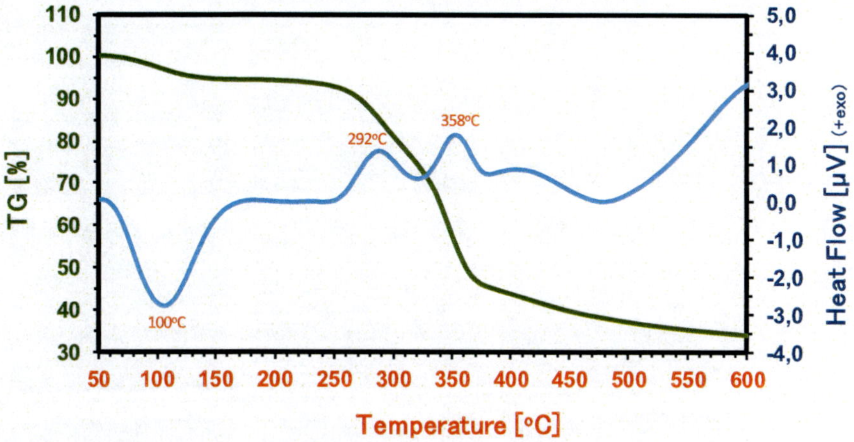

Fig. 58.3 DSC/TGA patterns of RH green sample before pyrolysis

Figure 58.5 shows the SEM images of RH mixing with KOH at different amount such as 0 wt% (Fig. 58.5a-d-g), 20 wt% (Fig. 58.5b-e-h) and 40 wt% (Fig. 58.5c-f-k) with different magnification such as 500, 5.000, and 10.000X. Upon pyrolysis, the RH change morphology from spherical-like shape (Fig. 58.5a-b-c) to plate-like shape (Fig. 58.5g-h-k) and expose the open pore due to the forming of burning gas as pyrolysis process.

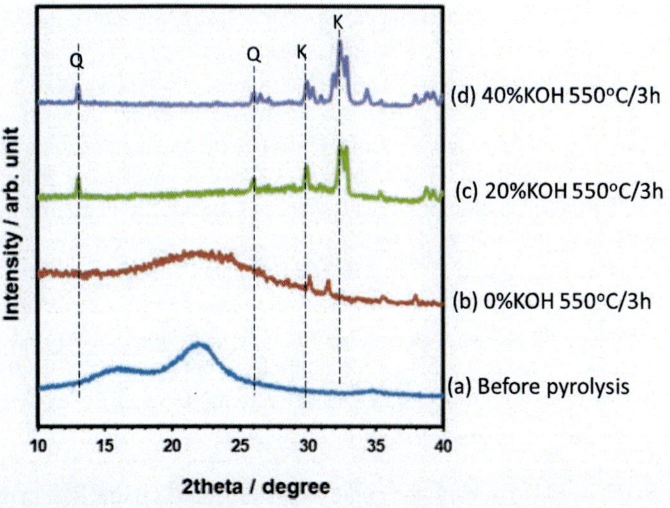

Fig. 58.4 XRD patterns of starting materials: **a** RH raw materials and biochar at different KOH content; **b** 0% KOH-derived biochar; **c** 20% KOH-derived biochar; and **d** 40% KOH-derived biochar. The pyrolysis condition was 550 °C for 3 h to obtain the biochar. Code: Q-Quartz (SiO_2) PDF#46-1045; K-Potassium Carbonate Hydrate ($K_2CO_3.1.5H_2O$) PDF#11-0655

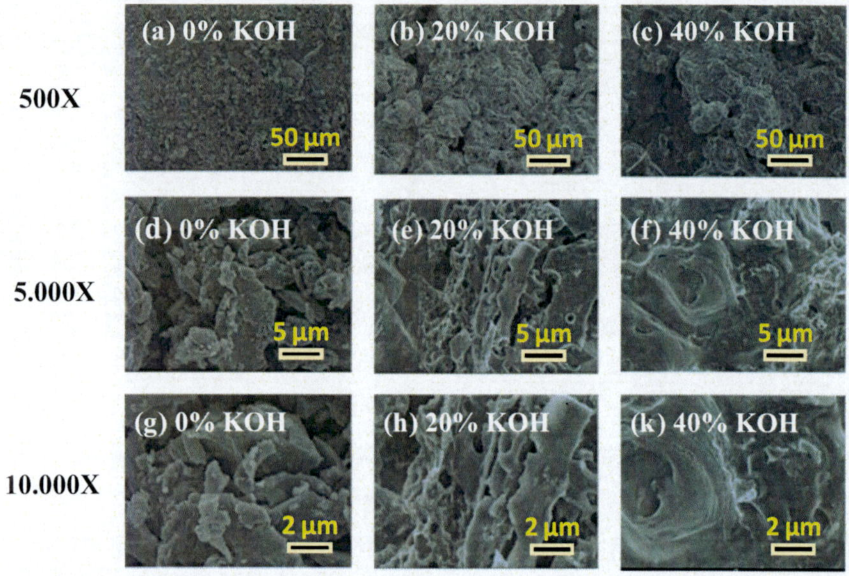

Fig. 58.5 SEM images of RH mixing with KOH at different amount such as 0 (**a-d-g**), 20 (**b-e-h**), and 40 wt% (**c-f-k**) with different magnification such as 500, 5.000, and 10.000X

Figure 58.6 shows the EDS images of RH/KOH with different amount of KOH: (a) 0 wt% KOH; (b) 20 wt% KOH; and (c) 40 wt% KOH. The C and K mapping element also was shown in these images. Upon increasing the KOH, the KOH particles distribution also increases as shown in Fig. 58.6(a-b-c) K-K layer. Figure 58.7 shows the EDS spectra of RH/KOH with different amount of KOH: (a) 0 wt% KOH; (b) 20 wt% KOH; and (c) 40 wt% KOH. Upon pyrolysis with 0 KOH (Fig. 58.7a), the peak of K could not be observed. However, when increasing the amount of KOH, the peak of K could be obtained at energy of 3.3 keV, respectively, as indicated in Fig. 58.7b, c. Based on the peak intensity of C, O, Si, and K peak, the amount of these elements were given in Table 58.2.

Apparently that the amount of C in RH/KOH biochar decrease with the increasing of KOH. As mentioned in the introduction, the present of KOH accelerates the binding of cellulose materials, resulting in easy pellet forming. It was obviously that the residual KOH and $K_2CO_3.1.5H_2O$ could block the pore size of RH/KOH-derived biochar, thus our research group suggest to activate the biochar by acidic treatment such as HCl acidic treatment to obtain the highly porous environmental treatment materials.

This research needs to put into the context of experiment conditions. In pyrolyzed technique, it is important to identify the pyrolyzed condition such as pyrolyzed temperature and pyrolyzed duration. By DSC/TG characterization technique, we can identify the pyrolyzed condition at 550 °C for 3 h when mixing with alkaline activator KOH. The main purpose of using KOH is to do chemical reaction with C-based cellulose during the pyrolysis process, thus capture C inside the biochar

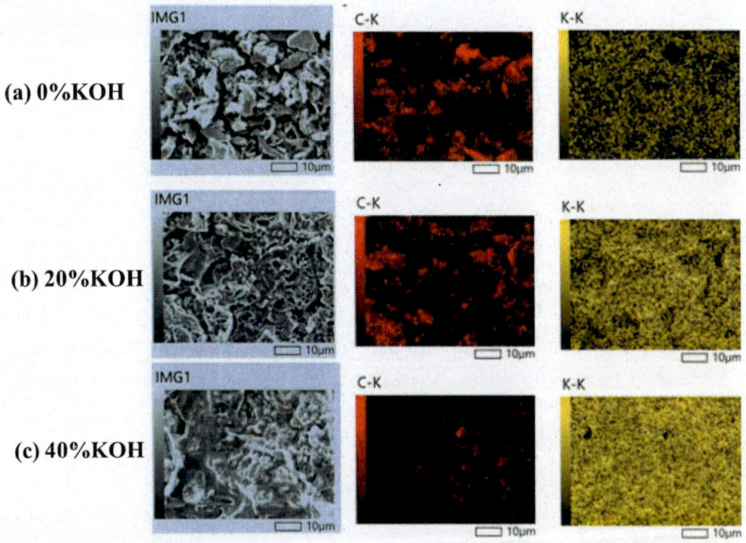

Fig. 58.6 EDS images of RH/KOH with different amount of KOH: **a** 0 wt% KOH; **b** 20 wt% KOH and **c** 40 wt% KOH. The C and K mapping element also was shown in these images

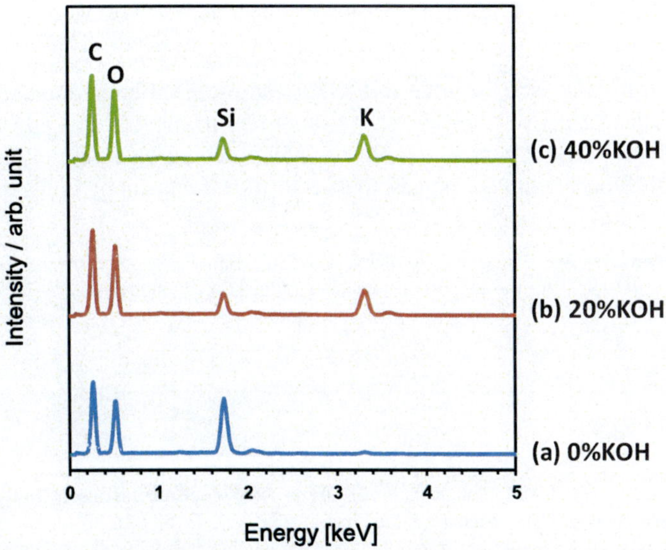

Fig. 58.7 EDS spectra of RH/KOH with different amount of KOH: **a** 0 KOH; **b** 20 KOH; and **c** 40 KOH. The C, O, Si, and K mapping element also were shown in these images

Table 58.2 Amount of C, O, Si, and K in biochar at different KOH content

Materials	Elemental analysis			
	C	O	Si	K
0% KOH	49.96	29.95	20.51	2.21
20% KOH	30.10	36.36	6.46	26.79
40% KOH	14.29	42.67	1.58	41.33

production to obtain the C-zero-emission release. Based on the XRD, SEM/EDX finding, we could observe the C capture inside the RH/KOH biochar vary from 14.29 to 49.96 wt%, depend on the KOH used. The next experiment on the activation of RH/KOH-derived biochar was under investigating to produce the activated carbon (FSB-AC) used as solid refuse fuel.

58.4 Conclusion

The aim of the study is to find the effect of adding KOH into rice hull, thus to fabrication the RH/KOH-based biochar toward the reduce CO_2 emission. The result shows that:

- Different mixing ratio between RH/KOH leading to different amount of fixed C in RH/KOH-derived biochar vary from 14.29 to 49.96% of C. In addition, the residua KOH as well as the $K_2CO_3.1.5H_2O$ could block the inside porous RH/KOH biochar, thus we suggest to do activation treatment by acidic treatment.

Acknowledgements We would like to thank Ho Chi Minh City University of Technology (HCMUT), VNU-HCM for the support of time and facilities for this study.

References

1. P.T. Kien, L.H. Tai, L.H.T. Anh, Effects of KOH on some biochar components from pyrolyzed fine sugar bagasse (FSB). Vietnam J. Sci. Technol. **59**(6A), 43–50 (2021)
2. Z. Amir, K. Sardar, Popular wood and sugarcane bagasse biochars reduced uptake of chromium and lead by lettuce from mine-contaminated soil. Environ. Pollut. **263**(Part A), 114446 (2020). https://doi.org/10.1016/j.envpol.2020.114446
3. A. Paulo, M. Cerqueira, Upgrading from batch to continuous flow process for the pyrolysis of sugarcane bagasse: structural characterization of the biochars produced. J. Environ. Manage. **285**, 112145 (2021). https://doi.org/10.1016/j.jenvman.2021.112145
4. M. Yongfei, Q. Yong, Adsorptive removal of imidacloprid by potassium hydroxide activated magnetic sugarcane bagasse biochar: adsorption efficiency, mechanism and regeneration. J. Clean. Prod. **292**, 126005 (2021). https://doi.org/10.1016/j.jclepro.2021.126005

5. A. Nahid, R. Fayez, Sugarcane bagasse biochar modulates metal and salinity stresses on microbial functions and enzyme activities in saline co-contaminated soils. Appl. Soil. Ecol. **167**, 104043 (2021). https://doi.org/10.1016/j.apsoil.2021.104043
6. N. Chengrong, Y. Xing, Impact of sugarcane bagasse-derived biochar on heavy metal availability and microbial activity: a field study. Chemosphere **200**, 274–282 (2018). https://doi.org/10.1016/j.chemosphere.2018.02.134
7. C.D. Sheng, Estimating the higher heating value of biomass fuels from basic analysis data. Biomass Bioenerg. **28**, 499–507 (2005). https://doi.org/10.1016/j.biombioe.2004.11.008
8. N.H. Thang, N.H.L. Ngoc, V.T.N. Uyen, P.T. Kien, Novel materials synthesized from red mud, bagasse, and bentonite for gas treatment by CO_2 absorption. MATEC **207**, 1–4 (2018)
9. H.T. Nguyen, V.H.Q. Pham, DT. Phong, N.Y. Nguyen, N.H. Uyen, T.K. Pham, Utilization of red mud and rice husk ash for synthesizing lightweight heat resistant geopolymer–based materials. Int. J. Eng. Sci. Invent. **7**(2), 10–15 (2018)
10. N.H. Thang, P.T. Kien, M.M.B. Abdullah, Lightweight heat resistant geopolymer-based materials synthesized from red mud and rice husk ash using sodium silicate solution as alkaline activator. MATEC **97**, 1–6 (2017)
11. M.A.B. Promentilla, N.H. Thang, P.T. Kien, H. Hinode, F.T. Bacani, S.M. Gallardo, Optimizing ternary-blended geopolymers with multi-response surface analysis. Waste Biomass Valorization **7**(29), 1–11 (2016). https://doi.org/10.1007/s12649-016-9490-8
12. P.T. Kien, H.D. Phu, N.V. Tam, N.V.K. Thanh, N.H. Thang, H. Hinode, Synthesize of calcium silicate hydrate (CSH) from vietnam rice hush. J. Sci. Technol. Food **12**, 73–80 (2017)
13. P.T. Kien, T.P.Q Nguyen, N. H. Thang, Effect of hydrothermal treatment temperature on morphology of obtained calcium silicate from rice husk ash. J. Sci. Technol. Dev. **19**(K6), 69–75 (2016)

Chapter 59
Research on Mechanical Properties and Anti-vibration Performances of NR/EPDM/Carbon Black Rubber Compounds

Quoc Phu Phan, Thi Ngoc Diem Huynh, Xuan Viet Cao, Thi Thai Ha La, and Dai Phu Huynh

Abstract Rotating machines operated at high speeds frequently create vibrations, which greatly reduce their accuracies and lifetimes. Besides, the vibrations generate noise pollutions, and therefore, affect the human brain and nervous system. To minimize the effects of the vibrations from machines, natural rubber (NR) has been commonly used for isolating the vibration sources thanks to its high elasticity and viscosity. However, several drawbacks in heat and ozone resistances may significantly decrease the lifespan of NR. As a solution to overcome this issue, ethylene propylene diene monomer rubber (EPDM), a high heat- and weather-resistant elastomer, and carbon black were selected and mixed with NR to form the NR/EPDM/carbon black compound. In this research, the mechanical properties and anti-vibration performances of the NR/EPDM/carbon black compounds were demonstrated through the tensile and compression tests. Furthermore, the EPDM and carbon black filler effects on the mechanical and anti-vibration performances of the synthesized compounds were presented. As a result, the rubber compound (composed of 40 phr of NR and 60 phr of EPDM) filled with 30 phr of carbon black showed the high mechanical properties with the anti-vibration efficiency of ~90%.

Keywords EPDM rubber · Natural rubber · Anti-vibration pad

Q. P. Phan (✉) · T. N. D. Huynh · X. V. Cao · T. T. H. La · D. P. Huynh
Department of Polymer Materials, Faculty of Materials Technology, Ho Chi Minh City University of Technology (HCMUT), 268 Ly Thuong Kiet Street, District 10, Ho Chi Minh City, Vietnam
e-mail: pqphu@hcmut.edu.vn

D. P. Huynh
e-mail: hdphu@hcmut.edu.vn

Vietnam National University Ho Chi Minh City (VNUHCM), Linh Trung Ward, Thu Duc City, Ho Chi Minh City, Vietnam

D. P. Huynh
Polymer Research Center, Ho Chi Minh City University of Technology (HCMUT), 268 Ly Thuong Kiet Street, District 10, Ho Chi Minh City, Vietnam

M. A. A. Mohd Salleh et al. (eds.), *Proceedings of the Green Materials and Electronic Packaging Interconnect Technology Symposium*, Springer Proceedings in Physics 289,
https://doi.org/10.1007/978-981-19-9267-4_59

59.1 Introduction

Rotating machines that work at high speed have enabled an appearance of vibrations, making noise, reducing their accuracy, and shortening their lifetime. In 1961, the first book about shock, vibration, anti-vibration theories, and measurements was introduced by Harris and Pierol [1]. Up to date, scientists and researchers have gained lots of research on minimizing the vibrations from machines using various methods such as controlling their natural frequency, decreasing their vibration's amplitude, or using passive systems such as springs, dampers, and rubber mounts. As one of the most effective, economical, and simplest methods, the uses of rubber materials help limit and even eliminate the vibrations during the machine's operations. Due to its elastic and viscous properties, natural rubber (NR) can disperse the vibration energy from machines to its structure leading to the reduction of the vibration phenomenon. As results, various numerical studies not only in theory and simulation but also in experiments about the effect of stiffness, creep, and viscoelastic behaviors of rubbers on the anti-vibration properties have been demonstrated [2–4]. Based on that account, the uses of the diverse rubber compounds that ensure both mechanical properties and good anti-vibration performances have been also researched [5–7]. Besides that, some blended rubber compounds of NR mixed with various synthetic rubbers were also focused on vibration resistance applications [8–10]. As one of the tricky requirements, the lifetime of anti-vibration rubber over the working time of machine is required. After a period of research, NR has been showing good properties of damping vibrations, however, the usage of NR also has its drawbacks in heat and ozone resistances which decrease its uses [11, 12]. To overcome the low environmental stability of NR, there have been various researches focusing on the combinations of NR with other synthetic rubber types, which have good weather resistance, heat resistance, and ozone resistance, for increasing the lifespan of the anti-vibration products. As results, there are lots of research have illustrated that the combinations between the synthetic rubbers with NR presented improvements in the mechanical and physical properties of rubber compounds [13–15]. In this research, the compounds of NR with ethylene propylene diene monomer rubber (EPDM) and carbon black filler were evaluated for the possibility of application in the anti-vibration area. During evaluation the mechanical properties and vibration resistances of the various NR/EPDM/carbon black rubber amounts, the optimal composition was found for effective damping elastomeric isolators.

59.2 Methodology

59.2.1 Materials

Natural rubber (SVR 3L), ethylene propylene diene monomer (EPDM) rubber, stearic acid, 2,2,4-trimethyl-1,2-dihydrquinoline (TMQ), phenolic 2,6-di-tert-butyl-*p*-Cresol (BHT), zinc oxide, paraffin oil (P150), carbon black (N330), 2–mercapto-benzothiazole (MBT), disulfur tetramethylthiuram (TMTD), and sulfur were bought from Hiep Phat Hung Services Trading Co., Ltd, Vietnam. The samples were fabricated with the recipes for the vulcanization systems with the compositions in parts per hundred of rubber (phr), including: SVR 3L of 20–40 phr, EPDM of 60–80 phr, stearic acid of 2.0 phr, N330 of 30–50 phr, TMQ and TMTD of 1.0 phr, ZnO of 5.0 phr, P150 of 2.5 phr, BHT of 0.5 phr, MBT and sulfur of 1.5 phr.

All ingredients except MBT, TMTD, and sulfur were mixed on an internal mixer (MX500-D75L90, Thailand) with the rotor speed at 60 rpm at 60 °C for 10 min following with the mixing with the remaining vulcanization materials using the two-roll mill (Deposes, France) to produce compounded rubbers. The rubber compounds were vulcanized at 150 °C under a pressure of 50 kgf with the curing time was identified from rheographs.

59.2.2 Methods

Cure characteristics: The curing and the thermal aging processes of the rubber compounds were measured using the Oscillating Disk Rheometer (ODR described in the ASTM D2084) at a temperature of 150 °C with pressure at 2.0 MPa applied for 30 min.

Hardness test: The hardness values were determined using the ASTM D2240 standard using the Shore A Durometer Hardness tester. All the tested samples were spherical pieces with a thickness of 6 mm and stored at room temperature for 72 h before measuring.

Tensile test: The static mechanical properties of the fabricated samples were tested using a Testometric M350–10CT machine. The ASTM D412 standard was followed in the measurements with the crosshead speed at 500 mm/min. Besides that, the tearing strength of the synthesized samples was also evaluated with the same measurement condition.

Compression test: The compression properties of the materials were evaluated using a Testometric M350–10CT machine with the crosshead speed was 2 mm/min according to the ASTM D695 standard. All samples were fabricated in cylinder shapes of approximately 29.0 mm in diameter and 12.5 mm in height sizes. Due to the compression test process, the deflection and the force were determined to calculate the stiffness (K), the natural frequency (f_{n}), the forced frequency (f_d), the transmission coefficient (T), and the anti-vibration performance (H) according to some of the formulas below [1].

$$K = F/d \tag{59.1}$$

$$f_n = 3.13 \times (K/W)^{1/2} \tag{59.2}$$

$$f_d = f_{\min}/60 \tag{59.3}$$

$$T = \left((f_d/f_n)^2 - 1\right)^{-1} \tag{59.4}$$

$$H = (1 - T) \times 100(\%) \tag{59.5}$$

where: F is the force corresponding to the static deflection (N), d is the static deflection (mm), W is the design load (15 kg), and $f_{\min}$ is the Duress frequency (60 Hz).

Compressive residual deformation test: The rubber cylinder samples of 29.0 mm diameter and 12.5 mm height sizes were compressed to a distance of 9.5 mm at room temperature for 72 h for determining the deformation properties following the ASTM D395 standard. The deformation of the sample was calculated below:

$$C_B = (t_o - t_i)/(t_o - t_n) \times 100(\%) \tag{59.6}$$

where: t_o is the original specimen thickness (mm), t_i is the specimen thickness after testing (mm) and t_n is the spacer thickness during the test (mm). The t_i values were measured after stabilizing the compressed samples at room temperature for 30 min.

59.3 Result and Discussions

59.3.1 *Effect of EPDM/NR Ratios on Mechanical and Anti-vibration Performances*

The rheological properties of the EPDM/NR/carbon black rubber compounds were presented in Fig. 59.1. It can be seen that the change in the ratio amounts of the EPDM and NR affects slightly on the pre-vulcanization time (T_{10}), however, it takes the significant effect on the optimal vulcanization time (T_{90}). The higher NR amount is replaced by the EPDM rubber in the mixed compound, the longer curing time is required. Besides that, the maximum torque gives an idea about the crosslink density and the degree of reinforcement of the EPDM rubber in the matrix. The increase in maximum torque value is due to the increase of the restriction to the NR chain

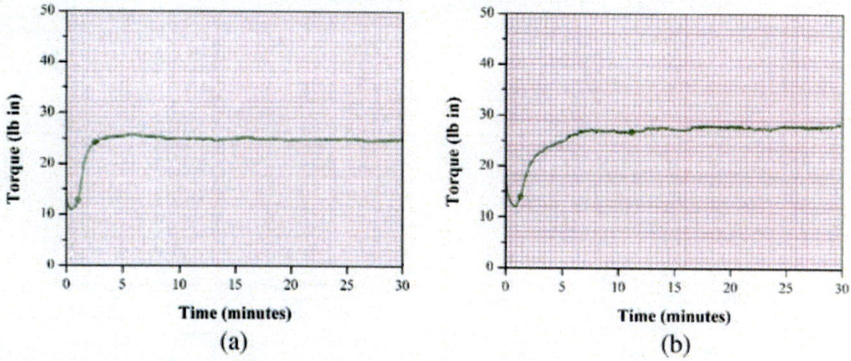

Fig. 59.1 Rheographs of the prepared rubber compounds with the different EPDM/NR ratios of 6/4 (**a**) and 7/3 (**b**) (carbon black composition of 40 phr)

mobility and viscosity caused by the EPDM rubber. Besides that, the changes in the torque values after curing in the measurement also indicate the thermal aging process. The sample contained with 60 phr EPDM rubber indicates the slight reduction of the torque level due to the thermal aging of NR at 150 °C. The positive effect of the EPDM rubber in the fabricated compound is illustrated from the stable line of the sample with the lower amount of NR.

Figure 59.2a indicates that the higher EPDM amounts contained in the samples lead to the increase in the hardness values. As results, the fabricated samples present suitable hardness for the rubber pad using in the anti-vibration application. Originally, the mixing of the EPDM rubber into the NR matrix is difficult and exhibits the representation of the phase interfaces. The present of the phase interface leads to the lower tearing strength of the samples from 15.847 N/mm to 12.788 N/mm when the amounts of the EPDM rubber is increased from 60 to 70 phr, respectively. Contrastingly, the sample contained the enormous EPDM amount shows the enhancement of the tearing strength at 15.444 N/mm (Fig. 59.2b).

In addition, the stress at peak and Young's modulus properties of the rubber compounds were characterized via tensile measurement (Fig. 59.3a). It can be seen that the increase of the EPDM amount leading to the increase of Young's modulus and the decrease of the stress at peak values. Particularly, the highest amount of the EPDM rubber was used in the sample indicating the highest values of the Young's modulus of 2.105 N/mm^2. However, the highest stress at peak is achieved by the sample contained with 60 phr EPDM which values at 4.895 N/mm^2. Furthermore, the compression test results of the rubber compounds with different ratios of the EPDM and NR were also evaluated. Regarding the testing results, the anti-vibration abilities of the samples were also calculated and presented in Table 59.1. It can be seen that the rubber compound containing 60 phr of EPDM in the rubber matrix exhibits the anti-vibration efficiency of 81.44% which is higher than the rubber compounds used with higher EPDM levels. Meanwhile, the anti-vibration property of the rubber compound is embedded with elastic and viscous properties. As shown in Fig. 59.3b,

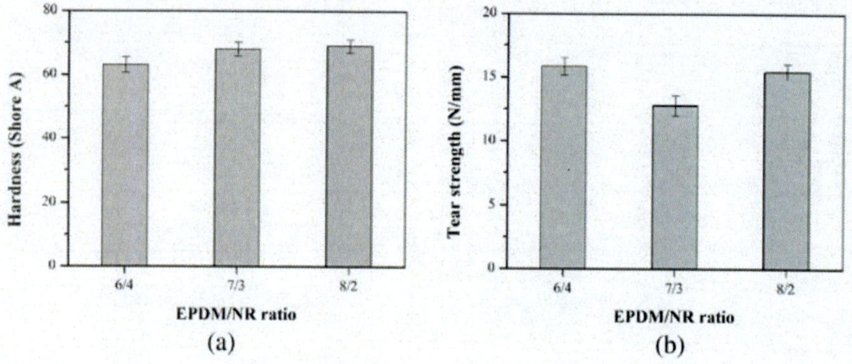

Fig. 59.2 Hardness Shore A (**a**) and tearing strength (**b**) of the synthesized samples with different EPDM/NR ratios

the increase of the EPDM amount leads to the increase of the Young's modulus and deformation. The lower Young's module indicates the more flexible of the rubber chain in the sample which supports the reduction of the vibration. Besides, the higher deformation indicates a higher shearing ability of the rubber chains in the compound which causes the irreversible molecular chain slippage. The shearing ability of the EPDM part is higher than that of the NR part due to the lower cross-linkages formed between the EPDM chains.

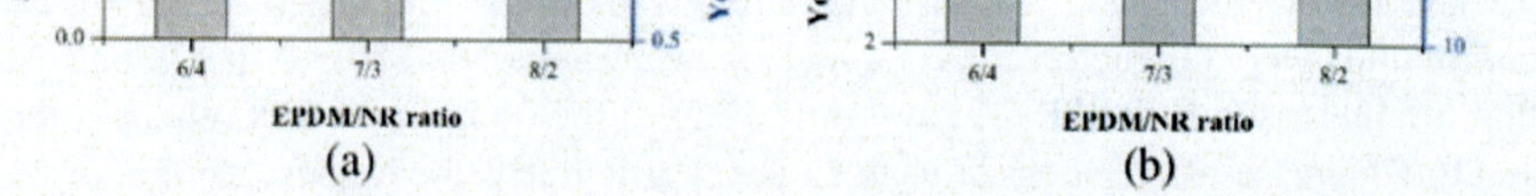

Fig. 59.3 Stress at peak and Young's modulus of the tensile test (**a**) and Young's modulus and deformation of the compression test (**b**) with different EPDM/NR ratios

Table 59.1 Compression measurement data of the various ratios of EPDM/NR

Sample	Force at 0.1 mm (N)	Stiffness K (N/mm)	Natural frequency (Hz)	Transmission coefficient T	Anti–vibration performance (%)
6/4	33.34	333.4	23.74	0.19	81.44
7/3	36.38	363.8	24.80	0.21	79.40
8/2	49.85	498.5	29.03	0.31	69.44

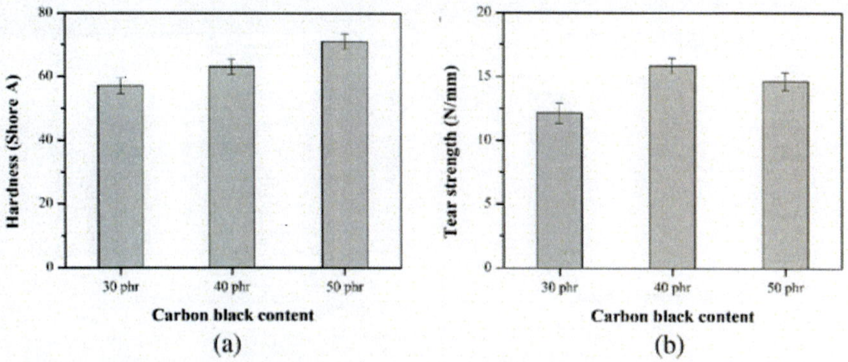

Fig. 59.4 Hardness shore A (**a**) and tearing strength (**b**) of the samples with different carbon black contents

59.3.2 Effect of Carbon Black Contents on Mechanical and Anti-vibration Performances

In this study, the carbon black filler amounts were varied from 30 to 50 phr when the ratio of the EPDM/NR was fixed at 6/4. Figure 59.4a shows that the sample contained with 50 phr carbon black filler shows the highest value of 71 Shore A. The high content of carbon black is used leading an increase in hardness. The tear strength of the fabricated samples was reduced from 15.847 to 14.646 N/mm when the carbon black amounts were increased from 40 to 50 phr, respectively, (Fig. 59.4b). Generally, the addition of the carbon black into the rubber compounds will enhance the tear strength due to the connection formation between rubbers and fillers. However, the overuse leads to the aggregate of carbon black particles in the rubber compound which affects the strength of the sample.

In the tensile test results, the sample contained 40 phr carbon black filler presents the highest stress at peak value of 4.895 N/mm^2 (Fig. 59.5a). The reason for the increase of the stress at peak is related to the formation of the interaction between filler and rubber during the measurement. Particularly, the sample contained with 50 phr of carbon black filler presents the highest Young's modulus at 1.981 N/mm^2. According to Payne's law, the increase of Young's modulus is related to the increase in flow restriction in the compounds caused by the higher adsorption of the rubber chains on the surface of the carbon black particles due to the increase of the filler amounts. Besides that, the compression test was also used for the rubber compounds with different amounts of carbon black.

Table 59.2 shows the changes in the anti-vibration performance values of the fabricated rubber samples. The best anti-vibration performance is found with the sample containing 30 phr carbon black which values at 92.61%. When the carbon black filler in the rubber compound is increased to 40 phr, the anti-vibration characteristic decreases to 81.44% due to the forming of more bonding between the filler and rubber. Oppositely, the slight enhancement of the anti-vibration performance

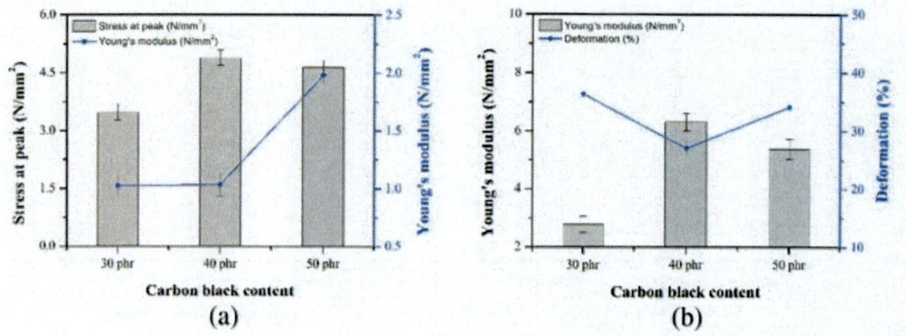

Fig. 59.5 Stress at peak and Young's modulus of the tensile test (**a**) and Young's modulus and deformation of the compression test (**b**) with different carbon black contents

Table 59.2 Compression measurement data of the various contents of carbon black

Sample	Force at 0.1 mm (N)	Stiffness K (N/mm)	Natural frequency (Hz)	Transmission coefficient T	Anti-vibration performance (%)
30 phr	14.66	146.6	15.74	0.07	92.61
40 phr	33.34	333.4	23.74	0.19	81.44
50 phr	28.43	284.3	21.92	0.15	84.59

from the sample containing of 50 phr carbon black occur due to the vibration adsorption ability of the filler part. Figure 59.5b indicates that the sample containing 30 phr carbon black produces the lowest Young's modulus and the highest deformation results. The lowest Young's module together with the highest deformation indicates not only the most flexible of the rubber chain but also the biggest molecular chain slippage in the compound. The samples with higher amounts the carbon black showed higher Young's modulus values due to the formation of more interaction between the fillers and rubbers chains. As can be seen in Fig. 59.5b, the deformation of sample containing 50 phr carbon black filler is higher than that of the sample with 40 phr carbon black filler. That mean due to the increase of the filler amounts, the excess carbon black might lower the elastic properties of the prepared sample.

59.4 Conclusion

The obtained results indicated the positive effect of using the NR/EPDM/carbon black rubber compound. By replacing 60 phr of NR with EPDM, the thermal aging process from the synthesized sample was controlled which indicates the potential for the long life using of the rubber pad. Besides that, the carbon black amount was also identified which presented the high mechanical properties as well as the effective anti-vibration abilities. In comparison with the requirements of the anti-vibration rubber pads, the sample with the recipe of using about 30 phr carbon black achieved the suitable hardness and the high anti-vibration performance values at 57 Shore A and 92.61%, respectively. Based on the experiment results, the NR/EPDM/carbon black rubber compound illustrates the high applicability for the anti-vibration products.

Acknowledgements We acknowledge Ho Chi Minh City University of Technology (HCMUT), VNU-HCM for supporting this study.

References

1. C.M. Harris, C.E. Crede, *Shock and Vibration Handbook Vol I Basic Theory and Measurements* (McGraw-Hill Book Company Incorporated, 1961)
2. D. Leng, K. Xu, L. Qin, Y. Ma, G. Liu, A hyper-elastic creep approach and characterization analysis for rubber vibration systems. Polymers (Basel) **11**(6), 988 (2019). https://doi.org/10.3390/polym11060988
3. S. Santhosh, V. Velmurugan, V. Paramasivam, S. Thanikaikarasan, Experimental investigation and comparative analysis of rubber engine mount vibration and noise characteristics. Mater. Today Proc. **21**, 638–642 (2020). https://doi.org/10.1016/j.matpr.2019.06.730
4. F. Carleo, E. Barbieri, R. Whear, J. Busfield, Limitations of viscoelastic constitutive models for carbon-black reinforced rubber in medium dynamic strains and medium strain rates. Polymers (Basel) **10**(9), 988 (2018). https://doi.org/10.3390/polym10090988
5. B.S. Kumar, P.R.M. Raju, C. Ratnam, Dynamic mechanical properties of magnetorheological polyurethane elastomers in comparison to existing antivibration rubbers. Adv. Mater. Process. Technol., 1–15 (2022). https://doi.org/10.1080/2374068X.2022.2036449
6. P.Q. Phu, N.K. Tien, Effects of several factors on anti–vibration ability of EPDM rubber. Vietnam J. Sci. Technol. **55**(1B), 202–207 (2017). https://doi.org/10.15625/2525-2518/55/1B/12109
7. N.M. Shukri, A.R. Azura, Effect of different types of rubbers and carbon black grades on curing characteristics and dynamic properties for anti-vibration application. IOP Conf. Ser.: Mater. Sci. Eng. **548**(1), 12002 (2019). https://doi.org/10.1088/1757-899X/548/1/012002/meta
8. S. Akyüz, M.E. Darı, Y.E. Esiyok, M.A. Ermeydan, Effects of NR/SBR ratio on mechanical properties and artificial mechanical performance of anti-vibration bushings. Iran. Polym. J. **30**(12), 1317–1328 (2021). https://doi.org/10.1007/s13726-021-00971-4
9. M. Banić, D. Stamenković, A. Miltenović, D. Jovanović, M. Tica, Procedure for the selection of rubber compound in rubber-metal springs for vibration isolation. Polymers (Basel) **12**(8), 1737 (2020). https://doi.org/10.3390/polym12081737
10. N. kadum Abd Ali, M.M. Farhan, A.S. Moosa, The effect of shape factor on the operation periods of anti-vibration rubber. Ind. Eng. Lett. **7**(1), 31–38 (2017)
11. F. Grasland, L. Chazeau, J.-M. Chenal, About thermo-oxidative ageing at moderate temperature of conventionally vulcanized natural rubber. Polym. Degrad. Stab. **161**, 74–84 (2019). https://doi.org/10.1016/j.polymdegradstab.2018.12.029
12. N.T. Thuong, Y. Yamamoto, P.T. Nghia, S. Kawahara, Analysis of damage in commercial natural rubber through NMR spectroscopy. Polym. Degrad. Stab. **123**, 155–161 (2016). https://doi.org/10.1016/j.polymdegradstab.2015.11.025
13. W. Wang et al., Enhanced thermal stability of gamma radiation vulcanized polybutadiene rubber (PBR)/nature rubber (NR) blends with sulfur added. Mater. Lett. **186**, 186–188 (2017). https://doi.org/10.1016/j.matlet.2016.09.128

14. Z. Mao, W. Wang, G. Mao, Improving the thermal aging resistance of γ-vulcanized polybutadiene rubber (BR)/nature rubber (NR) blends with sulfur added. Adv. Mater. Sci. Eng. **2021** (2021). https://doi.org/10.1155/2021/5782539
15. H. Nabil, H. Ismail, Enhancing the thermal stability of natural rubber/recycled ethylene–propylene–diene rubber blends by means of introducing pre-vulcanised ethylene–propylene–diene rubber and electron beam irradiation. Mater. Des. **56**, 1057–1067 (2014). https://doi.org/10.1016/j.matdes.2013.12.020

Chapter 60
A Review: Preliminary Study on Corrosion Behavior of Aluminum Metal Matrix Composites (MMCs) Reinforced with Strontium (Sr)

R. Rosmamuhamadani, N. F. M. Razali, N. N. A. Basir, S. Osman, M. M. Mahat, S. M. Yahaya, and M. N. Zakaria

Abstract Aluminum metal matrix composites (MMC) are potential materials for many applications due to their outstanding physical and mechanical properties. Reinforcements in the metallic matrix enhance the qualities of stiffness, specific strength, wear, creep, and fatigue when compared to conventional engineering materials. The discussion covered the corrosion behavior of strontium (Sr) reinforced with aluminum–magnesium and aluminum–copper. Utilizing electrochemical studies such electrochemical impedance spectra, linear polarization tests, open-circuit potential, and potentiodynamic tests, the corrosion resistance of the Al-MMC composites created in this study was examined. The linear polarization resistance (LPR) method of the Gamry Reference Potentiostat is used to evaluate the rate of corrosion of different Al-MMCs. Sr is frequently utilized in the casting industry to modify hypoeutectic aluminum–silicon (Al–Si) alloys' eutectic silicon. Sr has been found to increase the amount and severity of casting faults, particularly porosity and oxide inclusions, according to the reviews that have been done. This study made an effort to assess the effects of the reinforcing material Sr used in the synthesis of Al-MMC on the materials' corrosion behavior.

R. Rosmamuhamadani (✉) · N. F. M. Razali · N. N. A. Basir · S. Osman · M. M. Mahat
Material Science and Technology Program, Faculty of Applied Sciences, Universiti Teknologi MARA, 40450 Shah Alam, Selangor, Malaysia
e-mail: rosma614@uitm.edu.my

S. M. Yahaya
Applied Chemistry Program, Faculty of Applied Sciences, Universiti Teknologi MARA, 40450 Shah Alam, Selangor, Malaysia

M. N. Zakaria
Eco-Technology Program, Faculty of Applied Sciences, Universiti Teknologi MARA, 40450 Shah Alam, Malaysia

Selangor School of Materials and Mineral Resources Engineering, Universiti Sains Malaysia, 14300 Nibong Tebal, Penang, Malaysia

M. A. A. Mohd Salleh et al. (eds.), *Proceedings of the Green Materials and Electronic Packaging Interconnect Technology Symposium*, Springer Proceedings in Physics 289,
https://doi.org/10.1007/978-981-19-9267-4_60

Keywords Aluminum metal matrix composite · Corrosion behavior · Strontium · Reinforcing material

60.1 Introduction

Metals that have been strengthened by additional metal, ceramic, or biological elements, continuous or irregular fibers, whiskers, or particles are known as MMCs. They are made by mixing reinforcing materials throughout a metal matrix. The qualities of the base metal, such as strength, stiffness, conductivity, and so forth, are frequently increased by the use of reinforcements. MMC is produced using a variety of techniques, including co-spray deposition, powder metallurgy, stir casting, and vacuum hot pressing. The degree of porosity, particle dispersion, and matrix/reinforcement interfacial characteristics are all controlled during MMC manufacture, and they all have an impact on the MMC's behavior processes [1].

Al-MMC is an improved version of MMC that can meet the most recent requirements for cutting-edge technological applications in the field of engineering. The convenience of standard manufacturing techniques, an improvement in mechanical qualities, and a reduction in production costs all meet these needs. These materials have caught the interest of numerous industries due to their various advantages. In addition to being strong, they are lightweight and have a lower density than their monolithic cousins. In order to facilitate heat treatment and boost strength at room temperature and higher, Cu and Mg are frequently added. Strontium has also been found to have the potential to be a successful grain refiner for magnesium alloys [2].

There is a lot of aluminum in the crust of the earth. It is also the most prevalent metal and the third most prevalent chemical element. Aluminum alloys are frequently used as the main matrix component in composite materials, because a wide range of technical applications could employ it as a favored alternative material. Aluminum and its alloys have received the greatest attention as base metals in metal matrix composites in the modern world. They are frequently used in applications that need weight carrying, such as bridge decks, naval vessels, and light railways. Al alloy can be combined with other materials including magnesium, titanium, and zinc to create composite materials. Due to their unique mix of mechanical and physical characteristics, including chemical and corrosion resistance, low weight, and mechanical strength, these materials are produced [3].

The good qualities of aluminum also include its low density, excellent castability, high stiffness, high strength, extraordinary low thermal coefficient of expansion, and excellent tribological properties. These materials are used in a variety of industries, including aerospace, automotive, biotechnology, electrochemistry, electronics, maritime, nuclear, and thermal. They offer strong mechanical qualities and are corrosion- and wear-resistant. When compared to other MMCs, the production of Al-MMC is also less expensive [4].

When aluminum is exposed to air and water, a thick oxide coating forms due to the metal's thermodynamically reactive nature. The generated oxide layer shields the metal from corrosion, while reactive substances like chlorides and sulfates in aqueous solutions cause pitting corrosion. When anodized, reinforced parts dissolve more quickly than aluminum does, creating a heterogeneous layer that prevents corrosion. According to the available literature, aluminum's microstructural and macrostructural morphologies significantly affect corrosion resistance [5].

Based on the literature review, the performance of stretchable conductive ink is very important in stretchable electronic applications. Many works have been done on the mechanical examination of different conductive inks to study their behavior under tensile or compression loadings. However, very limited research has been conducted on the effect of cyclic loading on conductive polymers. The main objective of this research was to study the behavior of proposed silver-based conductive ink under cyclic tensile loading. The ink was tested using two modes: stress-controlled and strain-controlled mechanisms. Lastly, the surface morphology and element composition of the ink after cyclic loading were analyzed using finite element scanning electron microscopy (FESEM) and energy disruptive X-ray (EDX) spectroscopy, respectively.

60.2 Result and Discussions

60.2.1 Aluminum–Magnesium (Al–Mg)

Less flammable and lighter than conventional aluminum and magnesium alloys are aluminum–magnesium (Al–Mg) alloys. Al–Mg casting alloys offer a wide range of applications due to their exceptional mechanical qualities, particularly in the automotive and shipbuilding industries.

The two oxides that are present in Al–Mg alloys' oxide layer have essentially diametrically opposing, if not contradicting, characteristics. Magnesia (MgO), which is a porous oxide, has an RPB ratio of 0.81 compared to alumina (Al_2O_3), which has an RPB ratio of 1.283. Due to magnesium's higher coefficient of diffusion than aluminums, MgO rather than alumina develops at the metal/oxide interface [3].

Magnesium makes up 14.9 wt% at 451 °C and 1.7 wt% at room temperature in solid form, making it one of the two most soluble elements in aluminum (the other being zinc). The open-circuit potential of aluminum in solid solution is unaffected by magnesium. By reducing the cathodic process, it considerably increases the resistance of aluminum to corrosion [6].

60.2.2 Aluminum–Copper (Al–Cu)

It has long been known that copper weakens 99.99% of aluminum's ability to resist corrosion. According to [6], copper may have an impact on the oxide film's quality.

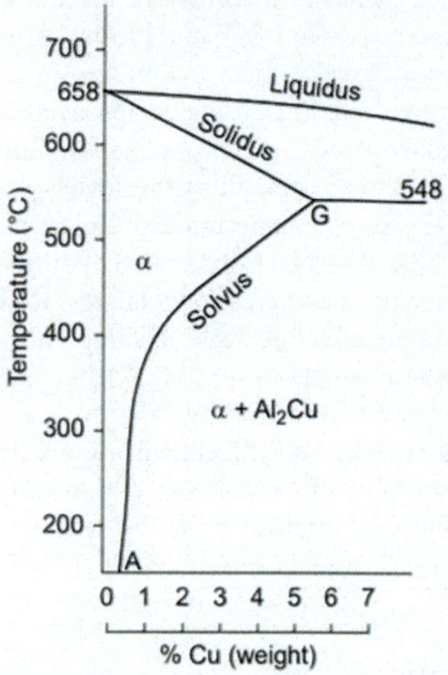

Fig. 60.1 Binary diagram of Al–Cu [3]

Copper addition to aluminum has two results. The crystal nucleation density of g-Al_2O_3 alumina decreases from 3.109 on aluminum with 0.09 wt% copper to 9.106 wt% with 4 wt% copper, irrespective of temperature. Second, there was an increase in the rate of amorphous alumina synthesis between g-Al_2O_3 crystals.

As long as copper is in solid solution and presents in concentrations less than 0.20 wt%, it has no impact on corrosion resistance. By delaying the onset and propagation of metastable pits, copper additions, in general, decrease the likelihood of stable pit formation. The presence of Al–Cu intermetallics or redeposited copper, not solid copper, is what causes structural heterogeneities and cathodic corrosion sites, despite the fact that this may seem to go against technical requirements [6].

Al–Cu alloy binary diagram is shown in Fig. 60.1. It demonstrated that copper has a detrimental effect on the quality of the oxide film. This explains why the corrosion resistance of copper-containing wrought alloys (2XXX and 7XXX series) and casting alloys (4XXX series) is restricted [3].

60.2.3 Effect of Strontium (Sr)

It was anticipated that the creation of compounds containing Sr would be very beneficial in terms of improving the higher temperature properties. Sr significantly increased the tensile strength and elongation at high temperatures. At high temperatures, strontium markedly improved the tensile strength and elongation. Nowadays, strontium is the most often used modifying agent in aluminum foundries due to its much higher modification capability than sodium (Na). Thus, it has become standard practice to add 0.015–0.025 wt% Sr to Al–Si alloys in order to improve mechanical properties [7]. Silicon-based casting alloys (4xx.x series) benefit from the eutectic's refinement by adding a few hundred ppm of strontium. The best alloy characteristic for predicting % porosity, largest pore size, and greatest pore area was found to be strontium [8].

As pore nucleation sites, oxide inclusions in aluminum melt perform better when Sr is present, it has been shown. The amount of Sr oxides in the solidified structure determines the variation in porosity characteristics. Porosity frequently lowers mechanical properties, producing castings of poor quality. Strontium additions also affect how finely the grains of alloys used in aluminum foundries are refined. Strontium is insoluble in aluminum, where it forms the intermetallic Al4Sr. Strontium has little impact on the corrosion resistance of aluminum, according to the electrochemical characteristics of binary alloys Al–Sr with Sr varying from 0.03 to 2.45 wt% [7].

The corrosion behavior of the MMCs with and without 0.01 wt% Sr was assessed using a potentiodynamic polarization test in a 0.5M H_2SO_4 solution, as shown in Fig. 60.2. As observed in the graph, the composite with the addition of Sr has a higher probability for failure than the basal composite (0 wt% Sr). Table 60.1 displays the corrosion potential (E_{corr}, VSCE), corrosion rate, and corrosion current density (i_{corr}) of samples taken from the polarization curves. The corrosion potential of the base composite, which is 30 mV nobler than the composite with Sr, was −1.067 V versus SCE. As a negative impact on corrosion resistance, the inclusion of Sr reduced the size of the Mg_2Si particle and increased its density, resulting in more Al/Mg_2Si interfaces. The Sr-infused composites had a lower corrosion current density (1.8 A/cm^2) than base composites (0.58 A/cm^2) [7].

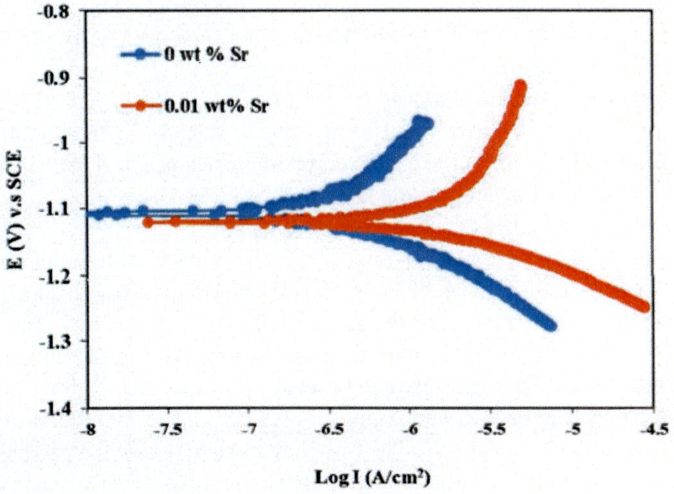

Fig. 60.2 Potentiodynamic polarization test in 0.5M H_2SO_4 [7]

Table 60.1 Electrochemical parameters of the Al–20Mg_2Si MMCs in H_2SO_4 attained from the polarization test [7]

Sr (wt%)	E_{corr} (V)	I_{corr} (A cm^{-2}) (10^6)	CR (mm y^{-1}) (10^{-3})
0	−1.067	0.58	0.2
0.01	−1.096	1.8	0.8

60.2.4 Corrosion Behavior of Al-MMCs

The existence of heterogeneities, such as the reinforcement phase, microcracks, porosity, secondary phase precipitates, and interaction products, is directly related to the corrosion behavior of MMCs. Porosity, the precipitation of intermetallic phases inside the matrix, high dislocation densities at the particle–matrix interfaces, the presence of an interfacial reaction product, and the electrical conductivity of the reinforcement are factors that affect the corrosion of composites [5].

60.2.5 Long-Term Exposure Studies

The requirements of sample preparation for this test involved the samples before and after being located in test cabinet during exposure and the position of samples in the test cabinet. In corrosion test done by using salt spray fog, samples were cut into the required dimension according to ASTM B-117, 2011, standard practice for operating salt spray (fog) apparatus.

The weight loss of an Al–6 wt% Cu alloy reinforced with TiB_2 and exposed to a salt environment for one month is shown in Fig. 60.3. The samples were tested in a cabinet using a 5% NaCl solution. Figure 60.1 shows an Al–6 wt% Cu alloy with 3 wt%. In comparison to Al–6 wt% Cu with 6 wt% TiB_2 and Al–6 wt% Cu alloy as cast, TiB_2 reduced weight by the least amount. Al–6 wt% Cu alloy with 3 wt% TiB_2 exhibited high corrosion resistance due to its capacity to naturally form a corrosion oxide film on the surface. The corrosion resistance was further improved due to the formation of a stable oxide layer as a result of passivation of the surface during corrosion testing [9].

Figure 60.4 depicts the corrosion rates of an Al–6 wt% Cu alloy reinforced with TiB_2 over the course of 30 days or 720 h in a 5% NaCl solution. According to the findings, the corrosion rates vary with varied TiB_2 particle contents. The findings indicate that when compared to other compositions in Al–6 wt% Cu alloy, Al–6 wt% Cu with 3 wt% TiB_2 exhibits good corrosion properties of alloy. Al–6 wt% Cu alloy in 5% NaCl solution and Al–6 wt% Cu composites exhibited the lowest corrosion rate values, respectively. The results demonstrate that the corrosion rates dropped proportionately, notably for the composition of 3 wt% TiB_2 content within 720 h, when the TiB_2 content particles were added to Al–6 wt% Cu alloy. In this study, using more TiB_2 caused the corrosion rate of Al–Cu–TiB_2 composites to increase. It has been demonstrated that the rate of corrosion increases with exposure

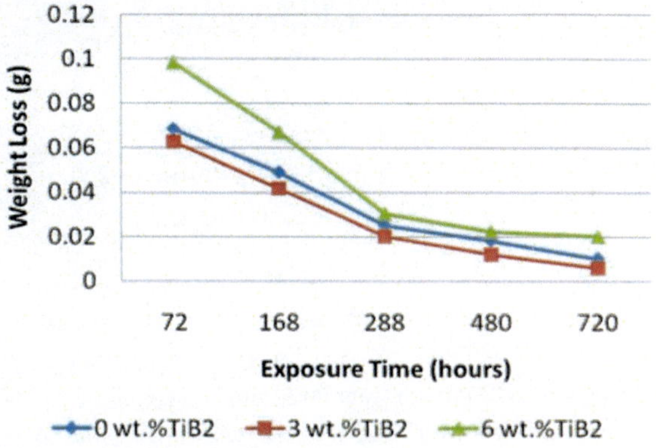

Fig. 60.3 Weight loss of Al–6 wt% Cu with 3 and 6 wt% TiB_2 in 5% NaCl [9]

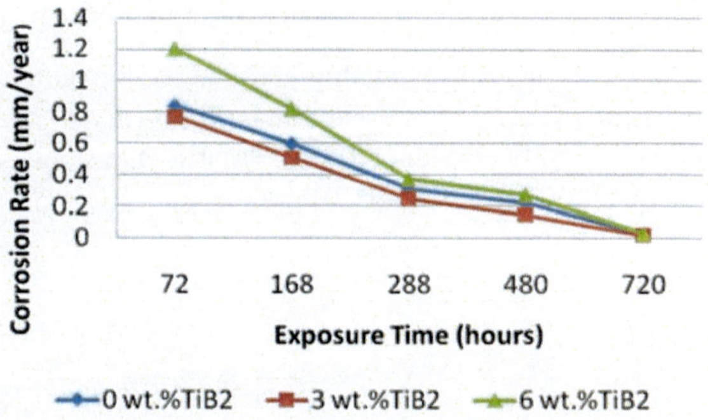

Fig. 60.4 Corrosion rate of Al–6 wt% Cu reinforced with TiB_2 in 5% NaCl measured by salt spray fog test [9]

time. As a result of the structure's porosity, a rise in TiB_2 composition will aid corrosion. According to Huang et al., reinforcement of in situ TiB_2 particles, which have unique properties like good matrix bonding, high conductivity, small size, and better oxidation resistance, may impede the initiation and growth of oxidation film and disrupt the film continuity, making it difficult to form a good protective anodic film on the surface of composites.

60.2.6 Potentiodynamic: Linear Polarization Resistance

The polarization resistance of a material is usually assessed using the LPR technique, an electrochemical process that aids in determining the degree of corrosion. The fluid's capacity to cause corrosion is correlated with its electrical conductivity. This method creates the link between the electrochemical potential and the current of charged electrons [9]. It typically consists of two or more electrodes, which, when subjected to erosion, produce polarization resistance, which, in turn, determines the degree of corrosion [10]. Typically, current travels between the counter electrode

Table 60.2 Electrochemical parameters of the MMCs attained from the potentiodynamic test [7]

Element	E_{corr} (mV)	I_{corr} (A cm^{-2}) (10^6)	CR (mm y^{-1}) (10^{-3})
Al–Cu	−385	7.29	6.67
Al–Cu–3ZrB_2	−290	1.52	2.85
Al–Cu–6ZrB_2	−321	1.81	0.36

and the electrode constructed of the substance under study. Calculating polarization resistance (PR) involves figuring out the electrochemical potential and current's slope. This PR can be translated into a corrosion rate using the Stern–Geary equation (Angst and Büchler 2015), according to this source.

$$R_p = B/i_{\text{corr}}, \tag{60.1}$$

where B is the proportionality constant for the particular corrosion system and i_{corr} is its corrosion current density (in units of μA cm^{-2}). Tafel plot is used to determine proportionality constant (B) by measuring the slopes of anode (β_a) and the cathode (β_b) empirically (Table 60.2).

$$B = \beta_a\beta_b/2.3(\beta_a + \beta_b), \tag{60.2}$$

$$R_p = \beta_a\beta_b/2.3(\beta_a + \beta_b)i_{\text{corr}}. \tag{60.3}$$

The corrosion barrier in Al–Cu is effectively altered by the addition of ZrB_2. This behavior may be related to the advantageous effect of ZrB_2 inclusion on grain size refinement. It has been found that materials with finer grains behave more conveniently in terms of corrosion and oxidation. In contrast to other compositions, the Al–Cu alloy with 6 wt% ZrB_2 demonstrated the best corrosion barrier, as shown in Fig. 60.5. Corrosion rate was 2.85×10^{-3} mm y^{-1} at 3 wt% ZrB_2, while it was 0.36×10^{-3} mm y^{-1} for Al–Cu at 6 wt% ZrB_2 [11].

60.3 Conclusions

We may draw the following conclusions from the review that has been done:

i. The addition of reinforcement such TiB_2, ZrB_2, and Sr to Al–matrix composites will minimize the wear rate of Al-MMCs.
ii. To determine the corrosion behavior, the Sr will add reinforcement to Al–matrix composites, particularly to Al–Mg and Al–Cu master alloys.

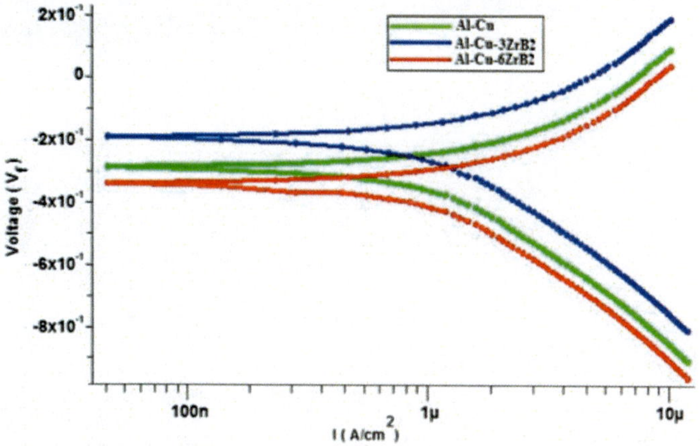

Fig. 60.5 Potential-dynamic polarization curves on Al–Cu alloy reinforced by ZrB_2 [11]

Acknowledgements The author acknowledge the Faculty of Applied Sciences for the laboratory facilities. Special thanks to who those contributed this project directly or indirectly.

References

1. A.K. Sharma, R. Bhandari, A. Aherwar, C. Pinca-Bretotean, Mater. Today: Proc. **27**, 1608–1612 (2020)
2. G.E. Byczynski, D.A. Cusinato, Int. J. Cast Metals Res. **14**(5), 315–324 (2002)
3. C. Vargel, in *Corrosion of Aluminium* (Elsevier, 2020), 91–111
4. B.V. Ramnath, C. Elanchezhian, R.M. Annamalai, S. Aravind, T. Sri, A. Atreya, V. Vignesh, C. Subramanian, Aluminium metal matrix composites. a review (2013)
5. S. Roseline and V. Paramasivam. *Journal of Alloys and Compounds 799,* 2019.
6. C. Vargel, in *Influence of Alloy Composition.* Corrosion of Aluminium (Elsevier, 2020), pp. 127–155
7. S. Derin, U. Aybarç, Y. Birol, Inter. Metalcast **11**, 688–695 (2017)
8. G.E. Byczynski, D.A. Cusinato, Int. J. Cast Metals Res. **14**(5), 315–324 (2002)
9. R. Rosmamuhamadani, M.K. Talari, S.M. Yahaya, S. Sulaiman, M.I.S. Ismail, M.A.A. Hanim, in *Corrosion Characterization of In-situ Titanium Diboride (TiB_2) Reinforced Aluminium-Copper (Al–Cu) Alloy by Two Methods: Salts Spray Fog and Linear Polarization Resistance (LPR).* Nano-SciTech 2017-AIP Conference Proceedings 1963 (2018)
10. M. Akhtar, M.I Lashari, M. Muzamil, M. Sattar, M.I. Shabir, S. Mohsin, M. Samiuddin, Revista de Metalurgia **57**(2) (2021)
11. U. Angst, M. Büchler, Mater. Corros. **66**(10), 1017–1028 (2014)
12. N.H. Mustafa, M.M. Mahat, S.M. Yahaya, R. Rosmamuhamadani, J. Phys.: Conf. Ser. **1349**(1) (2019)

Chapter 61
Characterization of Hybrid Composite Materials from Natural Fibres from Kenaf and Pineapple for Automotive Application

Raihanah Muhammad Navil and M. R. Isa

Abstract This study focus on behaviour of natural fibre from ground Kenaf and Pineapple for automotive interior application. The fabricated material was tested for mechanical and physical properties to define similarity of properties as suggestion of replacement with existing polymer used in automotive interior. The process of creating the composite material involves mercerization, manual fabrication and mechanical test. The data was collected and analysed, needs an improvement on strain and impacts strength to reach the standard. Based on tensile test done, the result of 100% Kenaf and 100% Pineapple has achieved required tensile strength for both dashboard and bumper. Next, strain result for all material fabricated composite shows 4 times higher than strain of dashboard; however, it is 80% lower for bumper. The density of the composite and bumper material shows large difference. The tensile result ranges from 29.5 to 3.5 MPa with percentage difference of 88.14%. According to hardness test done, 70% Kenaf with 30% Pineapple and 30% Kenaf with 70% Pineapple were the highest and lowest average reading obtained, respectively. The result ranges from 17.4 to 19.84 HRB, and the difference between the two material mentioned was 12.3%. Then the result of impact test obtained displays the highest and lowest were 100% Kenaf and 50% Kenaf with 50% Pineapple, accordingly. The range of impact test was from 0.70 to 0.30 J with percentage difference of 57.14%. From the three test done on the material, each material obtained different properties such 100% Kenaf has a good tensile strength and impact, 70% Kenaf is with 30% Pineapple excellent hardness despite poor tensile strength and impact, 50% Kenaf with 50% Pineapple was poor tensile strength and impact with excellent hardness, 30% Kenaf with 70% Pineapple was poor in every test done and lastly 100% Pineapple shows decent tensile strength and hardness but poor impact test. The properties of the material could be fit in other parts with the similar standards in shifting polymers to composite material made by Kenaf and Pineapple. A modification of fibre with an additive could improve the properties of the material, depending on desire properties which were shown in previous studies.

R. M. Navil · M. R. Isa (✉)
Department of Mechanical Engineering, College of Engineering, Univerisiti Tenaga Nasional, Jalan IKRAM-UNITEN, 43000 Kajang, Selangor, Malaysia
e-mail: mrashdan@uniten.edu.my

M. A. A. Mohd Salleh et al. (eds.), *Proceedings of the Green Materials and Electronic Packaging Interconnect Technology Symposium*, Springer Proceedings in Physics 289,
https://doi.org/10.1007/978-981-19-9267-4_61

Keywords Kenaf · Pineapple · Composite · Natural fibres

61.1 Introduction

Plastic polymer has been widely used and even owned by every household. It has been choice of every industry for the supremacy of anti-break compared to glass and high strength of tensile. The drawbacks of the material once it is obsolete are duration of decomposing the material. To encounter and as replacement of plastic, natural resources have been explored to be suitable candidate. Renewable sources are mainly living things that can be reproduced frequently in own nature such as animals and plants. The sources could be reproduced by breeding and plantation allow easy accessibility of the sources.

The natural fibre is extracted from renewable sources such as animals, plants and mineral. The type of natural fibre used in this paper are both from plants with main constituents of fibres are cellulose, hemicellulose and lignin. A unit of fibre is structured with region of cellulose connected to hemicellulose and lignin [1, 2]. The plant fibre group that was used was bast of Kenaf tree and leaves of a Pineapple (PALF). High elasticity, strength and physical properties of a fibre depend on level of cellulose in a fibre such as the PALF contains higher cellulose than kenaf. Study shows surface texture of the fibres that could be improved by removing non-cellulose content through mercerization process on plant natural fibre. The process could be done with several chemical solutions such as sodium hydroxide and potassium hydroxide.

Composite materials are created with combining two or more materials with different mechanical and physical properties. The result of merging different materials is together formed stronger properties compared to solitary material. Application of composite material in industries has been tested and compared for the strength to match and replace the polymers. The composite material is currently used widely in automotive interior application such as dashboard and bumper. Several parts of automotive have utilized Kenaf fibre as base composites for its lightweight properties and Pineapple fibre in textile industry (Fig. 61.1).

Production of composite materials of natural fibre has been used in engineering applications for the flexibility offered, substantial durability, corrosion resistance, reduction in weight which has directed researchers and industries to work in this area [3]. Theoretically, the composite shape, surface condition and durability are dominated by polymer matrix of the composite material [1]. However, recent research done addressed limitations addressed on fabrication of composite material. The low moisture resistance or high moisture absorption is owing to the presence of the hydroxyl group and fibre to act as hydrophilic material [1, 4, 5]. The results of water absorption by fibre will result in voids and swell on interface of the composite. The sequel to the formation of void is low mechanical properties and reduction in stability of the material [1]. Most of the fibre used was obtained from plants produced has limitation of tensile and flexural modulus. The value is below 100 MPa [5]. However,

Fig. 61.1 Composition of ground Kenaf and Pineapple fibre in composite material

the purpose of this paper is to provide the results for future reference. The composite is manually fabricated and tested to investigae the mechanical and physical properties.

61.2 Methodology

61.2.1 Alkali Treatment of Natural Fibre

Alkali treatment has been proven in several papers that the alkali helps in fibre surface improvement, fibre–matrix adhesion and removal of impurities. The result of surface roughness development is caused by disruption of hydrogen bond aids in proper bond of matrix material, higher surface exposure of cellulose and higher possible reaction sites These treatments will enhance tensile strength compared to untreated ones [1, 3, 4, 6]. Mercerization process removes constituent in fibre such as hemicellulose, lignin, pectin and impurities. The removal led to fibrillation which is formation of fibre process to create rough surfaces and enable interlocking of mechanical related to the matrix. The NaOH solution was prepared by using Eq. 61.1. Reaction of alkali treatment with fibre is shown in Eq. 61.2. The treatment reduces nature of hydrophilic in fibre or in another term and water absorption and provides better adhesion in composite matrix.

The fibre was prepared by immersing it into 6% concentration of NaOH solution with time of immersion in Table 61.1 [3, 4, 7]. The fibre then drained and soaked into water for 30 min to remove any access of NaOH solution [8]. The fibre is then ready for drying process with specified temperature and time in Table 61.1.

$$\%\text{Solution} = \frac{\text{mass of soluble (g)}}{\text{mass of solution (g)}} \times 100 \tag{61.1}$$

Table 61.1 Ideal mercerization procedures

Fibre	Sodium hydroxide concentration (NaOH %)	Immersion time (h)	Drying temperature (°C)	Drying time (h)	References
Kenaf	6	48	100	6	[9]
Pineapple	6	3	60	24	[3]

$$\text{Fibre} - \text{OH} + \text{NaOH} \rightarrow \text{Fibre} - \text{O} - \text{Na}^{+} + \text{H}_2\text{O} \tag{61.2}$$

61.2.2 Preparation of Fibre

After fibre has been mercerized, the fibre will be ground. The fibre was inserted through the chamber instead of the funnel due to the properties of the fibre that is lightweight unable to dive through the tunnel. The chamber contains three sections with different degree of grinder. The fibre was estimated about two index fingers sized for each cycle due to narrow space between the teeth of the grinder and the chamber. During the grinding process, the tunnel will be concealed with flat object to prevent the fibre from flying out of the funnel. The end product was channelled into a tub attached below the machine. The tub was attached with two secure clips on the left and right. After grinding process, the fibre was extracted from the tub by unbuckling the secure pins and placed in a closed container to prevent spill of ground fibre. Product of ground fibre is lightweight and similar to cotton physically.

61.2.3 Manual Fabrication

The mould preparation was done by placing the plastic cover on the mould and takes the shape of it. This is to prevent binding of composite with the metal plate and to ease removal of the material after hand lay-up (HLU) process. Then, the process started with placing a 500-ml beaker on the scale and weight the epoxy and hardener separately by referring to the ratio given along with the product. To achieve maximum strength of the matrix, it is essential to follow the proportions directed to complete chemical reaction. The resin is poured into non-plastic container to prevent the resin to melt during exothermic reaction. Then, Kenaf and Pineapple were set on the metal container according to the ratio set as in Table 61.2.

Afterwards, the epoxy was mixed by using flat stick to remove air bubbles in the resin for approximately 3 min and resin was poured onto layer of fibre. The roller was then used to distribute the resin onto the material uniformly and to remove air

Table 61.2 Ratio of fibre composition

Material	Sample no	Kenaf	Pineapple
Ratio	1	100: 0	
	2	70: 30	
	3	50: 50	
	4	30: 70	
	5	0: 100	

Table 61.3 Specimen specification

Type of test	Standards	Length (mm)	Width (mm)	Thickness (mm)	Amount of sample	Ref
Tensile	ASTM D3039	250	25	2.5	15	[10]
Hardness	ASTM E23	55	10	10	1	[11]
Impact	ASTM E23	55	10	10	1	[11]

bubbles. The material was left to cure at room temperature for 24–48 h. A amount of resin and fibres were calculated using Eq. 61.3. The value for weight of fibre obtained was doubled to achieve compact product of composite. The mass of material to be used was calculated using Eq. 61.4.

$$\text{Volume} = \text{Length} \times \text{Width} \times \text{Thickness} \tag{61.3}$$

$$\text{Mass} = \text{Density} \times \text{Volume} \tag{61.4}$$

The material was composed in two different sizes, rectangle flat plate and cuboid. The composite was then divided into desired size as shown in Table 61.3. The same specimen for hardness and impact test was used since the hardness test will only cause small indentation on the surface. Therefore, specimen was reused for impact test.

61.3 Result and Discussions

61.3.1 Tensile Test

The purpose of this test is to discover the mechanical properties of the composite fabricated. An axial load is applied until the specimen fractures to observe the point of rupture. The test provides information on tensile strength, yield strength and ductile. The result obtained from this test generates stress–strain graph. The dimension of the specimen was referred according to the ASTM standards for D3039, tensile properties

Table 61.4 Tensile specimen geometry recommendation

Fibre orientation	Width (mm)	Overall length (mm)	Thickness (mm)	Tab length (mm)	Tab thickness (mm)	Tab bevel angle (°)	Ref
Random discontinuous	25	250	2.5	Emery cloth	–	–	[10]

Table 61.5 Tensile strength of the samples

Fibre composition	Average stress (MPa)
100% Kenaf	17.14
70% Kenaf + 30% Pineapple	3.7
50% Kenaf + 50% Pineapple	2.32
30% Kenaf + 70% Pineapple	11.24
100% Pineapple	20.13

of polymer matrix composite materials. Table 61.4 shows tensile specimen geometry recommendations.

The tensile test was done using Zwick Roell Z100. The machine has feature to test load up to 100 kN. The machine was connected with the testXpert II to graph and record the tensile test. The suitable preload applied on sample was set to 0.1 MPa. Then, the speed of tensile modulus and test speed was set to 30 MPa/s and 10 mm/min, respectively. In total, there were five samples for each type. The first two were indicated as replica and the last three were real sample.

The software connected with the machine has generated graph for the test done. The graph shows linear graph which indicates that the material has a brittle behaviour. Each kind was prepared with three samples to record the highest, and average reading of stress was calculated in Table 61.5. The width, length and thickness of the samples are 25 mm, 250 mm and 2.5 mm, respectively.

Figure 61.2 indicates the highest stress of each sample against the elongation. The first sample contained 100% Kenaf with epoxy has the highest reading for tensile stress in green, 29.5 MPa. However, fifth sample with 100% Pineapple with epoxy in yellow obtained 28.8 MPa slightly lower reading than 100% Kenaf with the shortest elongation. Meanwhile, third sample with 50% Kenaf, 50% Pineapple with epoxy has the lowest reading of 3.38 MPa in orange with longest elongation among the samples tested. A low stress and linear graph are depicted plastic material on stress–strain graph.

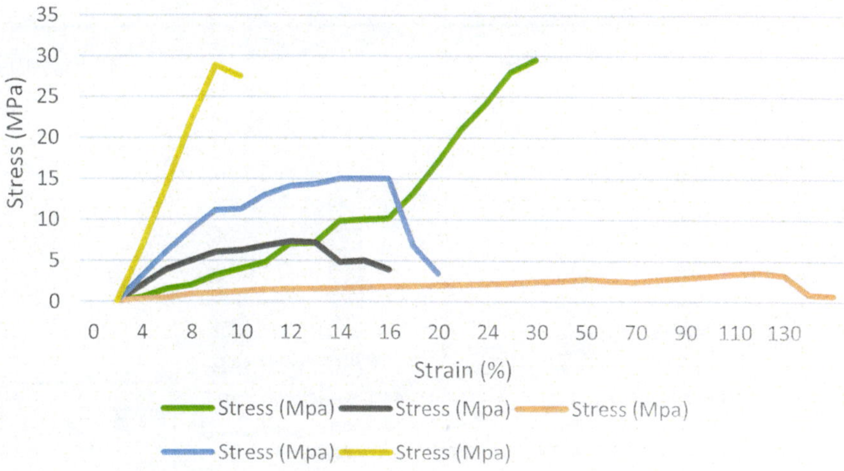

Fig. 61.2 Graph of stress against strain

61.3.2 Hardness Test

The hardness test is measuring the depth of indentation applied with load on the specimen. The load will be applied perpendicular to the surface. The type of stress created from this stress is monoaxial. Advantage of this test is the material properties that could be determined without breaking or destroying the specimen. However, the machine has limitation such as the indenter became sensitive when the results are almost fully load. The leftover limitation was not faced such thin samples which could lead to indenter puncture through the samples and irregular surface causing imprecise result. Results of hardness test were plotted into graph of fibre composition against hardness. The machine Zwick Roell Indentec ZHU was used for hardness test. The test was done by switching on the machine and placing the specimen on the plate. The disc of the machine was turned slowly in clockwise direction to elevate the specimen until it touches the indenter. Rapid rotation of disc will cause an error of reading by the machine as mentioned earlier, the indenter is sensitive towards reaching the results. While the indenter slightly penetrates through the material, the disc was turned gradually from the start due to high sensitivity of the indenter. The perimeter of indentation will be measured with bars indicated on the screen and notified by a beep sound. The specimen was then flipped over for different surface and point until five readings were taken from a specimen.

The hardness testing was done on different surface of the composite to obtain the average reading of hardness. The average reading was calculated to compare the difference. The highest reading recorded for hardness test was 70% Kenaf with 30% Pineapple indicated green. The sample with 30% Kenaf and 70% Pineapple has the lowest reading and is indicated with orange. However, the difference between the highest and lowest hardness was calculated about, 12.3% which were 70% Kenaf and

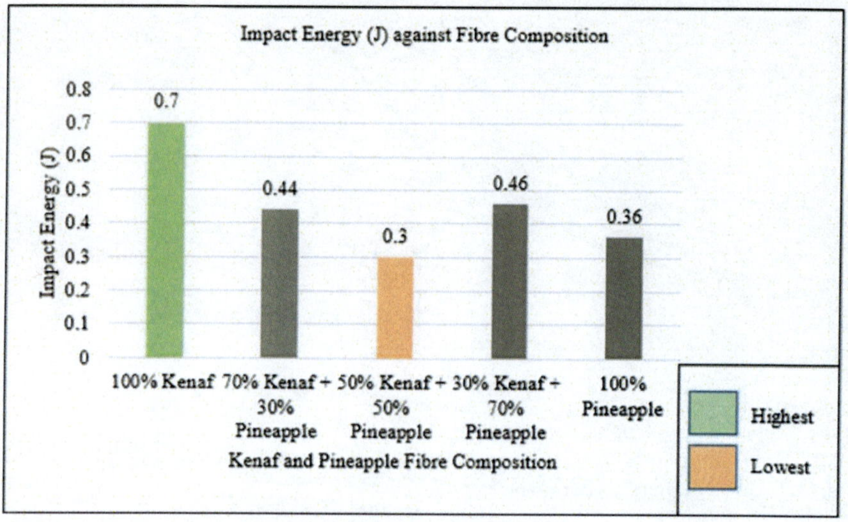

Fig. 61.3 Graph of hardness against the fibre composition of Kenaf and Pineapple

30% Pineapple with 19.84 and 30% Kenaf and 70% Pineapple with 17.4 respectively. The data collected was plotted in Fig. 61.3.

61.3.3 Impact Test

The impact test measures the energy absorbed by a material during fracture. The absorbed energy indicates the strength of the material and is used as method to observe the impact. The previous study states that the fibre orientation affects the fracture of composite of impact energy. The machine used in this test was Zwick Roell HIT50P. Before starting test, error test was done on the machine. The pendulum was released with no specimen positioned to observe the error displayed on the machine. Afterwards, the specimen was placed with v notch facing to the left or away from the pendulum. It was placed on the centre, and the pendulum was ensured to hit the v notch; otherwise, it will be considered as failed test due to inaccurate positioned of the sample. The pendulum was then lift upward to the right until it snaps to indicate the pendulum has been locked. The two-door will be shut for safety purposes to avoid the specimen from flying over after impact applied. The green button on left and right was pushed simultaneously to release the pendulum. The break was later then pulled slowly after pendulum hit the sample to stop the pendulum from swinging. The reading was recorded on the screen of the machine.

The impact machine that was tested before impact test was performed. From the data obtained, 100% Kenaf sample shows the highest impact energy. The difference between the highest and lowest impact energy was 48.6% which was 100% Kenaf

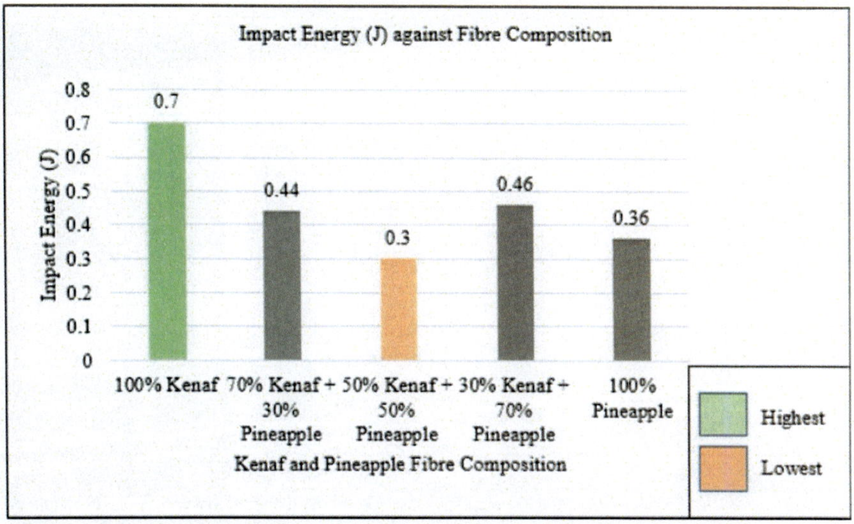

Fig. 61.4 Graph of impact energy against the fibre composition of Kenaf and Pineapple

with 0.7 J and 100% Pineapple with 0.36, respectively. The readings and graph of impact energy against the fibre composition were plotted in Fig. 61.4.

From the results obtained, discussion of the project was constructed to summarize the outcome of the project. The fabrication of composite material was done multiple trial due to faulty of specimen. The sample was incurred due to improper composition of resin, uneven absorption of resin, formation of porosity and uneven spread of fibres such in Fig. 61.5. The sample was put under the light examine the visible voids. The compactness of fibre was identified by the darker area which indicates compact and lighter area which indicates incompact such as in Fig. 61.6. Furthermore, few of the specimen was not fit for test due to physical form of the sample was flabby and fragile. The flabby part appears to be bloated with incurred epoxy. Therefore, some of the samples were made slightly thicker to straighten the sample for tensile test. This has caused weight of every composite to vary due to different densities of both fibre and thickness of the sample.

The process of mercerization has caused loss of weight on the fibres. The fibre was weight before and after the process as indicated in Table 61.6. Pineapple fibre shows larger weight reduction after the process compared to Kenaf with 65.38% and 30.77%, respectively.

The result obtained from the test performed was compared with dashboard and bumper beam in Table 61.7. The sample 100% Kenaf and 100% Pineapple reached desired value of tensile strength of dashboard and bumper beam indicated in green. However, the value of strain was far too small compared to polypropylene plastic. However, all the reached standard of strain of dashboard made from plastic. The difference of density could be seen and proven that natural fibre offers lightweight properties.

Fig. 61.5 Uneven absorption of resin and porosity formed on sample

Fig. 61.6 Compactness of fibres

Table 61.6 Weight of fibres affected by mercerization

Type of fibre	Weight before (g)	Weight after (g)	Weight lost difference (%)
Kenaf	39	27	30.77
Pineapple	26	9	65.38

Table 61.7 Comparison of sample with dashboard and bumper beam

Automotive part	Dashboard	Bumper beam	Sample 1	Sample 2	Sample 3	Sample 4	Sample 5
Material	Plastic	Polypropylene Plastic	100% Kenaf	70% Kenaf + 30% Pineapple	50% Kenaf + 50% Pineapple	30% Kenaf + 70% Pineapple	100% Pineapple
Tensile strength (MPa)	20–40	27.6–41.4	29.5	7.3	3.5	15	28.8
Strain (%)	2	100–600	24.5	14.2	131	18	9.2
Impact (kJ/m^2)	13.48	–	1.273	0.8	0.545	0.836	0.655
Density (g/mm^3)	–	890–910	0.00185	0.00283	0.00173	0.00166	0.00233
References	[9]	[12]	–	–	–	–	–

The impact value of the sample fabricated did not reached desired value. The composition of the samples could improve for future studies with additional of different natural fibre to enhance the strength of the sample.

61.4 Conclusions

As conclusion for the tensile, hardness and impact test performed, the data was recorded and tabulated in achieving the mechanical and physical properties of Kenaf and Pineapple composite material. The results for every test varied from one test to another. The range of tensile strength between the highest and the lowest was about 88.7%, such 100% Kenaf obtained 29.9 MPa and 50% Kenaf and 50% Pineapple obtained 3.38 MPa. Meanwhile, the average tensile calculated from three sample tested for each type was 100% Pineapple with 20.13 MPa highest tensile and 50% Kenaf and 50% Pineapple with 2.32 lowest tensile strength. The purpose of including average tensile value was due to irregular size of sample fabricated. The average hardness calculated indicates 70% Kenaf and 30% Pineapple obtained the maximum value and 30% Kenaf and 70% Pineapple with minimum value such as 19.84 HRB and 17.4 HRB, respectively. The difference between the readings was 12.3%. Then, the impact energy result shows 100% Kenaf with the highest and 50% Kenaf and 50% Pineapple with the lowest value of 0.7 and 0.44 J. The error value has been excluded from the results obtained. The physical properties of the sample were studied, and mercerization has caused weight reduction in fibre.

In short, the sample is not yet eligible to represent automotive parts for dashboard nor bumper beam. However, an improvement on elongation and impact strength of the sample could be done for future studies. In Malaysia, the thermal stability of a material should be taken into account with the climate condition of the country, hot and humid. Deformation of material made from thermoplastic is affected by high temperature while thermosetting is not. Therefore, an additional thermal test on samples will help fill in the void of information regarding the mechanical properties. Furthermore, additional of Halloysite Nanotubes (HNTs) with treated fibre as a study was done shows improvement in tensile properties [3]. An additional of different fibre might enhance the absorption of impact and strengthen the mechanical properties of composite [13].

Acknowledgements The authors gratefully acknowledge the support provided by the Universiti Tenaga Nasional through the Matching Grant under job number 2021006UNSRI. The authors also acknowledge Innovation and Research Management Centre (iRMC) of Universiti Tenaga Nasional (UNITEN).

References

1. D. Tholiban, I. Tharazi, A.B. Sulong, N. Muhammad, N.F. Ismail, M.K.F.M. Radzi, N.A.M. Radzuan, D. Hui, Kenaf fiber composites: a review on synthetic and biodegradable polymer matrix. Jurnal Kejuruteraan **31**, 65–76 (2019)
2. S.O. Amiandamhen, M. Meincken, L. Tyhoda, Natural fibre modification and its influence on fibre-matrix interfacial properties in biocomposite materials. Fibers Polym. **21**, 677–689 (2020)
3. B. Suresha, G. Hemanth, Effect of halloysite nanotubes on morphology and mechanical properties of alkali treated pineapple fiber reinforced epoxy composites. Proc. Mater. Today, 9047–9053 (2021)
4. V.L. Narayana, L.B. Rao, A brief review on the effect of alkali treatment on mechanical properties of various natural fibre reinforced polymer composites. Proc. Mater. Today, 1988–1994 (2021)
5. N.G. Shinde, D.M. Patel, in *A Short Review on Automobile Dashboard Materials.* IOP Conference Series: Materials Science and Engineering (2020)s
6. N. Chand, M. Fahim, in *Natural Fibers and their Composites.* Tribology of Natural Fiber Polymer Composites (2020), pp. 1–59
7. R. Verma, M. Shukla, D.K. Shukla, Effect of glass fibre hybridization on the water absorption and thickness of alkali treated Kenaf-epoxy composites. Proc. Mater. Today, 2093–2096 (2021)
8. N.F. Ismail, N.A.M. Radzuan, A.B. Sulong, N. Muhammad, C.H.C. Haron, The effect of alkali treatment on physical, mechanical and thermal properties of Kenaf fiber and polymer epoxy composites. Polymers **13** (2021)
9. M. Fitri, S. Mahzan, F. Anggara, The mechanical properties requirement for polymer composite automotive parts—a review. Int. J. Adv. Technol. Mech. Mechatron. Mater. **1**, 125–133 (2020)
10. American National Standard: Standard Test Method for Tensile Properties of Polymer Matrix Composite Materials (West Conshohocken, Pennsylvania, 2002)
11. American National Standard: Standard Test Methods for Notched Bar Impact Testing of Metallic Materials (West Conshohocken, Pennsylvania, 2007).
12. O. Osokoyo, An evaluation of polymer composites for car bumper beam. Int. J. Autom. Compos. **3**, 44–60 (2017)
13. Z. Liu, J. Lu, P. Zhu, Lightweight design of automotive composite bumper system using modified particle swarm optimizer. Compos. Struct. **140**, 630–643 (2016)

Chapter 62
Buffalo Reef Mesothermal Gold Mineralization Mineralogy and Geochemistry in Kuala Lipis, Pahang, Malaysia

S. H. Osman, A. Rahim, and W. M. W. Ibrahim

Abstract In Pahang, the East Coast of Peninsular Malaysia, Buffalo Reef remains active as a gold mine among the main gold deposits at the North of Selinsing Gold Mine. Mineralization of gold is possible when stibnite occurs in quartz veins. The main purpose of this study is to examine the geological condition governing the mineralization of ore and the origin of gold mineralization of Buffalo Reef. The XRD result shows the quartz stage (SiO_2) while quartz, sillimanite and stibnite are identified. Ten samples were tested with XRF geochemical studies showing SiO_2 contents of 24.98–70.67 wt%, and the overall $Na_2O + K_2O$ diagram versus SiO_2 alkaline diagram showed varies basaltic andesite, andesite and dacite series with favourable and random distribution. The deposits region consists of the thick Permian rock sequences especially tuff volcanic rock that is intertwined by clay (argillite, phyllite) and carbonate rocks. Mineralization of volcanic rock and pyroclastic tuff supporting and regulating the orogenic gold deposit under the Raub–Bentong framework. The region epizoic and mesonic with a mesothermal deposit characterized by the presence of gold, stibnite and arsenopyrite with the grade of gold is 4.01 g/t Au from AAS.

Keywords Mineralization · Deposit · Gold · Buffalo reef

A. Rahim (✉)
School of Civil Engineering, Faculty of Engineering, Universiti Teknologi Malaysia, 81310 Johor Bahru, Malaysia
e-mail: afikahrahim@utm.my

S. H. Osman
Fuel Cell Institute, Universiti Kebangsaan Malaysia, Bangi, Malaysia

W. M. W. Ibrahim
Faculty of Mechanical Engineering Technology, Universiti Malaysia Perlis, 026000 Arau, Perlis, Malaysia

M. A. A. Mohd Salleh et al. (eds.), *Proceedings of the Green Materials and Electronic Packaging Interconnect Technology Symposium*, Springer Proceedings in Physics 289,
https://doi.org/10.1007/978-981-19-9267-4_62

62.1 Introduction

Malaysia has a long background of extensive gold mining in the country, especially in the Central Belt of Peninsular Malaysia and a potentially attractive gold mining area. The Central "Gold" Belt represents 20 km wide N-S trend in gold mining localities, illustrating the substantial role of hydrothermal fluids in gold development in Malaysia's peninsular particularly in the North Pahang and Kelantan areas [1–4].

Most of the gold production seems to originate from Pahang to the North until the Central Belt of Kelantan. The geology of the Goldfield of the Central Belt demonstrates the significant role of hydrothermal fluids in the formation of gold deposits [4]. Gold mineralization in the Central Gold Belt is usually classified as a small mesothermal gold deposit due to its tectonic and geological location [5]. Mesothermal veins are generated in rocks known for their large size and depth at mild temperatures and pressures. In comparison, undeformed accretionary belts adjacent to continental magmatic arcs called "mesothermal" gold ores during compressive and transpression regime and throughout the upper crust [6]. Raub–Bentong Suture viewed the area of folded slates with vertical or isoclinal folds and small ophiolitic cells as an oceanic crust and sediment, producing an accretionary complex formed by east subduction [7]. The Raub–Bentong Suture Zone includes complex rocks, with oceanic ribbon-bedded cherts, argillite, turbidite rhythmites, melange, serpentinites and continental margin/shelf deposits [8].

In 2013, Makoundi et al. [7] have suggested a two-stage model for the formation of geology, geochemistry and metallogenesis of Selinsing gold deposit From his study to further his work, this study examined the geological situation regulating mineralization by characterizing the ore and the genesis of gold mineralization in Buffalo Reef. Accordingly, mineralogical and geochemical mineralization studies in the Buffalo Reef region aim to improve gold exploration in Malaysia.

62.2 Regional Geology

The study is located in the north-western of Pahang State of Malaysia Peninsular. The prospective area covers 2635.77 m^2 which is about 220 km to Kuala Lumpur and about 60 km north of the city of Raub (Fig. 62.1). The Buffalo Reef deposits are about 1 km east of the Raub–Bentong suture. Argillite and Permian calcareous is dominated eastward by conglomerates and Devonian era sandstones westward (Fig. 62.2). According to Naidu et al. [9], the regional low-grade metamorphism throughout the greenschist facies (locally up to amphibolite facies). The sediments were subsequently intruded by granite bodies of roughly Jurassic age. Argillite and calcareous Permian age are the dominant rock species in the Buffalo Reef, which are transformed into intrusive rocks by the following granite. Gold in the veins is usually connected with pyrite, arsenopyrite and stibnite.

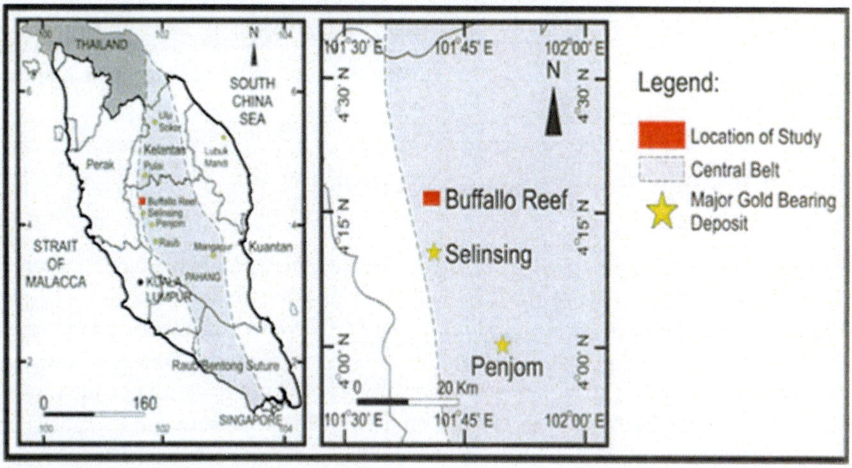

Fig. 62.1 Location of study area

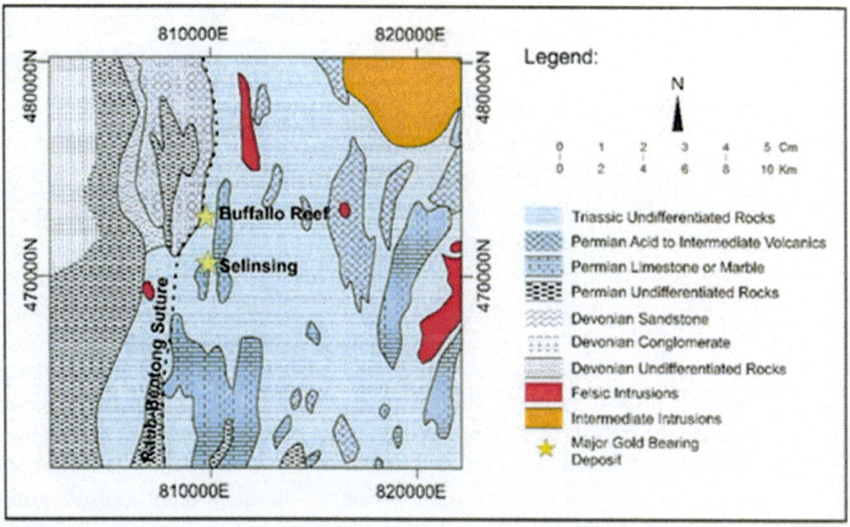

Fig. 62.2 Regional geology of the Selinsing–Buffalo Reef areas [9]

A conductive ink was formed by dissolving silver flakes into the base polymer PDMS-OH. Toluene was added into the mixture and allowed it to mix for 24 h at room temperature on a magnetic or mechanical stirrer at 220–280 rpm so that all silver particles dissolved in the binder. After that, organic solvent (D4) and cross-linking agent (ETMS) were added into the mixture and again stirred for 5–10 min. Finally, few drops of acetic acid and DBDTL were poured into the mixture to cure. The conductive ink mixture was then dispensed on a rectangular-shaped mould and allowed it to cure at room temperature for 24 h. Three samples of rectangular shape geometry (Fig. 62.1) were cut from the sheet and proceeded to the experimentation.

62.3 Lithology Recording and Analysis

The research techniques outlined in this study included fieldwork and lab work at School of Material and Mineral Resources, Universiti Sains Malaysia. For the fieldwork activity, a Global Positioning System (GPS) was used to provide relatively precise positions, generally from 3 to 10 m for mapping, lithology recording, and during the mapping. Ten key samples were drawn from drill holes which were collected. For the laboratory work, several laboratory analyses were undertaken at the School of Material and Mineral Resources, University Sains Malaysia. The analyses included (X-ray fluorescence) XRF and X-ray powder diffraction (XRD) analysis also for scanning electron microscopy SEM–EDX analysis, one sample was a success for atomic absorption spectrometry (AAS).

62.4 Result and Discussion

The geology of Buffalo Reef result from the analysis, it lies near the eastern side of the Raub–Bentong Suture and the primary granite situated just south of gold mineralization, intruders which share a comparable sting because of the widespread presence of low regional metamorphism of (greenschist to local amphibolite facies) marine classics, mainly consisting of pale to dark grey carbonated and calcareous and fine-grained schismatic sandstones from the Permian area [10]. Figure 62.3 shows the geological map of the sample based on the Selinsing Gold Mine Report (2016).

Gold mineralization is mostly confined to a marine sequence of clastic rock, generally striking in N-NW direction and dipping between 65° and 70° eastward. There are also irregular, fractured quartz-carbonate venations throughout the process. Quartz vein is mostly found parallel to the bedding. The main feature of the gold mineralization is the N-S shear zone in the Raub Group [11, 12].

The Bentong–Raub Suture is an important structure in SE Asia with a trending fault scheme NW–SE trending fault system. The sinistral strike-slip fault is parallel to the Raub–Bentong Suture. From Fig. 62.4, it shows the volcanic tuff (30%) is an older formation followed by volcanic lapilli tuff (5%), carbonaceous volcanic tuff (10%), limestone (30%), and the youngest is alluvium formation (25%). The percentages show the content of each rock type covered the geological map in Buffalo Reef. The mining process is very active in this area then the geological aspect is noted here because of the mining process all the overburden makes a new phase for the location. The alluvium is dominant in this area because of the weathering phenomena. The current pit limit always changes according to the mining process to extract the ore in Raub–Bentong Suture area.

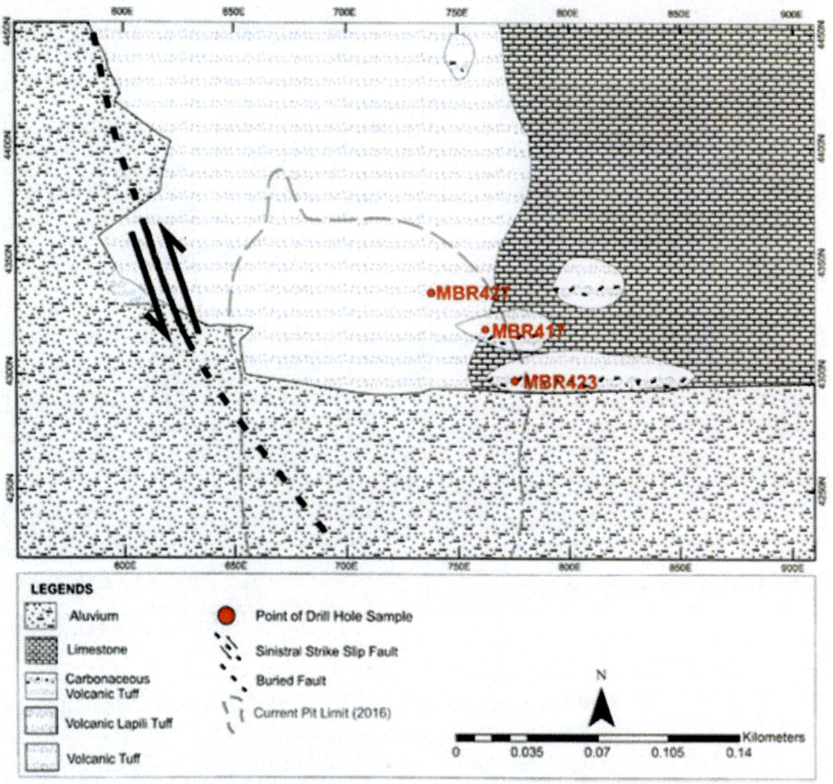

Fig. 62.3 Geological map of sample MBR417, MBR423 and MBR427 in Buffalo Reef mining area with plan view from 3D model of unpublished Report of Selinsing Gold Mine (2016)

Structurally, Raub–Bentong Suture interpreted the zone of folded slates with vertical or overturned isoclinal folds and minor ophiolitic bodies as representing the oceanic crust and sediments, forming an accretionary complex produced by eastwards subduction. The Raub–Bentong Suture zone includes accretionary complex rocks. The formation of sinister strike-slip fault is parallel to the Raub, Bentong Suture. The sinistral trend of the strike slips like the Raub–Bentong Suture where a collision structure overprint produced a significant N-S or NW-SW trend. Buffalo Reef is located near the eastern edge of the Buffalo Reef Suture Zone. The Buffalo Reef's gold mineralization is mainly confined to the marine clastic rock sequence which strike mainly in the north and south [13]. Table 62.1 shows the sequence according to the depth of the drill core. Most of the samples are located at volcanic tuff formation. Volcanic tuff is a compacted rock of volcanic ash ranging in size from fine to coarse gravel. Alluvium formation is not in these strata because of the geological aspect where the alluvium is on the surface. It just covers the surface. The geological column defined MBR417_1 dominant of volcanic tuff, MBR417_2 dominant of volcanic lapilli tuff, MBR417_3 dominant of volcanic tuff, MBR423_1

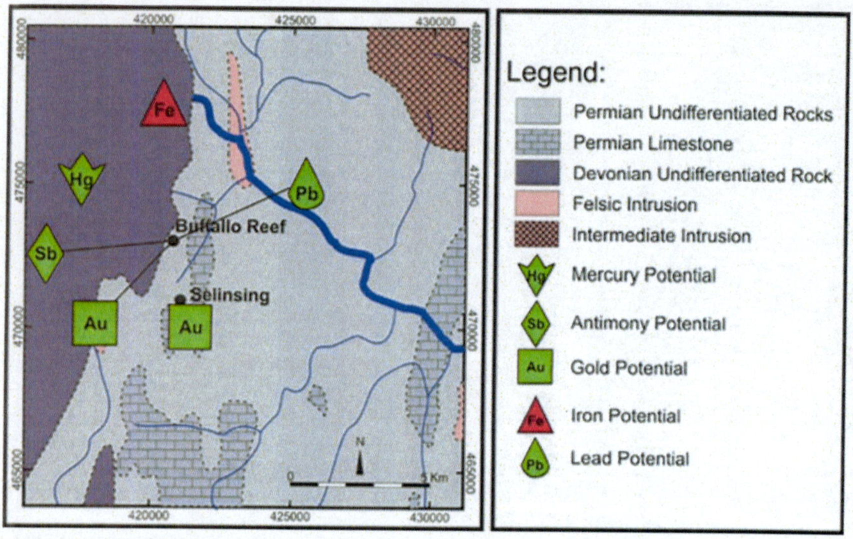

Fig. 62.4 Mineral distribution of Buffalo Reef

consists of carbonaceous rock alteration, MBR423_2 dominant of volcanic lapilli tuff, MBR423_3 dominant of volcanic tuff, MBR423_4 dominant of volcanic lapilli tuff, MBR427_1 dominant of volcanic lapilli tuff, MBR427_2 dominant of volcanic lapilli tuff, MBR427_3 dominant of volcanic lapilli tuff.

For the ore mineralization, the interpretation of ore mineral fabric and rock textures are important steps in understanding the sequence of events in the ore formation. Figure 62.5 shows the mineral distribution of the Buffalo Reef. The following minerals were found in the ore from the mineralized veins at Buffalo Reef area: quartz, stibnite, pyrite, arsenopyrite, calcite and gold [12]. Quartz is very abundant

Table 62.1 Geological column of samples

Sample	Geological column
MBR417_1	Volcanic tuff
MBR417_2	Volcanic lapilli tuff
MBR417_3	Volcanic tuff
MBR423_1	Carbonaceous rock alteration
MBR423_2	Volcanic lapilli tuff
MBR423_3	Volcanic tuff
MBR423_4	Volcanic lapilli tuff
MBR427_1	Volcanic lapilli tuff
MBR427_2	Volcanic lapilli tuff
MBR427_3	Volcanic lapilli tuff

in the ore-bearing faults. It generally occurs as massive milky quartz which is heavily fractured and shows undulose extinction [14].

Figure 62.6 petrography observations show that the stibnite sample has penetrated a crack dividing several large quartz grains. Polished section MBR417_2 shows in Fig. 62.6a and shows that milky quartz occurs as rounded grains in a matrix of massive stibnite. Stibnite occurs in a massive form and well-formed needle-like crystals of various sizes.

Quartz occurs as poor interlocking masses of well-formed quartz crystals as shown in Fig. 62.6b by polished section MBR427_2. The crystals are clear and do not show undulose extinction in the thin section. These well-formed crystals can occasionally be seen lining vugs in milky quartz. Pyrite in (Fig. 62.6c) from polished section sample MBR417_2 occurs in both crystalline and massive forms. Arsenopyrite in

Fig. 62.5 Geological column of rock types from drill holes of original mine

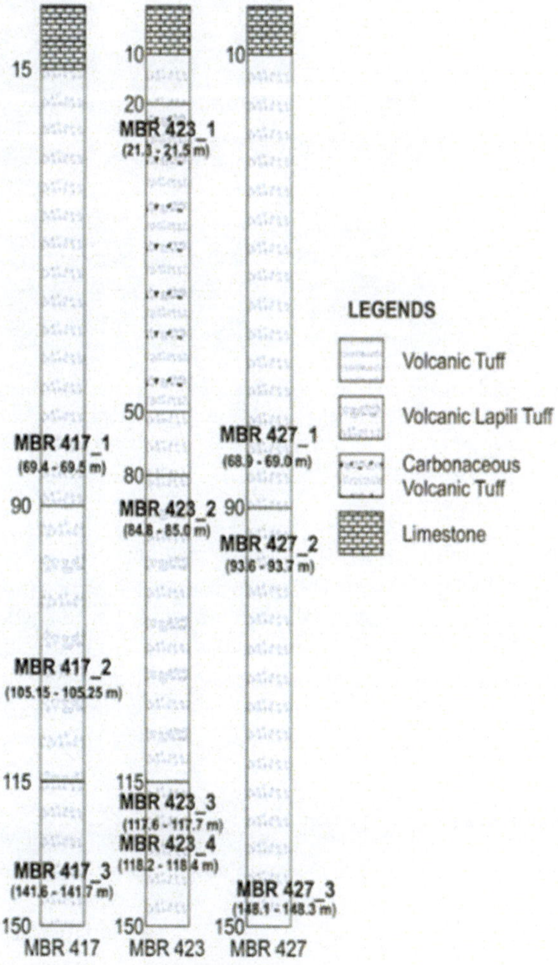

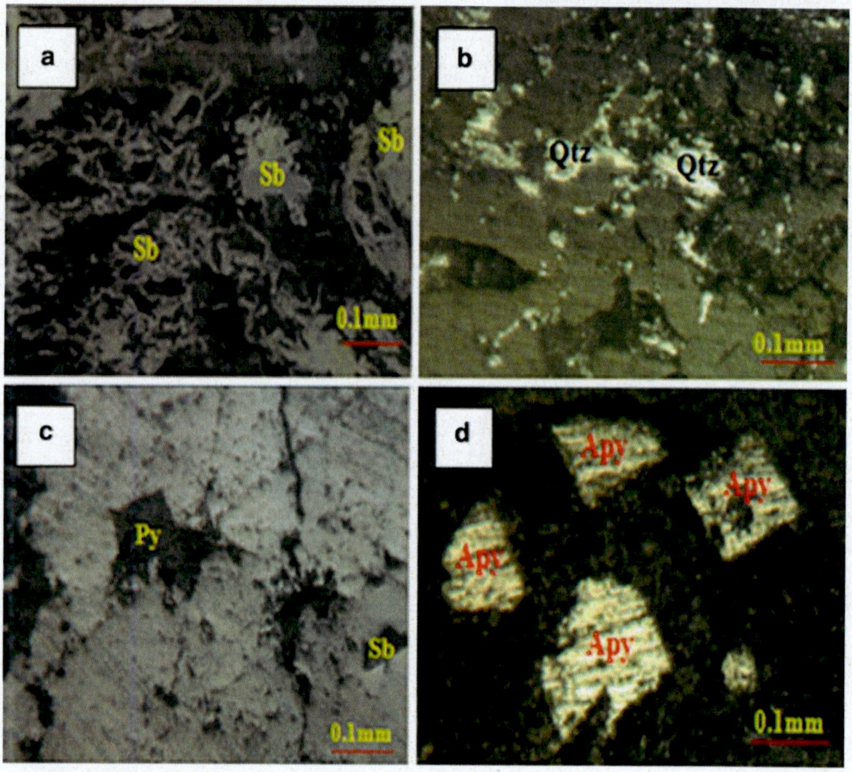

Fig. 62.6 Photomicrographs show **a** stibnite massive showing strange feather-shaped zonation. **b** Poorly interlocking crystals of polygonal quartz with even extinction. **c** Massive intergrowths of stibnite and pyrite. **d** Polygonal crystals of arsenopyrite in slate. Abbreviation: *Apy* Arsenopyrite, *Py* Pyrite, *Qtz* Quartz, *Sb* Stibnite

(Fig. 62.6d) from polished section sample MBR423_3 occurs in the massive quartz and stibnite found in the veins. It is in the same form as the previously described arsenopyrite in slate. Crystals of arsenopyrite isolated in stibnite.

Calcite Fig. 62.7 is minor constituent of these veins. Clear crystals are found in quartz fractures and often are found encasing other minerals such as pyrite. It also appears in stibnite fractures and in fact seems more common in combination with stibnite in the thin section sample MBR427_2.

Gold in Fig. 62.8 in polished section sample MBR423_2 is reported to have occurred in association with quartz and pyrite. As noted, free gold means gold found in loose particle or nugget in mining terms where only interbedded veins at depth are intercepted by the fissure veins. From the EDX analysis in Fig. 62.8, gold particle from the ground sample in MBR423_2 and MBR427_3 was spotted with amount 15.41 and 10.64 wt% while the highest is stibnite, 67.98 wt% shown by sample MBR417_2. The mineralogical analysis was carried out by EDX analysis at magnification used for sample between 500x and 5000x.

Fig. 62.7 Photomicrographs show calcite associated with the opaque mineral known as pyrite

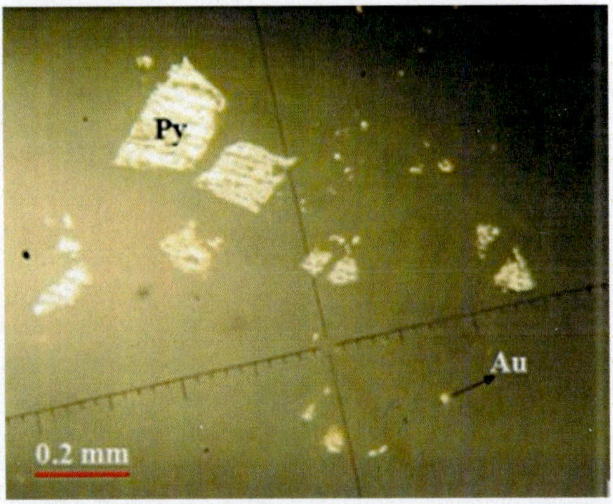

Fig. 62.8 Photomicrographs show gold very rare content with the pyrite

On the other hand, the rock sample shown in sample MBR423_2 shows the presence of associate minerals with the highest fraction of sulphur which is 26.93 wt%. Figure represents that the gold existed in the sample as finely disseminated in quartz–stibnite–ankerite–sillimanite. Besides, the existence of gold was not homogenously distributed in the formation and other mineral compounds can be determined.

Geochemical characteristics are based on the result from X-ray fluorescence (XRF) where 10 samples were analysed based on the major elements, $Na_2O + K_2O$ versus SiO_2, volcanic rock classification diagram and major elements versus SiO_2 where SiO_2 as the *x*-axis.

Table 62.2 Host rock associated with mineralization of research area

Sample	Host rock associated with mineralization
MBR417_2	Andesite series
MBR417_3	Andesite series
MBR423_3	Andesite series
MBR427_3	Andesite series
MBR417_1	Basaltic andesite series
MBR432_1	Basaltic andesite series
MBR423_2	Dacite series
MBR427_1	Dacite series
MBR423_4	Cannot be detect
MBR427_2	Cannot be detect

The association of basaltic andesite, andesite and dacite rocks is composed dominantly of ash-flow tuffs. MBR417_1 and MBR423_1 located basaltic andesite series. For sample MBR417_2, MBR417_3, MBR423_3 and MBR427_3 were in andesite series while the dacite series for sample MBR423_2 and MBR427_1. The range for SiO_2 is from 24.98 to 70.67wt%. From that, only 8 samples can be detected by the diagram which samples MBR417_1, MBR417_2, MBR423_1, MBR423_2, MBR423_3, MBR427_1 and MBR427_3 are being in rock type while the other 2 which MBRMBR423_4 and MBR427_2 are in ore type and cannot be detected because of alteration effect and silica content is less than 30% from the sample tabulate in Table 62.2. All the samples in the research area are altered to variable degrees and any interpretation on the relative abundances of major elements.

Chemical tests used to determine the proportion of metal elements (e.g. copper, plum, zinc, gold, silver and other or non-metallic components) in a sample such as fluoride, sulphur and phosphorus are commonly referred to as a test. For this purpose, various techniques may be used. Several analytical methods of the gold testing were used in this study, namely the digestion of acid by aqua regia. The analysis was necessary to reduce the free gold content for the determination of the lead grade. The use of the test methods with an efficacious combination of detection methods is important for gold analysis.

Fire assay is the gravimetric gold and silver analysis technique used for platinum group elements (PGEs) [15]. This technique involves mixing the gold ore sample with lead oxide (litharge) and other sodium carbonates, borax, silver nitrate and potassium nitrate fluxes. The gold content is 4.01 g/t is demonstrated by the results from the AAS test. Results of the sample MBR423_4 and MBR427_2 cannot be carried out with an atomic absorption spectroscopy (AAS) due to high content of sodium.

62.5 Conclusion

This work has purposely been carried out to determine the mineralogy and geochemistry of the Selinsing gold deposits at Selinsing Gold Mine. During the field and analytical portions of this study, the following conclusions were determined by rocks in the Buffalo Reef shear zone which have typically been altered to variable degrees by silica–sericite–pyrite. Gold is found in the veins which are usually linked with pyrite, arsenopyrite and stibnite. Temporarily, during brittle deformation, gold is coeval with stibnite. The presence of stibnite and arsenopyrite showed the transition epizonal (Au-Sb) and mesozonal (Au-As) occurs in the mesothermal gold deposit model. Based on the Central Gold Belt 2, gold quartz lodes that prove from single FA-AAS analysis shows a significant gold grade of 4.01 g/t. Therefore, for future works, geochemical, petrographically and dating methods are suggested to detect and measure elements based on the ionization of a sample.

Acknowledgements This work was supported by the Fundamental Research Grant Scheme (FRGS) awarded by the Ministry of Education of Malaysia entitle "Enhanced Coalbed Methane Recovery Processes by Coupling Fluid Flow of Mukah-Balingian Coal Deposits" with the referral no. R.J130000.7851.5F371.

This work was supported by the Universiti Teknologi Malaysia Encouragement Research Grant (UTMER) awarded by Universiti Teknologi Malaysia entitle "Diagnosis of Alkali-Aggregate Reaction–Polarizing Microscopy and SEMEDS Analysis" with the referral no. Q.J130000.3851.19J79.

References

1. K.S. Ariffin, N.J. Hewson, Gold-related sulfide mineralization and ore genesis of the Penjom gold deposit, Pahang, Malaysia. Resour. Geol. **57**(2), 149–169 (2007)
2. E.B. Yeap, Tin and gold mineralizations in Peninsular Malaysia and their relationships to the tectonic development. J. Southeast Asian Earth Sci. **8**(1–4), 329–348 (1993)
3. F.E. De et al., The geology and mineral resources of the neighbourhood of Raub, Pahang, Federated Malay States, with an account of the geology of The Raub Australian Gold Mine (1939)
4. B.W.D. Yardley, J.S. Cleverley, The role of metamorphic fluids in the formation of ore deposits Geol. Soc. Spec. Publ. **393**, 117–134 (2014)
5. K. Shah, K. Shah Ariffin, *Mesothermal Lode Gold Deposit Central Belt Peninsular Malaysia* (InTech, 2012), p. 32
6. D.I. Groves, R.J. Goldfarb, M. Gebre-Mariam, S.G. Hagemann, F. Robert, Orogenic gold deposits: a proposed classification in the context of their crustal distribution and relationship to other gold deposit types. Ore Geol. Rev. **13**(1–5), 7–27 (1998)
7. C. Makoundi, K. Zaw, R.R. Large, S. Meffre, C.K. Lai, T.G. Hoe, Geology, geochemistry and metallogenesis of the Selinsing gold deposit, central Malaysia. Gondwana Res. **26**(1), 241–261 (2014)
8. I. Metcalfe, The Bentong-Raub Suture zone. J. Asian Earth Sci. **18**(6), 691–712 (2000)
9. L.A. House, Monument Mining Ltd: Buffalo Reef Gold Deposit, Malaysia. NI43-101 Technical Report Project No. 2015
10. F.C.P. Spiller, I. Metcalfe, Palaeozoic and Mesozoic radiolarian biostratigraphy of Peninsula Malaysia. J. Geol. Geol. Surv. Vietnam Hanoi B **5**(6), 75–93 (1995)

11. K. Shah, Mesothermal lode gold deposit Central Belt Peninsular Malaysia. Earth Sci. (2012) (InTech)
12. P.J.J. Yeap, E B and Ng T F, Application of soil geochemistry to the detection of Sb-Au mineralization in the Buffalo Reef area, Kuala Medang, Pahang. Bull. Geol. Soc. Malaysia **35**, 123–133 (1994)
13. J. Pereira, Geology, mining and tailing characteristics of the Selinsing Gold Mine, Pahang: a preliminary study. War. Geol. **19**(2), 35–41 (1993)
14. T.M. Williams, N. Breward, A.J. Lumb, C. Curnmins, Trace element contamination around centres of gold mining: examples from peninsular Malaysia. Environ. Geochem. Health **16**(2), 90–90 (1994)
15. S. Parry, M. Asif, I.W. Sinclair, Radiochemical fire-assay for determination of the platinum group elements. J. Radioanal. Nucl. Chem. Artic. **123**(2), 593–606 (1988)

Chapter 63
A Recent Progress on Sustainable Construction Waste Management Using 3R (Reduce, Reuse, and Recycle) Approach in Malaysia

S. Abbirahmi, Mohamad Anuar Kamaruddin, Rasyidah Alrozi, and Mohd Mustafa Al Bakri Abdullah

Abstract Application of 3R (reduce, reuse, and recycle) is important to reduce dependability for dumping C&DW at landfills, a major environmental problem. Limited legal and enforcement measures have been implemented for construction waste management in Malaysia to ensure safety and quality construction while encouraging the use of the best methods in the industry. The fundamental goal of this research is the potential of recyclable materials and the C&DW management plans are assessed using SWOT analysis. The methodology focuses mainly on four parts, database selection and bibliometric search, refining, and sample selection key concepts, and structuring of the review. C&DW is becoming one of the significant waste sources, surpassing municipal solid waste. Concrete, heavy metal, plastic, and glass are among the C&D wastes highlighted. Reduce, reuse, recycle, compost, incinerate, and dump are the six stages of the waste disposal hierarchy, from low to high. Strengths, weaknesses, opportunities, and threats (SWOT) were analyzed for C&DW management. At the end of the research, a lot of information on the application and the benefits of the 3R method for C&DW are determined.

Keywords Construction and demolition waste · Recycling · Reuse · Reduce · Sustainable · SWOT

S. Abbirahmi · M. A. Kamaruddin (✉)
Environmental Technology Division, School of Industrial Technology, Universiti Sains Malaysia, 11800 Penang, Gelugor, Malaysia
e-mail: anuarkamaruddin@usm.my

M. A. Kamaruddin · M. M. Al Bakri Abdullah
Center of Excellence Geopolymer and Green Technology, Universiti Malaysia Perlis, 01000 Perlis, Malaysia

R. Alrozi
School of Chemical Engineering, College of Engineering, Universiti Teknologi MARA(UiTM), Cawangan Pulau Pinang Kampus, 13500 Permatang Pauh, Pulau Pinang, Malaysia

M. A. A. Mohd Salleh et al. (eds.), *Proceedings of the Green Materials and Electronic Packaging Interconnect Technology Symposium*, Springer Proceedings in Physics 289,
https://doi.org/10.1007/978-981-19-9267-4_63

Table 63.1 Laws under the authorities' jurisdiction are based on construction waste management

Authorities	Law/Act/Standard	Remarks
Construction Industry Development Board (CIDB)	CIDB Act 1994 (Act 520) • (Part I) Construction work • (Part IX) Enforcement and investigation	This defines the construction works, and the officer has the authority to examine the worksite
Ministry of Environment and Water	Environmental Quality Act 127, 1974	This is done to avoid, regulate contamination, and enhance the environment
Ministry of Works	Standard Specifications for Buildings Works (2014) *Section A* • 38 Waste Management • 44 Keeping the Site Tidy	The contractor should manage waste and keep the site clean and tidy condition by removing waste

63.1 Introduction

C&DW has been increasing with the growth of the economy and population. An expected total of 63,101.93 metric tonnes is produced all over Peninsular Malaysia for residential projects [1]. The premise of a low recycling ratio governs C&DW due to limited space at the construction site, and rising labor costs, making it difficult to use 3R in C&DW as well as dispose of it in a landfill. Proper waste management may save costs, minimize waste creation, and conserve natural resources [2]. In Malaysia, a few legal and enforcement measures for construction waste management have been implemented. Table 63.1 shows a few laws under the authorities' jurisdiction in handling construction waste management.

Present C&DW management heavily relies on landfill disposal gradually reduces the landfill life span and causes more emissions of harmful gases from C&DW disposal. The common harmful gas released contains 45–60% methane and 40–60% carbon dioxide. In general, C&DW consists of many recyclable materials that can be recovered. However, little research has been conducted on the possibility of recovering C&DW from waste creation at the source by using the 3R principle. Thus, the objective of this study is to evaluate the recyclable item potential from C&DW and analyze the strategy of C&DW management by using SWOT analysis.

63.2 Methodology

The aim of this work is to expand knowledge about construction and demolition waste by conducting a systematic literature review on C&DW behavior using a 3Rs approach. A four-stage plan is selected to distinguish the most relevant research

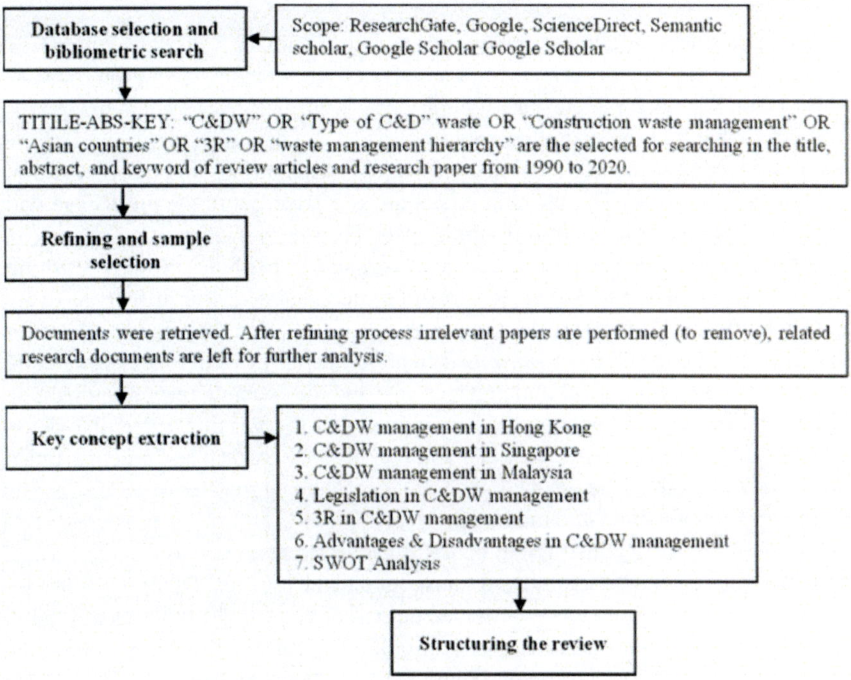

Fig. 63.1 Flowchart of the literature review methodology

papers, namely database selection and bibliometric search; refining and sample selection; key concept extraction; and structuring the review based on the main topics extracted. The literature review methodology is presented in Fig. 63.1.

63.3 Results and Discussion

63.3.1 Concrete

The pyramid of discarding alternatives categorizes environmental consequences into 6 levels, which range from peak to trough which is reduce, reuse, recycle, compost, incinerate, and landfill. The "3Rs" refers to the three basic tactics of reuse, recycling, and reduction. Construction and obliteration waste, including demolished concrete, bricks and masonry, wood, and many more accouterments similar as glass, sequestration, roofing, cables, pipes, jewels, and soil, makes up a considerable element of the overall waste [3].

Among the total construction and obliteration waste, concrete waste makes up about 50% of the total waste generated [3–5]. Overordering, potential harm during transit, impairment during installation, bad craftsmanship, as well as design revision are among the five major reasons for a massive concrete waste generation [6–8].

Based on Fig. 63.2, on-site construction waste is currently disposed of in landfills, while fresh materials are manufactured from rocks and delivered to the construction site to produce new concrete. The second stage is stripping, which involves clearing and leveling the rocks. The blasting procedure is carried out in the third stage with the help of blasting equipment and capital costs. The fourth stage, stockpiling, necessitates the hiring of one laborer at a wage of around $18 per hour. Primary crushing, magnetic separation, and secondary crushing are all part of the fifth stage of the crushing process. Primary crushers, secondary crushers, and shapers are all used in this process. As a result, the profit achieved at this step is the difference between the existing method's equipment maintenance cost and the concrete recycling method's expense of equipment maintenance. The sixth stage is the washing, screening, or air-sitting process that uses recycled water and fuel to settle the dust and other particles. Finally, in the stage of the finished product, completed aggregates of 7, 10, 20, and 75 mm are manufactured and marketed.

Based on Fig. 63.3 construction debris is discarded at a recycling center, where fresh materials are manufactured and transported to the job area. Dumping construction waste at a recycling factory provided numerous advantages in terms of landfill fees, landfill zone, transportation costs, air pollution, noise pollution, gas emission, and energy consumption. The second stage is stockpiling, which requires one laborer. Sorting is the third stage when pulverizers and excavators are used to estimate capital expenses. Primary crushing, magnetic separation, and secondary crushing operations are all part of the crushing process's fourth stage. The fifth stage is the manual removal procedure, which includes the removal of wood, paper, and plastics as well as crushed debris. The sixth stage is the washing, screening, or air-sitting process that uses recycled water and fuel to settle the dust and other particles. Finally, in the stage of the finished product, completed aggregates of 7, 10, 20, and 75 mm are manufactured and marketed (Table 63.2).

Overall, the concrete recycling method receives a positive net benefit of about $30,916,000 year^{-1} while the current method receives a negative net benefit of about—$44,076,000 year^{-1}.

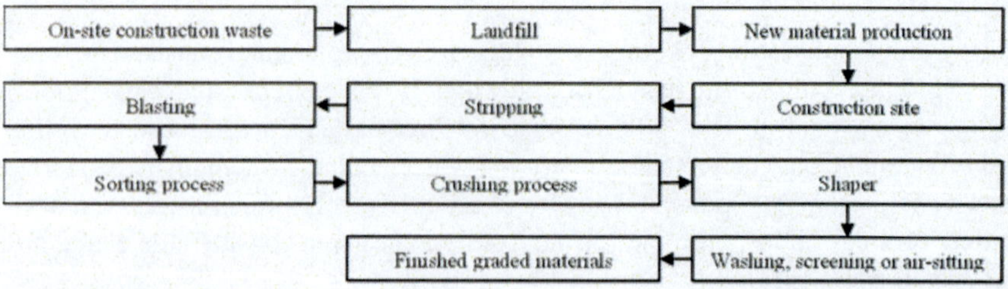

Fig. 63.2 Flowchart of the current practice

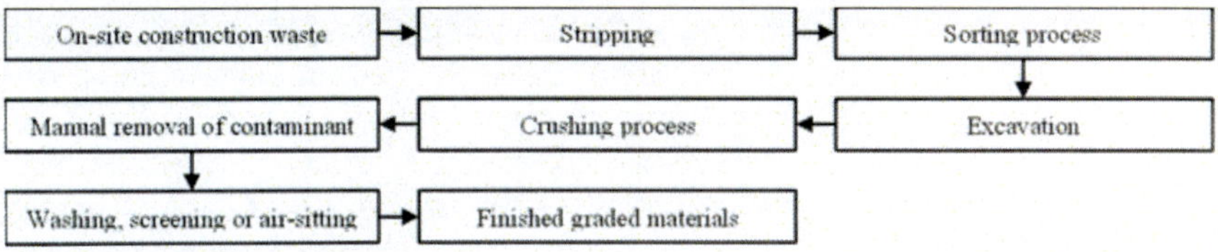

Fig. 63.3 Flowchart of the concrete recycling practice

Table 63.2 Comparison of prices of finished graded materials between the current and concrete recycling practices [9]

Price of finished graded materials (mm)	Price of finished graded materials ($)	
	Current practice	Concrete recycling practice
7	15	14.8
10	25	20
20	25	22
75	16	15.4

63.3.2 Heavy Metal

Sorted wood and metals could be recycled in China, although this accounts for very little around 10% (by weight) of the C&D waste created. Construction businesses and recyclers often separate only valuable components from construction sites, like metals as well as wood frames. In the case of asbestos removal, asbestos fibers can become airborne if asbestos is detected during crushing and screening, posing a health risk. To date, over sixty nations, along with the United States, the European Union, and Japan, who were the major users of asbestos throughout times of strong economic expansion, had prohibited its usage [10]. China, on the other hand, seems to be an outlier, continuing to manufacture and consume enormous quantities of asbestos due to bargainable cost, good efficiency, and quantity of asbestos mining areas as well as allied sectors [11]. The practice of recycling asphalt shingles is gaining popularity due to preservation of landfilling and the commercial viability of items that use repurposed asphalt shingles while the negative being the likely presence of asbestos [12]. As a result, a thorough examination of the asphalt shingle recycling process is required.

63.3.3 Plastic

Based on Fig. 63.4, there are two methods for collecting plastic garbage which plastics are collected after they reach the municipal trash stream, whereas plastics are collected before any subsequently enter the MSW. The majority of post-consumer garbage (disposals from homes, companies, and light industry sources) eventually

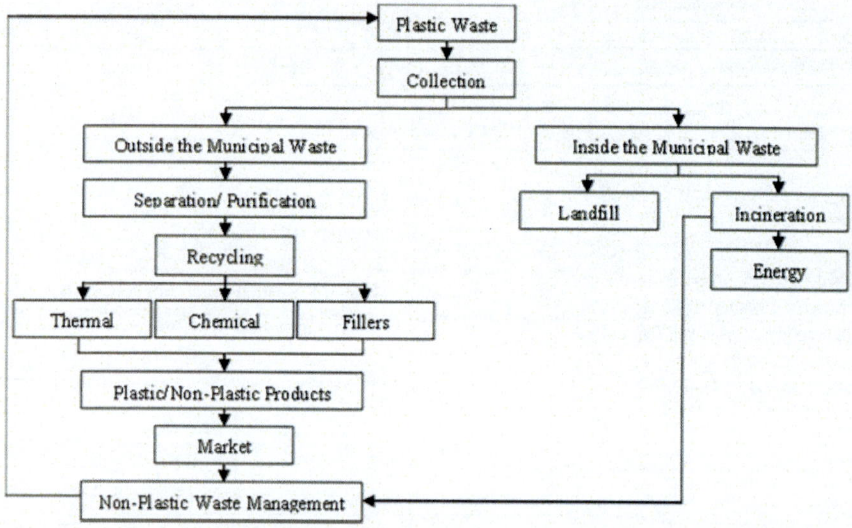

Fig. 63.4 Flowchart of the plastic waste management

winds up within municipal waste streams and is often highly tainted, making recycling economically impractical at present time [13]. As a result, garbage is either buried or burned to decrease size plus recuperate energy content. Plastic garbage gathered beyond MSW, on the other hand, is quite sanitary and it can be repurposed with minor isolation and purification. Because used plastic boules made of polyethylene terephthalate (PET) and high-density polyethylene (HDPE) could be recognized and compiled before entering the municipal waste stream and segregated by kind before getting handled, the infrastructure presently being built to gather plastic waste is primarily limited to these materials.

Landfilling plastics becomes a public concern because the substance degrades slowly but studies have demonstrated that plastic rubbish does not obstruct landfill workflow or contribute to landfill leachate toxicity [14]. Some people believe that degradable plastics should be used in landfills leading to investigation on biodegradation and photocatalytic degradation for commercial uses. However, for biodegradable polymers materials to disintegrate, they need light and air, as well as enough moisture and nutrients to support microbial activity which is not met in landfill conditions [14]. Moreover, the influence of biodegradable waste on the environment is uncertain [15].

The heat capacity of the plastic garbage could be recuperated through incineration which lowers the rubbish volume by 90 to 95 percent and offers an enticing origin of alternative energy, conserving naturally-born resources, and reducing the burden of international energy dependence. Modern technology has reduced public opposition by allowing incineration plants to run without hazardous emissions complying with the 1990 Clean Air Act Amendments [14]. Despite that, two kinds of ash created, fly ash and bottom ash must be disposed of and not accumulated due to the risk of groundwater and soil pollution from contaminated liquid containing toxic elements like lead and cadmium. Groundwater and soil protection methods, like lining landfills, could be costly and ineffective in terms of environmental protection. As a result, some research is being carried out to successfully stabilize and recycle incineration leftovers in building applications [16].

The 3 recycling methods of plastic waste are thermal reprocessing, chemical modification, and filler which are stated in Fig. 63.4. Thermal reprocessing is the process of a thermoplastic being heated to temperatures that are exceptionally high to make it flow. As the plastic cools, it gets transformed into a new product. Chemical modification or depolymerization of trash to reclaim its fundamental ingredients is a viable way of recovering plastics. Hydrolysis (chemical decomposition) and pyrolysis are the two methods for depolymerization. The filler method is where thermoplastic wastes are melted and co-extruded or co-injected into moldings with virgin resins. The virgin resins, with superior properties, are forced into the perimeter of the mold while the recycled plastics, with inferior properties, are injected into the center of the mold.

63.3.4 Glass

Recycling glass instead of dumping it in dumpsite can save 27% of energy and 37% of GHG emissions. In 1963, the first research on using waste glass as a construction material was completed [17]. The writers transformed discarded glass into glass chips that could be used to make wall panels. Following that, substantial research into the utilization of waste glass in construction materials was conducted. Figure 63.5 shows the materials in glass packaging that are saved from one tonne of glass recycling.

Coarse glass cullet used in cement mortar and concrete [19] causes the concrete's workability and strength to be harmed [20, 21] and the usage of discarded glass crushed into powder popularized. They consist of 70% SiO_2, 13% Na_2O, and 10% CaO and had an amorphous structure that allowed them to dissolve quickly high alkaline environment and functions in cement concrete as a reactive pozzolan [22, 23]. In other situations, GP was employed as a geopolymer predecessor to generate cement-free compounds [24] as well as a silica fume replacement or additive, resulting in significant cost reductions [25]. Fabrication of LWAs and PCMs that take advantage of glass's phase-changing capacity were fascinating applications for GP [26]. Below shows the comparison between concrete with GP and control concrete in Fig. 63.6

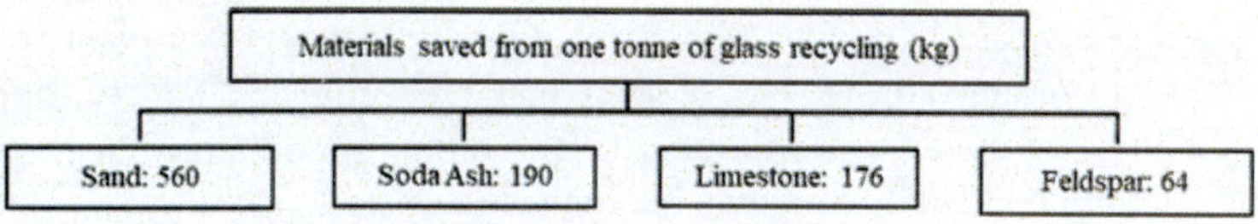

Fig. 63.5 Materials in glass packaging that is saved from one tonne of glass recycling [18]

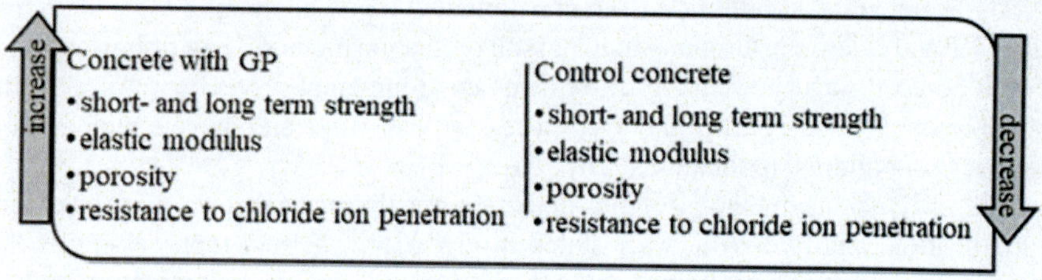

Fig. 63.6 Comparison between concrete with GP and control concrete [27]

Table 63.3 The result of replacing GP with specific material in the fabrication of blocks [27]

Material	GP (%)	Results
Dry cast block	10–25	Decrease in density Low compressive strength Increased water absorption
Other waste in the fabrication process of blocks (e.g., limestone sawdust)	Specific value was not mentioned	Lower water absorption Increased flexural strength

and the result of replacing GP with specific material in the fabrication of blocks in Table 63.3.

63.3.5 *Strength, Weakness, Opportunities, and Threat (SWOT) Analysis*

63.3.5.1 Strength Facilitating C&DW Management

The SWOT analysis was carried out to discover the positive and negative sides of C&DW, and to have a picture of how to improve and increase the quality of the management.

Strength_1 Reduce C&D Waste: 10–30% of C&DW occupies the total volume of landfills around the world which is 8 times municipal solid waste.

Strength_2 Reduce Landfill and More Land for Urban Development: In China, the landfills cover an area of 375 million m^2 equaling half of Singapore's total land area and will be reduced by 90%, with the optimum recycling scenario.

Strength_3 Increase Economic Benefits and Future Urban Development Funds: The rate per present value (R-PV) of scrap metal recycling in China is estimated at 201 billion US dollars and will turn to 401.6 billion US dollars if the recycling rate increases about 30% to include the four inert wastes.

Strength_4 Reduce Financial Burden: By reducing the number of landfills, the fund invested and used in the landfill sector can be reduced and reduces labor demand for the production and transport of virgin building supplies.
Strength_5 Reduce Sourcing of Raw Materials: Recycled materials can replace virgin materials in manufacturing brand-new buildings [28].

63.3.5.2 Weakness Impeding C&DW Management

Weakness_1 Need Specific Disposal Method: C&DW entails disposal requirements as it consists mostly of inert materials.
Weakness_2 Increase in Pollution: C&DW is not known as hazardous unless accumulated leading to negative environmental issues, such as dust produced in recycling facilities [29]. Furthermore, the growing popularity of recycling asphalt shingles was impeded by the potential environmental impacts of asbestos [12].
Weakness_3 Transfer of Contaminants: The production of new concrete from recycled aggregate can transfer contaminants from the waste to new products with reduced strength and unstable concrete [12].
Weakness_4 Lack of C&DW Management Data: Insufficient data in C&DW and environment consequences assessment leads to an ineffective C&DW categorization administration system.
Weakness_5 High Finances and Investment in Large-Scale Recycling Facility: A study carried out on recycling C&DW plants in the Netherlands concluded static and "on the move" recycling centers with second-hand tools have more economic feasibility and a higher turnover margin compared to centers with brand-new tools [30].

63.3.5.3 Opportunities Facilitating C&DW Management

Opportunities_1 New Job Available: The identification of products hazardous to health and the environment led to the development of more C&D workforces, building operators, cleaning services, and waste workers in the grouping, storing, and moving of waste.
Opportunities_2 More Concrete and Intensive Data: With the rising apprehension on C&DW, C&DW production approximations of other countries are useful for planning the current analysis effort and could assist academics, experts, and decision-makers in making more ecologically comprehensive supervision of C&DW.
Opportunities_3 Improved Law and Legislations: In the United States, each state has its specific policies on C&DW. For instance, C&D waste must be segregated from the waste stream and separated at sanitary landfills as required by the North Carolina Solid Waste Management Act of 1989.

Table 63.4 SWOT analysis

Internal conditions	External conditions
Strengths	*Opportunities*
S1: Reduce C&D waste	O1: New job available
S2: Reduce landfill and more land for urban development	O2: More concrete and intensive data
S3: Increase economic benefits and future urban development funds	O3: Improved law and legislations
S4: Reduce financial burden	
S5: Reduce sourcing of raw materials	
Weakness	*Threats*
W1: Need specific disposal method	T1: Cultural habits of illegal dumping
W2: Increase in pollution	T2: No proper legislation
W3: Transfer of contaminants	T3: Authorities power
W4: Lack of C&DW management data	
W5: High finances and investment in large-scale recycling facility	

63.3.5.4 Threat Impeding C&DW Management

Threat_1 Cultural Habits of Illegal Dumping: In developing and underdeveloped countries, it is not uncommon for C&DW to be transported to low-income neighboring provinces due to the inadequate dumping ground and high cost of land resources in developed metropolises.

Threat_2 No Proper Legislation: In China, there is no nationwide or local guideline to handle the removal of C&DW via dumping ground. Furthermore, there are no detailed guidelines on categorizing and secure-handling methods for active hazardous C&DW materials [31].

Threat_3 Authorities Power: In many China cities, due to the small retrieval value, material retrieval and reprocessing are not usually done by municipal officials or dumping ground operatives leading to foul play by construction authorities [32] (Table 63.4).

63.4 Conclusion

In conclusion, C&D Waste can undergo or practice the 3R waste minimalization method. Applying waste minimalization methods in the construction industry reduces the reliability or dependability of landfills and leads to a higher net gain of profit due to the 3R method consuming fewer expenses compared to the traditional method. Furthermore, 3R practice can be implemented in Malaysia since many parts of the country had already applied and have been practicing the method in the construction industry. Based on research from these past years by many researchers, it can be acknowledged that recyclable items are potentially inferior compared to the reference material if an appropriate ratio of materials is used in the production as per the research. Moreover, strength, weakness, opportunity, and threat (SWOT) were analyzed for the C&D waste management. Despite having threats and weaknesses most of the major problems can be solved and handled professionally.

Acknowledgements The authors would like to acknowledge Universiti Sains Malaysia (USM) for the opportunity of research work to be conducted. Special thanks to those who contributed to this project directly or indirectly.

References

1. K. Kupusamy, S. Nagapan, A.H. Abdullah, S. Kaliannan, S. Sohu, S. Subramaniam, H. Maniam, Construction waste estimation analysis in residential projects of Malaysia. Eng. Technol. Appl. Sci. Res. **9**, 4842–4845 (2019)
2. N.K. Bui, T.L. Nguyen, K.D. Phan, A.T. Nguyen, In *Legal Framework for Recycling the Domestic Solid Waste in Vietnam: Situation and Recommendation.* E3S Web of Conferences, 164, (2020)
3. S. Coventry, C. Woolveridge, S. Hillier, *The Reclaimed and Recycled Construction Materials Handbook* (1999)
4. H.S.P. Charles, F. Hendriks, SUSTAINABLE raw materials: construction and demolition waste (2000)
5. W. Li, Composition analysis of construction & demolition waste and enhancing waste reduction and recycling in construction industry in Hong Kong. (2002)
6. C.S. Poon, T.W. Ann, L.H. Ng, On-site sorting of construction and demolition waste in Hong Kong. Resour. Conserv. Recycl. **32**, 157–172 (2001)
7. L.Y. Shen, V.W.Y. Tam, Implementation of environmental management in the Hong Kong construction industry. Int. J. Project Manage. **20**, 535–543 (2002)
8. V.W.Y. Tam, C.M. Tam, Evaluations of existing waste recycling methods: a Hong Kong study. Build. Environ. **41**, 1649–1660 (2006)
9. Environmental Protection Agency, Construction and demolition waste, waste management and resource use opportunities. Queensland Government (2007)
10. S. Furuya, K. Takahashi, Experience of Japan in achieving a total ban on asbestos. Int. J. Environ. Res. Public Health **14**, 1261 (2017)
11. X. Qian, W. Zhou, K. Nakagami, An international comparative study on the regulation policy of asbestos usage in China. Ritsumeikan J. Asia Pacific Stud. **81** (2012)
12. T. Townsend, J. Powell, C. Xu, Environmental issues associated with asphalt shingle recycling. Construction Materials Recycling Association, US EPA Innovations Workgroup (2007)
13. T.R. Curlee, The economic feasibility of recycling: a case study of plastic wastes (1986)
14. H. Alter, The origins of municipal solid waste: II. Policy options for plastics waste management. Waste Manage. Res. **11**, 319–332 (1993)
15. P.P. Rumps, Plastic wastes: Management, control, recycling and disposal-pollution technology review no. 201. US Environmental Protection Agency, T. Randall Curlee, and Sujit Das, Noyes Data Corporation, Park Ridge, NJ,(1991) 479 pp. ISBN No.: 0-8155-1265-1. US List P. (1994)
16. J.J.J.M. Goumans, H.A. der Sloot, T.G. Aalbers, Waste materials in construction (1991)
17. F. Rajabipour, H. Maraghechi, G. Fischer, Investigating the alkali-silica reaction of recycled glass aggregates in concrete materials. J. Mater. Civ. Eng. **22**, 1201–1208 (2010)

18. Glass Packaging Institute, Why Recycle Glass? (2017). http://www.gpi.org/recycling/why-recycle-glass
19. K. Afshinnia, P.R. Rangaraju, Impact of combined use of ground glass powder and crushed glass aggregate on selected properties of Portland cement concrete. Constr. Build. Mater. **117**, 263–272 (2016)
20. E.E. Ali, S.H. Al-Tersawy, Recycled glass as a partial replacement for fine aggregate in self compacting concrete. Constr. Build. Mater. **35**, 785–791 (2012)
21. B. Taha, G. Nounu, Utilizing waste recycled glass as sand/cement replacement in concrete. J. Mater. Civ. Eng. **21**, 709–721 (2009)
22. S. Liu, G. Xie, S. Wang, Effect of curing temperature on hydration properties of waste glass powder in cement-based materials. J. Therm. Anal. Calorim. **119**, 47–55 (2015)
23. A.M. Matos, J. Sousa-Coutinho, Durability of mortar using waste glass powder as cement replacement. Constr. Build. Mater. **36**, 205–215 (2012)
24. R. Redden, N. Neithalath, Microstructure, strength, and moisture stability of alkali activated glass powder-based binders. Cement Concr. Compos. **45**, 46–56 (2014)
25. A. Omran, D. Harbec, A. Tagnit-Hamou, R. Gagne, Production of roller-compacted concrete using glass powder: field study. Constr. Build. Mater. **133**, 450–458 (2017)
26. V. Ducman, A. Mladenovič, J.S. Šuput, Lightweight aggregate based on waste glass and its alkali–silica reactivity. Cem. Concr. Res. **32**, 223–226 (2002)
27. A. Omran, A. Tagnit-Hamou, Performance of glass-powder concrete in field applications. Constr. Build. Mater. 109, 84–95 (2016)
28. N. Roussat, C. Dujet, J. Méhu, Choosing a sustainable demolition waste management strategy using multicriteria decision analysis. Waste Manage. **29**, 12–20 (2009)
29. I.M. Simion, M.E. Fortuna, A. Bonoli, M. Gavrilescu, Comparing environmental impacts of natural inert and recycled construction and demolition waste processing using LCA. J. Environ. Eng. Landsc. Manag. **21**, 273–287 (2013)
30. W. Zhao, R.B. Leeftink, V.S. Rotter, Evaluation of the economic feasibility for the recycling of construction and demolition waste in China—the case of Chongqing. Resour. Conserv. Recycl. **54**, 377–389 (2010)
31. H. Duan, D. Yu, J. Zuo, B. Yang, Y. Zhang, Y. Niu, Characterization of brominated flame retardants in construction and demolition waste components: HBCD and PBDEs. Sci. Total Environ. **572**, 77–85 (2016)
32. H. Duan, J. Wang, Q. Huang, Encouraging the environmentally sound management of C&D waste in China: an integrative review and research agenda. Renew. Sustain. Energy Rev. **43**, 611–620s (2015)

Chapter 64
Structural Mechanical Properties of Polyethylene Terephthalate (PET) Concrete Subjected to High Temperature

S. Beddu, A. J. H. Alansare, Z. Itam, Daud Mohamad, M. M. Zainoodin, and W. W. A. Zailani

Abstract Concrete is the most popular and used as building components around the world. One of the most crucial materials to form concrete is sand. Due to high volume in urbanization, this lead to mass concrete production. Consequently, the quantities of sand found in nature will reduce over time. On the other hand, large quantities of plastic are produced around the world, and due to daily consumption, most plastic is converted into waste. Using plastic inside concrete is one of the best ways to get rid of waste as well as reducing the amount of sand used in concrete production, which in turn helps to reduce environmental risks and preserve the environment. When designing a mixture of concrete containing plastic, the safety of the building must be considered, as the behavior and effect of plastic on concrete change at high temperatures. This study presents the effect of polyethylene terephthalate (PET) on fresh concrete as well as the mechanical properties of concrete containing PET at normal temperatures and high temperatures when using PET as fine aggregate.

Keywords Concrete · Polyethylene terephthalate (PET) · Fine aggregate

64.1 Introduction

The use of plastic in most industries led to an increase in the consumption of plastic, which led to an interest in plastic waste recycling. The problem of increasing and

S. Beddu (✉) · Z. Itam · D. Mohamad · M. M. Zainoodin
Department of Civil Engineering, Universiti Tenaga Nasional, Jalan Ikram-Uniten, 43000 Kajang, Selangor Darul Ehsan, Malaysia
e-mail: salmia@uniten.edu.my

A. J. H. Alansare
College of Graduate Studies, Universiti Tenaga Nasional, Jalan Ikram-Uniten, 43000 Kajang, Selangor Darul Ehsan, Malaysia

W. W. A. Zailani
School of Civil Engineering, College of Engineering, Universiti Teknologi MARA, 40450 Shah Alam, Selangor, Malaysia

M. A. A. Mohd Salleh et al. (eds.), *Proceedings of the Green Materials and Electronic Packaging Interconnect Technology Symposium*, Springer Proceedings in Physics 289,
https://doi.org/10.1007/978-981-19-9267-4_64

accumulating plastic waste is of great concern in the field of environmental protection, where large quantities of plastic waste are being sent to the landfill annually. Plastic includes chemical materials such as polyethylene terephthalate (PET), polypropylene (PP), polystyrene (PS), polycarbonate (PC), acrylonitrile butadiene styrene (ABS). PET is considered one of the most widespread polymers around the world and ranked third among polymers used in plastic industry, especially in food packaging and the manufacture of beverage bottles. PET covers approximately 16% of European plastic consumption [1]. Due to the biological nature of plastic, the PET decomposes slowly, which would take hundreds of years, resulting in plastic accumulation in landfills. Studies have tended to introduce PET into the concrete industry due to the operational life of the concrete that is longer than the operating life of the PET, where the fine or coarse aggregate is replaced with PET. Incorporating PET into the concrete will make it far from direct contact with the environment, Many researchers have studied the replacement of aggregate with plastic in concrete. Experiments have been conducted to understand the properties of concrete that contains plastic, and of the most important experiments are studying the tensile resistance and flexural resistance, studying the plastic proportions that can improve concrete properties, and studying the behavior of concrete containing plastic [2, 3].

Although concrete is a fire-resistant material due to its non-combustibility and low thermal conductivity coefficient (compared to steel), when exposure to high temperatures, a series of physical and chemical reactions occur leading to a deterioration of its mechanical properties, which leads to the heat reaching the inner layers of the concrete [4].

The temperatures affect the plastic component of the concrete due to the behavior change of plastic when exposed to high temperature. Most of the researchers used low plastic ratios that do not exceed 10% because increasing the percentage of replacing aggregates with plastic leads to the deterioration of the properties of concrete, It has been found that large quantities of plastic waste are not recycled, and the percentage of plastic waste remains greater than the rate of recycling; in 2018, in Europe out of 61.8 million tons produced, 9.4 million tons recycled [5, 6].

Plastic waste takes years to decompose due to the nature of plastic. Therefore, it occupies large areas of land for burial, and this has a negative impact on the environment, where plastic waste can restrict the movement of roots and hinder the movement of the ground water flow so it has become essential to recycle plastic waste and enter recycling in several fields, including construction industry, where environmentally friendly buildings are used, and the green buildings in which the waste is reused [7, 8].

Although the possibility of recycling plastic waste PET is important to environment, the studies on replacing aggregates inside concrete with PET is still limited. However, recent research shown the inclusion of PET not exceeding 10% is capable to avoid failure of concrete resistance. To make it possible using the concrete containing PET in the building sector, and increasing the recycling of PET waste inside the construction industry, it is necessary to study the properties of concrete containing PET. It is also essential to assess the compression-resistant after exposing the concrete containing PET to high temperatures, to investigate whether PET concrete can withstand high temperatures without collapse. Determining the effect of PET on the properties of concrete in high temperatures is complicated. And it is difficult to identify the main reason for the failure of concrete contains PET whether due to the behavior of PET under high temperatures or due to the behavior of the concrete itself [5].

The increase in the population has led to an increase in the demand for plastic industries; this is because plastic is used in many industries, which consequently led to an increase in PET waste. Among it, there are a lot of PET products that are used for a short time, such as plastic bags, and water bottles [1]. Therefore, scientists have become interested in recycling PET waste to be used in concrete production. However, large quantities of PET waste are not used in concrete; only 10% of the fine aggregates are replaced by PET inside the concrete. This is because increasing the rates of replacement or additives leads to a deterioration of the concretes properties such as resist compression and flexural strength [5].

This paper study to investigate the effectiveness of concrete containing polyethylene terephthalate (PET) when exposed to high temperatures. The effect of heat on the compressive strength of the concrete and the compressibility of concrete have been studied, as well as the effect of adding PET to concrete on compressive, flexural, and tensile strength. Studying whether the concrete containing PET will still be able to withstand pressure after being exposed to high temperatures is of great importance for the safety of the building occupants, as well as for economic aspect for determining the life span of the building after being exposed to high temperatures. Despite many studies related to the behavior of concrete containing PET, there are a limited study on the behavior of concrete containing PET under the influence of high temperatures, such as the heat caused by fires. Therefore, this research is study to determining the highest optimal ratio of replacement of the fine aggregate with PET that does not lead to significant deterioration of compressive strength of the concrete when exposed to high-temperature, studying the effects on the mechanical and chemical properties of the concrete, and exploring the compression resistance before and after being exposed to high temperatures.

64.2 Methodology

64.2.1 Materials and Fabrication of Ink

In this research, Ordinary Portland Cement (OPC) type I (ASTM-C150 2011) was used in concrete mixtures, and tap water was also used for the hydration process. Coarse aggregates were produced from gravel, and the type of fine aggregates were natural river sand. ASTM C33 standard (ASTM Standard C33 2003) was used to determine the gradients.

Fine aggregates are replaced by PET in different proportions (3%, 5%, 7%, 10%), and the mixture is designed using standard ACI 211.1 (reapproved 2009). 120 samples were cast, after curing the samples in water for a duration of 3, 7, 28, and 56 days.

30 samples were exposed to a temperature of 500 °C for 2 h, then a compressive strength test was used for samples after being exposed to the air to get cold for one day. Then a tensile strength test was used for 15 samples from the cylinder that have different percentages of PET in the concrete, and the other 15 samples from prisms contend different percentages of PET in the concrete were tested using the flexural strength test. These tests were done. The limits of material and the standards of the tests are as follows:

- Concrete mixture is designed depending on ACI 211.1–91 (reapproved 2009)
- W/C ratio: 0.55
- Fresh concrete slump: 25 mm–50 mm
- Size for PET that replaces with a fine aggregate: 2.36 mm
- Maximum size for coarse aggregates: 20 mm
- Gradation size of fine aggregates: between 5 and 0.075 mm
- Compressive strength test accordance ASTM C109/C109M: Cube size 100 × 100 × 100 mm
- Flexural strength test accordance ASTM C78 / C78M–18: Prism size 10 × 100 × 500 mm
- Tensile strength test in accordance with ASTM C496 / C496M–17: Cylindrical Concrete Specimens 150 mm diameter × 300 mm Height.

64.2.2 Design of Concrete Mix Design

One of the main goals when designing a concrete mix is to obtain an ideal concrete with high resistance. This is done by controlling the proportions of the main components of concrete such as cement, sand, gravel, and the water/cement ratio. Also, it was learned that improving the properties of concrete is possible when adding or replacing some of its components. Many researchers have sought to obtain the optimal proportions when replacing the fine aggregate with PET to get the best concrete mixture containing PET to achieve high-resistant concrete, more economical, and environmentally friendly. However, this is not an easy task as any replacement or addition requires many calculations and conducting many different experiments. In addition, determining the proportions should have the standard specifications. For instance, applying the premium method of the American Concrete Institute to determine the water content in kilograms per cubic meter of concrete is required to guide the workability of the mixture. Several researchers have designed the concrete using ACI specifications. Therefore, the fine aggregates were replaced by PET in certain proportions, and the size of the PET was conformed to have the specifications. The effect of PET on concrete after exposure to high temperatures was studied, as well as studying the different properties of the concrete containing PET. The concrete mixture was designed according to specifications ACI 211.1-91 where the properties of the constituent materials of the concrete mixture were considered, and the concrete mixture was designed through the following steps:

Step 1: Choice of the slump: The maximum and minimum slump must be specified. When the slump was chosen, it had a drop higher than 50 mm to avoid the separation of the concrete mixture.
Step 2: Choice of the maximum size of aggregates: The maximum aggregate used in this study was 20 mm.
Step 3: Estimate the water content and the air content: According to the slump and maximum size of aggregates, the water content is chosen. The water content was 190 kg/m^3, as shown in Table 64.1.
Step 4: Water/cement ratio selection: Concrete strength and durability are affected significantly by the water/cement ratio. According to ACI 211.1-91 and literature review, 0.55 w/c ratio was used
Step 5: Cement content calculation: According to ACI 211.1-91, the cement content in concrete mixture was calculated using the following Eq. 64.1:

$$\text{Cement Content} = \frac{\text{Water Content}}{\text{Water} - \text{Cement Ratio}} \tag{64.1}$$

Step 6: Coarse aggregates content calculation: According to Table 64.2, when the fineness modulus is 2.6 and the maximum fine aggregate is 20 mm, the bulk volume of oven-dry rodded coarse aggregate should be 0.64. The bulk density of the coarse aggregate was 1600 kg/m^3; therefore, the coarse aggregates' content was calculated according to Eq. (64.2), as following:

Table 64.1 Approximate mixing water and air content requirements for different slumps and nominal maximum sizes of aggregates (ACI 211.1-91)

Slump, mm	Water kg/m^3 of concrete for indicated nominal maximum size of aggregates							
	9.5	12.5	19	25	37.5	50	75	150
Non-air-entertained concrete								
25–50	207	199	190	179	166	154	130	113
75–100	228	216	205	193	181	169	145	124
150–175	243	228	216	202	190	178	160	–
Amount of entertained air, %	3	2.5	2	1.5	1	0.5	0.3	0.2
Air-entertained concrete								
25–50	181	175	168	160	150	142	122	107
75–100	202	193	184	175	165	157	133	119
150–175	216	205	197	184	174	166	154	–
Total air content % for								
Improvement of workability	4.5	4.0	3.5	3.5	2.5	2.0	1.5	1.0
moderate exposure	6.0	5.5	5.0	4.5	4.5	4.0	3.5	3.0
extreme exposure	7.5	7.0	6.0	6.0	5.5	5.0	4.5	4.0

Table 64.2 Bulk volume of course aggregate per unit volume of concrete (ACI 211.1-91)

Maximum size of aggregates, mm	Bulk volume of oven-dry rodded coarse aggregates per unit volume of concrete for fineness modulus of fine aggregates			
	2.4	2.6	2.8	3.0
9.5	0.50	0.48	0.46	0.44
12.5	0.59	0.57	0.55	0.53
20	0.66	0.64	0.62	0.60
25	0.71	0.69	0.67	0.65
37.5	0.75	0.73	0.71	0.69
50	0.78	0.76	0.74	0.72
75	0.82	0.80	0.75	0.76
150	0.87	0.85	0.83	0.81

$$\text{Coarse aggregates} = \text{coarse aggregates bulk volume} * \text{density} \tag{64.2}$$

Step 7: Fine aggregates content calculation: The volume of sand was obtained by summing the absolute volumes of water, cement, submerged air, and coarse aggregate, considering the division of each component by its specific weight. Then the total volumes were subtracted from 1. And then, the result was multiplied by the specific weight of fine aggregate to find the fine aggregate' content as shown in (64.3).

$$\text{Fine aggregates content} = \text{Fine aggregates volume} \times \text{fine aggregates SP} \tag{64.3}$$

Step 8: PET aggregates content calculation: The volume of PET (kg/m^3) was obtained by multiplying the required percentage of PET (3%, 5%, 7%, and 10%) by fine aggregate volumes (kg/m^3), as shown in (64.4)

$$\text{PET content} = \text{Fine aggregates volume} \times \text{required percentage of PET} \tag{64.4}$$

Step 9: Adjustments to mix proportions: Adjustments of mix proportions: The mass of all ingredients in kilograms per cubic meter of concrete are shown in Table 64.3.

Table 64.3 Concrete Mix Design of Coal Bottom Ash (CBA)

Samples	Cement (kg/m^3)	Water (kg/m^3)	Coarse Aggregate (kg/m^3)	Sand (kg/m^3)	PET (kg/m^3)	Replacement (%)
Mix 1	345.45	190	1024	737.88	0	0
Mix 2	345.45	190	1024	715.74	22.14	3
Mix 3	345.45	190	1024	700.99	36.89	5
Mix 4	345.45	190	1024	686.23	51.65	7
Mix 5	345.45	190	1024	664.09	73.79	10

64.3 Result and Discussions

64.3.1 Compressive Strength

The compressive strength test of concrete containing different percentage of PET (3%, 5%, 7%, 10%) was carried out after completing the concrete curing period for 3, 7, 28, and 56 days. Table 64.4 shows the average compressive strength for each concrete mix design.

At the end of the 28-day of concrete curing period, satisfactory failure results were obtained for the compressive strength test of the concrete containing 3% and 5% from PET; 34.65 N/mm^2 and 32.16 N/mm^2, respectively. However, in comparison with concrete that does not contain PET, it was noticed that compressive strength for concrete containing 3% of PET increased by 0.63%. While the compressive strength for concrete containing 5% decreased by 6.58%, and when concrete containing 7% and 10% of PET. The compressive strength was 29.96 N/mm^2 and 28.41 N/mm^2, respectively. And compared with concrete that does not contain PET, the decrease rates for concrete containing 7% and 10% were equal to 12.99% and 17. 49%, respectively.

At the end of the day 56 of concrete curing period, satisfactory failure results were obtained for the compressive strength test of the concrete containing 3%, 5%, 7%, and 10% from PET and 34.71 N/mm^2, 35.97 N/mm^2, 34.24 N/mm^2, and 33.99 N/mm^2,

Table 64.4 Average compressive strength results

Days mixture type	3 Days	7 Days	28 Days	56 Days
	Average compressive strength (N/mm^2)			
0%	25.61	30.50	34.43	36.94
3%	19.55	25.51	34.65	34.71
5%	21.66	26.93	32.16	35.97
7%	20.33	24.94	29.96	34.24
10%	19.34	20.49	28.41	33.99

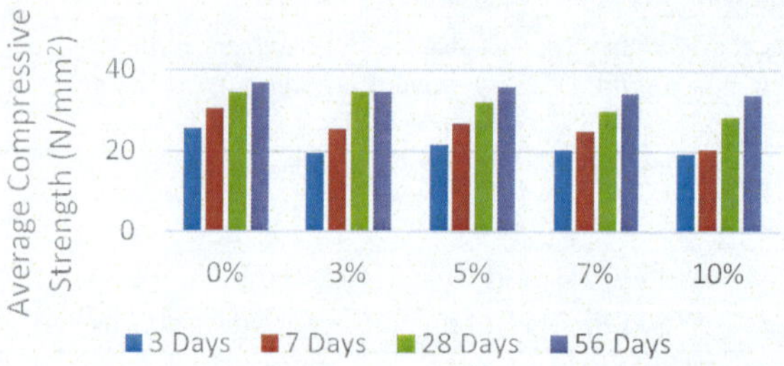

Fig. 64.1 Compressive strength

Fig. 64.2 Compressive strength failure behavior

respectively, but in comparison with concrete that does not contain PET, it was noticed that the compressive strength for concrete containing 5% of PET decreased by 2.63%. While for the concrete containing 3%, 7% and 10% of PET, a decrease in compressive strength was observed as 6.87%, 7.316%, and 7.99%, respectively, comparing with the concrete that does not contain PET. The time of the concrete treatment period was increased to obtain a higher resistance than the treated concrete to 28 days (Figs. 64.1, 64.2).

64.3.2 Split Tensile Strength

The split tensile strength test was performed after completing the 28-day treatment period. Table 64.5 and Fig. 64.3 show the split tensile strength results for all mixtures. It was observed that the split tensile strength of the concrete containing 3% PET increased by 4.046% comparing with the non-PET concrete. In regard to concrete containing 5%, 7%, and 10% of PET, it was found that the split tensile strength is equal to 1.7553 MPa, 1.5384 MPa, and 1.47126 MPa, respectively. And by comparing the results with concrete that does not contain PET, it was found that the split tensile strength comparison decreases by 1.154%, 13.36%, and 17.15%, respectively. And this decrease was attributed to the fact that PET cannot absorb water. Cement wetting

Table 64.5 Split tensile strength for concrete containing different percentage of PET

Percentage, %	Day	Average (Mpa)
0%	28	1.78
3%	28	1.85
5%	28	1.76
7%	28	1.54
10%	28	1.47

does not occur on surfaces of PET; therefore, replacing large quantities of fine aggregates with PET reduces tensile strength. And this is because of the use of a one size of PET (2.36 mm) while replacing it with different grades of fine aggregate (Fig. 64.4).

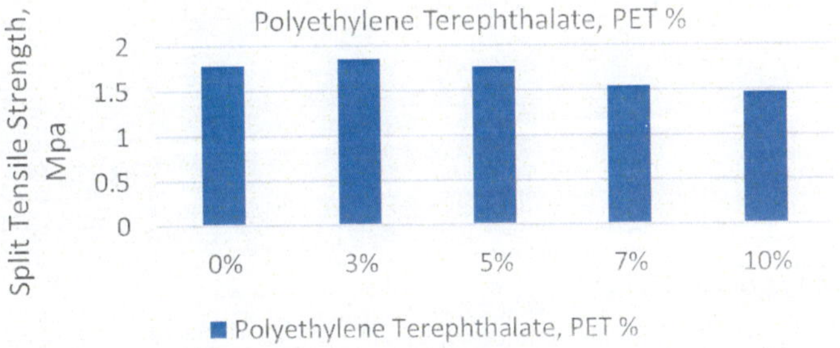

Fig. 64.3 Split tensile strength for concrete containing different percentages of PET

Fig. 64.4 Cylinder specimens' failure behavior in spilt tensile strength for concrete containing different percentages of PET

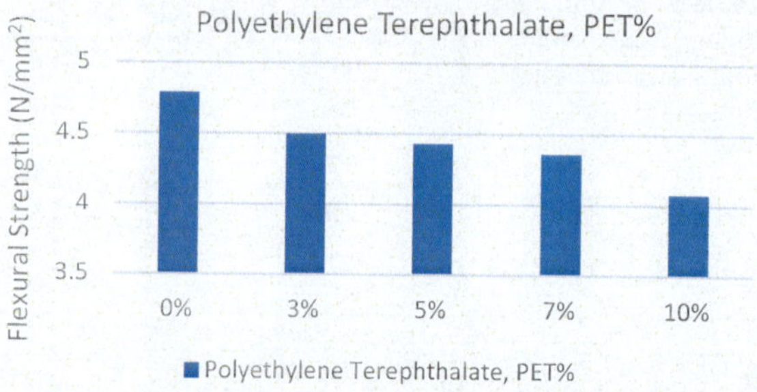

Fig. 64.5 Flexural strength for concrete containing different percentages of PET

Table 64.6 Flexural strength for concrete containing different percentage of PET

Percentage, %	Day	Average (Mpa)
0%	28	4.79
3%	28	4.50
5%	28	4.43
7%	28	4.36
10%	28	4.08

64.3.3 *Flexural Strength*

Results of flexural strength test performed according to ASTM C293/C293M after 28 days curing period are shown in Fig. 64.5, while Table 64.6 shows a general decrease in the flexural resistance of concrete containing PET of 3%, 5%, 7%, and 10%, where the results were found to be equal to 4.4961 N/mm^2, 4.43175 N/mm^2, 4.3635 N/mm^2, and 4.07685 N/mm^2, respectively. And when comparing with the concrete that does not contain PET, it was found that the flexural strength ratios for each concrete containing PET of 3%, 5%, 7%, and 10%, decreased to 6.067%, 7.41%, 8.837%, and 14.826%, respectively. In general, it was observed that the flexural resistance decreased when the percentage of PET increased, and the best flexural resistance was at a replacement rate of 3% of PET comparing with the rest of the percentages, but it is less than non-PET concrete. And this decrease was because the cement wetting does not occur on surfaces of PET attributed to the fact that PET cannot absorb water, therefore, replacing large quantities of fine aggregates with PET reduces flexural strength. Also, it is because of the use of one size of PET (2.36 mm) while replacing it with different grades of fine aggregate (Fig. 64.6).

64.4 Conclusion

This study summarizes the results obtained based on the experiments that were conducted. To achieve the goal of this research which was to know the effectiveness of waste plastic addition in concrete under high temperature. The results showed that

Fig. 64.6 Flexural strength specimen failure behavior for concrete containing different percentages of PET

the compressive strength of concrete containing different proportions of PET was affected when the temperature increased.

From previous and comparative results, it was concluded that the best replacement rate is 3%. It was observed that the compressive strength decreased as the percentage of PET in the concrete increased. The reason for the low compressive strength of the concrete is due to the release of free water and chemical water in the calcium silicate hydrate (CSH) during heating the samples, which in turn produces a fine surface cracks and internal stresses, followed by its interaction with moisture from the surrounding air during the cooling period. Small cracks occur because of the increase in volume associated with this process. Also, because the PET was transformed from a solid state to a gaseous state when exposed to a temperature of 500 °C, leaving behind gaps inside the concrete. This increases the porosity of the concrete, weakens the concrete resistance to compression, and distorts the outer surface of the concrete.

- For the resistance of **split tensile strength**, it was found that the best replacement rate for fine aggregates is 3% of PET, since the bending resistance increased comparing with concrete containing 0% of PET. For concrete containing 5%, 7%, and 10% of PET, it was found that the split tensile strength decreases, and this decrease was attributed to the fact that PET cannot absorb water.
- **Flexural strength** of concrete containing 3%, 5%, 7% and 10% have reduced compared to non-PET containing concrete. This is due to cement wetting does not occur on surfaces of PET, attributed the fact that PET cannot absorb water. Therefore, replacing large quantities of fine aggregates with PET reduces flexural strength, and because of the use of one size of PET (2.36 mm) while replacing it with different grades of fine aggregate.
- Based on the results of the **compressive strength** test of concrete containing different percentages of PET cured for 28 days and 56 days, exposed to 500°C for two hours, it was found that the compression strength for concrete containing

3%, 5%, 7%, and 10% of PET decreased, comparing with concrete containing 0% of PET.

Acknowledgements The authors are extremely grateful to UNITEN R&D for supporting this research under TNB SEEDING FUND 2021, through Project No: U-TG-RD-20-21 and (BOLD 2021) Grant J510050002/2022038. Special thanks to those who are contributed to this project directly or indirectly.

References

1. R. Nisticò, Polyethylene terephthalate (PET) in the packaging industry. Polym. Testing **90**, 106707 (2020)
2. R. Siddique, J. Khatib, I. Kaur, Use of recycled plastic in concrete: a review. Waste Manage. **28**(10), 1835–1852 (2008)
3. R. Sharma, P.P. Bansal, Use of different forms of waste plastic in concrete—a review. J. Clean. Prod. **112**, 473–482 (2016)
4. K. Venkatesh. Properties of concrete at elevated temperatures. Int. Sch. Res. Notices (2014)
5. A.M. Hameed, B.A.F. Ahmed, Employment the plastic waste to produce the lightweight concrete. Energy Procedia **157**, 30–38 (2019)
6. Plastics Europe, Plastics—the facts 2018. An analysis of European plastics production, demand, and waste data
7. Plásticos—Situación En 2019. Plastic Europe (2019). Available at: https://www.plasticseurope.org
8. R.H. Faraj, H.F.H. Ali, A.F.H. Sherwani, B.R. Hassan, H. Karim, Use of recycled plastic in self-compacting concrete: a comprehensive review on fresh and mechanical properties. J. Build. Eng. **30**, 101283 (2020)

Chapter 65
Structural Effect of Well-Graded Coal Bottom Ash in Improving Mechanical and Thermal Properties of Normal Concrete

S. Beddu, N. A. N. Basri, N. M. Kamal, Daud Mohamad, P. Sukhvinder, and W. W. A. Zailani

Abstract Concrete is one of the most widely and world's oldest used construction materials, due to its versatility, durability, sustainability, and economy. This study presents the way of improving normal coal bottom ash to well-graded coal bottom ash and to investigate the effects of workability, mechanical properties, and thermal properties of normal concrete when partially replaced with well-graded coal bottom ash with sand. It was observed that the workability of the concrete showed a slight increase with normal concrete containing well-graded coal bottom ash. The compressive strength of well-graded coal bottom ash achieved the designed mix design which was 40 Mpa at the ages of 3, 7, and 28 days compared to controlled concrete. The thermal conductivity of well-graded coal bottom ash concrete showed a decrease in value compared to controlled concrete. This research concluded that 80% of well-graded coal bottom ash concrete was the optimum percentage replacement which will help achieve the beneficial strength development.

Keywords Concrete · Sustainability · Coal bottom ash · Mechanical properties

65.1 Introduction

Thermal power plants produce coal bottom ash, which makes about 15–25% of the total amount of coal ash generated. It is a form of coal combustion residue made up

S. Beddu (✉) · N. M. Kamal · D. Mohamad · P. Sukhvinder
Department of Civil Engineering, Universiti Tenaga Nasional, Jalan Ikram-Uniten, 43000 Kajang, Selangor Darul Ehsan, Malaysia
e-mail: salmia@uniten.edu.my

N. A. N. Basri
College of Graduate Studies, Universiti Tenaga Nasional, Jalan Ikram-Uniten, 43000 Kajang, Selangor Darul Ehsan, Malaysia

W. W. A. Zailani
School of Civil Engineering, College of Engineering, Universiti Teknologi MARA, 40450 Shah Alam, Selangor, Malaysia

M. A. A. Mohd Salleh et al. (eds.), *Proceedings of the Green Materials and Electronic Packaging Interconnect Technology Symposium*, Springer Proceedings in Physics 289,
https://doi.org/10.1007/978-981-19-9267-4_65

of heavy particles that settle at the bottom of the combustion chamber. Workability of concrete is affected by incorporating bottom ash as a component of the mixing process. Researchers who have included oven-dried coal bottom ash (CBA) into concrete have discovered that the drop of the finished product diminishes as the ash content rises. Like this, the slump flow and passing ability of concrete reduces when 10–30 percent bottom ash is added to a high-workability concrete mix [1, 2]. T water is stored in the voids of the oven-dried CBA structure, which lowers slump. Ash's ability to absorb liquids and its abundance of open pores are to cause for the water loss that was intended to promote particle mobility in the mixture. As a result, the frictional force rises and the workability of the mix decreases.

The physical characteristics of ash, such as its interlocking behaviors, uneven form, and coarse surface roughness, further contribute to the mix's decreased workability. Combining coal bottom ash with particles smaller than 300 m lowers workability because coal bottom ash has a larger surface area than sand. On the other hand, other studies claim that bottom ash makes the mixture more workable. Two years later, the same researchers reported a similar observation.

The literature suggests that utilizing bottom ash as a partial sand substitution impacts the concrete's ability to support loads. Numerous research has shown that strengthening concrete by adding a little amount of bottom ash. The strength of regular concrete is increased by adding 10% coal bottom ash. When compared to other methods, using bottom ash that was 15% smaller in size (passing 600 m) produced the mix's perfect content for the maximum strength [1]. Studies show that coal bottom ash may be used as a binder, fine aggregate, or coarse aggregate in concrete, depending on the particle size. When coal bottom ash is used as a fine aggregate in place of sand, concrete strength, and pore structure characteristics develop most consistently [3]. Coal bottom ash is often well-graded and has a particle size distribution resembling river sand. Its particles can interact with one another. Coal bottom ash is lighter and more brittle than unprocessed river sand. When loaded or compacted, low-specific-gravity coal bottom ash's porous structure quickly breaks down. Coal bottom ash created from high sulfur and low rank coal is dense and not very permeable. The coal bottom ash properties are affected by the degree of pulverization of coal, the starting temperature in the furnace, and the furnace type [4]. According to Kumar et al. using up to 20% bottom ash boosted the concrete's compressive strength [1]. Sing and Siddique discovered that replacing fine bottom ash did not have a significant impact on compressive strength [5].

65.2 Methodology

65.2.1 Materials

Four different mix designs of well-graded coal bottom ash as sand replacement in concrete, with percentages of 0%, 35%, 40%, and 80%, were designed for the experimental investigation. According to Table 65.1 classification of concrete mix designs. Following that, they combined cement, gravel, and water with the mix designs 1, 2, 3, and 4 as given in Table 65.2. All mixture was then put to the test using the ASTM C143 standard slump test, ASTM C109 standard compressive strength test, and ASTM C518 standard thermal conductivity test.

As a crucial element in concrete, Portland cement is the variety of cement that is used the most frequently worldwide. A combination of limestone, clay, or shale is heated and ground into a gray powder to create this intricate formation component. In this study, ordinary Portland cement (OPC) Type I that satisfies ASTM C150, the Standard Specification for Portland Cement, will be utilized. In addition to water and Portland cement, aggregates are inert granular materials like gravel, sand, or crushed stone that are utilized in concrete. Two different forms of aggregate make up 60–75% of the total volume of concrete. Normal operation of a sieve with a 4.75 mm aperture produces fine aggregates. As the fine aggregate, river sand that satisfies ASTM C33 (ASTM, 2055i) standards will be applied.

Twelve samples for each mixture were prepared by slicing into dimensions of 100 × 100 × 100 mm cube samples will be used to cast. For each mixture, different proportions will be utilized, and the cubes will undergo wet curing at room temperature. Concrete cubes that had been cured for 3, 7, and 28 days were examined for compressive strength, with three cubes tested for each age. During the test, an axial load will be applied to the cube sample, which will gradually increase until the cube sample can no longer withstand it.

Fresh concrete is cast in 100 mm × 100 mm × 100 mm concrete cube as part of sample preparation. Wet curing took place for 28 days on the concrete cube. The concrete cube is then cored into cylinders using the coring machine. After coring, the sample is then sliced using a concrete slicing machine into small cylinder with 100 mm thickness. It is necessary to remove air gaps between the specimen's surface

Table 65.1 Classification of concrete mix design

Mix design No.	Sand (%)	Coal bottom ash (%)
1.0	100	0
2.0	65	35
3.0	60	40
4.0	20	80

Table 65.2 Mix design 1.0, Mix design 2.0, Mix design 3.0 and Mix design 4.0

Mix design	Cement (kg)	Sand (Kg)	Coal bottom ash (Kg)	Gravel (Kg)	Water (Kg)
1.0	5.15	6.23	–	17.7	2.6
2.0	5.15	4.05	2.18	17.7	2.6
3.0	5.15	3.73	2.50	17.7	2.6
4.0	5.15	1.23	5.00	17.7	2.6

and the apparatus's plates to acquire accurate results. The samples were polished with a rough surface grinder first to achieve flat parallel surfaces, and then with a fine grinder to achieve smooth finishes. The thermal conductivity of concrete is measured with the FOX50 heat flow meter, which operates on WinTherm 50 software.

65.3 Result and Discussions

65.3.1 Compressive Strength

The results obtained for Slump test were tabulated in Table 65.3. The lowest slump obtained was 45 mm which was from the controlled concrete mix where there was no replacement of well-graded coal bottom ash in the mixture. Furthermore, three other mix design with partial replacement of sand with well-graded coal bottom ash showed an increase in workability which is 55 and 56 mm compared to 0% mix design. Even though well-graded coal bottom ash has an interlocking capability and the absorbing property of ash, which is rich in empty pores, which can cause water loss that was supposed to aid particle movement in the mix, it still showed an increase in workability compared to controlled concrete with no replacement of well-graded coal bottom ash.

Figure 65.1 shows the compressive strength of different days which are 3, 7, and 28 days. Compressive strength of 3 days age was used as an early indication for 28 days strength to determine if any action should be taken if the strength values obtained from the test did not achieve the desired strength. Compressive strength for 7 and 28 days shows an increasing trend for the mix design with replacement of mixing sand with well-graded coal bottom ash. However, it shows slightly decreasing trend compared to 0% of sand replacement. Mix design 4.0 with 80% replacement of well graded coal bottom ash as sand replacement with 37.45 Mpa has the closest targeted compressive strength for this test was 40 Mpa which was achieved by Mix design 1.0 which is the controlled concrete with 0% replacement of well graded coal bottom ash with 40.78 Mpa. Nevertheless, mix design 3.0 with 40% replacement managed to achieve a strength of 31.90 Mpa followed by Mix design 2.0 with 35% replacement with 25.58 Mpa, respectively.

Table 65.3 Slump test for well-graded coal bottom ash concrete

Well-Graded coal bottom ash (%)	Slump (mm)	Workability	Type of slump
0	45	Medium	True
35	55	Medium	True
40	56	Medium	True
80	56	Medium	True

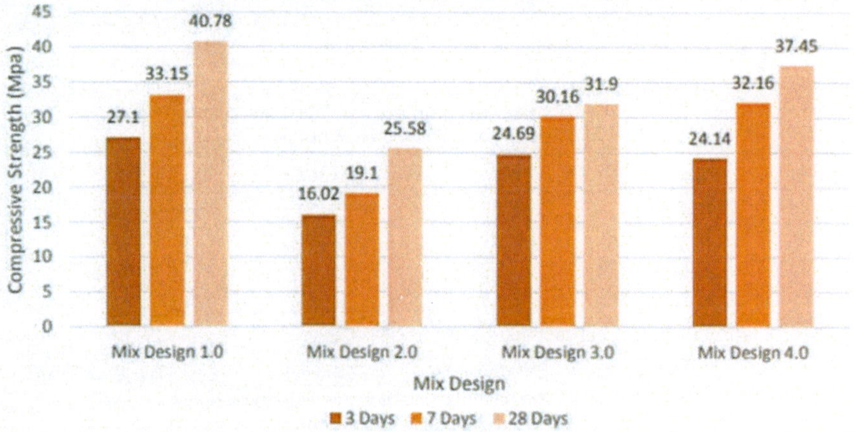

Fig. 65.1 Compressive strength test for 3, 7, and 28 days

The experiment results are consistent with previous research where certain percentage of well-graded coal bottom ash replacement with sand showed decrease in compressive strength while others showed an increase in compressive strength. The data also showed that the proportions utilized in the control mix were the correct proportions for producing a strong reinforced concrete, and that the concrete created was strong enough to be used in structural projects. As a result, Mix design 2.0 should not be used in any projects. The optimal percentage is 80% as it showed a positive on replacing well-graded coal bottom ash in concrete.

65.3.2 Thermal Conductivity

Concrete's thermal conductivity is influenced by the density, kind of cementitious material, aggregate type, temperature, and moisture level. Concrete's mechanical and thermal properties are significantly impacted by the presence of voids in it. Concrete conducts heat via conduction because of chemical vibrations and energy transfer by free electrons. It has been shown that standard concrete has a thermal conductivity of 2.25 W/mK. Changes in each component will have an impact on the concrete's thermal properties; aggregate makes up around 60% of the concrete's volume.

The thermal conductivity of normal concrete with replacement of sand with well graded coal bottom ash showed a slight drop in the thermal conductivity compared to controlled concrete. As shown in Table 65.6, 40% replacement of well graded coal bottom ash showed a slight increase in thermal conductivity compared to controlled concrete by 1.8%. This slight increase can be caused by moisture content, density, and temperature. Based on the results obtained, it can be discussed that the optimal percentage of replacement of well-graded coal bottom ash with sand as fine aggregate is 80% since it had lower thermal conductivity compared to controlled concrete. Heat

Table 65.4 Results for thermal conductivity test of 0% CBA

Sample name	Thickness (mm)	Thermal conductivity sample + rubber (W/mK)	Thermal conductivity sample (W/mK)	Average (W/mK)
0% NC 1	11.30	0.9647	2.1367	1.9157
0% NC 2	10.87	0.9755	0.9755	
0% NC 3	11.13	1.0330	2.6351	

Table 65.5 Results for thermal conductivity test of 35% CBA

Sample name	Thickness (mm)	Thermal conductivity sample + rubber (W/mK)	Thermal conductivity sample (W/mK)	Average (W/mK)
35% NC 1	11.18	0.9002	1.8312	0.6589
35% NC 2	13.26	1.0370	0.0629	
35% NC 3	11.13	0.9172	0.0826	

conductivity was also found to be highest in lightweight concrete that did not contain any bottom ash (Tables 65.4, 65.5 and 65.7).

Table 65.6 Results for thermal conductivity test of 40% CBA

Sample name	Thickness (mm)	Thermal conductivity sample + rubber (W/mK)	Thermal conductivity sample (W/mK)	Average (W/mK)
40% NC 1	11.02	0.9622	2.2001	1.9517
40% NC 2	11.43	0.9207	1.8798	
40% NC 3	12.09	0.9225	1.7736	

Table 65.7 Results for thermal conductivity test of 80% CBA

Sample name	Thickness (mm)	Thermal conductivity sample + rubber (W/mK)	Thermal conductivity sample (W/mK)	Average (W/mK)
80% NC 1	12.47	0.9177	1.7003	1.6029
80% NC 2	12.78	0.9283	1.7012	
80% NC 3	12.78	0.8429	1.4073	

65.4 Conclusion

In this paper, the effect of using well graded coal bottom ash on concrete as a sand replacement on normal concrete workability was successfully conducted. The conclusion that can be made based on the results obtained in the experiment on the slump test, it is showed that the presence of well graded coal bottom ash as partial replacement for sand has effect on the fresh concrete workability. The optimal percentage chosen was 80%. A comparison was done between concrete with 0% well-graded bottom ash content and concrete with up to 80% well-graded bottom ash content. Concrete with 80% replacement showed a slight increase in the workability compared to 0%. Both of this concrete mix had medium workability, but 80% replacement concrete showed slightly higher value than 0%. Based on the results obtained, for improvements of mechanical properties by using well-graded coal bottom ash as a sand replacement in normal concrete it is known that 80% of replacement has shown to be the optimal percentage. It can be concluded that 80 percent is very effective to be used as a replacement as it has the closest compressive strength when compared to controlled concrete of 40 Mpa. The compressive strength obtained from 80% replacement was 37.45 Mpa. Finally, thermal properties of normal concrete containing well-graded coal bottom ash as sand replacement were successfully conducted. According to the results obtained, well-graded coal bottom ash as partial replacement showed a significant decrease in thermal conductivity compared to controlled concrete which was 0%. The most effective replacement was found to be 80% with k value of 1.6029, making it the best choice of replacement.

Acknowledgements The authors are thankful to Universiti Tenaga Nasional, Malaysia for providing financial support (BOLD 2021) grant J510050002/2022038 and TNB Seeding Fund (U-TG-RD-2021) for this study

References

1. K. Muthusamy, M.H. Rasid, G.A. Jokhio, A.M. Albshir Budiea, M.W. Hussin, J. Mirza, Coal bottom ash as sand replacement in concrete: a review. Constr. Build. Mater. **236**, 117507 (2020)
2. M. Rafieizonooz, J. Mirza, M.R. Salim, M.W. Hussin, E. Khankhaje, Investigation of coal bottom ash and fly ash in concrete AEW3Q2S replacement for sand and cement. Constr. Build. Mater. **116**, 15–24 (2016)
3. Y.H. Kim, H.Y. Kim, K.H. Yang, J.S. Ha, Effect of concrete unit weight on the mechanical properties of bottom ash aggregate concrete. Constr. Build. Mater. **273**, 121998 (2021)
4. M. Singh, R. Siddique, Effect of coal bottom ash as partial replacement of sand on workability and strength properties of concrete. J. Clean. Prod. **112**, 620–630 (2016)
5. M. Singh, R. Siddique, Effect of coal bottom ash as partial replacement of sand on properties of concrete. Resour. Conserv. Recycl. **72**, 20–32 (2013)

Chapter 66
The Effects of Well-Graded Bottom Ash in Improving Mechanical and Thermal Properties of High-Strength Concrete

S. Beddu, M. S. M. Sufian, N. M. Zahari, Daud Mohamad, N. A. N. Basri, H. M. Yee, and W. W. A. Zailani

Abstract Well-graded bottom ash exhibited an improvement on mechanical and thermal properties of high-strength concrete as a fine aggregate replacement in concrete. The well-graded coal bottom ash was used by 25%, 50%, 75%, and 100% in a set of six concrete samples trial mix designs including normal mix. Existence of bottom ash affected the workability for high-strength concrete where the workability reduced as the well-graded bottom ash percentages increase in concrete. Other than that, the compressive strength test of high-strength concrete also showed an improvement with the optimum variable at 50% bottom ash replacement. Besides, by increasing the well-graded bottom ash, the thermal conductivity on high-strength concrete was reduced. Lastly, well-graded bottom ash was one of the best replacements for fine aggregate in improving the mechanical and thermal properties of high-strength concrete.

Keywords Well-graded bottom ash · Compressive strength · Thermal conductivity · Mechanical properties

---

S. Beddu (✉) · M. S. M. Sufian · N. M. Zahari · D. Mohamad
Department of Civil Engineering, Universiti Tenaga Nasional, Jalan Ikram-Uniten, 43000 Kajang, Selangor Darul Ehsan, Malaysia
e-mail: salmia@uniten.edu.my

N. A. N. Basri
College of Graduate Studies, Universiti Tenaga Nasional, Jalan Ikram-Uniten, 43000 Kajang, Selangor Darul Ehsan, Malaysia

H. M. Yee
School of Civil Engineering, University Teknologi MARA (UiTM) Cawangan Pulau Pinang, Kampus Permatang Pauh, 13500 Permatang Pauh, Pulau Pinang, Malaysia

W. W. A. Zailani
School of Civil Engineering, College of Engineering, Universiti Teknologi MARA, 40450 Shah Alam, Selangor, Malaysia

M. A. A. Mohd Salleh et al. (eds.), *Proceedings of the Green Materials and Electronic Packaging Interconnect Technology Symposium*, Springer Proceedings in Physics 289,
https://doi.org/10.1007/978-981-19-9267-4_66

66.1 Introduction

Cement, coarse aggregate, fine aggregate, and water are all essential components of concrete, a material used extensively in construction and infrastructure. According to Singh and Siddique [1], concrete is an almost indispensable component in the creation of high-rise buildings and subsequent construction projects, and river sand is the principal raw resource used in the manufacture of concrete as fine aggregate. Besides, river sand deposits are depleting over time in India. Binder, aggregate, and water are the three main constituents of concrete, and these three components are the basis for most analyses and designs. To meet the growing demand for concrete, more quantities of nature substance such as fine and coarse aggregates must be gathered from the environment.

When it comes to the creation of the high-strength concrete, the most efficient use of the fundamental element that are employed in the production of normal-strength concrete is really required. Concrete manufacturers who specialized in the high-strength concrete are familiar with the factors that determine compressive strength as well as how to differentiate those aspects to achieve the appropriate compressive strength. Beyond selecting high-quality Portland cement, manufacturers optimize aggregates before optimizing the combination of other components which including to consider replacing natural sand with coal bottom ash waste. Based on study by Kim and Lee [2], the results demonstrated that using untreated bottom ash particles for internal curing could reduce or eliminate autogenous shrinkage of high-strength concrete.

In addition, aggregates for high-strength concrete are chosen based on a variety of factors, including the strength of the aggregate, its size, how well the cement paste adheres to it, and the aggregate's surface quality. Based on Rafieizonooz et al. [3], coal bottom ash (CBA) amounts in Malaysia's electricity production are approximately 1.7 million tons. A well-graded material, coal bottom ash has an equivalent particle size distribution to the river sand (fine aggregates) and is commonly used in the construction industry [3]. Its particles have the capabilities to interlock with one another. When compared to sand of the river, coal bottom ash is much lightweight and extra fragile. Because of its little specific gravity, coal bottom ash has a penetrable surface that becomes more solid when loaded or compacted. For the manufacturing of coal bottom ash, which is impermeable and has a high density, greater sulfur coal and small rank coal are employed [4]. A coal furnace's ability to produce coal bottom ash depends on several factors, including the amount of pulverization, furnace temperature, and kind of furnace.

Furthermore, in production of concrete, currently waste material properties are being studied by past researchers around the world to consider replacing the fine and coarse aggregate as well as cement. Because of the amount of solid waste that being produced increases daily, the intention of using waste products in concrete goes beyond simple economics to address environmental concerns as well. Using waste derived aggregates such as coal bottom ash in concrete can result in high potential of reliability gains in terms of concrete's sustainability. Overall, coal bottom ash is essential because it has a size distribution like fine aggregates for example natural sand. In addition, because of the scarcity of natural aggregates, such as fine aggregates, the usage of ash from the waste plants is gaining traction as an alternative.

66.2 Methodology

The focus of this research is about the investigation of the implementation of coal bottom ash as the fine aggregate substitution in the manufacturing of concrete. The well-graded coal bottom ash will be used in a set of six concrete samples trial mix designs include normal mix and will be evaluated the potential for well-graded coal bottom ash itself as a half part replacement of sand in concrete by 25%, 50%, 75%, and 100%, in order to achieve the optimum results which will be to identify whether the well-graded coal bottom ash can influence or effects in enhancing the mechanical properties and thermal properties of high-strength concrete.

In addition, there are another several tests used on the concrete samples mix design to test its mechanical and thermal properties. Roughly, to analyze the mechanical properties of the four samples of concrete mix design, the concrete slump test and concrete cube compressive strength test were conducted. This four-sample mix concrete design with influence of bottom ash will be used for concrete slump test, and this test is to test the fresh concrete workability. Then, the prepared concrete previously will be cast to 100 mm × 100 mm × 100 mm cube specimens, and concrete cube compressive strength test was carried out. This compressive strength test will be carried out at age 3 days, 7 days, and 28 days.

Moreover, thermal conductivity test was carried out to identify the improvement of thermal properties on high-strength concrete from the well-graded coal bottom ash. For the thermal conductivity test, three identical cylinders with 50 mm diameter and 100 mm height will be prepared for each percentage variable of CBA. For this test, the conductivity test will be measured at ages of 28 days. Besides, all the mix design sample will be kept inside a water tank to cure before being preserved in oven at 105 °C for almost 24 h to eliminate any excess moisture. For both type of properties that will be the measure during the research, most of the tests will be conducted in the Civil Engineering Laboratory, UNITEN.

66.2.1 Materials

66.2.1.1 High-Strength Cement (HSC)

Even though higher strengths have been achieved and use, high-strength concrete (HSC) can be defined as concrete that has a predefined characteristic cube strength that ranges between 60 and 100 N/mm^2, depending on the use of the material. The

Fig. 66.1 Different particle sized of coal bottom ash

optimal proportion of water to cement ratio should fall in the range of 0.30–0.35 or below. Thus, in this project, the water–cement ratio of 0.28 was used.

66.2.1.2 Well-Graded Coal Bottom Ash

Bottom ash is a sort of incombustible waste that is left over after combustion takes place in a power plant, boiler, furnace, or incinerator. The proportion of coal bottom ash that is created when coal is burned ranges somewhere between 25 and 30%. The dimension of the particles that make up bottom ash can range from 10 to 0.75 mm and can be anywhere from the size of fine gravel to fine sand. The proportion of normal bottom ash which must be dried in an oven for 24 h at a temperature of 100 °C to meet the requirement for manufacturing well-graded bottom ash. The technology that was used to manufacture well-graded fine aggregates (SW) in accordance with the ASTM D2487 standard was utilized to produce well-graded bottom ash with D10, D30, D60, Cu, and Cc as the dimensions (Fig. 66.1 and Table 66.1).

66.2.1.3 Coarse Aggregates

Coarse aggregates are big and lumpy things placed in concrete that usually found in nature or quarries. Their shape and strength affect the concrete in many ways.

Table 66.1 ASTM C33/C33M well-graded fine aggregates specification

Sieve sizes (mm)	Retaining mass (g)	Cumulative mass (g)	Cumulative (%)	Passing (%)	ASTM C33 limits	
					Minimum	Maximum
9.50	0.00	0.00	0	100	100	100
5.00	20.20	20.20	2	98	95	100
2.36	129.30	149.50	15	85	80	100
1.18	198.37	347.87	35	65	50	85
0.60	198.37	546.24	55	45	25	60
0.30	198.37	744.61	75	25	5	30
0.15	150.20	894.81	90	10	0	10
0.075	49.84	944.65	95	5	0	5
< 0.075	49.84	994.50	100	0	–	–

The size of the coarse aggregates effects the concrete firmness and determines how much fine aggregates are required to make a concrete mixture. Coarse aggregates are nonetheless strongly impacted by the dry shrinkage as well as other deformations. Based on Fig. 66.2, the greatest size of the coarse aggregates can be approximately from 4.75 mm to 20 mm which they were sieved and passed through a sieve with a size of 20 mm and retained on a sieve with a size of 4.75 mm. Besides, aggregates of a coarser size have been washed with water to remove dirt and other impurities and then dried in the sun.

Fig. 66.2 Crush limestone coarse aggregates

Fig. 66.3 Sieved and dried fine aggregates (sand)

66.2.1.4 Fine Aggregates

Fine aggregate is used to fill the gaps between larger aggregates. Fine aggregates are one of the most vital components and ingredients that can have a considerable influence on the arrangement of the concrete mix with a range of 60–70% of the volume of the concrete. The ASTM C33/C33M (Specification, 1999) states that a minimum size of 4.75 mm for fine aggregates is required before they can be used in a building project. Because of this, it was possible for all the fine aggregates to move freely through a combined total of 4.75 mm sieves and then dried in an oven at a temperature of 100 °C for a period of 24 h or approximately near one whole day as shown in Fig. 66.3.

66.2.1.5 Mixing Water

Water is the most important ingredient that mixed with cement to make a paste that holds the aggregates together and to harden the concrete. Concrete's workability, permeability, compressive strength, and the durability of hardened concrete are all controlled by the amount of water that is present in the mixture. Water-to-cement ratio plays the most crucial role of producing high quality concrete. According to standards set by ASTM C1602/C1602M-18, regular and pure tap water used in this experimental works. Water used must be free from any chemicals components that may harm the concrete.

66.2.1.6 Design of Experiment Setup

Concrete mechanical and thermal properties are two different characteristics that being observed and analyzed in this project. In terms of mechanical properties study, it could be in two varies situation of concrete condition, fresh concrete, or hardened concrete. However, for the thermal characteristic study, it only involves the hardened condition of concrete. Testing on hardened concrete for quality control is essential for the construction of structures because hardened concrete increases in strength over time. There are a variety of tests that may be used to examine the qualities of the hardened concrete. In this study, one major types of mechanical property testing were carried out: a compressive strength test (ADR Touch Control Pro 2000 Auto Compression Machine). An evaluation of the thermal characteristics of the hardened concrete using the thermal conductivity test (FOX 50 Machine).

66.3 Result and Discussions

66.3.1 Particle Size Distribution

Since determining the characteristics of bottom ash and the particle distribution ratios of fine aggregates rely on the size and weight of the particles, it is very important to have a clear understanding of the particle size distribution. Utilizing a sieve as a part of the process that results in the granular distribution of materials is among the most common approaches. During this inquiry, a filtering analysis was carried out so that the totals could be differentiated based on the size of the screen. First, in accordance with ASTM C33, the percentage of aggregate that must be present for each sieve is determined. Subsequently, the percentage of aggregate which must be represented for each specimen is utilized in the computation of the mass carried.

66.3.1.1 Normal Bottom Ash

The results of the sieve analysis test of normal bottom ash are presented in Table 66.2. The percentage of material that passed through sieves with opening diameters of 9.5 mm, 5 mm, and 2.36 mm were 99.8%, 97.1%, and 74.60%, respectively. On the other hand, the percentage of material that passed through sieves of 0.18 mm, 0.6 mm, 0.3 mm, and 0.15 mm was, respectively, 55.62%, 39.30%, 28.00%, and 14.43%. Based on the result obtained, there was less material in that range, which indicates that there were more materials with a size greater than 0.15 mm. Table 66.3 indicates that the normal bottom ash does not meet the criteria for "well-graded bottom ash" because its Cc value is less than 1. Therefore, these outcomes shown that the bottom ash overall category for particle size distribution falls as SP, poorly graded fine aggregates, or sand (ASTM D-2487-98) since the Cu is greater than 6 but the Cc value less than 1.

Table 66.2 Normal bottom ash sieve analysis

Sieve sizes (mm)	Retaining mass (g)	Cumulative mass (g)	Cumulative (%)	Passing (%)	ASTM C33 limits	
					Minimum	Maximum
9.50	2.00	2.00	0.20	99.80	100	100
5.00	27.40	29.40	2.90	97.10	95	100
2.36	223.00	252.40	25.40	74.60	80	100
1.18	189.20	441.60	44.38	55.62	50	85
0.60	162.50	604.10	60.70	39.30	25	60
0.30	112.10	716.20	72.00	28.00	5	30
0.15	139.50	855.70	85.57	14.43	0	10
0.075	76.80	932.50	93.76	6.24	0	5
< 0.075	62.00	994.50	100.00	0.00	–	–

Table 66.3 Uniformity coefficient and coefficient of gradation of normal bottom ash

D10	D30	D60	Cu	Cc
0.11	0.36	1.42	13.50	0.88

66.3.1.2 Well-Graded Bottom Ash

The results of the sieve analysis test of well-graded bottom ash are presented in Table 66.4. The percentage of material that passed through sieves with opening diameters of 9.5 mm, 5 mm, and 2.36 mm were 100.0%, 98.0%, and 85.0%, respectively. On the other hand, the percentage of material that passed through sieves of 1.18 mm, 0.6 mm, 0.3 mm, and 0.15 mm was, respectively, 65.0%, 45.0%, 25.0%, and 10.0%. Based on the results obtained as in Table 66.4, all the passing percentage of well-graded bottom ash for all sieve opening sizes, all of them were significantly within the range of passing limit as minimum and maximum limit based on ASTM C33 limits. Moreover, Table 66.5 indicates that this well-graded bottom ash sieve analysis does really meet the standard criteria for "well-graded bottom ash" because its Cc value is greater than 1. Since its Cc value is greater than 1 and the Cu value is greater than 6, this bottom ash aggregates could be simply simplified or categorized as SW, well-graded sand as in ASTM D-2487-98 standards.

66.3.1.3 Coarse Aggregate

Test for the sieve analysis of coarse aggregates carried out in accordance with ASTM C136. The findings of the sieve analysis of coarse aggregate are displayed in Table 66.6, respectively. The fineness modulus of coarse aggregates is 5.46, and the large bulk of aggregates fall between the sieve size of 4.75 and 20 mm. The percentage of

Table 66.4 Well-graded bottom ash sieve analysis

Sieve sizes (mm)	Retaining mass (g)	Cumulative mass (g)	Cumulative (%)	Passing (%)	ASTM C33 limits	
					Minimum	Maximum
9.50	0.00	0.00	0.00	100.00	100	100
5.00	20.20	20.20	2.00	98.00	95	100
2.36	129.30	149.50	15.00	85.00	80	100
1.18	198.37	347.87	35.00	65.00	50	85
0.60	198.37	546.24	55.00	45.00	25	60
0.30	198.37	744.61	75.00	25.00	5	30
0.15	150.20	894.81	90.00	10.00	0	10
0.075	49.84	944.65	95.00	5.00	0	5
< 0.075	49.84	994.50	100.00	0.00	–	–

Table 66.5 Uniformity coefficient and coefficient of gradation of well-graded CBA

D10	**D30**	**D60**	**Cu**	**Cc**
0.15	0.38	1.04	6.93	1.00

Table 66.6 Coarse aggregates sieve analysis

Sieve opening (mm)	Retaining mass (g)	Cumulative mass (g)	Cumulative (%)	Passing (%)
80.00	0.0	0.0	0.0	100
40.00	0.0	0.0	0.0	100
20.00	0.0	0.0	0.0	100
10.00	2.6	2.6	52	48
4.75	2.1	4.7	94	6
2.36	0.3	5.0	100	0
1.18	0.0	5.0	100	0
0.06	0.0	5.0	100	0
0.03	0.0	5.0	100	0

material that passed through a sieve with a thickness of 20 mm, 10 mm, and 4.75 mm, respectively, was 100%, 48%, and 6%.

66.3.2 Specific Gravity and Water Absorption

The water absorption test analyzes the moisture retention capacity of the coarse and fine aggregates, while the specific gravity test of aggregates is done to quantify the strength or quality of the material. Aggregates are tested for their capacity to hold

Table 66.7 Specific gravity and water absorption

Properties	Bottom ash	Fine aggregates	Coarse aggregates
Bulk specific gravity	1.35	2.58	2.40
Apparent specific gravity	1.45	2.29	2.66
Saturated specific gravity	1.62	2.56	–
Water absorption	16.26	1.73	0.23

water. To obtain accurate results, a specific gravity test must be carried out. This test is used to determine the specific gravity of coarse and fine aggregate. The size of the aggregate that is classified as coarse aggregate is greater than 5 mm, while the size of the aggregate that is defined as fine aggregate is between 5 and 75 mm. The specific gravity and absorption of aggregates were evaluated using the ASTM C128-15 standard for fine aggregates and the ASTM C127-15 standard for coarse aggregates. As can be seen in Table 66.7, the water absorption of bottom ash was significantly higher when compared to that of fine particles. Bottom ash has a specific gravity of 1.8, while fine aggregates have a specific gravity of 2.6. This implies that ash is a lighter substance when compared with fine aggregates.

66.3.3 Slump Test

The slump test is the method that is used the most frequently to measure the consistency of concrete. This test can be performed in a laboratory setting or on the actual construction site. It is not an appropriate method for working with concrete that is either very wet or very dry. Based on this project study, the obtained slump values that defines the fresh concrete workability characteristics with different percentages implementation of well-graded bottom ash are being stated as in Table 66.8. According to Fig. 66.4 and Table 66.8 also, the slump value for CM, BAM25, BAM50, BAM75 are 135 mm, 110 mm, 60 mm, and 45 mm, respectively, which shows reducing slump value as the influence of the bottom ash in concrete increase. Regarding the study of bottom ash on high-strength concrete, when using 100% of coal bottom ash as concrete influences, the slump flow values of fresh concrete were found to be reduced by as much as one hundred millimeters, as stated by Kim and Lee [2].

The following factors may have contributed to the reduction in the slump flow value; for instance, coal bottom ash particles had a more complex shape and coarser surface than regular aggregate particles. Additionally, it was found that the adhesion between rougher particles was greater than that which occurred when regular aggregate was used. This increased friction force inhibited the flow properties of fresh

Table 66.8 Slump value

Concrete mix design	Slump value (mm)
CM	135
BAM25	110
BAM50	60
BAM75	45

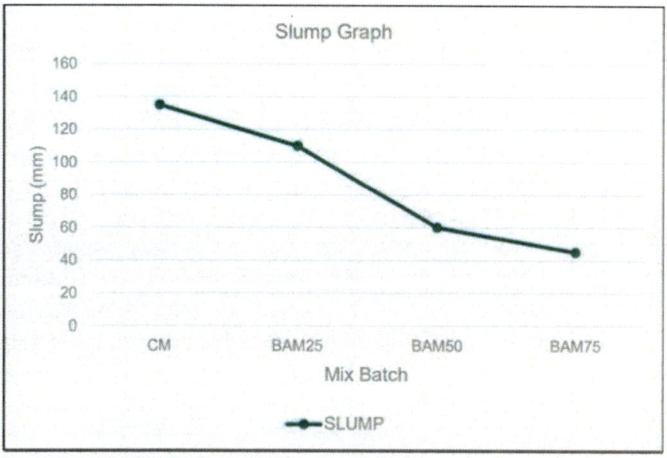

Fig. 66.4 Slump graph values versus different bottom ash percentage

concrete and caused the slump values to decline. Furthermore, the amount of water and cement paste used for lubrication between aggregates was reduced since some of the water and cement paste would be sucked into the bottom ash particles. This resulted in a reduction in the total amount of water and cement paste used [2].

66.3.4 *Compressive Strength*

The compressive strength test is a type of mechanical examination that determines the maximum degree of compressive force that a substance can withstand before it cracks. In addition, the compressive strength of every mixture of this project study was evaluated in period of 3, 7, and 28 days of curing after been casted in accordance with the ASTM C109 and C109M standards The involvement of well-graded bottom ash presents a positive influence on the concrete production and on its compressive strength. For this project study, as in Table 66.9 also, it shows that the value of compressive strength, almost for all the mixtures showing a gradually increment along with the increasing curing periods. In terms of each mix, the compressive strength of the concrete cubes at 28 days for the control mix (CM), 25% bottom ash

Table 66.9 Compressive strength value

Concrete mix design	Average compressive strength @ 3 days age (N/mm^2)	Average compressive strength @ 7 days age (N/mm^2)	Average compressive strength @ 28 days age (N/mm^2)
CM	53.90	55.30	62.33
BAM25	55.26	56.20	68.90
BAM50	55.80	56.83	70.70
BAM75	39.65	40.10	48.97

mix (BAM25), 50% bottom ash mix (BAM50), 75% bottom ash mix (BAM75) are 62.33 N/mm^2, 68.90 N/mm^2, 70.70 N/mm^2, and 48.97 N/mm^2, respectively.

The increment of the compressive strength occurs along with the increasing bottom ash influence in the concrete. However, the compressive strength of the concrete cubes started to reduce when the 75% bottom ash implemented on the concrete mix. Relatively shows that for this project, the compressive strength increment of all mixture stops at BAM50. This was due to the factor that the BAM50 mix reach the maximum of the compressive strength compared to the other mixes. This can be explained more detail as in Fig. 66.5 which shows that the compressive strength graphs increase direct proportionally from CM to BAM50 of all curing ages until it started to drop the strength at the BAM75 mix.

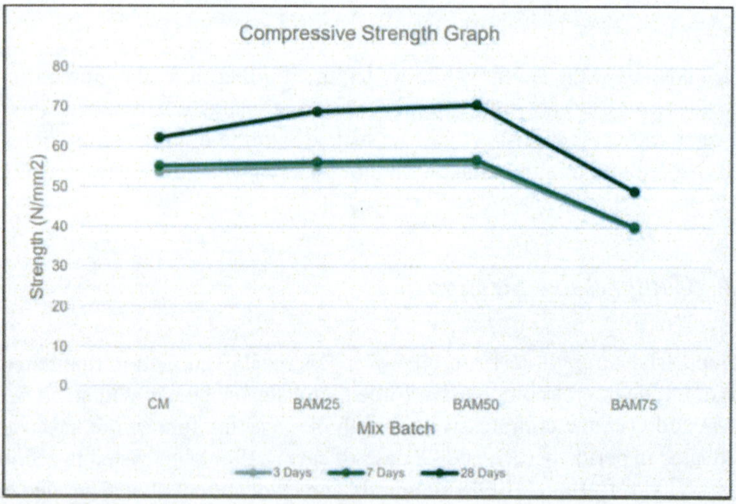

Fig. 66.5 Compressive strength

66.3.5 Thermal Conductivity

The FOX 50 heat flow meter machine was used to execute the experiments for thermal conductivity, and the results were compared to the ASTM C1784 standard. During the thermal conductivity test, the temperature was maintained at a steady state was 28 °C. Furthermore, this thermal conductivity test has been carried out with assist of rubber sheet. The use of the rubber sheet helped to ensure that there would be no inaccuracies brought on by the presence of an air gap between the specimens and the air surface. In addition to this, the rubber sheet contributed to an improvement in the contact between the surface of the specimens and the device. As can be seen in Fig. 66.6, the thermal conductivity of the concrete decreased when well-graded bottom ash was used as a replacement for fine particles in the concrete mix. This was determined by testing the thermal conductivity of the concrete.

Additionally, based on Table 66.10, the thermal conductivity of this concrete with influence of CBA reduces from 2.5604 W/mK of CM to 0.3047 W/mK of BAM75. This shows that the relative relation between the reducing factor of the thermal conductivity is because of the porosity material condition of the coal bottom ash itself. Apparently, based on the study by the [5], the thermal conductivity of the pervious concrete that was created with 25% of IBA replacement was the maximum; however, the thermal conductivity dropped with the corresponding increase in IBA substitution, and the thermal conductivity of the pervious concrete that included 100% of IBA replacement was 0.61 W/mK which is very low. As mentioned, IBA and CBA is a porous material which has a much lower thermal conductivity than normal granite aggregates [6]. Thus, the least amount of heat or thermal conductivity was lost through the permeable concrete that was made with 100% IBA/CBA and usually this type of concrete useful in making of less heat conductivity partition wall.

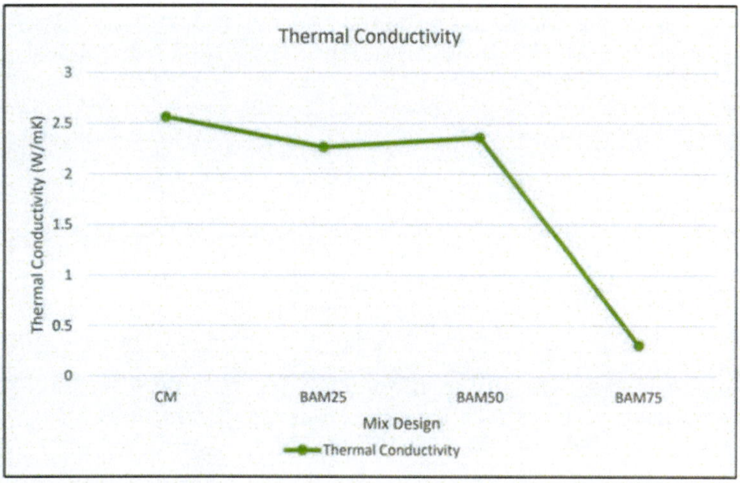

Fig. 66.6 Thermal conductivity

Table 66.10 Thermal conductivity value

Concrete mix design	Thermal conductivity samples + rubber (W/mK)	Thermal conductivity samples (W/mK)
CM	1.2913	2.5604
BAM25	1.084	2.2638
BAM50	1.1825	2.3550
BAM75	0.2924	0.3047

66.4 Conclusion

The workability effects for high-strength concrete due to the bottom ash existence for this project has successfully determined whereas along the increment of well-graded bottom ash percentages increase in concrete, the workability of the high-strength concrete reduced. Through the determination of the compressive strength test, it can be said that well-graded bottom ash brought a variety of effects including the improvement of the compressive strength of the high-strength concrete. Along with the positive increment of this strength, it possesses together with some limitations. This can be proved where the compressive strength started to diminish and dropped at the 75% replacement of the bottom ash as a fine aggregate in the concrete and the maximum bottom ash variable was 50% bottom ash replacement. Lastly, from the thermal conductivity test based on plenty of samples, the thermal conductivity has already been well studied. The analyzing of the reducing thermal conductivity along the increment of well-graded bottom ash on high-strength concrete is one of the reasons. This occurs due to several important factors including the material characteristics of the bottom ash which high in porosity and low thermal conductivity.

Acknowledgements The authors are thankful to Universiti Tenaga Nasional, Malaysia for providing financial support (BOLD 2021) grant J510050002/2022038 and TNB Seeding Fund (U-TG-RD-2021) for this study.

References

1. M. Singh, R. Siddique, Constr. Build. Mater. **68**, 39–48 (2014)
2. H.K. Kim, H.K. Lee, J. Civ. Eng. **19**(6), 1802–1818 (2015)
3. M. Rafieizonooz, J. Mirza, M.R. Salim, M.W. Hussin, E. Khankhaje, Constr. Build. Mater. **116**, 15–24 (2016)
4. M. Singh, R. Siddique, J. Clean. Prod. **91**, 269–278 (2015)
5. P. Shen, H. Zheng, J. Lu, C.S. Poon, Recourses. Conserv. Recycl. **175**, 105736 (2021)
6. W.J. Cho, S. Kwon, J.W. Choi, Eng. Geol. **107**(3–4), 167–171 (2009)

Chapter 67
The Effects of Well-Graded Bottom Ash in Improving Mechanical and Thermal Properties of Lightweight Concrete

S. Beddu, H. Ramachenran, Daud Mohamad, T. S. A. Manan, N. A. N. Basri, and H. M. Yee

Abstract The aim of this study is to determine the effects of well-graded coal bottom ash (CBA) on concrete as sand replacement on lightweight concrete in terms of mechanical and thermal properties. Coal bottom ash (CBA) is one of the by-products produced during energy production process in power plant. However, due to high demand electricity in Malaysia, this leads to excess production of CBA and this led to several environmental problems; therefore, to overcome these problems, the utilization of CBA needs to be done. Four different percentages of CBA are used as fine aggregate replacement which are 0, 5, 10, and 15%. In this study, several tests have been done during the utilization of CBA as sand replacement in lightweight concrete which are slump, density, compressive, modulus elasticity, and thermal conductivity test. The results indicated that density recorded was around 1600 kg/m3 which is lower than typical lightweight density (1750 kg/m3). 15% of CBA as sand replacement recorded highest Compressive strength with 17 MPa while 0% is the contra (12 MPa) on 28th days curing. On 28th days of curing, LWC with 15% CBA as sand replacement has the highest value (1.3 GPa) and compared to others. 15% CBA was recorded the lowest thermal conductivity value which is 0.5 W/m.K while the highest 0.7 W/m.K (0% CBA). Overall, CBA can be classified as one of the suitable materials as sand replacement to produce lightweight concrete as it promotes good compression strength as well as lowering the *K* value.

S. Beddu (✉) · H. Ramachenran · D. Mohamad
Department of Civil Engineering, Universiti Tenaga Nasional, Jalan IKRAM-UNITEN, 43000 Kajang, Selangor Darul Ehsan, Malaysia
e-mail: Salmia@uniten.edu.my

N. A. N. Basri
School College of Graduate Studies, Universiti Tenaga Nasional, Jalan IKRAM-UNITEN, 43000 Kajang, Selangor Darul Ehsan, Malaysia

H. M. Yee
School of Engineering, College of Engineering, Universiti Teknologi MARA Cawangan Pulau Pinang, Kampus Permatang Pauh, 13500 Permatang Pauh, Pulau Pinang, Malaysia

T. S. A. Manan
Institute of Tropical Biodiversity and Sustainable Development, Universiti Malaysia Terengganu (UMT), 21030 Kuala Nerus, Terengganu Darul Iman, Malaysia

M. A. A. Mohd Salleh et al. (eds.), *Proceedings of the Green Materials and Electronic Packaging Interconnect Technology Symposium*, Springer Proceedings in Physics 289,
https://doi.org/10.1007/978-981-19-9267-4_67

Keywords Coal bottom ash (CBA) · Thermal conductivity · Compressive strength · Lightweight

67.1 Introduction

United Nations recorded that the current population of Malaysian in 2022 is more than 33 million, a 1.09% increase from 2021. With these strong and continuous increments of population, this led to higher demand for electrical power consumptions. Malaysian government gave private investors permission to build power plants all throughout the nation to meet demand. The greatest energy was produced by thermal stations, which produced 119 TWh of Malaysia's total 134 GWh of electricity, while hydropower plants produced 9.0 TWh as of 2013 [1].

In Malaysia, coal has been identified as an important source in electric generation, cement production, and more [2]. The total consumption of coal can see increase gradually. It can be seen from Fig. 67.1, the consumption of coal from year 2005 to 2020 increased from 12.4 to 36 million tons. This increment led to higher amounts of CO_2 released as shown in Fig. 67.2.

According to Ref. [4], coal is identified as combustible black rock consisting mainly of carbon. Coal is formed from the remains of plants, by a process called coalification. The remains of dead plants were buried in low oxygen environments, i.e., swamp-type environment to prevent them from completely decompose and take up to millions of years to form coal. As mentioned in the previous study, higher demand for electricity led to higher production of coal ash and high amounts of industrial waste produced from power plant stations. Various types of coal ash are produced during coal combustion process, for instance, fly ash, bottom ash, boiler slag, and clinker [5]. However, a major by-product produced in power plant is fly ash and bottom ash.

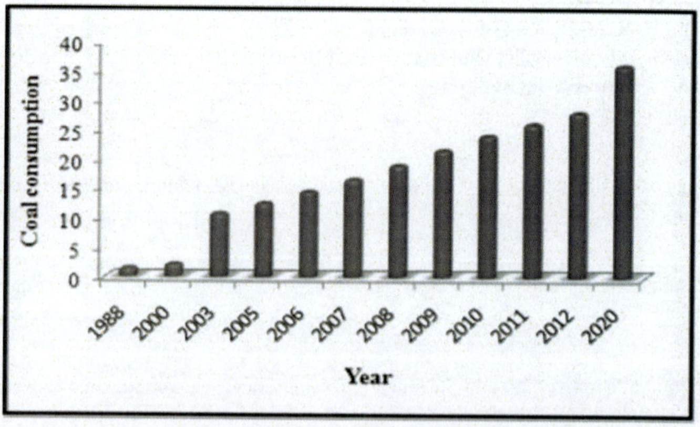

Fig. 67.1 Coal consumption for electricity generation in Malaysia (Million tons) from 1988 to 2020 [3]

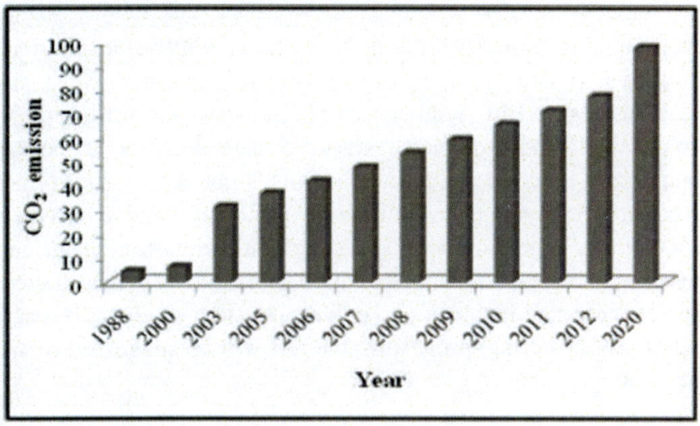

Fig. 67.2 CO_2 emission from coal combustion in year 1988 to 2020 (Million tons) [3]

Due to a lack of available dumping space, coal ash, such as fly ash and bottom ash, is still processed as garbage and dumped in impoundment ponds, silos, or landfills, which has led to environmental issues such as the contamination of ground and surface water. The Environmental Protection Agency (EPA) claims that residing close to a coal ash disposal site can raise the risk of cancer and other illnesses among nearby residents. It may also have an impact on drinking water from a well where arsenic contamination puts people at risk for cancer.

With these problems, a comprehensive action plan needs to be taken. Recently, many studies shown CBA bring many benefits as construction materials due to it low density, low cost and suitable as sand replacement in concrete manufacturing by increasing compressive, flexural, and splitting tensile test. More, the usage of CBA as fine aggregate replacement reduces the mass density of concrete which make it lighter than normal concrete and better performance. With these outstanding properties, CBA makes themselves suitable for many applications, especially in earthquake-prone area [6, 7].

Coal ash for concrete adds a new dimension to concrete mix composition. Even though CBA is known as one of the best materials to improve concrete properties, no proposal has been made for the manufacture of well-graded bottom ash and its impact on the mechanical and thermal qualities of lightweight concrete. Thus, for this study the effects of well-graded bottom ash in lightweight concrete properties with concrete mix.

67.2 Methodology

The purpose of this study is to determine the effects of well-graded CBA as sand replacement in lightweight concretes' workability, mechanical properties such as compressive, modulus elasticity, and thermal conductivity value. Various percentages of CBA (0, 5, 10 and 15%) used as sand replacement with constant values of cement, coarse aggregate, LECA, and water will be used in the mix design of building materials to observe its performance against mechanical as well as thermal properties. To obtain more accurate and reliable results, strict quality control needs to be taken. Therefore, by applying the existing code of standards and requirements for the collection of concrete-producing products and mixing proportions, spurious findings and false results can be avoided. Compressive and tensile strength are determined by using compression testing machine and universal testing machine, respectively. On the other hand, as for thermal properties, the test will be conducted by using FOX 50 instrumentation.

67.2.1 Materials

67.2.1.1 Ordinary Portland Cement (OPC)

Poly Portland Cement is the most widest type of cement used as major constituents in concrete production. Ordinary compound composed of complex chemical components such as lime (CaO), silica (SiO_3), alumina (Al_2O_3), and iron (Fe_2O_3); this fine gray powder acts as binding materials when mixed with water or any suitable solutions and hardened with time. Type I ordinary Portland cement (OPC) that meets ASTM Standard Specification for Portland Cement (ASTM C150) was used during experiments (Fig. 67.3).

Fig. 67.3 Ordinary Portland cement (OPC)

Fig. 67.4 Tap water

67.2.1.2 Water

Water, as well as cement and aggregates, is used as a combining agent in the production of concrete. The water to use for trying to mix is clean tap water that meets ASTM C1602/C1602 M specifications (Fig. 67.4).

67.2.1.3 Well-Graded Bottom Ash

The fine aggregate used is bottom ash that meets the requirements of ASTM D2487. CBA used in this study was obtained from Kapar Power Plant and then, dried in the oven for 24 h at 100 °C to remove excess moisture and prepare for sieving step. Coal Bottom Ash (CBA) was seived from the largest opening size of sieve 9.5 mm to smallest 0.075 mm and separated into distinct pans according to their size (Fig. 67.5).

Fig. 67.5 Well-graded coal bottom ash

Fig. 67.6 Lightweight expanded clay aggregate (LECA)

67.2.1.4 Coarse Aggregate

To produce lightweight concrete, lightweight expanded clay aggregate or known as LECA is one of the suitable materials used as coarse aggregate. LECA was made by heating clay up to 1200 °C in rotary kiln. LECA is eco-friendly nature-based waste product, indestructible, non-combustible, and impervious to attack by dry-rot, wet-rot, and insects [8]. A 15 mm nominal size aggregate (LECA) will be used as the coarse aggregate which conforms to the requirements ASTM C33 (ASTM, 2055i) (Fig. 67.6).

67.2.1.5 Design of Experiment Set up

Numerous lightweight concrete formulations exist, depending on the coarse aggregate used and the required strength of the finished product. In this study, the alternative way was chosen to prepare the mix proportion table. The water: cement: fine aggregates: coarse aggregates ratio that was used is 0.6: 2: 2: 1. Before the concrete mix, materials were prepared. The bottom ash was collected from Kapar Power Plant. The collected bottom ash was sieved and dried in the oven for 24 h exactly (Table 67.1).

Table 67.1 Percentage of CBA and mix design

Sample	Water (%)	Cement (%)	Coarse aggregate (LECA) (%)	Fine aggregate (Sand) (%)	Coal Bottom Ash (CBA) (%)
1	19	32.5	16	32.5	0
2	19	32.5	16	27.5	5
3	19	32.5	16	22.5	10
4	19	32.5	16	17.5	15

67.3 Result and Discussions

67.3.1 Slump Test

Test parameters such as bottom ash as fine aggregate replacement ratio on fresh and hardened concrete were studied. Slump test was carried out for each percentage mixture. The results are shown below. The lowest slump value reading was from 15% bottom ash replacement. This is high particle friction between particles of bottom ash when its replaced sand. The highest slump value reading was taken from 0% bottom ash replacement. This is due to no foreign particles that are changing the properties or mixture of a typical lightweight concrete mixture. As we can see in the table, 0% and 5% slump is giving is medium workability, where it can be considered in the usage of normal reinforced concrete placed with vibration. As for the 10% and 15% slump is giving us low workability where it is considered for the usage of foundations with light reinforcement.

67.3.2 Compressive Test

A compressive strength test reveals how much compressive stress a material can tolerate before breaking. The major comparison for this research was between 0% replacement of bottom ash and the optimal proportion of 15% replacement of bottom ash. The results showed an excellent indication of the projected compressive strength, with an average of 9 Mpa for 0% and 12 Mpa for 15% on the third day. The average value for the seventh day findings was 10 Mpa for 0% and 13 Mpa for 15%. And on the 28th day, the average compressive strength value was 12 Mpa for 0% and 17 Mpa for 15%. This study found that replacing fine aggregate with CBA had a positive impact on concrete strength. This was proven in Ref. [9]. The delayed hydration and pozzolanic activity of coal bottom ash led to a significant improvement in the compressive strength of bottom ash concrete mixes after 28 days of curing. The reduction in water–cement ratio have improved the compression strength of concrete (Fig. 67.7).

67.3.3 Modulus Elasticity

The modulus of elasticity, also known as Young's Modulus, is a measurement that determines how quickly strain changes in response to changes in stress. Regarding this test, the primary comparison that was carried out was between having no replacement of bottom ash at all and having the replacement of bottom ash at the optimal proportion of 15%. It was noted that the reading on the third day for both percentages is greater when compared to the reading on the seventh day; however, on the 28th

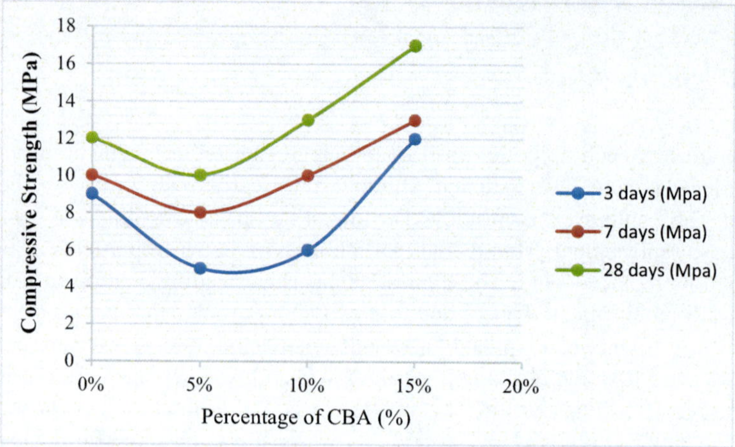

Fig. 67.7 Compressive strength

day, the reading for both 0 and 15% marginally climbs up when compared to the seventh day. This might be due to the gaps in days between measurements because the 14th day's findings were not obtained. Additionally, lightweight concrete tends to be fragile, and the use of bottom ash, which is itself fragile and porous, might have an impact on the values of the lightweight concrete (Fig. 67.8).

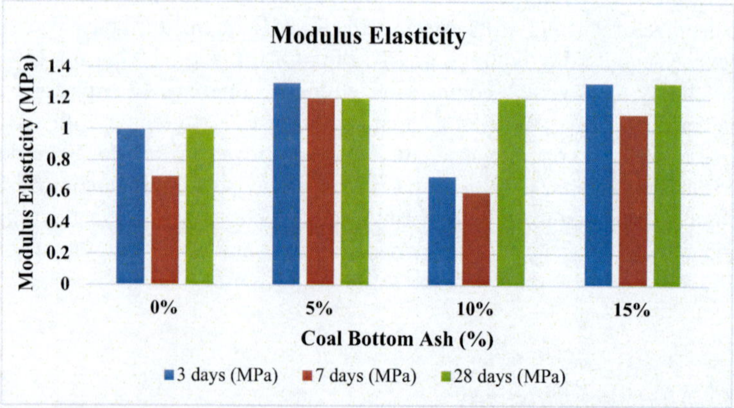

Fig. 67.8 Modulus elasticity

67.3.4 Thermal Conductivity

The FOX50 heat flow meter was used to test the thermal conductivity of mortar. The test procedure was carried out in line with ISO 8301 and ASTM C1784-13. Rubber was used throughout the measurement to assist in removing errors caused by an air gap between the specimen's surface and the apparatus' plate. Rubber also assisted in avoiding inadequate contact between the specimen's surface and the apparatus plate, eliminating thermal contact resistance. After the test, the resulting *k* values were rectified using Eq. (67.1):

$$\lambda\text{accurate} = \frac{(d\text{sample} + \text{rubber} - \text{drubber})}{\left[(d\text{sample} + \text{rubber}/\lambda\text{sample} + \text{rubber}) - (d\text{rubber}/\lambda\text{rubber})\right]} \tag{67.1}$$

The upper and lower plates had a 20 °C temperature differential during the thermal conductivity test, which was conducted at a mean temperature of 28 °C. It is clear from the Fig. 67.9 that the replacement fine aggregate generally decreased *k* values. The upper and lower plates had a 20 °C temperature differential during the thermal conductivity test, which was conducted at a mean temperature of 28 °C. It is clear from the figure below that the replacement fine aggregate generally decreased *k* values. However, the reduction in *k* value was not considerable. The lowest *k* value obtained in this study is 0.5 W/m.K when 20% of CBA was used as sand replacement. On the other hand, thermal conductivity of all mixture's series is 0.7 W/m.K for control mix, 0.6 W/m.K for 5%, 0.51% for 15%, respectively. This can be attributed to many reasons, for instance, the increasing in CBA dosage and inclusion of LECA causing density to drop more hence lowering the *k* values [10, 11].

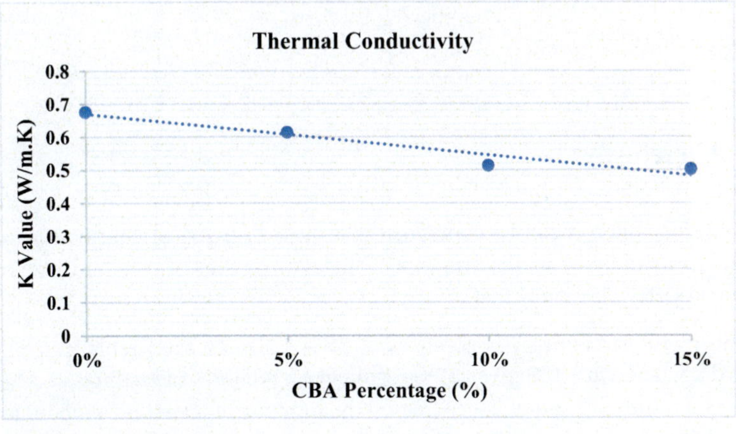

Fig. 67.9 Thermal conductivity

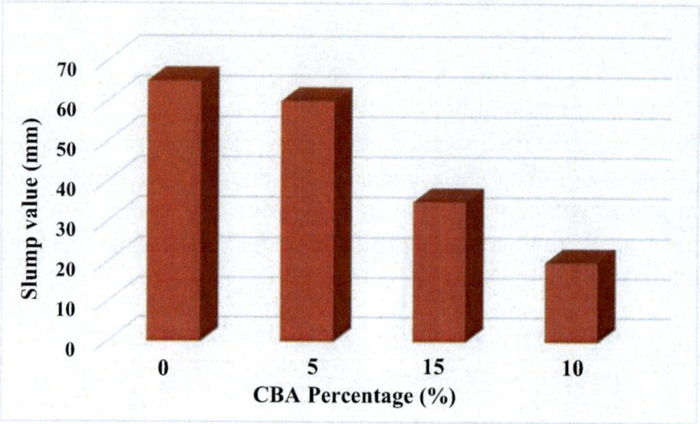

Fig. 67.10 Slump test

67.3.5 Slump Test

Test parameters such as bottom ash as fine aggregate replacement ratio on fresh and hardened concretes were studied. Slump test was carried out for each percentage mixture. The lowest slump value reading was from 15% bottom ash replacement. This is high particle friction between particles of bottom ash when its replaced sand. The highest slump value reading was taken from 0% bottom ash replacement. This is due to no foreign particles that are changing the properties or mixture of a typical lightweight concrete mixture. As we can see in the table, 0% and 5% slump is giving medium workability where it can be considered in the usage of normal reinforced concrete placed with vibration. As for the 10% and 15% slump is giving us low workability where it be considered for the usage of foundations with light reinforcement (Fig. 67.10).

67.4 Conclusion

Based on this study, it can be concluded that the utilization of coal bottom ash as sand replacement influenced thermo-mechanical properties of concrete. The results for each property also show that:

- Compressive strength of concrete with 15% CBA as sand replacement gives 17 MPa value while 0% gives 12% reading on 28th days of testing.
- From this study, 15% of CBA increased the lightweight concrete's modulus elasticity by 23% as compared to 0% of CBA but there are no significant changes in 5% and 10% CBA used.
- The 15% of CBA have recorded lowest thermal conductivity value which is 0.5 W/m.K compared to other percentage.

Acknowledgements The authors are thankful to Universiti Tenaga Nasional, Malaysia, for providing financial support (BOLD 2021) grant J510050002/2022038 and TNB Seeding Fund (U-TG-RD-2021) for this study.

References

1. M.S.N. Samsudin, M.M. Rahman, M.A. Wahid, Power generation sources in Malaysia: status and prospects for sustainable development. J. Adv. Rev. Sci. Res. **25**(1), 11–28 (2016)
2. A.U. Abubakar, K.S. Baharudin, I. Technology, *Potential use of Malaysian Thermal Power Plants*, vol. 3, no. 2, pp. 25–37 (2012)
3. Martunus, M. Othman, R. Zakaria, W. Fernando, Co 2 emission and carbon capture for coal fired power plants In Malaysia and Indonesia. Int. Conf. Environ. (Icenv 2008), No. May 2014 (2008)
4. J.H.L. Voncken, *Geology of Coal Deposits of South Limburg, The Netherlands*, No. November (2020)
5. N.I.R. Ramzi, S. Shahidan, M.Z. Maarof, N. Ali, Physical and Chemical Properties of Coal Bottom Ash (Cba) from Tanjung Bin Power Plant. Iop Conf. Ser. Mater. Sci. Eng. **160**(1) (2016). https://doi.org/10.1088/1757-899x/160/1/012056
6. F. Monika, H. Prayuda, M.D. Cahyati, E.N. Augustin, H.A. Rahman, A.D. Prasintasari, Engineering properties of concrete made with coal bottom ash as sustainable construction materials. Civ. Eng. J. **8**(1), 181–194 (2022). https://doi.org/10.28991/Cej-2022-08-01-014
7. P. Aggarwal, Y. Aggarwal, S. M. Gupta, Effect of bottom ash as replacement of fine aggregates in concrete. Asian J. Civ. Eng. (Building Housing) **8**(1), 49–62 (2007)
8. A. Zukri, R. Nazir, K.N.M. Said, H. Moayedi, Physical and mechanical properties of lightweight expanded clay aggregate (Leca). Matec Web Conf. **250** (2018). https://doi.org/10.1051/Mateconf/201825001016
9. M. Singh, R. Siddique, Effect of coal bottom ash as partial replacement of sand on workability and strength properties of concrete. J. Clean. Prod. **112**, 620–630 (2016). https://doi.org/10.1016/J.Jclepro.2015.08.001.
10. I.H. Yang, J. Park, K.C. Kim, S.W. Yoo, A comparative study on the thermal conductivity of concrete with coal bottom ash under different drying conditions. Adv. Civ. Eng. **2021** (2021). https://doi.org/10.1155/2021/7449298.
11. P. Torkittikul, T. Nochaiya, W. Wongkeo, And A. Chaipanich, Utilization of coal bottom ash to improve thermal insulation of construction material. J. Mater. Cycles Waste Manag. **19**(1), 305–317 (2017). https://doi.org/10.1007/s10163-015-0419-2

Chapter 68
Smart Attendance System Using Face Recognition

Muhammad Hazirul Bin Badrul, Suzi Seroja Sarnin, Mohd Nor Mad Tan, Azlina Idris, Nani Fadzlina Naim, and Ros Shilawani S. Abdul Kadir

Abstract In the past, a password was required to identify someone. Passwords are becoming obsolete because they can be easily cracked and used to access accounts. Passwords have been replaced with biometrics because they offer more security and are more difficult to hack. Biometric facial recognition is widely employed in our environment. An individual can be recognized by their face traits thanks to this sophisticated, automatic, and practical identification method. A digital camera is used to take the face image, a computer is used for processing and analysis, and an output device is used to show the identification outcome. This project's goal is to use a camera to take attendance while simultaneously identifying students' faces and entering the data into a database. The records of student attendance will be shown on a web user interface, and an authorized individual will be able to alter or delete the data. There will be no need to manually update the pupils' attendance because the data will be updated automatically when the face is recognized. When the face is recognized, it will also display the student's name and ID number. Because of COVID-19, it is necessary to record a person's temperature. As a result, this system keeps track of the students' temperatures, and the data is automatically uploaded after the face is recognized, and the temperature is scanned. This has led a quick and simple biometric system for recording students' attendance in the class.

M. H. B. Badrul · S. S. Sarnin (✉) · M. N. M. Tan · A. Idris · N. F. Naim · R. S. S. A. Kadir
School of Electrical Engineering, College of Engineering, Universiti Teknologi MARA, 40450 Shah Alam, Selangor, Malaysia
e-mail: suzis045@uitm.edu.my

M. N. M. Tan
e-mail: Mohdn829@uitm.edu.my

A. Idris
e-mail: azlina831@uitm.edu.my

N. F. Naim
e-mail: nanifadzlina@uitm.edu.my

R. S. S. A. Kadir
e-mail: ros885@uitm.edu.my

M. A. A. Mohd Salleh et al. (eds.), *Proceedings of the Green Materials and Electronic Packaging Interconnect Technology Symposium*, Springer Proceedings in Physics 289,
https://doi.org/10.1007/978-981-19-9267-4_68

Keywords Microcontroller · ESP32-CAM · Wi-Fi connection

68.1 Introduction

The paper "Smart Attendance System Using Face Recognition" is proposed to ease the management to handle the attendance of the students. The web server that held the attendance database records can be accessed remotely from a distance place by connecting the device within the same network or it can also be accessed remotely by setting up the router. The attendance of a student when entering the school can be recorded automatically where a camera will be used to recognize the student's face. The temperature of a student will automatically be scanned when the camera detects a face. The system will check the face of the student where if the student's face is recognized, the student's attendance will be recorded into the system. Otherwise, the system will display that the face is not recognized.

This system can prevent inaccuracy in recording attendance data where teachers or staff in school might incorrectly record the student attendance. The attendance of the student will be recorded into a database where it can be managed through the web user interface. The web user interface offers the display of the student's data such as student's ID, students name, class, and others. Thus, it will also display the student's attendance with the selections of day and class. The student's data such as a new student can be added to the system using the features added into the web user interface. Furthermore, the data of the student and the attendance can be modified and deleted depending on user input. The data of the student attendance will be transmitted by using a Wi-Fi connection through the MQTT broker. The data transmitted by the microcontroller which is ESP32-CAM will be stored in the database of the server. Thus, an SMS notification will be sent to the parents if their child not punctual and not present at school after a certain time.

68.2 Methodology

ESP32-CAM is the main microcontroller that will be used to communicate with the Raspberry Pi which acts as a server through the MQTT protocol. ESP32-CAM has the capability to attach and use a camera where this microcontroller will be used to detect and recognize a face. Figures 68.1 and 68.2 show the flowchart for the face recognition system for hardware and software. Initially, ESP32-CAM, ESP32, and Raspberry Pi will be booted up, and the ESP32-CAM will start the camera stream. Raspberry Pi will check either all of the students' attendance is defined on the present date or not. If the attendance is not defined, it will start to create new attendance for all students for that day and set it as absent where the attendance will be updated later after the student starts to scan the face on the face recognition system. Then, ESP32-CAM will start to detect a face using the camera where if a face is detected, a box frame will appear on the camera display which indicates that a human face is detected. Then, ESP32-CAM will serially communicate with ESP32 to turn on the temperature sensor to start measuring the temperature and display the output value to the LCD. ESP32-CAM will start face recognition upon finish in scanning the temperature and retrieve the temperature value from ESP32.

After recognizing a student's face, it will send the face identity number which is the student's id to the Raspberry Pi through the MQTT broker. Raspberry Pi will query the student's information from the database and set the student's attendance

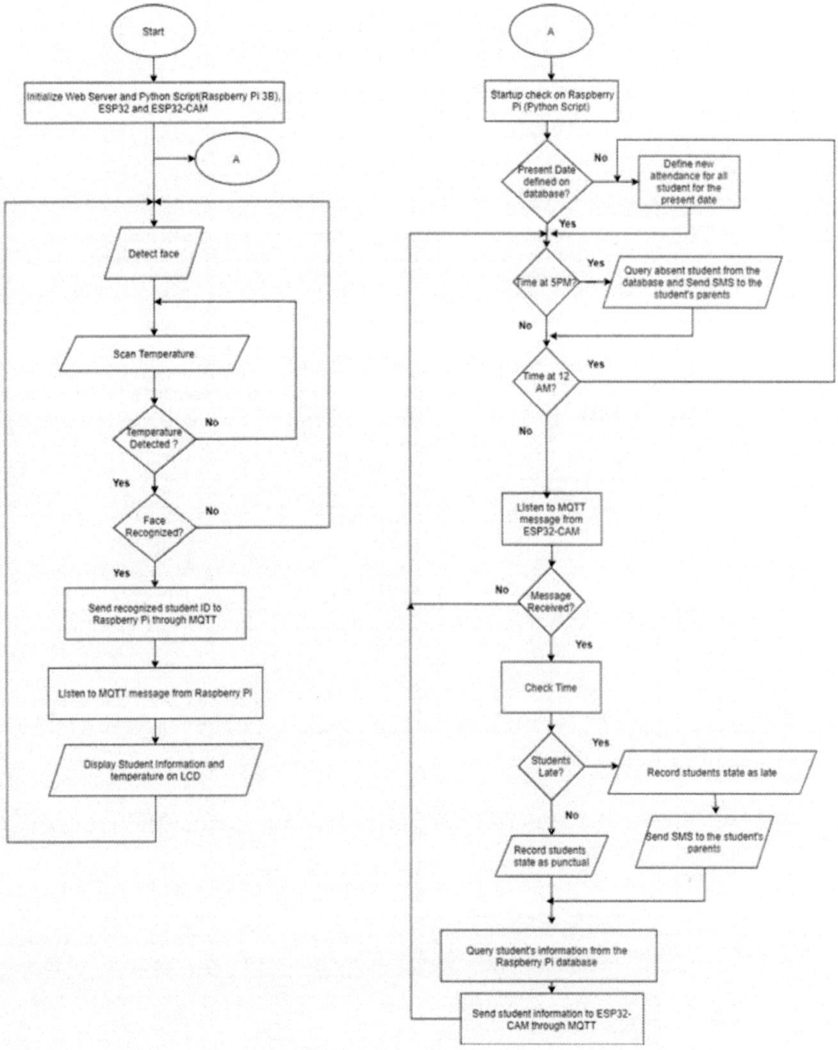

Fig. 68.1 Flowchart of the hardware system

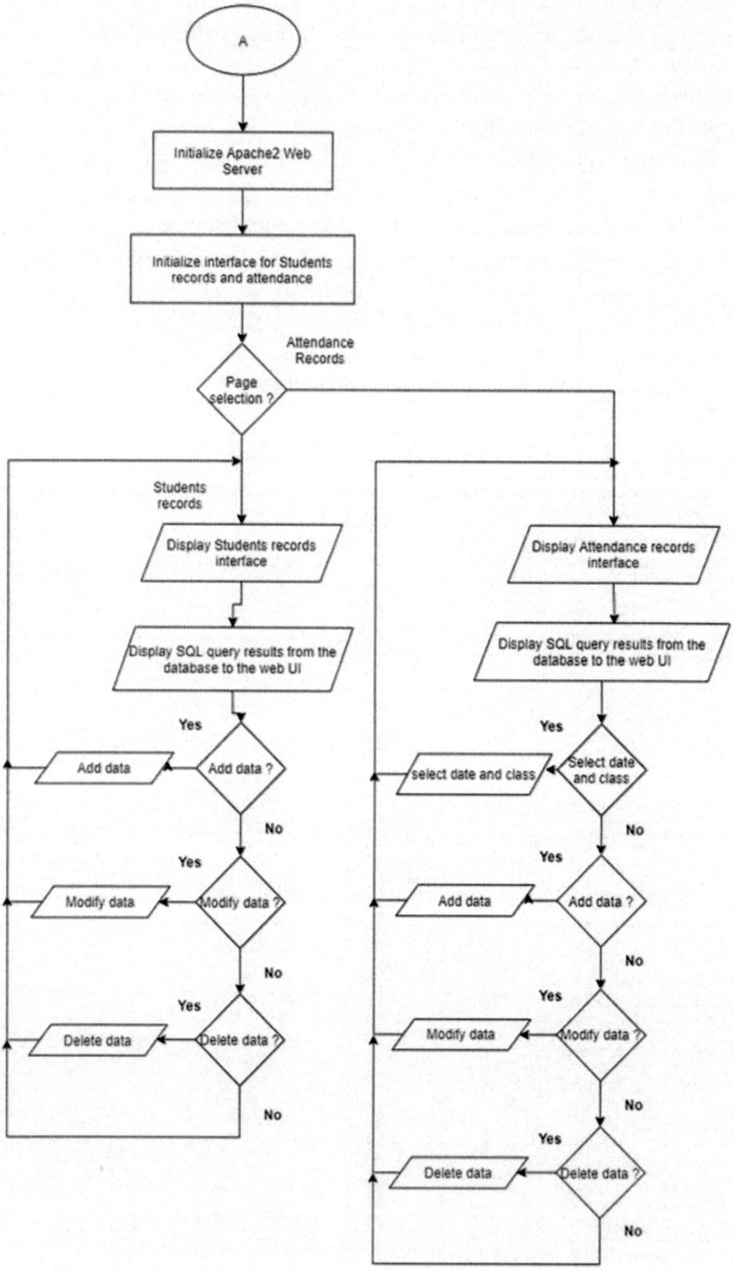

Fig. 68.2 Flowchart of the software system

for that day. Then, it will send the student's information such as the student's name to ESP32-Cam which will then serially send the information to ESP32 to display the data on the LCD. The transmission and retrieval of data through the MQTT broker will be conducted wirelessly using Wi-Fi protocol. Next, the temperature sensor will be used to scan the temperature of a student where the reading of the sensor will be transmitted to the microcontroller using an IIC serial connection. Furthermore, both of the student information such as name with the temperature reading will be displayed on the LCD.

Raspberry Pi 3B will be used as a server where it acts as a cloud server that can be used for web server and database storage. The ingoing and outgoing message or data will be channeled through the MQTT broker using Python with the subscription of the topic set. A combination of Python code will be created to handle the MQTT connection, SQL database query, and SMS protocol. A web user interface will be stored in the webserver where the Apache2 web server will be used to display the interface. The web user interface will consist of the students' records and attendance records where features such as adding, modifying, and deleting data will be added in the interface. Lastly, the web user interface will display the queries of the data in the database using SQLite based on the user input selection.

68.2.1 Temperature Measurement

MLX90614 is a non-contact infrared temperature sensor that can be used to scan the target object temperature without needing to have physical contact with it. The sensor was first tested by applying the codes to reads the temperature where the output results were observed on the serial monitor of Arduino IDE.

68.2.2 Displaying Data on LCD

Displaying data on LCD is a very straightforward process where minimal lines of code were created. A single function was being used to display the data on the LCD. The function has two parameters where it will be called in loop function where the data will be passed to the function. The initializations of the LCD were set on the setup function where this LCD used IIC or I2C serial data communication to interact with the microcontroller ESP32.

68.2.3 Face Recognition Using ESP32-CAM

Camera code and library were obtained from the ESP32 camera web server example from the Arduino. The ESP32-CAM was configured to have persistent storage where by default, the saved image for face recognition will vanish every time the ESP32-CAM reset or restart.

Persistent storage will enable the ESP32-CAM image to be stored internally without having loss of image data issues when restarting ESP32-CAM. The modifications of the ESP32-CAM camera library were done where the partitions of ESP32-CAM memory will be set to enable the persistent storage [1–3]. The library was modified to enable storing the face with a name that will be set as the student's id.

68.2.4 Web User Interface with Database Development

The web user interface will be used to display the records of the students. The development of this project requires HTML, CSS, JS, PHP, and SQL (SQLite3) language to be used to create the web user interface with database features [4–8].

Firstly, a minimal display of the web user interface was created by using HTML and CSS where a table containing a header and data list was initialized. The initial web user interface extension was .html where it will be needed to be renamed into .php to enable the usage of the PHP script. A welcome page was created which will list all of the page lists where all of the pages consist of a side panel which will help in navigating to another page. Four pages of web user interface were created to handle the attendance system.

68.2.5 Forwarding Live Camera Stream

ESP32-CAM camera server cannot be started automatically where an HTTP connection with a "GET" request needs to be done to start the camera. A Python library and script were used to repeat and extend the video stream of MJPEG from ESP32-CAM and encounter this problem. The Python's script and library named "MJPEG Relay" from GitHub was developed by a developer with an author name of "OliverF" where this library will allow more clients or web browsers to connect to the video stream of ESP32-CAM [9, 10]. The camera streams can be accessed on the web user interface where it will allow more connections to streams the camera through the web user interface.

68.2.6 Handling Face Recognition Data

Student's face needs to be enrolled and captured to enable the face recognition feature. The administrator will need to enter the student's id that is defined on the database where the student's face will be enrolled through the camera stream on the web user interface using an HTTP connection. Raspberry Pi will capture the student's face

Table 68.1 Hardware serial communication pin connections

Hardware serial communication pin connections			
ESP32-CAM		ESP32	
Communication pin	Location	Communication pin	Location
TX	14	RX	16
RX	15	TX	17

upon finish in enrolling the student's face that will store the image list and display it on the web user interface. A software package which is FFmpeg was used to enable the features of taking a snapshot of the ESP32-CAM stream. Thus, delete features were created to enable the function of deleting a student's face or delete all student's face. Codes for ESP32-CAM were also set to handle storing the enrolled face image with a name where student id was used as a name to ease the process of recognizing student's face with information.

68.2.7 Serial Communication Between ESP32 and ESP32-CAM

ESP32-CAM has limitations in terms of the availability of GPIO pins where most of the pins were internally used for the camera usage. The limitation of this microcontroller causes additional peripherals such as LCD and temperature sensors to be connected to another microcontroller which is ESP32. Table 68.1 shows the physical connection for the ESP32 and ESP32-CAM pin.

68.2.8 Python Script on Raspberry Pi

The main Python code which is "core.py" will be used to listen to the activity of the ESP32-CAM such as the state of face recognition, face detection, scanned temperature, recognized face, and others [11–13]. The other Pythons' code will be used separately to send commands to ESP32-CAM to enroll the student's face, delete the student's face, and delete all student data on ESP32-CAM. The main Python code which is "core.py" will retrieve the recognized face from ESP32-CAM and query the student's information on the Raspberry Pi database which will then submit the data back to the ESP32-CAM to display it on the LCD. The Python script will also handle the SMS message depending on the student attendance status.

68.3 Result and Discussion

Based on the result of this project, it can be seen that this project contains the improvement from the previous project [14, 15]. The project from the technical paper [15] provides an attendance system using RFID with SMS functionality, but the project did not automatically submit the SMS messages where it needs to be done manually by inserting a contact number and use a button to send the SMS. Another technical paper [14] provides a face recognition attendance system, but it does not have the user interface to handle the attendance. Therefore, the improvement from the previous report was improved in this project where face recognition attendance system with automatic SMS submission, and it also provides a friendly web user interface for the authorized user to handle the attendance system.

68.3.1 Hardware Development

ESP32 and ESP32-CAM microcontrollers were mounted into a single donut board where the temperature sensor, IIC LCD, LED, and buzzer were also connected on that board as can be seen in Fig. 68.3. A container was modified to allow all of the components on that board to be fitted into it. The container will be the enclosure for this project where the zero board and the components were mounted by using a hot glue gun to make it hold on the container. The container was later wrapped with a fabric to cover the container and gives a better view of the project. The modified container will be used for recognizing a face, scanning temperature, display information on the LCD, indicates the recognition state using LED light, and create a sound using the buzzer. Then, the container will communicate with Raspberry Pi for receiving and transmitting data. Raspberry Pi will wirelessly listen for the command from the modified container, process the data, and transmit information back to the modified container.

Fig. 68.3 Hardware components connected to donut board

68.3.2 Face Recognition System

Face recognition system is the developed hardware system which is the modified container that will be used to scan the students' faces as can be seen in Fig. 68.4. Initially, the red LED on the system will be turned on and the other LED will be turned off to indicate that the system is not detecting any face. This system will firstly detect either there is a face detected or not where it will start the temperature sensor if a face is detected. Yellow LED will be turned on after a face is detected, and the buzzer will beep for once by the system while other LED will be turned on. The system will record the student's temperature where the LCD will display the temperature measurement of the students.

The face recognition will be enabled upon finishing in measuring temperature. If the detected face is not recognized, red and yellow led will be turned on and the buzzer will beep for three seconds to indicate that the face is not recognized. The LCD will display that the face is not recognized. The system will reset and restart to its initial condition. If the face is recognized by the system, a green LED will light up while another LED will be turned off and the buzzer will beep twice to indicate that the face is recognized. Then, the student's id that is saved with the student's face by the system will be sent to the server which is the Raspberry Pi for querying the student's information such as the student's name. Then, it will send the data to the face recognition system to display the information on the LCD.

On the initial startup or boot, where the script will check the database to identify either the present date or the same day date is defined on the database or not. If the present date is defined on the database, then it will skip the startup routine which is updating all of the students' attendance as not present on the database. All of the attendance statuses will be automatically be set as not present by Raspberry Pi every day at 12 AM where the attendance will be updated when the students start to scan their face on the face recognition system. The script in the Raspberry Pi

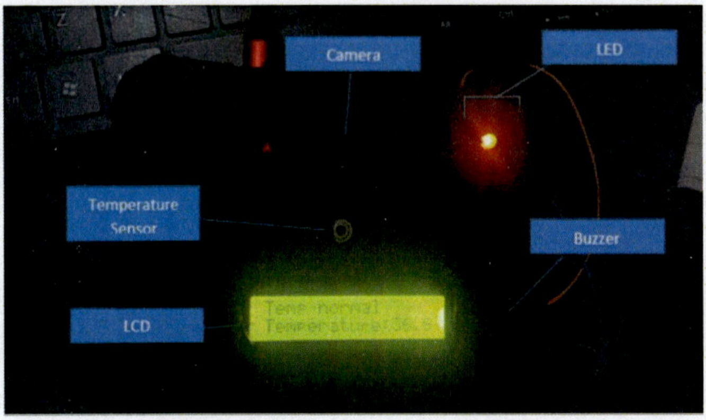

Fig. 68.4 Mounted hardware on the container

will automatically set the students' attendance based on time such as if the student's clock-in after 8 AM, the student's attendance status will be marked as late. Then, an SMS service using "Vonage" will be used to send a message to the student's parent to inform that their children are late. If the students not present or absent on that day, a message will be sent to their parents to inform them that their children were absent on that day at 5 PM. Raspberry Pi will handle the web user interface display where the first page will display the menu of the system which consists of student management, attendance management, data records, and test developments of the system. A side panel was created for each of the pages to enable the user to select pages within the same page. The web user interface will communicate with the back-end server services such as PHP and Python to process the input data from the web user interface.

68.3.3 Student Management Page and Attendance Management Page

Student management will be the page that displays the student information such as student id, student name, class, and parent's contact number as can be seen in Fig. 68.5. Three features can be used on this page such as adding a student, modify a student, and delete a student in the database. The administrator or user needs to define all of the students in the system by using the add student features which consists of student id, student name, class, and parent's contact number. Then, student lists will be displayed on the display box of the student management page Attendance management will be the page that displays the students' attendance where all of the attendance statuses will be automatically be set as not present by Raspberry Pi every day at 12 AM. This page will display the students' attendance information such as the student id, class, status, clock-in time, clock-out time, temperature when clock-in, and temperature when clock-out. The student attendance status will be updated when the students scan their faces on the face recognition system. The students' status consists of absent ("0"), punctual ("1"), late ("2"), and not clocked out ("3") where all of this status will be handled by the Raspberry Pi using Python code which will automatically modify the database based on the time. There are three features included on this page such as add an attendance, modify an attendance, and delete an attendance.

68.3.4 Data Records Page

Data records page will consist of the camera stream, student's image list, and other features such as enroll student's image and delete student's image as can be seen in

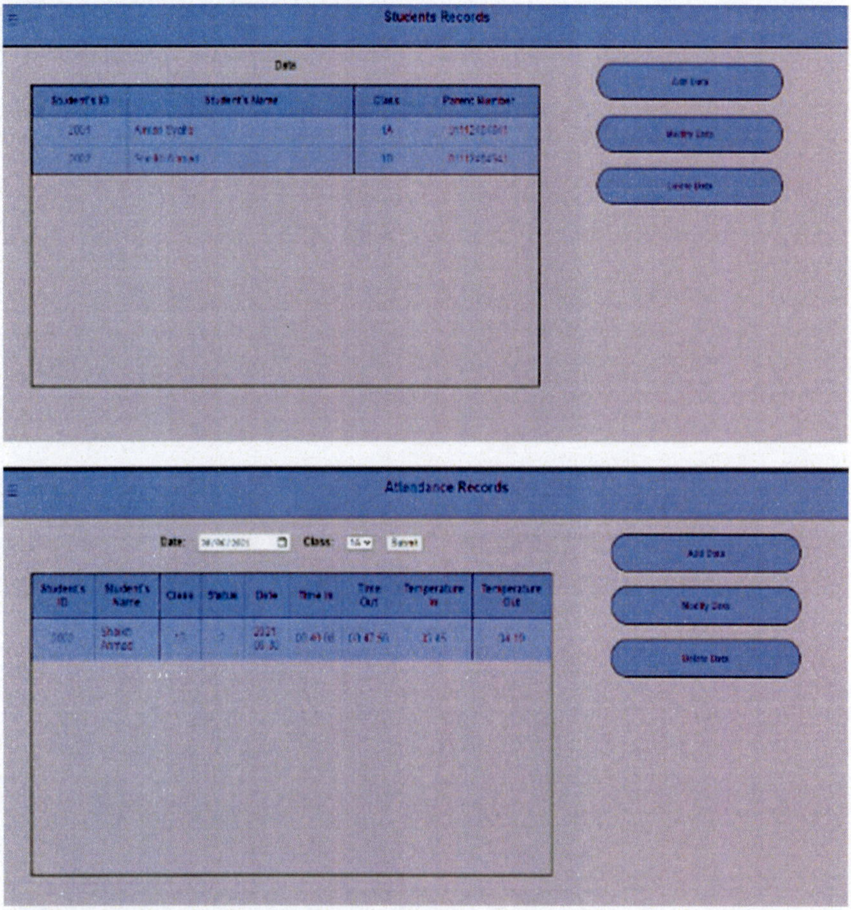

Fig. 68.5 Student management page display and attendance management page display

Fig. 68.6. In order to allow the students' faces to be recognized by the face recognition system, the students' faces need to be enrolled through this page.

This will update the student's information on the database which is column "registered" which will set that the student's face is registered. The camera stream will allow monitoring the incoming students on the face recognition system and enrolling face on the camera. Delete student will be used to remove the student's face on the defined list of the allowable face to be recognized. Then, all of the defined students' faces will be displayed on the page.

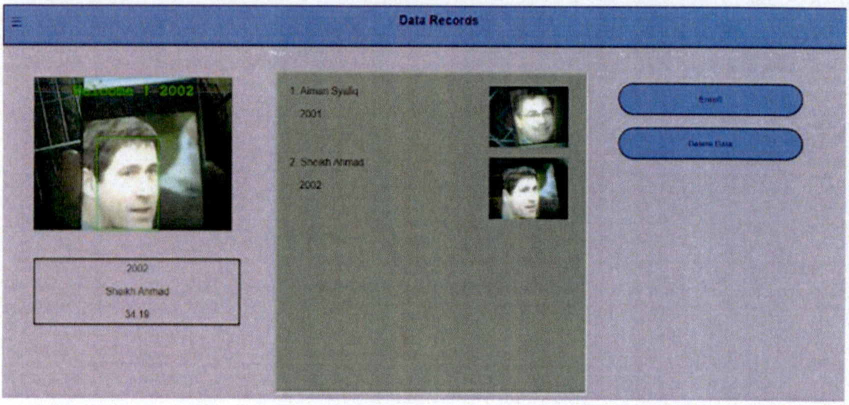

Fig. 68.6 Data records page display

68.3.5 *Test Development Page*

Test developments page will be the page for testing purposes where it consists of a box that will display logs and errors and buttons for testing the face recognition system as shown in Fig. 68.7. Firstly, any logs or errors will be displayed on the box where the internal error that occurs from the codes will be saved on a text file and displayed on the web user interface. Next, there are button features that can be used to restart ESP32 and ESP32-CAM, check the ESP32-CAM alive status, check face recognition state, and test the SMS feature.

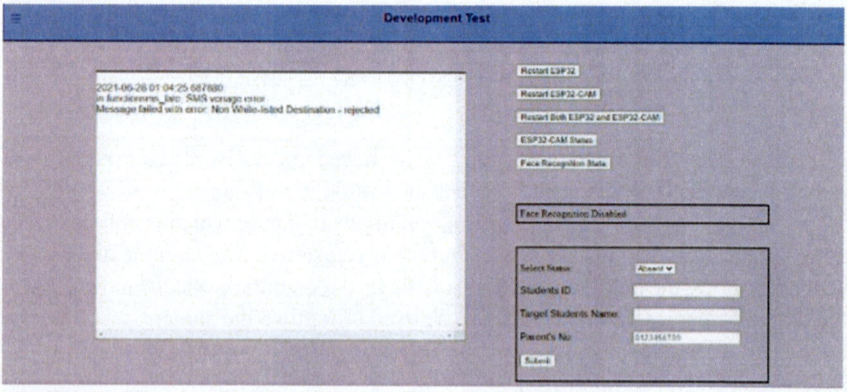

Fig. 68.7 Test development page display

68.4 Conclusion and Recommendation

Based on the final software and hardware developments, the attendance system was successfully developed. The student's face can be recognized by the system, and the student's information can be displayed on the LCD upon recognizing the face. The student's face was able to be stored in the ESP32 memory with a name where the system can communicate with Raspberry Pi as a server. The Python script developed can manage the communication between ESP32-CAM and Raspberry Pi through MQTT. The Python script can query, update, and delete data on the database based on user input from the web user interface and automatically managed by the face recognition system. Thus, the system can automatically submit the SMS message to the student's parent.

This project could be improved for future works where the face recognition could be handled by Raspberry Pi as a server. The ESP32-CAM could be used as a camera stream while Raspberry Pi will handle the face recognition processing. Furthermore, a board with more GPIO pin interface could be used to replace the serial communication between ESP32 and ESP32-CAM such as Raspberry Pi Zero. Raspberry Pi Zero is small in size where it could handle a camera interface, and it also has more GPIO pins which can also be expanded by adding an expansion board to it.

Acknowledgements The authors would like to thanks to Universiti Teknologi MARA (UiTM) and Research Management Centre (RMC) Shah Alam, Selangor, Malaysia for the financial support of this research under grant project code: 600-RMC/MYRA5/3/LESTARI (097/2020).

References

1. Robotzero, Esp32cam Access Control, *GitHub* (2021). https://github.com/robotzero1/esp32cam-access-control. Accessed 15 May 2021
2. Expressif, fr_flash.h, *GitHub* (2020). https://github.com/espressif/esp-face/blob/master/face_recognition/include/fr_flash.h. Accessed 15 May 2021
3. Espressif, fr_flash.c, *GitHub* (2020). https://github.com/espressif/esp-face/blob/master/face_recognition/fr_flash.c. Accessed 15 May 2021
4. S. Prettyman, *Learn PHP 7: Object Oriented Modular Programming Using HTML5, CSS3, JavaScript, XML, JSON, and MySQL* (Apress L. P, Berkeley, CA, 2016)
5. P. Hong, *QR Code Practical web design : learn the fundamentals of web design with HTML5, CSS3, Bootstrap, jQuery, and Vue.js.* (Packt, Birmingham, Mumbai, 2018)
6. D. Powers, *PHP Solutions Dynamic Web Design Made Easy* (Apress, Berkeley, CA, 2010)
7. J. Lengstorf, *PHP for Absolute Beginners* (Apress, Berkeley, CA, 2010)
8. Jit, Q. Shen, M. Ghazaly, Mariam, Development and analysis of face recognition system on a mobile robot environment. J, Mech. Eng. (JMechE) **15**(2), 169–189. ISSN 1823-5514; 2550-164X (2018)
9. OliverF, MJPEG Relay, *GitHub* (2019). https://github.com/OliverF/mjpeg-relay. Accessed 15 May 2021
10. Hdavid0510, MJPEG Relay (commit), *GitHub* (2021). https://github.com/OliverF/mjpeg-relay/archive/2bd4e93107b5a57c5d633fd22861fd995b97cd8d.zip. Accessed 15 May 2021

11. T. Cox, *Getting Started with Python for the Internet of Things: Leverage the Full Potential of Python to Prototype and Build IoT Projects Using the Paspberry pi.* (Packt, Birmingham, England, Mumbai, 2019)
12. T. Hall, J.-P. Stacey, *Python 3 for Absolute Beginners* (Apress, Berkeley, CA, 2009)
13. P. Membrey, D. Hows, *Learn Raspberry Pi with Linux* (Apress, Imprint, Apress, Berkeley, CA, 2013)
14. A. Md Sajid, S. Pronob, M. Ahmad Tamim, A.A. Abu Musa, U. Jia, *Face Recognition and RFID Verified Attendance System* (2018)
15. N.S. Soumil, A. Abdelshakour, *IoT Based Smart Attendance System (SAS) Using RFID* (2019)

Chapter 69
Modulus Resilient Analysis of Flexible Pavement AC–WC and AC–BC Using Asphalt Modification PG.70 from Marshall Testing Results

T. Tasman, B. H. Setiadji, A. Situmorang, and D. Mudjono

Abstract The quality of asphalt concrete that meets technical specifications as an asphalt pavement construction material is the main consideration for an engineer, so that the design life of the road can be achieved. Currently, the use of modified type II PG70 asphalt mixture is considered to be a relevant choice for road conditions in areas with a tropical climate, especially in hot temperature areas. Asphalt concrete is a construction material that has flexible behavior during traffic loads. One of the behaviors in question is the resilient modulus (Mr), the selection of aggregate gradations and the type of asphalt that meets certain requirements which greatly affects the quality of asphalt concrete. Method is using static loading with the Marshall test tool. The value of the modulus of elasticity can be measured with the Marshall parameter. The maximum resilience modulus value for the asphalt concrete mixture AC–WC = 67,678 Mpa and AC–BC = 72,732 Mpa. The application of the Marshall test is still feasible to use in determining the resilient modulus, although there are other, more detailed methods in finding the resilient modulus of a material. The resilient modulus analysis carried out by the author aims to determine the value of the resilience modulus (Mr) on the AC–WC and AC–BC pavement layers as an evaluation of the results of the design planning (Detail Engineering Design) of road pavements.

Keywords AC–WC and AC–BC · Asphalt modification type II PG70 · Loading static · ITS · Modulus resilient

T. Tasman (✉) · B. H. Setiadji
Doctoral Program of Civil Engineering, Faculty Engineering, Diponegoro University, Semarang 50275, Indonesia
e-mail: tasimahmad67@gmail.com

A. Situmorang
Departement of Civil Engineering, Faculty of Engineering, Semarang University, Semarang 50196, Indonesia

T. Tasman · D. Mudjono
Consultant of Pavement/Material Engineering, PT Aria Jasa Reksatama, Surabaya, Indonesia

M. A. A. Mohd Salleh et al. (eds.), *Proceedings of the Green Materials and Electronic Packaging Interconnect Technology Symposium*, Springer Proceedings in Physics 289,
https://doi.org/10.1007/978-981-19-9267-4_69

69.1 Introduction

The improved handling of the road structure is intended for the convenience of vehicular traffic. The damage that appears directly occurs in the surface layer, among the causes is the durability and stiffness of the asphalt pavement that does not meet the design life. Another problem is the use of asphalt mixture materials and the mixing process as well as the implementation of work in the field. Asphalt concrete technology is currently very developed, and it is proven that many studies and research have been carried out to answer the above problems.

Analysis of the behavior of the asphalt pavement layer is an important part of the studies that have been carried out in recent years. The following are some research studies on the behavior of asphalt pavement under different loading conditions through experimental and computational techniques focusing on stiffness modulus. Reference [1] analyzed the resilient modulus on AC–WC and AC–BC asphalt concrete pavement layers using asphalt penetration 60/70 based on Marshall test. Resilient modulus is one of the parameters determining the strength of the asphalt mixture, which is influenced by static loading and plastic deformation. Reference [2] evaluates the resilient behavior of flexible pavement in asphalt layers using asphalt pen 60/70. The test method uses one pulse loading, namely haversine, which represents the form of loading given to the binder layer, and an indirect tensile test is carried out to determine the resilient modulus using a universal testing machine.

Introduced an indirect tensile test method of stiffness modulus as the main property of asphalt mixtures closely related to permanent deformation. In the summary of findings by the Technical Committee of RILEM 206 Advanced Testing and Characterization of Asphalt Materials, the interlayer bonding of the pavement layers was seriously considered. The effect of sample diameter, test temperature, test velocity, stress build-up and aging on the shear bond strength was evaluated by shear test [3]. Reference [4] investigated compressive modulus of high modulus asphalt mixture using indirect tensile test and compared with complex modulus. In addition, the impact of wave loading on the test results is explored. Indirect tensile tests were carried out according to EN 12,697-26 standard procedure.

Reference [5] evaluated the application of indirect tensile stiffness modulus test as an alternative to direct tension modulus test. The results showed a good correlation between the two test methods. Based on the measuring instrument limit, achieving the target strain is difficult, especially for high modulus mixtures. The alternative test is used for cold mixtures with some sample damage during the test. In another study, [6] compared an indirect tension modulus test performed on two different mixtures using six machines with a direct tensile test. MR, measured according to EN 12,697-26 Annex C (IT-CY) standard procedure, was compared with estimates of complex modulus and direct tensile modulus. The indirect tensile test is a reliable, fast and low-cost method and is recommended for the design of semi-coarse asphalt mixtures. However, for high modulus mixtures, the test procedure requires some adjustments.

Currently, the pavement design method is mechanical and empirical, and flexible pavement is designed based on the influence of different temperature and loading conditions. One of the most important characteristics of asphalt in the pavement design procedure is the modulus of the asphalt layer. This parameter is the elasticity behavior of asphalt material under various loading and traffic conditions. In addition, the theory of elastic and finite element multilayer using models for the determination of asphalt pavement thickness and analysis of pavement structures has been highly developed.

Preservation work for the Pejagan–Prupuk–Tegal–Bts. Banyumas/Brebes-Wangon road, Central Java, Indonesia, for the 2021–2022 fiscal year can be used as an example of an analysis of the calculation of the resilience modulus on the AC–WC and AC–BC pavement layers based on the results of the Marshall test. Previous studies have shown that the determination of the modulus in the asphalt layer using the dynamic modulus instead of the resilient modulus (Mr) makes some responses in the pavement negligible, which can be effective in predicting pavement performance [7, 8]. Based on the reasons above, the author uses Mr in the analysis of the flexible pavement design.

69.2 Methodology

69.2.1 Aggregates

The aggregate used in the preparation of the sample test must meet the physical properties in accordance with the technical specifications rev.2 page 6–39 (2018), Reference [9] the maximum nominal aggregate size of the mixture is 19.1 mm for asphalt concrete surface layer (AC–WC), while for the binder layer (AC–BC) is 25.4 mm. The source of aggregate material is from the Asphalt Mixing Plant (AMP) PT. SPP, Jladri-Kebumen, Central Java. As a mixture of asphalt concrete, the required aggregate is divided into 2 (two) types of aggregate, namely coarse aggregate with conditions that are retained on filter no. 8 or a sieve size of 2.38 mm and fine aggregate with a size that passes sieve no. 8 (2.38 mm) and retained on filter no. 200 (0.075 mm). Prior to the gradation test, the material was subjected to an abrasion test using a Los Angeles tool using the SNI 06-2417-2008 method with a result of 19.3%. Furthermore, a gradation test is carried out to determine the composition of the aggregate mixture used. The results of the aggregate gradation test are shown in Fig. 69.1 and properties of aggregate in Table 69.1.

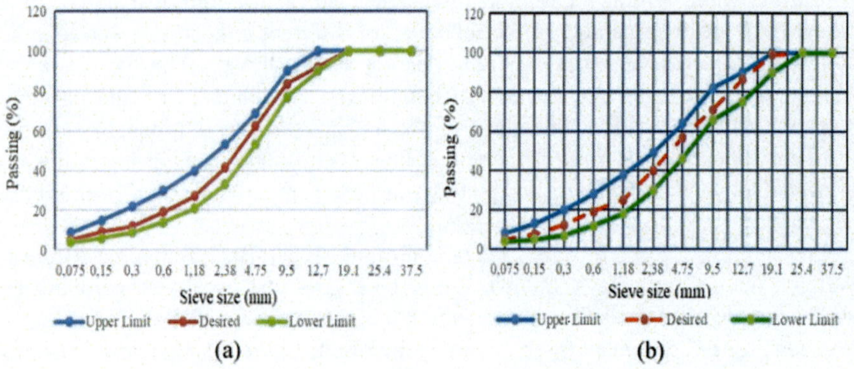

Fig. 69.1 Gradation aggregate of the Bina Marga (2018 rev.2) for asphalt concrete AC–WC (**a**) and AC–BC (**b**)

Table 69.1 Physical properties of aggregate

Test	Method	Value	Standard requirement
Los Angeles abrasion (%)	SNI 06-2417-2008	19.30%	Max 30 (%)
Conservation of aggregate form in solution	SNI 3407:2008		
Particle-specific gravity		2.69	
Material pass sieve no. 200	SNI ASTM C117: 2012		Max 1%

69.2.2 Bitumen

The asphalt mixture sample was prepared using modified asphalt type II PG70 which in the technical specifications of Bina Marga is called asphalt type II (technical specification rev.2 page 6–41, 2018) [9]. Before asphalt is used as a mixture of asphalt concrete for the pavement layer, an asphalt test is carried out at the DKI Jakarta-West Java National Road Implementation Center which aims to determine the physical properties of asphalt according to the technical requirements set. The following are the asphalt test results as shown in Table 69.2.

69.3 Design of Experiment Setup

69.3.1 Mixture Design

The design of the asphalt concrete mix was achieved according to the Marshall method specified in the Ref. [10]. The optimum asphalt content must be based on laboratory and field tests as stated in the working mix plan (JMF). Mixing is done

Table 69.2 Results of test on asphalt type II asphalt modification PG 70

Test	Method	Value	Standard requirement
Penetration 25 °C (0,1 mm)	SNI 2456-2011	57.90	–
Softening point (°C)	SNI 2434:2011	70.50	–
Ductility 25 °C (cm)	SNI 2432:2011	113.67	–
Flash point (°C)	SNI 2433:2011	321.83	≥ 230
Solubility in trichloroethylene (%)	SNI 06-2348-1991	99.89	≥ 99
Specific gravity (gr/ml)	SNI 2441:2011	1.031	–
Storage stability (softening point difference) (°C)	ASTM D 5976-87	1.75	≤ 2.2
Kinematic viscosity at temperature 135 °C (Cst)	AASHTO T316-13	1853.93	≤ 3000
DSR original binder fail temperature (°C)	AASHTO T315-12	77.70	Min 70
Residue testing of TFOT (SNI-06-2440-1991) or RTFOT (SNI-03-6835-2002) results:			
Weight loss (TFOT), (%)	SNI 06-6835-2002	0.032	≤ 0.8
Penetration at 25 °C after RTFOT (%)	SNI 2456:2011	69.78	≥ 54
Ductility after RTFOT (cm)	SNI 2432:2011	97.33	≥ 50
Temperature that produces dynamic shear (G * sinδ) at oscillations of 10 rad/sec 2.2 kPa (°C)	AASHTO T 315-12	74,10	Min 70
Fresh asphalt residue after PAV at a temperature of 100 °C and a pressure of 2.1 Mpa:			
Temperature that produces dynamic shear (G * sinδ) at oscillations of 10 rad/second 5000 kPa (°C)	AASHTO T 315-12	20.90	Max 31

at AMP PT. SPP is used for the work of preservation of Pejagan–Prupuk–Tegal–Boundary Banyumas/Brebes-Wangon, Central Java, of fiscal year 2021–2022. The test objects made were a mixture of concrete asphalt wearing course (AC–WC) and concrete asphalt bearing course (AC–BC), with a cylindrical shape with a diameter of 101.60 mm (4 in) and a thickness of 40 mm and 60 mm, respectively, six test objects. The temperature of the asphalt mixture is set at 170 °C, while for compaction, the temperature is 95–145 °C. Compaction is carried out with the number of collisions 2 × 75 times per plane. Furthermore, the test object is soaked in water for 24 h. After that, an indirect tensile test was carried out at a temperature of 60 °C using the Marshall test to determine the value of stability and flow.

Table 69.3 Results of recapitulation Marshall test of asphalt concrete (AC–WC and AC–BC)

y		AC–WC		AC–BC	
		Value	Specification	Value	Specification
Optimum asphalt content	%	6.01	–	5.70	–
VIM Marshall	%	4.26	3.0–5.0	4.43	3–5
VIM refusal	%	2.99	Min 2	2.87	Min 2
VMA	%	16.17	Min 15	15.68	Min 14
VFB	%	73.68	Min 65	71.75	Min 65
Stability	Kg	1361	Min 1000	1482	Min 1000
Flow	Mm	2.87	2.0–4.0	2.86	2–4
Density	Gr/cc	2.332	–	2.342	–
Marshall quotient	KN/mm	475.0		518.1	
Stability Marshall 24 h, 60 °C	%	91.47	Min 90	95.30	
Effective asphalt content	%	5.31		4.99	
Rasio Partikel pass no. 200 of effective asphalt content	%	0.95	0.6–1.6	0.92	

69.3.2 Marshall Test

Marshall testing was carried out to determine stability, plastic fatigue (flow), volume weight (density), percent voids in the mixture (VIM), percent voids filled with asphalt (VFB), percent voids between aggregate grains (VMA); and, Marshall quotient (MQ) is a measure of the resistance of the test object to deformation. The test method was carried out according to SNI 06-2489-1991, while in the calculation to determine the value of ITS and Mr. using the ASTM D4123-04 method [11]. The following are the Marshall test results of asphalt concrete (AC–WC and AC–BC) as shown in Table 69.3.

69.3.3 Indirect Tensile Strength (ITS)

Indirect tensile strength is a method to determine the tensile strength of asphalt concrete that occurs in the field. The asphalt pavement layer must experience two loadings, namely compressive loads and tensile loads. It is very difficult to get the tensile force loading that occurs in the field, so the appropriate method to determine the tensile strength in asphalt concrete is to use the laboratory indirect tensile strength method. The value of indirect tensile strength can be calculated using Eq. (69.1) below,

$$\text{ITS} = \frac{2 \times P}{\pi \times d \times h} \tag{69.1}$$

where ITS is an indirect tensile strength value (N/mm^2)essary, *P* stability value (N), *h* test object height (mm), and *d* test object diameter (mm). Indirect tensile test is a property of asphalt tensile strength failure test that is useful in predicting potential crack conditions. A good pavement layer is able to withstand the maximum load so that cracks can be prevented. The test object for the indirect tensile test is cylindrical in shape which is subjected to compression loading with two pressure plates.

69.3.4 Modulus Resilient (Mr)

The resilience modulus (Mr) is a measure of the stiffness of a material, which is an estimate of the modulus of elasticity (*E*). Reference [6] the modulus of elasticity is the stress divided by the strain using a slowly applied load. Meanwhile, the resiliency modulus is the stress divided by the strain for the load carried out quickly in accordance with that experienced by the pavement. The magnitude of the applied load in the Mr test should be selected as part of the indirect tensile strength (ITS) of the asphalt mixture [12, 13]. The static ITS of a given specimen was determined using the procedure specified in Ref. [6]. The resilience modulus which is also called the elasticity of a material in the indirect tension method can be calculated using Eq. (69.2),

$$\mathrm{Mr} = \frac{P \times (\mu + 0.27)}{t \times \Delta H} \tag{69.2}$$

where Mr is the resilient modulus (Mpa), *P* repeated load (N), is the Poisson ratio, *t* is the specimen thickness (mm) and *H* is the horizontal deformation (mm) [14]. The Poisson ratio in Eq. (69.2) is generally set at 0.35. Table 69.4 presents the result of indirect tensile strength (ITS) and resilient modulus (Mr).

Table 69.4 Results of indirect tensile strength (ITS) and modulus resilient (Mr)

No	AC–WC asphalt content 6.01%		AC–BC asphalt content 5.7%	
	ITS (Kpa)	Mr (Mpa)	ITS (Kpa)	Mr (Mpa)
1	2.244	75.280	2.346	78.702
2	2.167	75.310	2.549	88.601
3	2.167	72.714	2.422	81.268
4	2.040	64.021	2.346	73.624
5	1.989	64.501	2.295	74.424
6	1.989	64.501	2.295	74.424

69.4 Result and Discussions

The results of the Marshall test of the specimens for AC–WC and AC–BC asphalt concrete on each specimen showed different variations. Indirect tensile strength (ITS) and resilient modulus (Mr) values are calculated by Eqs. (69.1) and (69.2). Table 69.3 presents the average value of the Marshall test on each asphalt mixture. From the analysis, it is known that the indirect tensile test (ITS) and resilience modulus (Mr) for the asphalt concrete mixture (AC–WC) are maximum at 2244 kPa and 75,280 Mpa, while for AC–BC are 2549 kPa and 88,601 Mpa. Table 69.4 shows the varying values of ITS and Mr; the average value of ITS and Mr for the asphalt concrete mixture AC–WC and AC–BC is the value of ITS 1991 kPa, Mr 63,777 Mpa and ITS 2278 kPa, Mr 72,733 Mpa. In the AC–WC asphalt mixture, there are similarities in ITS values but different Mr; this condition occurs in test objects no. 2 = 3, while in AC–BC, only test objects no 1 = 4. The factor that affects the similarity is the stability value experienced the same.

The relationship between ITS and Mr values on AC–WC and AC–BC pavements shows that the magnitude of the tensile force value that occurs means that the resilient modulus is also large. Test object numbers 5 and 6 tend to be low, namely 1989 kPa with a resilient modulus of 64,501 Mpa compared to specimen number 4 of 64,021 Mpa; this is influenced by the amount of deformation experienced. The value of R = 0.9068 for AC–WC and R = 0.9142 for AC–BC as shown in Fig. 69.2. The results of the next analysis show that the Mr design value for AC–WC and AC–BC is 63777 Mpa and 72,733 Mpa, indicating that the resilient modulus of AC–BC is greater than that of AC–WC, and it is influenced by the difference in the maximum size of aggregate in the asphalt concrete mixture. And, on both layers, the pavement is in accordance with the Mr in the initial design, so that the performance is expected to match the service life.

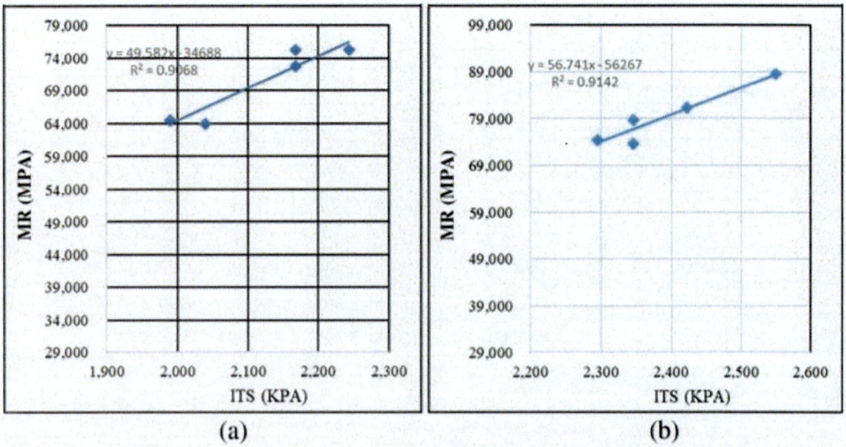

Fig. 69.2 Relation of the ITS and Mr of AC–WC (**a**) and AC–BC (**b**)

69.5 Conclusion

Modulus resilient is a parameter that determines the strength of the asphalt mixture, influenced by static loading and plastic deformation. And, the high value of indirect tensile strength generally indicates the value of the resilient modulus, so it can be said that the asphalt mixture has high strength. The ITS value in the AC–WC asphalt mixture is close to the ITS value in the AC–BC asphalt mixture, but has a different resilient modulus value. This is due to the static loading (Marshall stability) experienced by each test object which is the same but the deformation value is different. In the asphalt concrete mixing process, AC–WC uses an effective asphalt content of 6.01% for all test objects, while AC–BC is 5.70%. And, maximum resilient modulus value for asphalt mixture AC–WC = 75,310 Mpa and AC–BC = 88,601 Mpa. The results of the resilient modulus analysis prove that the resilience modulus of the AC–WC asphalt mixture is smaller than that of AC–BC, so that the compaction or field trial results can be used as a reference for implementation in the field. And, stability and flow test results in the Marshall test greatly affect the resilient modulus value. The application of the Marshall test is still feasible to use in determining the resilient modulus, although there are other, more detailed methods in finding the resilient modulus of a material.

Acknowledgements The authors would like to thank the National Road Planning and Supervision Work Unit and the Regional National Road Implementation Unit for Region 1.3 Central Java Province, The National Road Implementation Center Central Java—DIY Yogyakarta, The Directorate General of Highways, The Ministry of Public Works and Public Housing. Special thanks to those who contributed to this project directly or indirectly.

References

1. Munirwansyah, S.M. Saleh, Y. Hayati, T. Tasman, *Modulus resilientanalysis of flexible pavement AC–WC and AC–BC using asphalt penetration 60/70 from Marshall testing results*, no. 8, pp. 69–74 (2016). http://www.icesred.unsyiah.ac.id/proceedings/8_Munirwansyah_et_al_2016_69-74.pdf
2. G. Shafabakhsh, A. Tanakizadeh, Evaluation of resilient behavior of flexible pavement asphalt layers. Mater. Struct. Constr. **49**(7), 2829–2840 (2016). https://doi.org/10.1617/s11527-015-0689-0
3. F. Canestrari, *Advanced Testing and Characterization*, vol. 2005 (2005)
4. E. Chailleux, C. De La Roche, J.M. Piau, Modeling of complex modulus of bituminous mixtures measured in tension/compression to estimate secant modulus in indirect tensile test. Mater. Struct. Constr. **44**(3), 641–657 (2011). https://doi.org/10.1617/s11527-010-9655-z
5. X. Carbonneau, Y. Le Gal, P. Bense, Evaluation of the indirect tensile stiffness modulus test (2003)
6. X. Carbonneau, Y. Le Gal, Comparative test on indirect tension modulus test. Adv. Test. Charact. Bitum. Mater., 1211–1220 (2009). https://doi.org/10.1201/9780203092989.ch117
7. M. Ameri, M. Malakouti, P. Malekzadeh, Quasi-static analysis of multilayered domains with viscoelastic layer using incremental-layerwise finite element method. Mech. Time Depend. Mater. **18**(1), 275–291 (2014). https://doi.org/10.1007/s11043-013-9227-z

8. F. Zhou, E. Fernando, T. Scullion, Report FHWA/TX-10/0-5798-2: Development, calibration, and validation of performance prediction models for the texas m-e flexible pavement design system, vol. 7, no. 2, p. 216 (2010)
9. Kementerian Pekerjaan Umum dan Perumahan Rakyat, "Spesifikasi Umum 2018 rev.2, Seksi 6.3," no. Revisi 2, p. 19620221 (2018)
10. D. ASSHTO TP31-96, Wahington, American Association of State Highway and Transportation Officials, Washington, DC (1996)
11. A.T. Dachlan, M. Sjahdanulirwan, Kajian Pengaruh Modulus Resilien Dan Kepadatan (a study of effect the resilient modulus and refusal density of hot mix asphalt pavement on the strength and durability). J. Pusjatan (2012)
12. M. Elseifi et al., Accuracy of current complex modulus selection procedure from vehicular load pulse: NCHRP project 1–37A mechanistic-empirical pavement design guide. Transp. Res. Rec. **2087**, 81–90 (2008). https://doi.org/10.3141/2087-09
13. Asphalt Institute "The asphalt binder handbook—MS-26", p. 223 (2011)
14. W. C. ASTM D7369-11, ASTM International, "Standard Test Method for Determining the Resilient Modulus of Bituminous Mixtures by Indirect Tension Test", no. 2, (2011). https://doi.org/10.1520/D7369-11.2

Chapter 70
Liquid Phase Synthesis of Na_3SbS_4 Solid Electrolyte

Tran Anh Tu and Nguyen Huu Huy Phuc

Abstract We report the synthesis of tetragonal Na_3SbS_4 solid electrolyte by wet-chemical method. XRD revealed the peak splitting at $2\theta = 30.1$ and $35.1°$, which might origin from the defects-containing structure. A fairly high ionic conductivity of 6.0×10^{-4} Scm^{-1} at 25 °C was obtained. Molding pressure dependence of ionic conductivity of t-Na_3SbS_4 solid electrolyte was observed.

Keywords Na_3SbS_4 · Solid-state battery · Liquid phase synthesis

70.1 Introduction

Rechargeable sodium ion batteries (SIBs) show great potential as stationary energy storage systems because of the high abundance of sodium and their suitable redox potential ($E_{Na^+/Na} = -2.71$ V vs. SHE), which is just above that of lithium ($E_{Li^+/Li} = -3.04$ V vs. SHE) [1]. The discovery of Na^+ superionic conductor β-Al_2O_3 opened the way to Na–S cells and Na–$NiCl_2$ ZEBRA (Zero Emission Batteries Research Activities), which are being applied in megawatt-scale grid storage and electrical vehicles (EVs) at high temperature (about 300 °C). Therefore, finding a suitable solid electrolyte (SE) to realize SIBs at room temperature is necessary.

The search for sulfide SEs began with the development of cubic Na_3PS_4, which had a conductivity of 4.6×10^{-4} Scm^{-1} at room temperature (RT) [2]. A solid solution $0.94Na_3PS_4 - 0.06Na_4SiS_4$ showed improved ionic conductivity, 7.4×10^{-4} Scm^{-1} at RT [3]. A cubic Na_3PSe_4 exhibited high ionic conductivity of 1.16×10^{-3} Scm^{-1} at RT [4]. The large unit cell of cubic Na_3PSe_4 and high polarizability of Se were the origins of high ionic conductivity at RT and low activation energy compared with cubic Na_3PS_4.

T. A. Tu · N. H. H. Phuc (✉)
Faculty of Materials Technology, Ho Chi Minh City University of Technology (HCMUT), 268 Ly Thuong Kiet Str., Dist. 10, Ho Chi Minh City, Vietnam
e-mail: nhhphuc@hcmut.edu.vn

Vietnam National University, Linh Trung Ward, Thu Duc Dist., Ho Chi Minh City, Vietnam

M. A. A. Mohd Salleh et al. (eds.), *Proceedings of the Green Materials and Electronic Packaging Interconnect Technology Symposium*, Springer Proceedings in Physics 289,
https://doi.org/10.1007/978-981-19-9267-4_70

In the last five years, Na_3SbS_4 took the attention because of its high ionic conductivity at RT and ease of preparation. $Na_3SbS_4{\cdot}9H_2O$ (also known as Schlippe's salt) is available in nature, and its thermal treatment results in Na_3SbS_4 [5, 6]. There are two common crystal structures of Na_3SbS_4: cubic and tetragonal phases [7]. Structural analysis of the tetragonal phase showed a space group P-42_1c with the lattice parameter a, $b = 7.1597(5)$ Å and $c = 7.2906(6)$ Å. The lattice parameters of cubic Na_3SbS_4 were $a = b = c = 0.71910$ (7) nm. The tetragonal phase is stable at RT and changes to cubic phase at elevated temperature, from 100 to 130 °C.

There are three common methods to produce Na_3SbS_4: solid-state reaction at high temperature, liquid phase synthesis, and high-energy ball-milling [5, 7, 8]. t-Na_3SbS_4 is the dominant phase in the products received from solid-state reaction and liquid phase methods. c-Na_3SbS_4 is the main phase in the product obtained by high-energy ball-milling methods. Liquid phase synthesis is widely used to prepare Na_3SbS_4 and electrode composites because of the formation of ultimate contact among active materials and Na_3SbS_4 solid electrolytes [9, 10]. In this study, we report the synthesis of Na_3SbS_4 from an aqueous solution and the dependence of ionic conductivity on molding pressure.

70.2 Methodology

The following reaction occurred during synthesis procedure:

$$3Na_2S + Sb_2S_3 + 2S \rightarrow 2Na_3SbS_4$$

Na_3SbS_4's precursor samples were characterized using thermogravimetry—differential thermal analysis (TG–DTA; EVO II, Rigaku) and X-ray diffraction (XRD). TG–DTA was performed under an N_2 flow with a temperature ramp of 5 K min^{-1}. Powder XRD measurements were performed on a Rigaku Ultima IV diffractometer. The X-ray beam was generated by CuKα radiation (40 kV, 30 mA).

The temperature dependence of the total conductivity of the prepared samples was investigated using alternating-current impedance spectroscopy (SI 1260, Solatron) from 1 MHz to 10 Hz under a dry Ar flow. The samples for impedance measurements were prepared by uniaxial pressing each sample into pellets with a diameter of approximately 10.0 mm at different pressure at RT. Each prepared pellet was placed in a holder with two stainless steel rods used as blocking electrodes. The cell was then placed in a glass tube under a dry N_2 stream for temperature-dependent measurements. The temperature was gradually increased from room to 170 °C in various increments and held at each temperature for one hour prior to the impedance measurements (Fig. 70.1).

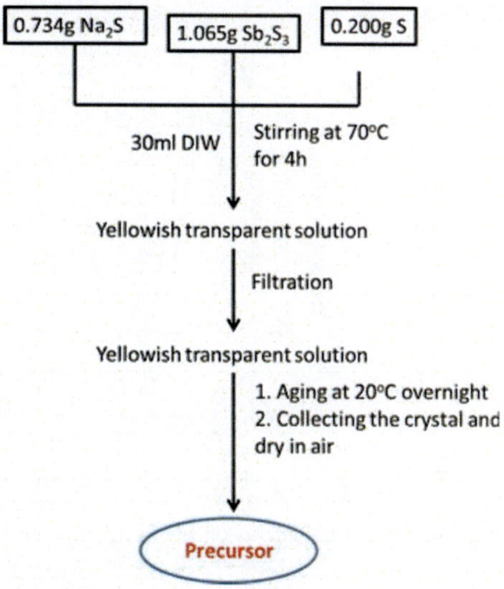

Fig. 70.1 Na_3SbS_4's precursor preparation procedure

70.3 Result and Discussions

Figure 70.2 shows the XRD patterns of the samples after air-dry for 3 and 24 h. JCPDS card number 00-043-0442 of $Na_3SbS_4{\cdot}9H_2O$ (Schlippe's salt) was also illustrated for comparison. The sample after air-dry for 3 h exhibited two intense peaks at $2\theta = 14.9$ and 31.0°, which are unassignable. The peaks' position in the XRD pattern of the sample after air-dry for 24 h was almost identical to JCPDS card number 00-043-0442 of $Na_3SbS_4{\cdot}9H_2O$. The differences in peak intensity ratios might arise from non-identical crystalline orientation.

Figure 70.3 shows the TG–DTA curves of the sample after air-dry for 24 h. TG curve exhibited a mass loss of about 33.7%, and the weight change finished at about 100 °C. The DTA curve exhibited only one endothermic peak associated

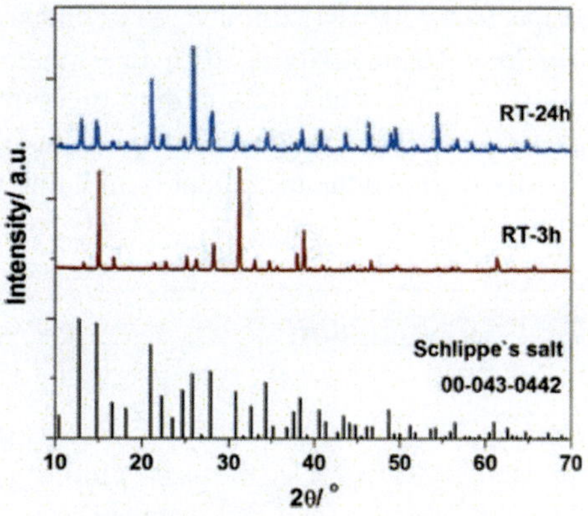

Fig. 70.2 XRD patterns of the precursors and standard $Na_3SbS_4{\cdot}9H_2O$ (Schlippe's salt)

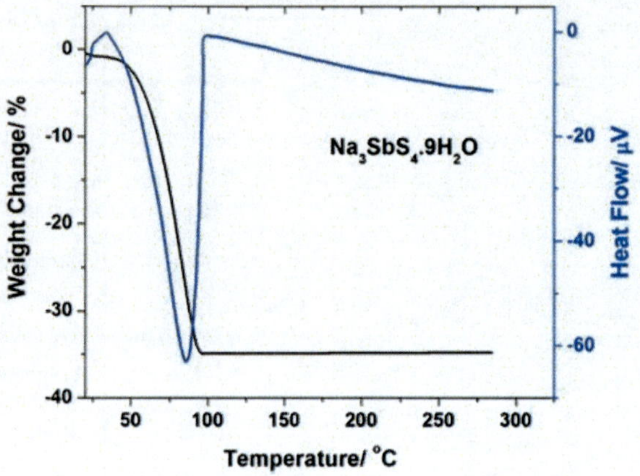

Fig. 70.3 TG–DTA curves of the precursor after air-dry for 24 h

with the weight loss in the TG curve. Therefore, water evaporation was the cause of the changes in TG–DTA curves. From the weight change, the sample formula was calculated to be $Na_3SbS_4{\cdot}9H_2O$. This result agreed with the XRD pattern. All the samples were heat treated at 100 °C under reduced pressure for 1 h prior to other characterization experiments because the mass loss in the TG curve ended before 100 °C.

Figure 70.4 illustrates the XRD pattern of the heat-treated sample and JCPDS card number 00-035-1016 of t-Na_3SbS_4 for comparison. Peak position in the sample's pattern was almost similar to the standard t-Na_3SbS_4. However, peaks at $2\theta = 30.1$ and 35.1° were split into two different peaks. The peak splitting observed in this study was in excellent agreement with t-Na_3SbS_4 prepared by wet-chemical method reported so far [9]. The peak splitting in t-Na_3SbS_4 was also observed in vacancy-contained t-Na_3SbS_4 prepared by high-temperature solid-state reaction [8].

The temperature dependences of ionic conductivity of the samples which were pelletized under different uniaxial pressure were illustrated in Fig. 70.5. The sample prepared at 330 MPa exhibited ionic conductivity at 25 °C of about 2.1×10^{-4} Scm^{-1}. The ionic conductivity at 25 °C (σ_{25}) increased to 2.5×10^{-4}, 3.2×10^{-4}, and 6.0×10^{-4} Scm^{-1} when the molding pressure increased to 403, 477, and 550 MPa, respectively. The change of σ_{25} might indicate the improvement of grain boundary resistivity due to the increment of molding pressure.

70.4 Conclusion

A t-Na_3SbS_4 solid electrolyte was successfully prepared by wet-chemical method in this study. XRD revealed the peak splitting at $2\theta = 30.1$ and 35.1°, which might origin from the defects-containing structure. A fairly high ionic conductivity of 6.0×10^{-4}

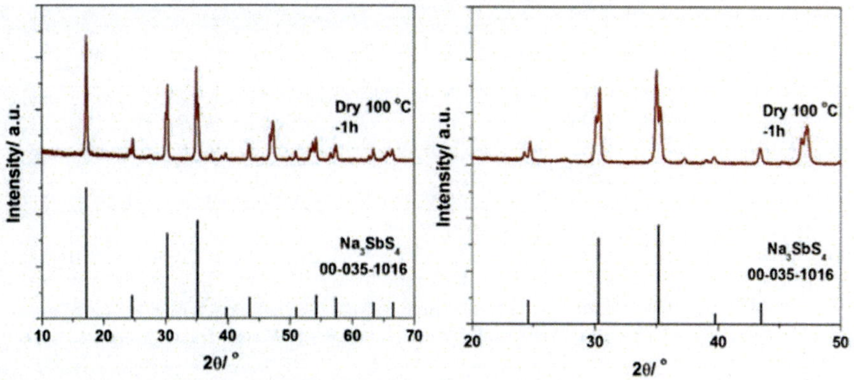

Fig. 70.4 XRD patterns of the precursor after dried at 100 °C for 1 h under reduced pressure, the tetragonal Na_3SbS_4, and their magnification (right)

Fig. 70.5 Temperature dependence of ionic conductivity of the samples which were pelletized under different uniaxial pressure

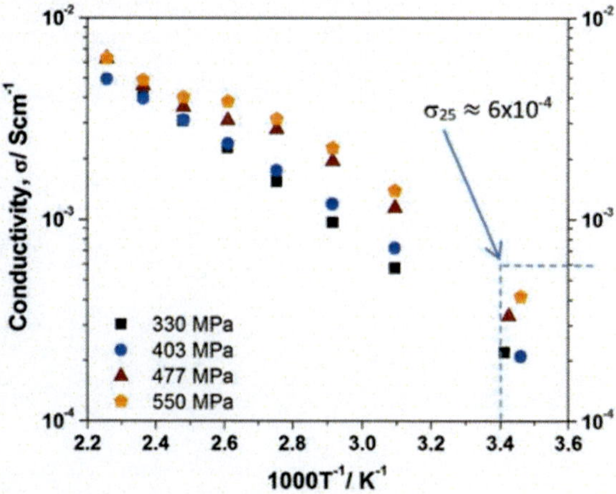

Scm^{-1} at 25 °C was obtained. Molding pressure dependence of ionic conductivity of t-Na_3SbS_4 solid electrolyte was observed.

Acknowledgements We acknowledge Ho Chi Minh City University of Technology (HCMUT), VNU-HCM, for supporting this study.

References

1. D. Buchholz, C. Vaalma, M. Weil, S. Passerini, Nat. Rev. Mater. **3**, 18013 (2018)
2. A. Hayashi, K. Noi, A. Sakuda, M. Tatsumisago, Nat. Commun. **3**, 856 (2012)
3. N. Tanibata, K. Noi, A. Hayashi, M. Tatsumisago, RSC Adv. **4**, 17120 (2014)

4. L. Zhang, K. Yang, J. Mi, L. Lu, L. Zhao, L. Wang, Y. Li, H. Zeng, Adv. Energy Mater. **5**, 1501294 (2015)
5. A. Banerjee, K.H. Park, J.W. Heo, Y.J. Nam, C.K. Moon, S.M. Oh, S.T. Hong, Y.S. Jung, Angew. Chem. **55**, 9634–9638 (2016)
6. H. Wang, Y. Chen, Z.D. Hood, G. Sahu, A.S. Pandian, J.K. Keum, K. An, C. Liang, Angew. Chem. **55**, 8551–8555 (2016)
7. D. Zhang, X. Cao, D. Xu, N. Wang, C. Yu, W. Hu, X. Yan, J. Mi, B. Wen, L. Wang, L. Zhang, Electrochim. Acta **259**, 100–109 (2018)
8. L. Zhang, D. Zhang, K. Yang, X. Yan, L. Wang, J. Mi, B. Xu, Y. Li, Adv. Sci. **3**, 1600089 (2016)
9. K.H.P.T.W. Kim, Y.E. Choi, J.Y. Lee, Y.S. Jung, J. Mater. Chem. A **6**, 840–844 (2018)
10. H. Gamo, N.H.H. Phuc, R. Matsuda, H. Muto, A. Matsuda, Mater. Today Energy **13**, 45–49 (2019)

Chapter 71
Using Geopolymer Technology to Fabricate Spray Fire Resistance Material (SFRM)

W. H. Lee, C. Y. Huang, and T. W. Cheng

Abstract High-rise buildings usually use steel construction as the main structure because of their lightweight, toughness, and superior earthquake resistance. However, one of the disadvantages of steel construction is its fire resistance ability. Therefore, protecting steel constructions from the reduction of strength affected by high temperatures has always been an important issue. Geopolymer materials have many great properties, such as high mechanical properties, acid/alkali resistance, low permeability, low carbon emissions, and fire resistance. Therefore, this study uses geopolymer materials to develop spray-applied fire-resistive materials (SFRMs) and compares the fire resistance performance between geopolymer SFRM and the commercial product. The experiment results showed that the thickness of the SFRM had the most significant effect on the improvement of fire resistance performance. In general, the fire resistance period of geopolymer SFRM with a thickness of one centimeter can reach more than 70 min, and the fire resistance ability is better than commercial SFRM. The mechanical properties and tensile strength of the geopolymer SFRM also meet the specification requirements of SFRM. According to the results, geopolymer material is expected to be used as SFRM and be practically used in building structures.

71.1 Introduction

With the development of Taiwanese cities, population density has increased, and so did the number of floors in buildings. High-rise buildings often use steel construction as the main structure because of their lightweight, toughness, and superior earthquake resistance. However, one of the disadvantages of steel construction is that its fire resistance is not as good as that of reinforced concrete or steel-reinforced concrete.

W. H. Lee (✉) · C. Y. Huang · T. W. Cheng
Institute of Mineral Resources Engineering, National Taipei University of Technology, 1, Sec. 3, Zhongxiao E. Rd., Taipei 10608, Taiwan, R.O.C.
e-mail: whlee@ntut.edu.tw

M. A. A. Mohd Salleh et al. (eds.), *Proceedings of the Green Materials and Electronic Packaging Interconnect Technology Symposium*, Springer Proceedings in Physics 289,
https://doi.org/10.1007/978-981-19-9267-4_71

Therefore, protecting steel constructions from the reduction of strength affected by high temperatures has always been an important issue.

Geopolymer materials have been developed for more than 30 years. They have received the attention of various international research institutions due to their simple production equipment, fabrication process, and other great engineering properties [1–7]. The structure of geopolymer material is similar to the amorphous or semi-crystalline structure of the zeolite, and it has good fire resistance [8] and thermal stability. Therefore, this research uses geopolymer materials to develop spray-applied fire-resistive materials (SFRMs) and compares the fire resistance performance between geopolymer SFRM and the commercial one.

In this study, metakaolin, GGBF slag, coal fly ash, and potassium-based alkaline solution were used to form a geopolymer slurry, in which vermiculite [9] and sodium percarbonate (foaming agent) were added to increase the porosity of the material and reduce thermal conductivity to increase fire resistance performance and heat insulation capacity. This research aims to develop geopolymer-based SFRM and clarify the correlation between the geopolymer composition and fire resistance performance.

71.2 Experimental

The raw materials for geopolymer synthesis were metakaolin (MK), ground granulated blast furnace slag (GGBFS), and coal fly ash (CFA). The metakaolin was produced by Pei Long Enterprise Co., Ltd. The GGBFS was produced by China Steel Corporation. The CFA was produced by Hsinta Power Plant. The vermiculite was produced by Nan Hai Gardencing Co. All the chemical composition of the starting materials were analyzed by X-ray fluorescence (XRF) and were shown in Table 71.1. Wollastonite and sodium percarbonate were also added as functional materials. Wollastonite has fiber-like crystal and features flame retardant, thermal insulation, and thermal stability. It can also improve the toughness of geopolymers [10]. Sodium percarbonate was used as a foaming agent to increase the porosity in geopolymers and thus improve the thermal insulation. Alkali solution was prepared by mixing KOH, potassium silicate, and sodium aluminum oxide. The concentration of KOH was controlled at 6, 8, and 10 M. The SiO_2/Na_2O molar ratio was controlled at 1.28. The SiO_2/Al_2O_3 molar ratio was controlled at 50. The experimental procedure and test parameters are shown in Fig. 71.1 and Table 71.2. This research also compared the fire resistance performance of geopolymer SFRM with the commercial one. The commercial SFRM is produced by Isolatek. The main components include gypsum, lime, vermiculite, and perlite. The fire resistance test in this research follows ISO 834–1: Fire resistance tests–elements of building construction. Therefore, the heating curve of the furnace is followed by Eq. (71.1).

$$T = 345 \times \log(8t + 1) + 20 \tag{71.1}$$

Table 71.1 Raw materials chemical composition

wt%	Composition						
	SiO_2	CaO	Al_2O_3	TiO_2	Fe_2O_3	Other	L.O.I.
Metakaolin	55.2	–	39.6	2.4	0.6	0.1	1.5
GGBFS	27.5	57.6	10.8	1.1	0.5	2.5	1.1
CFA	63.0	5.8	17.3	1.7	7.9	4.3	3.0
Vermiculite	40.0	5.9	11.2	–	23.3	19.6	4.4

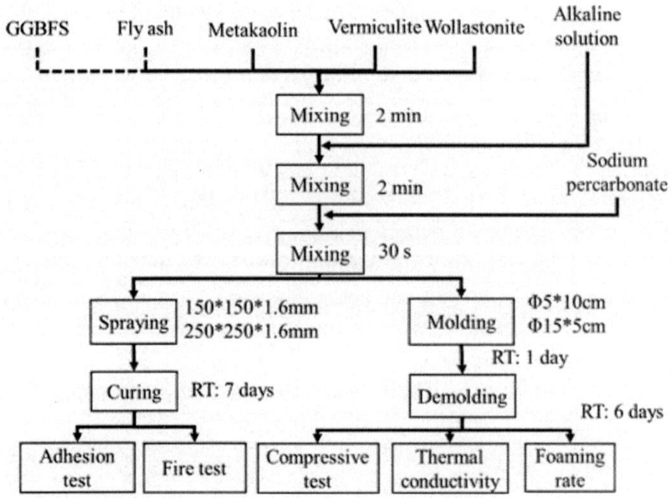

Fig. 71.1 Overall experiment procedures

Table 71.2 Experiment parameters

Experiment No.	GGBFS	CFA	MK	Vermiculite: Binder	KOH(M)	L/S	Addition	
							W	NC
M	–	–	100	1:0.8	6	1.21	10	5
S25	25	–	75					
FA25	–	25	75					
AS6	–	25	75		*6*	*1.18*		
AS8					*8*			
AS10					*10*			
AS6-1.0 cm					6	1.18		
AS6-2.0 cm								
AS6-3.0 cm								

CFA Coal fly ash; *MK* Metakaolin; *L/S* Liquid/solid ratio; *W* Wollastonite; *NC* Sodium percarbonate

71.3 Results and Discussions

71.3.1 Effect of Binder Ratio

The effect of the binder ratio on mortar viscosity, foaming rate, and thermal conductivity is illustrated in Fig. 71.2. The results showed that the foaming rate was highly affected by mortar viscosity except for the CFA replacement group. The foaming rate rises when the viscosity is higher. The CFA replacement group was different, however. The foaming rate increased sharply even though the viscosity is lower. This phenomenon is caused by decreasing friction force between particles led by the sphere shape of the CFA particle. Thus, the thermal conductivity became much lower when CFA partially replaced metakaolin.

The results of the fire resistance test are illustrated in Fig. 71.3. The final temperature of the reverse side of M, S25, and FA25 is 481, 486, and 461 °C, respectively. From here, it can observe that the thermal conductivity shows a high correlation with the foaming rate, and the foaming rate is highly affected by the binder ratio. Thus, we concluded that using CFA partially replacing metakaolin can reach optimal fire resistance performance. Furthermore, when comparing the fire resistance performance of geopolymer SFRM with the commercial one, the performance of geopolymer SFRM is far better than the commercial one because the commercial one tends to spall when exposed to high temperatures for a period of time.

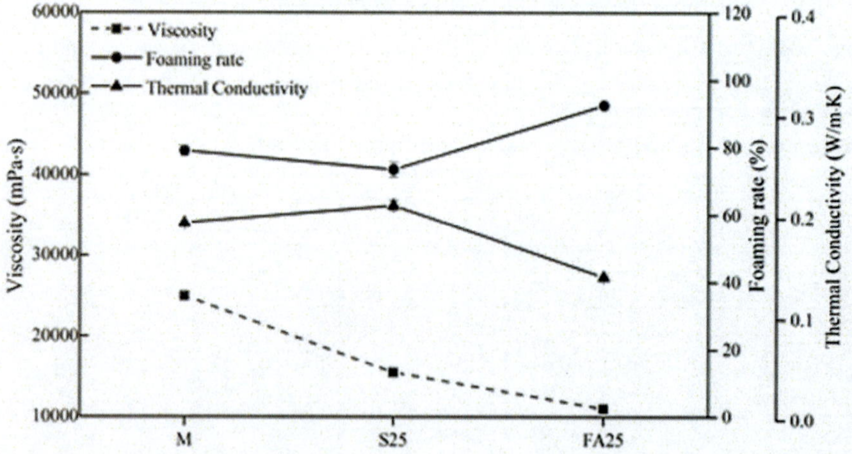

Fig. 71.2 Effect of binder ratio on mortar viscosity, foaming rate, and thermal conductivity (alkali solution $SiO_2/Na_2O = 1.28$, $SiO_2/Al_2O_3 = 50$, $L/S = 1.21$)

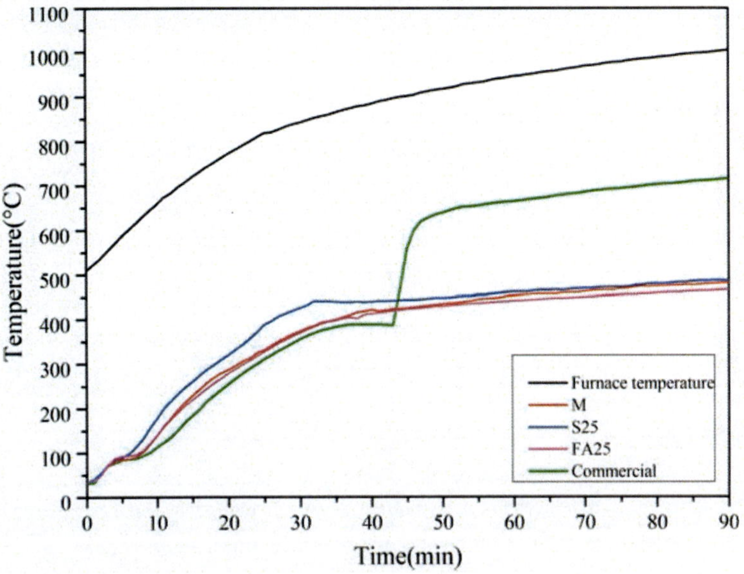

Fig. 71.3 Effect of binder ratio on fire resistance performance (thickness 1 cm)

71.3.2 Effect of KOH Concentration

The effect of KOH concentration on mortar viscosity, foaming rate, and thermal conductivity is illustrated in Fig. 71.4. The results showed that the mortar viscosity was also affected by KOH concentration. A higher concentration of KOH can dissolve more silicone gel and aluminum gel, leading to a higher mortar viscosity and increasing the foaming rate slightly. However, the thermal conductivity increased while the KOH concentration increased. The reason is that a higher concentration of KOH could improve the geopolymeric reaction, and the structure of the geopolymer will be densified.

The results of fire resistance test are illustrated in Fig. 71.5. The final temperature of the reverse side of AS6 and AS8 is around 440 °C. However, the final temperature of AS10 is 475 °C, which is higher than AS6 and AS8. These results indicated that a higher concentration of KOH will densify the geopolymer structure and lead to poor fire resistance performance. Therefore, the optimal selection of KOH concentration will be 6 M.

71.3.3 Effect of Coating Thickness

This section extended the experiment in previous, enhancing the thickness of the coating to 2 and 3 cm, then comparing their fire resistance performance. The results

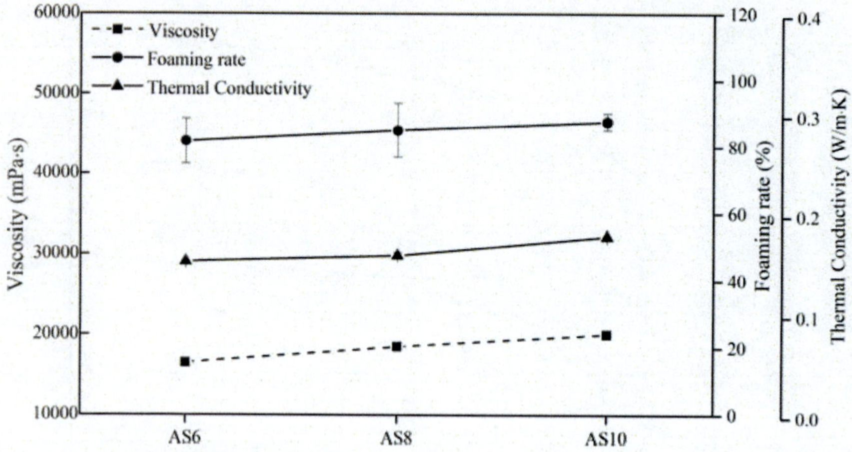

Fig. 71.4 Effect of KOH concentration on mortar viscosity, foaming rate, and thermal conductivity (alkali solution $SiO_2/Na_2O = 1.28$, $SiO_2/Al_2O_3 = 50$, L/S = 1.18)

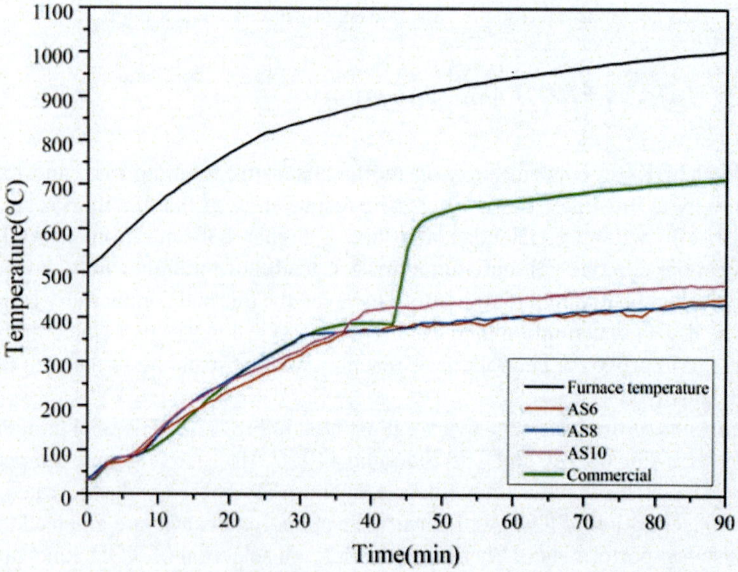

Fig. 71.5 Effect of KOH concentration on fire resistance performance (1 cm)

of the fire resistance test of geopolymer SFRM are illustrated in Fig. 71.6. The results of fire resistance test of commercial SFRM are shown in Fig. 71.7. The results in Fig. 71.6 indicate that the temperature of the reverse side of AS6-1.0 cm, AS6-2.0 cm, and AS6-3.0 cm is 468, 410, and 335 °C, respectively. This means for each centimeter increase in thickness, the temperature of the reverse side decreases by 15%.

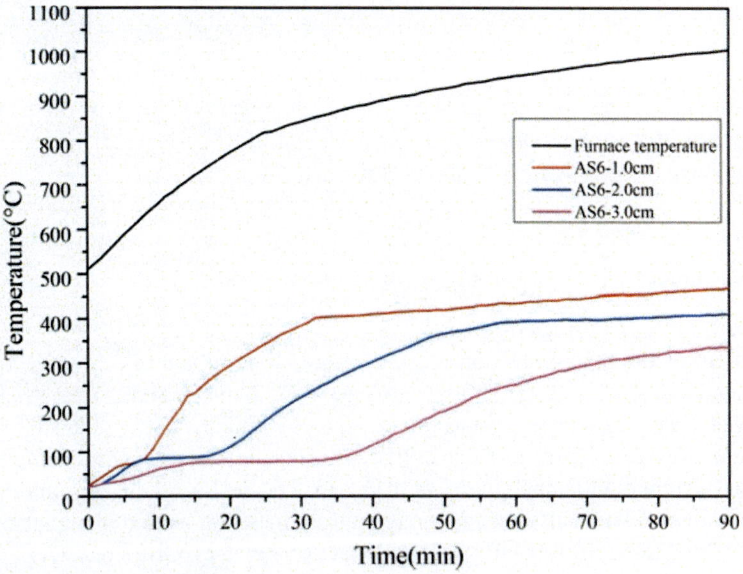

Fig. 71.6 Effect of coating thickness on fire resistance performance (geopolymer)

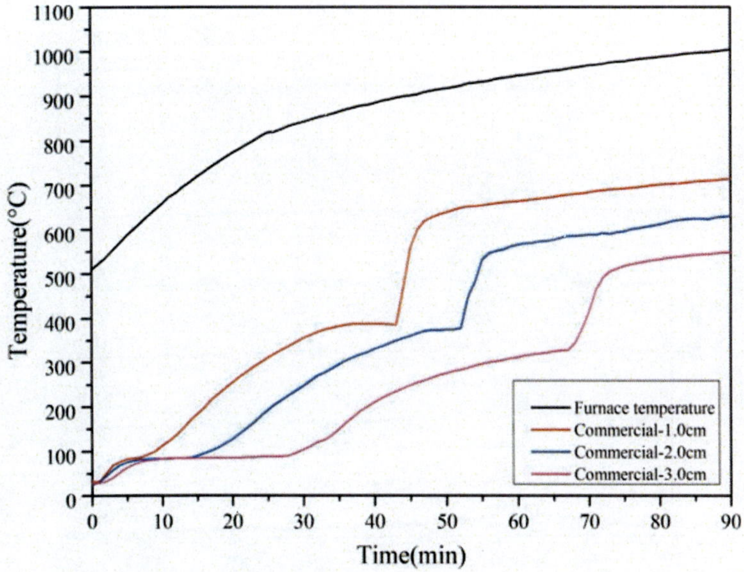

Fig. 71.7 Effect of coating thickness on fire resistance performance (commercial)

On the other hand, the results in Fig. 71.7 also indicate that increasing the coating thickness will effectively improve the fire resistance performance. However, it is obvious that all the commercial specimens experienced a rapid rise in temperature. This is due to the spalling of the commercial SFRM when experiencing high temperatures for an extended period of time.

71.3.4 Micro-structure Analysis

The effect of the binder ratio on crystal structure before and after the fire test as illustrated in Fig. 71.8. Before the fire test, the wollastonite or quartz phase could be observed. However, leucite crystals formed in FA25 while $KAlSiO_4$ or kalsilite crystals formed in S25. This result suggests that adjusting the binder proportion might also lead to different high-temperature crystal phases. Leucite is a relatively stable crystal phase compared with kalsilite, and it might afford better protection in high temperatures. ^{29}Si NMR analysis results indicated that using either GGBFS or CFA partially replacing metakaolin could improve the geopolymeric reaction, and therefore, the $Q^4{}_4$ and $Q^4{}_3$ bonding was increased. After fire test, most of the microstructure shifted to $Q^4{}_2$ bonding because of the crystallization process (Fig. 71.9).

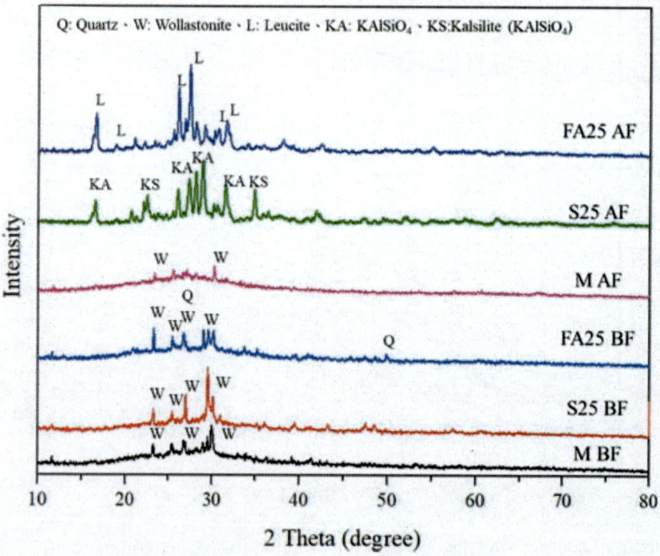

Fig. 71.8 XRD analysis on different binder ratio. BF: before fire test; AF: after fire test (alkali solution $SiO_2/Na_2O = 1.28$, $SiO_2/Al_2O_3 = 50$, $L/S = 1.21$)

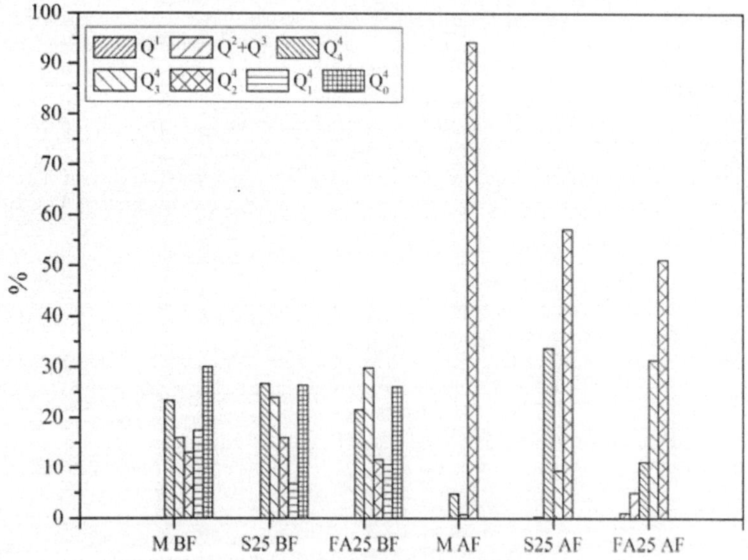

Fig. 71.9 ^{29}Si NMR analysis on different binder ratios. BF: before fire test; AF: after fire test (alkali solution $SiO_2/Na_2O = 1.28$, $SiO_2/Al_2O_3 = 50$, L/S = 1.21)

71.4 Conclusion

This research focuses on the effect of binder ratio, the concentration of KOH, and the coating thickness of geopolymer SFRM to fire resistance performance. It can conclude from the experiment results that are partially replacing metakaolin with CFA (25%) could increase the foaming rate and improve the fire resistance performance. Moreover, it can also enhance the geopolymer structure and change the crystal phase at high temperatures, as proven by XRD and NMR analyses. However, a higher concentration of KOH would lower the fire resistance performance. Thus, the optimal composition in this research is AS6. On the other hand, we can also increase the SFRM coating thickness to improve the fire resistance performance depending on the demand for fire resistance time.

References

1. T.A. Aiken, J. Kwasny, W. Sha, M.S. Soutsos, Effect of slag content and activator dosage on the resistance of fly ash geopolymer binders to sulfuric acid attack. Cem. Concr. Res. **111**, 23–40 (2018)
2. J. Aliques-Granero, M.T. Tognonvi, A. Tagnit-Hamou, Durability study of AAMs: sulfate attack resistance. Constr. Build. Mater. **229**, 117110 (2019)
3. M. Lahoti, K.H. Tan, E.H. Yang, A critical review of geopolymer properties for structural fire-resistance applications. Constr. Build. Mater. **221,** 514–526 (2019)

4. Y. Liu, P. Su, M. Li, Z. You, M. Zhao, Review on evolution and evaluation of asphalt pavement structures and materials. Sci. Dir. **7**, 573–599 (2020)
5. S.K. Shill, S. Al-Deen, M. Ashraf, W. Hutchison, Resistance of fly ash based geopolymer mortar to both chemicals and high thermal cycles simultaneously. Constr. Build. Mater. **239**, 117886 (2020)
6. M. Vafaei, A. Allahverdi, P. Dong, N. Bassim, Acid attack on geopolymer cement mortar based on waste-glass powder and calcium aluminate cement at mild concentration **193**, 363–372 (2018)
7. P. Cong, Y. Cheng, Advances in geopolymer materials: A comprehensive review. J. Traffic Transp. Eng. **8**, 283–314 (2021)
8. J. Davidovits, Fire proof geopolymeric cements. Géopolymère '99 Proceedings, pp. 165–170 (1999)
9. V. Medri, E. Papa, M. Mazzocchi, L. Laghi, M. Morganti, J. Francisconi, E. Landi, Production and characterization of lightweight vermiculite/geopolymer-based panels. Mater. Des. **85**, 266–274 (2015)
10. H. Li, X. Peng, J. Li, L. Li, D. Hu, Y. Xiang, L. Han, Z. Xu, Preparation and fireproofing performance of the wollastonite-metakaolin-based geopolymer foams. Mater. Lett. **307**, 131020 (2022)

Chapter 72
Determination of Municipal Solid Waste Composition, Generation Rate and Its Recyclable Potential in Penang, Malaysia—A Statistical Approach

Yuganantheni K. Marippan, Mohamad Anuar Kamaruddin, Rasyidah Alrozi, and Mohd Mustafa Albakri Abdullah

Abstract The municipal solid waste (MSW) generation rate in Malaysia is increasing annually due to population growth, industrialization, advancement in technology, economic growth, and lifestyle. Landfilling is the main method of solid waste management, while recycling is still underrated in Malaysia. Penang state has initiated a "Waste Segregation at Source" (WSAS) campaign in 2016 to reduce MSW disposal in Pulau Burung landfill to cultivate recycling habits among Penangites. However, there are no studies being conducted to measure the effectiveness of this campaign. Hence, this study was conducted to determine the effectiveness of the WSAS campaign before (2011 until 2015) and after (2016 until 2020) implementation. The objectives of the study are to evaluate MSW generation, the recyclable potential based on the composition, and its relationship in Penang by using statistical analysis and the SWOT-AHP model from 2011 until 2020. Secondary data were obtained from Penang Island City Council (MBPP), Seberang Perai City Council (MBSP), and journals. SPSS was used to carry out the statistical analysis by using ANOVA one way, correlation, and regression. The analytic hierarchy process (AHP) model was conducted through three series of questionnaire to obtain expert opinion and mathematical calculation to determine the degree of SWOT on MSW generation and recycling in Penang. The outcome of this study showed that both the MSW generation and the recycling potential increased in the state for the past 10 years. Besides, the effectiveness of WSAS showed positive results in terms of recycling rates before (2011–2015) and after (2016–2020) the campaign was implemented.

Y. K. Marippan · M. A. Kamaruddin (✉)
Environmental Technology Division, School of Industrial Technology, Universiti Sains Malaysia, 11800, Gelugor, Penang, Malaysia
e-mail: anuarkamaruddin@usm.my

M. A. Kamaruddin · M. M. A. Abdullah
Center of Excellence Geopolymer and Green Technology, Universiti Malaysia Perlis, 01000 Perlis, Malaysia

R. Alrozi
School of Chemical Engineering, College of Engineering, Universiti Teknologi MARA(UiTM), Cawangan Pulau Pinang Kampus, 13500 Permatang Pauh, Pulau Pinang, Malaysia

M. A. A. Mohd Salleh et al. (eds.), *Proceedings of the Green Materials and Electronic Packaging Interconnect Technology Symposium*, Springer Proceedings in Physics 289,
https://doi.org/10.1007/978-981-19-9267-4_72

Keywords Solid waste · Generation rate · Sustainable · Statistical analysis

72.1 Introduction

Solid waste generation has the greatest environmental impact without appropriate management, which poses a significant concern for local authorities globally, especially in most of the developing countries [1, 2]. Population, urbanization, and economic growth have resulted in the accumulation of solid waste, which could reduce the land availability for waste disposal. 1.3 billion metric tons of MSW are generated globally per year, and it is expected to rise to approximately 2.2 billion tons by 2025 [3]. The world population is expected to increase by more than 37.3 million by 2030 [4]. This statement is particularly evident in urban areas, especially in Africa and Asia, which result in increased MSW generation [3]. Remarkable economic growth over the past 20 years and urbanization in Malaysia are expected to raise the population to 34.4 million by 2030 [4, 5]. The per capita MSW generation among Malaysians is higher than other Asian countries such as Myanmar, Vietnam, Laos, Indonesia, and the Philippines because of development in socioeconomic status and lifestyle, which changed the composition of MSW to become more complex. Approximately, 80% of MSW disposed of at landfills is recyclable [6]. Recycling is one of the sustainable solid waste management strategies to reduce disposal and transport costs, increase the lifespan of a landfill, and reduce degradation of the environment [2]. Malaysia is still lacking in the implementation of recycling. Most people don't separate their organic and inorganic waste at home, so they end up with a mix of both.

Malaysia's major mode of MSW management is landfilling, which is done by most of the states. Due to rapid development and land constraints, the Penang state government is in the process of transitioning from its current landfilling approach to a more sustainable solution. The Waste Segregation at Source (WSAS) approach involves the process of separating solid waste at source into recyclable waste and general waste. The goals are to reduce pollution, maintain ecological balance, increase recycling rates, reduce waste generated at landfill, extend the life of the Pulau Burung landfill (which disposes of approximately 1800 tons of municipal and non-hazardous industrial solid waste each year), and reduce costs [7]. According to the Penang Green Council official Web site on WSAS, one Penangite can generate about 1.1 kg of MSW daily, which ends up in Pulau Burung Landfill, with approximately 60% of recyclable waste to be recovered through committed segregation. There are no studies conducted to measure the effectiveness of the WSAS campaign organized by the state government. Therefore, this study is proposed to gather historical data for waste management, which can be used as a benchmark to measure the effectiveness of WSAS in Penang before (2011–2015) and after (2016–2020) the implementation of the campaign. Thus, this study aims to evaluate MSW generation, the recyclable potential, and its relationship in Penang by using statistics and the SWOT-AHP model from 2011 until 2020.

72.2 Methodology

Pulau Pinang or mostly recognized with the name of Penang state is comprised of Penang Island and Seberang Perai. The local authorities responsible for solid waste management in Penang Island is MBPP and Seberang Perai is MBSP. It is mandatory for collected waste from Penang Island to be sent to Batu Maung Transfer Station operated by a private contractor (Eurasia Express Sdn. Bhd.) before disposal at Pulau Burung landfill. Pulau Burung is the only approved MSW landfill in the state, which is in the Byram Reserves, Nibong Tebal, with an area of 62.4 ha. Currently, 33 ha are operating, which can receive 1800 tons of MSW and non-hazardous industrial solid waste daily. 600 tons of solid waste come from Penang Island itself [7]. Besides, there are also 20 hectares of "landfill" disposal sites at Jelutong Solid Waste Transfer Station that receive waste mainly of construction and demolition materials, garden waste, and other bulky waste materials. The collection areas for MBSP are the northern, central, and southern zones. Ampang Jajar Transfer Station was a landfill until 2001, which was then converted to a transfer station upon reaching capacity. Wastes from the Northern and Central zones are transported to Ampang Jajar transfer station to be compacted before being disposed of at Pulau Burung, while wastes from the Southern zone will be transferred directly to the landfill.

The MSW generation and recycling rate were based on secondary data obtained from municipalities, while the composition was evaluated based on previous studies. ANOVA one way, correlation, and regression were evaluated by using SPSS to determine the significance between waste generation and recycling potential. Expert opinion from both municipalities was taken into consideration for the SWOT-AHP method via three series of questionnaires. Previous research on SWOT-AHP utilized the opinions of a group of 4 to 40 experts [8–12]. AHP is used to determine relative priorities based on pairwise comparisons by assigning a number from a comparison scale to represent the relative importance of the items per SWOT group [8, 13]. SWOT factors from a previous study were used as the template to prepare Questionnaire 1. It is suggested to have no more than 10 items for each SWOT factor [11]. Hence, the questionnaire was constructed in the "Four-by-Four SWOT Matrix." A 24 pairwise comparison was carried out via Questionnaire 2 to determine the weight of each factor based on Saaty's 9 value scale of relative importance (Table 72.1). The consistency ratio (CR) (Eq. 72.1) should be less than 0.1 for a consistent survey with accurate results. Equation 72.2, Tables 72.1 and 72.2 illustrate the necessary information required to compute CR. The final questionnaire was created based on six pairwise comparisons with the highest criteria weight, to evaluate the overall priority by multiplying the factor weight by the group weight. Figure 72.1 shows the summary of the methodology used in this study.

$$\text{Consistency Ratio (CR)} = (\text{Consistency Index (CI)})/(\text{Random Index (RI)}) \quad (72.1)$$

$$\text{Consistency Index (CI)} = ((\lambda \max - n))/((n - 1)) \quad (72.2)$$

Table 72.1 Saaty's 9 value scale of relative importance [8, 13]

Numerical value	Explanation
1	Importance was given equally to two criteria
3	Importance was given moderately favoring one criterion over another
5	Importance was given strongly favoring one criterion over another
7	Importance was given very strongly favoring one criterion over another
9	Extreme or absolute importance was given to one criterion compared to another
2, 4, 6, 8	Intermediate values between two criteria which require a compromise

Table 72.2 RI for various matrix sizes (*n*) [13]

n	1	2	3	4	5	6	7	8	9
RI	0.00	0.00	0.58	0.90	1.12	1.24	1.32	1.41	1.45

72.3 Results and Discussion

Figure 72.2 and Table 72.3 show that the MSW generation in Penang state was similar with a small fluctuation from 2011 until 2016. There is a slight increase in waste generation in 2017 and 2020. Penang is leading in economic activity, especially manufacturing, services, and construction, besides Selangor and Johor in Peninsular Malaysia. Thus, rapid economic growth, rural–urban migration, and lifestyle changes experienced in Penang state have resulted in an almost steady increase in MSW generation with a slight fluctuation [14]. In conclusion, most Asian countries' MSW generation is influenced by population, locality area, and economic activities [7, 14, 15].

The recycling rate in Penang shows a gradual increase. Approximately, 26% of MSW generated in Penang can be recycled. The recyclable waste materials are plastic, paper, metal, and tetra pak. Plastic waste dominates the recyclable fraction as it is widely used in daily life. The implementation of "No Styrofoam Campaign" is one of the reasons behind the accumulation of plastic and paper waste in Penang, as the state government banned polystyrene food containers [16]. Penang's recycling rate is based on cooperation between recycling businesses and local authorities [17]. A simple survey was conducted by Moh and Manaf in 2017 to study source separation and recycling. The result showed that most Malaysians agree that lack of space at home to place recyclables is one of the reasons for non-recycling. The number of high-rise buildings in Penang has the greatest percentage compared to other types of landed houses. This indirectly supports that Penangites have fewer opportunities to carry out source separation and recycling at home due to a lack of space. Recycling plays an important role in reducing dependency on Pulau Burung Sanitary Landfill as it is the second largest category of waste in Penang.

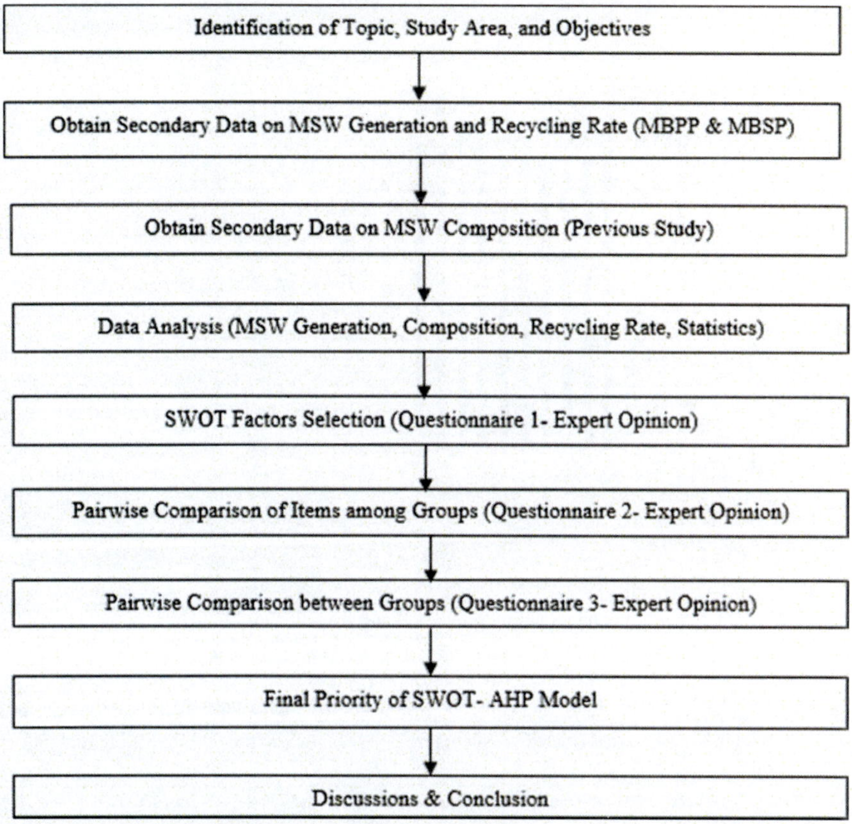

Fig. 72.1 Summary of methodology

ANOVA One Way was used to prove that there are two different variables that have a significant relationship with each other. In this case, data analysis shows a significant relationship between MSW generations and when the P-value is 0.05. A correlation coefficient is used to determine the strength of the linear relationship between two variables [1]. At a 95% confidence interval, there is a significant relationship between MSW generation and recycling when the beta value (b value) shows a positive relationship, which indicates that when MSW increases, recycling also increases. Also, the beta value shows how much the amount of trash produced affects the overall recycling rate in each city and state. The recycling rate depends about 86% ($R^2 = 0.858$) on MSW generation in Penang.

In general, information on waste composition is lacking due to unknown waste stream pathways in the environment, various MSW management, and a limited social survey [2]. Data on MSW composition cannot be obtained as the municipalities do not carry out waste characterization. Thus, the MSW composition of the state was obtained by referring to the MSW composition percentage in Malaysia. Based on Table 72.4 and Fig. 72.3, food or organic waste was the largest component in

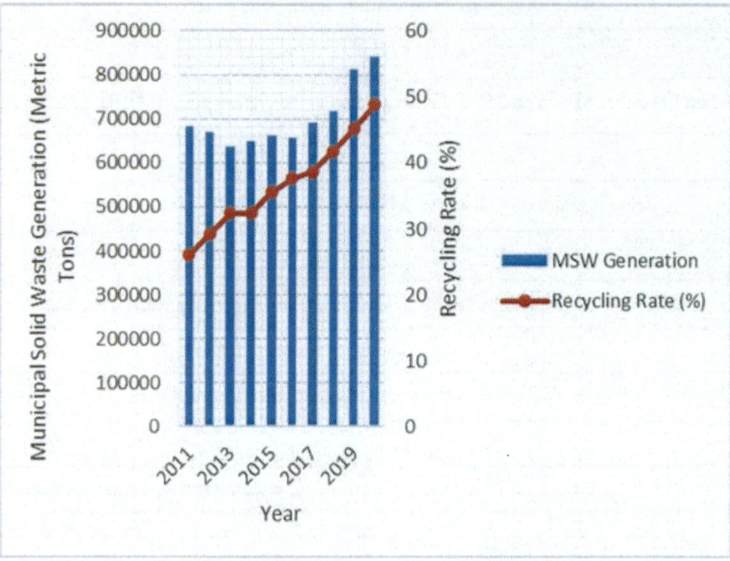

Fig. 72.2 MSW generation and recycling rate in Penang from 2011 until 2020

Table 72.3 MSW generation and recycling rate in Penang from 2011 until 2020

Year	Total waste generated (metric tons)	Recycling per year (metric tons)	Recycling rate (%)
2011	683,678.00	178,669.01	26.13
2012	671,566.06	196,464.78	29.25
2013	637,349.08	206,673.80	32.43
2014	648,809.15	210,363.17	32.42
2015	662,416.77	235,658.70	35.58
2016	657,473.55	247,457.83	37.64
2017	690,340.38	266,369.63	38.59
2018	718,436.22	300,232.27	41.79
2019	810,731.61	366,082.47	45.15
2020	842,041.63	411,613.39	48.88

the MSW at 44.5%. The other major components were plastic (13.2%), diapers (12.1%), paper (8.5%), garden waste (5.8%), glass (3.3%), textiles (3.1%), metal (2.7%), rubber (1.8%), tetra pak (1.6%), wood + peep/husk (1.4%), and hazardous household waste (1.3%). Others, accounting for 0.5%, and leather, accounting for 0.4%, were considered minor and insignificant wastes. Thus, the MSW composition in Penang consists of 51.7% of biodegradable or organic waste, 26.0% of recyclable waste, and the remaining non-biodegradable or inorganic waste.

Table 72.4 MSW composition in Penang from 2011 until 2020

MSW composition	Percentage of composition (%)	Composition based on year (metric tons)									
		2011	2012	2013	2014	2015	2016	2017	2018	2019	2020
Organic or food waste	44.5	304,236.71	298,846.90	283,620.34	288,720.07	294,775.46	292,575.73	307,201.47	319,704.12	360,775.57	374,708.53
Plastic	13.2	90,245.50	88,646.72	84,130.08	85,642.81	87,439.01	86,786.51	91,124.93	94,833.58	107,016.57	111,149.50
Paper	8.5	58,112.63	57,083.12	54,174.67	55,148.78	56,305.43	55,885.25	58,678.93	61,067.08	68,912.19	71,573.54
Diaper	12.1	82,725.04	81,259.49	77,119.24	78,505.91	38,420.17	79,554.30	83,531.19	86,930.78	98,098.52	101,887.04
Garden waste	5.8	39,653.32	38,950.83	36,966.25	37,630.93	38,420.17	38,133.47	40,039.74	41,669.30	47,022.43	48,838.41
Glass	3.3	22,561.37	22,161.68	21,032.52	21,410.70	21,859.75	21,696.63	22,781.23	23,708.40	26,754.14	27,787.37
Textiles	3.1	21,194.02	20,818.55	19,757.82	20,113.08	20,534.92	20,381.68	21,400.55	22,271.52	25,132.68	26,103.29
Metal	2.7	18,459.31	18,132.28	17,208.43	17,517.85	17,885.25	17,751.79	18,639.19	19,397.78	21,889.75	22,735.12
Rubber	1.8	12,306.20	12,088.19	11,472.28	11,678.56	11,923.50	11,834.52	12,426.13	12,931.85	14,593.17	15,156.75
Tetra Pak	1.6	10,938.85	10,745.06	10,197.59	10,380.95	10,598.67	10,519.58	11,045.45	11,494.98	12,971.71	13,472.67
Wood + peel/husk	1.4	9571.49	9401.92	8922.89	9083.33	9273.83	9204.63	9664.77	10,058.11	11,350.24	11,788.58
Hazardous household waste	1.3	8887.81	8730.36	8285.54	8434.52	8611.42	8547.16	8974.42	9339.67	10,539.51	10,946.54
Leather	0.4	2734.71	2686.26	2549.40	2761.36	2649.67	2629.89	2761.36	2873.74	3242.93	3368.17
Others	0.5	3418.39	3357.83	3186.75	3244.05	3312.08	3287.37	3451.70	3592.18	4053.66	4210.21

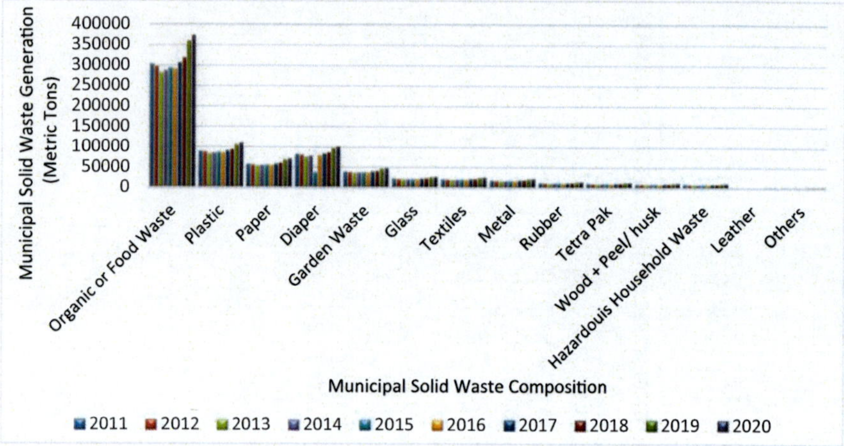

Fig. 72.3 MSW composition in Penang from 2011 until 2020

Information on waste composition helps in selecting the possible options for sustainable waste management methods [2]. Based on the results, the MSW stream in Penang is dominated by food waste. Generally, food waste is composed mainly of organic matter. Burying food waste in landfill results in the emission of greenhouse gases such as methane when decomposition poses serious harm to the atmosphere without proper treatment [18]. Incineration is not possible in Malaysia with the presence of food waste due to high moisture content, temperature, tropical climate, and pressure build-up in waste that is sent to landfills. Besides, this also results in the production of leachate [5, 19–21]. The presence of moisture in MSW would affect the incineration operation as it lowers the calorific value, resulting in a reduction in energy production [21]. Composting is an indirect method of recycling biodegradable waste in an enclosed space [19]. The product can be used as a fertilizer for crops. MBPP has a pilot project on circular economy in Batu Maung Transfer Station to convert food waste to compost and green energy (biogas). This effort can reduce about 68,000 tons of CO_2/year and fossil fuel for electricity by about 11,000 tons/year, reducing the dependency on inorganic fertilizer and carbon dioxide emissions from about 100 tons of food waste per day. In Japan, Korea, and most European countries, food waste will be segregated from MSW for it to be easily degraded [22]. In the end, composting can get rid of 51.7% of Penang's biodegradable waste, which is made up of food waste, garden waste, and wood + peep/husk waste.

Developed Asian countries such as Japan and Korea, as well as European countries such as Germany and Denmark, have ranked landfilling as the least preferred method, as those countries' main waste disposal treatment is incineration [23]. MSW without waste segregation results in harmful emissions of gas and leachate because of biological, chemical, and physical degradation [23]. Thus, incineration can be used as an alternative to landfilling as it has been proven to be an effective approach in reducing the dependency on landfilling, the quantity of MSW, and energy recovery.

In Singapore, incineration is chosen as the waste management option after recyclable materials have been recovered. This method is applicable in Malaysia due to similar waste characteristics and climate between both countries [14]. However, for the successful implementation of incineration, an accurate estimation of future MSW generation and composition based on demographic area and economic indicators is required. So, burning MSW could be one way to deal with the remaining 22.3% of inorganic or non-biodegradable waste in Penang, but only if strict waste sorting rules are put in place to reduce the amount of trash that ends up in landfills.

According to Table 72.5, efficient waste collection and disposal at landfills has the highest percentage of 65% in the Strength group. The SWOT factors or items with highest criteria weight were utilized in the final questionnaire to determine the weight of each SWOT groups. The SWOT factors or items with highest criteria weight were utilized in the final questionnaire to determine the weight of each SWOT groups. The final weight of each SWOT factor or item was determined by multiplying criteria weight of each factor or items (which were determined through Questionnaire 2) with criteria weights of each group (which were determined through Questionnaire 3). Penang Island's waste collection is divided into eight zones: Tanjung Tokong, Pulau Tikus, Padang Kota Lama, Jelutong, Gelugor, Air Itam, Batu Maung with Bayan Baru, and Balik Pulau with Teluk Kumbar. MSW collection is the responsibility of separate contractors in each zone. Seberang Perai is divided into 8 localities, which are Kepala Batas, Tasek Gelugor, Bagan, Permatang Pauh, Bukit Mertajam, Batu Kawan, and Nibong Tebal. Due to the vast area covered, these localities were further divided into separate zones (about 36 zones). Generally, domestic waste will be collected from each housing unit three times per week; compactor bins and Ro-Ro bins at roadsides daily; and garden waste every weekend depending on the schedule of each contractor at both municipalities. About 42% of the experts from both municipalities agree that poor culture or lack of willingness for waste segregation is due to a lack of environmental awareness among local people [24]. When awareness is not translated into action, public participation in waste segregation and recycling remains low among Malaysians [25]. Furthermore, a lack of space in households to carry out recycling activities contributes to a lack of willingness for waste segregation [26]. Experts believe that a strong regulatory framework will allow them to increase recycling rather than landfilling by about 30% in the Opportunity group [27]. Penang is exempted from The Household Solid Waste and Public Cleansing Management Act 2007 (Act 672) but adheres to the WSAS Policy, which was implemented in 2016. Bans on plastic straws, polystyrene, and plastic bags (on Mondays and Wednesdays) have also been implemented in Penang successfully over the years. "Rapid economic growth and changing lifestyles increased resource consumption, which in turn increased waste generation," is the most voted factor among the threat group, accounting for approximately 48%. Generally, Penang is undergoing economic and population growth, together with non-Malaysian residents who have settled down in the state. From 2007 to 2017, the major types of economic activity in Penang, such as services, manufacturing, construction, agriculture, and mining and quarrying, all contributed to an increase in the Gross Domestic Product.

Table 72.5 Final priority of SWOT factors and groups

SWOT group	SWOT factor	SWOT factor weight	SWOT group weight	Overall priority
Strength	**S1**: Active community participation	0.1628	0.5854	0.0953
	S2: Efficient waste collection and disposal at landfills	0.6474		**0.3790**
	S3: Municipality carry out recycling activities such as award and training programs	0.1400		0.0820
	S4: Strong human resource to meet the demand of MSW management	0.0498		0.0292
Weakness	**W1**: Poor culture or lack of willingness for waste segregation due to lack of environmental awareness of local people	0.4208	0.1472	**0.0620**
	W2: High dependence on landfilling	0.2290		0.0337
	W3: Highly comingled waste	0.2969		0.0437
	W4: Lack of financial support from government	0.0533		0.0078
Opportunity	**O1**: Location of Penang is good for attracting investment from outside foreign and internal organization	0.2538	0.2028	0.0515
	O2: Better engagement of private sectors in solid waste management by changing economic structures and policies	0.2117		0.0429

(continued)

Table 72.5 (continued)

SWOT group	SWOT factor	SWOT factor weight	SWOT group weight	Overall priority
	O3: Strong regulatory framework provides the basis for increasing recycling instead of landfilling	0.2950		**0.0598**
	O4: Participation in waste prevention activities	0.2395		0.0486
Threat	**T1**: Increased cases of illegal dumping due to high costs and poor enforcement of law	0.2899	0.0646	0.0187
	T2: Rapid economic growth and changing lifestyle increased consumption of resources which in turn increasing waste generation	0.4819		**0.0311**
	T3: Increasing population due to migration from other part of country resulting increased waste generation	0.1755		0.0113
	T4: Environmental degradation due to continued resource acquisition and landfilling	0.0527		0.0034

72.4 Conclusion

The MSW generation in Penang was evaluated from 2011 until 2020. Generally, MSW generation in the state remained fluctuant from 2011 until 2016 and increased from 2017 until 2020. Organic waste constituted the largest proportion of waste generated in the state, followed by recyclables and finally inorganic waste. Thus, composting can be carried out to remove the organic fraction in the first place, followed by recycling. Inorganic waste can be incinerated, while any waste that cannot be treated at all should be disposed of in a landfill. This would reduce the dependency on landfilling, reduce environmental emissions and degradation, and create a new waste management system in Penang which has more benefits than the overdependence on landfilling. The recyclable potential of MSW generated in

Penang from 2011 until 2020 was evaluated. The recyclable potential in the state showed an increasing trend from 2011 until 2020. The relationship was evaluated by using statistical analysis and the SWOT-AHP model. The recycling rate in Penang increased with an increase in MSW generation from 2011 until 2020. ANOVA One Way, correlation, and regression showed significant relationships between MSW generation and recycling potential.

Acknowledgements The authors would like to acknowledge Universiti Sains Malaysia (USM) for the opportunity of research work to be conducted. The authors would like to also acknowledge Penang Island City Council (MBPP) and Seberang Perai City Council (MBSP) for providing the secondary data required for data analysis. Special thanks to those who contributed to this project directly or indirectly.

References

1. M. Noufal, L. Yuanyuan, Z. Maalla, S. Adipah, J. Environ. Public Health, 1–15 (2019)
2. Y.C. Moh, L.A. Manaf, Resources. Conserv. Recycl. **82**, 50–61 (2014)
3. K. Kawai, T. Tasaki, J. Mater. Cycles Waste Manag. **18**, 1–13 (2016)
4. N.I.A. Rahman, M.A. Khoiry, S. Rahim, Basri, Int. J. Disaster Recovery Bus. Continuity **11**, 2230–2242 (2020)
5. S.A.S.A. Kadir, C.Y. Yin, M.R. Sulaiman, X. Chen, X.M. El-Harbawi, Renew. Sustain. Energy Rev. **24**, 181–186 (2013)
6. A. Malakahmad, M.S. Abualqumboz, S.R.M. Kutty, T.J. Abunama, Waste Manage. **70**, 282–292 (2017)
7. M.A. Kamaruddin, M.S Yusoff, H.A. Aziz, R. Alrozi, Current status of Pulau Burung Sanitary Landfill Leachate, Penang Malaysia, in *AIP Conference Proceedings* (2016), pp. 1–6
8. A. Gorener, K. Toker, K. Ulucay, Application of combined SWOT and AHP: a case study for a manufacturing firm, in *Procedia-Social and Behavioral Sciences*, pp. 1525–1534 (2012)
9. I.D. Adamo, P.M. Falcone, M. Gastaldi, P. Morone, Energy Policy **138**, 1–15 (2020)
10. H.M.E. Haque, S. Dhakal, S.M.G. Mostafa, Energy Policy **137**, 1–12 (2020)
11. Y. Lee, Y.J. Kim, M.C. Lee, Int. J. Hydrogen Energy **46**, 17597–17607 (2021)
12. N. Pournabi, S. Janatrostami, A. Ashrafzadeh, K. Mohammadi, Land Use Policy **107**, 1–9 (2021)
13. A. Shahnazari, H. Pourdej, M.D. Kharage, Biocatalysts Agric. Biotechnol. **32**, 1–7 (2021)
14. A. Johari, H. Alkali, H. Hashim, S.I. Ahmed, R. Mat, Mod. Appl. Sci. **8**, 37–49 (2014)
15. R.L. Verma, G. Borongan, M. Memon, Municipal solid waste management in Ho Chi Minh City, Vietnam, current practices ad future recommendation, in *International Conference on Solid Waste Management, 5IconSWM 2015*, pp. 127–139 (2016)
16. F. Giritto, Waste Manage. **41**, 1–3 (2015)
17. J.A.P. Olivera, J. Environ. Manage. **233**, 481–499 (2019)
18. M.A.S. Samah, M.Y. Ishak, K.B.A. Hamid, J. Environ. Stud. **24**, 1969–1982 (2015)
19. O.C. Aja, H.H. Al-Kayiem, J. Mater. Cycles Waste Manage. **16**, 693–710 (2014)
20. S.T. Tan, H. Hashim, J.S. Lim, W.S. Ho, C.T. Lee, J. Yan, Appl. Energy **138**, 797–804 (2014)
21. S.T. Tan, W.S. Ho, H. Hashim, Lee, M.R. Taib, C.T. Ho, Energy Convers. Manage. **102**, 111–120 (2015)
22. P. Kumaran, D. Hephzibah, R. Sivasankari, N. Saifuddin, A.H. Shamsuddin, Renew. Sustain. Energy Rev. **56**, 929–940 (2016)
23. S.N.S. Ismail, L.A. Manaf, J. Toxicol. Environ. Health Sci. **5**, 86–96 (2013)

24. M.A. Kamaruddin, M.S. Yusoff, L.M. Rui, A.M. Isa, M.H. Zawawi, R. Alrozi, Environ. Sci. Pollut. Res. **24**(35), 26988–27020 (2017)
25. P.S.M. Davadoss, P. Agamuthu, S.B. Mehran, C. Santha, S.H. Fauziah, Waste Manage. **119**, 135–144 (2021)
26. Y. Moh, L.A. Manaf, Resources. Conserv. Recycl. **116**, 1–14 (2017)
27. M.A. Kamaruddin, M.S. Yusoff, N. Ibrahim, M.H. Zawawi, in *AIP Conference Proceedings*, vol. 1835, no. 1 (AIP Publishing LLC, 2017), p. 020031

Chapter 73
The Effect of Heating Temperature on Coercivity Force of Superparamagnetic Zinc Nickel Ferrite Nanoparticles

Luong Thi Quynh Anh

Abstract This study is toward the fabrication of superparamagnetic particles applied to biomedical. A superparamagnetic zinc nickel ferrite $Zn_{0.8}Ni_{0.2}Fe_2O_4$ nanoparticles were synthesized by hydrothermal method (0.1 M precursors concentration), then heated at different temperatures 100, 120, 140, 160, 180 °C for 6 h in the autoclave. The effect of ferrite heating temperature on nanoparticle size and magnetic properties of ferrite was studied. Then, superparamagnetic nanoparticles were analyzed by EDS, XRD, TEM, and VSM to test their microstructure, chemical compositions, morphology, and magnetic properties. The TEM results confirm that nanoparticle sizes ranged from 3.5 to 8 nm when increases heating temperatures. Their magnetic properties were measured by a vibrating sample magnetometer (VSM) showed that the coercive force H_c increased from 2.05 to 8.67 Oe when heating temperatures increase from 100 to 180 °C.

Keywords Zinc nickel ferrite nanoparticles · Superparamagnetic ferrites · Coercivity force

73.1 Introduction

Zinc–nickel (Zn–Ni) ferrite nanoparticles are one of the most interesting magnetic nanomaterials nowadays because of their potential for important biotechnological applications in cancer treatment, medical diagnosis, and MRI contrast enhancement [1–5]. Superparamagnetism is a phenomenon observed when the crystallite size is smaller than 10 nm, the coercivity, and the remanence reach zero [5].

L. T. Q. Anh (✉)
Department of Metallurgy and Alloys Materials, Faculty of Materials Technology, Ho Chi Minh City University of Technology (HCMUT), 268 Ly Thuong Kiet Street, District 10, Ho Chi Minh City, Vietnam
e-mail: ltqanh@hcmut.edu.vn

Vietnam National University Ho Chi Minh City, Linh Trung Ward, Thu Duc District, Ho Chi Minh City, Vietnam

M. A. A. Mohd Salleh et al. (eds.), *Proceedings of the Green Materials and Electronic Packaging Interconnect Technology Symposium*, Springer Proceedings in Physics 289,
https://doi.org/10.1007/978-981-19-9267-4_73

Various methods are used for the synthesis of ferrite nanoparticles such as co-precipitation, sol–gel technique, and hydrothermal method [6–10]. Compared with other methods, hydrothermal synthesis seems to be the most promising because of its simplicity, productivity and keeps small nanoparticles size by low heating temperature than other methods [11]. The size, shape, and agglomeration of magnetic nanoparticles influence their applicability; good control over these features is essential, so, we need to add the solvents or surface agents, which it is capable of meeting the strict requirements mentioned earlier. Oleic acid is commonly used for the preparation of hydrophobic nanoparticles. As a surfactant, the oleic acid provides the steric stabilization of the nanoparticles against the Van der Waals and magnetic attractive interactions and thereby prevents their agglomeration [12–15].

In this paper, nanoparticles of $Zn_{0.8}Ni_{0.2}Fe_2O_4$ ferrite were first synthesized by chemical co-precipitation, then modified by oleic acid, and finally heated at different temperatures: 100, 120, 140, 160, and 180 °C in 6 h in autoclave heating—hydrothermal method. This study aims to understand how the heating temperature influences nanoparticle size and the coercivity force of ferrite.

73.2 Materials and Experimental

73.2.1 Materials

Ferric chloride ($FeCl_3 \cdot 6H_2O$), zinc chloride ($ZnCl_2$), nickel chloride ($NiCl_2 \cdot 6H_2O$), and 1-pentanol ($C_5H_{12}O$) were purchased from Merck (Germany). Ethanol and *n*-hexane (C_6H_{14}) were obtained from Xilong Scientifics (China), and oleic acid ($C_{18}H_{34}O_2$) was purchased from Sigma Aldrich (United States).

73.2.2 Experimental Processing

The pure salts were dissolved in doubly distilled water to produce a solution with a concentration of 0.1 M. The solution was then heated to 60 °C under stirring at 500 rpm. Subsequently, metal (Fe, Zn, and Ni) nano-hydroxides were formed by adding a 25% ammonia solution to the above solution under constant stirring. The ammonia solution was added until the solution pH reached 8.5. To prevent particle agglomeration, a 27 ml solution of oleic acid in 1-pentanol (concentration = 0.25 M) was added to the above mixture. To generate magnetic properties, the mixture was heated in an autoclave for 6 h at different temperatures of 100, 120, 140, 160, and 180 °C (hydrothermal heating). Finally, the mixture was dried at 100 °C for 4 h, and the brown powder products were washed five times by re-dispersing in *n*-hexane followed by precipitation with ethanol to remove the impurities. Figure 73.1 shows a schematic of the ferrite synthesized process.

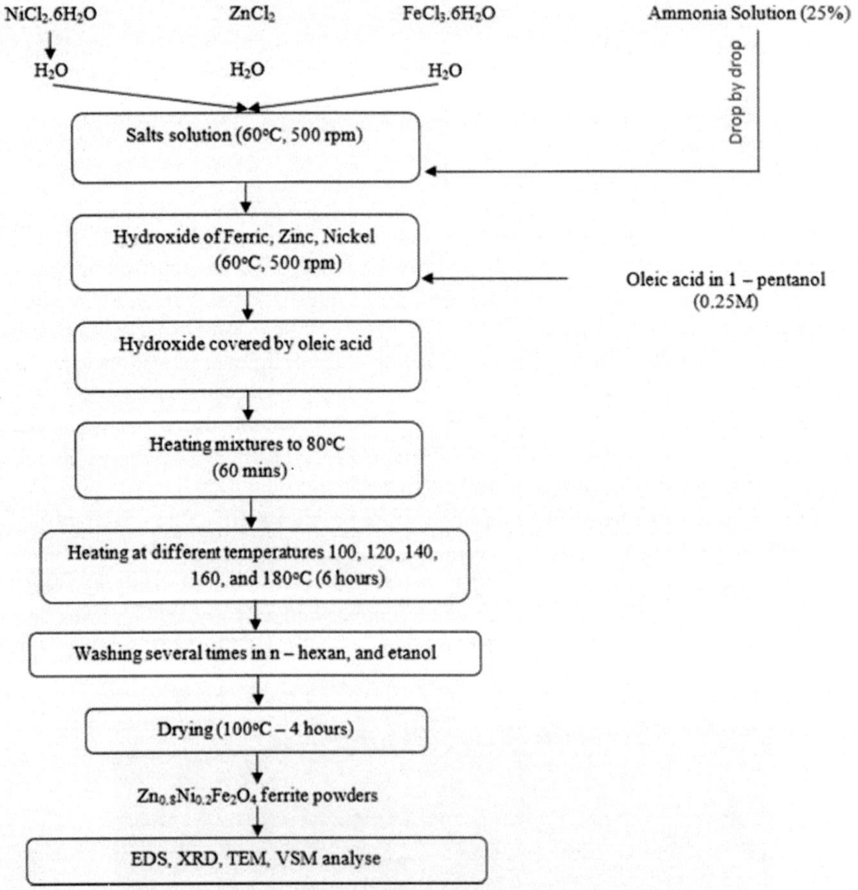

Fig. 73.1 Schematic of $Zn_{0.8}Ni_{0.2}Fe_2O_4$ ferrite synthesized process

73.2.3 Methods Analysis

The ferrite structure and morphology were analyzed by X-ray diffraction (XRD) and transmission electron microscopy (TEM). The chemical composition of ferrite was determined by energy-dispersive X-ray spectroscopy (EDS). Finally, the magnetic properties of the synthesized ferrite were evaluated using a vibrating sample magnetometer (VSM).

73.3 Result and Discussions

73.3.1 Chemical Composition, Microstructure, Morphology Results

The chemical composition of the ferrite samples is determined by the EDS method. The EDS spectrum of the sample heated at 140 °C (Fig. 73.2) confirms the presence of Fe, Zn, Ni, and O elements in Fig. 73.2. Table 73.1 compares the chemical composition of the sample heated at 140 °C to the theoretical composition, and Table 73.1 indicates that the difference in composition was negligible and suitable for the $Zn_{0.8}Ni_{0.2}Fe_2O_4$ formula.

The XRD spectrum of samples in Fig. 73.3 provides the crystallinity of $Zn_{0.8}Ni_{0.2}Fe_2O_4$ nanoparticles. Figure 73.3 shows the diffraction patterns of the samples heated at temperatures of 100, 120, 140, 160, and 180 °C. The peaks in the XRD pattern of the heated samples correspond to the (220), (311), (400), (422), (511), and (440) peaks in the standard powder diffraction spectrum of Zn–Ni ferrite [16]. The spectrum of the sample heated at 100 and 120 °C showed only two low-intensity peaks, indicating that this sample did not completely crystallize. Thus, the high heating temperature leads to high complete crystallinity (140, 160, and 180 °C).

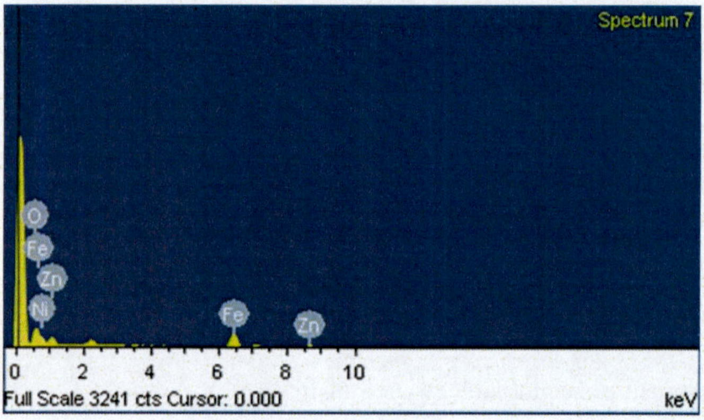

Fig. 73.2 EDS spectrum of $Zn_{0.8}Ni_{0.2}Fe_2O_4$ nanoparticle

Table 73.1 Chemical composition of $Zn_{0.8}Ni_{0.2}Fe_2O_4$ sample heated at 140 °C

Elements	Samples weight (%)	Theoretical weight (%)
O K	25.20	26.71
Fe K	45.99	46.74
Ni K	5.01	4.84
Zn K	23.81	21.70

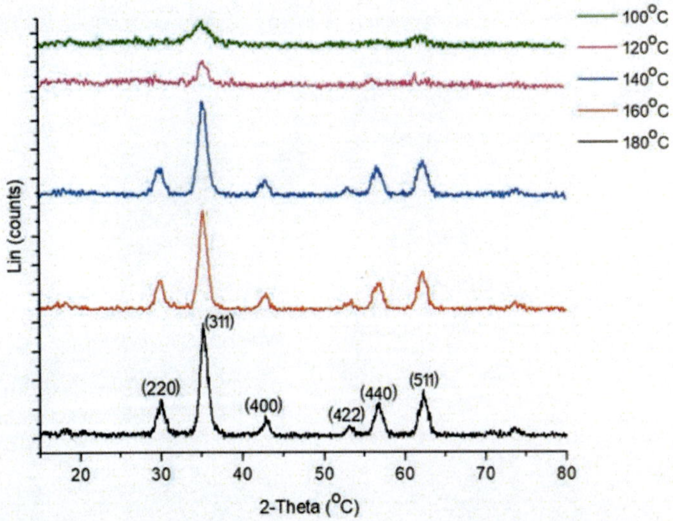

Fig. 73.3 XRD patterns of $Zn_{0.8}Ni_{0.2}Fe_2O_4$ samples heated at different temperatures

73.3.2 Morphological Analysis

The TEM images of samples heated at 180, 160, 140, 120, and 100 °C temperatures are shown in Fig. 73.4. Figure 73.3 shows the ferrite nanoparticle of all samples exhibited nearly spherical particle morphology and a particle size range of 3.5–8 nm (using Image J software) because of the growth of crystallite nanoparticles of ferrites with temperature, the average nanoparticle be seen in Table 73.2.

In Table 73.2, we can conclude that heating temperature significantly influences nanoparticle size. The average size increases from 3.5 to 8 nm when the heating temperature increases. However, all of the nanoparticle sizes are below 10 nm, so, all samples reach a superparamagnetism state.

73.3.3 The Effect of Heating Temperature on Coercivity Force of $Zn_{0.8}Ni_{0.2}Fe_2O_4$ Ferrite

The magnetization curves of the $Zn_{0.8}Ni_{0.2}Fe_2O_4$ ferrite nanoparticles and a line graph of the heating temperature–coercivity force relationship are displayed in Fig. 73.5. All samples showed the remanence value (M_r) is zero approximately (in Table 73.3). So, the lower coercivity force leads to a superparamagnetism state easily.

Figure 73.5 also shows that the heating temperature increases from 100 to 180 °C; magnetization saturation M_s increases from 10.62 to 27.11 emu/gr; this depends on

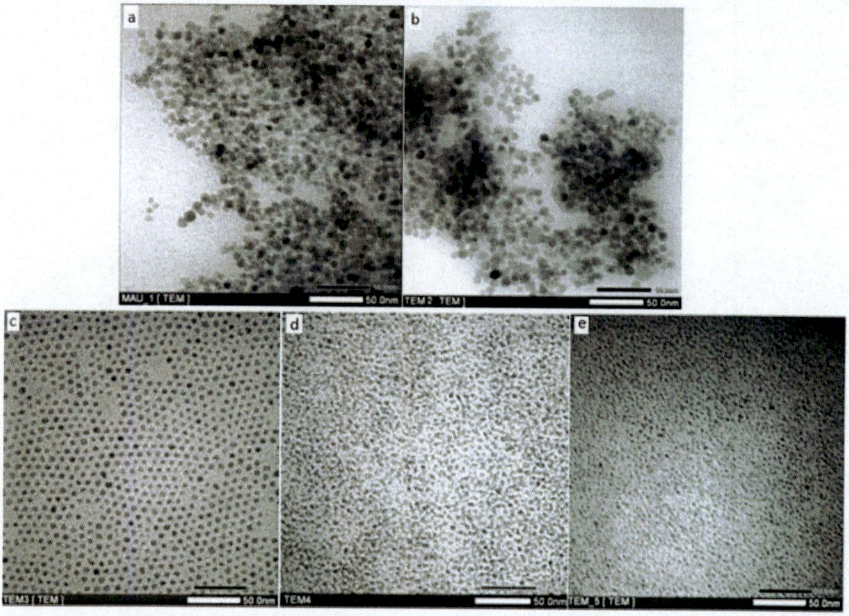

Fig. 73.4 TEM images of samples heated at: **a** 180 °C, **b** 160 °C, **c** 140 °C, **d** 120 °C, and **e** 100 °C

Table 73.2 Average nanoparticle size at different heating temperatures

Heating temperatures (°C)	Average nanoparticle size (nm)
100	3.5 ± 0.1
120	5 ± 0.2
140	6 ± 0.1
160	7.5 ± 0.2
180	8 ± 0.1

the increase in the proportion of magnetic energy with respect to crystal volume as the nanoparticle size increases, and the corresponding coercivity force increases from 2.05 to 8.67 Oe (Table 73.3). The lower the heating temperature, the lower the coercivity force obtained.

73.4 Conclusion

This study aims to know how the effect of the heating temperature of zinc nickel ferrite nanoparticles on coercivity force, the main parameter strong impact on the superparamagnetism state. The results show the main conclusions:

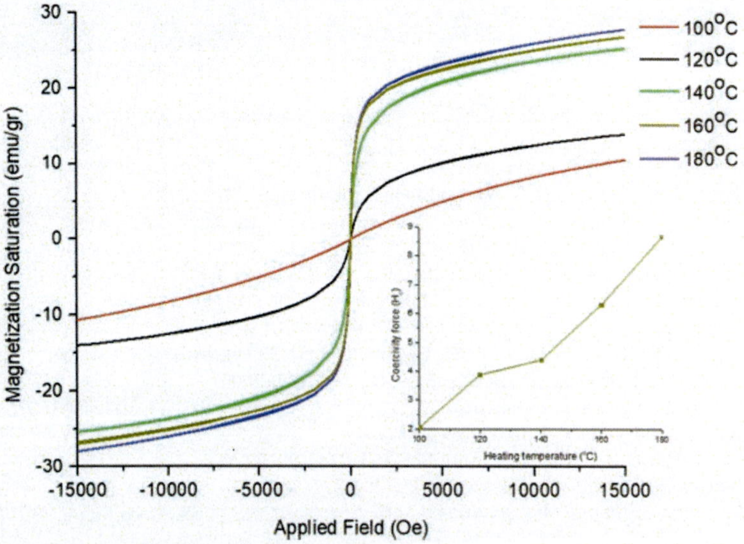

Fig. 73.5 Magnetization curves of $Zn_{0.8}Ni_{0.2}Fe_2O_4$ ferrite crystallized at different temperatures and a line graph of heating temperature–coercivity force relationship

Table 73.3 Magnetic properties of all samples

Heating temperatures (°C)	Coercivity force (Oe)	Magnetization saturation (emu/gr)	Remanence (emu/gr)
100	2.05	27.11	≈ 0
120	3.88	26.81	
140	4.39	25.39	
160	6.31	14.02	
180	8.67	10.62	

(a) By using the XRD method, the results show the suitable heating temperature should be from 140 to 180 °C in 6 h.

(b) TEM result confirms that the average size strongly depends on heating temperature; all samples obtain the average nanoparticle size below 10 nm, ensuring the superparamagnetism state.

Testing magnetic properties displayed the strongly depend of coercivity force on heating temperature, the high heating temperature, the high coercivity fore. To ensure the magnetism state of ferrite nanoparticles, the smaller heating temperature is, the better coercivity force archived.

Acknowledgements We acknowledge the support of time and facilities from Ho Chi Minh City University of Technology (HCMUT), VNU-HCM for this study.

References

1. A. Figuerola, R.D. Corato, L. Manna, T. Pellegrino, Pharmacol. Res. **62**, 1126–1143 (2010)
2. J.H. Lee, et al., Nat. Med. *13*, 95–99 (2007)
3. E. Fantechi, et al., ACS nano **8**, 4705-4719 (2014)
4. F. Hyder, S. Manjura Hoque, Contrast Media Mol. Imaging 2017 (2017)
5. M. Benz, *Superparamagnetism: Theory and Applications* (Research Gate, 2012)
6. S. Kumar, et al., Sci. Res. **5**, 145–151 (2013)
7. K. Bhattacharjee, C.K Ghosh, J. Nanopart. Res. **13**, 739-750 (2011)
8. S. Jovanovich, et al., J. Phys. Chem. C **118**, 13844–13856 (2014)
9. U.D. Israf, et al., Aust. J. Basic Appl. Sci. **7**, 154-162 (2013)
10. P.B.C. Rao, S.P. Setty, Int. J. Eng. Sci. Technol. **2**, 3351-3354 (2010)
11. W. Suchanek, R. Riman, Adv. Sci. Technol. **45**, 184–193 (2006)
12. A. Repko, D. Nižňanský, J. Poltierová-Vejpravová, J. Nanopart. Res. **13**, 5021–5031 (2011)
13. N. Wu, et al., Nano Lett. **4**, 383-386, (2004)
14. L. Zhang, R. He, H.C. Gu, Appl. Surf. Sci. **253**, 2611-2617 (2006)
15. R. Tadmor, et al., Langmuir **16**, 9117-9120 (2000)
16. D. Jiles, *Introduction to Mmagnetism and Magnetic Materials* (USA, Boca Raton, 1994)

Chapter 74
Investigation on the Effect of Electrospinning Parameters: Voltage and Flow Rate on PVDF Fiber

Ammar Athallah Budiarto, Sharifah Shahnaz Syed Bakar, Shahrizam Saad, and Shuhaida Yahud

Abstract Polyvinylidene fluoride (PVDF) is a high purity thermoplastic fluoropolymer and high-capacity polymer that is employed in the electronics, space, and aeronautics industries. Electrospinning parameters have a considerable impact on the morphology, crystallinity, and electrical conductivity of PVDF fiber. Voltage and flow rate are the electrospinning parameters investigated in this study. As a result, the goal of this study is to investigate the effects of various voltages and flow rates on PVDF electrospun fiber. Electrospinning is the simplest and low-cost method to produce PVDF fibers by dissolving PVDF in solvent *N*, *N*-Dimethylformamide (DMF). About 15 wt% PVDF solution was used. Electrospinning parameters used were varied applied voltages 15 and 20 kV, flow rates 0.8, 1.0, 1.2, and 1.4 ml/h. Tip-to-collector distance was fixed at 15 cm. Scanning electron microscope (SEM), four-point probe, and X-ray diffraction (XRD) were performed to determine the morphology, electrical conductivity, and crystallinity phase of electrospun PVDF fibers. The SEM analysis concluded that bead formation occurs at higher flow rate and higher voltage. With higher flow rates and lower applied voltage, the fiber's thickness increases. XRD analysis shows that shift peak is observed among all fibers due to short electrospinning time which leads to insufficient thickness of electrospun mat, which affects the mechanical properties of fibers and causes peak shift. Four-point probe analysis concluded that the average conductance increases sharply as flow rate increases and the average resistance decreases. Also, as applied voltage increases, average conductance decreases and the average resistance increases.

Keywords Electrospinning · PVDF · Fibers · Applied voltage · Flow rate

---

A. A. Budiarto · S. S. Syed Bakar (✉) · S. Saad
Faculty of Chemical Engineering and Technology, Universiti Malaysia Perlis (UniMAP), 01000 Kangar, Perlis, Malaysia
e-mail: shahnaz@unimap.edu.my

S. Yahud
Faculty of Electronic Engineering Technology, Universiti Malaysia Perlis, Pauh Putra Campus, 02600 Arau, Perlis, Malaysia

M. A. A. Mohd Salleh et al. (eds.), *Proceedings of the Green Materials and Electronic Packaging Interconnect Technology Symposium*, Springer Proceedings in Physics 289,
https://doi.org/10.1007/978-981-19-9267-4_74

74.1 Introduction

Polyvinylidene fluoride (PVDF) is a fluoropolymer made up of the monomer units CF_2–CH_2 [1]. It is also a polymorphic, semi-crystalline polymer that exhibits at least five crystal phases under various processing circumstances. Those main crystalline structures are α, β, γ, δ, and ε with different polarities, i.e., α-phase is a non-polar, whereas the β-phase has polar properties. Those each five phases can be employed for a specific application [2].

The β-phase, which has a net crystallographic dipole moment orientated essentially in the direction normal to the molecular chain, is currently the most important PVDF polymorph, and it is responsible for the improved piezoelectric and ferroelectric activity [3]. These piezoelectric materials can create electrical charges in response to direct piezoelectric effect and mechanical stimulation; they can also be deformed reversibly by an electric stimulus (reverse piezoelectric effect). In fact, under mechanical stress, piezoelectric materials can create an electrical field across their limits, or vice versa [4].

Electrospinning parameters have a considerable impact on the morphology of the fibers produced as a result of electrospinning, and by manipulating these parameters properly, nanofibers of desired morphology and diameters can be formed. In electrospinning, the applied voltage is thought to be crucial in launching the electrospinning jet. The impact of applied voltage on fiber shape and diameter is disputed. The fiber morphology, initial polymer droplet shape, polymer jet trajectory, and Taylor cone formation are all influenced by the flow rate [1]. In this study, the phase, electrical properties, and the morphology of PVDF at different flow rates and applied voltages were investigated.

74.2 Methodology

74.2.1 Materials

N, *N*-Dimethylformamide (DMF) was used as a solvent which was supplied from Fisher Scientific Company to dissolve polyvinylidene fluoride (PVDF) in the form of powder that was supplied from Sigma-Aldrich Company.

74.2.2 Methods

About 15 wt% of PVDF powder which is 0.8329 g was dissolved into 5 ml of DMF solution to obtain a PVDF solution. The mixing process was prepared by using magnetic stirrer for 2 h until its fully homogeneous and transparent polymer solution.

The solution was covered with aluminum foil before it was used for electrospinning to prevent evaporation.

The electrospinning process was conducted after the preparation of PVDF solution (see Fig. 74.1). The PVDF solution was transferred to a 5 mL syringe with a 0.8 mm inner diameter syringe nozzle, which was placed horizontally on the injection holder at the syringe pump. Electrospinning process was conducted under specific electrospinning parameters as shown in Table 74.1. A roller metal plate which was wrapped with aluminum foil that acted as the fiber's collector was aligned perpendicular to the syringe nozzle's tip. To ensure the needle stays level on the surface, the syringe was scraped and cleaned with sandpaper. To transmit voltage from the power source, the crocodile clip was fastened on the needle. The pump machine's power and voltage supply are turned on, allowing PVDF to flow from the needle's tip. The PVDF fibers were collected from the aluminum foil.

Fig. 74.1 Electrospinning process setup

Table 74.1 Electrospinning parameters for all eight samples

Applied voltage (kV)	Flow rate (ml/h)	TCD (cm)
15	0.8	15
	1.0	15
	1.2	15
	1.4	15
20	0.8	15
	1.0	15
	1.2	15
	1.4	15

74.2.3 Characterization

Morphology of PVDF fibers was investigated by scanning electron microscope (SEM), Hitachi TM3000. The samples were prepared in size of 1 cm × 1 cm. Shimadzu's model XRD-600 X-ray diffraction (XRD) analysis was used to determine the phase of PVDF fibers. The composition and phase of PVDF and PVDF fibers were identified. The 2θ scan range is between 5° and 30°, and the scan rate was 0.02°/s. The electrical conductivity of PVDF fibers was assessed using four-point probe technique. Keithley 6221 AC/DC source and Keithley 2182 A nanovoltmeter were used to measure the conductivity. The samples were tested using a 100 V voltage.

74.3 Results and Discussion

74.3.1 Morphological Study

As shown in Fig. 74.2, PVDF electrospun fiber at 1.4 ml/h (sample d) showed beaded structure with inconsistent fiber diameter and contains more beaded fibers than the other samples. Electrospun PVDF fiber at 0.8 ml/h (sample a) showed the least beads' fiber formation as compared to other samples. Electrospun PVDF fiber at 1.2 ml/h (sample c) and PVDF fiber at 1.4 ml/h (sample d) contain beaded fibers. PVDF solution prepared in this experiment has a weight percentage of 15 wt%. Low solution viscosity will cause electrospraying rather than bead-producing electrospinning to occur [5].

Fiber at 20 kV with 1.4 ml/h (sample d) showed beaded structure with inconsistent fiber diameter. It is also observed that it contains more beaded fibers than the remaining samples at 20 kV. Meanwhile, electrospun PVDF fiber with 1.0 ml/h (sample b) showed the least beads' fiber formation as compared to other samples. Because of the increased columbic forces in the jet and the stronger electric field caused by the higher voltage of the fiber, the solution stretches more, which results in smaller beads and faster solvent evaporation from the fibers [1].

74.3.2 Phase Determination

Figure 74.3a shows the diffraction angle of different flow rates of PVDF at 15 kV which was enlightened by XRD patterns observed from 10° to 40°. It is observed that sample (ii) (15 kV/1.0 ml/h) exhibits most intense diffraction peak among the other samples at 22.42°, followed by sample (iii) (15 kV/1.2 ml/h) at 22.76°, sample (i) (15 kV/0.8 ml/h) at 21.73°, and sample (iv) (15 kV/1.4 ml/h) at 23.28°, while Fig. 74.3b shows the diffraction angle of different flow rates of PVDF at 20 kV

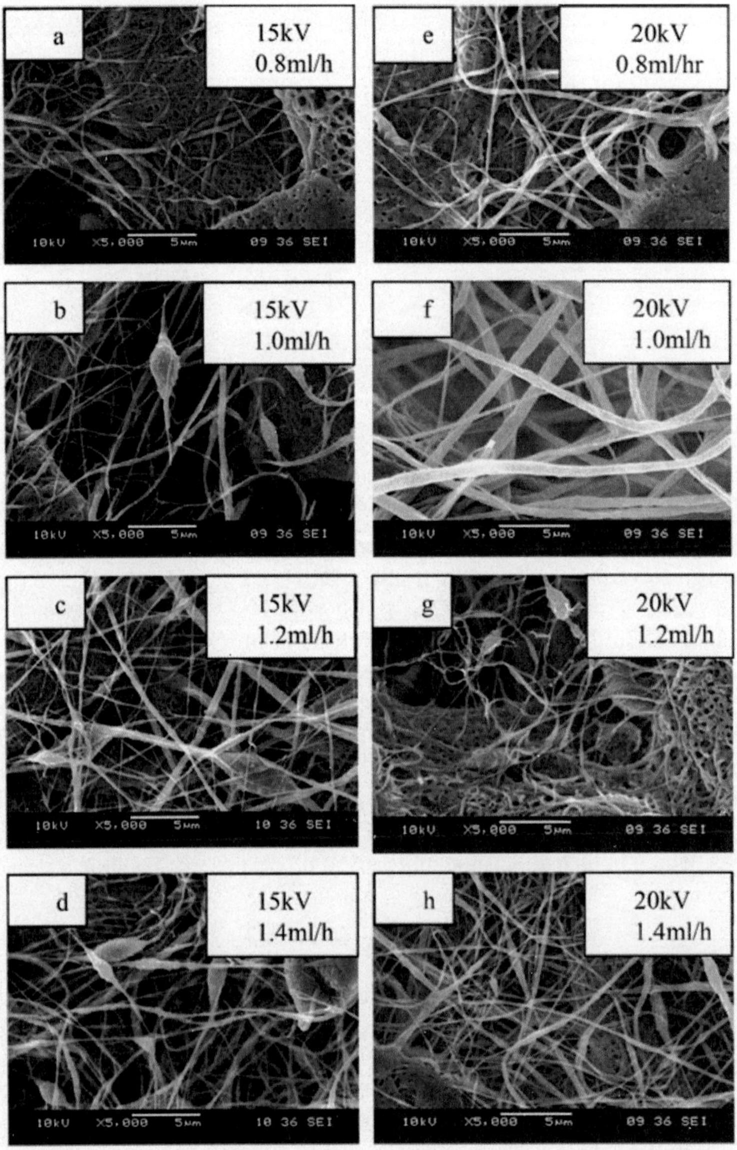

Fig. 74.2 SEM image of PVDF electrospun fibers at different applied voltages and flow rates

which was enlightened by XRD patterns observed from 10° to 40°. It is observed that sample (vi) (20 kV/1.0 ml/h) exhibits most intense diffraction peak among the other samples at 22.53°, followed by sample (v) (20 kV/0.8 ml/h) at 21.71°, sample (vii) (20 kV/1.2 ml/h) at 22.27°, and sample (viii) (20 kV/1.4 ml/h) at 21.69°.

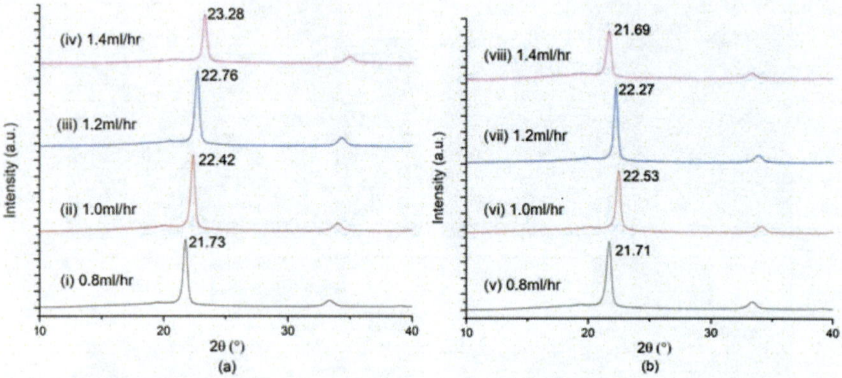

Fig. 74.3 XRD patterns of electrospun PVDF fiber at different flow rates and applied voltages **a** 15 kV and **b** 20 kV

The samples should have shown up an intense peak around 20.26° indicating β-phase [6]. Nevertheless, from the result of XRD in Fig. 74.3, it is shown that the peak showed up at range of 21.71°–23.28° instead. It is suspected that the main cause for shifted peak in this XRD analysis is because insufficient electrospun thickness affects the mechanical properties that are resulting in a peak shift in XRD analysis [7]. According to previous research, the samples' lattice strain and crystallite size are the main factor to the displaced peak in this XRD analysis [8]. Shorter fiber–fiber junctions as a result of the samples' decreased Young's modulus performance disturb with the fibers' capacity to reorganize themselves in response to rising strain [9]. As a result, it can be stated that the electrospun samples fail to meet the optimal thickness, which in turn affects the mechanical properties that result in a peak shift in XRD analysis.

74.3.3 Electrical Properties

Figure 74.4 shows average conductance of samples at different applied voltages and flow rates which are measured using four-point probe. It is observed that sample (a) (1.4 ml/h) has the lowest average conductance which is 0.0683 S. In contrast, sample (d) (0.8 ml/h) has the highest average conductance which is 0.0708 S.

Figure 74.5 shows average resistance of samples at different applied voltages and flow rates which are measured using four-point probe. It is observed that sample (e) (1.4 ml/h) has the highest average resistance which is 25.2797 Ω. In contrast, sample (h) (0.8 ml/h) has the lowest average resistance which is 21.9637 Ω.

The applied voltage difference can be seen also by comparing Figs. 74.4 and 74.5. Samples (a) and (e) with the same flow rate but different applied voltages showed significant changes of average conductance and resistance. At 0.8 ml/h with 15 kV applied voltage, the average conductance and average resistance are 25.2797 Ω and

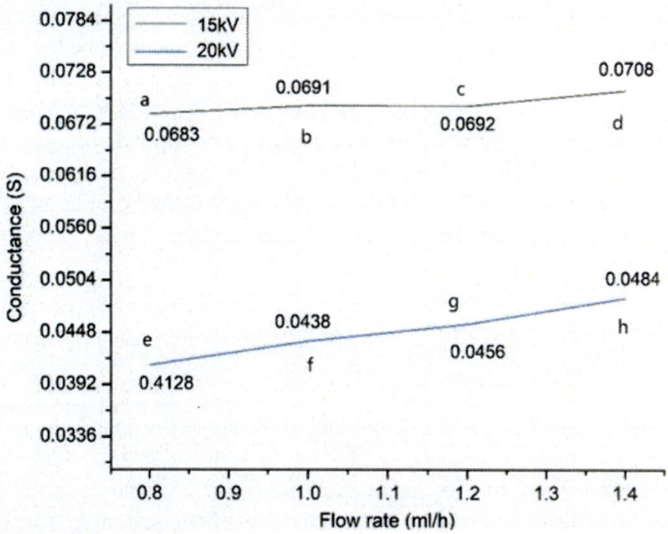

Fig. 74.4 Average conductance of different flow rate and applied voltage samples using four-point probe measurement

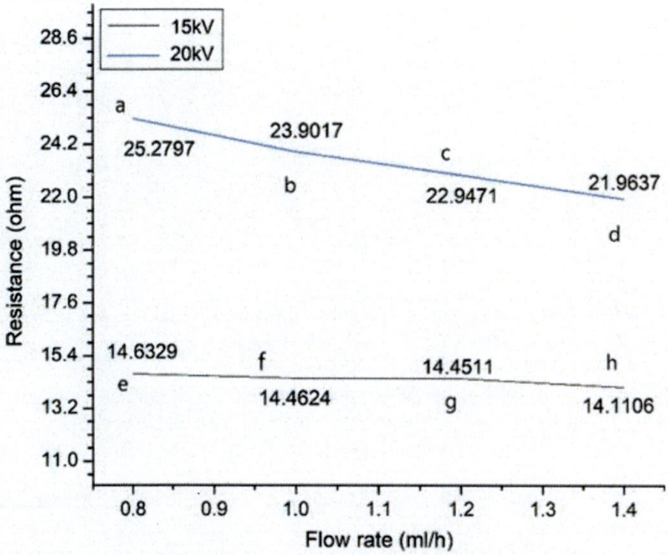

Fig. 74.5 Average resistance of different flow rate and applied voltage samples using four-point probe measurement

0.0412, while at same flow rate with 20 kV, the average resistance and lowest average conductance are 14.6392 Ω and 0.0683 S, respectively. It shows that the higher the voltage, the lower average conductance and the higher the average resistance which applied to all samples with same flow rate but different applied voltages. Based on data from Figs. 74.4 and 74.5, sample (e) with 15 kV applied voltage and 1.4 ml/h flow rate showed a better electrical conductivity than the other samples as it achieves lowest average resistance and highest average conductance value.

74.4 Conclusion

In conclusion, the effect of applied voltage and flow rate on the phase, electrical properties, and morphology of electrospun PVDF fibers were evaluated. Due to inadequate PVDF concentration and insufficient stirring time, all PVDF fibers showed beaded fibers. Flow rate and applied voltage have significant effect on morphology of PVDF electrospun fiber. Bead formation occurs at higher flow rate, and as applied voltage increases, the bead formation also increases. Furthermore, the electrospun samples fail to meet the optimal thickness, which in turn affects the mechanical properties that result in a peak shift in XRD analysis. The average conductance increases sharply as flow rate increases and the average resistance decreases. Also, as applied voltage increases, average conductance decreases and the average resistance increases.

References

1. K.G. Tamberg, *Effect of Electrospinning Conditions on Morphology of PVDF Fibers* (Tallinn Univ. Technol., Estonia, 2014)
2. M. Nasir, H. Matsumoto, T. Danno, M. Minagawa, T. Irisawa, M. Shioya, A. Tanioka, Control of diameter, morphol-ogy, and structure of PVDF nanofiber fabricated by elec-trospray deposition. J. Polym. Sci. Part B Polym. Phys. **44**(5), 779 (2006)
3. K.S. Ramadan, D. Sameoto, S. Evoy, A review of piezoelectric polymers as functional materials for electrome-chanical transducers. Smart Mater. Struct. **23**(3), 033001 (2014)
4. A.J. Lovinger, Developments in crystalline polymers. J. Polym. Sci. Part C Polym. Lett. **20**(10), 559–560 (1982)
5. N. Shehata, E. Elnabawy, M. Abdelkader, A. Hassanin, M. Sa-lah, R. Nair, S. Ahmad Bhat, Static-aligned Pi-ezoelectric poly (vinylidene fluoride) electrospun nano-fibers/MWCNT composite membrane: facile method. Polymers **10**(9), 965 (2018). https://doi.org/10.3390/polym10090965
6. R. Gregorio Jr., Determination of the α, β, and γ crystal-line phases of poly (vinylidene fluoride) films prepared at dif-ferent conditions. J. Appl. Polym. Sci. **100**(4), 3272–3279 (2006)
7. J.W. Lee, S.S. Shahnaz, A.N. Hidayah, S. Yahud, N. Samat, Effect of carbon nanotube loading on electrical properties of electrospun polyvinylidene fluoride (PVDF) fiber. J. Phys. Conf. Series, vol. 2080, No. 1 (IOP Publishing, 2021), p. 012015

8. S.K. Mishra, H. Roy, A.K. Lohar, S.K. Samanta, S. Tiwari, K. Dutta, A comparative assessment of crystallite size and lattice strain in differently cast A356 aluminium alloy, in *IOP Conference Series: Materials Science and Engineering*, vol. 75, no. 1 (IOP Publishing, 2015), p. 012001
9. A.A. Conte, K. Sun, X. Hu, V.Z. Beachley, Effects of fiber density and strain rate on the mechanical properties of electrospun polycaprolactone nanofiber mats. Front. Chem. **8**, 610 (2020)

Chapter 75
Synthesis of Zeolite from Rice Husk Ash

Faizul Che Pa and Nurul Nazihah Mohamad Nasir

Abstract Rice husk ash (RHA) is used as the starting material for the synthesis of zeolites. In this study, alkaline hydrothermal treatment has been chosen for the zeolite conversion. Sodium hydroxide solution is used as the activation for the synthesis of zeolites. Two parameters affecting the hydrothermal synthesis of zeolites were investigated, including sodium hydroxide concentration (1 and 2 M) and reaction temperature (60, 80, and 100 °C). The produced samples' chemical composition and crystalline phase were investigated using XRF and XRD. The result showed that the zeolite obtained from 2 M of NaOH solution under 80 °C of reaction temperature was gismondine.

Keywords Rice husk ash · Zeolite · Hydrothermal

75.1 Introduction

Zeolites are a hydrated crystalline aluminosilicate with open three-dimensional (3D) framework structures, comprising a $SiO_4{}^{4-}$ and $AlO_4{}^{5-}$ tetrahedral. The tetrahedral is linked to each other by sharing oxygen atoms to form regular intracrystalline cavities and channels of atomic dimensions. About 40 naturally occurring zeolite species have been recorded, and more than 100 species have been synthesized [1]. Zeolites are made up of silica (SiO_2) and alumina (Al_2O_3). Several by-products contain high silica content that can be used to produce zeolites such as rice husk ash, fly ash, kaolinite, oil palm ash, and others [2]. Gismondine is a mineral in the zeolite family. It is a hydrated calcium aluminosilicate with the formula $Ca_2Al_4Si_4O_{16} \cdot 9H_2O$. Microcrystalline zeolites of the gismondine family are often reported in alkali-activated and blended cement systems.

F. Che Pa (✉) · N. N. Mohamad Nasir
Faculty of Chemical Engineering and Technology, Universiti Malaysia Perlis, 02600 Arau, Perlis, Malaysia
e-mail: faizul@unimap.edu.my

M. A. A. Mohd Salleh et al. (eds.), *Proceedings of the Green Materials and Electronic Packaging Interconnect Technology Symposium*, Springer Proceedings in Physics 289,
https://doi.org/10.1007/978-981-19-9267-4_75

In this research, rice husk ash (RHA) was used as the raw material for the conversion of zeolites. RHA was abundantly produced in rice-producing countries. Current rice production in the world is about 700 million tons. RHA is obtained from the incineration of rice husk. Every 100 kg of rice husks burnt will produce about 25 kg of RHA [3]. RHA contains high silica content, about 92% [4]. This makes RHA can be used as an alternative source of silica to produce zeolites due to its high silica content.

The objectives of this research are to investigate the effect of sodium hydroxide (NaOH) concentration and the reaction temperature on the production of zeolites and to characterize the zeolite obtained by using X-ray fluorescence (XRF) and X-ray diffraction (XRD).

75.2 Methodology

75.2.1 *Materials*

Rice husk ash (RHA) is a byproduct produced from the incinerated rice husk obtained from a local rice mill. RHA was ground and sieved to obtain 150 μm particle size powder. Table 75.1 shows the chemical composition of RHA which was obtained using XRF.

75.2.2 *Hydrothermal Treatment Method*

In this study, sodium hydroxide (NaOH) was used as an alkali source for zeolite synthesis using the hydrothermal treatment method. In each test, 10 g of dry sample (8 g of RHA + 2 g of kaolin) was mixed and stirred in NaOH solution to obtain a homogeneous mixture. The mixture was then aged and crystallized for 5 h in a shaking water bath. The synthesis was performed at different NaOH concentrations of 1 and 2 M under different reaction temperatures of 60, 80, and 100 °C.

Table 75.1 Chemical composition of RHA

Oxides	Rice hush ask
SiO_2	92.4
Al_2O_3	–
Fe_2O_3	2.26
CaO	1.96
TiO_2	0.07
K_2O	2.06
MgO	1.25

After the alkaline hydrothermal treatment was performed, the solid particles were separated from the alkaline solution by filtration and washed several times until all traces of excess alkali are removed. Then, dried in the oven overnight at 60 °C and stored in desiccators for further characterization. RHA and the product were characterized by using X-ray fluorescence (XRF), X-ray diffraction (XRD), and scanning electron microscopy (SEM).

75.3 Results and Discussion

75.3.1 Alkaline Hydrothermal Treatment

The precursor elements of the alkaline hydrothermal conversion into zeolites were based on the silicon (Si) and aluminum (Al) content in the raw materials. The XRF result in Table 75.1 shows the absence of aluminum oxide (Al_2O_3) in the rice husk ash (RHA). Kaolin is added into the RHA to add in some concentration of Al_2O_3, so that zeolites synthesis can be performed. As stated by Atta et al. [5] in their report, nearly, half of the kaolin is made up of Al_2O_3 (44.8%), 50.9% of SiO_2, and other compounds that are present in trace amounts.

75.3.2 Synthesis of Zeolite

Figure 75.1 shows the XRD pattern of zeolite produced under a reaction temperature of 80 °C, NaOH concentration of 2 M, and RHA to NaOH ratio of 1:5. It is observed that there are a total of three crystalline zeolite phases that are present in the produced zeolite. The crystalline zeolite phases are gismondine ($CaAl_2Si_2O_8 \cdot 4H_2O$), gismondine (tetramethylammonium exchanged), syn ($(CH_3)_4NAlSi_3O_8$), and zeolite K-F, (Na) ($Na_5Al_5Si_5O_{20} \cdot 9H_2O$).

Figure 75.2 shows the surface morphology of the zeolite produced. Zeolite has a microporous structure. The size of zeolite produced is ranging from 15 to 20 μm.

75.3.3 Effect of Reaction Temperature

The effect of reaction temperature on the end products after 5 h of reaction time has been investigated. The experiment has been done at a NaOH concentration of 1 M with RHA to NaOH ratio of 1:5. Table 75.2 shows the chemical composition of the end product after hydrothermal treatment at different reaction temperatures.

The minimum total concentration of SiO_2 and Al_2O_3 is at the reaction temperature of 60 °C which is at only 64.4%. Whereas at the reaction temperature of 80 °C, the total concentration of SiO_2 and Al_2O_3 is at maximum (70.5%). Increasing the temperature up to 100 °C will change the zeolites formed [6].

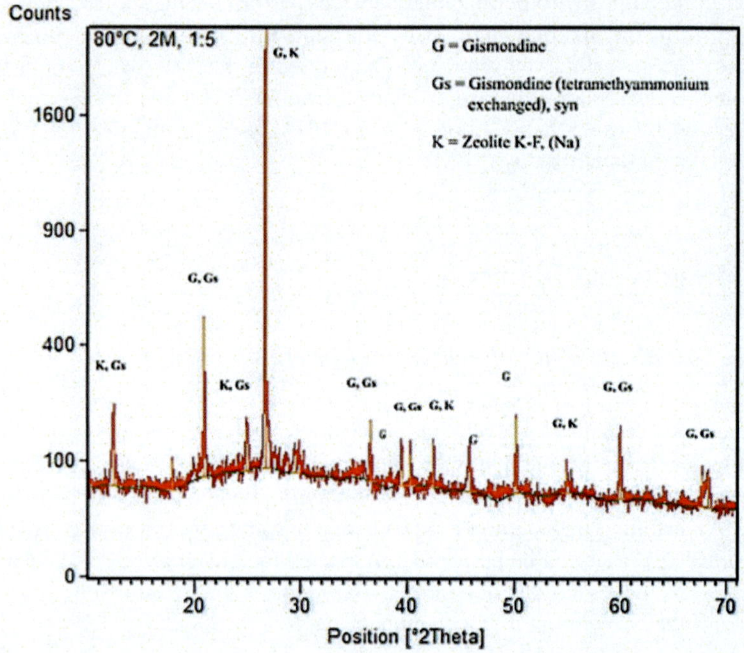

Fig. 75.1 XRD pattern of zeolite under reaction temperature of 80 °C, NaOH concentration of 2 M, and RHA/NaOH ratio of 1:5

Figure 75.3 illustrates the XRD spectrum of zeolite under different reaction temperature that was conducted at a NaOH concentration of 1 M and a ratio of RHA to NaOH of 1:5 for 5 h of crystallization time. As seen in the XRD pattern, the gismondine and zeolite K-F crystals are observed to grow the most at the temperature of 80 °C. The main peak of gismondine is observed to be located at $2\Theta = 26.71°$.

75.3.4 Effect of NaOH Concentration

The effect of NaOH concentration on the end products after 5 h of reaction time has been investigated. The experiment has been done under a reaction temperature of 80 °C with RHA to NaOH ratio of 1:5. The chemical composition of the end product after hydrothermal treatment using different NaOH concentrations is shown in Table 75.3.

Table 75.3 clearly stated that when NaOH concentration increase, the total concentration of SiO_2 and Al_2O_3 also increases. Increasing the concentration of NaOH can also increase the concentration of hydroxide (OH^-) ions. The presence of OH^- ion in the alkali solution helped in extracting the silica from the RHA. The sodium (Na^+) ion that is present in the alkali solution will contribute to the crystallization of the

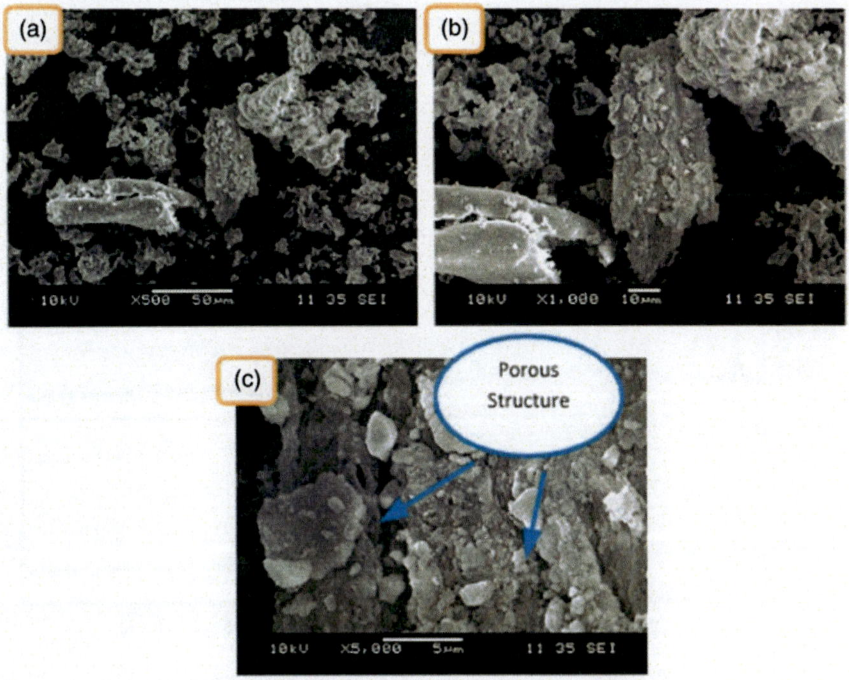

Fig. 75.2 SEM micrograph of zeolite produced at various magnifications

Table 75.2 Chemical composition of samples after alkaline hydrothermal treatment at various reaction temperatures

Oxides	60 °C	Concentration (%) 80 °C	100 °C
SiO_2	59.9	59.2	57.9
Al_2O_3	4.5	11.3	10.8
MgO	5.10	4.5	4.0
K_2O	4.49	4.09	3.64
CaO	19.9	16.3	15.7
TiO_2	1.06	0.89	0.93
Fe_2O_3	4.24	3.12	3.38

zeolite produced [7]. The higher the NaOH concentration, the higher the amount of zeolites that can be produced.

Figure 75.4 shows the XRD pattern of zeolites produced at a hydrothermal temperature of 80 °C and RHA to NaOH ratio of 1:5 using different NaOH concentrations. From the peak, it can be observed that the quartz was slowly dissolved in the alkali medium, which is NaOH, and was crystallized to form zeolites. The quartz was found as the minor phase, while gismondine was present as a major phase. Increasing the NaOH concentration can ease quartz dissolution [7]. Quartz will be fully dissolved and form zeolite gismondine at a higher NaOH concentration.

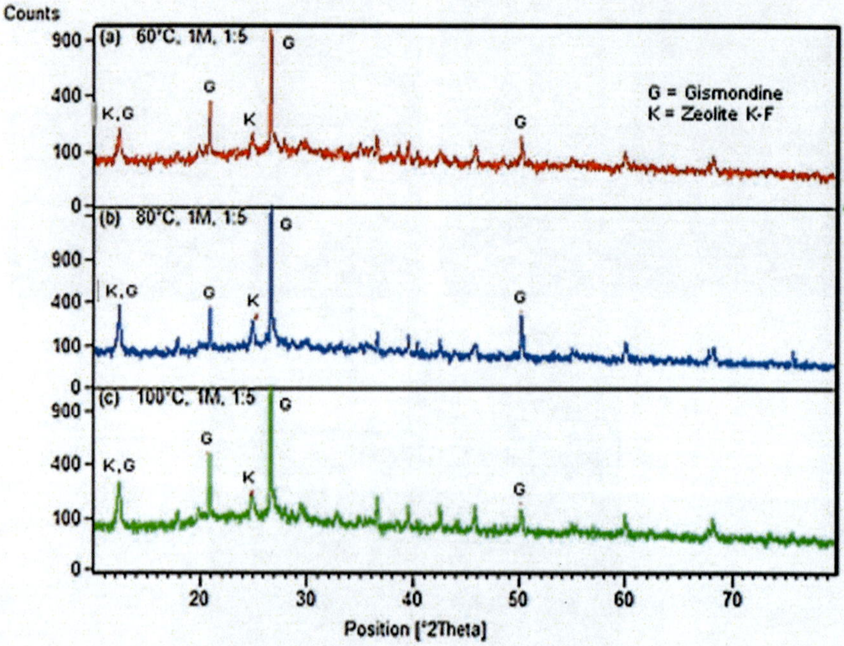

Fig. 75.3 **a, b, c** XRD patterns of zeolite prepared at various hydrothermal temperatures conducted at a 1 M NaOH solution and treated RHA: NaOH ratio of 1:5 for 5 h

Table 75.3 Chemical composition of samples after hydrothermal treatment with various NaOH concentrations

Compound	Concentration (%)	
	1 M	2 M
SiO_2	59.2	63.1
Al_2O_3	11.3	10.1
MgO	4.5	4.0
K_2O	4.09	4.27
CaO	16.3	13.4
TiO_2	0.885	0.84
Fe_2O_3	3.12	3.28

75.4 Conclusion

The optimum condition in producing zeolites using RHA via alkaline hydrothermal treatment is by using a reaction temperature of 80 °C and NaOH concentration of 2 M at a reaction time of 5 h. The total concentration of SiO_2 and Al_2O_3 is also at the optimum point. Zeolite gismondine was produced at the end of the experiment after alkaline hydrothermal treatment of RHA using 2 M of concentration of NaOH solution under a reaction temperature of 80 °C.

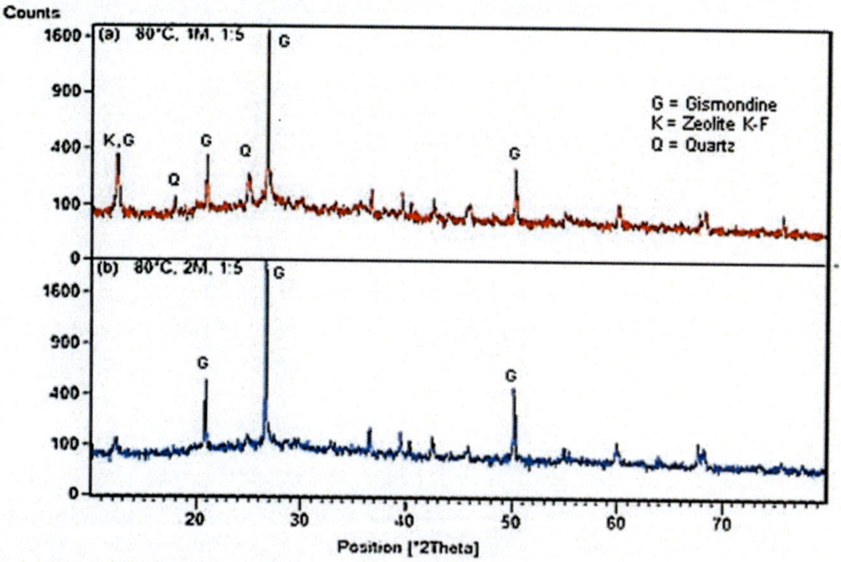

Fig. 75.4 a, b XRD patterns zeolites at different NaOH concentrations were conducted at a hydrothermal temperature of 80 °C and treated RHA: NaOH ratio of 1:5 for 5 h

Acknowledgements The authors would like to thank Universiti Malaysia Perlis, especially the Faculty of Chemical Engineering and Technology for allowing us to do this study.

References

1. T. Wajima, O. Kiguchi, K. Sugawara, T. Sugawara, Synthesis of zeolite—a using silica from rice husk ash. J. Chem. Eng. Jpn **42**(SUPPL. 1), 61–66 (2009)
2. T. Abdullahi, Z. Harun, M.H.D. Othman, A review on sustainable synthesis of zeolite from kaolinite resources via hydrothermal process. Adv. Powder Technol. **28**(8), 1827–1840 (2017)
3. B. Singh, *Rice Husk Ash* (Elsevier Ltd, 2018)
4. C.R. Melo et al., Production of zeolite from rice husk ash. Mater. Sci. Forum **798–799**, 617–621 (2014)
5. A.Y. Atta, B.Y. Jibril, B.O. Aderemi, S.S. Ade, Preparation of analcime from local kaolin and rice husk ash. Appl. Clay Sci. **61**, 8–13 (2012)
6. K. Yamni, N. Tijani, A. Aouidate, Preparation of zeolites by different bases with 1.5 of initial ratio Si/Al: the effect of crystallization temperature. IOSR J. Appl. Chem. **8**(12), 63–66 (2015)
7. F.C. Pa, A. Chik, Effect of NaOH concentration and reaction time on zeolite synthesized from treated oil, *Effect of NaOH Concentration and Reaction Time on Zeolite Synthesized from Treated Oil Palm Ash*, vol. 020214 (2018)

Chapter 76
Tribological Investigation of 2D Ti_3C_2 MXene Via Microwave-Assisted Hydrothermal Synthesis as Additives for Different Lubrications

H. A. Zaharin, M. J. Ghazali, A. Numan, M. Khalid, N. Thachnatharen, F. Ezzah, and A. K. Rasheed

Abstract In this study, Ti_3C_2 MXene additives were synthesised by using a microwave-assisted hydrothermal technique which was dispersed in two different oil samples (marine engine oil and hydraulic oil). An X-ray diffraction (XRD) was conducted to validate the formation of MXene. The tribological properties of Ti_3C_2 MXene were investigated by using a 4-ball tribotester. The results revealed a reduction in the coefficient of friction (COF) by 4.6 and 26.2% in MXene/marine oil and MXene/hydraulic oil, respectively, compared to the base oils. Improvement of the wear scar diameter (WSD) by 7.3 and 8.1% in MXene/marine oil and MXene/hydraulic oil, respectively, were also observed. This showed that the Ti_3C_2 MXene additives improved the tribological performance by forming a protective film which avoids direct contact between the sliding surfaces, resulting in a significant COF and wear reduction.

H. A. Zaharin · M. J. Ghazali (✉)
Department of Mechanical and Manufacturing Engineering, Faculty of Engineering and Built Environment, Universiti Kebangsaan Malaysia, 43600 Bangi, Selangor, Malaysia
e-mail: mariyam@ukm.edu.my

A. Numan · M. Khalid
Graphene and Advanced 2D Materials Research Group (GAMRG), School of Engineering and Technology, Sunway University, No. 5, Jalan Universiti, Bandar Sunway, 47500 Petaling Jaya, Selangor, Malaysia

N. Thachnatharen
Faculty of Defence Science and Technology, National Defence University of Malaysia, Kuala Lumpur, Malaysia

F. Ezzah
Department of Chemical and Environmental Engineering, Malaysia-Japan International Institute of Technology (MJIIT), Universiti Teknologi Malaysia, Jalan Sultan Yahya Petra, 54100 Kuala Lumpur, Malaysia

A. K. Rasheed
Department of New Energy Science and Engineering, School of Energy and Chemical Engineering, Xiamen University Malaysia (XMUM), Jalan Sunsuria, Bandar Sunsuria, 43900 Sepang, Selangor, Malaysia

M. A. A. Mohd Salleh et al. (eds.), *Proceedings of the Green Materials and Electronic Packaging Interconnect Technology Symposium*, Springer Proceedings in Physics 289,
https://doi.org/10.1007/978-981-19-9267-4_76

Keywords 2D · Ti_3C_2 · MXene · Microwave synthesis · COF · Additives

76.1 Introduction

In 2011, MXenes were discovered and joined the two-dimensional materials' family. MXenes were derived from the precursor MAX phase through selective etching. The first etching strategy involved the usage of a high concentration of hazardous hydrofluoric acid, HF [1]. Different approaches like hydrothermal heating [2], solvothermal heating [3], and electrochemical etching [4] with various kinds of etchant [4–6] have successfully produced the 2D MXene without involving direct handling of HF. On the other hand, these methods demand higher etching time and higher temperatures to complete the MAX phase exfoliation. Furthermore, none of the studies focuses on speeding up the MXene etching procedure. Our recent work successfully achieved a high-quality MXene by combining microwave and hydrothermal methods [7].

Microwave-assisted hydrothermal synthesis method works by combining hydrothermal and microwave mechanisms, which fully utilise the benefits of water heat and microwave irradiation [8]. The heat penetration improved with the assistance of microwave. Unlike conduction heating by conventional hydrothermal, microwave-assisted hydrothermal offers homogeneous heating, where heat distributes evenly to the sample at a time, hence improving the chemical reaction speed, enhancing the heating process, and reducing the reaction time [7–9]. Other advantages of the microwave-assisted hydrothermal approach include high yield and efficiency, consistent distribution of small particles, excellent phase purity, and energy- and cost-effectiveness [9].

Lately, increasing numbers of reports regarding the two-dimensional Ti_3C_2 MXene are being used to lubricate a system attributed to the low shear strength, high surface area, and excellent surface chemical stability [10–13]. Yet again, no studies have been reported regarding the tribological performance of Ti_3C_2 prepared via the microwave-assisted hydrothermal route.

Thus, this work was proposed to investigate the tribological properties of 2D Ti_3C_2 MXene as additives in various lubrications. Highly exfoliated MXene was prepared via microwave-assisted hydrothermal route with LiF/HCl in-situ method. The as-prepared 2D Ti_3C_2 MXene was dispersed in different oil samples as the additive, and the COF and WSD were determined using a four-ball tribometer. This study provides a new approach to synthesising MXenes, that can be useful for future tribology research and application.

76.2 Experimental

76.2.1 Preparation of MXENE Using Microwave-Assisted Hydrothermal Synthesis

The synthesis procedure was adopted from our previous work [7]. MAX phase, Ti_3AlC_2 400 mesh, purchased from Xiamen Tob New Energy Technology Co., LTD, was used as a precursor. Ten ml of 2.5 M HCl (Merck) and 6 M LiF (Sigma-Aldrich) were mixed to provide an in-situ environment for MXene synthesis. About 0.5 g MAX phase powder was slowly immersed into LiF/HCl solution. The mixture was stirred for 30 min and was sonicated for 30 min. The mixture was then placed in a microwave-assisted hydrothermal FlexiWAVE Microwave Synthesis Platform for synthesis at 40 °C for 30 min. After the reaction was completed, the mixture was washed with DI water and ethanol several times until MXene colloidal was attained. The colloidal was sonicated for 1 h in a water bath sonicator in argon condition. In order to obtain the Ti_3C_2 MXene powder, the supernatant from the colloidal was frozen and dried using a freeze-dryer.

76.2.2 Characterisation

The elemental composition of 2D Ti_3C_2 established using microwave-assisted hydrothermal synthesis was characterised utilising BRUKER D8 Advance X-ray diffractometer (XRD).

76.2.3 Preparation of MXENE Nanolubricants

About 0.005 wt% of nanolubricant samples were prepared by dispersing as-synthesised Ti_3C_2 powder nanoparticles' samples in marine engine oil and hydraulic oil. The samples were subjected to homogenisation at 7000 rpm for 10 min using a homogeniser and direct ultrasonication using a water bath sonicator for 30 min to ensure uniform dispersion.

76.2.4 Tribological Testing

The tribological testing using a four-ball tribometer (DUCOM) was demonstrated as per ASTM D 4172 (temperature: 75 °C, speed: 1200 rpm, weight: 40 kg, duration: 3600 s) to investigate the friction and wear properties of Ti_3C_2 nanolubricant in different oil samples. Before the experiment started, all equipment was cleaned with ethanol to prevent contamination. The rotating ball was fixed, pushing other stationary three balls immersed in the nanolubricant. The terminal data processor provided the COF of nanolubricant at the end of the experiment. At the same time, the WSD was measured using the image acquisition system.

76.3 Results and Discussion

76.3.1 XRD Analysis

Figure 76.1 shows the XRD analysis of the MAX phase (ICDD No: 00-052-0875) before and after etching using microwave-assisted hydrothermal synthesis. The formation of 2D Ti_3C_2 MXene is displayed in the XRD pattern [1, 7]. The intense (002) peak detected at 9.6° in the MAX phase broadened and shifted towards a lower angle, 6.6°. The result demonstrates that the Ti–Al bond was disrupted, which caused an increase in the c lattice parameter [3]. Our result suggests that by using a microwave-assisted hydrothermal method, the Ti_3AlC_2 MAX phase can be etched to form Ti_3C_2 MXene.

76.3.2 Friction and Wear Analysis

Figure 76.2a exhibits the COF of different oil samples with and without the 0.005 wt% Ti_3C_2 MXene added as an additive. The COF of pure marine oil and hydraulic oil was 0.1252 and 0.0861, respectively. The reduction of COF for both samples was observed after adding 0.005 wt% of Ti_3C_2 MXene into the oil samples. The COF of marine oil with 0.005 wt% Ti_3C_2 MXene was reduced by 4.6% compared to pure marine oil, while COF of hydraulic oil with 0.005 wt% Ti_3C_2 MXene was decreased

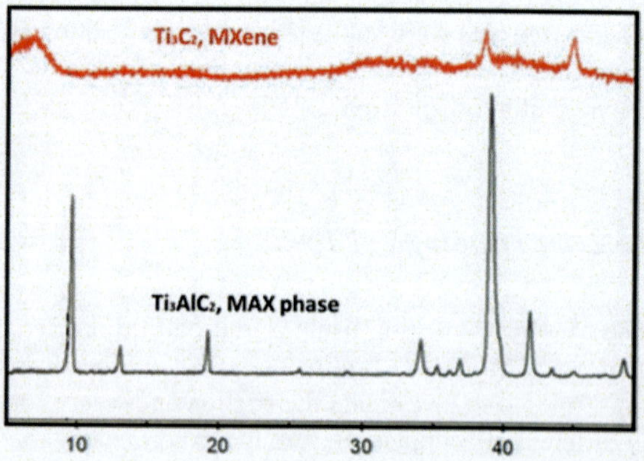

Fig. 76.1 XRD pattern of MAX phase and microwave-assisted hydrothermal synthesis Ti_3C_2 MXene

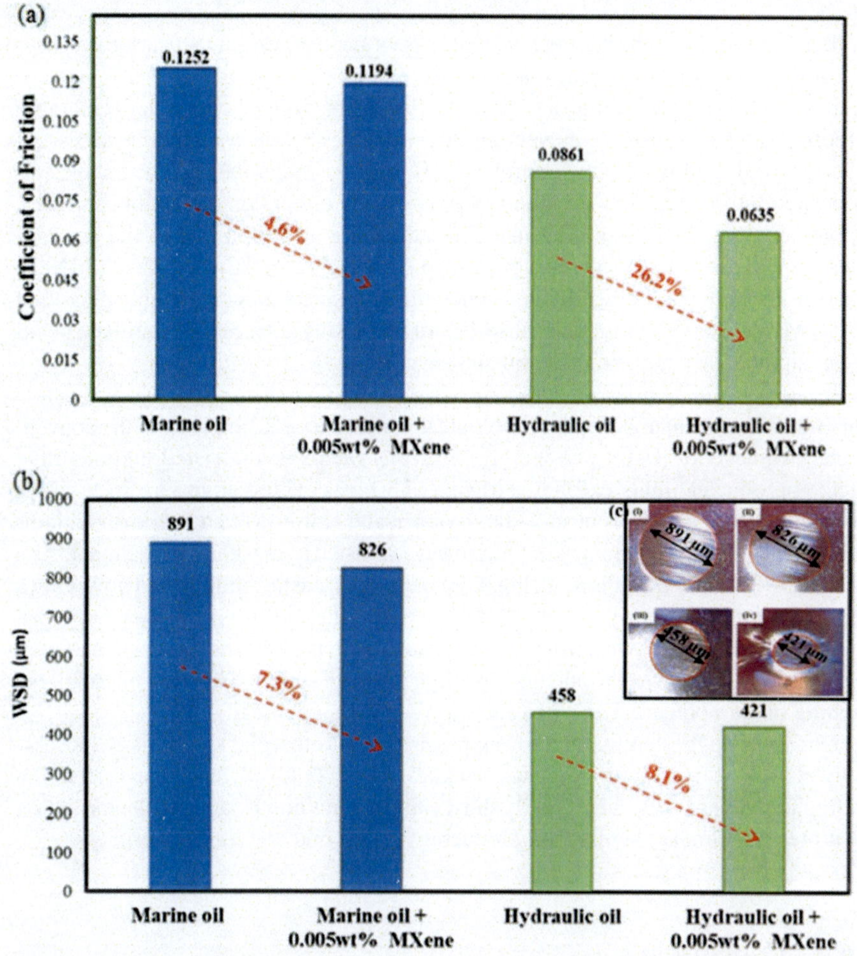

Fig. 76.2 Reduction of **a** COF and **b** average WSD of different oil samples with MXene additive. **c** Microscopic images of WSD with (i) marine oil, (ii) marine oil + 0.005 wt% MXene, (iii) hydraulic oil, and (iv) hydraulic oil + 0.005 wt% MXene

by 26.2% compared to pure hydraulic oil. The reduction of COF with the addition of 2D Ti_3C_2 MXene is attributed to the formation of a protective layer, also known as a tribofilm. The mechanism of formation of tribofilm was subjected to the constant force supplied during the rubbing process, which induced the building of 2D Ti_3C_2 MXene tribofilm. The tribofilm prevents direct contact between the sliding surfaces.

From Fig. 76.2a, it was observed that the average WSD was reduced by 7.3 and 8.1%, respectively, when 0.005 wt% MXene was added to marine oil and hydraulic oil as an additive compared to the base oil. The reduction in COF was directly proportional to the decrease in WSD. The microscopic images reveal microscale scratches (valleys and ridges). The sliding between valleys and hills on contact surfaces causes friction to develop. In agreement with the previous study, adding Ti_3C_2 MXene to base oil assists in filling the spaces between the valleys and hills, building a tribofilm, and successively preventing direct contact between mating surfaces. As a result, COF and wear rates improved [13]. In Fig. 76.2c, the MXene nanosheets deposited into the scratch's region cause "mending effect" on the sliding surfaces, minimising the WSD size. The mending effect was occurred due to the arrangement of MXene nanosheets during the high shearing process, where Ti_3C_2 MXene nanosheets were parallelly arranged under high pressure and force. Highly pressured Ti_3C_2 MXene nanosheets tend to absorb onto the mating surfaces and facilitate lower wear, significantly reducing friction and wear [13, 14].

In addition, the synthesis method to prepare 2D Ti_3C_2 MXene was also responsible for enhancing the nanolubricant tribological performance. Compared to the conventional synthesis of Ti_3C_2 MXene [15, 16], this microwave-assisted hydrothermal method produced highly exfoliated MXene. This was evidenced in Fig. 76.1, where an intense (002) peak was observed to broaden and shift toward a lower angle, indicating high interlayer spacing after microwave-assisted hydrothermal treatment. As a result, low Van der Waals is reducing the shear forces, easing sliding, and providing a lubricating effect. As the outcome, MXene added as the additive in oil samples helped in reducing friction between friction pairs. Besides that, since this method dealt with the heat from microwave irradiation and high pressure supplied by hydrothermal, the etching time to prepare Ti_3C_2 MXene was reduced dramatically.

Our work suggested that microwave-assisted hydrothermal synthesis is successfully etched Al from the MAX phase and developed Ti_3C_2 MXene nanosheets. The individual MXene nanosheets easily shear and slip, thus contributing to the formation of protective film, are significantly reducing friction and increasing lubricity.

76.4 Conclusion

In summary, the microwave-assisted hydrothermal synthesis is confirmed to successfully etch the MAX phase and form Ti_3C_2 MXene at 40 °C for 30 min. About 0.005 wt% of as-synthesised 2D Ti_3C_2 MXene added into different oil samples can effectively enhance the tribological performance of the lubricant. The tribological analysis demonstrated that Ti_3C_2 MXene prepared using this method can easily shear and improve lubricity by forming a protective film and avoiding direct contact between sliding surfaces, resulting in significant COF and wear reduction. Furthermore, mending effect contributed from 2D Ti_3C_2 MXene nanosheets as the additives in lubricant facilitated in significant reduction of friction and wear. Compared to other MXene synthesis methods, MXene produced through a microwave-assisted hydrothermal approach is safer, saves time and energy, and contributes to excellent tribological properties.

Acknowledgements The research work was funded by the Ministry of Higher Education and Universiti Kebangsaan Malaysia (FRGS/1/2018/TK03/UKM/02/8). The authors would like to acknowledge the invaluable assistance and support from Graphene and Advanced 2D Materials Research Group (GAMRG) members, Sunway University. The authors also acknowledged the Centre for Research and Instrumentation Management (CRIM) for the analytical services.

References

1. M. Naguib et al., Two-dimensional nanocrystals produced by exfoliation of Ti3AlC2. Adv. Mater. **23**(37), 4248–4253 (2011). https://doi.org/10.1002/adma.201102306
2. L. Wang et al., Synthesis and electrochemical performance of Ti3C2Tx with hydrothermal process. Electron. Mater. Lett. **12**(5), 702–710 (2016). https://doi.org/10.1007/s13391-016-6088-z
3. C. Peng et al., A hydrothermal etching route to synthesis of 2D MXene (Ti3C2, Nb2C): Enhanced exfoliation and improved adsorption performance. Ceram. Int. **44**(15), 18886–18893 (2018). https://doi.org/10.1016/j.ceramint.2018.07.124
4. S.-Y. Pang et al., Universal strategy for HF-free facile and rapid synthesis of two-dimensional MXenes as multifunctional energy materials. J. Am. Chem. Soc. **141**(24), 9610–9616 (2019). https://doi.org/10.1021/jacs.9b02578
5. J. Halim et al., Transparent conductive two-dimensional titanium carbide epitaxial thin films. Chem. Mater. **26**(7), 2374–2381 (2014). https://doi.org/10.1021/cm500641a
6. M. Ghidiu, M.R. Lukatskaya, M.-Q. Zhao, Y. Gogotsi, M.W. Barsoum, Conductive two-dimensional titanium carbide 'clay' with high volumetric capacitance. Nature **516**(7529), 78–81 (2014). https://doi.org/10.1038/nature13970
7. Numan et al., Microwave-assisted rapid MAX phase etching and delamination: a paradigm shift in MXene synthesis. Mater. Chem. Phys. **288**, 126429 (2022). https://doi.org/10.1016/j.matchemphys.2022.126429
8. G. Yang, S.J. Park, Conventional and microwave hydrothermal synthesis and application of functional materials: a review. Materials (Basel) **12**(7) (2019). https://doi.org/10.3390/ma12071177
9. O. Gerard, A. Numan, S. Krishnan, M. Khalid, R. Subramaniam, R. Kasi, A review on the recent advances in binder-free electrodes for electrochemical energy storage application. J. Energy Storage **50**, p. 104283 (2022). https://doi.org/10.1016/j.est.2022.104283
10. H.T. Nguyen, K.-H. Chung, Assessment of tribological properties of Ti3C2 as a water-based lubricant additive. Materials (Basel) **13**(23), 5545 (2020). https://doi.org/10.3390/ma13235545
11. Z. Ji et al., Mechanical and tribological properties of nanocomposites incorporated with two-dimensional materials. Friction **8**(5), 813–846 (2020). https://doi.org/10.1007/s40544-020-0401-4
12. J. Chen, W. Zhao, Simple method for preparing nanometer thick Ti3C2TX sheets towards highly efficient lubrication and wear resistance. Tribol. Int. **153**, 106598 (2021). https://doi.org/10.1016/j.triboint.2020.106598
13. M. Xue et al., Preparation of TiO2/Ti3C2Tx hybrid nanocomposites and their tribological properties as base oil lubricant additives. RSC Adv. **7**(8), 4312–4319 (2017). https://doi.org/10.1039/c6ra27653a

14. X. Zhang et al., Preparation and tribological properties of Ti3C2(OH)2 nanosheets as additives in base oil. RSC Adv. **5**(4), 2762–2767 (2015). https://doi.org/10.1039/c4ra13800g
15. O. Mashtalir, M. Naguib, B. Dyatkin, Y. Gogotsi, M.W. Barsoum, Kinetics of aluminum extraction from Ti3AlC2 in hydrofluoric acid. Mater. Chem. Phys. **139**(1), 147–152 (2013). https://doi.org/10.1016/j.matchemphys.2013.01.008
16. O. Mashtalir et al., Intercalation and delamination of layered carbides and carbonitrides. Nat. Commun. **4**(1), 1716 (2013). https://doi.org/10.1038/ncomms2664

Chapter 77
Influence of the Part Orientation and Type of Used Photopolymer Resin on Surface Roughness in the Process of Digital Light Processing Technology

Jan Milde, Jozef Peterka, Marcel Kuruc, Jakub Hrbal, and Patrik Dobrovszky

Abstract The presented scientific article deals with the influence of the orientation of the model and used photopolymer resins during additive manufacturing using the Digital Light Processing (DLP) method on the resulting surface roughness of the models. The examined component is a cuboid with dimensions of 14 mm × 7 mm × 28 mm. A total of five orientations of the component in the *Y* axis (0°, 30°, 45°, 60°, 90°) were used in this experiment. The examined samples were produced on a Zortrax Inkspire DLP 3D printer, where three parts were produced for each orientation and material. The surface roughness parameters (Ra, Rz, Rq) were measured on the printed samples using a Mitutoyo SJ-210 roughness meter. In this experiment, a total of 30 samples were produced, i.e., three samples for each of the orientations for each material used. Considering the analysis of literary studies, we can conclude that the area of surface roughness in the process of DLP technology was addressed only by a minimum of authors. Regarding research on the influence of model orientation on surface roughness in the DLP technology process, the authors Hanon and Zsidai in [1] report that the orientation of the part model has a significant effect on surface roughness. In their experiment, they prove that the best orientation of the model in relation to the least surface roughness is at the 0° and 90° orientations, just as the authors of this article point out.

Keywords Surface roughness · DLP · Additive manufacturing · Resin

J. Milde (✉) · J. Peterka · M. Kuruc · J. Hrbal · P. Dobrovszky
Faculty of Materials Science and Technology in Trnava, Slovak University of Technology in Bratislava, Jána Bottu 2781/25, 917 24 Trnava, Slovakia
e-mail: jan.milde@stuba.sk

M. A. A. Mohd Salleh et al. (eds.), *Proceedings of the Green Materials and Electronic Packaging Interconnect Technology Symposium*, Springer Proceedings in Physics 289,
https://doi.org/10.1007/978-981-19-9267-4_77

77.1 Introduction

Modern additive manufacturing (AM) dates to the mid-1980s with the invent of Stereolithography. Several AM processes emerged in the late 1980s and early 1990s [2, 3]. The impact and use of AM continue to grow in both commercial and scientific activities. AM is the process to make real objects from 3D model data, usually, layer by layer [4]. It is also known as rapid manufacturing [5] or rapid prototyping [6]. Unlike conventional manufacturing techniques such as milling, drilling, or turning, which produce products by removing materials from a block of raw material, AM creates the final shape by adding materials. AM has the ability to efficiently use raw materials and produce minimal waste while achieving satisfactory geometric accuracy [4, 5]. According to ISO/ASTM 52900 [7], AM can be divided into seven groups: material jetting, material extrusion, binder jetting, directed energy deposition, powder bed fusion, sheet lamination, and vat photopolymerization. The light-based AM methods include Digital Light Processing (DLP) printing, laser-assisted printing, Selective Laser Melting (SLM), and Selective Laser Sintering (SLS) [8–10]. The technique vat photopolymerization is first described in 1986 by Charles [11]. DLP and Stereolithography (SLA) can be grouped into vat photopolymerization-based AM approaches. The SLA-printing process is based on light exposure carried out by scanning a concentrated ultraviolet laser beam on the surface of the photopolymer. The DLP-based printing process also relies on the photopolymerization process, with the difference that every single layer is polymerized all-at-once by projecting a single digital image. When compared, both technologies use the same resins and create model layer by layer. The main difference is that while SLA uses a laser to cure the resin, DLP uses ultraviolet light instead [12–14]. DLP technology uses controlled square micromirrors with an edge length of approximately 16 μm. The light source is controlled optically either onto the build platform, which resides in a translucent vat of photopolymer or onto a diffuse surface (absorber). This is made possible by tilting the individual micromirrors in the unit, which are triggered by forces exerted by electrostatic fields [15, 16]. DLP uses a platform that sits in a vat containing a liquid epoxy resin or an acrylate resin. During the additive process, the platform sits just below the surface of the resin and an elevator is used to incrementally lower the platform after each layer of the photosensitive polymer has been exposed to ultraviolet light by the highly focused laser beam. The laser scans the predefined area according to the slice information and cures the resin in a defined penetration depth. Using ultraviolet light instead of the laser allows the DLP printer to illuminate the whole layer at once and that reduces building time. While DLP offers faster building time, SLA on the other hand can produce parts with better quality, since the resolution in SLA is defined by minimum laser spot size as opposed to minimum pixel size in DLP [17–19]. A schematic representation of SLA and DLP techniques is shown in Fig. 77.1.

Several authors already focused on the surface quality in the process of DLP. The authors Beghari [20], Barone [21], Dikova [22], and Keßler [23] were involved in the research of various parameters of DLP technology and their subsequent influence

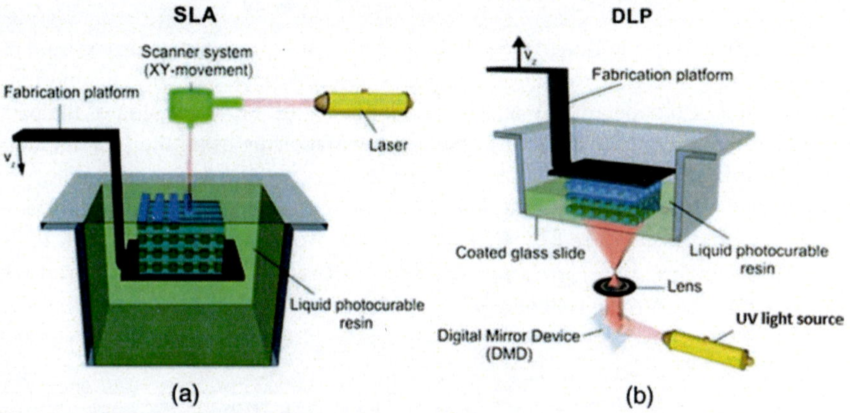

Fig. 77.1 Schematic representation of SLA and DLP technique. **a** SLA technology, **b** DLP technology [16]

on the resulting surface roughness. They even dealt with the comparison of two technologies, mainly DLP and FDM from the point of view of surface roughness. As samples, these authors chose models from the field of dentistry. However, after studying these literatures, we can conclude that even though in their research these authors focused on the influence of various parameters of DLP technology and their subsequent influence on the surface roughness, none of them dealt with the orientation of the model and the influence of the polymer itself expect of M. M. Hanon and L. Zsidai. These authors in [1] focused on the orientation of the model on the resulting surface roughness, where they declared in this publication that the best orientation in the point of view of lowest surface roughness is the orientation at 0° and 90° and they reported that orientation has significant influence on surface roughness in the process of DLP technology. Milde et al. [24] investigated the influence of the used photopolymers on the surface roughness and dimensional accuracy of the sample in the production process using DLP technology. After studying the literature in the field of the influence of model orientation on surface roughness, we can conclude that this area has been devoted to a minimum of literature, and therefore, the authors of this article decided to penetrate better and deeper into this issue. Part orientation is an important parameter in the planning of a DLP process as it directly governs productivity, part quality, and cost of manufacturing. From previous research, it was confirmed that the orientation of the model significantly affects the quality and roughness of the surface, and the authors decided to complete an experiment focusing on the quality of DLP because this technology is expected to have the best surface quality and the lowest surface roughness. In this experiment, we also focused on the comparison of two materials of different types, the first, which is used for ordinary prints, and the second, much more significant and important for the authors and other users, material from the castable category. Photopolymer is very often used in the jewelry industry, and therefore, it is necessary to know and set the best possible production parameters in order to achieve the best possible surface quality and the lowest roughness. In addition to the jewelry industry, this material can also be used in general casting, where the roughness and quality of the surface also play an important role. The aim of this paper is to analyze the influence of the orientation of the part and used photopolymer resin during additive manufacturing using the DLP method on the resulting surface roughness of the models.

The article is intended to bring better knowledge to achieve increased surface quality of samples produced by DLP technology. Authors of this article in their previous researches [25–27] focused on influence of various parameters and various AM techniques of surface roughness.

77.2 Methodology and Geometrical Model of Roughness of DLP Process

In this experiment, we printed a total of 30 prints, three prints for each orientation for each resin. Two various resins were used. The first used resin was *Zortrax Basic White/Ivory*, a common resin for general usage. The second used resin was *Bluecast X10* which is a castable resin, widely used in jewelry production. Samples printed from *Basic White/ivory* resin were cured in a UV chamber Formlabs Cure in the post-processing phase. Curing took 30 min at 40 °C. Samples produced from castable resin were not subjected to post-processing as that step is not recommended by manufacturer of resin. Support structures used were tree-type structures sprouting only from raft. Process parameters used for manufacturing of samples can be seen in Table 77.1.

Surface tester Mitutoyo SJ 210 was used to measure the surface roughness of all samples produced from both resins. The roughness was measured at a distance of 2.4 mm three times in three different areas for each sample to eliminate discrepancies and random errors. The authors measured three surface roughness parameters in total (Ra, Rz, and Rq).

Table 77.1 Manufacturing process parameters

Resin	Zortrax basic white/ivory	Bluecast X10
Layer thickness (mm)	0.05	0.05
Layer exposure time (s)	8	11.5
Raft height (mm)	1	1
Bottom layers	3	5
Bottom layer exposure time (s)	80	110
Z-lift distance (mm)	5	5
Print time	3 h 38 min	3 h 59 min

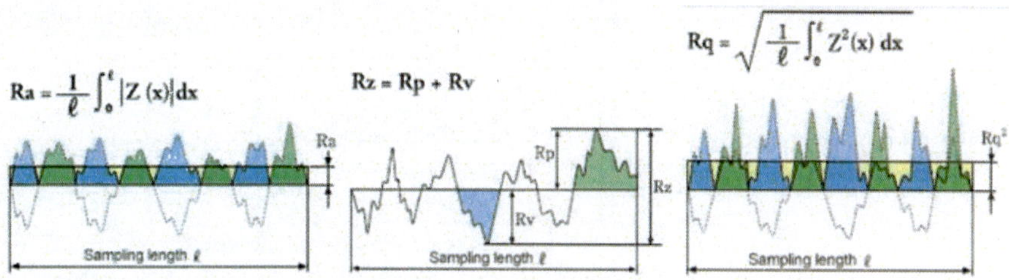

Fig. 77.2 Visual definition of the most used surface roughness parameters [30]

The quality of the product's surface, especially its texture, is determined by the surface roughness of the product. The surface roughness is the most often measured by devices as surface testers and profilometers, which utilize contact principle. Besides contact methods of surface roughness measuring, there are also contactless methods, which utilize, e.g., laser or atomic forces [28, 29]. There are known many surface roughness parameters (linear parameters, planar parameters, bearing ratio, skewness, kurtosis, etc.). Parameters Ra, Rz, and Rq are the most often used ones of them all. Their definition is visualized in Fig. 77.2.

The quality of the newly formed additive surface is affected by the layer thickness, the additive head step, the pixel size, or the size of the laser beam. Here, you can see an analogy with the technology of material removal—machining with machining parameters: feed rate [31–33] and tool diameter size (tip radius) [34–36]. These machining parameters also affect the accuracy of the achieved dimensions [37, 38].

For the theoretical calculation of the roughness of the DLP additive surface, we used analytical pixel analysis. We used the following theoretical assumptions for our analysis:

- We do not consider the inaccuracy of pixel dimensions,
- we do not consider deformation of the material during cooling,
- we do not consider the oscillations and vibrations of the 3D display in any direction (X, Y, Z).

Our geometric model is based on the arrangement of pixels around the theoretical profile (shape). The squares represent voxels with a height Lt and with the dimensions of the base in the XY plane with the size of pixels of the same resolution. Gray squares represent the material after curing. White squares represent uncured material. Figure 77.3 shows a geometric model of the DLP profile with examples of the influence of the layer thickness Lt for different angles of profile inclination $\alpha =$ 0°, 30°, 45°, 60°, 90°.

Where Rz is the theoretical maximum height of the irregularities of the newly formed additive DLP surface, Lt is layer thickness, and PS_x is the size of the pixel in the *X*-axis direction.

We can say that the surface roughness in this case is a function of the layer thickness Lt, the pixel size $Ps_{x,}$ and the angle of inclination of the profile:

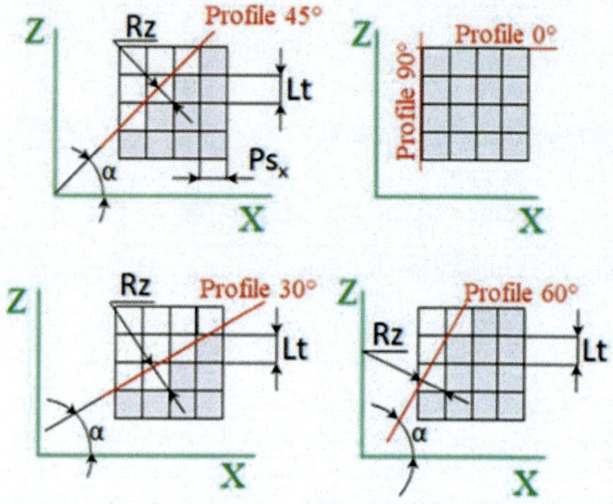

Fig. 77.3 Geometric model of DLP profile, influence of layer thickness for different angles of profile inclination α [24]

$$\mathrm{Rz} = f(\mathrm{Lt}, \mathrm{Ps}_x, \alpha). \tag{77.1}$$

The direction of probe movement has great impact on the results during the measuring of the surface roughness. The direction, in which the roughness is the greatest, is often the most important one. For products of AM, it is usually vertical direction—direction perpendicular on the created layers [39].

77.3 Results

In this experiment, three samples were manufactured for each orientation and for each of the resins used (Basic Ivory and Bluecast X10), so we manufactured 30 specimens in total. All surface roughness parameters (Ra, Rq, and Rz) were measured on each sample three times in a row for better statistical evaluation. Example of all measurements for each sample, orientation, and material is shown in Table 77.2. *WIB* means White/Ivory Basic, *AM* means arithmetical mean, *BCX10* means Bluecast X10.

Authors created the arithmetical mean of the three measurements for each sample, which authors then made the total arithmetic mean of all three samples for each orientation and for both materials. This evaluation is shown in Table 77.3. Due to the fact that the roughness parameters Rz and Rq showed the same dependencies as the Ra parameter, only the roughness parameter Ra was shown in the form of a table and in the form of a graph in the article.

On the graphs in Fig. 77.4, we can see the values of roughness Ra, Rq, and Rz for individual orientations and for both used photopolymers. According to this graph, we can say that the orientation of the part clearly affects the roughness of the surface as well as the material used. In Fig. 77.5, we can see a detail of the examined

Table 77.2 Example of all measurements for orientation 30°

30°	Ra (μm)				Rq (μm)				Rz (μm)			
	#1	#2	#3	#AM	#1	#2	#3	#AM	#1	#2	#3	#AM
WIB 1	9.47	8.10	9.97	9.18	11.44	9.80	12.02	11.09	46.93	39.21	48.37	44.84
WIB 2	10.34	8.57	9.11	9.34	12.33	10.35	10.89	11.19	47.76	41.49	43.15	44.13
WIB 3	9.35	8.68	10.25	9.43	11.19	10.50	12.28	11.32	45.01	42.24	49.60	45.61
WIB AM	9.32				11.20				44.86			

Table 77.3 Ra versus orientation (both materials)

Orientation (°)	0	30	45	60	90
Ra (basic white/ivory) (μm)	0.461	9.321	8.001	9.006	5.024
Ra (Bluecast X10) (μm)	1.671	4.198	3.589	6.042	0.981

surface magnified 37× using the CoolingTech Digital MicroScope microscope as an additional evaluation. The difference in surface roughness is highly visible.

According to the results obtained from this experimental research focused on the orientation in the DLP technology process, we can unequivocally claim that the orientation in the DLP process affects the resulting surface roughness for all three

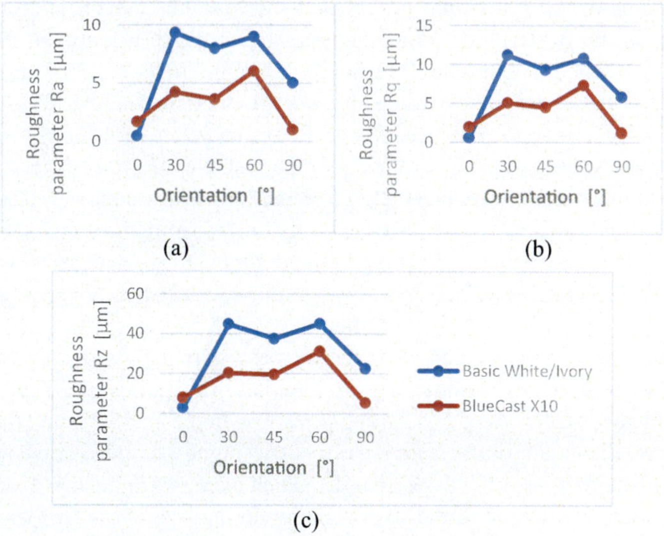

Fig. 77.4 Roughness versus orientation (both materials): **a** Ra, **b** Rq, **c** Rz

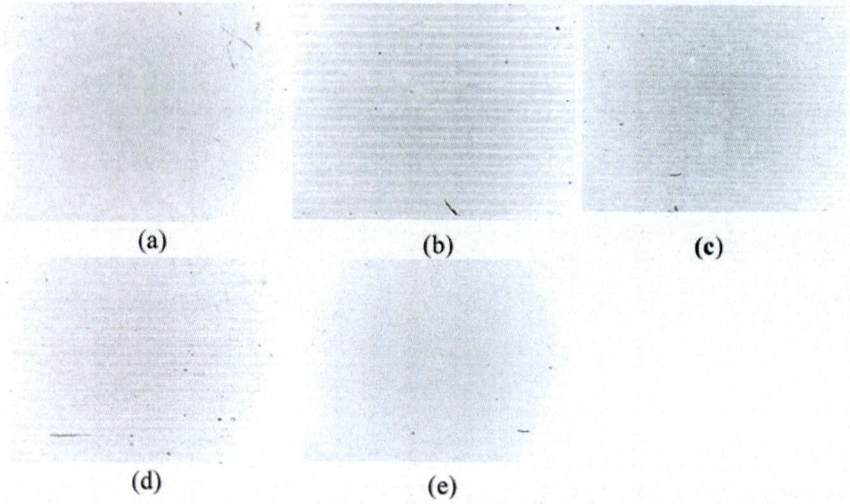

Fig. 77.5 Detail of surface at orientation: **a** 0°, **b** 30°, **c** 45°, **d** 60°, **e** 90°

investigated roughness parameters (Ra, Rq, and Rz). These dependencies can be seen on the graphs in Fig. 77.4. Based on the achieved results, we can also claim that also type of photopolymer used affects the roughness of the surface. In this section, it could be argued whether this difference in the two materials was not also caused by the post-processing phase, where the samples made of material from the Basic category were subsequently cured in the Formlabs Cure UV chamber for 30 min at a temperature of 40 °C. The graphs in Fig. 77.4 show that the most negligible surface roughness for the Basic Ivory/White material was at the 0° orientation, i.e., at the basic position, where the measured surface was parallel to the level of the liquid. In this case, the surface roughness values Ra were 0.461 μm. This was caused due to the fact that there was no staircase effect [40] at this position of the test specimen, and thus, the roughness was the smallest. For the castable material Bluecast X10, the roughness values were the smallest at 90° orientations with values of 0.981 μm. We consider this finding to be accidental because the surface roughness at the 0° position was very close to this value. According to the findings, the surface roughness value depended on the staircase effect. It is also possible to see that at 0° orientations, the surface roughness values were also smaller. In general, we can say that at 0° and 90° orientations, the roughness values were the smallest for both investigated photopolymers. The highest values of roughness appeared at orientations of 30° and 60° for both investigated materials. Surface roughness was highest at 30° and 60° orientations because the nature of the staircase effect was highest at these orientations. We can imagine the staircase effect as a triangle, which increases or decreases depending on the orientation and therefore also depending on the size of the hypotenuse, i.e., the longest side of the triangle opposite to the right angle. The staircase effect is a phenomenon associated with additive manufacturing, where layer marks are clearly visible on the surface of parts, creating the impression of a staircase. In the graphs, we can see a decrease at 45° and then a sudden increase in the roughness value. At the 45° orientation, the character of the staircase effect was smaller than at the mentioned two orientations. From a statistical point of view, it would be most advantageous to carry out this experiment with several orientations of the model at a scale of 1°, in order to penetrate more deeply into the issue of the influence of the orientation of the model on the surface roughness of the parts and to better understand the dependence of this parameter. The surface roughness parameters Rq and Rz showed the same dependencies as Ra, and these parameters are shown in Fig. 77.4b, c. In the case of comparing two investigated resins, the surface roughness values were smaller for the castable material Bluecast X10.

The different surface roughness values for the two investigated materials arose for three main reasons. The first and most important reason is that the samples made of Basic White Ivory were post-cured with UV light in the Formlabs Cure UV chamber in the post-processing phase, while the specimens made of castable X10 were not post-cured in the post-processing phase. The second reason is that each material used has different technical parameters. Among the parameters that can most cause these differences in roughness, the authors consider viscosity, density, glass transition temperature, and shrinkage value indicated by the manufacturer. The third reason that these differences in both materials could have arisen is that some production parameters were set differently in the software during the production itself. These parameters, which were set differently, include the exposure time of the layer, the number of underlying layers, and the exposure time of the underlying layer.

77.4 Conclusion

The authors consider the results of this experiment to be important and scientifically significant also because the quality and roughness of the surface are important for DLP technologies, and especially, for casting resin models. As for the achieved results, based on the conducted experiment, we can claim that the highest value of roughness was at the orientation of 30° for the White/Ivory Basic material and at 60° for the X10 material. The smallest roughness value was at 0° for the White/Ivory Basic material and 90° for the X10 material. In the future, it would be appropriate to focus on other parameters of the DLP technology, such as exposure time, layer thickness, and consequently, their influence on the resulting surface roughness. The authors also point out the importance of the influence of different photopolymers on the resulting surface roughness.

Acknowledgements The article was written with the support of the project VEGA 1/0019/20.

References

1. M. Hanon, L. Zsidai, Sliding surface structure comparison of 3D printed polymers using FDM and DLP technologies. In *IOP Conference Series: Materials Science and Engineering*. (IOP Publishing, 2020), p. 012015
2. D.L. Bourell, D.W. Rosen, M.C. Leu, The roadmap for additive manufacturing and its impact. 3D Print Addit. Manuf. **1**(1), 6–9 (2014). https://doi.org/10.1089/3dp.2013.0002
3. N. Guo, M.C. Leu, Additive manufacturing: technology, applications and research needs. Front. Mech. Eng. **8**(3), 215–243 (2013). https://doi.org/10.1007/s11465-013-0248-8
4. J. Jiang, X. Xu, J. Stringer, Support structures for additive manufacturing: a review. J. Manuf. Mater. Process. **2**(4) (2018). https://doi.org/10.3390/jmmp2040064
5. G.N. Levy, R. Schindel, J.-P. Kruth, Rapid manufacturing and rapid tooling with layer manufacturing (LM) technologies, state of the art and future perspectives. CIRP Ann. **52**(2), 589–609 (2003). https://doi.org/10.1016/S0007-8506(07)60206-6
6. J.-P. Kruth, M.C. Leu, T. Nakagawa, Progress in additive manufacturing and rapid prototyping. CIRP Ann. **47**(2), 525–540 (1998). https://doi.org/10.1016/S0007-8506(07)63240-5
7. ISO Aditive Manufacturing—General Principles—Terminology; ISOASTM 52900
8. T.D. Ngo, A. Kashani, G. Imbalzano, K.T.Q. Nguyen, D. Hui, Additive manufacturing (3D Printing): A review of materials, methods, applications and challenges. Compos. Part B: Eng. 143, 172–196 (2018). https://doi.org/10.1016/j.compositesb.2018.02.012
9. S.C. Ligon, R. Liska, J. Stampfl, M. Gurr, R. Mulhaupt, Polymers for 3D printing and customized additive manufacturing. Chem. Rev. **117**(15), 10212–10290 (2017). https://doi.org/10.1021/acs.chemrev.7b00074
10. J. Zhang, Q. Hu, S. Wang, J. Tao, M. Gou, Digital light processing based three-dimensional printing for medical applications. Int. J. Bioprint. **6**(1), 242 (2019). https://doi.org/10.18063/ijb.v6i1.242. PMID: 32782984; PMCID: PMC7415858
11. Ch.W. Hull, Ch.W. Lewis, R.S. Freed, S.T. Spence, W.A. Vinson, D.R. Smalley, *Stereolithographic Curl Reduction,* (European Patent Office, 1989). EP 0362982 A3
12. H. Bikas, P. Stavropoulos, G. Chryssolouris, Additive manufacturing methods and modeling approaches: a critical review. Int. J. Adv. Manuf. Technol. **83**(1–4), 389–405 (2016). https://doi.org/10.1007/s00170-015-7576-2
13. J. Borrello, P. Nasser, J.C. Iatridis, K.D. Costa, 3D printing a mechanically-tunable acrylate resin on a commercial DLP-SLA printer. Addit. Manuf. **23**, 374–380 (2018). https://doi.org/10.1016/j.addma.2018.08.019
14. J. Gardan, Additive manufacturing technologies: state of the art and trends. Int. J. Prod. Res. **54**(10), 3118–3132 (2016). https://doi.org/10.1080/00207543.2015.1115909
15. J. Schweiger, D. Edelhoff, J.F. Güth, 3D printing in digital prosthetic dentistry: An overview of recent developments in additive manufacturing. J. Clin. Med. **10**(9), (2010). https://doi.org/10.3390/jcm10092010. PMID: 34067212; PMCID: PMC8125828
16. C. Feng, K. Zhang, R. He, G. Ding, M. Xia, X. Jin, Ch. Xie, Additive manufacturing of hydroxyapatite bioceramic scaffolds: Dispersion, digital light processing, sintering, mechanical properties, and biocompatibility. J. Adv. Ceram. **9**, 360–373 (2020). https://doi.org/10.1007/s40145-020-0375-8
17. T. Osswald, Understanding polymer processing. Process. Governing Eqn. (2017). ISBN: 978-1-56990-647-7
18. J. Schmidt, H. Elsayed, E. Bernardo, P. Colombo, Digital light processing of wollastonite-diopside glass-ceramic complex structures. J. Eur. Ceram. Soc. **38**(13), 4580–4584 (2018). https://doi.org/10.1016/j.jeurceramsoc.2018.06.004
19. M. Pagac, J. Hajnys, Q.-P. Ma, L. Jancar, J. Jansa, P. Stefek, J. Mesicek, A review of vat photopolymerization technology: Materials, applications, challenges, and future trends of 3D printing. Polymers (Basel) **13**(4), 1–20 (2021). https://doi.org/10.3390/polym13040598
20. A.B. Saed, A.H. Behravesh, S. Hasannia, S.A.A. Ardebili, B. Akhoundi, M. Pourghayoumi, Functionalized poly l-lactic acid synthesis and optimization of process parameters for 3D

printing of porous scaffolds via digital light processing (DLP) method. J. Manuf. Process. **56**, 550–561 (2020). https://doi.org/10.1016/j.jmapro.2020.04.076

21. S. Barone, P. Neri, A. Paoli, A.V. Razionale, F. Tamburrino, Development of a DLP 3D printer for orthodontic applications. Procedia Manuf. **38**(2019), 1017–1025 (2019). https://doi.org/10.1016/j.promfg.2020.01.187
22. T.D. Dikova, D.A. Dzhendov, D. Ivanov, K. Bliznakova, Dimensional accuracy and surface roughness of polymeric dental bridges produced by different 3D printing processes. Arch. Mater. Sci. Eng. 94(2), 65–75 (2018). https://doi.org/10.5604/01.3001.0012.8660
23. A. Keßler, M. Dosch, M. Reymus, M. Folwaczny, Influence of 3D- printing method, resin material, and sterilization on the accuracy of virtually designed surgical implant guides. J. Prosthet. Dent. (2021). https://doi.org/10.1016/j.prosdent.2020.08.038
24. J. Milde, J. Peterka, F. Jurina, P. Dobrovsky, J. Hrbal, P. Pitek, Influence of selected photopolymers on the resulting accuracy and surface roughness of the component in digital light processing technology. Ann. DAAAM Proc. Int. DAAAM Symp. **32**(1), 235–245 (2021). https://doi.org/10.2507/32nd.daaam.proceedings.034
25. **J. Milde, F. Jurina, J. Peterka, P. Dobrovszký, J. Hrbál, J. Martinovič**, Influence of part orientation on the surface roughness in the process of fused deposition modeling. KEM **896**, 29–37 (2021). https://doi.org/10.4028/www.scientific.net/kem.896.29
26. J. Milde, M. Kuruc, J. Hrbal, P. Dobrovszky, T. Machac, Research of scanning speed and orientation of the model to the resulting surface roughness in the process of Selective Laser Melting. In *10th International Conference On Recent Challenges In Engineering And Technology (ICRCET),* 22nd–23rd June 2022 (Bali, Indonesia, 2022). ISBN: 978-93-92105-60-9
27. J. Milde, J. Hrbal, J. Peterka, M. Kuruc, Research of layer thickness and selected thermoplastic materials and their influence on the surface roughness in the process of Fused Deposition Modeling technology. In *2nd International Conference on Innovative Research in Engineering and Technology 2022* (Hanoi, Vietnam, 2022). ISBN: 978-93-92105-65-4
28. S.H. Patil, R. Kulkarni, Surface roughness measurement based on singular value decomposition of objective speckle pattern. Opt. Lasers Eng. **150**, 106847 (2022). https://doi.org/10.1016/j.optlaseng.2021.106847
29. Ch.Y. Poon, B. Bhushan, Comparison of surface roughness measurements by stylus profiler, AFM and non-contact optical profiler. Wear **190**(1), 76–88 (1995). https://doi.org/10.1016/0043-1648(95)06697-7
30. Roughness (2D) parameter. [online], [cit. 2022-07-20]. Available on the Internet: http://www.olympus-ims.com/en/knowledge/metrology/roughness/2d_parameter/
31. J. Peterka, A new approach to calculating the arithmetical mean deviation of a profile during copy milling. Strojniski vestnik - J. Mech. Eng. **50**(12), 594–597 (2004)
32. M. Polakovic, I. Buransky, J. Peterka, Simulation concept for machined surface roughness and shape deviation prediction. In *Conf. 19th Intern. Sympos. of DAAAM,* Trnava, Slovakia, Oct 22–25, 2008 *Annals of DAAAM Proceedings of the 19th Int. DAAAM Symp. Book Series: Annals of DAAAM and Proceedings*, pp. 1089–1090
33. T. Vopat, J. Peterka, M. Kovac, I. Buransky, The wear measurement process of ball nose end mill in the copy milling operations. In *24th DAAAM International Symposium on Intelligent Manufacturing and Automation, 2013, Procedia Engineering*, vol. 69, pp. 1038–1047 (2014). https://doi.org/10.1016/j.proeng.2014.03.088
34. J. Peterka, P. Pokorny, Influence of the Lead Angle from the Vertical Axis Milling on Effective Radius of the Cutter. In *Conference: 7th International Congress of Precision Machining (ICPM 2013),* (Miskolc, HUNGARY, 2013) OCT 03–05, 2013 Sponsor(s): Int Visegrad Fund; SANDVIK Cromant. Precision Machining VII, Book Series: Key Engineering Materials, vol. 581, pp. 44–49
35. T. Vopat, J. Peterka, V. Simna, M. Kuruc, The influence of different types of copy milling on the surface roughness and tool life of end mills. In *Conference: 25th DAAAM International Symposium on Intelligent Manufacturing and Automation,* (Oldest technical univ, Vienna, AUSTRIA) Nov 26–29 (2014)

36. J. Peterka, P. Pokorny, S. Vaclav, B. Patoprsty, M. Vozar, Modification of cutting tools by drag finishing. MM Sci. J. 3822–3825 (2020). https://doi.org/10.17973/MMSJ.2020_03_2019130
37. J. Peterka, P. Pokorny, S. Vaclav, CAM strategies and surfaces accuracy. In *Conference: 19th International Symposium of the Danube-Adria-Association-for-Automation-and-Manufacturing,* (Trnava, Slovakia Date: OCT 22–25, 2008), *Annals of DAAAM for 2008 & Proceedings of the 19th International DAAAM Symposium, Book Series: Annals of DAAAM and Proceedings,* (pp. 1061–1062)
38. M. Beno, M. Zvoncan, M. Kovac, J. Peterka, Circular interpolation and positioning accuracy deviation measurement on five axis machine tools with different structures. Tehnicki vjesnik - Tech. Gazette 2(3) 479–484 (2013)
39. I. Misumi, R. Kizu, K. Sugawara, A. Hirai, S. Gonda, Extension of the probe-tip error evaluation for areal surface roughness measurements using metrological AFM. Measurem. Sens. **18**, 100092 (2021). https://doi.org/10.1016/j.measen.2021.100092
40. H. Wenbin, Y.T. Lee, G. Haiqing, A study of the staircase effect induced by material shrinkage in rapid prototyping. Rapid Prototyping J. **11**(2), 82–89 (2005). https://doi.org/10.1108/13552540510589449

Chapter 78
Investigation on the Mixing Ratio of Dimethylformamide (DMF) and Acetone Binary Solvent on the Electrospun Polyvinylidene Fluoride (PVDF) Fiber

Muhamad Haikal Abdullah, Sharifah Shahnaz Syed Bakar, and Shuhaida Yahud

Abstract Polyvinylidene Fluoride (PVDF) fibers were electrospun from PVDF solution with varied solvent mixing ratios of Dimethylformamide (DMF) and acetone in this work. The goal of this study is to see how the mixing ratio of DMF and acetone as solvent affects the fiber size, crystal structure, and electrical conductivity of PVDF fiber. In weight, the solvent mixing ratios are 100 DMF:0 acetone, 70 DMF:30 acetone, 50 DMF:50 acetone, 30 DMF:70 acetone, and 0 DMF:100 acetone. To be employed in this study, the PVDF solution was fixed at a concentration of 15%. The electrospinning machine parameters were likewise set to 20 kV, 10 cm tip-to-collector distance (TCD), and 1 mL/h flow rate. After the electrospinning process is completed, the sample will be kept for a short period of time in room temperature to allow the fiber to dry before being taken for characterization stages. Scanning Electron Microscopy (SEM), X-Ray Diffraction (XRD), and Four-Point Probe (FPP) were used to examine the fibers. The fiber size of electrospun PVDF fibers is shown by SEM; the crystal structure of the electrospun PVDF fiber is shown by XRD; and the electrical resistivity of the electrospun PVDF fiber is shown by FPP. According to the results of the experiment, one of the five samples is in powder form, while the other is in fiber form. Two of the four fibers produced contain beads, as determined by SEM. For XRD and FPP characterization, it was discovered that a sample with a high peak intensity of β phase produces the maximum electrical conductivity.

Keywords Electrospinning · Binary solvent · PVDF · DMF · Acetone

M. H. Abdullah · S. S. Syed Bakar (✉)
Faculty of Chemical Engineering and Technology, Universiti Malaysia Perlis, 02600 Arau , Perlis, Malaysia
e-mail: shahnaz@unimap.edu.my

S. Yahud
Faculty of Electronic Engineering Technology, Universiti Malaysia Perlis, Pauh Putra Campus, 02600 Arau, Perlis, Malaysia

M. A. A. Mohd Salleh et al. (eds.), *Proceedings of the Green Materials and Electronic Packaging Interconnect Technology Symposium*, Springer Proceedings in Physics 289,
https://doi.org/10.1007/978-981-19-9267-4_78

78.1 Introduction

Polymer nanofibers are a type of nanomaterial that has sparked a lot of interest in the recent decade due to its high surface-to-mass (or volume) ratio and distinctive properties. Polymer nanofibers have been manufactured using a number of processing methods in recent years, including drawing, template synthesis, phase separation, self-assembly, electrospinning, and so on. Electrospinning has gained popularity as a more straightforward and adaptable method of producing submicrometer or nanoscale fiber than previous procedures. Electrostatic force accelerates and lengthens the jet of polymer solution during the electrospinning process, allowing polymeric nanofibers to be formed. This is due to the interaction of the electric field with the fluid, which is governed by the polymer solution's characteristics.

Several polymers are utilized in the production of fibers in electrospinning applications, including Polyethylene Terephthalate (PET), Polyacrylonitrile (PAN), Polyethylene Oxide (PEO), Polybenzimidazole (PBI), and Polyvinylidene Fluoride (PVDF). Fluorinated polymers, such as Polyvinylidene Fluoride (PVDF) and its copolymers, are being researched for use in the fabrication of polymeric piezoelectric devices. PVDF has a wide range of properties, including widespread availability. Due to the existence of β phase, PVDF has a high purity, is lightweight and flexible, solvent resistant, and stable under strong electric fields.

To increase the piezoelectric properties of PVDF fiber, a solvent can be added during the PVDF solution manufacturing process. Dimethylformamide (DMF), which has a greater dielectric constant and electric conductivity as well as a lower vapor pressure, is frequently employed as a solvent additive. Meanwhile, due to its compatibility for biocompatible polymers, acetone is a non-toxic and sustainable solvent that has seen limited utilization in the electrospinning business. The simultaneous use of DMF and acetone as solvents results in a binary solvent system.

78.2 Methodology

78.2.1 Materials

Polyvinylidene Fluoride (PVDF) powder with average molecular weight of 534,000 and density of 1.74 g/mL was used without further purification. For the solvents, 99.8% purity of *N*, *N*-Dimethylformamide (DMF) and 99.86% purity of acetone were also used without further purification.

Table 78.1 Mixing ratio of solvent (DMF and acetone) in wt%

Sample name	Dimethylformamide (DMF) wt%	Acetone wt%
100D	100	0
70D30A	70	30
50D50A	50	50
30D70A	30	70
100A	0	100

78.2.2 Sample Preparation

In this study, mixing ratio of DMF and acetone as solvent was determined as shown in Table 78.1. DMF and acetone were mixed together and make use as binary solvent. Then, the mass of PVDF was calculated and mixed with solvent to prepare PVDF solution with concentration of 15 wt% for each sample.

78.2.3 Characterization

Scanning Electron Microscopy (SEM) (Model: JEOL JSM-6460LA) was employed to analyze the fiber size of electrospun PVDF fibers. The sample was placed on the sample holder and taped with double-sided tape, and the sample was coated with thin platinum layer by using sputter coater (model: JFC 1600). The crystal structure of electrospun PVDF fibers is analyzed by using X-Ray Diffraction (XRD) (model: XRD-6000, Shimadzu, Tokyo, Japan). The diffraction patterns were recorded using Cu Kα radiation over 2θ range of 10°–60°, a position-sensitive detector using a step size of 0.02° and scan rate of 10°/min. The analyzed data are then interpreted by using a software named HighScore Plus. The electrical conductivities of electrospun PVDF fiber were measured by using Four-Point Probe (FPP). This approach employs a linear array of four sharp probes. A known current (I) is injected between two probes, and the voltage drop (V) that results is measured using the other two probes.

78.3 Result and Discussion

78.3.1 Morphological Analysis

By naked eye observation, samples 100D, 70D30A, 50D50A and 30D70A present in fiber or sheet state with solvent mixing of 100 DMF:0 acetone, 70 DMF:30 acetone, 50 DMF:50 acetone, and 30 DMF:70 acetone, while sample 100A with solvent mixing of 0 DMF:100 acetone presents in powder state as shown in Fig. 78.1. While for samples 100D, 70D30A, 50D50A, and 30D70A, it was observed that there are likely liquid droplets on the surface of samples. Based on study done by Prasad G et al., low polymer concentration may lead to electrospraying instead of electrospinning. In the present study done by Prasad et al. also, they found out that polymer concentration of 18 wt% produced the most optimized conditions for electrospun PVDF fiber. So, the liquid droplets on surface of samples may due to the low polymer concentration [1]. While by SEM observation in Fig. 78.2, from four samples that present in sheet state, only samples 100D and 50D50A are free from beading, while the other two showed beaded fiber. All of the samples were taken at eight different points for its fiber diameter, and the average fiber diameter for samples 100D, 70D30A, 50D50A, and 30D70A is 0.5983 μm, 0.5323 μm, 0.7292 μm, and 3.2434 μm, respectively. According to Choi et al. experiment, adding volatile solvents evaporates the solvent entirely during the electrospinning process and lowers the structure of the beads, which is advantageous for increasing beta phase. A further benefit of acetone addition is a decrease in PVDF solution viscosity, which aids in stretching and beta-phase development. Furthermore, Choi et al. discovered that acetone may greatly lower the surface tension of the combined solvent, which helps to prevent the production of beads. As a result, the addition of the volatile solvent raises the beta phase [2]. Ghafari et al. conducted a detailed investigation of the DMF/acetone volume ratio's impact on morphologies (% of beads and nanofibers diameter) and beta phase. The quantity of beads decreases when the DMF/acetone volume ratio decreases. As a result, thorough solvent evaporation is advantageous in eradicating the beads which is contrasting to Gee's experiment, which found that increasing acetone level increased beta phase [3].

78.3.2 Crystallinity Analysis

From Fig. 78.3, the strongest diffraction peak of electrospun PVDF fiber for each parameter occurs at $2\theta = 22.55°$ for sample 100D, $2\theta = 23.1°$ for sample 70D30A, $2\theta = 22.43°$ for sample 50D50A, $2\theta = 23.08°$ for 30D70A, and $2\theta = 20.04°$ for sample 100A. According to research made by Liuxia Ruan et al., it confirms the peaks at 17.7°, 18.3°, 19.9° characteristic of α phase of PVDF and peaks at 18.5°, 19.2°, 20° characteristic of the γ phase. Meanwhile, the peak at $2\theta = 20.26°$ is related to the diffraction of β phase at (110) and (200) [4]. However, from the result of XRD in

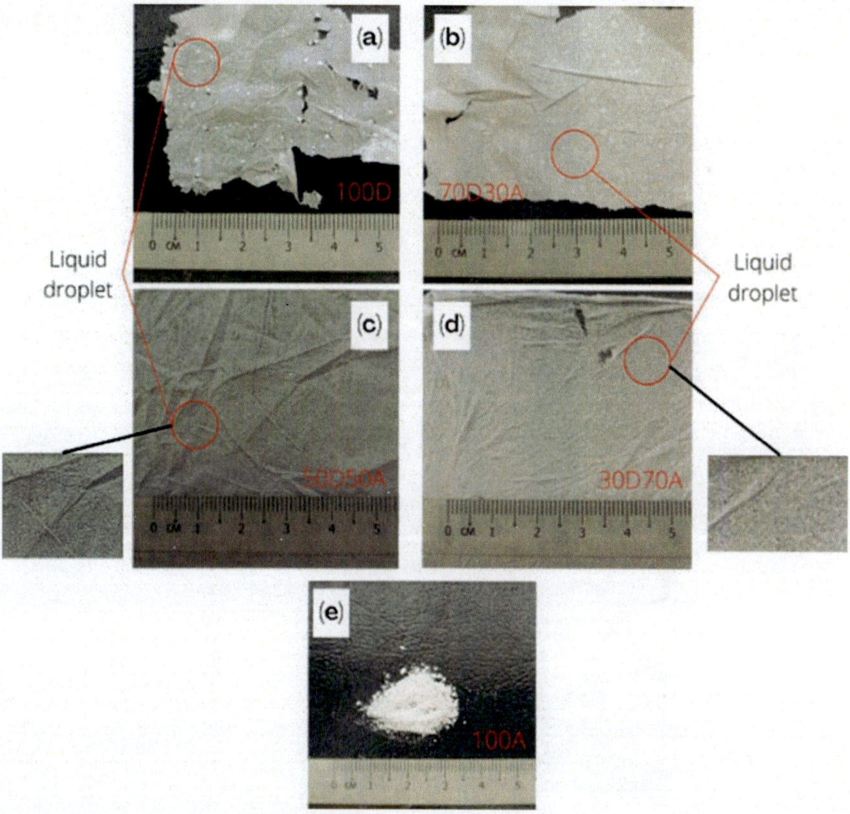

Fig. 78.1 Image of PVDF fiber surface **a** 100 DMF:0 acetone, **b** 70 DMF:30 acetone, **c** 50 DMF:50 acetone, **d** 30 DMF:70 acetone, and **e** 0DMF:100 acetone

Fig. 4.2, it was shown that the peak showed up at range of 22.43°–23.1° for samples A, B, C, and D which is crystalline, while for sample E which is amorphous, it shows peak at 20.04°. Sample 50D50A has the highest peak intensity which is 19400 a.u with peak location at 22.43°. So, sample 50D50A may have the highest percent of β phase and hence make it the most conductive sample among the others. Since the peak for sample 50D50A presents at location of 22.43°, the crystallite size and lattice strain effect of the sample are thought to be the primary causes of the displaced peak in the XRD study.

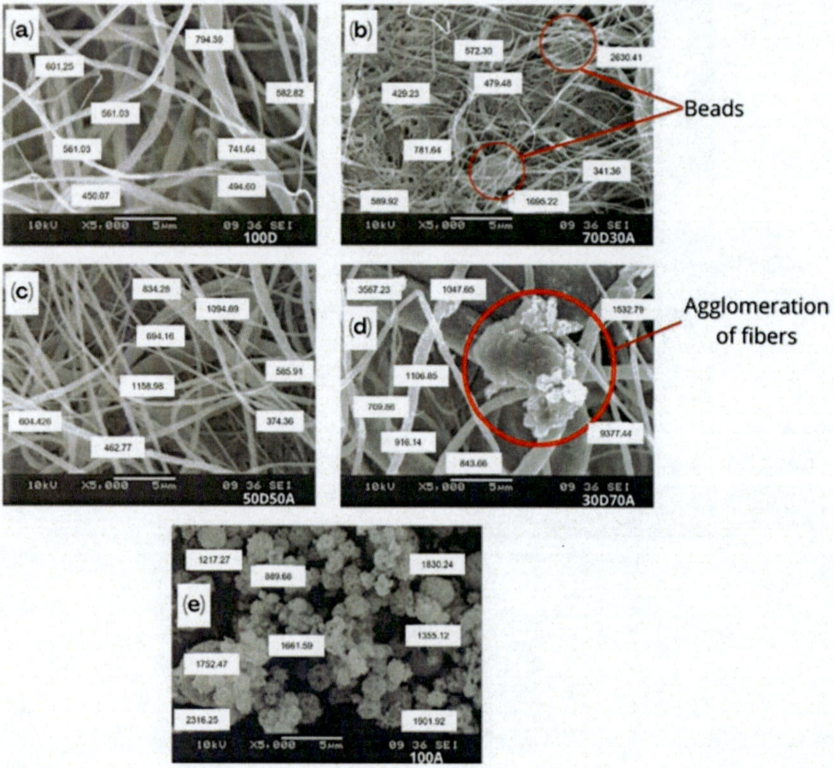

Fig. 78.2 SEM micrographs of electrospun PVDF fiber at ×5000 SEM magnification **a** 100 DMF:0 acetone, **b** 70 DMF:30 acetone, **c** 50 DMF:50 acetone, **d** 30 DMF:70 acetone, and **e** 0DMF:100 acetone

78.3.3 Electrical Conductivity Analysis

From the works done, all electrospun PVDF fibers have varying conductivity values with different parameters based on the FPP test results. As shown in Fig. 78.4, resistance and conductivity levels are related to one another. It has been observed that sample 50D50A has the highest electrical conductance and the lowest electrical resistance, while sample 30D70A has the lowest electrical conductance and the highest electrical resistance compared to the others. Based on XRD analysis, sample 50D50A has the highest peak intensity compared to the others as it contains the highest β phase which verify that high contain of β phase will affect in good electrical conductance in PVDF.

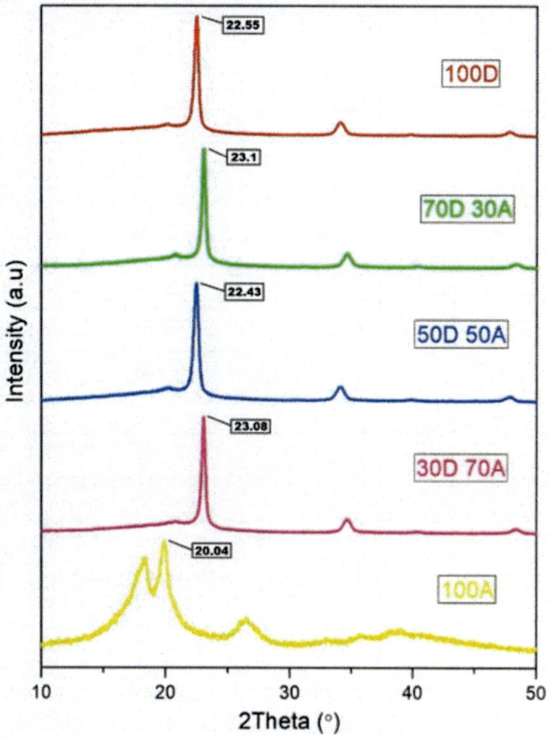

Fig. 78.3 XRD diffraction plots of five samples with different solvent DMF:acetone wt%: (100D) 100:0, (70D30A) 70:30, (50D50A) 50:50, (30D70A) 30:70, and (100A) 0:100

78.4 Summary

As a conclusion, electrospun PVDF fibers with desired properties can be obtained by controlling the electrospinning parameters. With the naked eye observation, it was observed that one sample mixed with 100% acetone presents in powder state, while the others are in sheet state. This could occur as acetone has a high solvent content, which makes it possible for the sample to solidify before it reaches the collector and results in a powdered condition. By observation of Scanning Electron Microscope (SEM), it was found that there is beading present on samples 70D30A and 30D70A. Sample 70D30A has the smallest average fiber diameter, followed by samples 100D, 50D50A, and 30D70A which is 0.5323 μm, 0.5984 μm, 0.7292 μm, and 3.2434 μm, respectively. Crystallinity of electrospun PVDF fiber was also analyzed by using X-Ray Diffraction (XRD) technique. From the analysis conducted by XRD, it was observed that the peaks did not show at expected peak which is at 20.26° but showed at 22.43°–23.1° for samples 100D, 70D30A, 50D50A, and 30D70A, while 20.04° for sample 100A instead. Moreover, the electrical properties of electrospun PVDF fibers were determined by using the Four-Point Probe (FPP). From the result, it shows that sample with solvent ratio of 50 DMF:50 acetone has the highest electrical conductance and the lowest electrical resistance, while sample with 100% DMF has

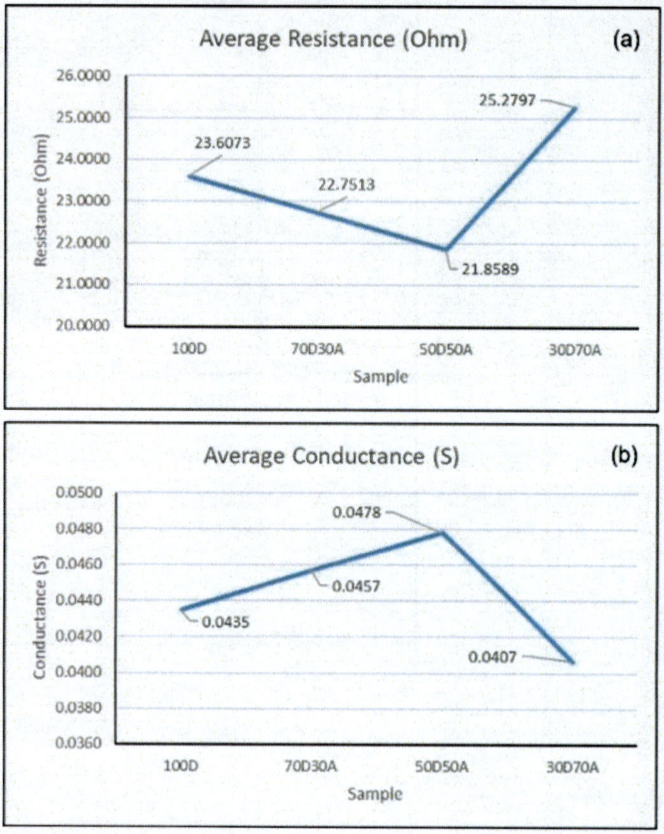

Fig. 78.4 Electrical conductivity analysis of electrospun PVDF fiber for **a** average resistance and **b** average conductance

the lowest electrical conductance and the highest electrical resistance compared to the others.

References

1. G. Prasad et al., Enhancement of solvent uptake in porous PVDF nanofibers derived by a water-mediated electrospinning technique. J. Mater. **7**(2), 244–253 (2021). https://doi.org/10.1016/j.jmat.2020.05.004
2. S.W. Choi, J.R. Kim, Y.R. Ahn, S.M. Jo, E.J. Cairns, Characterization of electrospun PVdF fiber-based polymer electrolytes. Chem. Mater. **19**(1), 104–115 (2007). https://doi.org/10.1021/cm060223+

3. E. Ghafari, X. Jiang, N. Lu, Surface morphology and beta-phase formation of single polyvinylidene fluoride (PVDF) composite nanofibers. Adv. Compos. Hybrid Mater. **1**(2), 332–340 (2018). https://doi.org/10.1007/s42114-017-0016-z
4. L. Ruan, X. Yao, Y. Chang, L. Zhou, G. Qin, X. Zhang, Properties and applications of the β phase poly (vinylidene fluoride). Polymers (Basel) **10**(3), 1–27 (2018). https://doi.org/10.3390/polym10030228

Chapter 79
Isothermal Oxidation Behavior of Ni-Based Fe–Ni–Cr Superalloys: Role and Effect of Nb Alloying Element

Noraziana Parimin, Esah Hamzah, and Astuty Amrin

Abstract The isothermal oxidation of heat-treated Ni-based Fe–Ni–Cr superalloy at 500 °C was studied to further investigate the role and effect of Nb alloying element during oxide scale formation. The Ni-based Fe–Ni–Cr superalloys underwent a heat treatment process at two different temperatures, that are, 950 °C and 1100 °C, for 3 h of soaking time, followed by water quenching. Heat-treated sample at 950 °C exhibited fine grain size, while heat-treated sample at 1100 °C exhibited coarse grain size. The heat-treated Ni-based Fe–Ni–Cr superalloys were subjected to an isothermal oxidation test at 500 °C for 500 h exposure. The oxidized heat-treated Ni-based Fe–Ni–Cr superalloys were analyzed in terms of oxidation kinetic, oxide phase formation and oxide surface morphology. The oxide phases were analyzed by using XRD technique. The surface morphology was examined by using SEM, FESEM and EDX techniques. Both oxidized heat-treated Ni-based Fe–Ni–Cr superalloys' samples followed a parabolic rate law, indicating a diffusion-controlled oxide growth mechanism. Several oxide phases were formed on the alloy surface after exposures at 500 h. The surface morphology of oxidized samples at 300 h exhibited the formation of continuous oxide scale with distributed overgrown Nb-rich oxides. Nb-rich oxides were formed on coarse-grained heat-treated alloys that developed excessively large Nb-rich oxides. This phenomenon will initiate a crack that propagates around the oxide.

Keywords Isothermal oxidation · Ni-based superalloys · Fe–Ni–Cr superalloys · Heat treatment

N. Parimin (✉)
Faculty of Chemical Engineering and Technology, Universiti Malaysia Perlis (UniMAP), 02600 Arau, Perlis, Malaysia
e-mail: noraziana@unimap.edu.my

E. Hamzah
Faculty of Mechanical Engineering, Universiti Teknologi Malaysia (UTM), 81310 Skudai, Johor, Malaysia

A. Amrin
Razak Faculty of Technology and Informatics, UTM Kuala Lumpur, 54100 Kuala Lumpur, Malaysia

M. A. A. Mohd Salleh et al. (eds.), *Proceedings of the Green Materials and Electronic Packaging Interconnect Technology Symposium*, Springer Proceedings in Physics 289,
https://doi.org/10.1007/978-981-19-9267-4_79

79.1 Introduction

Ni-based Fe–Ni–Cr superalloys are usually used in high-temperature conditions with various atmospheric conditions. At high-temperature applications, one of the utmost significant properties of alloys is the oxidation resistance at high-temperature condition [1]. In this research, the Fe–40Ni–24Cr superalloy is a type of Ni-based Fe–Ni–Cr superalloy, generally used in industrial heating operations and heat treatment applications [2]. Fe–40Ni–24Cr superalloy was developed to enhance the oxidation resistance at high-temperature condition with accumulation of numerous alloying elements to further enhance their properties. The major elements in this superalloy are Fe, Ni and Cr. Cr acts as an oxide former in the alloy system. Nb, Mn and Ti alloying elements are usually added to the alloy system to procedure sturdy precipitates at operational temperatures [3, 4]. The formation of protective chromia scale, Cr_2O_3, is characterized by a slow growth rate responsible for the protection of metal alloys in high-temperature applications and has outstanding resistance to oxidation in various harsh atmospheres [5]. Several researches associated with chromia establishing alloys report the establishment of Cr_2O_3 and $MnCr_2O_4$, which act as a barricade against supplementary oxidation and then offer superior resistance to oxidation [5–8]. Fe–40Ni–24Cr superalloy usually contains small amounts of Nb and Ti as alloying elements, which provides structural stability under severe service conditions. These alloys are strengthened by the precipitation of carbides such as MC carbides consisting of NbC, TiC and (Nb, Ti)C. However, under certain conditions, massive growth of oxide precipitates containing Nb and/or Ti elements will cause crack formation.

Extensive research has been conducted to enhance the oxidation resistance properties in Ni-based superalloy. These researches include grain refinement, surface treatment and thermomechanical treatment. The interesting methods were refining the grain structure of the alloy by heat treatment procedure. Therefore, grain refinement is a possible approach to enhance the oxidation resistance [9, 10]. Fine grain size alloys may have a significant effect on oxidation behavior, which can lower the oxidation rate. Fine grain size alloy consumes supplementary grain boundary area which provides numerous diffusion paths to boost the diffusion process. Therefore, a heat treatment process was performed in this work to change the grain size of the alloy, in order to produce a fine grain structure.

This alloy experiences very severe conditions at various temperatures and environments during service. Under these conditions, this alloy undergoes a change in the composition of the formation of an oxide layer due to oxidation at high temperatures. The fine grain size alloy obtained from the heat treatment process is expected to have a good outcome on the oxidation performance of this alloy. Consequently, it is important to know the role of alloyed Nb in the isothermal oxidation of Fe–24Ni–24Cr superalloy by examining the oxide scale morphology at high temperature.

79.2 Materials and Methods

79.2.1 *Materials*

The Fe–40Ni–24Cr superalloy is a subgroup of Ni-based Fe–Ni–Cr superalloys used in this research with a chemical composition measured in weight percent (wt%) of 40.45 Ni, 24.11 Cr, 0.05 C, 0.44 Nb, 0.44 Si, 0.70 Mn, 0.08 Al, 0.03 Ti and the rest Fe. The Fe–40Ni–24Cr superalloy is a commercial HAYNES® HR-120® alloy under UNS N08120 with standard ASTM-B-409; ASME-SB-409.

79.2.2 *Methods*

A sample was cut into dimensions of 10 mm × 10 mm × 3 mm. The cut samples underwent a heat treatment procedure at two temperatures of 950 and 1100 °C for 3 h of soaking time followed by water quenching. These heat treatment samples are registered as H950 and H1100. The grain size of heat-treated samples was calculated using linear intercepts' method of standard ASTM E112.

The grinding process was performed to the heat-treated samples. All samples were ground to a P600 surface finish, and the dimension of each sample was recorded. The weight of each sample was recorded before and after the oxidation test. The heat-treated samples were then undergoing an isothermal oxidation test at 500 °C for 500 h exposure. The sample was discontinuously exposed at 50, 100, 150, 200, 300, 400 and 500 h to determine the oxidation kinetics. The oxidized samples were characterized using X-ray diffraction (XRD) technique, scanning electron microscope (SEM) and field-emission scanning electron microscope (FESEM) furnished with energy-dispersive X-ray (EDX) spectrometer.

79.3 Results and Discussion

79.3.1 *Oxidation Kinetics*

Isothermal oxidation tests were performed on two heat-treated samples H950 and H1100. From the heat treatment results, the H950 sample produced a fine-grained structure, while the H1100 sample produced a coarse-grained structure. Fine-grained H950 and coarse-grained H1100 samples yielded average grain sizes of 27.27 μm and 36.31 μm, respectively. The oxidation kinetic graph was plotted based on the weight change per area as a function of time for H950 and H1100 samples. The oxidation kinetics presented in Fig. 79.1 exhibits a pattern of weight gain for both samples with H950 displaying a lower weight gain. At the commencement of the oxidation test,

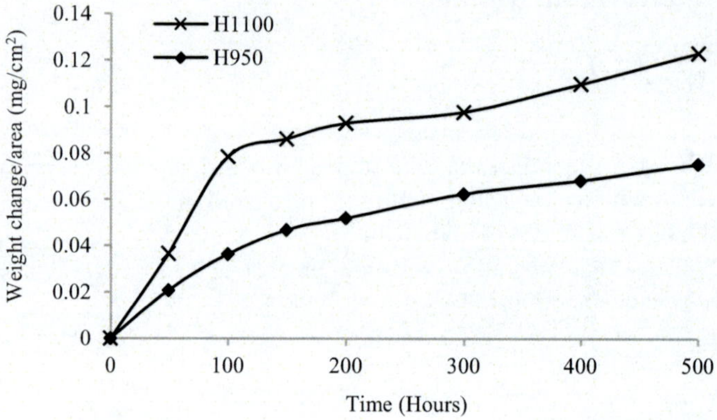

Fig. 79.1 Oxidation kinetics

the H950 sample oxidized relatively quickly up to 50 h, which is 0.0208 mg cm^{-2}. Extending the exposure time up to 500 h, the weight gain increased continuously up to 0.0753 mg cm^{-2}. On the other hand, the weight gain of the H1100 sample shows a more weight gain owed to the incessant growing of oxide scale. A sharp pattern of weight gain was recorded up to 100 h exposure, continuing with an increasing trend of weight gain. This sample recorded an increment pattern of weight gain that increased from 100 to 500 h from 0.0784 to 0.1232 mg cm^{-2}.

To identify the oxidation rate law, an analysis was done based on Eq. (79.1), where x refers to the weight change over surface area, t refers to time and m is a constant. Figure 79.2a shows a double log plot for oxidation rate analysis. The m values of 1, 2 and 3 denote linear, parabolic and cubic rates, respectively. The values of m for H950 and H1100 samples are 1.86 and 2.21, respectively, indicating a parabolic rate law. All plots obtained the appropriate R^2 results. The fitting parameter R^2 with a value approaching one indicates an increase in the consistency of the data with the curve.

The determination of the parabolic rate constant was identified based on Eq. (79.2) as shown in Fig. 79.2b, where K_p represents the parabolic rate constant. Parabolic rate constant, K_p value for H950 sample is lesser than H1100 sample, which is 3.12 $\times$ 10^{-9} mg^2 cm^{-4} s^{-1} and 7.11 $\times$ 10^{-9} mg^2 cm^{-4} s^{-1} correspondingly. The low parabolic rate constant of the H950 sample indicates good oxidation resistance with a low oxidation rate. All plots show the corresponding good fitting R^2 results.

$$\log x = (1/m) \log t + c, \tag{79.1}$$

$$x^2 = K_p t + c. \tag{79.2}$$

The low parabolic rate constant of sample H950 indicates that this sample obeys a diffusion-controlled oxide growth rate mechanism. It begins when the alloy is exposed to high-temperature conditions, and oxide nuclei are formed when metal

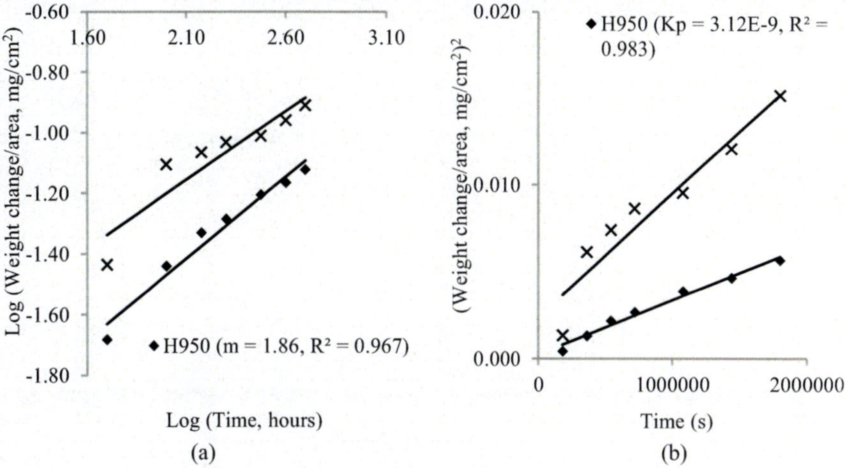

Fig. 79.2 **a** Oxidation rate curve and **b** oxidation rate constant

ions react with oxygen gas in the atmosphere. As the exposure time increases, a layer of oxide has encased the metal that serves as a protective barricade to the metal from the atmosphere. The high grain boundary region obtained from the fine-grained heat-treated H950 sample will serve as a diffusion pathway for metal ion diffusion to the outside which provides more passage than the coarse-grained heat-treated H1100 sample. This effect will simultaneously produce a speedy initial thin oxide layer on the metal alloy surface. Since the entire exterior of the metal alloy is shielded with an oxide layer, the diffusion process will slow down, further reducing the oxidation rate. This phenomenon is recorded by the H950 sample. A similar finding was reported in our previous study on Fe–33Ni–19Cr alloy, indicating that the fine-grained alloy exhibits a low parabolic rate constant [11, 12].

79.3.2 Oxide Scale Analysis

The establishment of oxide phases identified using the XRD technique for H950 sample is shown in Fig. 79.3, recorded the formation of various oxide phases consisting of Cr-rich oxides, Ti–rich oxides and Nb-rich oxides. Cr-rich oxides consist of corundum structure, which composes of Cr_2O_3 and $(Cr_{0.88}Ti_{0.12})_2O_3$, and also spinel structure, which composes of $MnCr_2O_4$, $FeCr_2O_4$ and $NiCr_2O_4$. Ti-rich oxides consist of TiO_2 and $(Ti_{0.97}Cr_{0.03})O_2$. In addition, NbO_2 was also detected. The development of $(Cr, Ti)_2O_3$ corresponds to $(Cr_{0.88}Ti_{0.12})_2O_3$ oxides, and Ti-rich oxides are believed to reduce Cr volatility due to the lowest Cr vapor pressure of the $(Cr, Ti)_2O_3$ oxide phase associated with Cr_2O_3 [12, 13]. Similar oxide phases were recorded for both samples.

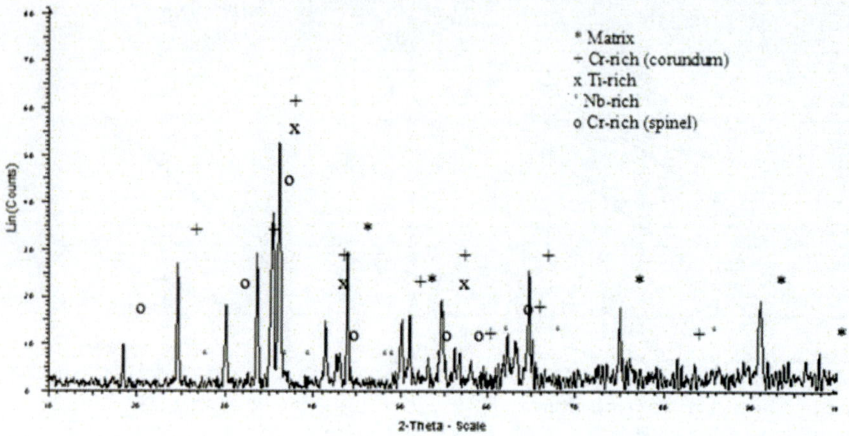

Fig. 79.3 XRD analysis of H950 sample

The establishment of the oxide scale for H950 sample after an exposure time of 300 h was characterized using SEM as shown in Fig. 79.4. Figure 79.4a of the low magnification SEM image displays an incessant oxide scale formed on the surface of the alloy with excessively large oxide particles distributed with detail view in image Fig. 79.4b. EDX elemental analysis on matrix A detected the major elements O, Fe, Ni and Cr, with minor elements Mn as shown in Fig. 79.4c. This analysis suggests that this region consists of a mixed oxide phase as analyzed using the XRD technique consisting of Cr-rich oxides such as Cr_2O_3, $MnCr_2O_4$, $FeCr_2O_4$ and/or $NiCr_2O_4$. Oversized oxide particles in region B with a size of about 10 μm consist of large amounts of O and Nb as analyzed using EDX analysis in Fig. 79.4d, most likely originating from NbO_2 oxide. This analysis is consistent with XRD phase analysis which detected the formation of NbO_2 oxide. In contrast, the SEM image of the H1100 sample in Fig. 79.5a, b shows a similar oxide morphology with continuous oxide scales with larger distributed oxide particle sizes. EDX elemental analysis of the C region (Fig. 79.5c) recorded the main elements O and Nb, detected from NbO_2 oxide. This sample shows a distribution of oxide particles similar to the H950 sample.

Figure 79.6a displays the FESEM image of the H950 sample after 500 h of exposure, displaying the development of isolated oxide particles. Alternatively, a high magnification image of the matrix area reveals an even oxide scale formed on the surface of the alloy showing a good protective layer as shown in Fig. 79.6b. EDX elemental analysis of oxide particles in area D (Fig. 79.6c) shows the detection of a large amount of element Nb and O, originating from the Nb-rich oxide as in line with XRD analysis which perceived the NbO_2 oxide particles.

The FESEM image of the H1100 sample in Fig. 79.7a reveals larger NbO_2 particle sizes that overgrow on the surface. This oxide induces cracks around the oxide particles as shown in Fig. 79.7b which is further analyzed by EDX analysis (Fig. 79.7c) which confirms the detection of Nb and O elements. The formation of cracks can reveal and disrupt the protection behavior of the oxide scale. The crack will expose

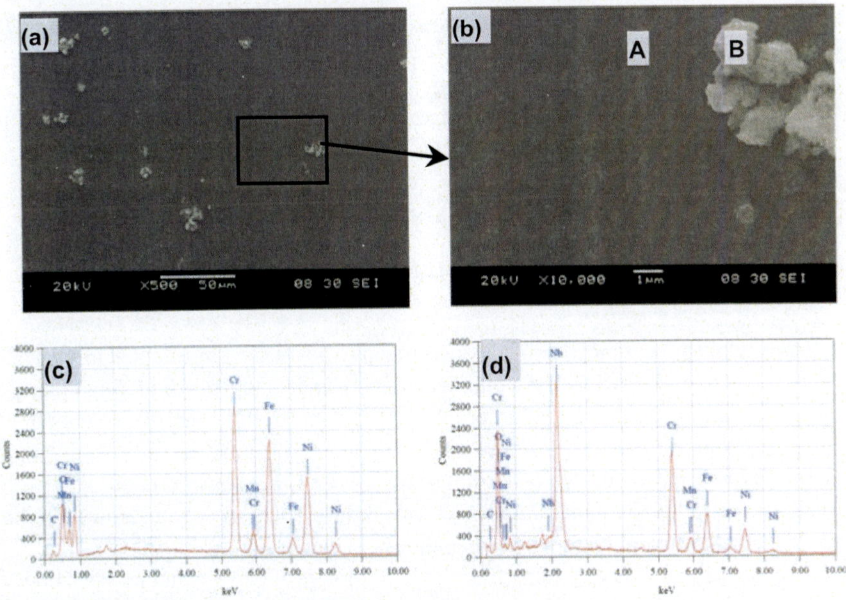

Fig. 79.4 **a**, **b** SEM images of H950 after exposure for 300 h, **c** EDX analysis in area A and **d** EDX analysis in area B

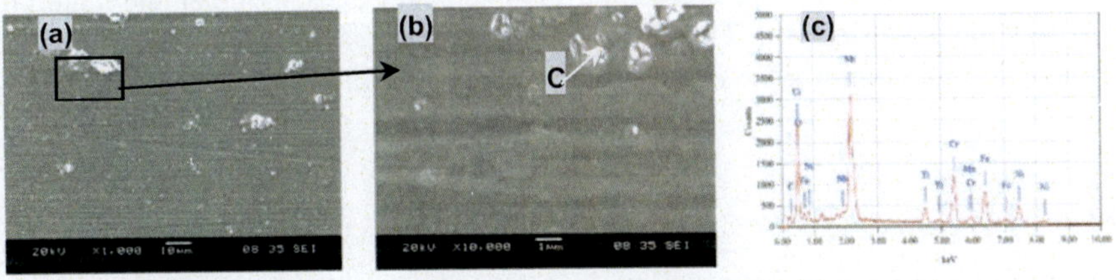

Fig. 79.5 **a**, **b** SEM images of H1100 after exposure for 300 h and **c** EDX analysis in area C

Fig. 79.6 **a**, **b** FESEM images of H950 after exposure for 500 h and **c** EDX analysis in area D

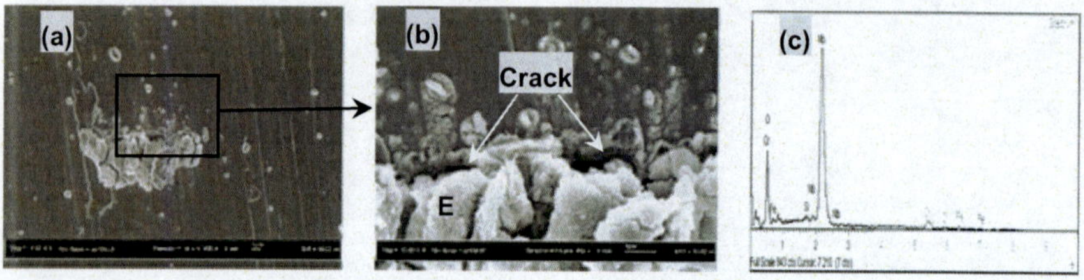

Fig. 79.7 **a**, **b** FESEM images of H1100 after exposure for 500 h and **c** EDX analysis in area E

the base metal causing more oxide scale to form to cover the unprotected area. This phenomenon is evidenced by the higher weight gain and increasing oxidation kinetics of this H1100 sample. A higher parabolic rate constant for this sample also indicates a higher oxidation rate.

Fine-grained H950 sample of Ni-based Fe–Ni–Cr alloy showed good oxidation resistance with a low oxidation rate which was desired in this research. However, the addition of Nb alloying elements in this alloy will cause the uncontrolled formation of NbO_2 oxide which is detrimental to the oxidation protective behavior, especially for coarse-grained H1100, although the additions of Nb alloying element are to boost the properties of the alloy. Nevertheless, disclosure of this alloy to high temperatures and extended times will rise the propensity to form Nb-rich oxides that will serve as favored sites for hole and/or crack development and further worsen the material into certain conditions [13]. Therefore, from this analysis, we found that grain size refinement will minimize the adverse effects of excessively large NbO_2 oxides.

79.4 Conclusion

Fine-grained H950 sample of Ni-based Fe–Ni–Cr alloy showed good oxidation resistance with lower weight gain and parabolic rate constant, indicating excellent oxidation rate. The oxide morphology of the H950 sample displays a uniform oxide scale with the establishment of isolated oxide particles prominent on the alloy surface. In contrast, the coarse-grained H1100 sample showed a high weight gain with a high parabolic rate constant, indicating a high oxidation rate. This phenomenon is consistent with the formation of cracks around excessively large oxide particles.

Acknowledgements The authors would like to acknowledge the support from the Fundamental Research Grant Scheme (FRGS) under a grant number of FRGS/1/2020/TK0/UNIMAP/02/43 from the Ministry of Higher Education Malaysia.

References

1. X. Wang, J.A. Szpunar, Effects of grain sizes on the oxidation behavior of Ni-based alloy 230 and N. J. Alloy. Compd. **752**, 40–52 (2018)
2. HAYNES International. HAYNES® HR-120® alloy. (H-3125D) (Haynes International, USA, 2008)
3. Y.X. Xu, J.T. Lu, X.W. Yang, J.B. Yan, W.Y. Li, Effect and role of alloyed Nb on the air. Corros. Sci. **127**, 10–20 (2017)
4. Y.X. Xu, J.T. Lu, W.Y. Li, X.W. Yang, Oxidation behavior of Nb-rich Ni–Cr–Fe alloys: role and effect of carbides precipitates. Corros. Sci. **140**, 252–259 (2018)
5. A. Col, V. Parry, C. Pascal, Oxidation of a Fe–18Cr–8Ni austenitic stainless steel at 850 °C in O_2: microstructure evolution during breakaway oxidation. Corros. Sci. **114**, 17–27 (2017)
6. T.D. Nguyen, J. Zhang, D.J. Young, Effect of Mn on oxide formation by Fe–Cr and Fe–Cr–Ni alloys in dry and wet CO_2 gases at 650 °C. Corros. Sci. **112**, 110–127 (2016)
7. D. Saber, I.S. Emam, R. Abdel-Karim, High temperature cyclic oxidation of Ni based superalloy at different temperature in air. J. Alloys Compd. **719**, 133–141 (2017)
8. C.N. Athreya, K. Deepak, D.I. Kim, B. Boer, S. Mandal, V.S. Sarma, Role of grain boundary engineered microstructure on high temperature stream oxidation behavior of Ni based superalloy alloy 617. J. Alloy. Compd. **778**(25), 224–233 (2019)
9. H.Z. Zheng, S.Q. Lu, Y. Huang, Influence of grain size on the oxidation behavior of $NbCr_2$ alloys at 950–1200 °C. Corros. Sci. **51**(2), 434–438 (2009)
10. T.S. Jo, S.H. Kim, D.G. Kim, J.Y. Park, Y.D. Kim, Effects of grain refinement on internal oxidation of alloy 617. J. Nucl. Mater. **402**(2–3), 162–166 (2010)
11. N. Parimin, E. Hamzah, Oxidation kinetics of Fe-Ni-Cr alloy at 900 °C. Mater. Sci. Forum **1010**, 58–64 (2020)
12. N. Parimin, E. Hamzah, N.A. Ahad, M.N. Derman, Isothermal oxidation behaviour of 800H Nickel-based alloy. J. Phys: Conf. Ser. **2169**, 012038 (2022)
13. L. Tan, X. Ren, K. Sridharan, T.R. Allen, Effect of shot-peening on the oxidation of alloy 800H exposed to supercritical water and cyclic oxidation. Corros. Sci. **50**(7), 2040–2046 (2008)

Chapter 80
Short Review: Corrosion Mitigation of AZ31 Magnesium Alloy by Superhydrophobic Coatings

Nur Fatihah Mohd Fadzil, Muhammad Salihin Zakaria, Razif Muhammed Nordin, Khairul Anwar Abdul Halim, and Lokman Hakim Ibrahim

Abstract Corrosion is a slow chemical process that protects metals from deterioration. This can be accomplished in a variety of ways, one of which is coating protection. All such metals and their objects are usually coated with a protective coating. Due to their ability to resist corrosion, these protective coatings can provide a better esthetic as well as a longer lifespan than uncoated ones. Superhydrophobic coatings provide various applications, especially, in corrosion protection due to their water-repellent properties, which can be reflected by measuring the degree of water contact angle. In this short review, different fabrication methods of the superhydrophobic coating were discovered to give good water contact angles and at the same time enhance the anticorrosion performance of the AZ31 Mg alloy. Tafel polarization curve as well as Electrochemical Impedance Spectroscopy is able to provide the necessary assessment of corrosion protection performance of AZ31 Mg alloy.

Keywords Superhydrophobic coating · Corrosion · AZ31 Mg alloy

N. F. M. Fadzil · M. S. Zakaria (✉) · K. A. A. Halim · L. H. Ibrahim
Faculty of Chemical Engineering and Technology, Universiti Malaysia Perlis (UniMAP), 02600 Arau, Malaysia
e-mail: salihin@unimap.edu.my

M. S. Zakaria · K. A. A. Halim
Biomedical and Nanotechnology Research Group, Center of Excellence Geopolymer and Green Technology (CEGeoGTech), Universiti Malaysia Perlis (UniMAP), 02600 Arau, Malaysia

R. M. Nordin
Department of Chemistry, Faculty of Applied Sciences, Universiti Teknologi MARA, Perlis Branch Arau Campus, 02600 Arau, Perlis, Malaysia

Green and Functional Polymer Research Group, Universiti Teknologi MARA, 40450 Shah Alam, Selangor, Malaysia

L. H. Ibrahim
Advanced Polymer Research Group, Center of Excellence Geopolymer and Green Technology (CEGeoGTech), Universiti Malaysia Perlis (UniMAP), 02600 Arau, Malaysia

M. A. A. Mohd Salleh et al. (eds.), *Proceedings of the Green Materials and Electronic Packaging Interconnect Technology Symposium*, Springer Proceedings in Physics 289,
https://doi.org/10.1007/978-981-19-9267-4_80

80.1 Introduction

Due to its biodegradability, biocompatibility, low density and good thermal conductivity as well as other mechanical qualities [1, 2], magnesium (Mg) alloy is often used in biomedical business, aerospace, electronics and automotive industries [1, 3, 4]. In nature, AZ31 Mg alloys are subjected to strong chemical activity and quickly generated a porous and non-protective oxide film, which, in turn, limits its widespread use [5]. Corrosion of AZ32 Mg alloy has become a major issue that can cause untimely flops of components, leading to financial losses, environmental pollution, along with injury or death.

Recently, various techniques and methods have been developed to reduce any damage caused by corrosion of AZ31 Mg alloy. The main key point for all methods is to enhance its corrosion performance. The most frequent methods for preventing this damage include oiling, painting and electroplating. However, superhydrophobic coating is a method that is being used efficiently because protective coatings are viable practical solutions for corrosion [6].

Superhydrophobic coatings have been the subject of several studies with their applications in anticorrosion [7, 8], separation of oil and water, anti-fogging [9], self-cleaning [10, 11] as well as antifouling [12, 13]. Materials coated with superhydrophobic coating seems to be extremely difficult to wet, displays a water contact angle of more than 150° and a sliding angle of less than 10° [2, 14]. It is due to the influence of hierarchical micro/nano-surface roughness and low surface energy material which will decrease the interaction between the coating and liquid [15].

Water contact angle (WCA) is a key metric for assessing superhydrophobic material's wettability. The study of WCA's effect variables has become a research hotspot for researchers all over the world. Young, Wenzel and Cassie–Baxter's equations determine the WCA of smooth and rough substances [16]. However, because no real surface is ideal, models and theories to depict the water contact angles on real surfaces have been established. Wenzel model and Cassie–Baxter model are two methods for describing the contact angle on a physical surface. The real surface, in contrast to the ideal, features chemical heterogeneity as well as surface roughness. The rough surface is taken into account in the Wenzel model, but chemical homogeneity is maintained.

80.2 Superhydrophobic Coating Limits the Interaction Between AZ31 Mg Alloy and Corrosive Species

Many corrosion resistance studies are typically performed on short-term trials in NaCl solutions, taking into account the likely time scale to which they should be exposed [17]. The influence of industrial pollutants [18], salt spray [19], oxidation [20], as well as moisture or environment [21] can all cause metal components to corrode. Therefore, the fabrication of anticorrosion coatings can render additional protection by acting as a barrier to prevent chemical compounds or corrosive chemicals from coming into contact with metal surfaces.

By minimizing AZ31 Mg alloy interactions with corrosive species like water and ions, increasing surface hydrophobicity reduces their corrosion rate. For all aqueous solutions of all pH levels, the Mg alloy is hydrophilic, with a value of WCA of less than 90°. Acids, bases, as well as salt solutions are corrosive liquids that spread rapidly over the Mg alloy surface and, as a result, erode the substrate. Surprisingly, the superhydrophobic coating will keep the coated Mg alloy from coming into direct contact with water. As a result, the development of a superhydrophobic coating is a viable strategy for protecting Mg alloys from corrosion [22].

80.2.1 Potentiodynamic Polarization Curve of Superhydrophobic Coating on AZ31 Mg Alloy

Potentiodynamic polarization curve (Tafel) is a one of the methods to assess the anticorrosion performance of superhydrophobic coating. This curve is to investigate the relationship between the potential of metal and current produced in the electrochemical cells. This method is highly being used to properly study the corrosion protection performance of superhydrophobic coating as it is able to provide specific information related to corrosion such as corrosion potential and corrosion current density (corrosion rate). Ideally, the higher the corrosion potential (more positive), the lower the value of corrosion current density and the better the corrosion protection performance.

In 2019, Jin et al. [5] reported an excellent performance of AZ31B Mg alloy coated with superhydrophobic magnesium hydroxide that was hydrothermally treated and grafted with stearic acid, resulting in maximum WCA of 159°. The fabrication technique is green and low cost as no harsh chemical and expensive equipment were being used. In their research, they discovered that the anticorrosion performance of AZ31B Mg alloy was improved due to the presence of lamellar magnesium hydroxide which leads to formation of uniform micro-nanostructure. Tafel polarization curve measurements in 3.5 wt% NaCl solution displayed the corrosion potential (E_{corr}) and corrosion current (i_{corr}) values (Table 80.1) for both bare sample and superhydrophobic AZ31B Mg alloy. The E_{corr} value of superhydrophobic AZ31B Mg alloy was enhanced toward positive direction, and i_{corr} values were lowered by two orders of magnitude, which indicates that the AZ31B Mg alloy was successfully protected.

Table 80.1 Corrosion potential (E_{corr}) and corrosion current density (i_{corr}) of bare and superhydrophobic AZ31B Mg alloy [5]

Samples	E_{corr} (V)	i_{corr} (A/cm^2)
Bare AZ31B Mg alloy	− 1.5220	1.6200×10^{-5}
Superhydrophobic AZ31B Mg alloy	− 1.3264	1.7253×10^{-7}

Table 80.2 Summary of a few studies on anticorrosion performance of superhydrophobic coating on AZ31 Mg alloy

Superhydrophobic coating materials	Highest WCA	E_{corr} (V)	i_{corr} (A/cm^2)	Reference
Stearic acid/CeO_2	> 158°	– 1.488	5.43×10^{-5}	[4]
MAO/PDMS-SiO_2	158.3 ± 2.5°	– 0.005	6.67×10^{-9}	[24]
MAO-PP	167.2 ± 0.8°	– 1.50 ± 0.14	$(7.21 \pm 2.23) \times 10^{-9}$	[25]
F_3-POE	170.2°	– 1.21	4.6×10^{-10}	[26]
$Mg(OH)_2$/DTMS	165.1 ± 2.1°	– 1.224	1.777×10^{-8}	[27]
$Mg(OH)_2$/PP	165.5 ± 3.6°	– 1.18	3.12×10^{-9}	[28]

Another recent study published in 2022 by Wang et al. [23] revealed good performance of superhydrophobic calcium phosphate (Ca-P)/chitosan (CS) composite coating on AZ31 Mg alloy. The coating was prepared by electrophoretic deposition of CS coating followed by immersion of prepared coating in Ca-P solution and low energy modification by stearic acid. The method was low cost, simple and resulted in highest WCA of 158.6° and sliding angle of 1.5°. The Ca-P/CS composite coating recorded a high E_{corr} at – 1.3092 V and i_{corr} at 1.6135×10^{-7} A/cm^2 compared to bare AZ31 Mg alloy. Those values indicate that superhydrophobic Ca-P/CS composites' coating offers more resistance to corrosion and is able to protect AZ31 Mg alloy. Table 80.2 summarizes the anticorrosion performance of superhydrophobic coating on AZ31 Mg alloy assessed by Tafel polarization curve by a few studies [4, 24–28].

80.2.2 Electrochemical Impedance Spectroscopy (EIS) of Superhydrophobic Coating on AZ31 Mg Alloy

EIS is a method that is usually being performed by using the standard three electrodes' system as shown in Fig. 80.1. The three electrodes are working electrode (WE), counter electrode (CE) and reference electrode (RE). Common examples of CE and RE are platinum plate and saturated calomel electrode (SCE), respectively. EIS will result in two types of plots, namely Nyquist plot and Bode plot. In the Nyquist plot, the charge transfer process causes a capacitive loop of AZ31 Mg alloy at high frequency, and the dissolution and pitting corrosion of Mg, as well as the relaxing of adsorbed intermediate products from the Mg surface, causes an inductive loop at low frequency. However, if the inductive loop is not present, indicating that the AZ31 Mg alloy may be effectively protected by the nanocomposite coating [26]. Bode plot is a plot of frequency response. The magnitude and phase of the system are displayed by the plot as the function of frequency. Usually, the superhydrophobic coating will exhibit high order of magnitude than the bare sample which indicates good corrosion resistance [4].

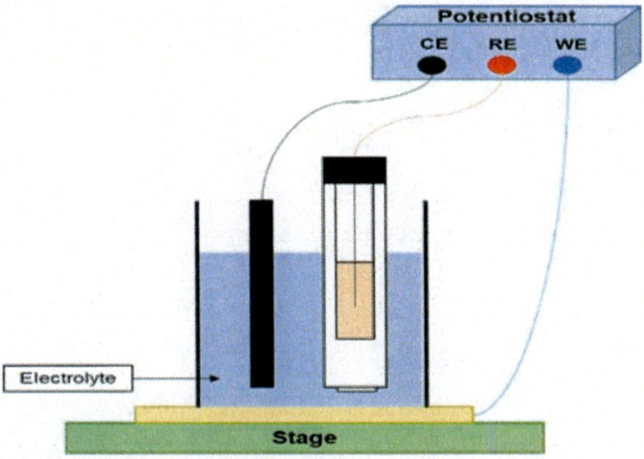

Fig. 80.1 Simple EIS setup

A study on AZ31 Mg alloy coated with ZnO/epoxy resin superhydrophobic coating was reported by Zhou et al. [29] in 2019. The coating was fabricated by chemical etching of AZ31 Mg alloy in hydrochloric acid solution, followed by spraying with ZnO seeds' solution and epoxy resin before placing the AZ31 Mg alloy in a reactor containing a ZnO growth solution, and finally modified with stearic acid solution, which resulted in good corrosion resistance, high wear resistance as well as excellent WCA of 163°. The corrosion resistance of superhydrophobic AZ31 Mg alloy coating is increased due to two main reasons: (1) the controlled mixing process of ZnO seeds to epoxy resin so that epoxy resin did not completely coated the ZnO seeds and (2) preparation of cluster-like ZnO rods on the surface had covered the epoxy resin, making surface modification easier. These two main reasons had created a strong barrier which further improved the corrosion protection performance of superhydrophobic coating.

Zhou et al. [29] performed the corrosion study by using EIS in 3.5 wt% NaCl solution. The EIS was tested with naked AZ31 Mg alloy and under three different $Zn(NO_3)_2$ growth concentrations, namely 2, 2.5 and 3 mmol/L. Compared to naked AZ31 Mg alloy, the superhydrophobic coating exhibited higher capacitance and resistance, which indicates that the coated AZ31 Mg alloy has excellent impermeability and inertness, leading to difficulty of chloride ions to reach the surface of AZ31 Mg alloy. The superhydrophobic coating prepared by growth concentration of 3 mmol/L recorded a low impedance ($|Z|$) compared to 2.5 mmol/L (optimum growth concentration). It was due to high concentration of growth solution as well as the size of cluster-like ZnO rods was bigger than the other two concentrations. The preparation of cluster-like ZnO rods on the surface structure of ZnO/epoxy resin coating led to formation of interdigited micro-nanostructure and the numerous air layers that prevent the corrosive species from diffusing.

EIS method was also being conducted by [23] to precisely evaluate the corrosion behavior of superhydrophobic Ca-P/CS composite coating and bare AZ31 Mg alloy, which displayed that superhydrophobic Ca-P/CS composite coating has better corrosion resistance compared to the bare AZ31 Mg alloy sample as the radius of the capacitive reactance arc in the Nyquist plot was larger and the $|Z|$ value was increased by more than one order of magnitude at low frequency. Furthermore, the value of charge transfer resistance of superhydrophobic coating (2.82×10^4 Ω.cm^2) was 22 times higher than bare AZ31 Mg alloy (1.28×10^3 Ω.cm^2), indicating that the superhydrophobic Ca-P/CS composite coating is able to protect AZ31 Mg alloy against corrosive species.

80.3 Conclusion

In conclusion, most of the superhydrophobic coatings on AZ31 Mg alloy resulted in excellent corrosion protection performance assessed by potentiodynamic polarization curve and EIS. From Tafel polarization curve, E_{corr} is changing in positive direction. The larger the i_{corr}, the greater the negative E_{corr}. A lower i_{corr} value is thought to signify a slower corrosion rate, but a greater E_{corr} value indicates a lower corrosion thermodynamic propensity. In EIS, data obtained from both the Nyquist plot and the Bode plot provided additional information to accurately analyze the corrosion behavior of AZ31 Mg alloy. At low frequency, the Nyquist plot revealed larger capacitance of superhydrophobic coating on AZ31 Mg alloy, while Bode plot displayed the high $|Z|$ value, illustrating that AZ31 Mg alloy coated with superhydrophobic coating was successfully being protected.

Acknowledgements The authors would like to acknowledge the support by the Ministry of Education of Malaysia through Fundamental Research Grant Scheme (FRGS/1/2020/TK0/UNIMAP/03/24).

References

1. J. Kuang, Z. Ba, Z. Li, Z. Wang,J. Qiu, The study on corrosion resistance of superhydrophobic coatings on magnesium. Appl. Surf. Sci. **501** (2020)
2. Y. Wang, Z. Gu, Y. Xin, N. Yuan, J. Ding, Facile formation of superhydrophobic nickel coating on magnesium alloy with improved corrosion resistance. Physicochem. Eng. Aspects **538**, 500–505 (2018)
3. T. Zheng, Y. Hu, F. Pan, Y. Zhang, A. Tang, Fabrication of corrosion-resistant superhydrophobiccoating on magnesium alloy by one-step electrodeposition method. J. Magn. Alloys **7** (2019)
4. X. Liu, T.C. Zhang, H. He, L. Ouyang, S. Yuan, A stearic acid/CeO_2 bilayer coating on AZ31B magnesium alloy with superhydrophobic and self-cleaning properties for corrosion inhibition. J. Alloys Compd. **834** (2020)
5. Q. Jin, G. Tian, J. Li, Y. Zhao, H. Yan, The study on corrosion resistance of superhydrophobic magnesium hydroxide coating on AZ31B magnesium alloy. Colloids Surf. A **577**, 8–16 (2019)
6. A. Jishnu, J. Jayan, A. Saritha, A.S. Sethulekshmi, G. Venu, Superhydrophobic graphene-based materials with self-cleaning and anticorrosion performance: an appraisal of neoteric advancement and future perspectives. Coll. Surf. A **606**, 1–19 (2020)

7. T. Ren, G. Tang, B. Yuan, Z. Yan, L. Ma, X. Huang, One-step fabrication of robust superhydrophobic coatings with corrosion resistance by a self-curing epoxy-resin-based adhesive. Surf. Coat. Technol. **380** (2019)
8. H. Yang, Y. Gao, W. Qin, J. Sun, Z. Huang, Y. Li, J. Sun, A robust superhydrophobic surface on AA3003 aluminum alloy with intermetallic phases in-situ pinning effect for corrosion protection. J. Alloys Compd. **898** (2021)
9. S. Parvate, P. Dixit, S. Chattopadhyay, Superhydrophobic surfaces: insight from theory and experiment. J. Phys. Chem. B **124**, 1323–1360 (2020)
10. S.S. Latthe, S.R. Sutar, V.S. Kodag, A.K. Bhosale, A.M. Kumar, Self-cleaning superhydrophobic coatings: potential industrial applications. Progr. Org. Coat. Self. **128**, 52–58 (2018)
11. P.C. Uzoma, F. Liu, L. Xu, Z. Zhang, E.H. Han, W. Ke, I.O. Arukalam, Superhydrophobicity, conductivity and anticorrosion of robust siloxane- acrylic coatings modified with graphene nanosheets. Prog. Org. Coat. **127**, 239–251 (2019)
12. X. Wu, F. Yang, J. Gan, W. Zhao, Y. Wu, A flower-like waterborne coating with self-cleaning, self-repairing properties for superhydrophobic applications. J. Market. Res. **14**, 1820–1829 (2019)
13. Z. Yin, D. Zhou, K. Liu, M. Li, X. Chen,M. Xue, S. Lei, J. Ou, Y. Lou, Z. Ren, An all-in-one bio-inspired superhydrophobic coating with mechanical/chemical/physical robustness. Coll. Surf. A: Physicochem. Eng. Aspects **643** (2022)
14. C. Ke,C. Zhang, X. Wu, Y. Jiang, Highly transparent and robust superhydrophobic coatings fabricated via a facile sol-gel process. Thin Solid Films **723** (2021)
15. T.P. Rasitha,S.C. Vanithakumar, D.N.G. Krishna, R.P. George, R. Srinivasan, J. Philip, Facile fabrication of robust superhydrophobic aluminum surfaces with enhanced corrosion protection and antifouling properties. Progr. Org. Coatings **162** (2022)
16. S. Ren, J. Chen, M. Jiang, S. Wang, Z. Wan, Y. Xie, L. Li, Colloids Surf. A **611**, 14 (2020)
17. M. Ferrari, A. Benedetti, F. Cirisano, Superhydrophobic coatings from recyclable materials for protection in a real sea environment. Coatings **9**(5) (2019)
18. W. Zhang, S. Fan, Y. Wang, X. Lang, G. Li, Preparation and performance of biomimetic superhydrophobic coating on X80 pipeline steel for inhibition of hydrate adhesion. Chem. Eng. J. **419** (2021)
19. N.A. Polyakov, I.G. Botryakova, V.G. Glukhov, G.V. Red'kina, ,Y.I. Kuznetsov, Formation and anticorrosion properties of superhydrophobic zinc coatings on steel. Chem. Eng. J. **421** (2021)
20. L. Li, J. Zhu, Z. Zeng, A sky-blue superhydrophobiccoating and applications. Progr. Org. Coat. **147** (2020)
21. B. Subeshan, R. Asmatulu, Corrosion mitigation of metals and alloys via superhydrophobic coatings with plasma surface and heat treatment processes. Eng. Fail. Anal. **139** (2022)
22. W. Yao, L. Wu, G. Huang, B. Jiang, A. Atrens, Superhydrophobic coatings for corrosion protection of magnesium alloy. J. Mater. Sci. Technol. **52**, 100–118 (2020)
23. L. Wang, X. Xiao, X. Yin, J. Wang, G. Zhu, S. Yu, X. Yang, Preparation of robust, self-cleaning and anti-corrosion superhydrophobic Ca-P/chitosan (CS) composite coating on AZ31 magnesium alloy. Surf. Coat. Technol. **432** (2022)
24. S. Song, H. Yan, Y. Huang, X. Fan, X. Cui, M. Zhu, Superhydrophobic composite coating for reliable corrosion protection of Mg alloy. Mater. Des. **215** (2022)
25. Z.Q. Zhang, L. Wang, R.C. Zeng, M.Q. Zeng, C.G. Lin, Z.L. Wang, Q. Zhang, Corrosion resistance and superhydrophobic of one-step polypropylene coating on anodized AZ31 Mg alloy. J. Magn. Alloys **9**(4), 1443–1257 (2021)
26. S. Wang, Y. Wang, G. Cao, J. Chen, Y. Zou, B. Yang, Y. Zhou, Highly reliable double-layer coatings on magnesium alloy surfaces for robust superhydrophobicity, chemical durability and electrical property. Ceram. Int. **47**(24), 35037–35047 (2021)
27. J.Z. Sheng, D.L. Cao, L.K. Xu, J.K. Tang, R.Q. Meng, D.L. Hong, Corrosion resistance of a superhydrophobic dodecyltrimethoxysilane coating on magnesium hydroxide-pretreated magnesium alloy AZ31 by electrodeposition. Coll. Surf. A: Physicochem. Eng. Aspects **625** (2021)

28. Z.Q. Zhang, R.C. Zeng, W. Yan, C.G., Lin, L. Wang, Z.L. Wang, D.C. Chen, Corrosion resistance of one-step superhydrophobic polypropylene coating on magnesium hydroxide-pretreated magnesium alloy AZ31. J. Alloys Compd. **821** (2020)
29. H. Zhou, R. Chen, Q. Liu, J. Liu, J. Yu, C. Wang, M. Zhang, P. Liu, J. Wang, Fabrication of ZnO/epoxy resin superhydrophbic coating on AZ31 Magnesium alloy. Chem. Eng. J. **368**, 261–272 (2019)

Chapter 81
Effect of Titanium Dioxide in Superhydrophobic Coating Using Expanded Polystyrene Foam and Palm Slag

Nur Zalifah Binti Hasmadi, Muhammad Salihin Zakaria, Razif Muhammed Nordin, Khairul Anwar Abdul Halim, Lokman Hakim Ibrahim, and Nur Fatihah Mohd Fadzil

Abstract A simple solvent casting method was proposed to prepare a superhydrophobic expanded polystyrene (EPS)/stearic acid modified palm slag (SA-PS)/titanium dioxide (TiO_2) coating with excellent water contact angle. The influence of components, such as mass ratio of palm slag to TiO_2 as well as the influence of TiO_2 on the morphology and wettability of coating, was evaluated. The morphology, tear behavior, surface roughness, and water contact angle were observed by scanning electron microscopy (SEM), scratch tester, 3D profilometer, and ImageJ software, respectively. The results displayed that the superhydrophobic coating exhibited highest WCA of 151.93° under the optimum condition formula. Additionally, the superhydrophobic EPS/SA-PS/TiO_2 showed non-wetting ability. Thus, the existence of a micro/nano-hierarchical structure of superhydrophobic EPS/SA-PS/TiO_2 may provide potential practical applications that are useful in many industries such as automotive or shipping.

N. Z. B. Hasmadi · M. S. Zakaria (✉) · K. A. A. Halim · L. H. Ibrahim · N. F. M. Fadzil
Faculty of Chemical Engineering and Technology, Universiti Malaysia Perlis (UniMAP), 02600 Arau, Malaysia
e-mail: salihin@unimap.edu.my

M. S. Zakaria · K. A. A. Halim
Biomedical and Nanotechnology Research Group, Center of Excellence Geopolymer and Green Technology (CEGeoGTech), Universiti Malaysia Perlis (UniMAP), 02600 Arau, Malaysia

R. M. Nordin
Department of Chemistry, Faculty of Applied Sciences, Perlis Branch Arau Campus, Universiti Teknologi MARA, 02600 Arau, Perlis, Malaysia

Green and Functional Polymer Research Group, Universiti Teknologi MARA, 40450 Shah Alam, Selangor, Malaysia

L. H. Ibrahim
Advanced Polymer Research Group, Center of Excellence Geopolymer and Green Technology (CEGeoGTech), Universiti Malaysia Perlis (UniMAP), 02600 Arau, Malaysia

M. A. A. Mohd Salleh et al. (eds.), *Proceedings of the Green Materials and Electronic Packaging Interconnect Technology Symposium*, Springer Proceedings in Physics 289,
https://doi.org/10.1007/978-981-19-9267-4_81

Keywords Superhydrophobic coating · Palm slag · Titanium dioxide · Expanded polystyrene foam · Water contact angle

81.1 Introduction

Nature serves as the source of inspiration for humans seeking to understand the mechanisms behind the development of novel superhydrophobic surfaces for everyday use. Many plant leaves have a constant self-cleaning phenomenon that allows them to live in unclean environments. Many natural species display superhydrophobicity behavior, such as lotus leaves, rice leaves, rose petals, cicada surface, water strider legs, mosquito eyes, and so on [1]. Superhydrophobic surfaces are surfaces that exhibited contact angles greater than 150° and hysteresis lower than 10° [2]. Subsequently, the development of superhydrophobic surfaces has drawn considerable attention due to its widespread application in creating surfaces with self-cleaning, anti-fogging, anti-biofouling, and other properties, and at the same time allows for actual industrial applications.

Malaysia is the second-largest producer of oil palm in the world, where is expected to be the main raw material for biofuel production in the country soon [3]. However, several risks have been identified, such as the increasing amount of solid waste generated from palm oil has contributed to the biggest problem for Malaysia in recent years. As a result of the occurrence, the issue of oil palm waste disposal has also become a hot topic and the main focus of the problems faced. There are two types of waste produced from palm oil, namely palm slag (PS) and palm ash (PA). PS is produced from palm oil waste such as empty fruit bunches (EFB), palm fiber, and palm oil shells. Both PS and PA were produced in palm oil mill industry to generate steam from the combustion process of solid wastes in combustion chamber. The results of the previous preview study found that waste from PS has no commercial value and can pollute the environment [4].

Some studies showed the preparation of superhydrophobic coating by incorporating inorganic filler with organic filler to create a hybrid (inorganic–organic) coating which in turn will elevate the coating's performance. For instance, [5] developed an inorganic–organic superhydrophobic coating by using polysiloxane double networks and fluorinated nanoparticles that withstood UV exposure for 54 h as well as remained its superhydrophobicity in alkaline and acidic environments. [6] also displayed an excellent UV resistance and corrosion protection performance of inorganic–organic superhydrophobic coating by incorporating fluorinated ethylene propylene with aluminum dihydrogen phosphate. Moreover, a low-cost hybrid superhydrophobic coating was successfully created by Yu et al. [7] using a mixture of methylsiloxane and alumina nanoparticles, which allowed the coating to absorb the oil almost instantly. These outstanding efforts had greatly improved the properties of superhydrophobic coating and opened many chances for further development. Herein, we used PS and nanoparticles titanium dioxide (TiO_2) to prepare a low-cost superhydrophobic EPS/SA-PS/TiO_2 coating by controlling the mass ratio of PS to

TiO_2. In line with this, to increase the adhesive strength of the coating on the substrate, the EPS has been selected as a matrix for PS powder which acts as a filler due to its low cost. Meanwhile, organic chemicals such as stearic acid have been used in these superhydrophobic coatings as additives for PS modification. At the same time, TiO_2 was chosen as a parameter to enhance the surface roughness of the coating and elevated it to achieve superhydrophobic coating. The presence of hydrophobic TiO_2 nanoparticles will assist in formation of micro/nano-hierarchical structure as well as to enhance the surface roughness to produce more air pockets.

81.2 Experimental

81.2.1 Materials and Methods

EPS is a recycled material that has been collected from public places around Perlis and has the density of 19.8 kg/m^3. PS (125 μm) was supplied by Felda Palm Industries Sdn Bhd, which is in Terengganu, Malaysia. Ethyl acetate (EtOAc), ethanol (EtOH), and stearic acid (SA) were purchased from Sigma-Aldrich®, Systerm®, and Fluka®Analytical, respectively. Solid TiO_2 nanoparticles was supplied by Huntsman with molecular weight of 19.87 g/mole.

The coating was fabricated by solvent casting technique. PS had undergone grinding process in planetary mill machine to further reduce the size of PS. 2 g of EPS was dissolved in EtOAc at room temperature and stirred for 5 min to allow the polymer solution to be completely homogeneous. At the same time, the designated amount of SA was mixed with EtOH under constant heat and stir. The SA solution was mixed for a few minutes before adding PS and nano-TiO_2 powder, and the solution was continued to be heated for 10 min until it mixed well. The above solution was mixed with the polymer solution under constant stirring, followed by ultrasonicated for 5 min under temperature of 60 °C. Finally, the solution was casted onto plate and the sample was left at room temperature for 24 h. This preparation coating step will be repeated for other samples by following the mass ratio of PS to TiO_2 as shown in Table 81.1.

Table 81.1 The mass ratio of PS to TiO_2

Sample	PS (g)	TiO_2 (g)
Sample 1	2	0
Sample 2	2	1
Sample 3	2	5
Sample 4	2	6
Sample 5	2	7

81.2.2 *Measurement and Characterization*

81.2.2.1 Water Contact Angle Measurement

The water contact angle (WCA) for this coating was measured using a dropper by dripping distilled water continuously over the sample surface. Then, by using a smartphone's external micro-lens (STK-L22, Huawei Y9 Prime 2019), images of water droplets formed on the sample surface were captured. Subsequently, contact angle data were analyzed using ImageJ software.

81.2.2.2 Microstructural Analysis

Microstructural analysis of the samples was analyzed using a scanning electron microscope (SEM) (JSM-6460LA, JEOL) and the operating energy voltage used is 10 kV. The sputter coater (Quorum, Q150RS) was used as a pre-preparation coating for this sample by applying a thin platinum coating on the sample surface to prevent the formation of electrostatic charges during the testing process.

81.2.2.3 Surface Roughness Analysis

Surface roughness analysis was performed by using a 3D surface profiler machine (Pemtron, Hawk 3D Wt-250). The scanning probe image processor (SPIP) tracking program has been used to evaluate all photos captured, and several statistical features were used to describe the topography.

81.2.2.4 Scratch Analysis

CSM Revetest Xpress was used to run this test. A Rockwell indenter-D247 consisting of a cone with a half-peak angle of 60° ending in a sphere with an end radius of 200 m was selected during machine testing and the scratch distance and indenter speed were 15 mm and 5 mm/min.

81.3 Results and Discussion

Based on Fig. 81.1, significant differences can be seen in the WCA values with and without the presence of TiO_2 in the coating. The WCA increases as the amount of TiO_2 increases. The initial WCA of sample 1 is 115.89° with no addition of TiO_2 and the highest WCA of 151.93° was showed by sample 4 with addition of 6 g TiO_2. However, the steady increase of WCA values suddenly dropped to 122.23° when

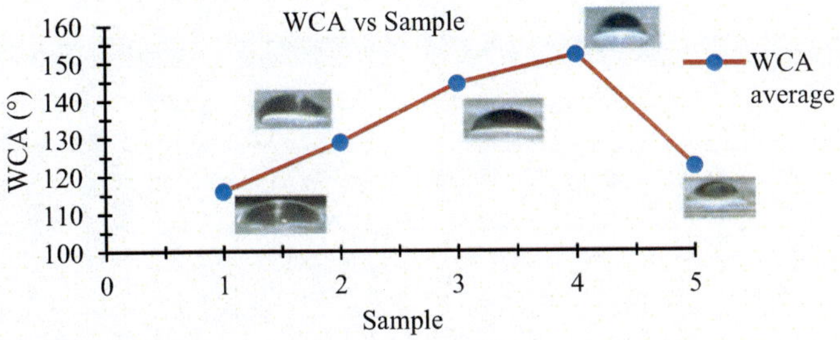

Fig. 81.1 The relationship between WCA and sample

the amount of TiO_2 increased to 7 g for sample 5. This is because the higher of the amount of TiO_2 the lower the adhesion of EPS to the particles, which means less air pockets were produced and thus, affecting the superhydrophobicity of sample 5.

The results of microstructural analysis obtained for all samples were shown in Fig. 81.2. Based on Fig. 81.2a, although the texture of the coating is smooth, but the surface of the coating was covered with air bubbles trapped inside it. This is due to the least adhesion of PS onto polymer matrix to create surface roughness, exposing the matrix to trap air during drying process. While Fig. 81.2b has shown that there is a change in terms of the results obtained finding that in the presence of 1 g of additional TiO_2, the texture of the coating changed, resulted in non-homogenous surface with the presence of white particles appear on the sample. In Fig. 81.2c, d, the surface of the coatings is uneven and highly jagged compared to surface of coating in Fig. 81.2e. The surface of coating is slightly smooth and least jagged, and the white particles is highly visible. This may be attributed to high content of TiO_2 nanoparticles, which hindered the formation of micro/nano-hierarchical structure and in turn reduces the adherence of fillers onto polymer matrix. The presence of jagged texture in both surfaces of Fig. 81.2c, d indicates the formation of micro/nano-hierarchical structure in the coatings. Sample 4 is determined as the optimum condition to fabricate EPS/SA-PS/TiO_2 superhydrophobic coating as the surface of coating has the uneven microstructure and highest jagged texture. Therefore, more air pockets will be generated and increases the contact angle.

A simple surface roughness analysis shows the roughness profile of sample 4 at different points. Figure 81.3 have been shown the placement of the points surface roughness in sample 4. Figure 81.4a illustrates the roughness profile of sample 4. It reveals the aggregation of particles, which are unevenly distributed on the scanned surface. Such surface topography occurs from uneven coating techniques causing such occurrences. Figure 81.4b shows the red color on the left and right of the coating has caused a large loss in the value of WCA. Patches on the surface layer suggest that impact uniformity plays a major role in superhydrophobic coatings [8]. Figure 81.4c shows such homogeneous surface topography is considered less responsible, especially for the observed superhydrophobic features [9]. Table 81.2 shows the analytical data of surface roughness that were taken from different points on

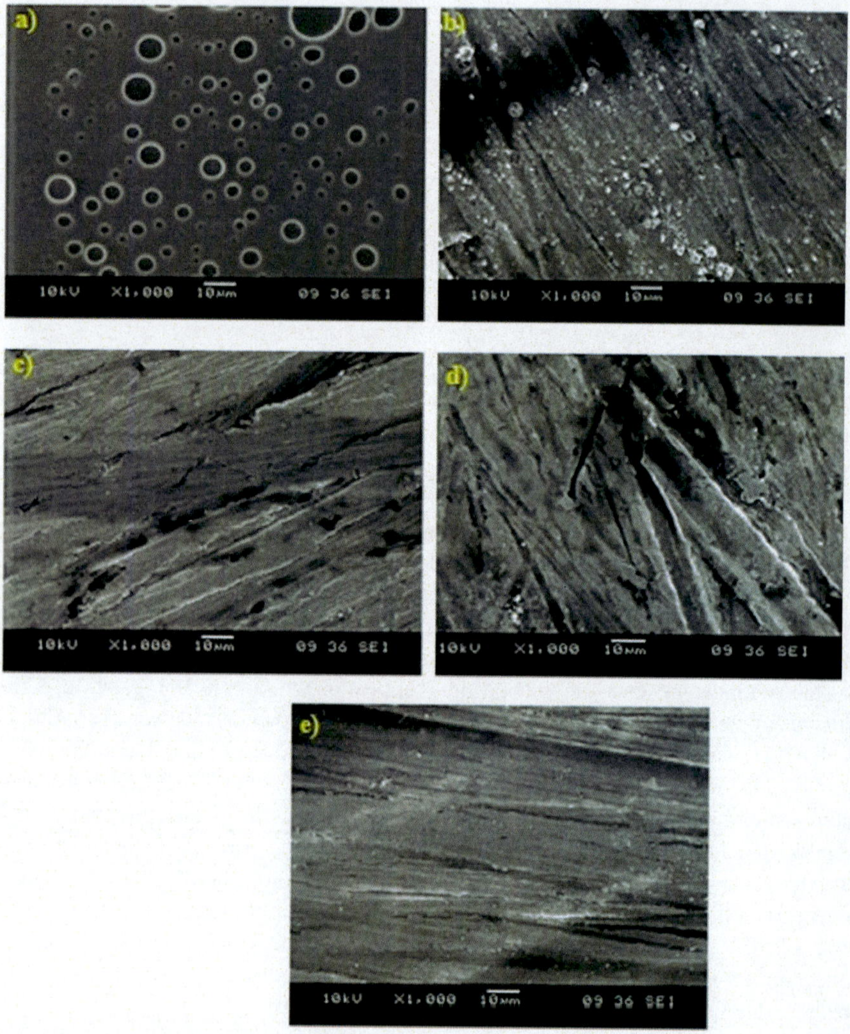

Fig. 81.2 SEM image of superhydrophobic coating on **a** sample 1 **b** sample 2 **c** sample 3 **d** sample 4, and **e** sample 5

sample 4. The surface roughness of the coating was described using mean roughness (Ra) and roughness mean root (Rq).

A simple surface roughness analysis shows the roughness profile of sample 4 at different points. Figure 81.3 have been shown the placement of the points surface roughness in sample 4 In Fig. 81.4a illustrates the roughness profile of sample 4. It reveals the aggregation of particles, which are unevenly distributed on the scanned surface. Such surface topography occurs from uneven coating techniques causing such occurrences. Figure 81.4b shows the red color on the left and right of the

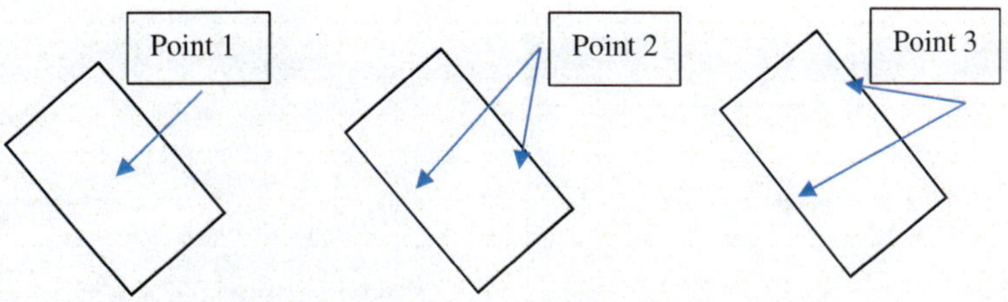

Fig. 81.3 The placement of the points surface roughness for sample 4

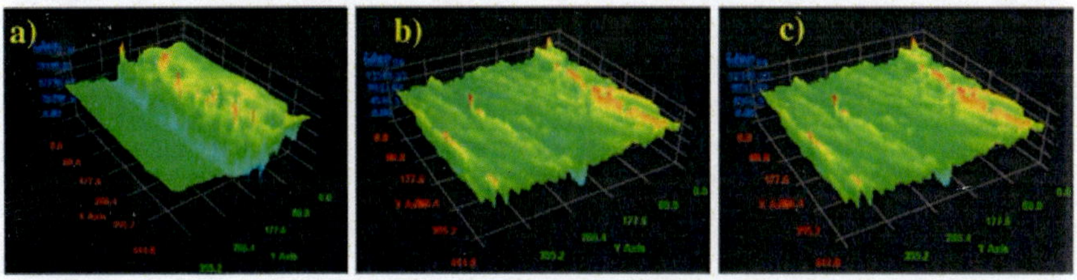

Fig. 81.4 The roughness profile of sample 4 at **a** point 1 **b** point 2 **c** point 3

Table 81.2 Ra and Rq values of three different points on sample 4

Sample 4	Ra (μm)	Rq (μm)
Point 1	8.73	10.53
Point 2	3.27	4.73
Point 3	1.30	1.80

coating has caused a large loss in the value of WCA. Patches on the surface layer, suggest that impact uniformity plays a major role in superhydrophobic coatings [8]. Figure 81.4c shows such homogeneous surface topography is considered less responsible, especially for the observed superhydrophobic features [9]. Table 81.2 shows the analytical data of surface roughness that were taken from different points on sample 4. The surface roughness of the coating was described using mean roughness (Ra) and roughness mean root (Rq).

Scratch tests were performed on sample 1 and sample 4 consisting of hydrophobic and superhydrophobic samples using a scratch and cantilever machine with a diamond tip, to see the resulting scratch effect, 30× and 50× optical microscope images were used to analyze in more detail the resulting scratch effect on the coating. Figure 81.5a, b show the results of scratches that have been done on sample 1 have found that there is little tear effect on the sample along the resulting scratch line and found that there is little distortion effect on the left and right of the scratch line produced during the scratch test process. Figure 81.5c, d have shown the effect of scratches produced on sample 4 which is the optimum superhydrophobic EPS/SA-PS/TiO_2 coating, it is found that the resulting tearing effect on the sample occurred uniformly compared to the tearing effect that occurred on the previous sample. At

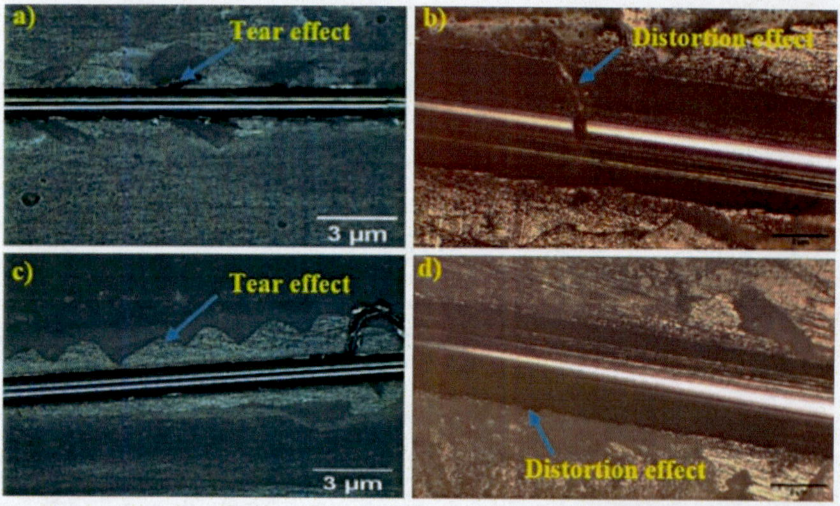

Fig. 81.5 Scratch analysis results for samples 1 and 4 using 50× magnification for (**a**) and (**c**), and 30× magnification for (**b**) and (**d**), respectively

the same time, through the 50× magnification image can also be observed the effect of distortion produced on this sample is a little compared to the previous sample. This can prove that this superhydrophobic sample has high mechanical resistance.

81.4 Conclusion

In conclusion, the superhydrophobic EPS/SA-PS/TiO_2 is successfully achieved by a simple and low-cost solvent casting technique. The coating resulted in high WCA and the optimum condition formula was achieved by controlling the mass ratio of PS to TiO_2. Furthermore, the presence of TiO_2 nanoparticles in fabricating helps to elevate the surface roughness of superhydrophobic EPS/SA-PS/TiO_2 as it contributed in creating a hierarchical micro-nanostructure along with PS. Such effective hierarchical micro-nanostructure contributes to maximum WCA of 151.93°. For future research, it is expected to fabricate a superhydrophobic coating that is highly durable and stable with various specific testing being conducted on it so that the ability of superhydrophobic coating will further explored.

Acknowledgements The authors would like to acknowledge the support by the Faculty of Chemical Engineering and Technology Universiti Malaysia Perlis (UniMAP) and Universiti Teknologi MARA (UiTM) Branch Perlis for providing all the necessary facilities for this study.

References

1. S.K. Sethi, U. Shankar, G. Manik, Fabrication and characterization of non-fluoro based transparent easy-clean coating formulations optimized from molecular dynamics simulation. Progr. Org. Coat. **136** (2019)
2. B. Bhushan, Y.C. Jung, Natural and biomimetic artificial surfaces for superhydrophobicity, self-cleaning, low adhesion, and drag reduction. Prog. Mater Sci. **56**(1), 1–108 (2011)
3. S. Dey, N.M. Reang, P.K. Das, M. Deb, A comprehensive study on prospects of economy, environment, and efficiency of palm oil biodiesel as a renewable fuel. J. Clean. Prod. **286** (2021)
4. N. Abdullah, F. Sulaim, *The Oil Palm Wastes in Malaysia*, 1st ed (InTech, Croatia, 2013)
5. X.Y. Wang, C. Zhang, S. Sun, M. Kalulu, L. Chen, X. Zhou, Y. Jiang, Durable superhydrophobic coating based on inorganic/organic double-network polysiloxane and functionalized nanoparticles. Coll. Surf. A **578** (2019)
6. Z. Liu, L. Ren, J. Jing, C. Wang, F. Liu, R. Yuan, M. Jiang, H. Wang, Fabrication of robust superhydrophobic organic-inorganic hybrid coating through a novel two-step phase separation method. Progr. Org. Coat. **157** (2021)
7. M. Yu, Z. Cui, F. Ge, H. Tian, X. Wang, Simple spray deposition of a hot water-repellent and oil-water separating superhydrophobic organic-inorganic hybrid coatings via methylsiloxane modification of hydrophilic nano-alumina. Prog. Org. Coat. **125**, 15–22 (2018)
8. A.B. Gurav, Q. Xu, S.S. Latthe, R.S. Vhatkar, S. Liu, H. Yoon, S.S. Yoon, Superhydrophobic coatings prepared from methyl-modified silica particles using simple dip-coating method. Ceram. Int. **41**(2), 3017–3023 (2015)
9. B. Bolvardi, J. Seyfi, I. Hejazi, M. Otadi, H.A. Khonakdar, A. Drechsler, M. Holzschuh, Assessment of morphology, topography and chemical composition of water-repellent films based on polystyrene/titanium dioxide nanocomposites. Appl. Surf. Sci. **396**, 616–624 (2017)

Chapter 82
Effect of Surface-Modified Copper Substrate by Photolithography On the Solderability of Lead-Free Solder Alloy

Nurul Aida Husna Mohd Mahayuddin, Juyana A. Wahab, Mohd Arif Anuar Mohd Salleh, and Siti Faqihah Roduan

Abstract Various modifications of copper substrates have been developed for soldering applications due to their promising properties. The present study aimed to investigate the effect of dimple micro-texture on a copper substrate surface on soldering. Photolithography technique was used to fabricate a dimple micro-texture onto copper substrate. After surface texturing, the copper substrate was reflowed with Sn–0.7Cu solder. The surface microstructural was determined using optical microscopy (OM), the 3D surface profiler, and visual tests. The spreading area of the Sn–0.7Cu solder and the copper substrate was also examined. In comparison with Sn-0.7Cu solder on textured surface, TD500 has larger spreading area as the diameter was large nearly equivalent to non-textured surface. However, Sn–0.7Cu textured substrate exhibits low spreading area than non-textured surface that is beneficial for solderability properties.

Keywords Surface texture · Photolithography · Lead-free solder · Sn–0.7Cu solder

82.1 Introduction

Surface texture technology has been widely used in manufacturing technologies. For instance, nature that has textured surfaces like lotus leaves, frog toes, and shark skin can be found and extensive research has been done to clarify their unique characteristics [1, 2]. It is proven that this textured surface can improve the properties of the material such as superhydrophobicity, friction via adhesion, and drag reduction [3]. In tribological systems, the textured surfaces serve to increase the materials' hydrophilicity, wettability, hardness [4, 5], and friction and wear resistance [6–8]. In

N. A. H. Mohd Mahayuddin · J. A. Wahab (✉) · M. A. A. Mohd Salleh · S. F. Roduan
Faculty of Chemical Engineering and Technology, Universiti Malaysia Perlis (UniMAP), Kompleks Pusat Pengajian Jejawi 2, Taman Muhibbah, 02600 Arau, Perlis, Malaysia
e-mail: juyana@unimap.edu.my

M. A. A. Mohd Salleh et al. (eds.), *Proceedings of the Green Materials and Electronic Packaging Interconnect Technology Symposium*, Springer Proceedings in Physics 289,
https://doi.org/10.1007/978-981-19-9267-4_82

1940s, the piston ring and bushing contact of diesel engines have undergone surface treatment to prevent seizures under the hot running circumstances, represent the first effective applications of surface texturing. Some of surface texturing methods such as micro-milling, hot embossing, ion beam etching, lithography, laser surface machining, diamond cutting, and laser surface machining [9–11] were used extensively. One of the surface texturing methods is the lithographic texturing technique, which uses the UV light in patterning to balance the final specified texture's quality and complexity at a low processing temperature. Due to its superior properties such as feature size resolution enhancement, processing dependability, alignment accuracy, and pattern replication flexibility, it has been broadly applied in nano-processing [12–14].

The fluidity of the molten solder is believed to be influenced by the surface roughness of the micro-textured copper substrate [15]. Surface roughness of the copper substrate is greatly influenced by the etching time of dimple micro-texture. Etching time will results in depth of the dimple micro-texture making the etching time one of the significant parameters in photolithography. However, there are no systematic studies which have been carried out regarding the effect of etching duration and the formation of dimple micro-textures on the wettability of Sn–0.7Cu lead-free solders. Thus, in order to improve the wettability and solderability of lead-free solders, surface-modified substrates with dimple micro-textures via photolithography technique were introduced in this work.

82.2 Methodology

In this work, a high-purity copper plate (99.9%) with dimensions of 15 mm × 15 mm × 1 mm was used as the substrate material, and Sn–0.7Cu was selected as the lead-free solder material. Prior to the fabrication of dimple micro-textures by photolithography process, the copper plate has been grinded and polished to create a smooth, mirror-like, and scratch-free surface. The photolithography technique of fabrication involved (i) photoresist coating, (ii) spin coating, (iii) soft baking, (iv) UV exposure, (v) development, (vi) etching the pattern into the copper substrate, and (vii) stripping. The diagram of photolithography process is presented in Fig. 82.1.

For surface treatment, the dimples with a diameter ranging from 100 to 500 μm were fabricated, with a 300 μm distance between each dimple. All sample designations are listed in Table 82.1. The covered part of the photomask design acted to form the dimples' micro-texture on the copper substrate. Positive photoresist (2 ml) was applied onto the copper substrate by spin coating technique at 850 rpm for 15 s, then increased to 3000 rpm for 30 s, and lastly reducing the speed to 0 rpm in 15 s. Later, the sample was soft baked within 90 s at 90 °C. The sample was then exposed to UV light for 120 s using the photolithography mask aligner.

To eliminate undesired parts, photoresist coating on the substrates' surface was removed using an RD-6 developer. Development times ranging between 30 and 150 s with the hard bake procedure taking place at 90 °C for 60 s. Before hard bake

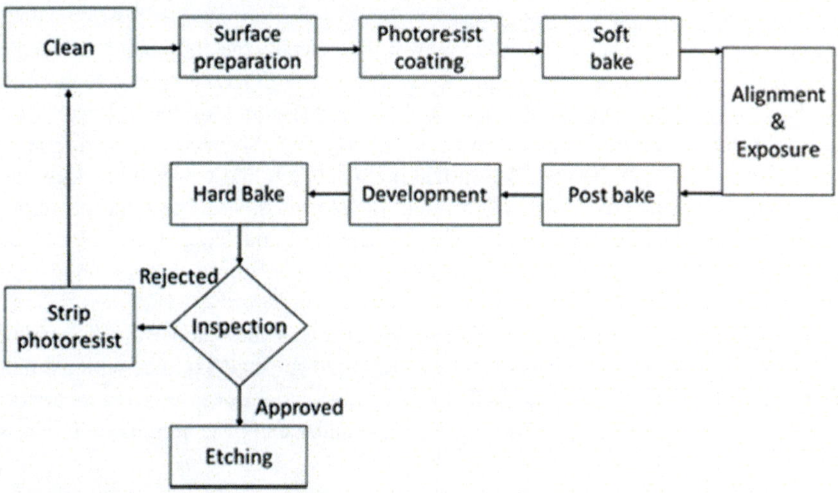

Fig. 82.1 Photolithography process

Table 82.1 Designation of the samples

Sample (dimple diameter) (μm)	Designation
100	T_{D100}
200	T_{D200}
300	T_{D300}
400	T_{D400}
500	T_{D500}

process, the copper substrates were etched three times using ferric chloride ($FeCl_3$) in the exposure time ranging between 15 and 45 min. The dimple micro-texture for the copper substrate was illustrated in Fig. 82.2.

Fig. 82.2 Schematic diagram of dimple micro-textured copper with the cross-sectional view

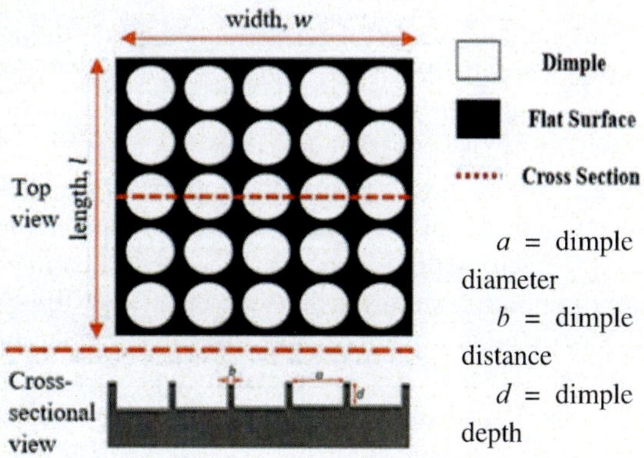

The dark area on the substrates' surface was protected by the photomask, while the bright region was exposed to UV light. X-ray diffraction (XRD) was used to identify the phases of the pure copper substrate via the X-pert High Score Plus software with the ICDD PDF-2 database using X-pert High Score Plus Software at a scan rate of 0.1° per min and 2θ of 20–80°. An optical microscope (OM) was used to view the images of the developed sample at different developed times. The dimple's depth during etching was measured using a 3D profiler. T_{D300} and T_{D500} were selected to proceed to the following soldering process and the non-textured sample functioned as the control in this study.

For soldering materials, the Sn–0.7Cu solder alloy was fabricated using the casting method. Then, 0.4 g of Sn–0.7Cu was soldered onto selected samples with a small amount of resin mildly activated (RMA) flux in the reflow oven. For characterization, the solder spreading areas were observed visually and measured using Image-J software.

82.3 Results and Discussions

82.3.1 Raw Material Phase Identifications

As shown in Fig. 82.3, the XRD diffraction pattern consisting the peaks of the pure copper substrate at angles 2θ of 43.254°, 50.375°, and 73.997°. This pattern matches with International Committee for Diffraction Data (ICDD) with reference patterns of (ICCD 00-003-1005). All patterns for this pure copper substrate were almost identical, indicating that the copper substrate was free from any impurities.

82.3.2 Fabrication of Dimple Micro-texture

There are three stages of formation condition during development process: underdeveloped, fully developed, and overdeveloped (Table 82.2). In the case of underdeveloped, the dimple micro-texture not yet formed on the copper substrate's surface successfully. As development time less than 60 s, the dimple micro-texture was hardly seen on the surface of the copper substrate. This was due to the insufficient time for the dimple to develop during this time range. Fully developed is when the dimple micro-pattern is successfully formed on the copper substrate's surface. The dimple micro-texture can be seen clearly on the surface of the copper substrate. It was found that the best time taken for dimple micro-texture to fully develop is between 100 and 120 s. Overdeveloped is when the dimple micro-texture disappears from the copper substrate's surface. This stage occurs when the development time is more than 150 s and within this stage, the dimple micro-texture has entirely vanished from the copper substrate's surface. It can be concluded that the best condition for

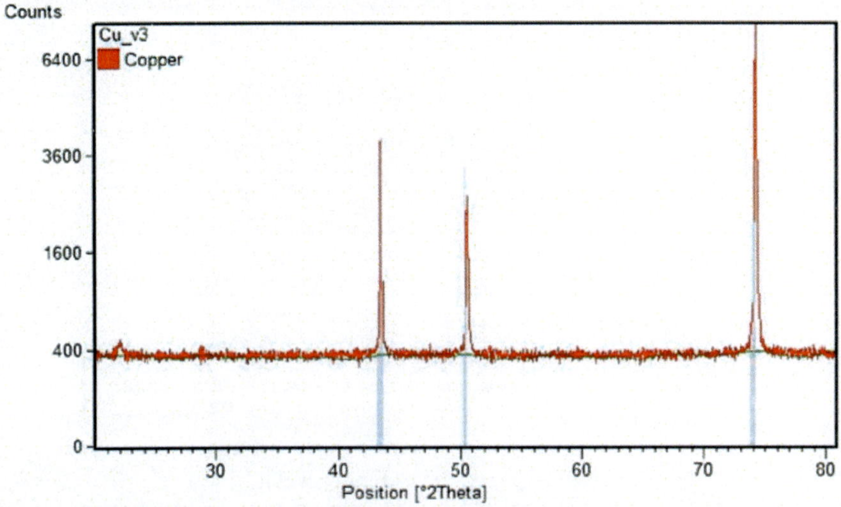

Fig. 82.3 XRD diffraction pattern of pure copper substrate

fabricating the dimple micro-textures was during fully developed stage (100 s to 120 s). When fully developed conditions were achieved, the samples proceeded to the etching process. The etching time at this stage was investigated in work.

The depth of dimple micro-texture for all samples was measured and analyzed. The wet etching method was used to transfer the dimple micro-texture through photolithography mask to the copper substrate as this method is simple and economical. Wet etching processes are usually isotropic, indicating that the etch rate is uniform in all directions. $FeCl_3$ is used as the chemical etchant for copper substrate in wet etching procedures due to these etchant's advantages including a high etch rate and ease of regeneration [16]. $FeCl_3$ reacts with the copper substrate with the presence of positive photoresist producing the dimple micro-texture on the copper substrate. As shown in Fig. 82.4, as the etching time increased, the depth of dimple micro-texture for all samples increased. Sample $T_{D500\mu m}$ shows the maximum depth of the dimple. All dimple micro-textured samples were achieving similar depth of dimple which is approximately 23 ± 5 μm depth of dimple when the etching time was at 45 min. The depth of dimple on the surface of the copper substrate is believed which can improved lead-free solder wettability by reducing the solder's contact area.

82.3.3 Spreading Characteristic of Sn–0.7Cu Solder

In order to study the performance of modified copper substrates for soldering applications, the spreading areas of the solidified solder for both textured and non-textured

Table 82.2 Development stages of photolithography

Development stages
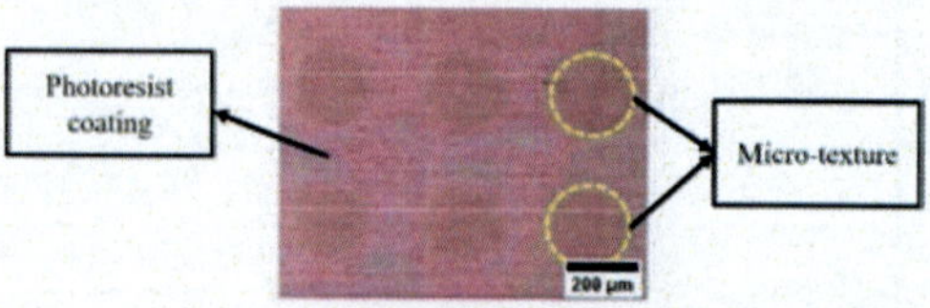 Underdeveloped
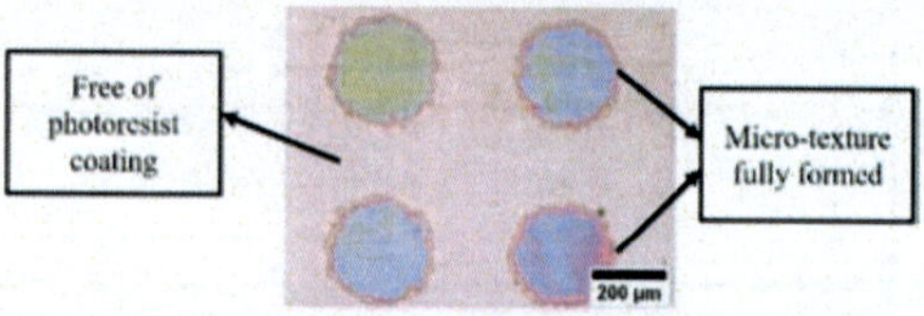 Fully developed
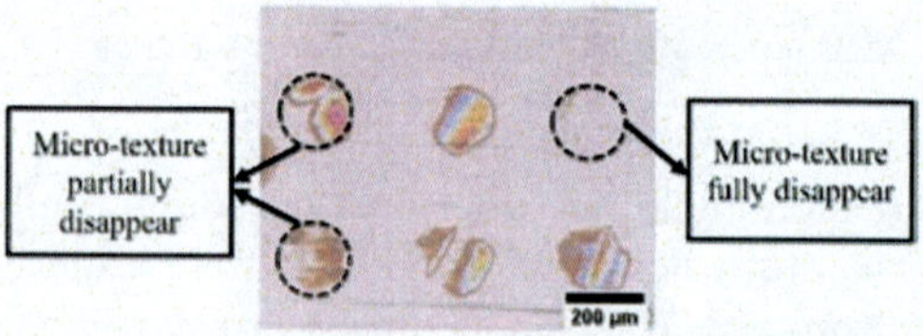 Overdeveloped

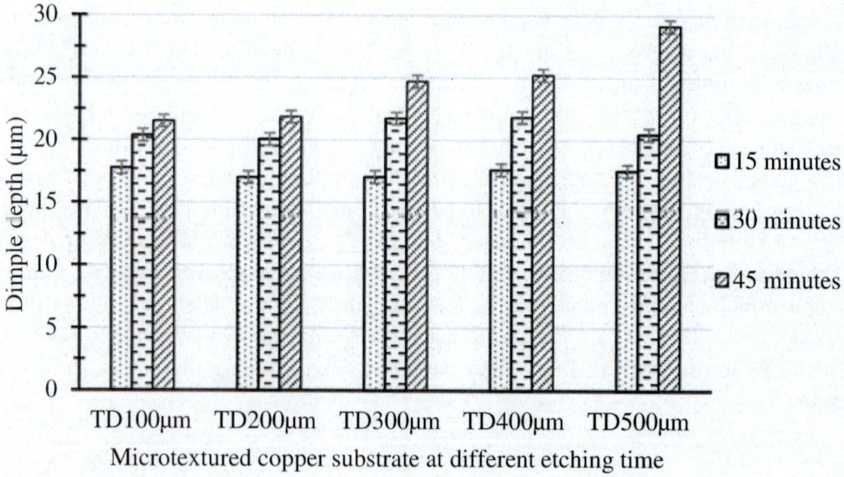

Fig. 82.4 Depth of micro-textured copper substrate at different etching times

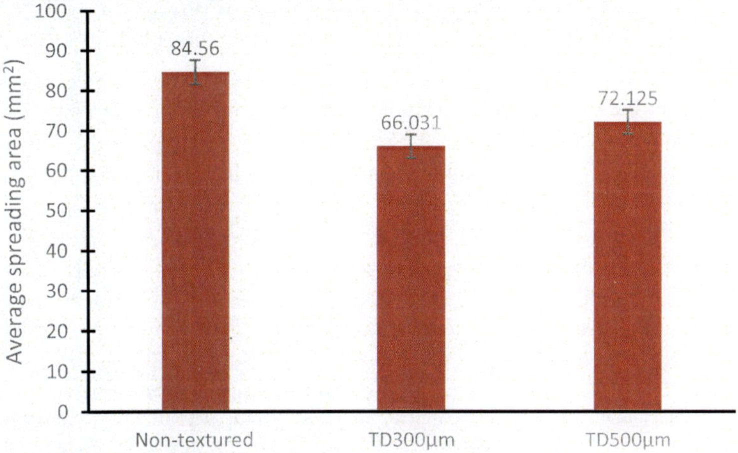

Fig. 82.5 Average spreading area of solder on copper substrate

substrates were measured and compared. The results from Fig. 82.5 demonstrated that as the dimple micro-textures was introduced to the copper substrate surface, the solder's spreading area was lower than the non-textured surface. The spreading areas of the T_{D300} and T_{D500} samples were roughly 66.031 mm^2 and 72.125 mm^2, respectively. This occurred due to the surface's dimple micro-texture, which prevented the molten solder from spreading. The trapping action of the flow of molten solder into the dimple can be explained to application of a Wenzel model with a constrained spreading region. Wenzel model is used to describe a surface in which droplets penetrate the asperities leading to high adhesive forces [17]. Compared to a textured surface, a greater driving force is required for molten solder to spread over a textured sample at a given time. However, T_{D500} has larger spreading area of solder on the copper substrate compared to T_{D300} as the diameter of T_{D500} was large nearly equivalent to the non-textured copper substrate.

On the other hand, the spreading area of non-textured was roughly 84.56 mm^2. The molten solder spreads easily across a large smooth, non-textured surface region. This was due to the non-textured surface having lower surface roughness than the textured surface. These findings are consistent with Satyanarayan et al. which found that the surface roughness is a significant factor in affecting the fluids wetting properties and molten solders and provides more interfacial area for liquid to spread [18, 19]. The substrate metal surface dissolves by molten solder during the solder flow process through diffusion. Substrate dissolution affects the flow and usually lowers the solder viscosity [20, 21].

82.4 Conclusion

In conclusion, the dimple micro-texture was successfully fabricated on the copper substrate's surface using the photolithography process. The fabrication of the dimple micro-texture on the copper substrate necessitates a fully developed state during the development steps, and the depth of the dimple of 23 ± 5 μm and 45 min seems to be an effective solution for etching time. Compared to the non-textured copper substrate, the samples T_{D300} and T_{D500} were reported to have smaller spreading areas. Overall, the solderability of Sn–0.7Cu Pb-free solder can be enhanced by the dimple micro-texture fabricated on a copper substrate.

Acknowledgements The authors would like to express their gratitude for the financial support provided by the Fundamental Research Grant Scheme (FRGS/1/2019/TK05/UNIMAP/03/3), which is funded by the Ministry of Education Malaysia (KPM). Also, thanks to the Faculty of Chemical Engineering Technology, Universiti Malaysia Perlis (UniMAP) and Nihon Superior Co. Ltd. providing facility guidance and support during this research endeavor. The author would also like to express his gratitude for the financial support in the form of a publication incentive grant from Universiti Malaysia Perlis (UniMAP).

References

1. H. Gao, Y. Liu, G. Wang, S. Li, Z. Han, L. Ren, Biomimetic metal surfaces inspired by lotus and reed leaves for manipulating microdroplets or fluids. Appl. Surf. Sci. **519**(Feb), 146052 (2020). https://doi.org/10.1016/j.apsusc.2020.146052
2. D.-G. Lee, Y.-J. Kim, Effects of biomimetic riblet configurations on the drag reduction. DEStech Trans. Environ. Energy Earth Sci. eccsd, 1–5 (2017). https://doi.org/10.12783/dteees/eccsd2016/5830
3. L. Zhang et al., Super-hydrophobicity and corrosion resistance of laser surface textured AISI 304 stainless steel decorated with Hexadecyltrimethoxysilane (HDTMS). Opt. Laser Technol. **127**(Feb), 106146 (2020). https://doi.org/10.1016/j.optlastec.2020.106146
4. Y.Y. Chen, J.G. Duh, B.S. Chiou, Effect of substrate surface roughness on the wettability of Sn–Bi solders. J. Mater. Sci.: Mater. Electron. **11**(4), 279–283 (2000). https://doi.org/10.1023/A:1008917530144
5. J.I. Ahuir-Torres, M.A. Arenas, W. Perrie, G. Dearden, J. de Damborenea, Surface texturing of aluminium alloy AA2024-T3 by picosecond laser: effect on wettability and corrosion properties. Surf. Coat. Technol. **321**, 279–291 (2017). https://doi.org/10.1016/j.surfcoat.2017.04.056
6. H. Yu, H. Deng, W. Huang, X. Wang, The effect of dimple shapes on friction of parallel surfaces. **225**, 693–703 (2015). https://doi.org/10.1177/1350650111406045
7. J. Zhang, J. Zhang, A. Rosenkranz, N. Suzuki, E. Shamoto, Frictional properties of surface textures fabricated on hardened steel by elliptical vibration diamond cutting. Precis. Eng. **59**(Feb), 66–72 (2019). https://doi.org/10.1016/j.precisioneng.2019.06.001
8. J. Hu, H. Xu, Friction and wear behavior analysis of the stainless steel surface fabricated by laser texturing underwater. Tribol. Int. **102** (2016). https://doi.org/10.1016/j.triboint.2016.06.001
9. H. Yue, J. Deng, D. Ge, X. Li, Y. Zhang, Effect of surface texturing on tribological performance of sliding guideway under boundary lubrication. J. Manuf. Process. **47**(Aug), 172–182 (2019). https://doi.org/10.1016/j.jmapro.2019.09.031

10. S.J. Won, H.S. Kim, Effects of laser parameters on morphological change and surface properties of aluminum alloy in masked laser surface texturing. J. Manuf. Process. **48**(Nov), 260–269 (2019). https://doi.org/10.1016/j.jmapro.2019.10.034
11. P. Pou et al., Laser texturing of stainless steel under different processing atmospheres: from superhydrophilic to superhydrophobic surfaces. Appl. Surf. Sci. **475** (2019). https://doi.org/10.1016/j.apsusc.2018.12.248
12. S. Zhu et al., Programmable pupil correction method for photolithography illumination system. Optik **208**(Nov 2019), 164072 (2020). https://doi.org/10.1016/j.ijleo.2019.164072
13. N.D. Pratiwi, M. Handayani, R. Suryana, O. Nakatsuka, Fabrication of porous silicon using photolithography and reactive ion etching (RIE). Mater. Today: Proc. **13**, 92–96 (2019). https://doi.org/10.1016/j.matpr.2019.03.194
14. R. Balma, K. Petsch, T. Kaya, Development of thin film photolithography process: patterning printed circuit boards (PCBs) and copper electroplating, in *ASEE NC & IL/IN Section Conference* (2011), pp. 1–9
15. H.C. Cheng, Z.X. Jiang, T.L. Chang, P.H. Chen, Roughness and wettability properties of plain and silica-coated copper surfaces textured with picosecond laser. Appl. Surf. Sci. **514**(Feb), 145918 (2020). https://doi.org/10.1016/j.apsusc.2020.145918
16. T.-S. Choi, D.W. Hess, Chemical etching and patterning of copper, silver, and gold films at low temperatures. ECS J. Solid State Sci. Technol. **4**(1), N3084–N3093 (2015). https://doi.org/10.1149/2.0111501jss
17. W.M. Sigmund, S.H. Hsu, Wenzel model, in *Encyclopedia of Membranes*, ed. by E. Drioli, L. Giorno (Springer, Berlin, Heidelberg, 2013). https://doi.org/10.1007/978-3-642-40872-4_1380-2
18. Satyanarayan, K.N. Prabhu, Reactive wetting, evolution of interfacial and bulk I MCs and their effect on mechanical properties of eutectic Sn–Cu solder alloy. Adv. Coll. Interf. Sci. **166**(1–2), 87–118 (2011). https://doi.org/10.1016/j.cis.2011.05.005
19. L.M. Vilhena, M. Sedlaček, B. Podgornik, J. Vižintin, A. Babnik, J. Možina, Surface texturing by pulsed Nd:YAG laser. Tribol. Int. **42**(10), 1496–1504 (2009). https://doi.org/10.1016/j.triboint.2009.06.003
20. P.T. Vianco, Y. Feng, Electronic packaging: solder mounting technologies, in *Reference Module in Materials Science and Materials Engineering* (Elsevier, 2016). https://doi.org/10.1016/b978-0-12-803581-8.10099-2
21. O. Çakır, Review of etchants for copper and its alloys in wet etching processes. Key Eng Mater **364–366**, 460–465 (2007). https://doi.org/10.4028/www.scientific.net/kem.364-366.460

Chapter 83
Effect of Wetting Characteristics of Dimpled Micro-textured Substrate on the Spreading Area of Sn–0.7Cu Solder Alloy

Siti Faqihah Roduan, Juyana A. Wahab, M. A. A. Mohd Salleh, Nurul Aida Husna Mohd Mahayuddin, M. H. Aiman, A. Q. Zaifuddin, and M. Ishak

Abstract This study aimed to investigate the effect of wetting characteristics of a dimpled micro-textured copper substrate on the spreading area of the Sn–0.7Cu solder alloy. A laser surface texturing (LST) technique was used to fabricate dimpled micro-textured on the copper substrate. Three samples were fabricated by varying the dimple spacing on the copper substrate such as 100, 140, and 180 μm. A 3D measuring laser microscope was used to study the surface roughness. The copper substrates' water contact angle measurement and surface roughness were studied for their wetting characteristics. In order to study the effect of wetting characteristics of the dimpled micro-textured copper substrate, spreading area measurement was conducted. The result showed that dimpled micro-textured introduces higher surface roughness and contact angle, increasing the hydrophobicity. Consequently, this reduces the spreading area of Sn–0.7Cu solder alloy.

Keywords Surface texturing · Surface roughness · Wettability · Water contact angle · Hydrophobicity · Sn–0.7Cu solder

83.1 Introduction

Surface texturing is a part of surface modification technologies to improve the surface behavior of materials [1]. Various surface texturing techniques have been developed over the years, such as lithography [2], reactive ion etching [3], and laser

S. F. Roduan · J. A. Wahab (✉) · M. A. A. Mohd Salleh · N. A. H. Mohd Mahayuddin
Faculty of Chemical Engineering and Technology, Universiti Malaysia Perlis (UniMAP), Kompleks Pusat Pengajian Jejawi 2, Taman Muhibbah, 02600 Arau, Perlis, Malaysia
e-mail: juyana@unimap.edu.my

M. H. Aiman · A. Q. Zaifuddin · M. Ishak
Faculty of Mechanical and Automotive Engineering Technology, Universiti Malaysia Pahang, 26600 Pekan, Pahang Darul Makmur, Malaysia

M. A. A. Mohd Salleh et al. (eds.), *Proceedings of the Green Materials and Electronic Packaging Interconnect Technology Symposium*, Springer Proceedings in Physics 289,
https://doi.org/10.1007/978-981-19-9267-4_83

surface texturing [4, 5]. Laser surface texturing (LST) has gained attention in surface texturing, mainly in tribological, brazing, and soldering. This process is applied to enhance tribological performance in tribology [1], biomedical [6], and optical fields as it offers excellent surface wettability [7].

Surface roughness measures the unevenness of a solid surface texture where the roughness will affect substrate wettability [8]. Wettability is defined as the ability of a liquid to maintain contact with a solid surface when it comes into contact, where water contact angle measurement is used to determine the wettability of a solid. When the measured water contact angle is below 90°, the wettability of the solid is said to have good wetting and is called as hydrophilic. Meanwhile, if the measured water contact angle is above 90°, the solid has poor wettability, and the terms used are hydrophobic [9].

In soldering applications, the condition of the substrate such as the wetting characteristics is essential to improve the solder joint's performance. The spreading area of solder alloy was affected by various factors, including surface roughness and solder alloy properties. Previous study has stated that the micro-textured substrate can improve the spreading and wetting behavior of molten lead-free solder alloy and the quality of soldering [10]. Hence, the dimpled micro-textured copper substrate was used as a substrate material in soldering to study the effect of wetting characteristics on the spreading area of the Sn–0.7Cu solder alloy.

83.2 Methodology

83.2.1 Raw Materials

A flat square-shaped pure copper plate with a purity of 99.9% and a dimension of 15 mm × 15 mm × 1 mm was used as a substrate, while Sn–0.7Cu ingots were used as solder material in this study. To remove any impurities and scratches, the copper substrate was ground and polished using different grit sizes of silicon carbide and then polished using 1 μm alumina polishing paste.

83.2.2 Fabrication of Dimple Micro-texture on Copper Substrate

A dimple micro-texture with an opening diameter of 50 μm was fabricated on the copper substrate using a laser surface texturing (LST) technique with a set of parameters shown in Table 83.1. The dimple micro-texture was fabricated by ranging the dimple spacing, which was 100, 140, and 180 μm.

Table 83.1 Laser surface texturing parameters

Parameter	Value
Name	Ytterbium Fiber Laser Marking Machine
Model	ML-MF-A01
Beam size (μm)	50
Wavelength (nm)	1064
Scanning speed (mm/s)	4000
Laser power (W)	70
Frequency (kHz)	20

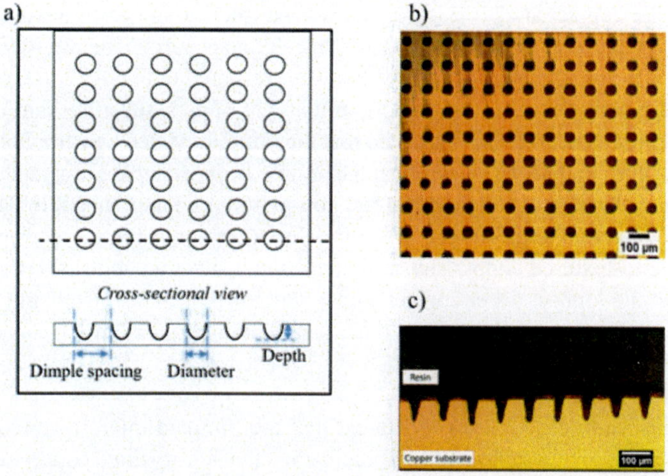

Fig. 83.1 **a** Schematic diagram of the copper substrate, **b** dimpled micro-texture on copper substrate, and **c** cross-sectional view of the copper substrate

Figure 83.1a shows the illustration of the dimple micro-textured and its cross-sectional view, which was developed, while Fig. 83.1b, c shows the dimpled micro-textured fabricated on the copper substrate. Water contact angle (WCA) measurement was carried out to examine the wettability of the sample after the laser surface texturing process. Then, the surface roughness of the dimpled micro-textured copper substrate was analyzed using Olympus OLS5000 3D measuring laser microscope.

83.2.3 *Fabrication of Sn–0.7Cu Solder Joint*

In preparing the solder material, Sn–0.7Cu alloy was prepared by melting the ingots at 350 °C in a solder pot. The ingot was then cold rolled to produce thin solder sheets. The thin solder sheet was cut and sectioned into approximately 0.4 g. The copper substrate was cleaned with acetone to remove any impurities before the solder joint fabrication. The 0.4 g of Sn–0.7Cu solder alloy was placed on the center of each copper substrate with a small amount of flux and then reflowed using a reflow oven (Model F4N). The image of the solder joint was captured by using optical microscope (OM), and the spreading area of the Sn–0.7Cu solder alloy was calculated by measuring the area of spreading on the copper substrate using Image-J software.

83.3 Results and Discussion

Water droplet contact angle measurement was carried out to determine the wetting properties of the copper substrate. The water contact angle for dimpled micro-textured and non-textured copper substrate was presented in Fig. 83.2. It was shown that both dimpled micro-textured and non-textured copper substrate samples have contact angles above 90°. This indicated that the sample showed a hydrophobic characteristic. Another observation was that the dimpled copper substrate samples exhibited higher water contact angles than the non-textured copper substrate. However, the water contact angle decreased when the dimple spacing increased. Among the dimpled micro-textured copper substrate, the copper substrate with 100 μm dimple spacing had the highest water contact angle, which was 129.85°, while the samples with 180 μm had the lowest water contact angle, which was 115.6°. Even so, the non-textured copper substrate recorded the lowest water contact angle among all samples, which was 111.55°.

From this finding, it can be concluded that the dimpled micro-textured copper substrate surfaces have better hydrophobicity than the non-textured copper substrate. A similar result was obtained in a study of superhydrophobic copper surface textured by laser for delayed icing phenomenon. It was revealed that the laser-textured surfaces were more hydrophobic than the polished substrate [11]. This was due to the larger surface area provided by the dimpled micro-textured, which tends to increase the

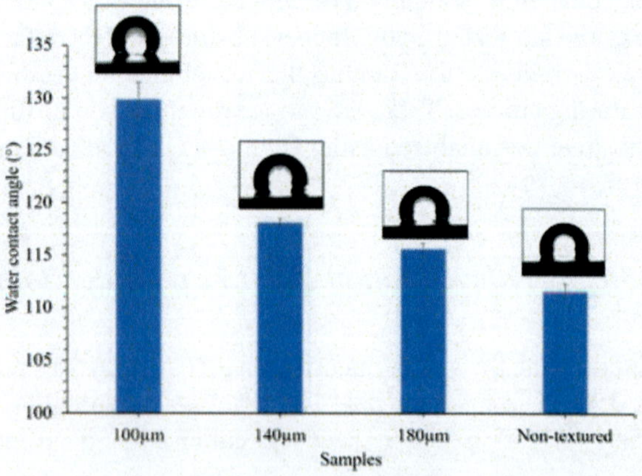

Fig. 83.2 Water contact angle of copper substrates

hydrophobicity [12]. This water contact angle can be related to the surface roughness of the copper substrate. A larger surface area leads to increased surface roughness. It was said that during a rough surface's wetting process, the droplets were suspended and did not fill the micro-texture on the surface and increased the hydrophobicity [13].

Figure 83.3 shows the surface roughness of dimpled micro-textured and non-textured copper substrates. It can be seen that the dimpled micro-textured copper substrates had higher surface roughness compared to the non-textured copper substrate. From the findings, it can be concluded that the samples with 100 μm dimple spacing had the highest surface roughness, which was found in the range of 6.675 μm, while the lowest surface roughness obtained among the micro-textured copper substrate was the samples with 180 μm dimple spacing, which was 4.384 μm. Nonetheless, the non-textured copper substrates had the lowest surface roughness compared to all samples, which was 0.074 μm. The surface roughness of dimpled micro-textured was increased from the laser surface texturing process. As the laser beam was in contact with the copper substrate, the substrate absorbed the heat energy resulting in thermalization and the substrate's rising temperature, which led to material removal from the substrate surface. This process was called as laser ablation. As the material from the surface was removed, more micro-textured was produced, changing the surface's texture and raising the surface roughness of the substrate [14, 15].

It can be seen that the copper surface roughness was directly proportional to the water contact angle, which means that high surface roughness had a higher water contact angle. This was due to the dimples that directly increased the surface roughness and surface area. According to Tran et al. [16], surfaces with higher surface roughness had more air pockets to trap the air. This process leads to an increased contact angle and thus has higher hydrophobicity [16].

To investigate the performance of dimpled micro-textured copper substrates in interconnect applications, the micro-textured copper was used as a substrate material for soldering. The ability of molten solder on the micro-textured copper substrate was examined. Figure 83.4 shows the recorded spreading area of Sn–0.7Cu on

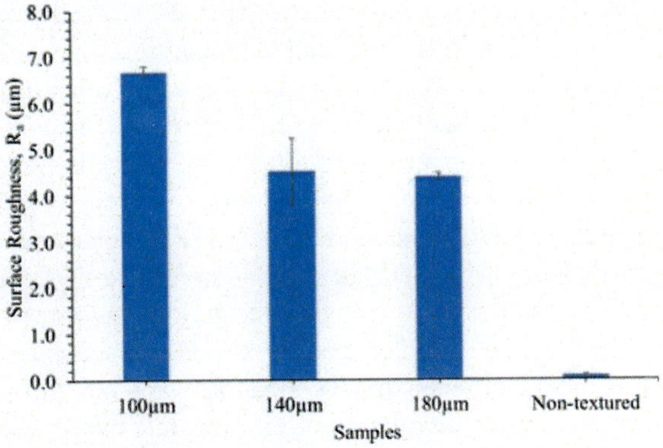

Fig. 83.3 Surface roughness of copper substrates

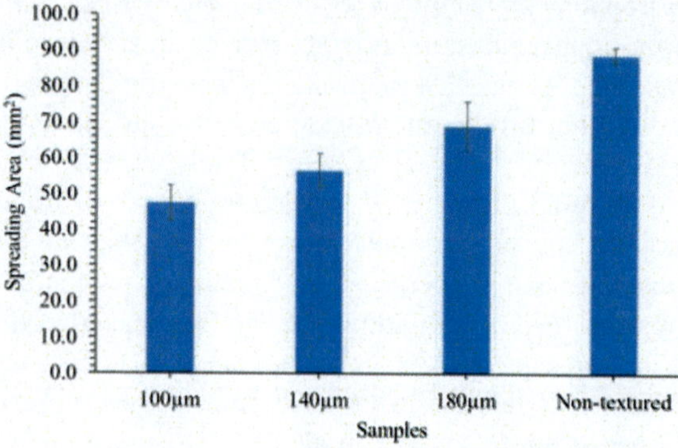

Fig. 83.4 Spreading area of the solder joint

dimpled micro-textured and non-textured copper substrates. It can be seen that the spreading area for the dimpled micro-textured copper substrate was smaller than the non-textured copper substrate. Among dimpled micro-textured copper substrates, the largest spreading area was recorded for samples with 180 μm, which was 68.930 mm^2, meanwhile, the smallest spreading area was for the samples with 100 μm dimple spacing, which was 47.441 mm^2. However, the largest spreading area among all samples was obtained by the non-textured copper substrate, which was 89 mm^2. This was because the non-textured surface has a smooth surface and the lowest surface roughness, allowing the Sn–0.7Cu solder alloy to spread on the substrate surface easily. The dimpled copper substrate has a higher surface roughness that causes the solder alloy to overcome the cavities present on the substrate surface before it can fully spread, which leads to a smaller spreading area of the solder alloy.

According to Cheng et al. [17], the spreading area of SAC305/Ni-W-P/Cu joint systems decreased when the substrate surface roughness increased. The study also found that the bigger spreading area of SAC305 solder was obtained on a smooth surface compared to a rough surface [17]. During the spreading process, the molten solder alloys need to flow into cavities present on the surface before they can further spread on the substrate. Hence, the smooth surface usually has a larger spreading area and spreads faster due to the uniform asperities on the smooth surface [18].

Several processes were involved in the spreading of solder over the substrate which were fluid flow, heat flow, and chemical reactions [19]. The result of the chemical reaction during spreading was the formation of an intermetallic compound. During the spreading area of the solder, an intermetallic reaction occurred between the solder alloy and the copper substrate [20]. A diffusion process occurs during the solder flow, where the molten solder dissolves the substrate. The wetting and formation of the intermetallic compound were driven by interfacial energy, where the interfacial energy will lower when the compound formation occurs [21]. It is said that the flow of molten solder was affected by the dissolution process, reducing the viscosity of the solder [22]. The dimpled micro-textured on the copper substrate provided additional surface area, allowing more copper diffusion into the interface. As the diffusion area increased, more dissolution of the substrate happened. This process led to a slower flow of solder spreading on the copper substrate. When the solder flow becomes slower, the molten solder needs more time to flow into the cavities on the copper substrate, making the solder wet the substrate in a smaller spreading area.

83.4 Conclusion

In summary, the water contact angle and surface roughness of dimpled micro-textured copper substrate were studied, and its effect on the spreading area of Sn–0.7Cu solder alloy was investigated. It was revealed that the dimpled micro-textured copper substrate exhibits better hydrophobicity than the non-textured copper substrate. It was also observed that the dimpled copper substrate had higher surface roughness than the non-textured copper substrate. The relation between the water contact angles was found to be directly proportional to the surface roughness. This study also found that the samples with better hydrophobicity had a smaller spreading area of Sn–0.7Cu solder. This is due to the higher diffusion of copper into the interface that led to a higher dissolution process. This action influences the solder's viscosity and flows, making it slower and having a smaller spreading area.

Acknowledgements The authors gratefully acknowledge the financial support through the Ministry of Education (MoE), Malaysia, under the Fundamental Research Grant Scheme (FRGS/1/2019/TK05/UNIMAP/03/3) and Faculty of Chemical Engineering Technology, Universiti Malaysia Perlis (UniMAP) for supporting this research effort through materials and facilities. The authors would also acknowledge Nihon Superior Co. Ltd. for the support given throughout the research project. The author would like to acknowledge the financial support in the form of a publication incentive grant from Universiti Malaysia Perlis (UniMAP).

References

1. J.A. Wahab, M.J. Ghazali, W.M.W. Yusoff, Z. Sajuri, Enhancing material performance through laser surface texturing: a review. Trans. Inst. Met. Finish. **94**(4), 193–198 (2016). https://doi.org/10.1080/00202967.2016.1191141
2. M. Imran, A.S. Saragih, M.S.U. Sahar, T.J. Ko, Digital maskless lithography capabilities for surface texturing with biomachining. Int. J. Adv. Manuf. Technol. **89**(9–12), 3709–3719 (2017). https://doi.org/10.1007/s00170-016-9317-6
3. L. Zhang, A. Verma, H.G. Xing, D. Jena, Inductively-coupled-plasma reactive ion etching of single-crystal β-Ga_2O_3 (2017)
4. A. Riveiro, A.L.B. Maçon, J. del Val, R. Comesaña, J. Pou, Laser surface texturing of polymers for biomedical applications. Front. Phys. **5**(Feb) (2018). https://doi.org/10.3389/fphy.2018.00016
5. J.A. Wahab, M.J. Ghazali, Erosion resistance of laser textured plasma-sprayed Al_2O_3—13% TiO_2 coatings on mild steel. Wear **432–433**(June), 202937 (2019). https://doi.org/10.1016/j.wear.2019.202937
6. A.G. Demir, P. Maressa, B. Previtali, Fibre laser texturing for surface functionalization. Phys. Procedia **41**, 759–768 (2013). https://doi.org/10.1016/j.phpro.2013.03.145
7. F.A. Müller, C. Kunz, S. Gräf, Bio-inspired functional surfaces based on laser-induced periodic surface structures. Materials (Basel) **9**(6) (2016). https://doi.org/10.3390/ma9060476

8. A.T. Kittu, R. Bulut, J. Puckette, Effects of surface roughness on contact angle measurements on a limestone aggregate. Geotech. Spec. Publ. **249 GSP**, 1–8 (2014). https://doi.org/10.1061/9780784478493.001
9. G. Agrawal, Y.S. Negi, S. Pradhan, M. Dash, S.K. Samal, Wettability and contact angle of polymeric biomaterials, in *Characterization of Polymeric Biomaterials* (Elsevier, 2017), pp. 57–81
10. H. Chen, J. Peng, L. Fu, X. Wang, Y. Xie, Solder wetting behavior enhancement via laser-textured surface microcosmic topography. Appl. Surf. Sci. **368**, 208–215 (2016). https://doi.org/10.1016/j.apsusc.2016.01.167
11. J. Li, Y. Zhou, W. Wang, C. Xu, L. Ren, Superhydrophobic copper surface textured by laser for delayed icing phenomenon. Langmuir **36**(5), 1075–1082 (2020). https://doi.org/10.1021/acs.langmuir.9b02273
12. P. Bizi-Bandoki, S. Valette, E. Audouard, S. Benayoun, Time dependency of the hydrophilicity and hydrophobicity of metallic alloys subjected to femtosecond laser irradiations. Appl. Surf. Sci. **273**, 399–407 (2013). https://doi.org/10.1016/j.apsusc.2013.02.054
13. C. Li, J. Zhang, J. Han, B. Yao, A numerical solution to the effects of surface roughness on water–coal contact angle. Sci. Rep. **11**(1), 1–12 (2021). https://doi.org/10.1038/s41598-020-80729-9
14. A.J. Sulaiman, M.H. Aiman, M.M. Quazi, M. Ishak, T. Ariga, Enhancement of mechanical properties of Copper brazed by laser surface modification. IOP Conf. Ser. Mater. Sci. Eng. **788**(1) (2020). https://doi.org/10.1088/1757-899X/788/1/012012
15. A.Q. Zaifuddin, F. Zulhilmi, M.H. Aiman, M.M. Quazi, M. Ishak, Enhancement of laser heating process by laser surface modification on Titanium alloy. **15**(3), 8310–8318 (2021)
16. P.A. Tran, T.J. Webster, Understanding the wetting properties of nanostructured selenium coatings: the role of nanostructured surface roughness and air-pocket formation. Int. J. Nanomed. **8**, 2001–2009 (2013). https://doi.org/10.2147/IJN.S42970
17. J. Cheng, X. Hu, S. Li, Effects of the surface roughness on wetting properties and interfacial reactions between SAC305 solder and Cu substrate with Ni–W–P coating. J. Mater. Sci. Mater. Electron. **31**(18), 15086–15096 (2020). https://doi.org/10.1007/s10854-020-04072-5
18. Satyanarayan, K.N. Prabhu, Spreading behavior and evolution of IMCs during reactive wetting of SAC solders on smooth and rough copper substrates. J. Electron. Mater. **42**(8), 2696–2707 (2013). https://doi.org/10.1007/s11664-013-2619-7
19. J.A. Warren, W.J. Boettinger, A.R. Roosen, Modeling reactive wetting. Acta Mater. **46**(9), 3247–3264 (1998). https://doi.org/10.1016/S1359-6454(97)00487-4
20. F.G. Yost, F.M. Hosking, D.R. Frear, Introduction: the mechanics of solder alloy wetting and spreading, in *The Mechanics Solder Alloy Wetting Spreading* (1993), pp. 1–7. https://doi.org/10.1007/978-1-4684-1440-0_1
21. M. Abtew, G. Selvaduray, Lead-free solders in microelectronics. Mater. Sci. Eng. R Reports **27**(5), 95–141 (2000). https://doi.org/10.1016/S0927-796X(00)00010-3
22. A. Imc, N. Imcs, Wetting and joint formation, in *A Guide to Lead-free Solder* (2007), pp. 53–78. https://doi.org/10.1007/978-1-84628-310-9_3

Chapter 84
Investigating the Effect of Calcium Alloying and Electrolyte Medium on the Corrosion Behavior of AZ31 Mg Alloy

Umer Masood Chaudry, Ameeq Farooq, Ho Seon Ahn, Kotiba Hamad, and Tea-Sung Jun

Abstract Magnesium (Mg) has the lowest density among all the structural materials (Al, Fe); however, the diverse applications of Mg have been hampered by its intrinsic brittleness at room temperature and high corrosion susceptibility. Hence, enormous efforts have been made to design the highly ductile corrosion resistant Mg by alloying or other techniques to catapult its utilization. AZ31 is the most commercially used Mg alloy in the transportation sector, yet its high corrosive nature poses a significant disadvantage among other materials. In this study, the effect of calcium on microstructure and corrosion characteristics of AZ31 were investigated. Microstructure was characterized by electron probe microanalysis, scanning electron microscopy and electron back-scattered diffraction. For investigating the corrosion behavior, open circuit potential, potentiodynamic polarization and electrochemical impedance spectroscopy were analyzed in NaCl and Na_2SO_4 electrolyte. The microstructure revealed the formation of $(Mg, Al)_2Ca$ in AZ31–0.5Ca. The corrosion rates were observed to be significantly reduced for AZ31–0.5Ca as compared to AZ31 (74.53, 67.33 and 29.01, 49.56 mpy for AZ31 and AZ31–0.5Ca in NaCl and Na_2SO_4 electrolyte, respectively). Improved corrosion resistance was attributed to segregation of aluminum with calcium to form $(Mg, Al)_2Ca$ which reduced β–Mg_{17}–Al_{12} area fraction.

Keywords Magnesium · AZ31–0.5Ca · Electron back-scattered diffraction · Potentiodynamic polarization · Electrochemical impedance spectroscopy

U. M. Chaudry · H. S. Ahn · T.-S. Jun (✉)
Department of Mechanical Engineering, Incheon National University, Incheon 22012, Republic of Korea
e-mail: t.jun@inu.ac.kr

A. Farooq
Department of Metallurgy and Materials Engineering, Corrosion Control Research Cell, University of the Punjab, Lahore, Pakistan

K. Hamad
School of Advanced Materials Science and Engineering, Sungkyunkwan University, Suwon 16419, Republic of Korea

M. A. A. Mohd Salleh et al. (eds.), *Proceedings of the Green Materials and Electronic Packaging Interconnect Technology Symposium*, Springer Proceedings in Physics 289,
https://doi.org/10.1007/978-981-19-9267-4_84

84.1 Introduction

Magnesium (Mg), as the lightest metallic material is considered as next generation structural material due to its high specific strength and good recyclability [1–5]. However, the intrinsic low ductility and high corrosion susceptibility of Mg has hampered its widespread applications [6–8]. Hence, in recent years, Mg have been researched intensively to uncover and resolve the key issues associated with Mg which are persistent obstacle in its utilization [9–11]. Secondly, the significantly high corrosion susceptibility of Mg and its alloys as compared to conventional construction materials (e.g., Fe, Al) also pose pertinent threat. Accordingly, designing highly ductile Mg alloy and sufficient corrosion resistance can catapult and revolutionize the automotive and aerospace sector. The corrosion properties of Mg can be enhanced by developing high corrosion resistant materials, implying surface coatings, inducing electrochemical protection or using corrosion inhibitors. Previously, Ca has been frequently reported in lieu of improvement in mechanical properties of pure Mg and Mg alloys [12, 13]. Umer et al. critically investigate the effect of Ca addition on the room temperature performance of AZ31 Mg alloy [14]. The detailed analysis using electron back-scattered diffraction and viscoplastic self-consistent modeling confirmed the activation of non-basal slip systems in Ca-added AZ31 Mg alloy. Moreover, few studies also investigated the effect of Ca on the corrosion characteristics of Mg. Yang et al. studied the microstructure, mechanical performance and corrosion resistance of ZM21 Mg alloy with various Ca additions (0.1–1.6 wt%) [15]. The results revealed that the mechanical properties and corrosion resistance were significantly improved on Ca addition, which was attributed to the grain refinement and formation of intermetallic particle ($Ca_2Mg_6Zn_3$). Moreover, the optimal Ca concentration was reported to be 0.7 wt% beyond which the properties further deteriorate. Jiang et al. investigated the corrosion properties of Mg–Al–Mn alloy with different Ca additions (0.2–4 wt%) [16]. Microstructural and phase analysis revealed the formation of Al_2Ca, $(Mg, Al)_2Ca$ and Mg_2Ca with Ca addition. Potentiodynamic polarization showed that the alloy with 2 wt% Ca exhibited the highest corrosion resistance because of discontinuous distribution of $(Mg, Al)_2Ca$ phase in the matrix, which preferentially corrode to protect the α-Mg matrix. It is pertinent to mention that AZ series (Mg–Al–Zn) has been extensively utilized in various sectors owning to good machineability and adequate mechanical properties. The dispersed β-phase ($Mg_{17}Al_{12}$) in the matrix results in high strength in AZ31 due to precipitation hardening. However, β-phase has been reported to be susceptible for the corrosion and lead to formation of galvanic cell which adversely effects the corrosion resistance. Hence, immense efforts have been made to control the morphology and distribution of deleterious β-phase by alloy design with suitable alloying addition. Chen et al. investigated the corrosion characteristics of AZ31 on simultaneous additions of strontium and yttrium [17]. Remarkable improvement in corrosion resistance of AZ31–0.5Sr–1.5Y was recorded, which was attributed to the breakdown of continuous β-phase to the Al_2Y and Al_3Y segregated in the matrix and along the grain boundaries. Since, rare earths are extremely scarce and costly so it will be interesting to investigate the effect of Ca addition to the corrosion characteristics of AZ31 Mg alloy. Moreover, the corrosion susceptibility of both materials in different electrolyte (NaCl and Na_2SO_4) was also studied.

84.2 Materials and Methods

AZ31 and AZ31–0.5Ca sheets were received from POSCO Co (South Korea). Both materials were fabricated by the twin roll strip casting. Both were rolled, heat treated followed by water quenching to receive 1 mm thick sheet with homogeneous microstructure. For microstructural characterization, scanning electron microscopy (SEM), electron probe microanalyzer (EPMA) and electron back-scattered diffraction (EBSD) were employed to investigate the presence of precipitates and texture analysis [18]. For electrochemical measurements, specimens were wire cut in the circular form with area of 2 cm^2. The samples were prepared by grinding to SiC grit 2000. The test was carried out at 30 °C in two different electrolytes (NaCl and Na_2SO_4) at 0.15 M concentration to investigate the influence of different electrolyte on the corrosion susceptibility. The test was carried out by using three electrode system (1000E, Gamry USA). The samples act as working electrode, Ag/AgCl acts as a counter electrode and graphite rod acts as a reference electrode. The open circuit potential (OCP) was determined for 1 h after stabilizing the sample in the electrolyte. Electrochemical impedance spectroscopy (EIS) was performed to investigate the corrosion behavior at (100 kHz to 10 MHz) and potentiodynamic polarization curves were obtained by polarizing the sample from − 2.0 to 1.5 V with a scan rate of 5 mV/s.

84.3 Results and Discussion

EBSD based detailed microstructural characterization was conducted to investigate the microstructure of both materials. The initial microstructural features of AZ31 and AZ31–0.5Ca in terms of inverse pole figure (IPF), pole figure (PF) and grain size distribution characterized by EBSD are presented in Fig. 84.1a. It can be seen from the IPF that both the alloys consist of twin free equiaxed microstructure owing to dynamic recrystallization during primary processing with almost similar grain sizes. Grain size distribution showed that the average grain size was 7 μm and 8 μm for AZ31 and AZ31–0.5Ca, respectively. Moreover, the texture analysis showed that AZ31 has typical strong basal texture (texture intensity ~13.4 mrd) where most of the grains were oriented along *c*-axis, while on the other hand, AZ31–0.5Ca showed the significantly weak texture (texture intensity ~5.2 mrd) with some peaks that can be witnessed to spread along transverse direction. It has already been reported in the literature that the Ca addition can significantly weaken the basal texture of Mg alloy. The solute segregation and intermetallic precipitates can induce particle stimulated nucleation (PSN) by grain boundary pinning effect, reducing the grain boundary energy and hampering the grain growth into favorable basal orientations resulting in grains with random orientations and weakened texture. Risheng et al. investigated the influence of multiple substitutional elements on the qualitative and quantitative segregation behavior, growth kinetics and recrystallization behavior of Mg. The results revealed that the co-segregation of Zn, Al and Ca can significantly hamper the grain growth, decreasing the grain boundary mobility and resulting and restricts the preferential growth of recrystallized grains with basal texture [19, 20]. EMPA was conducted on both materials to investigate the distribution of elements in the Mg matrix. Figure 84.1b, c present the detailed EPMA maps of AZ31 and AZ31–0.5Ca illustrating the distribution of Mg, Al, Zn and Ca. The EPMA maps of AZ31 revealed that the Al was co-existed with Mg as Mg17Al12 particles and was distributed throughout the matrix.

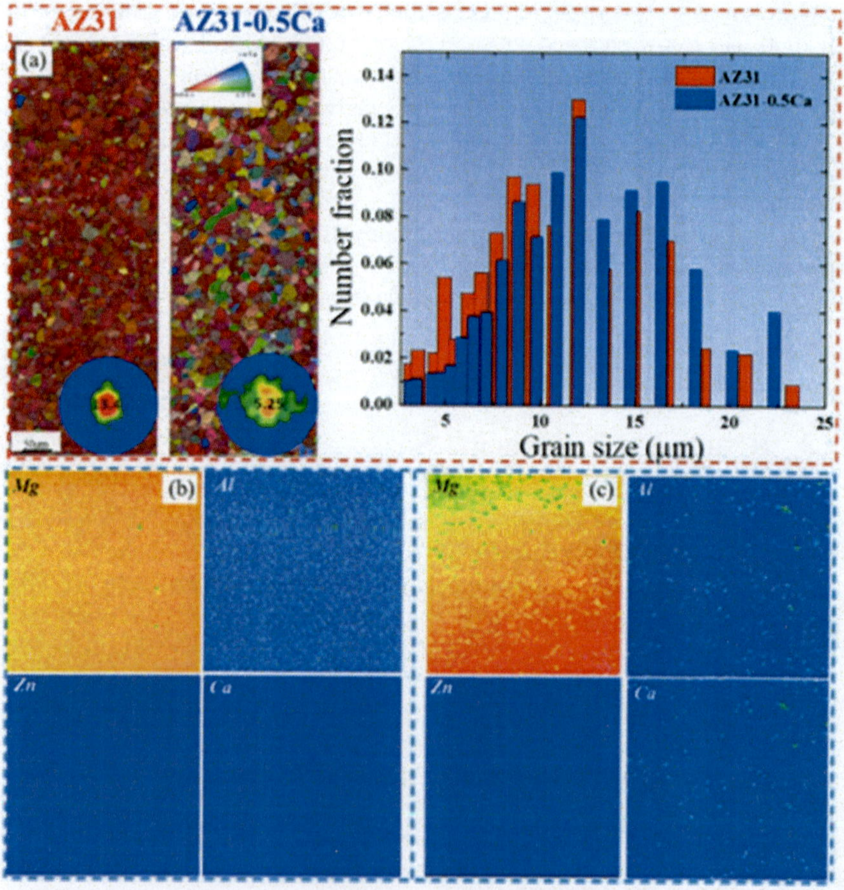

Fig. 84.1 **a** EBSD IPF maps, PFs and grain size distribution, **b**, **c** EPMA maps for the major elements in AZ31 and AZ31–0.5Ca

Moreover, Zn was found to be homogeneously distributed throughout the area. On the other hand, EPMA maps of AZ31–0.5Ca (Fig. 84.1c) revealed the co-segregation of the Al and Ca in the matrix, where white particles discerned in the matrix can be seen. This, in turn, will reduce the overall Mg17Al12 phase fraction in the AZ31–0.5Ca due to limited concentration of Al available to form the deleterious Mg17Al12 phase and consequently will enhance the corrosion resistance in AZ31–0.5Ca.

To characterize the precipitate observed in the EPMA maps of AZ31–0.5Ca XRD and SEM/EDS analysis was carried out (Fig. 84.2). XRD analysis illustrated in Fig. 84.2a revealed the emergence of new peaks in AZ31–0.5Ca, which were related to the formation of precipitates. Moreover, SEM analysis showed a large number of white precipitates dispersed throughout the matrix (Fig. 84.2b). EDS revealed that the precipitate structure composed of Mg (40 wt%), Al (40 wt%) and Ca (18 wt%). The received results are consistent with the previous studies, where formation of (Mg, Al)$_2$Ca precipitates were reported in Ca-added AZ31. As mentioned above, the ability of Ca to segregate with Al to form (Mg, Al)$_2$Ca precipitate will consequently result in the decreased fraction of $Mg_{17}Al_{12}$ phase as compared to AZ31. Furthermore, the effect of Ca addition on the corrosion characteristics of AZ31 was thoroughly investigated.

Figure 84.3a shows the Open Circuit Potential (OCP) analysis of both materials. The potential of the AZ31/AZ31–0.5 Ca surface was measured with respect to Ag/AgCl reference electrode in 0.15 M NaCl and Na_2SO_4 electrolyte at 30 °C. Different electrolytes were used to investigate the effect of alloying and nature of electrolyte on the corrosion susceptibility. The results show that the AZ31 was more active as compared to AZ31–0.5 Ca in both the electrolytes as its reduction potential was − 1.501 $V_{vs\ Ag/AgCl}$ and − 1.446 $V_{vs\ Ag/AgCl}$ in 0.15 M NaCl and Na_2SO_4 (Fig. 84.3a). As the Cl^- anions are comparatively more aggressive as compared to SO_4^{2-} anions, so lower reduction potential was recorded for both the alloys in NaCl electrolyte in comparison to Na_2SO_4. To analyze the surface mechanism, electrochemical impedance spectroscopy (EIS) was employed on both the specimens. The Nyquist spectra of AZ31 and AZ31–0.5 Ca in 0.15 M NaCl and Na_2SO_4 electrolyte

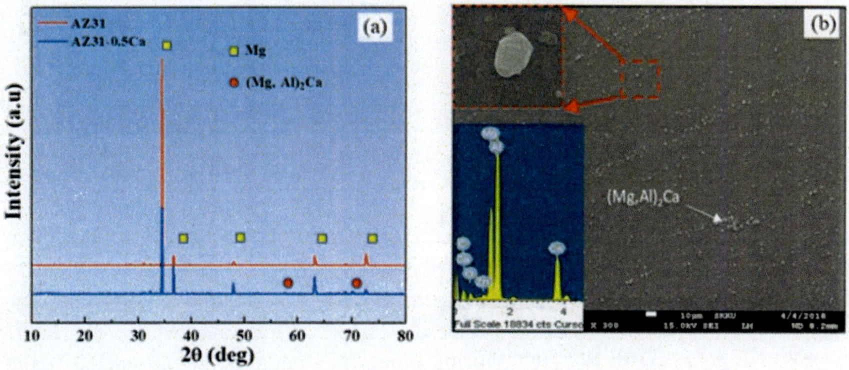

Fig. 84.2 **a** XRD of AZ31 and AZ31–0.5Ca, **b** SEM/EDS analysis of AZ31–0.5Ca

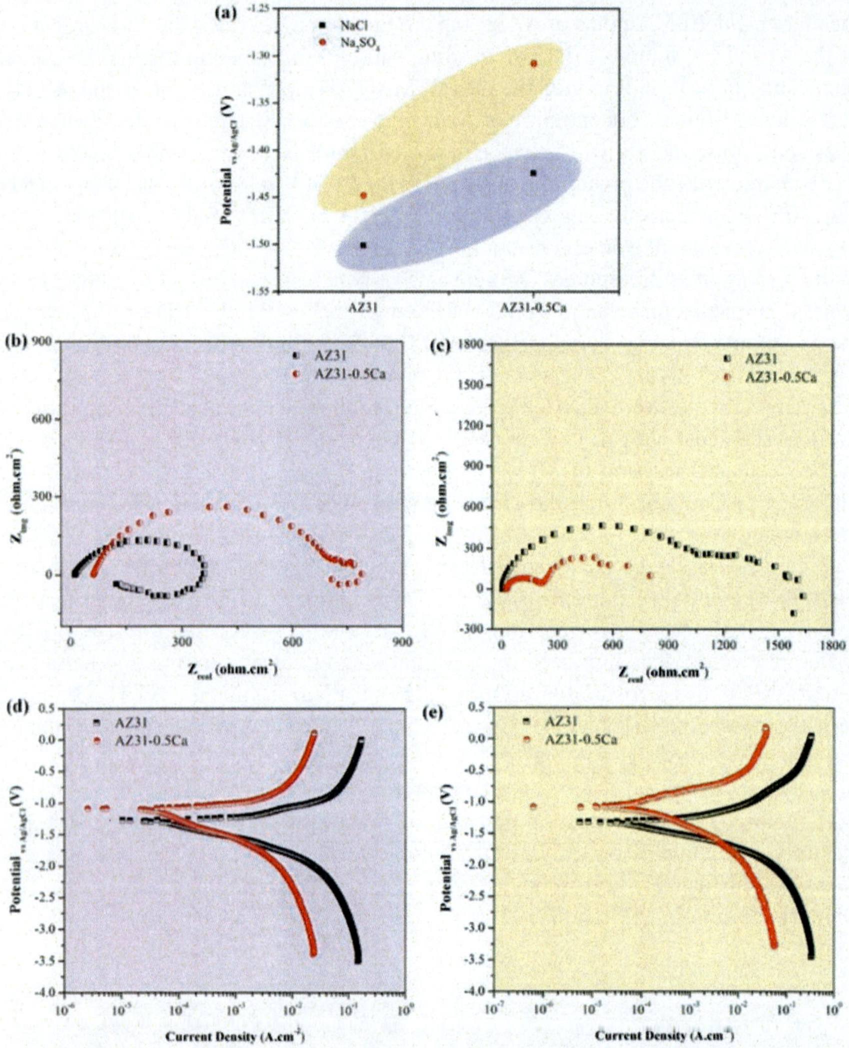

Fig. 84.3 Electrochemical analysis **a** open circuit potential (OCP), Nyquist spectra **b** 0.15 M NaCl **c** 0.15 M Na_2SO_4 and potentiodynamic polarization curves **d** 0.15 M NaCl **e** 0.15 M Na_2SO_4 at 30 °C

at 30 °C are illustrated in Fig. 84.3b, c. The spectra show that both dissolution and adsorption mechanism take place on both magnesium alloy surfaces. The results show the dissolution/capacitive loop at the high frequency and inductive loop at the low frequency. The charge transfer resistance is 343 $\Omega.cm^2$ for AZ31 while 732 $\Omega.cm^2$ for AZ31–0.5Ca in 0.15 M NaCl. During Potentiodynamic polarization (PDP), the samples of AZ31 with and without Ca were analyzed in NaCl and in Na_2SO_4 with 0.15 M concentration at 30 °C temperature.

Table 84.1 Kinetics parameters from the potentiodynamic polarization curves after Tafel fitting

	β_a (mV/decade)	β_c (mV/decade)	E_{corr} ($V_{vs\ Ag/AgCl}$)	i_{corr} (μA/cm^2)	Corrosion rate (mpy)
0.15 M NaCl					
AZ31	89.80	212.7	− 1.320	85.40	74.53
AZ31–0.5Ca	58.30	349.2	− 1.100	33.40	29.01
0.15 M Na_2SO_4					
AZ31	117.1	155.4	− 1.370	76.90	67.33
AZ31–0.5Ca	207.9	266.0	− 1.050	57.10	49.56

The results obtained from potentiodynamic polarization curves are listed in Table 84.1. It can be seen from the result values that the AZ31 with 0.5Ca showed good resistance to corrosion as it has less corrosion rate in both solutions at 30 °C. The potentiodynamic curves of the AZ31 and AZ31–0.5Ca in chloride solution and in sulphate solution at 0.15 M concentration are plotted as shown in Fig. 84.3d, e. In case of NaCl electrolyte, the chloride ions start to attack the native protective layer formed on the magnesium surface and ultimately breaks the layer and resulted in the pitting of the sample. While in case of sulphate electrolyte, the magnesium hydroxide ($Mg(OH)_2$) film that forms on the magnesium when immersed in aqueous solution reacts with the sulphate ions present in the electrolyte and formed magnesium sulphate ($MgSO_4$).

84.4 Conclusion

In conclusion, the addition of Ca significantly enhanced the corrosion resistance of AZ31 magnesium alloy. The microstructure investigations revealed the formation of precipitate dispersed in the Mg matrix, which was confirmed to be $(Mg, Al)_2Ca$ by EDS. Moreover, the potentiodynamic polarization measurements in NaCl and Na_2SO_4 electrolyte showed remarkable improvement in AZ31–0.5Ca, where corrosion rates of 74.53 mpy, 67.33 mpy and 29.01 mpy, 49.56 mpy were recorded for AZ31 and AZ31–0.5Ca in NaCl and Na_2SO_4. The excellent improvement was attributed to the formation of $(Mg, Al)_2Ca$ in AZ31–0.5Ca which decreased the overall amount of Al to form β-phase ($Mg_{17}Al_{12}$), which act as a micro galvanic cell and accelerate the corrosion rate.

Acknowledgements This work was supported by the National Research Foundation (NRF) of Korea grant funded by the Korean Government (MSIT) (No. 2020R1C1C1004434).

References

1. U. Masood Chaudry, S. Tekumalla, M. Gupta, T.-S. Jun, K. Hamad, Designing highly ductile magnesium alloys: current status and future challenges. Crit. Rev. Solid State Mater. Sci., 1–88 (2021)
2. G. Han, H.-K. Park, H.-K. Kim, T.-S. Jun, Local and global deformation behaviour in rolled pure magnesium sheets at room temperature under different strain rates. Mater. Sci. Eng., A **762**, 138110 (2019)
3. S.W. Lee, G. Han, T.-S. Jun, S.H. Park, Effects of initial texture on deformation behavior during cold rolling and static recrystallization during subsequent annealing of AZ31 alloy. J. Mater. Sci. Technol. **66**, 139–149 (2021)
4. U.M. Chaudry, Y. Noh, G. Han, R. Jaafreh, T.-S. Jun, K. Hamad, Effect of CaO on structure and properties of AZ61 magnesium alloy. Mater. Sci. Eng.: A., 143189 (2022)
5. D. Lee, B. Kim, S. Lee, S.-M. Baek, J.C. Kim, H.-T. Son et al., Enhanced corrosion resistance of Mg–Sn–Zn–Al alloy by Y microalloying. Scripta Mater. **163**, 125–129 (2019)
6. U.M. Chaudry, K. Hamad, Effect of calcium on the superplastic behavior of AZ31 magnesium alloy. Mater. Sci. Eng.: A., 140874 (2021)
7. G. Han, Y. Noh, U.M. Chaudry, S.H. Park, K. Hamad, T.-S. Jun, {10–12} extension twinning activity and compression behavior of pure Mg and Mg-0.5 Ca alloy at cryogenic temperature. Mater. Sci. Eng.: A. **831**, 142189 (2022)
8. A. Malik, U.M. Chaudry, T. Yan, J. Long, C. Li, Y. Wang, Achieving higher dynamic mechanical response by adjusting texture through twinning in a ZK61 Mg alloy. J. Alloys Compd., 163755 (2022)
9. K.-H. Kim, J.B. Jeon, N.J. Kim, B.-J. Lee, Role of yttrium in activation of < c+ a> slip in magnesium: an atomistic approach. Scripta Mater. **108**, 104–108 (2015)
10. A. Atrens, G.L. Song, M. Liu, Z. Shi, F. Cao, M.S. Dargusch, Review of recent developments in the field of magnesium corrosion. Adv. Eng. Mater. **17**, 400–453 (2015)
11. E. Lukyanova, N. Martynenko, V. Serebryany, A. Belyakov, L. Rokhlin, S. Dobatkin et al., Structure and mechanical and corrosion properties of a magnesium Mg–Y–Nd–Zr alloy after high pressure torsion. Russ. Metall. (Metally). **2017**, 912–921 (2017)
12. U. Masood Chaudry, K. Hamad, J.-G. Kim, A further improvement in the room-temperature formability of magnesium alloy sheets by pre-stretching. Materials. **13**, 2633 (2020)
13. U. Masood Chaudry, T. Hoo Kim, S. Duck Park, Y. Sik Kim, K. Hamad, J.-G. Kim, On the high formability of AZ31–0.5 Ca magnesium alloy. Materials **11**, 2201 (2018)
14. U. Masood Chaudry, K. Hamad, J.-G. Kim, Ca-induced plasticity in magnesium alloy: EBSD measurements and VPSC calculations. Curr. Comput.-Aided Drug Des. **10**, 67 (2020)
15. J. Yang, J. Peng, M. Li, E.A. Nyberg, F.-S. Pan, Effects of Ca addition on the mechanical properties and corrosion behavior of ZM21 wrought alloys. Acta Metall. Sin. (English Letters). **30**, 53–65 (2017)
16. J. Yang, J. Peng, E.A. Nyberg, F.-S. Pan, Effect of Ca addition on the corrosion behavior of Mg–Al–Mn alloy. Appl. Surf. Sci. **369**, 92–100 (2016)
17. C. Gang, X.-D. Peng, P.-G. Fan, W.-D. Xie, Q.-Y. Wei, M. Hong et al., Effects of Sr and Y on microstructure and corrosion resistance of AZ31 magnesium alloy. Trans. Nonferr. Metals Soc. China. **21**, 725–731 (2011)
18. U.M. Chaudry, A. Farooq, K. bin Tayyab, A. Malik, M. Kamran, J.-G. Kim et al., Corrosion behavior of AZ31 magnesium alloy with calcium addition. Corr. Sci. **199**, 110205 (2022)
19. R. Pei, Y. Zou, D. Wei, T. Al-Samman, Grain boundary co-segregation in magnesium alloys with multiple substitutional elements. Acta Mater. **208**, 116749 (2021)
20. A. Malik, Y. Wang, H. Cheng, F. Nazeer, M.A. Khan, Microstructure evolution of Mg–Zn–Zr magnesium alloy against soft steel core projectile. J. Mater. Sci. Technol. **79**, 46–61 (2021)

Chapter 85
Effect of Pulse Current Amplitude on Corrosion Protection of Mild Steel in the Atmospheric Environment

W. M. H. Wan Ahmad, S. H. Salleh, S. R. Shamsudin, and Mahalaksmi Gunasilan

Abstract Atmospheric metal corrosion affects the appearance and life span of iron-based metals. There are several methods for dealing with atmospheric corrosion, but none are perfectly ideal. Pulse Current Prevention (PCP) is a new method for atmospheric corrosion protection. Mild steel samples were studied using PCP for 14 days in a semi-sealed container. The pulse current is 50 Hz, 50% duty cycle, with different amplitudes of 2, 5, and 7 V. Amplitude effectiveness in the PCP technique is determined by morphological observation and image analysis technique. According to the findings, the use of pulsed current can prevent atmospheric corrosion on the steel sample. The amplitude of pulsed current has a major impact on atmospheric corrosion protection. The amplitude of 2 V is inadequate to create the necessary negative charges for steel surface protection. In the meantime, a 7 V amplitude produces over-protection, which exacerbates atmospheric corrosion deterioration. The meaningful finding is that 5 V is found to be a fairly good amplitude parameter for atmospheric corrosion prevention. Nonetheless, it needs to be investigated further to give optimal protection.

Keywords Atmospheric corrosion · Pulse current · Amplitudes · Corrosion protection

---

W. M. H. Wan Ahmad (✉) · S. H. Salleh
Faculty of Chemical Engineering and Technology, Universiti Malaysia Perlis, 02600 Arau, Perlis, Malaysia
e-mail: haqqi@studentmail.unimap.edu.my

S. R. Shamsudin · M. Gunasilan
Faculty of Mechanical Engineering and Technology, Universiti Malaysia Perlis, 02600 Arau, Perlis, Malaysia

S. H. Salleh
Centre of Excellence Geopolymer and Green Technology, Universiti Malaysia Perlis, 02600 Arau, Perlis, Malaysia

M. A. A. Mohd Salleh et al. (eds.), *Proceedings of the Green Materials and Electronic Packaging Interconnect Technology Symposium*, Springer Proceedings in Physics 289,
https://doi.org/10.1007/978-981-19-9267-4_85

85.1 Introduction

Corrosion is an electrochemical process that happens when metal is exposed to a harsh environment [1]. Atmospheric corrosion is the deterioration of materials caused by air and airborne contaminants. It is an electrochemical process that requires an electrolyte (which in this case is moisture), which could be precipitation, rain, dew, humidity, or snow melting [2]. Atmospheric corrosion occurs when the relative humidity of the atmosphere is higher than the equilibrium relative humidity of any saturated solution on the metal surface (usually NaCl solution, which is ubiquitous). Methods for preventing atmospheric corrosion typically include cleaning and dust protection, drying and cleaning the atmosphere and preventing condensation, the use of corrosion inhibitors, protective coatings (wrapping, temporary protective materials, paint films, metal surface coatings, polymers or vitreous enamel, conversion coating), and sacrificial coating [3].

The use of the pulsed current in the prevention of corrosion is gaining popularity among researchers. The DC feed current is transformed into offset AC by means of pulse current protection. On the exposed metal surface, a skin effect mechanism is achieved in the presence of a pulsed current supply. The charge flows mostly on the metal's surface due to the skin effect, which is the phenomenon described by Maxwell's equation when a signal with a high frequency is present in the conductor. According to Macdonald et al. [4], the electrical signal frequency is high enough that the skin effect, in which the current (charge flow) is concentrated at the conductor's surface, may play an important role in the protective mechanism.

The development of a device that employs pulse current to prevent corrosion under insulation against the automotive body is at the forefront of cutting-edge technology that is garnering increasing attention from consumers. According to Xu and Zhou [5], pulse current cathodic protection offers more uniform current density, greater protection depth, and a better protection effect than conventional cathodic protection. However, there are many pulses current variables such as amplitude, frequency, duty cycle, and pulse shape, which have no clear influence on the protection effect. Automotive manufacturers throughout the world were the first to fund research and development on corrosion prevention systems utilizing the pulsed current protection (PCP) method. Recent investigations include the one undertaken by Macdonald et al. [4]. They have developed an electromagnetic device that, according to their claims, functions as an antenna by activating the galvanic zinc layer to prevent corrosion of steel components on the automobile body. In the meanwhile, Camp [6] created a device that employs the capacitor idea via capacitive coupling. In this technology, Camp demonstrated that a negative charge on the metal beneath the paint layer may prevent corrosion under insulation on the automotive body paint layer. As both methods are capable of protecting exposed metals from corrosion caused by scratching, it is believed that they can also be manipulated to prevent atmospheric corrosion. This article presents the effect of pulse current amplitude parameters on exposed mild steel in an atmospheric environment.

Table 85.1 Material composition of mild steel samples

Element	C	Mn	P	S	Si	Fe
Sample (%)	0.164	0.679	0.012	0.011	0.167	98.70
AISI 1016	0.12–0.18	0.60–0.90	$\leq$ 0.050	$\leq$ 0.050	–	98.13–99.58

Table 85.2 Parameters setting of pulse current applied to the steel samples

Setting amplitude (V)	DC offset (V)	Total AC (μA)
2	– 0.5	23.8
5	– 1.25	75.4
7	– 1.75	120.9

85.2 Methodology

Mild steel is used as a sample to be protected against corrosion. The chemical composition of the steel was examined using optical emission spectroscopy, OES (Q8 Magellan, Bruker) revealed it complies with AISI 1016 type steel (Table 85.1). AISI 1016 steel is known to be very prone to corrosion. An AISI 1016 steel plate was first cut to dimensions 17 × 160 × 3 mm for experimental purposes. After that, the surface of the sample was ground with 80-to-600-grit silica carbide paper. The ground sample was then painted with a polymer-based paint, leaving a 1 cm^2 exposed area.

The steel sample was suspended in a semi-sealed container containing a 3.5% NaCl solution. The pulse current supply to the metal sample was generated through a custom-made signal generator with 5 V voltage output and 5Ω resistance. The applied amplitudes were varied, i.e., 2, 5, and 7 V. DC offset, frequency, and duty cycle were fixed at – 1 V, 50 Hz, and 50% respectively. A multimeter was used to measure the alternating current (AC) density. Table 85.2 shows the parameters setting of pulse current applied to the steel samples. The experiment was conducted for 14 days.

The surface morphology was visually inspected using a stereomicroscope (RL-M3T, Relife) to determine the presence of corrosion reaction (or rust) on the sample surfaces. The captured images were analyzed by ImageJ software.

85.3 Results and Discussion

Macroscopic inspection of exposed regions determines the efficacy of the PCP method in confirming the existence of corrosion or rust on mild steel. Figure 85.1a depicts the surface of a steel sample for atmospheric corrosion and Fig. 85.1b–d samples for the PCP experiment. Figure 85.1a depicts the surface of a steel sample subjected to atmospheric corrosion in the absence of any corrosion

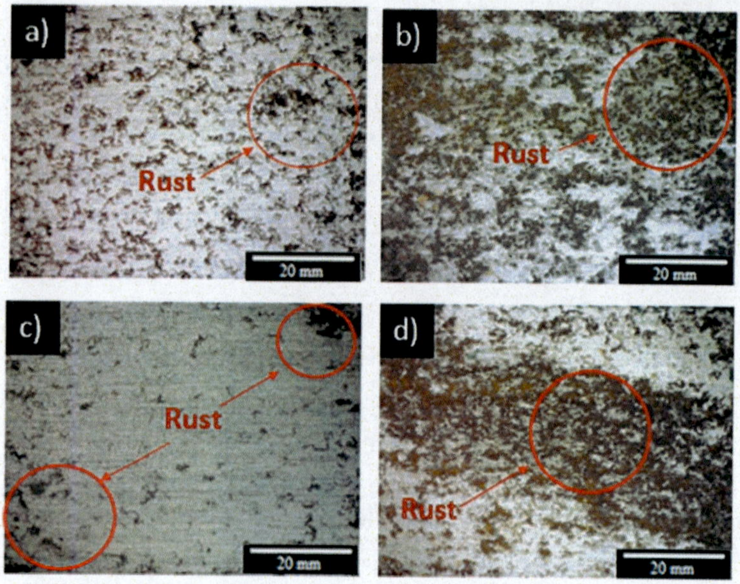

Fig. 85.1 Stereomicroscopic images of mild steel surface showing **a** atmospheric corrosion, and samples under PCP experiment subjected to different pulse current amplitudes, namely **b** 2 V, **c** 5 V, and **d** 7 V

prevention. While and Fig. 85.1b–d shows the surface of the sample after 14 days under the PCP experiment. Figure 85.1b displays the surface of the steel sample with amplitudes of 2 V. It exhibits a prominent rust formation on the sample's surface. With the lowest pulse current (23.8 μA), it is unable to adequately cover the mild steel sample against corrosion reaction. The pulsed charges on the sample surface are believed to have insufficient to prevent the oxidation of Fe atoms to Fe^{2+} ions, as seen by the brownish color. In this case, the corrosion protection on the metal surface is classified as under-protection.

Figure 85.1c, on the other hand, does not exhibit a very pronounced corrosion reaction. On the surface, there are barely a few dark patches. With an amplitude of 5 V, the pulse current output is 75.4 μA, and this value may be further fine-tuned to achieve the most optimal settings. In contrast, in Fig. 85.1d, the rusty brownish color is mostly dispersed over the sample's surface as a result of the formation of iron oxide during the oxidation process. Consequently, the 7 V pulse current amplitude utilized in this experiment does not contribute to the 14-day atmospheric corrosion prevention. Even though the 7 V amplitude resulted in a greater pulse current (120.92 μA), the dispersion of brownish rust on the sample seemed to be rather high. The use of such high currents is expected to create overprotection on the steel sample surface. Overprotection can lead to the formation of alkaline conditions on the metallic surface when there is insufficient water flow to return the water to its natural pH. It results in accelerated corrosion by creating more alkaline conditions and leads to the rapid deterioration of metal surfaces exposed to water vapor.

Figure 85.2 shows the image analysis performed on the same image in Fig. 85.1. All the samples were labeled with 'W' which represent a white or protected area and 'B' for a black or rust area. A large percentage of white color (W) characterizes better corrosion protection. In general, the image analysis results in Fig. 85.2 correspond to the stereomicroscope observation in Fig. 85.1. The sample surface exposed to atmospheric corrosion exhibited a white coloration of 70.87% (Fig. 85.2a). It is referred to as a borderline to the other samples' surface. The sample charged with an amplitude of 2 V has a lower (67.54%) than the borderline percentage. This demonstrates that at the amplitude of 2 V, the charges produced by pulse current feeding are unable to protect the steel surface against atmospheric corrosion. The sample treated to a pulse current amplitude of 5 V exhibited the largest percentage of white surfaces (84.76%). However, a thorough examination reveals that the protection is not so perfect, as 15.24% of the surface is covered by black, indicating that iron atoms have been oxidized to iron oxide. However, with this finding, the 5 V amplitude might be utilized as a reference for further investigations.

While for the sample with an amplitude of 7 V, it can be seen that the percentage of white color on the surface is quite low (77.31%), but still passes the borderline. As mentioned before, a pulsed current with an amplitude of 7 V causes overprotection which accelerates the corrosion reaction on metal surfaces exposed to water vapor.

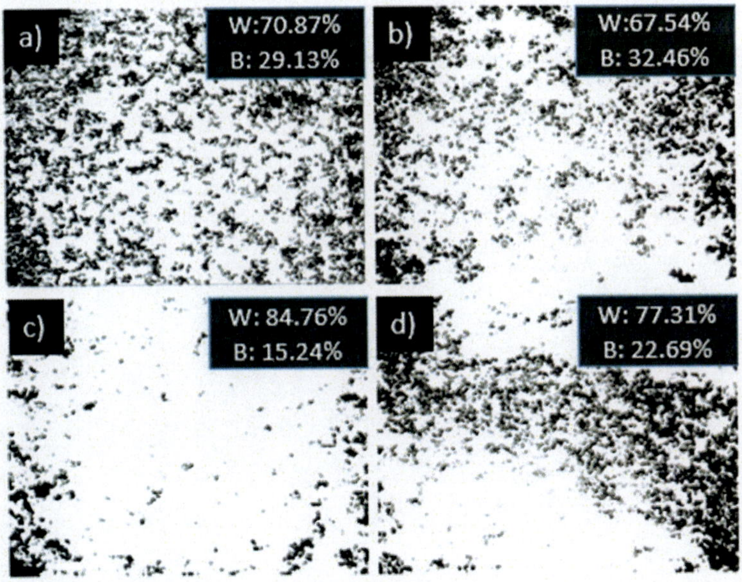

Fig. 85.2 Image analysis of mild steel surface showing **a** atmospheric corrosion and samples under PCP experiment subjected to different pulse current amplitudes, namely **b** 2 V, **c** 5 V, and **d** 7 V

85.4 Conclusion

The current investigation revealed that the application of pulsed current can prevent atmospheric corrosion on the surface of a steel sample. The amplitude of pulsed current has a significant effect in protecting against atmospheric corrosion. 2 V amplitude is insufficient to generate necessary negative charges to protect the steel surface. In the meantime, a 7 V amplitude generates overprotection, which exacerbates atmospheric corrosion degradation. Finally, 5 V was discovered to be a pretty excellent amplitude parameter for atmospheric corrosion prevention; however, it must still be examined further to provide the optimum protection.

Acknowledgements The author would like to acknowledge the support from the Fundamental Research Grant Scheme (FRGS) under a grant number of FRGS/1/2020/TK0/UNIMAP/02/85 from the Ministry of Higher Learning Malaysia.

References

1. A.S.H. Makhlouf, H. Victor, E. Muñoz, Corrosion and protection of the metallic structures in the petroleum industry due to corrosion and the techniques for protection (Elsevier Ltd, 2018)
2. Z. Ahmad, Atmospheric corrosion in principles of corrosion engineering and corrosion control (Elsevier, 2006)
3. V. Kucera, E. Mattsson, Atmospheric corrosion in Corrosion mechanisms (CRC Press, 2020), pp. 211–284
4. D.D. Macdonald, M. Lewis, J. McLafferty, E. Maya-Visuet, R. Peek, Electromagnetic induction corrosion control technology (EICCT). Mater. Corros. **69**(4), 436–446 (2018)
5. X.Q. Xu, H.B. Zhou, Study on the influence of pulse current cathodic protection parameters of oil well casing, in *Advances in Materials Science and Engineering* (2019)
6. W. Camp, Electronic corrosion protection device. U.S. Patent No. 10,982,338 (U.S. Patent and Trademark Office, Washington, DC, 2021)

Chapter 86
Effect of Surface-Treated Filler on the Wettability of Composite Solder: Short Review

Wan Hafizah Mohd Saufee, Wai Keong Leong, Ahmad Azmin Mohammad, and Muhammad Firdaus Mohd Nazeri

Abstract This paper evaluates and analyzes the effect of surface-treated filler on the wettability of composite solder. The basic concepts, key variables, arrangement, and most recent advances on wettability effect of selected surface treatment of ceramic-reinforced composite lead-free solder will be covered. The wettability characteristics of solders serve as the connection between the ceramic support and were found to be significantly improved by the ceramic filler with surface treatment as it minimizes the agglomeration caused by density differences and the Van der Waals attraction force. After the surface modification procedure, it was also possible to see the good dispersion of ceramic reinforcement in the solder as it assists filler in overcoming the Van der Waals attraction on the particle. Based on ultrasonic, electroless plating, and pyrolysis treatment that have been reviewed, the wettability of the composite solder is observed to be better than the untreated or initial pure solder without filler as surface treatment helps filler molecules effectively confine the mobility of the grain boundaries due to excellent interfacial bonding with Zener pinning effect. In solder, where mechanical aspect is one of the most important properties, substantial advancements in mechanical qualities have partially resulted from surface treatment of the ceramic reinforcement. Study on filler surface treatment has beneficially influences the improvement of the ceramic reinforcement-solder interfacial response and opens prospects for more specific future research lines and composite solder application opportunities as the response from surface treatment greatly impact the solder's wettability and its reliability as lead-free solder (LFS) product that can substitute the traditional lead-based solder.

Keywords Surface treatment · Filler · Lead-free solder · Composite solder

W. H. M. Saufee · W. K. Leong · M. F. M. Nazeri (✉)
Center of Excellence Geopolymer and Green Technology (CEGeoGTech), Universiti Malaysia Perlis (UniMAP), 02600 Arau, Perlis, Malaysia
e-mail: firdausnazeri@unimap.edu.my

A. A. Mohammad
Advance Soldering Material Group, School of Materials and Mineral Resources Engineering, Universiti Sains Malaysia (USM), 14300 Nibong Tebal, Pulau Pinang, Malaysia

M. A. A. Mohd Salleh et al. (eds.), *Proceedings of the Green Materials and Electronic Packaging Interconnect Technology Symposium*, Springer Proceedings in Physics 289,
https://doi.org/10.1007/978-981-19-9267-4_86

86.1 Introduction

Solder is one of the main components in electronic device that serves as the foundation for electronic devices connection. A fusible metal alloy called solder is utilized to permanently connect metal workpieces together. Solder offers mechanical, thermal, and electrical connectivity in electronics installations as a linking substance. The reliability of a solder joint, which has been essential to the assembly's total functionality, depends on the function and quality of the solder. The connection pad and silicon die are connected electrically and mechanically via the solder. Additionally, it acts as a conduit for the semiconductor's heat to be dissipated [1]. Years have passed since the invention of tin–lead or stannum–plumbum (Sn–Pb) solder, where this solder is reliable and is the general solder used by many electronic assembly factories.

However, owing to Pb's hazardous effects, rising environment legislation was focused on eliminating its use in electrical circuitry, particularly in the EU and Japan [1]. Leads obvious harmful effects as well as Pb-containing substances are regarded as ecosystem concerns that not only will harm the environment but it is a poison to human body as well [2]. Many researchers have now aware of the impact of lead usage to the environment and slowly changing to, LFS thus sparking interest in LFS research. In comparison with lead-contain solders LFS shows greater melting range, that shows significant impact across both mechanical and electrical characteristics [3]. The fundamental issue with LFS is it is typically overly steep, low, or wide separated liquidus and solidus temperatures. Additionally, it is necessary to improve the wettability to ensure good bonding is achieved. To combat the problem, reinforcement is introduced to the mix to increase the positive effect of the solder, alloying or ceramic element can be added in mind of this effect. Metallic processes have traditionally been strengthened by the addition of a fine dispersion of second-phase nanoparticles. The particles can be directly deposited using powder metallurgy processes [4] or using various other method in incorporating the reinforcement into the solder. Metal matrix composites were indeed configurations of two or more chemically inert components that create a new composite structure to improved material properties. For non-metallic components, most frequently ceramics like silicon carbide (SiC) [5], aluminum oxide (Al_2O_3) [6], graphite [7], or geopolymer [8], may be used as reinforcing materials [9]. These materials have several surface treatments option such as ultrasonic treatment, acid treatment, electroless plating, pyrolysis, etc.

Yet, filler alone is not entirely enough as agglomeration can happen due to the density differences between solder and ceramic material [10]. The majority of soldering fillers, indeed, cannot be wetted upon ceramics interfaces since they offer predominantly covalent or ionic bonding, have a relatively steady electron configuration, and seem to be chemically inert [11]. On molecules that are smaller than 1 mm, the Van der Waals force predominates, and Van der Waals pull is rivaled by the Coulomb force of repulsion. The Van der Waals force of attraction overcomes the electrostatic resistance for an interparticle spacing as small as 1 nm. Because the potential energy of repulsion reduces with decreasing particle size, aggregation is more likely. Another crucial factor is the Brownian motion of the nanostructure particles in the metal matrix composites. Brownian motion completely governs the kinetics of agglomeration for small particulates resulting in agglomeration of filler when incorporated into LFS [12]. The filler surface treatment is, therefore, the most crucial element in affecting wettability. When agglomeration happen after the filler has been incorporated in the composite solder, distribution of the properties for the solder would be different in different part of the solder making the composition and properties varies. This will cause segregation of properties in the solder and effect the wettability of the solder.

Surface treatment can enhance filler property and become beneficial in the mechanical properties namely wettability of the solder that led to better application. Wettability aspect such as wetting angle, surface energy, and surface tension as well as intermetallic compound (IMC) formation are the components that will be emphasized to judge the mechanical properties advancement. In this work, recent advances on the wettability improvement of ultrasonic, electroless plating, and pyrolysis surface-treated ceramic filler in LFS are compiled and discussed. Ultrasonic, electroless plating, and pyrolysis have been discussed in other work as an efficient and most utilized technique of surface treatment on the composite solder influencing the reduction of agglomeration of ceramic filler prompted author to select these methods of surface treatment [10]. The basic understanding of each technique, main parameter, and setup used and case studies is accumulated to improve the understanding on how surface treatment can enhance the wettability.

86.2 Wettability Measurement

As a suitable metallic bond to develop through the soldering operation, the molten solder's capability to flow and spread is crucial. When talking about soldering procedures, the word "wetting" is frequently mentioned. Wetting is another term that is typically used to describe the spreading phenomena. Wetting is the capability of one material, typically a liquid, to flow across other material, typically a solid, according to its meaning [13]. The contact angle developed at the intersection of a solid and liquid during a specific setting is used to gauge the degree of wetting, spreading area was used for wettability measurement as well. Generally, a system is thought of being wet if somehow the wetting or contact angle is around 0°–90°, whereas it is non-wetting if it ranges within 90°–180°. A good wetting angle has high surface energy and ranges from 0° to 30° for optimal wetting and, 30° to 40° for good wetting [14]. Wetting angle of a material can be determined by using J-Image software. The Young-Dupre equation states that the contact angle (γ) can be calculated from the equilibrium of surface tensions at the junction [1] (Fig. 86.1).

$$\text{Young-Dupre Equation}: \gamma_s = \gamma_{si} + \gamma_l \tag{86.1}$$

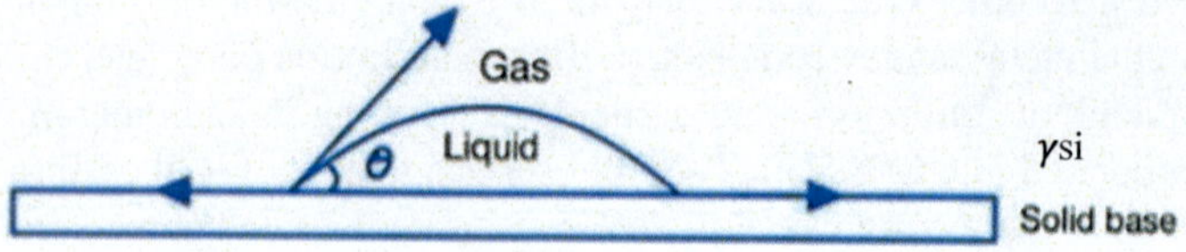

Fig. 86.1 Wetting angle environment

Lin et al. have reported about 63Sn–37Pb contact angles that are observed to have a dispersion characteristic and ranging between 13° and 32°, to match the low contact angle LFS must have either equal or superior surface angle than lead base solder [15]. Among the LFS studied, tin–bismuth–copper (Sn–10Bi–0.8Cu) has the lowest wetting angle at 32° while tin–bismuth-argentum (Sn–48Bi–3.4Ag) valued at ± 33° [1], tin-zinc (Sn–9Zn) at ± 54°, (Sn–58Bi) at ± 43° [16] and eutectic tin-argentum-copper (Sn3.0–Ag–0.5Cu) (SAC305) is ± 45° [17]. The effectiveness of wettability can be measured using several technique including spreading area [18], spreading angle [19], surface tension [1], wetting force, and wetting time [20]. Spreading area and spreading angle are inversely related to each other where when spreading area is high, spreading angle will be low and vice versa. This relationship can be applied for wetting time and wetting force as well where low wetting time and high wetting angle are optimum. The desired respond of wettability test is high spreading area, low contact angle, low wetting time, and high wetting force. However, it is challenging to compare the effectiveness of the alloys because the data regarding wetting angles for LFS alloys generally is somewhat inconsistent. This could be owing to the fact that each researcher uses different measurements of temperature, Cu substrate preparation, applied fluxes, as well as other research factors.

86.3 Wettability of Ultrasonicated Filler in Composite Solder

A new burgeoning fabrication method for creating metals, alloys, and composites is ultrasonic treatment. In that kind of process, an electric flow is converted into sound waveforms via piezoelectric transducer. These so-called ultrasonic vibrations are sent into a pool of molten metal either by transmitter or sonotrode that is submerged [21]. As a result, the liquid metal experiences longitudinal tension oscillations involving alternate periods of rarefaction and compression. The surface energy of ceramic materials with molten solder is high. The filler's mass can be countered by the total upwards force of surface tension, as well as the nanoparticle floats toward the top of the melt pool upon addition. Therefore, using the kinetics of outside forces is the main principle to get around this issue. External assistance is used to propel filler into the melt, creating a dynamic environment which traps the particles [22]. In ultrasonic treatment, as the melted solder is immediately influenced by the ultrasonic frequencies. The output voltage and frequency level are negatively associated proportionally. For non-destructive examination and health-care assessment, low-intensity, high-frequency ultrasound in megahertz is frequently utilized because it does not change the condition of the medium by which it passes. These sound waves have a frequency that is typically between 18 and 20 kHz [23]. The deagglomeration as well as dispersal of nanoparticles could benefit from the shock pulses and micro-jets produced from cavitation bubbles. Additionally, the sono-capillary effect (a term used to describe the wetting of ceramic nanoparticles with molten metal) is said to be triggered by micro-jets produced while cavitation undergoes [20]. Furthermore, scientists claim that dispersing nanoparticles in molten metals using ultrasound is among the most efficient ways to do surface treatment [24].

Liquid-state ultrasonic treatment was reported to effectively distribute 0.05–0.4 wt%. zinc oxide (ZnO) nanoparticles in Sn–57Bi (Sn–Bi) [25]. The nanoparticles were dispersed using a piezoelectric ultrasonic transmitter linked to an SS304 sonotrode with a 3 cm diameter, 20 kHz frequency, and 70% amplitude. When composite solder treated with ultrasonic treatment, Sn–Bi solders with treated filler show spreading ratio for monolithic and ZnO nanoparticles on a Cu substrate. On the Cu substrate, monolithic Sn-Bi solder spreads at a 70.7% average rate. The axisymmetric Laplace–Young equation in the lubrication limit reads the lowering wetting angle if the drop may be considered to be thin [26]. The average spreading ratio rises to 72.6% and 74.4% with introduction of 0.05 and 0.1 wt% of ZnO nanoparticles, respectively, and decline beyond 0.1 wt%. The spreading ratio for Sn-Bi solder has been enhanced, according to spreading tests, by the inclusion of ZnO nanoparticles up to 0.2 wt%. Higher adding quantities, unfortunately, deteriorated the spreading ratio [25].

Table 86.1 summarizes the effect of stir casting and ultrasonic technique on zero cross time (ZTC) and maximum wetting force (F_{max}) that are comparable to wetting time and wetting force for SAC-ZnO with different wt% values of (0.05, 0.1, and 0.2 wt%). According to this study, ultrasonic has proven to have shorter wetting time and higher wetting force until 0.1 wt% and declined onward. Both stir cast and ultrasonic show better wettability compared to initial SAC with untreated ZnO however, ultrasonic shows a better result proving its efficiency [20].

For the Sn–9Zn with addition of aluminum nitride (AlN) ceramic, it is observed that without ultrasonication, wetting is impossible, and once the solder has solidified, the foundation component will disengage from it [27]. Implying that Sn–9Zn eutectic metal could wet the AlN with the help of ultrasonication under low temperature, the joint was created without any evident flaws after only a brief 5 s of ultrasonic-assisted wetting time. It was explained that ultrasound cavitation introduces high pressure that allow the arrangement of AlN and improves the wettability.

Table 86.1 Wetting time and wetting force for SAC 305–ZnO initial and SAC 305–ZnO with different ZnO weight percent [20]

SAC composition		ZCT (zero cross time), (s)	F_{max} (Maximum wetting force), (mN)	Wetting time comparing to initial (%)	Wetting force comparing to initial (%)
SAC 305–ZnO (initial)		0.91	6.32		
SAC 305–0.05ZnO	Stir cast	0.78	6.48	14.2	2.5
	Ultrasonic	0.61	6.57	32.9	3.7
SAC 305–0.1ZnO	Stir cast	0.87	6.45	4.3	2.0
	Ultrasonic	0.53	6.91	41.7	9.3
SAC 305–0.2ZnO	Stir cast	1.32	6.13	− 45.2	− 3
	Ultrasonic	0.83	6.32	8.7	0

86.4 Wettability of Electroless Plated Filler in Composite Solder

Electroless plating is a non-galvanic type plating technique, sometimes referred to as chemical or auto-catalytic plating, entails several parallel chemical processes taking place in an aqueous medium without any of the assistance of external electric energy. Electroless nickel plating is among the most popular electroless plating technique. Due to the rate of deposition relies on the separation between the anode and the cathodic item, traditional electroplating will result in inconsistent thicknesses therefore changes to electroless plating. In the as-plated form, electroless nickel plating results in an amorphous layer. Despite the size or form of plated surface, the deposit is nearly evenly distributed in thickness because it is not reliant; on current distribution, it is crucial because it provides uniform coating thicknesses on products with odd or complicated shapes [28]. A reducing agent, often sodium hypophosphite, releases hydrogen during the process, which is then oxidized to produce a negative charge upon that part's surface [29]. Electroless coating can be divided into three main categories which including alloy, metallic, and composite coating. Metal alloys are deposited upon the substrate via the electroless alloy plating process. With suitable context, electroless metallic (pure nickel) coating is a reliable coating technique that could offer a workable substitute for standard electroplating while the co-deposition of composite substances in the alloy coating is known as electroless composite plating. In order to achieve co-deposition of both the solid materials as well as Ni–P matrix, electroless composite coatings are often made by adding solid particles to the standard electroless nickel phosphorous plating solution. This kind of coating has recently received a lot of attention when creating composite coatings. Among the often-utilized solid particles are polytetrafluoroethylene (PTFE), ZrO_2, Al_2O_3, SiC, and carbon (C) [30]. Electroless composite coating is found use in many studies of composite solder as it is a reliable technique of metallic distribution onto ceramic composite metallic elements such as nickel, gold, and palladium were used. Electroless plating can make use of any metallic element on top of the ceramic filler;

it can be applied on its own as well as with pretreatment before electroless plating surface treatment. Ultrasonic cleansing and acid pretreatment are commonly used coupled with the electroless plating. Deposits of electroless nickel offer a range of unique qualities. The amount of phosphorus (P) alloyed in the deposit affects these properties. Within 1 to 14% per weight of phosphorus, the physical, electrical, and surface characteristics change as phosphorus content rises [31].

Study for SAC with Ni-modified graphene nanosheets (GNSs) has been reported in several reports [17, 18, 32, 33]. Electroless plating for SAC was achieved by following three processes to create the Ni–GNS in this study: ultrasonic dispersion, sensitization, and activation with acid pretreatment, and electroless Ni-plating. The adsorption concept has a significant impact upon that morphology of alloys throughout the solidification process, therefore, the mechanical characteristics would be improved with finer microstructure [17]. Wetting angle for SAC with Ni–GNS was different for each weight percentage of Ni–GNS initial value of 37.5° for SAC followed by ± 32.7° for both 0.05 and 0.1 wt% to 32.6° for 0.2 wt%. The differences of wetting angle for each percentage are not big but with addition of Ni–GNS the contact angle is lower and average thickness of IMC is higher. However, the contact angle is steadily decreasing from ± 37.5° for pure SAC to ± 31.8° for SAC–0.2 Ni–GNS [33].

It was found that with varying Ni–GNSs composition added into SAC (0, 0.01, 0.05, 0.10 and 0.15 wt% of GNSs). The solder's distributing area on the Cu substrate first shows increment before decreasing, 0 wt% Ni–GNS has spreading area of ± 45 mm^2 and rose to ± 52 mm^2 at 0.05 wt% and falls to ± 44 mm^2, at 0.15 wt%. At 0.01 to 0.05 wt% GNSs in the solder matrix where the solder's wettability is at its best as the amount of GNSs in the solder risen. During the soldering process, interfacial interactions regulate the solder's wettability. Ni–GNSs, the interface active phase in the composite solder, would adsorb upon that surface of Ag_3Sn as its ratio increases, lowering the surface energy of the Ag_3Sn crystal plane [18]. While in study that involves SAC and Ni-MWCNTs (Ni-modified multiwalled carbon nanotubes), sample preparation for surface treatment is similar to [17] where three steps made up the majority of the procedure: ultrasonic dispersion, pretreatment, and electroless Ni modification process, with different percentage of Ni-CNTs wt% (0, 0.1, and 0.2) that has been reflowed for 5 min and aged for 0 to 360 h. The thickness of the IMC layer is largest at 360 h but decreases as the filler incorporation increases [32]. Found that even when the material and filler used are identical results can vary based on researcher technique and parameter making comparison difficult.

The silicon carbide treated using nickel (Ni–SiC) has been used in SAC, according to multiple publications [5, 34, 35]. In the study where SAC mixed with integration of different concentration (0.5, 1.0, 1.5 wt%) of nickel-coated Ni–SiC filler and obtain its wetting contact angle, the lead-free solder composite's contact angles shrank as SiC and Ni–SiC content rose. More precisely, the composite solder containing 1.5 wt% SiC and Ni–SiC reinforcement had a lower contact angle than pure SAC305, lowering from 38.8° to 23.2° and 18.1° correspondingly. These findings demonstrate that adding modest amounts of SiC and Ni–SiC particles to the solder matrix enhances the wettability of composite solder. This happens because Ni–SiC particles clump together during soldering at its interaction in between flux as well as the molten lead-free solder, which lowers the interfacial force. Consequently, it can be said that the addition of Ni–SiC lower the contact angle across the lead-free solder plus flux, which then in effect enhances the wettability of the lead-free solder composites [5, 34].

Mechanically, homogeneous mixing of the pre-weighted SiC and Ni–SiC reinforcement (0, 1, and 2 wt%) and SAC solder paste was performed with three different pre-treatment techniques used to activate the SiC's surface were conducted: surface oxidation, acidic pre-treatment, and $PdCl_2$ techniques were not successful. Author then changes the technique to acid pretreatment and electroless plating of Ni. Upon every specimen, the layer thickness was averaged after 30 SEM measurements. Pure SAC had an intermetallic layer (IML) thickness that was 1.85 μm on average. In the SAC-2% SiC solder composite, the thinnest IML produced shows reduction of 30% from initial SAC. In general, it can be said that Ni coating favorably influences the production of IML. In SiC composites with Ni coating, the contact angle found significantly reduced. The SiC joint with 1% Ni coating had the lowest recorded contact angle at 10% increases change. Since SiC significantly changed the droplet's shape, the standard deviation of the contact angle in the 2 wt% Ni–SiC sample at positive 20% was considerable. This might be because there is greater number of Ni–SiC, that makes the melt viscous and prevents melt flow [35].

Analyzed are SAC with inclusion of graphene nanoplatelets (GNPs) that have various concentrations of 0.05, 0.1, and 0.2 wt%. In respect of wetting angle, the wettability of solder nanocomposites is contrasted with that of the SAC solder alloy. In this sense, improved wettability is correlated with a smaller wetting angle. The findings show that GNPs have a positive impact since they consistently reduce the wetting angle when their content is increased in all circumstances. By including 0.2 wt% of GNPs, a maximum wetting angle drop of roughly 5 degrees is attainable. The integration of GNPs as well as the presence of a Ni coating causes the wetting angle to show a consistent pattern toward decline. Pure SAC has a high wetting angle at $\pm$ 37° and gradually decreasing with incorporation of 0.2 wt% GNPs to $\pm$ 32° while SAC with addition of Ni-GNPs shows consistent decrease pattern from $\pm$ 36° at 0.05 wt% to $\pm$ 33° at 0.2 wt% [19].

86.5 Wettability of Pyrolysis Filler in Composite Solder

Pyrolysis surface-treated filler would help in incorporating the positive impact to the lead-free solder. In the process of pyrolysis, biomass is broken down by heat in the exclusion of oxygen to produce charcoal, liquid, and gaseous byproducts. The three subclasses of pyrolysis are traditional pyrolysis, fast pyrolysis, as well as flash pyrolysis [36]. All organic (carbon-based) compound may be subjected to pyrolysis, a thermochemical process. Throughout this process, substance is heated to a high temperature while also undergoing chemical and physical segregation. Due to the finite thermal durability of material's chemical bonds, that enables heat to destroy them, breakdown occurs. Pyrolysis is typically used to transform carbon-based materials, such as polysilazane, into ceramics or carbon aerogels. It able to increase the interfacial adherence of the reinforcing materials to the matrix phase. Based on the effects of pyrolysis surface alterations, several advantageous effects have been documented, including a rise in thermal conductivity, electrical, flame retardant, and water proofing, as well as convenience in customizing optoelectronic characteristics [37]. According to Bauer et al., the following are some of the primary elements that have been shown to have a significant impact on the quality of surface modification achieved by pyrolysis processes: precursor system, degree of crosslinking, pyrolysis temperature, and environment are the first three factors and finally, humidity in the surroundings. The most effective pyrolysis was 300 °C in nitrogen environment [38].

To operate the tetra-needle-like zinc oxide (T-ZnO) in this investigation, pyrolysis approach premised on self-assembly was carried out as surface treatment on the filler that was used with SAC solder. Exterior alterations pyrolysis of T-ZnO using a self-assembly-based approach were accomplished satisfactorily. Upon the surface of T-ZnO, NiO nanoparticles were consistently adhered [39]. Ball milling and pyrolysis were used to modify the surface of zirconium dioxide (ZrO_2) in another study. With ultrasonic stirring, NiO-altered ZrO_2 (NiO–ZrO_2) was introduced onto the solder substrate. The microstructure of composite solders can be improved by adding NiO–ZrO_2 in the proper proportions of 0–0.3 mass % [24].

86.6 Conclusion

Solder is an important element, especially in the electronic application. When poor mechanical properties solder can be enhanced and developed into an excellent product, research community will be disclosed to something new and yet to be found. These can give a positive impact to our developer as well as industry. With the advancement of surface-treated filler for composite solder many have expect increasing reliability of the lead-free solder product. Surface treatment such as ultrasonic, electroless plating and pyrolysis were used in many studies and shows positive impact in term of mechanical characteristic. While ultrasound and pyrolysis are a good technique, the most used technique is electroless plating where the metallic elements such as nickel and argentum were used to coat the ceramic filler. The ability of Ni and Ag element to carry the non-wettable material that was used as the filler in the composite solder is the most desirable factor of using this method. Therefore, many studies have been conducted to observe the effect of electroless plating across various composite solder as well as across various ceramic filler. In ultrasonic, studies have varied from solder used and filler reinforcement used, making direct comparison tough, but ZnO has been found in several studies as filler deemed to be the most utilized. Electroless plating, however, has a pretty set pattern where most used SAC solder with varied filler, most Ni-plating were done after ultrasonic treatment and acid pre-treatment to ensure successful coating. The best result is shown with study of SAC with Ni–SiC filler with contact angle of 18.1° with incorporation of

1.5 wt%. Pyrolysis surface treatment has shown good wetting result. All three techniques have its own pros and cons making the technique selection difficult and need to be dependent on material selected as well as parameter.

Acknowledgements The authors appreciate the financial support provided by the Ministry of Higher Education Malaysia through FRGS grant scheme (FRGS/1/2019/TK05/UNIMAP/02/5).

References

1. M. Abtew, G. Selvaduray, Lead-free solders in microelectronics. **27**(5–6), 95–141 (2000)
2. K. Zeng, K.-N. Tu, Six cases of reliability study of Pb-free solder joints in electronic packaging technology. **38**(2), 55–105 (2002)
3. M.K. Pal, G. Gergely, D. Koncz-Horváth, Z. Gácsi, Investigation of microstructure and wetting behavior of Sn–3.0 Ag–0.5 Cu (SAC305) lead-free solder with additions of 1.0 wt% SiC on copper substrate. **128**, 106991 (2021)
4. W. Yang, H. Araki, C. Tang, S. Thaveethavorn, A. Kohyama, H. Suzuki, T. Noda, Single-crystal SiC nanowires with a thin carbon coating for stronger and tougher ceramic composites. **17**(12), 1519–1523 (2005)
5. M.K. Pal, G. Gergely, D. Koncz-Horváth, Z.J.P.M. Gácsi, M. Ceramics, Investigation of the electroless nickel plated sic particles in sac305 solder matrix. **58**(9), 529–537 (2020)
6. W.-q. Xing, X.-ye. Yu, H. Li, L. Ma, W. Zuo, P. Dong, W.-x. Wang, M. Ding, Effect of nano Al_2O_3 additions on the interfacial behavior and mechanical properties of eutectic Sn–9Zn solder on low temperature wetting and soldering of 6061 aluminum alloys. **695**, 574–582 (2017)
7. Z. Yang, W. Zhou, P. Wu, Effects of Ni-coated carbon nanotubes addition on the electromigration of Sn–Ag–Cu solder joints. **581**, 202–205 (2013)
8. N.S.M. Zaimi, M.A.A.M. Salleh, A.V. Sandu, M.M.A.B. Abdullah, N. Saud, S.Z.A. Rahim, P. Vizureanu, R.M. Said, M.I.I. Ramli, Performance of Sn–3.0 Ag–0.5 Cu composite solder with kaolin geopolymer ceramic reinforcement on microstructure and mechanical properties under isothermal ageing. **14**(4), 776 (2021)
9. D.K. Das, P.C. Mishra, S. Singh, S. Pattanaik, Fabrication and heat treatment of ceramic-reinforced aluminium matrix composites-a review. **9**(1), 1–15 (2014)
10. L.W. Keong, A.A. Mohamad, M.F.M. Nazeri, Surface modifications on ceramic reinforcement for tin-based composite solders, in *Recent Progress in Lead-Free Solder Technology*. (Springer, 2022), pp.53–75
11. J.A. Fernie, R. Drew, K. Knowles, Joining of engineering ceramics. **54**(5), 283–331 (2009)
12. A. Basak, A. Pramanik, M.N. Islam, V. Anandakrishnan, Challenges and recent developments on nanoparticle-reinforced metal matrix composites. **2015**, 349–367 (2015)
13. T. Zhao, L. Jiang, Contact angle measurement of natural materials. **161**, 324–330 (2018)
14. A.M. Erer, S. Oguz, Y. Türen, Influence of bismuth (Bi) addition on wetting characteristics of Sn–3Ag–0.5Cu solder alloy on Cu substrate. **21**(6), 1159–1163 (2018)
15. C.-T. Lin, K.-L. Lin, Contact angle of 63Sn–37Pb and Pb-free solder on Cu plating. **214**(1–4), 243–258 (2003)
16. T. Matsumoto, K. Nogi, Wetting in soldering and microelectronics. **38**, 251–273 (2008)
17. G. Chen, F. Wu, W. Xia, H. Liu, C. Liu. Effect of Ni-coated graphene on the performance of SAC305 solder, in *2014 15th International Conference on Electronic Packaging Technology* (IEEE, 2014)
18. H. Wang, K. Zhang, M. Zhang, Fabrication and properties of Ni-modified graphene nanosheets reinforced Sn–Ag–Cu composite solder. **781**, 761–772 (2019)

19. R. Sayyadi, F. Khodabakhshi, N.S. Javid, G. Khatibi, Influence of graphene content and nickel decoration on the microstructural and mechanical characteristics of the Cu/Sn–Ag–Cu/Cu soldered joint. **9**(4), 8953–8970 (2020)
20. S.H. Rajendran, D.H. Cho, J.P. Jung, Comparative study on the wettability and thermal aging characteristics of SAC 305 nanocomposite solder fabricated by stir-casting and ultrasonic treatment. **31**, 103814 (2022)
21. D. Meroni, R. Djellabi, M. Ashokkumar, C.L. Bianchi, D.C. Boffito, Sonoprocessing: from concepts to large-scale reactors. **122**(3), 3219–3258 (2021)
22. T. Paul, C. Zhang, B. Boesl, A. Agarwal, Analytical review of reinforcement addition techniques during ultrasonic casting of metal matrix composites. **22**(10), 2000524 (2020)
23. L.H. Thompson, L. Doraiswamy, Sonochemistry: science and engineering. **38**(4), 1215–1249 (1999)
24. F. Huo, Z. Jin, D. Le Han, J. Li, K. Zhang, and H. Nishikawa, Novel interface regulation of Sn1.0–Ag–0.5Cu composite solders reinforced with modified ZrO_2: microstructure and mechanical properties. **125**, 157–170 (2022)
25. S.H. Rajendran, H. Kang, J.P. Jung, Ultrasonic-assisted dispersion of ZnO nanoparticles to Sn–Bi solder: a study on microstructure, spreading, and mechanical properties. **30**(5), 3167–3172 (2021)
26. L.Y.-M. Yeo, Wetting and spreading, in *Encyclopedia of Microfluidics & Nanofluidics*. (Springer, 2008), pp.2186–2196
27. S. Chen, Z. Xu, Z. Li, Z. Ma, M. Zhang, Y. Lu, J. Yan, Ultrasonic-assisted wetting and soldering of AlN ceramic by using a nonactive solder (Sn_9Zn) in air. **48**(2), 1898–1907 (2022)
28. J. Sudagar, J. Lian, W. Sha, Electroless nickel, alloy, composite and nano coatings–a critical review. **571**, 183–204 (2013)
29. C. Loto, Electroless nickel plating–a review. **8**(2), 177–186 (2016)
30. S. Zhang, K. Han, L. Cheng, The effect of SiC particles added in electroless Ni–P plating solution on the properties of composite coatings. **202**(12), 2807–2812 (2008)
31. K.H. Krishnan, S. John, K. Srinivasan, J. Praveen, M. Ganesan, P. Kavimani, An overall aspect of electroless Ni-P depositions—a review article. **37**(6), 1917–1926 (2006)
32. H. Wang, X. Hu, X. Jiang, Effects of Ni modified MWCNTs on the microstructural evolution and shear strength of Sn–3.0 Ag–0.5 Cu composite solder joints. **163**, 110287 (2020)
33. G. Chen, F. Wu, C. Liu, V.V. Silberschmidt, Y.C. Chan, Microstructures and properties of new Sn–Ag–Cu lead-free solder reinforced with Ni-coated graphene nanosheets. **656**, 500–509 (2016)
34. M.K. Pal, G. Gergely, D. Koncz-Horváth, Z. Gácsi, Characterization of the interface between ceramics reinforcement and lead-free solder matrix. **20**, 100576 (2020)
35. Z. Gyökér, G. Gergely, D.K. Horváth, E. Bodnár, Z. Gácsi, Role of reinforcement surface treatment on the $SnAg_3Cu_{0.5}$ microelectronic joints. **475**, 982–985 (2019)
36. A. Demirbas, G. Arin, An overview of biomass pyrolysis. **24**(5), 471–482 (2002)
37. J. Wie, J. Kim, Thermal properties of binary filler hybrid composite with graphene oxide and pyrolyzed silicon-coated boron nitride. **12**(11), 2553 (2020)
38. F. Bauer, U. Decker, A. Dierdorf, H. Ernst, R. Heller, H. Liebe, R. Mehnert, Preparation of moisture curable polysilazane coatings: Part I. Elucidation of low temperature curing kinetics by FT-IR spectroscopy. **53**(3), 183–190 (2005)
39. F. Huo, K. Zhang, H. Nishikawa, Surface modification of tetra-needle like ZnO (T-ZnO) and characterization of interface between Sn1.0Ag0.5Cu and NiO decorated T-ZnO, in *2021 International Conference on Electronics Packaging (ICEP)* (IEEE, 2021)